Springer Monographs in Mathematics

T0215630

Springer
New York
Berlin
Heidelberg
Barcelona
Hong Kong
London
Milan
Paris
Singapore
Tokyo

Yisong Yang

Solitons in Field Theory
and Nonlinear Analysis

Springer

Yisong Yang
Department of Applied Mathematics
  and Physics
Polytechnic University
Brooklyn, NY 11201
USA
yyang@magnus.poly.edu

Mathematics Subject Classification (2000): 35JXX, 58GXX, 81E10, 53C80

Library of Congress Cataloging-in-Publication Data
Yang, Yisong.
    Solitons in field theory and nonlinear analysis / Yisong Yang.
      p. cm. — (Springer monographs in mathematics)
    Includes bibliographical references and index.

      1. Solitons.   2. Field theory (Physics)   I. Title.
    (Springer-Verlag New York, Inc.)
    QA1 .A647
    [QC174.26.W28]
    510 s—dc21
    [531'.1133]   00-067919

Printed on acid-free paper.

Photocomposed copy prepared from the author's LaTeX files.

9 8 7 6 5 4 3 2 1

ISBN 978-1-4419-2919-8

Springer-Verlag New York Berlin Heidelberg
A member of BertelsmannSpringer Science+Business Media GmbH

For Sheng
Peter, Anna, and Julia

Preface

There are many interesting and challenging problems in the area of classical field theory. This area has attracted the attention of algebraists, geometers, and topologists in the past and has begun to attract more analysts. Analytically, classical field theory offers all types of differential equation problems which come from the two basic sets of equations in physics describing fundamental interactions, namely, the Yang–Mills equations governing electromagnetic, weak, and strong forces, reflecting internal symmetry, and the Einstein equations governing gravity, reflecting external symmetry. Naturally, a combination of these two sets of equations would lead to a theory which couples both symmetries and unifies all forces, at the classical level. This book is a monograph on the analysis and solution of the nonlinear static equations arising in classical field theory.

It is well known that many important physical phenomena are the consequences of various levels of symmetry breakings, internal or external, or both. These phenomena are manifested through the presence of locally concentrated solutions of the corresponding governing equations, giving rise to physical entities such as electric point charges, gravitational blackholes, cosmic strings, superconducting vortices, monopoles, dyons, and instantons. The study of these types of solutions, commonly referred to as solitons due to their particle-like behavior in interactions, except blackholes, is the subject of this book.

There are two approaches in the study of differential equations of field theory. The first one is to find closed-form solutions. Such an approach works only for a narrow category of problems known as integrable equations, and, in each individual case, the solution often depends heavily on an ingenious construction. The second one, which will be the main focus of this book, is to investigate the solutions using tools from modern nonlinear analysis, an approach initiated by A. Jaffe and C. H. Taubes in their study of the Ginzburg–Landau vortices and Yang–Mills monopoles (*Vortices and Monopoles*, Birkhäuser, 1980).

The book is divided into 12 chapters. In Chapter 1, we present a short introduction to classical field theory, emphasizing the basic concepts and terminology that will be encountered in subsequent chapters. In Chapters 2–12, we present the subject work of the book, namely, solitons as locally concentrated static solutions of field equations and nonlinear functional analysis. In the last section of each of these chapters, we propose some open problems.

The main purpose of Chapter 1 is to provide a quick (in 40 or so pages) and self-contained mathematical introduction to classical field theory. We start from the canonical description of the Newtonian mechanics and the motion of a charged particle in an electromagnetic field. As a consequence, we will see the natural need of a gauge field when quantum mechanical motion is considered via the Schrödinger equation. We then present special relativity and its action principle formulation, which gives birth to the Born–Infeld theory, as will be seen in Chapter 12. We also use special relativity to derive the Klein–Gordon wave equations and the Maxwell equations. After this, we study the important role of symmetry and prove Noether's theorem. In particular, we shall see the origins of some important physical quantities such as energy, momentum, charges, and currents. We next present gauge field theory, in particular, the Yang–Mills theory, as a consequence of maintaining local internal symmetry. Related notions, such as symmetry-breaking, the Goldstone particles, and the Higgs mechanism, will be discussed. Finally, we derive the Einstein equations of general relativity and their simplest gravitational implications. In particular, we explain the origins of the metric energy-momentum tensor and the cosmological constant.

In Chapter 2, we start our study of field equations from the 'most integrable' problem: the nonlinear sigma model and its extension by B. J. Schroers containing a gauge field. We first review the elegant explicit solution by A. A. Belavin and A. M. Polyakov of the classical sigma model. We then present the gauged sigma model of Schroers and state what we know about it. The interesting thing is that, although the solutions are topological and stratified energetically as the Belavin–Polyakov solutions, their magnetic fluxes are continuous. We shall see that the governing equation of the gauged sigma model cannot be integrated explicitly and a rigorous understanding of it requires nonlinear analysis based on the weighted Sobolev spaces.

In Chapter 3, we present an existence theory for the self-dual Yang–Mills instantons in all $4m$ Euclidean dimensions. The celebrated Hodge theorem states that, on a compact oriented manifold, each de Rham cohomology class can be represented by a harmonic form. In the Yang–Mills theory, there is a beautiful parallel statement: each second Chern–Pontryagin class on S^4 can be represented by a family of self-dual or anti-self-dual instantons. The purpose of this chapter is to obtain a general representation theorem in all S^{4m}, $m = 1, 2, \cdots$, settings. We first review the unit charge instantons in 4 dimensions by G. 't Hooft (and A. M. Polyakov). As a preparation for E. Witten's charge N solutions, we present the Liouville equation and its explicit solution. We then introduce Witten's solution in 4 dimensions, which motivates our general approach in all $4m$ dimensions. We next review the $4m$-dimensional Yang–Mills theory of D. H. Tchrakian (the 8-dimensional case was also due to B. Grossman, T. W. Kephart, and J. D. Stasheff) and use a dimensional reduction technique to arrive at a system of 2-dimensional equations generalizing Witten's equations. This system will further be reduced into a quasilinear elliptic equation over the Poincaré half-plane and solved using the calculus of variations and a limiting argument.

In Chapter 4, we introduce the generalized Abelian Higgs equations, governing an arbitrary number of complex Higgs fields through electromagnetic interactions. These equations are discovered by B. J. Schroers in his study of linear sigma models and contain as special cases the equations recently found in the electroweak theory with double Higgs fields by G. Bimonte and G. Lozano and a supersymmetric electroweak theory by J. D. Edelstein and C. Nunez. Using the Cholesky decomposition theorem, we shall obtain a complete understanding of these equations defined either on a closed surface or the full plane. When the vacuum symmetry is partially broken, we give some nonexistence results.

In Chapter 5, we start our study of the Chern–Simons equations from the Abelian case. The Chern–Simons theory generally refers to a wide category of field-theoretical models in one temporal and two spatial dimensions that contain a Chern–Simons term in their action densities. These models are relevant in several important problems in condensed matter physics such as high-temperature superconductivity and quantum and fractional Hall effect. In their full generality, the Chern–Simons models are very difficult to analyze and only numerical simulations are possible. However, since the seminal work of J. Hong, Y. Kim, and P.-Y. Pac and R. Jackiw and E. J. Weinberg on the discovery of the self-dual Abelian Chern–Simons equations, considerable progress has been made on the solutions of various simplified models along the line of these self-dual equations, Abelian and non-Abelian, non-relativistic and relativistic. This chapter presents a complete picture of our rigorous understanding of the Abelian self-dual equations: topological and nontopological solutions, quantized and continuous charges and fluxes, existence, nonexistence, and degeneracy (nonuniqueness) of spatially periodic solutions.

In Chapter 6, we study the non-Abelian Chern–Simons equations. In order to study these equations, we need a minimum grasp of the classification theory of the Lie algebras. Thus we first present a self-contained review on some basic notions such as the Cartan–Weyl bases, root vectors, and Cartan matrices. We next consider the self-dual reduction of G. Dunne, R. Jackiw, S.-Y. Pi, and C. Trugenberger for the non-Abelian gauged Schrödinger equations for which the gauge fields obey a Chern–Simons dynamics and the coupled system is non-relativistic. We show how this system may be reduced into a Toda system, with a Cartan matrix as its coefficient matrix. We then present the solution of the Toda system due to A. N. Leznov in the case that the gauge group is $SU(N)$ and write down the explicit solution for the original non-relativistic Chern–Simons equations. After this we begin our study of the non-Abelian relativistic Chern–Simons equations. We shall prove the existence of topological solutions for a more general nonlinear elliptic system for which the coefficient matrix is not necessarily a Cartan matrix. We shall also discuss several illustrative examples.

In Chapter 7, we present a series of existence theorems for electroweak vortices. It is well known that the electroweak theory does not allow vortex-like solutions in the usual sense due to the vacuum structure of the theory. More precisely, vortices in the Abelian Higgs or the Ginzburg–Landau theory occur at the zeros of the Higgs field as topological defects and are thus viewed as the Higgs particle condensed vortices but there can be no finite-energy Higgs particle condensed vortex solutions in the electroweak theory. However, J. Ambjorn and P. Olesen found in their joint work that spatially periodic electroweak vortices occur as a result of the W-particle condensation. This problem has many new features, both physical and mathematical. We shall first present our solution to a simplified system describing the interaction of the W-particles with the weak gauge field. We then introduce the work of Ambjorn–Olesen on the W-condensed vortex equations arising from the classical Weinberg–Salam electroweak theory and state our existence theorem. The Campbell–Hausdorff formula will be a crucial tool in the proof that the spatial periodicity conditions under the original non-Abelian gauge group and under the Abelian gauge group in the unitary gauge are equivalent. Our mathematical analysis of the problem will be based on a multiply constrained variational principle. Finally we present a complete existence theory for the multivortex equations discovered by G. Bimonte and G. Lozano in their study of the two-Higgs electroweak theory.

In Chapter 8, we present existence theorems for electrically and magnetically charged static solutions, known as dyons, in the Georgi–Glashow theory and in the Weinberg–Salam theory. We first review the fundamental idea of P. A. M. Dirac on electromagnetic duality and the existence of a magnetic monopole in the Maxwell theory. We will not elaborate on the original derivation of the charge quantization condition of Dirac based on considering the quantum-mechanical motion of an electric charge in the field of a magnetic monopole but will use directly the fiber bundle devise due to T. T. Wu and C. N. Yang to arrive at the same conclusion. We then present the argument of J. Schwinger for the existence of dyons in the

Maxwell theory and state Schwinger's extended charge quantization formula. We next introduce the work of B. Julia and A. Zee on the existence of dyons in the simplest non-Abelian gauge field theory, the Georgi–Glashow theory. The physical significance of such solutions is that, unlike the Dirac monopoles and Schwinger dyons, the Julia–Zee dyons carry finite energies. We will first present the explicit dyon solutions due to E. B. Bogomol'nyi, M. K. Prasad, and C. M. Sommerfield known as the BPS solutions. Away from the BPS limit, the equations cannot be solved explicitly. In fact, the existence of electricity leads us to a complicated system of nonlinear equations that can only be solved through finding critical points of an indefinite action functional. Recently, Y. M. Cho and D. Maison suggested that dyons, of infinite energy like the Dirac monopoles, exist in the Weinberg–Salam theory. Mathematically, the existence problem of these Weinberg–Salam or Cho–Maison dyons is the same as that of the Julia–Zee dyons in non-BPS limit: the solution depends on the optimization of an indefinite action functional and requires new techniques. In this chapter, we show how to solve these problems involving indefinite functionals. These techniques will have powerful applications to other problems of similar structure.

In Chapter 9, we concentrate on the radially symmetric solutions of a nonlinear scalar equation with a single Dirac source term. We shall use a dynamical system approach to study the reduced ordinary differential equation. The results obtained for this equation may be used to achieve a profound understanding of many field equation problems of the same nonlinearity. For example, for the Abelian Chern–Simons equation, we will use the results to prove that the radially symmetric topological solution is unique and the charges of nontopological solutions fill up an explicitly determined open interval, of any given vortex number; for the cosmic string problem, we will derive a necessary and sufficient condition for the existence of symmetric finite-energy N-string solutions over $\mathbb{R}^2$ and S^2.

In Chapter 10, we study cosmic strings as static solutions of the coupled Einstein and Yang–Mills field equations. It is well accepted that the universe has undergone a series of phase transitions characterized by a sequence of spontaneous symmetry-breakings which can be described by quantum field theory models of various gauge groups. Cosmic strings appear as mixed states due to a broken symmetry which give rise to a multi-centered display of energy and curvature and may serve as seeds for matter accretion for galaxy formation in the early universe, as described in the work of T. W. B. Kibble and A. Vilenkin. Since the problem involves the Einstein equations, a rigorous mathematical construction of such solutions in general is extremely hard, or in fact, impossible. In their independent studies, B. Linet, and A. Comtet and G. W. Gibbons found that the coupled Einstein and Abelian Higgs equations allow a self-dual reduction as in the case of the Abelian Higgs theory without gravity and they pointed out that one might obtain multi-centered string solutions along the line of the work of Jaffe–Taubes. In the main body of this chapter, we present a fairly complete understanding of these multi-centered cosmic string solutions. In particular, we show that there are striking new surprises due to the presence of gravity. For example, we prove that the inverse of Newton's gravitational constant places an explicit upper limit for the total string number. In the later part of this chapter, we combine the ideas of Linet, Comtet–Gibbons, and Ambjorn–Olesen to investigate the existence of multi-centered, electroweak, cosmic strings in the coupled Einstein and Weinberg–Salam equations. We shall see that consistency requires a uniquely determined positive cosmological constant. We will begin this chapter with a brief discussion of some basic notions such as string-induced energy and curvature concentration, deficit angle, and conical geometry.

In Chapter 11, we consider a field theory that allows the coexistence of static vortices and anti-vortices, or strings and anti-strings, of opposite magnetic behavior, both local and global. This theory originates from the gauged sigma model of B. J. Schroers with a broken symmetry and has numerous interesting properties. The magnetic fluxes generated from opposite vortices or strings annihilate each other but the energies simply add up as do so for particles. Gravitationally, strings and anti-strings make identical contributions to the total curvature and are equally responsible for the geodesic completeness of the induced metric. Hence, vortices and anti-vortices, or strings and anti-strings, are indistinguishable and there is a perfect symmetry between them. However, the presence of a weak external field can break such a symmetry which triggers the dominance of one of the two types of vortices or strings. Mathematically, this theory introduces a new topological invariant in field theory, the Thom class. A by-product is that these vortices and anti-vortices may be used to construct maps with all possible half-integer 'degrees' defined as topological integrals. As in the Abelian Higgs theory case, the existence of such strings and anti-strings implies a vanishing cosmological constant.

In Chapter 12, we study the solutions of the geometric (nonlinear) theory of electromagnetism of M. Born and L. Infeld which was introduced to accommodate a finite-energy point electric charge modelling the electron and has become one of the major focuses of recent research activities of field theoreticians due to its relevance in superstrings and supermembranes. Mathematically, the Born–Infeld theory is closely related to the minimal surface type problems and presents new opportunities and structure for analysts. We begin this chapter with a short introduction to the Born–Infeld theory and show how the theory allows the existence of finite-energy point charges, electrical or magnetical. We then discuss the electrostatic and magnetostatic problems and relate them to the minimal surface equations and the Bernstein theorems. We shall also obtain a generalized Bernstein problem expressed in terms of differential forms. We next study the Born–Infeld wave equations and show that there is no more Derrick's theorem type constraint on the spatial dimensions for the static problem. Finally we obtain multiple strings or vortices for the Born–Infeld theory coupled with a Higgs field, originally proposed in the work of K. Shiraishi and S. Hirenzaki. In particular, we show that the Born–Infeld parameter plays an important role for the behavior of solutions, both locally and globally.

Bibliographical notes on the development in other topic areas will also be made at appropriate places in the book.

In developing the subjects presented in this work, I have benefited from helpful communications, conversations, and in several cases collaborations with many mathematicians and physicists: S. Adler, J. Ambjorn, H. Brezis, L. A. Caffarelli, X. Chen, Y. M. Cho, G. Dunne, Weinan E, A. Friedman, Y. Z. Guo, S. Hastings, D. Hoffman, R. Jackiw, H. T. Ku, G. P. Li, F. H. Lin, J. B. Mcleod, E. Miller, P. Olesen, B. J. Schroers, L. M. Sibner, R. J. Sibner, T. Spencer, J. Spruck, G. Tarantello, D. H. Tchrakian, E. J. Weinberg, E. Witten, D. Yang. In particular, I wish to thank J. Spruck for initiating my interest in this area and for some fruitful joint work. Finally, I am grateful to my parents, Zhaoqi Yang and Hua Han, and my brothers, Wei Yang and Jin Yang, for their unwavering encouragement and support over the years.

I hope that this book will be useful to both mathematicians and theoretical physicists, especially those interested in nonlinear analysis and its applications.

Brooklyn, New York   Yisong Yang
July, 2000

Contents

Notation and Convention

The signature of an $(n+1)$-dimensional Minkowski spacetime is always $(+-\cdots-)$. The $(n+1)$-dimensional flat Minkowski spacetime is denoted by $\mathbb{R}^{n,1}$ and is equipped with the inner product

$$xy = x^0 y^0 - x^1 y^1 - \cdots - x^n y^n,$$

where $x = (x^0, x^1, \cdots, x^n), y = (y^0, y^1, \cdots, y^n) \in \mathbb{R}^{n,1}$ are spacetime vectors.

Unless otherwise stated, we always use the Greek letters α, β, μ, ν to denote the spacetime indices,

$$\alpha, \beta, \mu, \nu = 0, 1, 2, \cdots, n,$$

and the Latin letters i, j, k to denote the space indices,

$$i, j, k = 1, 2, \cdots, n.$$

The standard summation convention over repeated indices will be observed. For example,

$$a_i b_i = \sum_{i=1}^n a_i b_i, \quad a_i b^i = \sum_{i=1}^n a_i b^i, \quad a_i^2 = \sum_{i=1}^n a_i^2, \quad |b^i|^2 = \sum_{i=1}^n |b^i|^2.$$

Similarly,

$$F_{jk}^2 = \sum_{j,k=1}^n F_{jk}^2.$$

The roman letter e is reserved to denote the base of the natural logarithmic system and the italic letter e is reserved to denote an irrelevant physical coupling constant such as the charge of a positron ($-e$ will then be the charge of an electron), except within some special statement environment such as a theorem where italic type is used throughout, which should not cause confusion. Likewise, the roman letter i denotes the imaginary unit $\sqrt{-1}$ and the italic letter i is an integer-valued index. We use the roman type abbreviation supp to denote the support of a function. However, within some special statement environment such as a theorem, this abbreviation will also be printed in italic type.

The letter C will be used to denote a positive constant which may assume different values at different places.

For a complex number c, we use $\bar{c}$ to denote its conjugate. For a complex matrix A, we use $A^\dagger$ to denote its Hermitian conjugate, which consists of a matrix transposition and a complex conjugation.

When a system consists of several equations or relations, we often number the system by labelling its last equation or relation.

Although n denotes an integer, the symbol $\partial/\partial n$ stands for the outward normal differentiation on the boundary of a domain.

The symbol $W^{k,p}$ denotes the Sobolev space of functions whose distributional derivatives up to the kth order are all in the space L^p.

By convention, various matrix Lie algebras are denoted by lowercase letters. For example, the Lie algebras of the Lie groups $SO(N)$ and $SU(N)$ are denoted by $so(N)$ and $su(N)$, respectively.

The notation for various derivatives is as follows,

$$\partial_\mu = \frac{\partial}{\partial x^\mu}, \quad \partial_\pm = \partial_1 \pm i\partial_2, \quad \partial = \frac{1}{2}(\partial_1 - i\partial_2), \quad \bar{\partial} = \frac{1}{2}(\partial_1 + i\partial_2).$$

Besides, with the complex variable $z = x^1 + ix^2$, we always understand that $\partial_z = \partial/\partial z = \partial, \partial_{\bar{z}} = \partial/\partial\bar{z} = \bar{\partial}$. Thus, for any function f that only has partial derivatives with respect to x^1 and x^2, the quantities $\partial_z f = \partial f/\partial z$ and $\partial_{\bar{z}} = \partial f/\partial\bar{z}$ are well defined.

Vectors and tensors are often simply denoted by their general components, respectively, following physics literature. For example, it is understood that

$$A_\mu \equiv (A_\mu) = (A_0, A_1, A_2, A_3), \quad g_{\mu\nu} \equiv (g_{\mu\nu}).$$

In a volume of this scope, it is inevitable to have a letter carry different but standard meanings in different contexts, although such a multiple usage of letters has kept to a minimum. Here are some examples: r may stand for the radial variable or the rank of a Lie group; δ may stand for a small positive number or variation of a functional, δ_p stands for the Dirac distribution concentrated at the point p, and δ_{ij} is the Kronecker symbol; g may stand for a coupling constant, a metric tensor or its determinant, or a function.

1

Primer of Field Theory

The purpose of this chapter is to present a very concise introduction to the basic concepts and terminology of field theory which will be encountered in the rest of the book. We shall begin in §1.1 with a discussion of classical mechanics in view of a variational formulation and then consider the Schrödinger equation in quantum mechanics. In §1.2 we present special relativity and relativistic wave equations. In particular, we derive the Maxwell equations for electromagnetism. In §1.3 we study the role of continuous symmetry in field theory and prove Noether's theorem. In particular, we consolidate the notion of energy and momenta and introduce the concept of conserved charges and currents. In §1.4 we present the Abelian gauge field theory and related concepts such as symmetry-breaking and the Higgs mechanism. In §1.5 we discuss in general the non-Abelian gauge field theory. In §1.6 we establish Einstein's equations of gravitation and discuss some simplest consequences in cosmology.

1.1 Mechanics and Fields

In this section, we first discuss classical mechanics in terms of Lagrange's variational formulation, which serves as a foundational starting point for field theory in general. We next consider the motion of a charged particle in an electromagnetic field. We then introduce the non-relativistic Schödinger equations governing the quantum-mechanical motion of such a particle.

In particular, we will observe the natural need of using gauge-covariant derivatives and gauge fields.

1.1.1 Action principle in classical mechanics

Consider the motion of a point particle of mass m and coordinates $(q^i) = q$ in a potential field $V(q, t)$ described by Newtonian mechanics. The equations of motion are

$$m\ddot{q}^i = -\frac{\partial V}{\partial q^i}, \quad i = 1, 2, \cdots, n, \tag{1.1.1}$$

where $\cdot$ denotes time derivative. Since $-\nabla V = (-\partial V / \partial q^i)$ defines the direction along which the potential energy V decreases most rapidly, the equation (1.1.1) says that the particle is accelerated along the direction of the flow of the steepest descent of V.

With the *Lagrangian function*

$$L(q, \dot{q}, t) = \frac{1}{2} m \sum_{i=1}^{n} (\dot{q}^i)^2 - V(q, t), \tag{1.1.2}$$

(1.1.1) are simply the *Euler–Lagrange equations* of the action

$$\int_{t_1}^{t_2} L(q(t), \dot{q}(t), t)\, \mathrm{d}t \tag{1.1.3}$$

over the admissible space of trajectories $\{q(t)\,|\,t_1 < t < t_2\}$ starting and terminating at fixed points at $t = t_1$ and $t = t_2$, respectively.

The *Hamiltonian* function or *energy* at any time t is the sum of kinetic and potential energies given by

$$H = \frac{1}{2} m \sum_{i=1}^{n} (\dot{q}^i)^2 + V(q, t) = m \sum_{i=1}^{n} (\dot{q}^i)^2 - L. \tag{1.1.4}$$

Introduce the *momentum vector* $p = (p_i)$,

$$p_i = m\dot{q}^i = \frac{\partial L}{\partial \dot{q}^i}, \quad i = 1, 2, \cdots, n. \tag{1.1.5}$$

Then, in view of (1.1.4), H is defined by

$$H(q, p, t) = \sum_{i=1}^{n} p_i \dot{q}^i - L(q, \dot{q}, t) \tag{1.1.6}$$

and the equations of motion, (1.1.1), are an Hamiltonian system,

$$\dot{q}^i = \frac{\partial H}{\partial p_i}, \quad \dot{p}_i = -\frac{\partial H}{\partial q^i}, \quad i = 1, 2, \cdots, n. \tag{1.1.7}$$

The above formulations may be extend to the case when L is an arbitrary function of $q, \dot{q}$, and t. The equations of motion are the Euler–Lagrange equations of (1.1.3),

$$\frac{\mathrm{d}}{\mathrm{d}t}\left(\frac{\partial L}{\partial \dot{q}^i}\right) = \frac{\partial L}{\partial q^i}, \quad i = 1, 2, \cdots, n. \tag{1.1.8}$$

In order to make a similar Hamiltonian formulation, we are motivated from (1.1.5) to introduce the *generalized momentum vector* $p = (p_i)$ by

$$p_i = \frac{\partial L}{\partial \dot{q}^i}, \quad i = 1, 2, \cdots, n. \tag{1.1.9}$$

We still use (1.1.6) to define the corresponding Hamiltonian function H. A direct calculation shows that (1.1.8) are now equivalent to the Hamiltonian system (1.1.7).

We note that an important property of an Hamiltonian function is that it is independent of the variables $\dot{q}^i$ $(i = 1, 2, \cdots, n)$. In fact, from the definition of the generalized momentum vector given by (1.1.9), we have

$$\frac{\partial H}{\partial \dot{q}^i} = p_i - \frac{\partial L}{\partial \dot{q}^i} = 0, \quad i = 1, 2, \cdots, n.$$

This fact justifies our notation of $H(p, q, t)$ in (1.1.6) instead of $H(p, q, \dot{q}, t)$.

Let F be an arbitrary function depending on p_i, q^i $(i = 1, 2, \cdots, n)$ and time t. We see that F varies its value along a trajectory of the equations of motion, (1.1.7), according to

$$\begin{aligned}
\frac{\mathrm{d}F}{\mathrm{d}t} &= \frac{\partial F}{\partial t} + \frac{\partial F}{\partial q^i}\dot{q}^i + \frac{\partial F}{\partial p_i}\dot{p}_i \\
&= \frac{\partial F}{\partial t} + \frac{\partial F}{\partial q^i}\frac{\partial H}{\partial p_i} - \frac{\partial F}{\partial p_i}\frac{\partial H}{\partial q^i},
\end{aligned}$$

where and in the sequel we observe the summation convention over repeated indices. Thus, we are motivated to use the *Poisson bracket* $\{\cdot, \cdot\}$,

$$\{f, g\} = \frac{\partial f}{\partial p_i}\frac{\partial g}{\partial q^i} - \frac{\partial f}{\partial q^i}\frac{\partial g}{\partial p_i},$$

to rewrite the rate of change of F with respect to time t as

$$\frac{\mathrm{d}F}{\mathrm{d}t} = \frac{\partial F}{\partial t} + \{H, F\}. \tag{1.1.10}$$

In particular, when the Hamiltonian H does not depend on time t explicitly, $H = H(p, q)$, then (1.1.10) implies that

$$\frac{\mathrm{d}H}{\mathrm{d}t} = 0, \tag{1.1.11}$$

which gives the fact that energy is conserved and the mechanical system is thus called *conservative*.

It will be useful to 'complexify' our formulation of classical mechanics. We introduce the complex variables

$$u_i = \frac{1}{\sqrt{2}}(q^i + \mathrm{i}p_i), \quad i = 1, 2, \cdots, n, \quad \mathrm{i} = \sqrt{-1}.$$

Then the Hamiltonian function H depends only on $u = (u_i)$ and its complex conjugate $\bar{u} = (\bar{u}_i)$,

$$H = H(u, \bar{u}, t).$$

Hence, in terms of the differential operators,

$$\frac{\partial}{\partial u_i} = \frac{\sqrt{2}}{2}\left(\frac{\partial}{\partial q^i} - \mathrm{i}\frac{\partial}{\partial p_i}\right), \quad \frac{\partial}{\partial \bar{u}_i} = \frac{\sqrt{2}}{2}\left(\frac{\partial}{\partial q^i} + \mathrm{i}\frac{\partial}{\partial p_i}\right),$$

the Hamiltonian system (1.1.7) takes the concise form

$$\mathrm{i}\dot{u}_i = \frac{\partial H}{\partial \bar{u}_i}, \quad i = 1, 2, \cdots, n. \tag{1.1.12}$$

Again, let F be a function depending on $u, \bar{u}$, and t. Then (1.1.12) gives us

$$\begin{aligned}
\frac{\mathrm{d}F}{\mathrm{d}t} &= \frac{\partial F}{\partial t} + \frac{\partial F}{\partial u_i}\dot{u}_i + \frac{\partial F}{\partial \bar{u}_i}\dot{\bar{u}}_i \\
&= \frac{\partial F}{\partial t} - \mathrm{i}\frac{\partial F}{\partial u_i}\frac{\partial H}{\partial \bar{u}_i} + \mathrm{i}\frac{\partial F}{\partial \bar{u}_i}\frac{\partial H}{\partial u_i}.
\end{aligned}$$

With the notation

$$\{f, g\} = \frac{\partial f}{\partial u_i}\frac{\partial g}{\partial \bar{u}_i} - \frac{\partial f}{\partial \bar{u}_i}\frac{\partial g}{\partial u_i}$$

for the Poisson bracket, we have

$$\frac{\mathrm{d}F}{\mathrm{d}t} = \frac{\partial F}{\partial t} + \mathrm{i}\{H, F\}. \tag{1.1.13}$$

In particular, the complexified Hamiltonian system (1.1.12) becomes

$$\dot{u}_i = \mathrm{i}\{H, u_i\}, \quad i = 1, 2, \cdots, n, \tag{1.1.14}$$

which is in a close resemblance of the *Schrödinger equation*, in the *Heisenberg representation*, in quantum mechanics.

1.1.2 Charged particle in electromagnetic field

Consider a point particle of mass m and electric charge $-Q$ moving in an electric field $\mathbf{E}$ and a magnetic field $\mathbf{B}$, in addition to a potential field V. The equation of motion is

$$m\ddot{\mathbf{x}} = -Q(\mathbf{E} + \dot{\mathbf{x}} \times \mathbf{B}) - \nabla V, \qquad (1.1.15)$$

where $\mathbf{x} = (x^1, x^2, x^3)$ gives the location of the particle in space, $-Q\mathbf{E}$ is the electric force and $-Q\dot{\mathbf{x}} \times \mathbf{B}$ is the *Lorentz force* of the magnetic field $\mathbf{B}$ exerted on the particle of velocity $\dot{\mathbf{x}}$.

We can represent $\mathbf{B}$ and $\mathbf{E}$ by a vector potential $\mathbf{A}$ and a scalar potential Ψ as follows,

$$\mathbf{B} = \nabla \times \mathbf{A},$$

$$\mathbf{E} = -\nabla\Psi - \frac{\partial \mathbf{A}}{\partial t}.$$

Consequently, using $-A_i$ $(i = 1, 2, 3)$ to denote the components of the vector field $\mathbf{A}$ and $\mathbf{y} = (y_i) = m\dot{\mathbf{x}} = m(\dot{x}^i)$ to denote the mechanical momentum vector, the equation (1.1.15) becomes

$$
\begin{aligned}
\dot{y}_i &= Q\left(\frac{\partial \Psi}{\partial x^i} - \frac{\partial A_i}{\partial t}\right) + Q\dot{x}^j\left(\frac{\partial A_j}{\partial x^i} - \frac{\partial A_i}{\partial x^j}\right) - \frac{\partial V}{\partial x^i} \\
&= -Q\frac{dA_i}{dt} + Q\frac{\partial \Psi}{\partial x^i} + Q\dot{x}^j\frac{\partial A_j}{\partial x^i} - \frac{\partial V}{\partial x^i},
\end{aligned}
$$

which may be recast into the form

$$\frac{d}{dt}(y_i + QA_i) = \frac{\partial}{\partial x^i}(Q\Psi + Q\dot{x}^j A_j - V)$$

or

$$\frac{d}{dt}\left(\frac{\partial L}{\partial \dot{x}^i}\right) = \frac{\partial L}{\partial x^i}, \quad i = 1, 2, 3 \qquad (1.1.16)$$

if we define the function L to be

$$
\begin{aligned}
L(\mathbf{x}, \dot{\mathbf{x}}, t) &= \frac{1}{2}m(\dot{x}^i)^2 + Q\Psi + Q\dot{x}^i A_i - V \\
&= \frac{1}{2}m\dot{\mathbf{x}}^2 + Q\Psi(\mathbf{x}, t) + Q\dot{\mathbf{x}} \cdot \mathbf{A}(\mathbf{x}, t) - V(\mathbf{x}, t). \quad (1.1.17)
\end{aligned}
$$

In other words, the formula (1.1.17) gives us the Lagrangian function of the problem. It is interesting to note that the momentum vector has a correction due to the presence of the electromagnetic field through the vector potential $\mathbf{A}$,

$$p_i = \frac{\partial L}{\partial \dot{x}^i} = \dot{y}_i + QA_i, \quad i = 1, 2, 3. \qquad (1.1.18)$$

Hence the Hamiltonian function becomes

$$
\begin{aligned}
H &= p_i \dot{x}^i - L = \frac{1}{2m}\dot{y_i}^2 - Q\Psi + V \\
&= \frac{1}{2m}(p_i - QA_i)^2 - Q\Psi + V.
\end{aligned}
\tag{1.1.19}
$$

Finally, if we use $A = (A_\mu)$ $(\mu = 0, 1, 2, 3)$ to denote a vector with four components, $A = -(\Psi, \mathbf{A})$, and define the electromagnetic field tensor $F_{\mu\nu}$ by

$$
F_{\mu\nu} = \partial_\mu A_\nu - \partial_\nu A_\mu,
$$

then we have

$$
\begin{aligned}
\mathbf{B} &= (B^i), \quad B^i = -\frac{1}{2}\epsilon^{ijk}F_{jk}, \quad i, j, k = 1, 2, 3, \\
\mathbf{E} &= (E^i), \quad E^i = F_{0i}, \quad i = 1, 2, 3.
\end{aligned}
$$

In particular, the Hamiltonian function (1.1.19) takes the form

$$
H = \frac{1}{2m}(p_i - QA_i)^2 - QA_0 + V.
\tag{1.1.20}
$$

1.1.3 Schrödinger equation via first quantization

As usual, we use $\hbar$ $(= 1.054 \times 10^{-27}$ erg sec$)$ to denote the *Dirac constant* ($h = 2\pi\hbar$ is called the *Plank constant*) which is of the dimension energy $\times$ time. The first step in the process of *quantization* from classical mechanics to quantum mechanics is to make the correspondence

$$
t \to t, \quad E \to i\hbar\frac{\partial}{\partial t}, \quad q^i \to q^i \quad p_i \mapsto -i\hbar\frac{\partial}{\partial q^i}, \quad i = 1, 2, \cdots, n
\tag{1.1.21}
$$

from classical energy E and momentum vector $p = (p_i)$ to time and space derivatives. The second step is to use the correspondence (1.1.21) to translate a relation such as (1.1.4),

$$
E = H(p, q, t) = \frac{1}{2m}p_i^2 + V,
\tag{1.1.22}
$$

into an *operator equation* and understand that such an equation governs a complex-valued wave function ψ of the particle described,

$$
i\frac{\partial\psi}{\partial t} = -\frac{\hbar}{2m}\Delta\psi + V\psi,
\tag{1.1.23}
$$

which is a simplest Schrödinger equation. Without loss of generality, we will scale $\hbar$ to 1 to save notation. Hence we see that (1.1.23) has the following Hamiltonian formulation

$$
i\frac{\partial\psi}{\partial t} = \frac{\delta H}{\delta\bar{\psi}},
\tag{1.1.24}
$$

where δ denotes the Fréchet derivative and H is defined by

$$H = \int_{\mathbb{R}^3} \left\{ \frac{1}{2m} |\nabla\psi|^2 + V|\psi|^2 \right\} dx, \qquad (1.1.25)$$

which is naturally identified with the energy of the particle. In fact, with the wave function ψ and the energy operator $\hat{E}$

$$\hat{E} = -\frac{1}{2m}\Delta + V \qquad (1.1.26)$$

(see (1.1.22)), we know that the expectation value of the energy is

$$E = \int_{\mathbb{R}^3} \overline{\psi}\hat{E}\psi \, dx = \int_{\mathbb{R}^3} \left\{ -\overline{\psi}\frac{1}{2m}\Delta\psi + V|\psi|^2 \right\} dx, \qquad (1.1.27)$$

which coincides with (1.1.25) when boundary terms are neglected in integration by parts.

We note that the Schrödinger equation (1.1.24) is analogous to its classical version, (1.1.12), which is a Hamiltonian system.

We now turn our attention to the case when electromagnetic interaction is present.

From the Hamiltonian (1.1.20) and the correspondence (1.1.21), we have

$$i\frac{\partial\psi}{\partial t} = -\frac{1}{2m}(\partial_i - iQA_i)^2\psi - QA_0\psi + V\psi. \qquad (1.1.28)$$

Thus, if we introduce the gauge-covariant derivatives

$$D_\mu\psi = \partial_\mu\psi - iQA_\mu\psi, \quad \mu = 0, 1, 2, 3, \qquad (1.1.29)$$

then the *gauged Schrödinger equation* (1.1.28) assumes an elegant form,

$$iD_0\psi = \frac{1}{2m}D_i^2\psi + V\psi. \qquad (1.1.30)$$

Note that (1.1.28) or (1.1.30) is semi-quantum mechanical in the sense that the point particle of mass m is treated quantum mechanically by the Schrödinger equation but the electromagnetic field is a classical field (through the coupling of the vector potential A_μ).

We show also that (1.1.28) has a Hamiltonian formation as before.

Again, in view of the relation (1.1.20) and the correspondence (1.1.21), we have the energy operator

$$\hat{E} = -\frac{1}{2m}D_i^2 + (V - QA_0). \qquad (1.1.31)$$

As a consequence, we have by using $\partial_i(\overline{\psi}D_i\psi) = |D_i\psi|^2 + \overline{\psi}D_i^2\psi$ and neglect boundary integrals at infinity the following result,

$$\begin{aligned}
E &= \int_{\mathbb{R}^3} \overline{\psi}\hat{E}\psi \, dx = \int_{\mathbb{R}^3} \overline{\psi}\left\{ -\frac{1}{2m}D_i^2\psi + (V - QA_0)\psi \right\} dx \\
&= \int_{\mathbb{R}^3} \left\{ \frac{1}{2m}|D_i\psi|^2 + (V - QA_0)|\psi|^2 \right\} dx. \qquad (1.1.32)
\end{aligned}$$

As before, we may use (1.1.32) as an Hamiltonian, H, to arrive at the Schrödinger equation

$$i\frac{\partial \psi}{\partial t} = \frac{\delta H}{\delta \overline{\psi}} = -\frac{1}{2m}(\partial_i - iQA_i)^2\psi - QA_0\psi + V\psi. \qquad (1.1.33)$$

It is interesting to observe that, since $A = (A_\mu)$ is taken as a background field, A_0 is viewed as a part of the potential density. Another formally equivalent path to arrive at the same equation is to use the Hamiltonian

$$H = \int_{\mathbb{R}^3} \left\{ \frac{1}{2m}|D_i\psi|^2 + V|\psi|^2 \right\} dx \qquad (1.1.34)$$

and replace the ordinary temporal derivative ∂_t in (1.1.33) by the gauge-covariant derivative, D_0, and write down the Hamiltonian equation

$$iD_0\psi = \frac{\delta H}{\delta \overline{\psi}}, \qquad (1.1.35)$$

which is exactly (1.1.30).

When the electromagnetic vector potential $A = (A_\mu)$ is not treated merely as a background field, its governing equation, which involves also the wave function ψ, must be considered. In this situation, we start from the Hamiltonian (1.1.34) and regard the equations of motion for A as constraints.

1.2   Relativistic Dynamics and Electromagnetism

In this section we first introduce the Minkowski spacetime and special relativity. The relativistic energy-momentum relation leads us to the simplest quantum field equation – the Klein–Gordon wave equation. From this wave equation, we derive the simplest gauge field equations – the Maxwell equations for electromagnetism.

1.2.1   *Minkowski spacetime and relativistic mechanics*

We use the vector $x = (x^\mu) = (x^0, x^1, x^2, x^3)$ to denote a point in the *Minkowski spacetime* $\mathbb{R}^{3,1}$ where $x^0 = ct$ with c being the vacuum speed of light and t the time variable and $\mathbf{x} = (x^i) = (x^1, x^2, x^3)$ determines the position coordinates of the point in the Euclidean space $\mathbb{R}^3$. The *Minkowski metric* of $\mathbb{R}^{3,1}$ is

$$g = (g_{\mu\nu}) = \begin{pmatrix} 1 & 0 & 0 & 0 \\ 0 & -1 & 0 & 0 \\ 0 & 0 & -1 & 0 \\ 0 & 0 & 0 & -1 \end{pmatrix}. \qquad (1.2.1)$$

The inverse of g is denoted by $g^{-1} = (g^{\mu\nu})$. Then $g^{-1} = g$. We always use g or g^{-1} to lower or raise indices,

$$x_\mu = g_{\mu\nu}x^\nu, \quad x^\mu = g^{\mu\nu}x_\nu.$$

Note that $x_0 = x^0$ but $(x_i) = -(x^i)$. The coordinates x_μ of x are called covariant and x^μ, contravariant.

Observing the *summation convention* over repeated indices again, we can define the scalar product for a pair of points x and y in $\mathbb{R}^{3,1}$ by using the metric 2-tensor g,

$$xy = g_{\mu\nu}x^\mu y^\nu = x^\mu y_\mu = x^0 y^0 - x^1 y^1 - x^2 y^2 - x^3 y^3 = g^{\mu\nu}x_\mu y_\nu. \quad (1.2.2)$$

We call a vector x in $\mathbb{R}^{3,1}$ time-like, light-like (or null), or space-like if $x^2 = xx$ is positive, zero, or negative.

We now consider the motion of a massive point particle. In Newtonian mechanics, time t is absolute (t is invariant in all inertial frames) and trajectories of motion are parametrized by t. In *special relativity*, time t does not occupy such a unique position anymore and is replaced by a new variable called the *proper time* τ, which is related to the spacetime variables x^μ by the formula

$$c^2 d\tau^2 = ds^2 = g_{\mu\nu}dx^\mu dx^\nu = (dx^0)^2 - (dx^1)^2 - (dx^2)^2 - (dx^3)^2. \quad (1.2.3)$$

Since the right-hand side of (1.2.3) is of the dimension (length)^2, the variable τ is of the dimension (time) as well.

Let the trajectory of the motion of a point particle carrying (invariant or rest) mass m be given by its coordinate functions of proper time τ, $x = x(\tau)$ or $x^\mu = x^\mu(\tau)$. Since the position of this particle is defined by four coordinate functions in $\mathbb{R}^{3,1}$, we introduce a four-coordinate velocity vector u accordingly by the expression

$$u(\tau) = \frac{dx(\tau)}{d\tau}, \quad \text{or in components,} \quad u^\mu(\tau) = \frac{dx^\mu(\tau)}{d\tau}. \quad (1.2.4)$$

It is interesting to see that u is a time-like vector under the Minkowski metric,

$$u^2 = g_{\mu\nu}\frac{dx^\mu}{d\tau}\frac{dx^\nu}{d\tau} = \frac{ds^2}{d\tau^2} = c^2. \quad (1.2.5)$$

In particular, $dt/d\tau \geq 1$ and by the inverse function theorem it is always possible to use the conventional time variable t in a fixed frame to parametrize the motion instead of using the invariant parameter τ for all frames. Hence, if we denote by $\mathbf{v}$ the ordinary velocity vector of the moving particle,

$$\mathbf{v} = (v^i) = \frac{d\mathbf{x}}{dt} = \frac{d}{dt}(x^i), \quad (1.2.6)$$

then the velocity four-vector u has the components

$$u^0 = \frac{dx^0}{d\tau} = c\frac{dt}{d\tau}, \quad \mathbf{u} = (u^i) = \frac{d}{d\tau}(x^i) = \frac{u^0}{c}\frac{d}{dt}(x^i) = \frac{u^0}{c}\mathbf{v}. \qquad (1.2.7)$$

The ordinary speed of the particle is denoted by v, $v = |\mathbf{v}|$. Inserting (1.2.5) into (1.2.7), we have

$$(u^0)^2\left(1 - \frac{v^2}{c^2}\right) = c^2. \qquad (1.2.8)$$

In particular we have the important conclusion that the speed of the particle will never exceed that of light, $v < c$. From (1.2.7) and (1.2.8), we obtain

$$\frac{dt}{d\tau} = \frac{u^0}{c} = \frac{1}{\sqrt{1 - \frac{v^2}{c^2}}} = \frac{1}{\sqrt{1 - \beta^2}} = \gamma. \qquad (1.2.9)$$

In summary, the four-velocity vector u takes the form

$$u = (c\gamma, \gamma\mathbf{v}), \qquad (1.2.10)$$

which suggests that for the motion in four-dimensional Minkowski spacetime we can define the four-component 'momentum' vector

$$p = mu; \quad p_0 = \gamma mc, \quad \mathbf{p} = (p_i) = \gamma m\mathbf{v}. \qquad (1.2.11)$$

It is interesting to note that the relativistic momentum three-vector $\mathbf{p}$ reduces to that of Newtonian mechanics when the speed of light is infinite.

With the above definition of the momentum four-vector and motivated by classical mechanics, we can write down now the following relativistic equation of motion subject to an applied 'force' four-vector K,

$$\frac{dp}{d\tau} = \frac{d}{d\tau}(mu) = K, \quad K = (K_0, K_1, K_2, K_3) = (K_0, \mathbf{K}). \qquad (1.2.12)$$

To study the structure of (1.2.12), we first observe by using (1.2.5) that

$$u\frac{du}{d\tau} = \frac{d}{d\tau}\left(\frac{1}{2}u^2\right) = 0.$$

Consequently, in view of (1.2.12), we have $Ku = 0$. Namely,

$$K_0 u^0 = K_i u^i = \mathbf{K} \cdot \mathbf{u}. \qquad (1.2.13)$$

Furthermore, inserting (1.2.10) into (1.2.12), we arrive at

$$\frac{d}{dt}(mu^0) = \frac{d\tau}{dt}K_0 = \frac{K_0}{\gamma}, \qquad (1.2.14)$$

$$\frac{d}{dt}(mu) = \frac{dp}{dt} = \frac{d}{dt}(m\gamma\mathbf{v}) = \frac{d\tau}{dt}\mathbf{K} = \frac{1}{\gamma}\mathbf{K}. \qquad (1.2.15)$$

Equation (1.2.15) suggests that we identify $\mathbf{K}/\gamma$ with the ordinary force three-vector $\mathbf{F}$,

$$\mathbf{K} = \gamma \mathbf{F}. \tag{1.2.16}$$

Thus, with the velocity-dependent effective inertial mass

$$m_{\text{eff}} = m\gamma = \frac{m}{\sqrt{1 - \frac{v^2}{c^2}}}, \tag{1.2.17}$$

the equation (1.2.15) looks the same as Newton's second law of motion of classical mechanics.

Substituting (1.2.16) into (1.2.13), we obtain the temporal component of the force four-vector,

$$K_0 = \frac{\gamma}{c} \mathbf{F} \cdot \mathbf{v}. \tag{1.2.18}$$

Using (1.2.18) in (1.2.14), we find that

$$\frac{d}{dt}(m_{\text{eff}}c^2) = \frac{d}{dt}(mc^2\gamma) = \mathbf{F} \cdot \mathbf{v}. \tag{1.2.19}$$

The right-hand side of (1.2.19) of course determines the rate of work the applied force does on the moving particle. Thus the left-hand side is the rate of energy change of the particle. Consequently we arrive at the amazing total *energy formula* of Einstein known to everyone,

$$E = m_{\text{eff}}c^2 = m\gamma c^2. \tag{1.2.20}$$

In view of (1.2.20), the temporal component p_0 of the momentum four-vector p is simply $p_0 = E/c$. Hence

$$p = \left(\frac{E}{c}, \mathbf{p}\right), \tag{1.2.21}$$

which is also often called the energy-momentum vector. Besides, using (1.2.5) again, we obtain the relativistic energy-momentum relation

$$E^2 = \mathbf{p}^2 c^2 + m^2 c^4. \tag{1.2.22}$$

There is a simple relation between the two governing equations, (1.2.14) and (1.2.15): the energy equation (1.2.14) where K_0 satisfies (1.2.13) is a consequence of the momentum equation (1.2.15) subject to the constraint (1.2.5) or

$$u^0 = \sqrt{c^2 + |\mathbf{u}|^2}. \tag{1.2.23}$$

As a consequence, we may formulate the following *classical* action principle for the motion of a relativistic particle with the spatial coordinates $(x^i) = \mathbf{x}$,

$$L = \int_{t_1}^{t_2} \left\{ mc^2 \left(1 - \sqrt{1 - \frac{1}{c^2}\left(\frac{d\mathbf{x}}{dt}\right)^2}\right) + \mathbf{F} \cdot \mathbf{x} \right\} dt, \tag{1.2.24}$$

with the understanding that the temporal component functions u^0 and K_0 are to be determined through (1.2.23) and (1.2.13), respectively. Here $\mathbf{F}$ is independent of the spatial position vector $\mathbf{x}$. If $\mathbf{F}$ depends on $\mathbf{x}$, for example, if $\mathbf{F}$ is induced from a potential field $V : \mathbf{F} = -\nabla V$, then the term $\mathbf{F} \cdot \mathbf{x}$ in (1.2.24) should be replaced by $-V$.

1.2.2   Klein–Gordon fields

We now use the quantization scheme (1.1.21) in the *relativistic energy-momentum relation* (1.2.22) to arrive at the *Klein–Gordon wave equation*

$$\left(\frac{1}{c^2} \frac{\partial^2}{\partial t^2} - \Delta \right) \phi + \lambda \phi = 0, \tag{1.2.25}$$

where $\lambda = m^2 c^2 / \hbar$ and the 'wave' function ϕ may be real, complex, or vector-valued. This equation is invariant under the Lorentz transformations and is relativistic. In fact it can be rewritten intrinsically as

$$\partial^\mu \partial_\mu \phi + \lambda \phi = 0 \quad \text{or} \quad g^{\mu\nu} \partial_\mu \partial_\nu \phi + \lambda \phi = 0, \tag{1.2.26}$$

in terms of the coordinates (x^μ), where $\partial_\mu = \partial / \partial x^\mu$.

The original motivation for the introduction of the Klein–Gordon equation is to follow the success of the Schrödinger equation (1.1.23) for a non-relativistic particle to give a quantum-mechanical description of a relativistic particle. However, there are two difficulties. The first one is that a probabilistic interpretation of the wave function ϕ is no longer possible and the second one is the existence of solutions to (1.2.25) of negative energy originated from the energy-momentum relation (1.2.22) or

$$E = \pm \sqrt{\mathbf{p}^2 c^2 + mc^4}. \tag{1.2.27}$$

In fact, to obtain a correct interpretation of the equation (1.2.22), we need to study it in the context of quantum field theory. In such a situation (1.2.25) describes the field distribution of certain spin-0 particles so that ϕ is viewed as a quantum field and $|\phi|^2$ is proportional to the number of particles present.

1.2.3   Maxwell equations

In this subsection, we present a simple mathematical derivation of the *Maxwell equations*. For this purpose, we consider a sourceless electromagnetic vector field $\mathbf{F}$,

$$\nabla \cdot \mathbf{F} = 0. \tag{1.2.28}$$

From now on it will be convenient to adopt suitably normalized dimensions so that the speed of light becomes unit, $c = 1$. Since $\mathbf{F}$ propagates like light, it satisfies the wave equation

$$\left(\frac{\partial^2}{\partial t^2} - \Delta\right)\mathbf{F} = \mathbf{0},\tag{1.2.29}$$

which may also be obtained by inserting the operator realization scheme
(1.1.21) in the energy-momentum relation (1.2.22) and using the zero-mass
condition $m = 0$. In view of the identity $\text{curl}^2\mathbf{F} = \nabla(\nabla \cdot \mathbf{F}) - \Delta\mathbf{F}$ and
(1.2.28), we can rewrite (1.2.29) as

$$\left(\frac{\partial^2}{\partial t^2} + \text{curl}^2\right)\mathbf{F} = \mathbf{0}.\tag{1.2.30}$$

To solve (1.2.30), we use the imaginary number i to make the following
factorization,

$$\frac{\partial^2}{\partial t^2} + \text{curl}^2 = \left(\frac{\partial}{\partial t} - i\,\text{curl}\right)\left(\frac{\partial}{\partial t} + i\,\text{curl}\right) \equiv D^*D.\tag{1.2.31}$$

In order to use (1.2.31) to solve (1.2.30), it is necessary to assume that
the vector field $\mathbf{F}$ is also complex-valued. Since $(D^*D)\mathbf{F} = D^*(D\mathbf{F})$, it is
seen that the electromagnetic wave equation (1.2.30) may be reduced to
the first-order factor equation

$$D\mathbf{F} = \left(\frac{\partial}{\partial t} + i\,\text{curl}\right)\mathbf{F} = \mathbf{0},\tag{1.2.32}$$

and any solution of (1.2.32) is also a solution of (1.2.30). Consequently, we
may concentrate on (1.2.32).

Let $\mathbf{E}$ and $\mathbf{B}$ be two real-valued vector fields which are the real and
imaginary parts of the complex-valued vector field $\mathbf{F}$,

$$\mathbf{F} = \mathbf{E} + i\,\mathbf{B}.\tag{1.2.33}$$

Inserting (1.2.33) into (1.2.32) and using (1.2.28), we see that $\mathbf{E}$ and $\mathbf{B}$
satisfy

$$\frac{\partial\mathbf{E}}{\partial t} = \text{curl}\,\mathbf{B}, \quad \frac{\partial\mathbf{B}}{\partial t} = -\text{curl}\,\mathbf{E}, \quad \nabla\cdot\mathbf{E} = 0, \quad \nabla\cdot\mathbf{B} = 0,\tag{1.2.34}$$

which are the vacuum Maxwell equations and $\mathbf{E}$ and $\mathbf{B}$ may be identified
with the electric and magnetic fields.

We now consider the presence of source terms. Let the *dielectric coefficient* and *permeability* of the medium be denoted by ϵ and μ. Then the
electric displacement $\mathbf{D}$ and *magnetic intensity* $\mathbf{H}$ are defined by

$$\mathbf{D} = \epsilon\mathbf{E}, \quad \mathbf{B} = \mu\mathbf{H}.\tag{1.2.35}$$

In the presence of an electric charge distribution given by a density function, $\rho = \rho(x)$, we have

$$\nabla \cdot \mathbf{D} = \rho. \tag{1.2.36}$$

Integrating (1.2.36) over any given bounded domain Ω in $\mathbb{R}^3$, we have

$$q = \int_\Omega \rho \, dx = \int_{\partial\Omega} \mathbf{D} \cdot d\mathbf{S}, \tag{1.2.37}$$

which determines the total electric charge contained in the domain Ω at the present time. On the other hand, the rate of increase of charge in Ω is balanced by the current (density) $\mathbf{j}$ flowing into Ω through its surface,

$$\frac{dq}{dt} = \int_\Omega \frac{\partial\rho}{\partial t} \, dx = -\int_{\partial\Omega} \mathbf{j} \cdot d\mathbf{S}. \tag{1.2.38}$$

In view of (1.2.37) and (1.2.38), we have

$$\int_{\partial\Omega} \left(\frac{\partial\mathbf{D}}{\partial t} + \mathbf{j} \right) \cdot d\mathbf{S} = 0. \tag{1.2.39}$$

Since Ω is arbitrary, the divergence theorem implies that the vector field

$$\frac{\partial\mathbf{D}}{\partial t} + \mathbf{j}$$

is solenoidal and there is a vector field $\mathbf{M}$ such that

$$\frac{\partial\mathbf{D}}{\partial t} + \mathbf{j} = \operatorname{curl} \mathbf{M}. \tag{1.2.40}$$

Recall that the dielectric coefficient ϵ and permeability μ satisfy

$$\epsilon\mu = \frac{1}{c^2} = 1. \tag{1.2.41}$$

Using (1.2.35) and (1.2.41), setting $\mathbf{j} = \mathbf{0}$ in (1.2.40), and comparing it with the first equation in (1.2.34), we are led to the conclusion $\mathbf{M} = \mathbf{H}$. Consequently we arrive at the following fundamental equations of Maxwell in presence of sources,

$$\frac{\partial\mathbf{D}}{\partial t} + \mathbf{j} = \operatorname{curl} \mathbf{H}, \qquad \nabla \cdot \mathbf{D} = \rho, \tag{1.2.42}$$

$$\frac{\partial\mathbf{B}}{\partial t} = -\operatorname{curl} \mathbf{E}, \qquad \nabla \cdot \mathbf{B} = 0. \tag{1.2.43}$$

In view of (1.2.35) and (1.2.41) again, we may also rewrite (1.2.42) as

$$\frac{\partial\mathbf{E}}{\partial t} + \mathbf{j} = \operatorname{curl} \mathbf{B}, \qquad \nabla \cdot \mathbf{E} = \rho, \tag{1.2.44}$$

where we have used the rescaling $\rho \mapsto \epsilon\rho, \mathbf{j} \mapsto \epsilon\mathbf{j}$ for the charge and current densities.

In view of the quantum realization (1.1.21) and wave-particle duality, the Maxwell fields are already quantum fields. This is why the quantization of the Maxwell fields, or more generally, gauge fields, is often called the *second quantization*.

1.3   Scalar Fields and Symmetry

We first present a variational formulation for scalar fields and write down their equations of motion. We then state and prove Noether's theorem and introduce the associated energy-momentum tensor, conserved charges, and currents. We also state the Derrick theorem concerning the existence of static solutions.

1.3.1   Variational formalism

Consider a general Lagrange action functional

$$S = \int \mathcal{L}(x; u, Du) \, \mathrm{d}x, \tag{1.3.1}$$

where $x = (x^\mu) = (x^0, x^1, \cdots, x^n)$ stands for the coordinates in an $(n+1)$-dimensional spacetime, Euclidean or Minkowskian,

$$u = (u^a(x)) = (u^1(x), u^2(x), \cdots, u^m(x))$$

is an m-component scalar field, $Du = (\partial_0 u, \partial_1 u, \cdots, \partial_n u)$, and $\partial_\mu = \partial/\partial x^\mu$. It is easily seen that the Euler–Lagrange (variational) equations of (1.3.1), or $\delta S = 0$, are

$$\frac{\partial \mathcal{L}}{\partial u^a} - \frac{\partial}{\partial x^\mu} \left\{ \frac{\partial \mathcal{L}}{\partial(\partial_\mu u^a)} \right\} = 0, \qquad a = 1, 2, \cdots, m. \tag{1.3.2}$$

For example, the Klein–Gordon equation (1.2.26) may be obtained from the Lagrange density

$$\mathcal{L} = \frac{1}{2}\partial^\mu u \cdot \partial_\mu u - \frac{1}{2}\lambda u^2 \quad \text{or} \quad \mathcal{L} = \frac{1}{2}\partial^\mu u \cdot \partial_\mu \bar{u} - \frac{1}{2}\lambda u\bar{u}, \tag{1.3.3}$$

depending on whether u is real or complex.

Comparing this formalism with the discussion of the Lagrange mechanics in §1.1, we see that $(u^a(x))$ is analogous to the generalized coordinates (q_i) there. In particular, the Euler–Lagrange equations of motion (1.3.2) resemble (1.1.8). Therefore, in view of (1.1.5), we are motivated to introduce the following *canonical momentum density*,

$$\pi_a(x) = \frac{\partial \mathcal{L}}{\partial(\partial_0 u^a)}, \qquad a = 1, 2, \cdots, m, \tag{1.3.4}$$

where x^0 is identified with the time variable t. Hence, in analogy with (1.1.6), we can write our Hamiltonian (energy) density as

$$\mathcal{H} = \pi_a \partial_0 u^a - \mathcal{L}. \tag{1.3.5}$$

Integrating (1.3.5) over the domain $\mathbb{R}^n$ of the spatial variables, x^i, $i = 1, 2, \cdots, n$, we have the total energy

$$E = H = \int_{\mathbb{R}^n} \mathcal{H} \, dx. \qquad (1.3.6)$$

With this definition, denoting ∂_0 by ∂_t, and the Fréchet functional derivative by δ, it can be checked that the equations of motion (1.3.2) take the form

$$\frac{\partial u^a}{\partial t} = \frac{\delta H}{\delta \pi_a}, \quad \frac{\partial \pi_a}{\partial t} = -\frac{\delta H}{\delta u^a}, \quad a = 1, 2, \cdots, m, \qquad (1.3.7)$$

which can be viewed as an infinite-dimensional extension of the system (1.1.7).

As a simple example, we see in view of (1.3.5) that the Klein–Gordon action density (1.3.3) has the positive definite energy density function

$$\mathcal{H} = \frac{1}{2}|\partial_t u|^2 + \frac{1}{2}|\nabla u|^2 + \frac{1}{2}\lambda|u|^2. \qquad (1.3.8)$$

1.3.2   Noether's theorem and conserved quantities

Suppose that spacetime and the field u are transformed by a Lie group G,

$$x \mapsto x', \quad u(x) \mapsto u'(x'). \qquad (1.3.9)$$

Note that x and x' refer to two sets of coordinates of the same spacetime point. *Noether's theorem* states that the equations of motion (1.3.2) have n conservation laws associated with the group G if the Lagrange action (1.3.1) evaluated over any spacetime domain is invariant under G, where n is the dimension of G. For example, energy conservation can be derived as a consequence of the translational invariance of (1.3.1) in the time variable,

$$t' = t + \varepsilon, \quad u'(x') = u(t', \mathbf{x}).$$

Hence Noether's theorem identifies conserved quantities by the symmetry of the system and is of fundamental importance.

If the action (1.3.1) is invariant under G, it has vanishing variation $\delta S = 0$ with

$$\delta S = \int \mathcal{L}(x'; u'(x'), D'u'(x')) \, dx' - \int \mathcal{L}(x; u(x), Du(x)) \, dx$$

$$= \int \{\mathcal{L}(x'; u'(x'), D'u'(x'))J(x'; x) - \mathcal{L}(x; u(x), Du(x))\} \, dx$$

$$\equiv \int \{(\delta\mathcal{L})J(x'; x) + \mathcal{L}(x; u(x), Du(x))(J(x'; x) - 1)\} \, dx, \quad (1.3.10)$$

where D' is the differentiation with respect to the variable x', J is the Jacobian of the transformation $x' = x'(x)$, and the integrals are carried out over an arbitrary spacetime domain. In order to evaluate (1.3.10), we use δx, $\delta u(x)$, and $\delta(\partial_\mu u(x))$ to denote the total variations,

$$\delta x = x' - x, \quad \delta u(x) = u'(x') - u(x), \quad \delta(\partial_\mu u(x)) = \partial'_\mu u'(x') - \partial_\mu u(x),$$

where $\partial'_\mu = \partial/\partial x'^\mu$. We also introduce the modified variation $\tilde{\delta}$ by the expression

$$\tilde{\delta} u = u'(x) - u(x). \tag{1.3.11}$$

It is clear from (1.3.11) that $\tilde{\delta}$ and ∂_μ commute,

$$\partial_\mu(\tilde{\delta} u(x)) = \tilde{\delta}(\partial_\mu u(x)). \tag{1.3.12}$$

However, the total variation δ does not satisfy such a property,

$$\begin{aligned}
\partial_\mu(\delta u(x)) &= \partial_\mu(u'(x')) - \partial_\mu(u(x)) \\
&= \{\partial'_\mu(u'(x')) - \partial_\mu(u(x))\} + \{\partial_\mu(u'(x')) - \partial'_\mu(u'(x'))\} \\
&= \delta(\partial_\mu u(x)) + \partial_\mu(\delta x^\nu)\partial'_\nu(u'(x')) \\
&= \delta(\partial_\mu u(x)) + \partial_\mu(\delta x^\nu)\partial_\nu(u(x)). \tag{1.3.13}
\end{aligned}$$

Furthermore, by expanding the determinant, we see that

$$J(x'; x) = 1 + \partial_\mu(\delta x^\mu) \tag{1.3.14}$$

up to the first order of the variation δ.

We now estimate $\delta\mathcal{L}$ up to the first order of variation to get

$$\begin{aligned}
\delta\mathcal{L} &= \mathcal{L}(x'; u'(x'), D'u'(x')) - \mathcal{L}(x; u(x), Du(x)) \\
&= \{\mathcal{L}(x; u'(x), Du'(x)) - \mathcal{L}(x; u(x), Du(x))\} \\
&\quad + \{\mathcal{L}(x'; u'(x'), D'u'(x')) - \mathcal{L}(x; u'(x), Du'(x))\} \\
&= \frac{\partial\mathcal{L}}{\partial u^a}\tilde{\delta} u^a + \frac{\partial\mathcal{L}}{\partial(\partial_\mu u^a)}\tilde{\delta}(\partial_\mu u^a) + \partial_\mu(\mathcal{L}(x; u(x), Du(x))(\delta x^\mu) \\
&= \partial_\mu\left(\frac{\partial\mathcal{L}}{\partial(\partial_\mu u^a)}\right)(\tilde{\delta} u^a) + \frac{\partial\mathcal{L}}{\partial(\partial_\mu u^a)}\partial_\mu(\tilde{\delta} u^a) + \partial_\mu(\mathcal{L})(\delta x^\mu),
\end{aligned}$$

where we have used the equations of motion (1.3.2) and the commutativity (1.3.12).

Inserting (1.3.14) and the above result into (1.3.10), we obtain

$$\delta S = \int \left\{\partial_\mu\left(\frac{\partial\mathcal{L}}{\partial(\partial_\mu u^a)}\tilde{\delta} u^a + \mathcal{L}\delta x^\mu\right)\right\} dx. \tag{1.3.15}$$

On the other hand, from (1.3.11), we have

$$\begin{aligned}
\tilde{\delta} u(x) &= (u'(x') - u(x)) - (u'(x') - u'(x)) \\
&= \delta u(x) - (\partial_\mu u'(x))\delta x^\mu \\
&= \delta u(x) - (\partial_\mu u(x))\delta x^\mu. \tag{1.3.16}
\end{aligned}$$

Substituting (1.3.16) into (1.3.15), we arrive at the following expression,

$$\delta S = \int \left\{ \partial_\mu \left(\frac{\partial \mathcal{L}}{\partial(\partial_\mu u^a)} (\delta u^a - \partial_\nu u^a \delta x^\nu) + \mathcal{L} \delta x^\mu \right) \right\} dx$$

$$= \int \left\{ \partial_\mu \left(\frac{\partial \mathcal{L}}{\partial(\partial_\mu u^a)} \delta u^a - \left[\frac{\partial \mathcal{L}}{\partial(\partial_\mu u^a)} \partial_\nu u^a - \delta^\mu_\nu \mathcal{L} \right] \delta x^\nu \right) \right\} dx. \tag{1.3.17}$$

Let the Lie algebra of G be parametrized by $\{\omega^s\}$. Then we can write δx^μ and δu as

$$\delta x^\mu = X^\mu_s \delta \omega^s, \quad \delta u^a = U^a_s \delta \omega^s. \tag{1.3.18}$$

Inserting (1.3.18) into (1.3.17), we find that

$$\delta S = \int \left\{ \partial_\mu \left(\frac{\partial \mathcal{L}}{\partial(\partial_\mu u^a)} U^a_s - T^\mu_\nu X^\nu_s \right) \right\} \delta \omega^s \, dx, \tag{1.3.19}$$

where T^ν_μ is the *energy-momentum tensor* defined by

$$T^\nu_\mu = \frac{\partial \mathcal{L}}{\partial(\partial_\nu u^a)} \partial_\mu u^a - \delta^\nu_\mu \mathcal{L}. \tag{1.3.20}$$

Using the metric tensor to raise or lower indices, we have the pair

$$T_{\mu\nu} = \frac{\partial \mathcal{L}}{\partial(\partial^\mu u^a)} \partial_\nu u^a - g_{\mu\nu} \mathcal{L},$$

$$T^{\mu\nu} = \frac{\partial \mathcal{L}}{\partial(\partial_\nu u^a)} \partial^\mu u^a - g^{\mu\nu} \mathcal{L}.$$

Since the variations $\delta \omega^s$ ($s = 1, 2, \cdots, n$) are arbitrary, (1.3.19) (with $\delta S = 0$) gives us

$$\int \partial_\mu J^\mu_s \, dx = 0, \quad s = 1, 2, \cdots, n, \tag{1.3.21}$$

where

$$J^\mu_s = \frac{\partial \mathcal{L}}{\partial(\partial_\mu u^a)} U^a_s - T^\mu_\nu X^\nu_s, \quad s = 1, 2, \cdots, n. \tag{1.3.22}$$

However, since the domain of integration in (1.3.21) is arbitrary, we conclude that

$$\partial_\mu J^\mu_s = 0, \quad s = 1, 2, \cdots, n \tag{1.3.23}$$

everywhere. These are the n conservation laws stated in Noether's theorem.

Consider the n *Noether current* four-vector defined in (1.3.22),

$$(J^\mu_s) = (\rho_s, \mathbf{j}_s), \quad s = 1, 2, \cdots, n,$$

where ρ_s is viewed as a *charge* density. Then we have the conservation of total charge q_s for each s,

$$q_s = \int_{\mathbb{R}^3} \rho_s \, dx \tag{1.3.24}$$

because (1.3.23) implies that

$$\frac{dq_s}{dt} = \int_{\mathbb{R}^3} \partial_0 \rho_s \, dx = \int_{\mathbb{R}^3} \nabla \cdot \mathbf{j}_s \, dx = 0. \tag{1.3.25}$$

One of the most importance situations is that the system is invariant under translations,

$$x'^\mu = x^\mu + \omega^\mu, \quad u'(x') = u(x). \tag{1.3.26}$$

Inserting (1.3.26) into (1.3.18) gives us $U_\mu^a = 0$ and $X_\mu^\nu = \delta_\mu^\nu$. Hence the Noether current gives rise to the energy-momentum tensor, $J_\nu^\mu = -T_\nu^\mu$, and the conservation laws

$$\partial_\mu T_\nu^\mu = 0, \quad \partial^\mu T_{\mu\nu} = 0; \quad \partial^\mu T_\mu^\nu = 0, \quad \partial_\mu T^{\mu\nu} = 0; \quad \nu = 0, 1, 2, 3.$$

In particular, by our discussion on special relativity (see (1.2.21)), we can immediately recognize that the conserved charges $\rho_\nu = J_\nu^0$ ($\nu = 0, 1, 2, 3$) correspond to the energy, E, and momentum vector, $\mathbf{P}$, because in view of (1.3.5) the component function T_0^0 is the Hamiltonian density,

$$(P_\nu) = (E, \mathbf{P}),$$

$$E - \int_{\mathbb{R}^3} T_0^0 \, dx,$$

$$P_i = \int_{\mathbb{R}^3} T_i^0 \, dx. \tag{1.3.27}$$

Of course, these quantities are constants.

1.3.3 Static solutions and Derrick's theorem

Following the discussion of §1.3.1, we study the field theory described by the Lagrange action density

$$\mathcal{L} = \frac{1}{2} \partial^\mu u \cdot \partial_\mu u - V(u) \tag{1.3.28}$$

over the Minkowski spacetime $\mathbb{R}^{n,1}$, where $V \geq 0$ is a general potential density function. It is easily seen that the equations of motion (1.3.2) for static solutions become

$$\Delta u = \frac{\delta V}{\delta u}, \tag{1.3.29}$$

which is the system of the Euler–Lagrange equations of the Hamilton energy

$$E(u) = \int_{\mathbb{R}^n} \left\{ \frac{1}{2} |\nabla u|^2 + V(u) \right\} dx. \tag{1.3.30}$$

Hence we are to consider finite-energy critical points of the functional (1.3.30).

Suppose that u is a critical point of (1.3.30). Then $u_\lambda(x) = u(\lambda x)$ is a critical point as well when $\lambda = 1$, which leads us to the assertion

$$\left\{ \frac{\mathrm{d}}{\mathrm{d}\lambda} E(u_\lambda) \right\}\bigg|_{\lambda=1} = 0. \tag{1.3.31}$$

On the other hand, if we use x_λ to denote λx and ∇_λ to denote the gradient operator with derivatives in terms of differentiation in x_λ, then

$$
\begin{aligned}
E(u_\lambda) &= \int_{\mathbb{R}^n} \left\{ \frac{1}{2} |\nabla u_\lambda|^2 + V(u_\lambda) \right\} \mathrm{d}x \\
&= \int_{\mathbb{R}^n} \left\{ \frac{1}{2} \lambda^2 |\nabla_\lambda u_\lambda|^2 + V(u_\lambda) \right\} \mathrm{d}x \\
&= \int_{\mathbb{R}^n} \left\{ \frac{1}{2} \lambda^2 |\nabla_\lambda u(x_\lambda)|^2 + V(u(x_\lambda)) \right\} \lambda^{-n} \mathrm{d}x_\lambda \\
&= \int_{\mathbb{R}^n} \left\{ \frac{1}{2} \lambda^{2-n} |\nabla u|^2 + \lambda^{-n} V(u) \right\} \mathrm{d}x.
\end{aligned} \tag{1.3.32}
$$

Combining (1.3.31) and (1.3.32), we obtain the identity

$$(2-n) \int_{\mathbb{R}^n} |\nabla u|^2 \, \mathrm{d}x = 2n \int_{\mathbb{R}^n} V(u) \, \mathrm{d}x. \tag{1.3.33}$$

Consequently, we see that there is no nontrivial solution if $n \geq 3$. This statement is known as the *Derrick theorem* [89]. Besides, the case $n = 2$ is interesting only in the absence of potential energy, $V = 0$. In this case the model is called the σ-model (harmonic maps). Finally, when $n = 1$, the potential density function V is not subject to any restriction and locally concentrated static solutions are often called kinks or domain walls.

1.4   Gauge Field Theory

In this section, we introduce gauge fields as a result of maintaining local internal symmetry. For simplicity, we shall concentrate on the Abelian case. We recover the Maxwell equations as gauge field equations. We discuss the notion of spontaneous symmetry breaking, the Goldstone particles, and the Higgs mechanism.

1.4.1   *Local symmetry and gauge fields*

To see the necessity for the introduction of gauge fields, we focus on a complex scalar field ϕ governed by the same Lagrange action density (1.3.28), namely,

$$\mathcal{L} = \frac{1}{2} \partial_\mu \phi \partial^\mu \overline{\phi} - V(|\phi|^2). \tag{1.4.1}$$

The Derrick theorem states that the equations of motion of (1.4.1) do not have nontrivial finite-energy static solutions when space dimension is larger than two, which rules out a possible use of the field ϕ to describe a system of particles in equilibrium and is a drawback. Another equally serious drawback comes from the internal symmetry of the model as we will now see.

It is clear that (1.4.1) is invariant under the following phase change for the field ϕ,

$$\phi(x) \mapsto e^{i\omega}\phi(x), \tag{1.4.2}$$

where ω is a real constant. Such a symmetry is called a *global* symmetry because it simply says that an everywhere identical phase shift for the field ϕ does not change anything. However, when this global symmetry is *enlarged* to a *local* one for which ω becomes a function of the spacetime coordinates, $\omega = \omega(x)$, we need to replace the ordinary derivatives by *covariant* derivatives of the form

$$D_\mu\phi = \partial_\mu\phi - ieA_\mu\phi, \quad \mu = 0, 1, 2, 3, \tag{1.4.3}$$

where $A = (A_\mu)$ is a vector field, and replace the action density (1.4.1) by

$$\mathcal{L} = \frac{1}{2}D_\mu\phi\overline{D^\mu\phi} - V(|\phi|^2). \tag{1.4.4}$$

In order that (1.4.4) is invariant under the required local (x-dependent) phase shift

$$\phi(x) \mapsto e^{i\omega(x)}\phi(x), \tag{1.4.5}$$

it suffices to make $D_\mu\phi$ change itself covariantly,

$$D_\mu\phi \mapsto e^{i\omega(x)}D_\mu\phi. \tag{1.4.6}$$

Using (1.4.3) in (1.4.6), we see that the vector field $A = (A_\mu)$ should obey the transformation rule

$$A_\mu \mapsto A_\mu + \frac{1}{e}\partial_\mu\omega. \tag{1.4.7}$$

The action density (1.4.4) is now invariant under the *gauge transformation* defined by (1.4.5) and (1.4.7). However, (1.4.4) does not contain any derivative terms of the gauge vector field A, which are necessary for dynamics and must be added in the model. Hence we are motivated to introduce suitable terms containing

$$F_{\mu\nu} = \partial_\mu A_\nu - \partial_\nu A_\mu, \quad \mu, \nu = 0, 1, 2, 3, \tag{1.4.8}$$

which are invariant under (1.4.7). Another motivation for the introduction of $F_{\mu\nu}$ is that it is a 'curvature 2-tensor', which measures the non-commutativity of the covariant derivatives defined in (1.4.3),

$$D_\mu D_\nu\phi - D_\nu D_\mu\phi = -ieF_{\mu\nu}\phi. \tag{1.4.9}$$

The simplest Lorentz scalar containing the field strength tensor $F_{\mu\nu}$ is of the form

$$F_{\mu\nu}F^{\mu\nu} = F_{\mu\nu}g^{\mu\mu'}g^{\nu\nu'}F_{\mu'\nu'}.$$

Adding the above (after a numerical rescaling) to (1.4.4), we arrive at the complete Lagrange action density

$$\mathcal{L} = -\frac{1}{4}F_{\mu\nu}F^{\mu\nu} + \frac{1}{2}D_\mu\phi\overline{D^\mu\phi} - V(|\phi|^2), \qquad (1.4.10)$$

which is invariant under the gauge and Lorentz transformations and contains all necessary dynamical terms for the fields ϕ and A_μ.

It is interesting to write down the Euler–Lagrange equations of (1.4.10),

$$D_\mu D^\mu \phi = 2V'(|\phi|^2)\phi, \qquad (1.4.11)$$
$$\partial_\nu F^{\mu\nu} = -J^\mu, \qquad (1.4.12)$$

where $V'(s) = dV/ds$ and

$$J^\mu = \frac{i}{2}e(\overline{\phi}D^\mu\phi - \phi\overline{D^\mu\phi}) \qquad (1.4.13)$$

is the Noether current associated with the continuous symmetry (1.4.2). In fact, the conservation law $\partial_\mu J^\mu = 0$ is an easy consequence of the equation (1.4.11).

Recall that $J^0 = \rho$ is a conserved charge density and $\mathbf{j} = (J^i)$ is a current density. We now identify the components of $F^{\mu\nu}$ with the electric field $\mathbf{E} = (E^i)$ and magnetic field $\mathbf{B} = (B^i)$ such that

$$F^{0i} = -E^i, \quad F^{ij} = -\varepsilon^{ijk}B^k, \qquad (1.4.14)$$

or in the matrix form,

$$(F^{\mu\nu}) = \begin{pmatrix} 0 & -E^1 & -E^2 & -E^3 \\ E^1 & 0 & -B^3 & B^2 \\ E^2 & B^3 & 0 & -B^1 \\ E^3 & -B^2 & B^1 & 0 \end{pmatrix}. \qquad (1.4.15)$$

Hence $F^{\mu\nu}$ is also called the electromagnetic field tensor. The $\mu = 0$ component of (1.4.12) is simply $\nabla \cdot \mathbf{E} = \rho$ which suggests that

$$\rho = \frac{i}{2}e(\overline{\phi}D^0\phi - \phi\overline{D^0\phi}) \qquad (1.4.16)$$

should be identified with the electric charge density. It is important to note from (1.4.16) that, if ϕ is real-valued, there is no electric charge and thus ϕ describes electrically *neutral* particles. The other spatial components, $\mu = 1, 2, 3$, correspond to $\partial\mathbf{E}/\partial t + \mathbf{j} = \text{curl } \mathbf{B}$. Hence we have recovered a part of the Maxwell equations, (1.2.44). On the other hand, by the definition (1.4.8), we have the *Bianchi identity*

$$\partial^\gamma F^{\mu\nu} + \partial^\mu F^{\nu\gamma} + \partial^\nu F^{\gamma\mu} = 0, \tag{1.4.17}$$

which may be recast into the form

$$\partial_\mu \tilde{F}^{\mu\nu} = 0, \tag{1.4.18}$$

where $\tilde{F}^{\mu\nu}$ is the dual tensor of $F^{\mu\nu}$ defined by

$$\tilde{F}^{\mu\nu} = \frac{1}{2}\varepsilon^{\mu\nu\alpha\beta} F_{\alpha\beta}. \tag{1.4.19}$$

In terms of the fields **E** and **B**, we have

$$(\tilde{F}^{\mu\nu}) = \begin{pmatrix} 0 & -B^1 & -B^2 & -B^3 \\ B^1 & 0 & E^3 & -E^2 \\ B^2 & -E^3 & 0 & E^1 \\ B^3 & E^2 & -E^1 & 0 \end{pmatrix}. \tag{1.4.20}$$

Inserting (1.4.20) into the identity (1.4.18), we obtain another part of the Maxwell equations, (1.2.43).

Hence (1.4.12) coincides with the Maxwell equations. In other words, the Maxwell equations can be introduced from the need of maintaining local symmetry through a gauge field. Classically, gauge field is a potential field and is not observable. However, quantum-mechanically, gauge field can be observed in topologically nontrivial situations. This phenomenon is known as the *Bohm–Aharonov effect*.

We can also represent the Maxwell equations by the covariant tensor field $F_{\mu\nu}$,

$$F_{\mu\nu} = \begin{pmatrix} 0 & E^1 & E^2 & E^3 \\ -E^1 & 0 & -B^3 & B^2 \\ -E^2 & B^3 & 0 & -B^1 \\ -E^3 & -B^2 & B^1 & 0 \end{pmatrix}. \tag{1.4.21}$$

Then, with the dual $\tilde{F}_{\mu\nu}$ defined by

$$\tilde{F}_{\mu\nu} = \frac{1}{2}\varepsilon_{\mu\nu\alpha\beta} F^{\alpha\beta}, \tag{1.4.22}$$

the Maxwell equations (1.2.43) and (1.2.44) take the form

$$\partial^\mu F_{\mu\nu} = -J_\nu, \quad \partial^\mu \tilde{F}_{\mu\nu} = 0. \tag{1.4.23}$$

We consider again the Derrick theorem over the Minkowski space $\mathbb{R}^{n,1}$. For a finite-energy static solution (ϕ, A) of (1.4.11) and (1.4.12) in the *temporal gauge*

$$A_0 = 0, \tag{1.4.24}$$

we may use the same argument as in §1.3.3 with $\phi_\lambda(x) = \phi(\lambda x)$ and $A_\lambda(x) = \lambda A(\lambda x)$ to arrive at the new identity

$$(4 - n) \int_{\mathbb{R}^n} F_{ij}^2 \, \mathrm{d}x + 2(2 - n) \int_{\mathbb{R}^n} |D_i \phi|^2 \, \mathrm{d}x = 4n \int_{\mathbb{R}^n} V(|\phi|^2) \, \mathrm{d}x \quad (1.4.25)$$

instead of (1.3.33) because the Hamiltonian density is now of the form

$$\mathcal{H} = \frac{1}{4} F_{ij}^2 + \frac{1}{2} |D_i \phi|^2 + V(|\phi|^2) \quad (1.4.26)$$

(summations over repeated indices $i, j = 1, 2, \cdots, n$ are made) so that the static solutions of (1.4.11) and (1.4.12) are critical points of the energy $E = \int_{\mathbb{R}^n} \mathcal{H} \, \mathrm{d}x$. Therefore, with a gauge field, the allowance of spatial dimensions is extended to $n \leq 4$. In particular, in the borderline case $n = 4$, the matter field coupling must be trivial, $\phi = 0$ and $V = 0$, and we are left with a pure gauge field model which is analogous to the harmonic map model without a gauge field in the case $n = 2$ discussed earlier.

In conclusion, we have seen that the local symmetry requirement makes it necessary to incorporate a gauge field which serves as a mediating field for interacting particles and increases spatial dimensions for the existence of static solutions.

1.4.2 Low temperature and spontaneous symmetry-breaking

We begin by considering the static solutions of the model (1.4.1) with temperature dependence. The simplest situation is that the potential density V is of the form

$$V(|\phi|^2) = \frac{1}{2} m^2(T) |\phi|^2 + \frac{1}{8} \lambda |\phi|^4, \quad (1.4.27)$$

where m^2 is a function of the temperature T, which is typically of the form

$$m^2(T) = a \left(\left[\frac{T}{T_c} \right]^2 - 1 \right), \quad (1.4.28)$$

$\lambda, a > 0$ are suitable parameters, and $T_c > 0$ is a critical temperature.

Vacuum solutions, or ground states, are the lowest energy static solutions. In high temperature, $T > T_c$, we have $m^2(T) > 0$ and, in view of the Klein–Gordon equation (1.2.25), the quantity $m(T) > 0$ is the mass of two real scalar particles represented by ϕ_1 and ϕ_2 where $\phi = \phi_1 + i\phi_2$ and higher order terms of ϕ_1 and ϕ_2 describe self-interactions. The only minimum of the Hamiltonian density

$$\mathcal{H} = \frac{1}{2} |\nabla \phi|^2 + \frac{1}{2} m^2(T) |\phi|^2 + \frac{1}{8} \lambda |\phi|^4 \quad (1.4.29)$$

is

$$\phi_{\mathrm{v}} = 0, \quad (1.4.30)$$

which is the unique vacuum state of the problem. This vacuum state is of course invariant under the $U(1)$-symmetry group (1.4.2).

In general, given a symmetry group, the Lagrangian density should be invariant if the vacuum state is already invariant, based on some consideration from quantum field theory. Such a statement is known as the *Coleman theorem* (the invariance of the vacuum state implies the invariance of the universe). If both the vacuum state and the Lagrangian density are invariant, we say that there is *exact* symmetry. If the vacuum state is non-invariant, the Lagrangian density may be non-invariant or invariant. In both cases, we say that the symmetry as a whole is broken. The former case is referred to as *explicit* symmetry-breaking and the latter case is referred to as *spontaneous* symmetry-breaking, which is one of the fundamental phenomena in low-temperature physics.

To see this, we assume that $T < T_c$. From (1.4.28) we have $m^2(T) < 0$ and we see that there is a phase transition: although $\phi = 0$ is still a solution but it is no longer stable. In fact the minimum of (1.4.29) is attained instead at any of the configurations

$$\phi_{v,\theta} = \phi_0 e^{i\theta}, \quad \phi_0 = \sqrt{\left(1 - \left[\frac{T}{T_c}\right]^2\right)\frac{2a}{\lambda}} > 0, \quad \theta \in \mathbb{R}, \qquad (1.4.31)$$

which give us a continuous family of distinct vacuum states (a circle) labelled by θ and quantum tunnelling from a state to another is impossible. Since for $\omega \neq 2k\pi$ (1.4.2) transforms any given vacuum state, $\phi_{v,\theta}$, to a different one, $\phi_{v,\theta+\omega}$, we observe the non-invariance of vacuum states although the Lagrangian density (1.4.1) is still invariant. Consequently, the symmetry is spontaneously broken. The quantity ϕ_0 measures the scale of the broken symmetry.

1.4.3 Goldstone particles and Higgs mechanism

We continue to consider the system at low temperature, $T < T_c$. Since $m^2(T) < 0$, it seems that we would have particles of imaginary mass. However, this is not the case as will be seen below.

In fact, the Lagrangian density (1.4.1) governs fluctuations around vacuum state. For $T > T_c$, the vacuum state is the zero state and $m(T)$ in (1.4.27) clearly defines mass. For $T < T_c$, we need to consider fluctuations around a given nonzero vacuum state, say ϕ_0, represented by two real scalar functions ϕ_1 and ϕ_2,

$$\phi(x) = \phi_0 + \phi_1(x) + i\phi_2(x). \qquad (1.4.32)$$

In this case, the minimum of (1.4.27) is strictly negative,

$$V(\phi_0^2) = \frac{1}{2}m^2(T)\phi_0^2 + \frac{1}{8}\lambda\phi_0^4 = -\frac{1}{8}\lambda\phi_0^4. \qquad (1.4.33)$$

Thus, in order to maintain finite energy in an unbounded space, we need to shift the potential energy density (1.4.27) by the quantity given in (1.4.33),

$$V \mapsto V + \frac{1}{8}\lambda\phi_0^4 = \frac{\lambda}{8}(|\phi|^2 - \phi_0^2)^2, \qquad (1.4.34)$$

and the new minimum energy level is zero. Inserting (1.4.32) into (1.4.1) and using the potential (1.4.34), we have the Lagrangian density

$$\begin{aligned} \mathcal{L} &= \frac{1}{2}\partial_\mu\phi_1\partial^\mu\phi_1 + \frac{1}{2}\partial_\mu\phi_2\partial^\mu\phi_2 - \frac{\lambda}{2}\phi_0^2\phi_1^2 \\ &\quad - \frac{\lambda}{8}(\phi_1^4 + \phi_2^4 + 2\phi_1^2\phi_2^2 + 4\phi_0\phi_1^3 + 4\phi_0\phi_1\phi_2^2) \end{aligned} \qquad (1.4.35)$$

governing the scalar fields ϕ_1 and ϕ_2 fluctuating around the vacuum state, $\phi_1 = 0, \phi_2 = 0$. The coefficient of ϕ_1^2 defines the mass of the ϕ_1-particles,

$$m_1 = \sqrt{\lambda}\phi_0 = \sqrt{2}|m(T)| > 0. \qquad (1.4.36)$$

However, since the ϕ_2^2 term is absent in (1.4.35) (the higher=order terms describe interactions), these ϕ_2-particles are massless and are called the *Goldstone* particles. Hence, spontaneous symmetry-breaking leads to the presence of the Goldstone particles, namely, particles of zero mass instead of particles of imaginary mass. This statement is known as the *Goldstone theorem*.

The Goldstone particles are massless, and hence, are curious. We see in the following that these particles may be removed from the system when gauge fields are switched on. For this purpose, we return to the locally invariant Lagrangian density (1.4.10) with the potential function V being defined by (1.4.34), i. e.,

$$\mathcal{L} = -\frac{1}{4}F_{\mu\nu}F^{\mu\nu} + \frac{1}{2}D_\mu\phi\overline{D^\mu\phi} - \frac{\lambda}{8}(|\phi|^2 - \phi_0^2)^2, \qquad (1.4.37)$$

and we consider fluctuations around the vacuum state

$$\phi_\mathrm{v} = \phi_0, \quad (A_\mu)_\mathrm{v} = 0. \qquad (1.4.38)$$

Using (1.4.32), we obtain the Lagrangian density for the interaction of the fields ϕ_1, ϕ_2, and A_μ as follows,

$$\mathcal{L} = -\frac{1}{4}F_{\mu\nu}F^{\mu\nu} + \frac{1}{2}\partial_\mu\phi_1\partial^\mu\phi_1 + \frac{1}{2}\partial_\mu\phi_2\partial^\mu\phi_2 + \frac{1}{2}e^2\phi_0^2 A_\mu A^\mu - \frac{\lambda}{2}\phi_0^2\phi_1^2 + \mathcal{L}_\mathrm{inter}, \qquad (1.4.39)$$

where $\mathcal{L}_\mathrm{inter}$ contains all off-diagonal interaction terms involving the mixed products of the fields ϕ_1, ϕ_2, A_μ, and their derivatives. Recall that the gauge transformation is defined by (1.4.5) and (1.4.7). Hence, ϕ_1, ϕ_2, A_μ transform themselves according to the rule

$$\phi_0 + \phi_1 + i\phi_2 \;\mapsto\; \phi_0 + \phi_1' + i\phi_2', \quad A_\mu \mapsto A_\mu',$$
$$\phi_1' = \phi_1 \cos\omega - \phi_2 \sin\omega + \phi_0(\cos\omega - 1),$$
$$\phi_2' = \phi_1 \sin\omega + \phi_2 \cos\omega + \phi_0 \sin\omega,$$
$$A_\mu' = A_\mu + \frac{1}{e}\partial_\mu\omega. \tag{1.4.40}$$

From the Lagrangian density (1.4.39), we see that ϕ_2 remains massless. Besides, the gauge field A_μ becomes massive (a mass of $e\phi_0$). However, using (1.4.40), we can find a suitable gauge transformation so that $\phi_2' = 0$. For example, we may choose

$$\omega = -\arctan\frac{\phi_2}{\phi_0 + \phi_1}.$$

If we use the phase function ω determined above in the transformation (1.4.40) and the new field variables ϕ_1', A_μ', and suppress the prime sign $'$, we see that (1.4.39) becomes

$$\mathcal{L} = -\frac{1}{4}F_{\mu\nu}F^{\mu\nu} + \frac{1}{2}\partial_\mu\phi_1\partial^\mu\phi_1 + \frac{1}{2}e^2\phi_0^2 A_\mu A^\mu - \frac{\lambda}{2}\phi_0^2\phi_1^2 + \mathcal{L}_{\text{inter}}, \tag{1.4.41}$$

where $\mathcal{L}_{\text{inter}}$ contains all off-diagonal interaction terms involving the mixed products of the fields ϕ_1, A_μ, and their derivatives. Thus we see that, in such a fixed gauge, we are left with a massive real scalar field and a massive gauge field and the massless Goldstone particle is eliminated. In other words, spontaneous breaking of a continuous symmetry does not lead to the appearance of a massless Goldstone particle but to the disappearance of a scalar field and the appearance of a massive gauge field. This statement is known as the *Higgs mechanism* and the massive scalar particles are called Higgs particles. In particular, the Lagrangian density (1.4.37) is commonly referred to as the *Abelian Higgs model* and the complex scalar field ϕ is called the *Higgs field*. In superconductivity, ϕ gives rise to density distribution of superconducting electron pairs known as the Cooper pairs and the fact that the electromagnetic field behaves like a massive field due to the Higgs mechanism is simply a consequence of the Meissner effect.

1.5   Yang–Mills Fields

In the last section, we discussed a gauge field theory with the Abelian group $U(1)$. The *Yang–Mills theory* is now a generic name for the gauge field theory with an arbitrary non-Abelian Lie group G. In this section, we present a short introduction to the Yang–Mills theory. For notational convenience, we shall concentrate on the specific case where G is either the orthogonal matrix group, $O(n)$, or unitary matrix group, $U(n)$, which is sufficient for all physical applications.

Let ϕ be a scalar field over the Minkowski spacetime and take values in either $\mathbb{R}^n$ or $\mathbb{C}^n$, which is the representation space of G (with $G = O(n)$ or $U(n)$). We use † to denote the Hermitian transpose or Hermitian conjugate in $\mathbb{R}^n$ or $\mathbb{C}^n$. Then $|\phi|^2 = \phi\phi^\dagger$. We may start from the Lagrangian density

$$\mathcal{L} = \frac{1}{2}(\partial_\mu\phi)(\partial^\mu\phi)^\dagger - V(|\phi|^2). \tag{1.5.1}$$

It is clear that (1.5.1) is invariant under the global symmetry group G,

$$\phi \mapsto \Omega\phi, \quad \Omega \in G. \tag{1.5.2}$$

However, as in the Abelian case, if the group element Ω is replaced by a local one depending on spacetime points,

$$\Omega = \Omega(x), \tag{1.5.3}$$

the invariance of (1.5.1) is no longer valid and a modification is to be devised. Thus we are again motivated to consider the derivative

$$D_\mu\phi = \partial_\mu\phi + A_\mu\phi, \tag{1.5.4}$$

where we naturally choose A_μ to be an element in the Lie algebra $\mathcal{G}$ of G which has an obvious representation over the space of ϕ. The dynamical term in (1.5.1) becomes

$$\frac{1}{2}(D_\mu\phi)(D^\mu\phi)^\dagger. \tag{1.5.5}$$

Of course, the invariance of (1.5.5) under the local transformation

$$\phi(x) \mapsto \phi'(x) = \Omega(x)\phi(x) \tag{1.5.6}$$

is ensured if $D_\mu\phi$ transforms itself covariantly according to

$$D_\mu\phi \mapsto D'_\mu\phi' = \partial_\mu\phi' + A'_\mu\phi' = \partial_\mu(\Omega\phi) + A'_\mu\Omega\phi = \Omega(D_\mu\phi). \tag{1.5.7}$$

Inserting (1.5.4) into (1.5.7) and comparing, we conclude that A_μ should obey the following rule of transformation,

$$A_\mu \mapsto A'_\mu = \Omega A_\mu \Omega^{-1} - (\partial_\mu\Omega)\Omega^{-1}. \tag{1.5.8}$$

It is easily examined that the $U(1)$ gauge field theory presented in the last section is contained here as a special case (the Lie algebra is the imaginary axis $i\mathbb{R}$). Thus, in general, the Lie algebra valued field A_μ is also a gauge field.

In order to include dynamics for the gauge field A_μ, we need to introduce invariant quadratic terms involving derivatives of A_μ. For this purpose, recall that there is a standard inner product over the space of $n \times n$ matrices, i.e.,

$$(A, B) = \text{Tr}(AB^\dagger).$$

For any $A \in \mathcal{G}$, since $A^\dagger = -A$, we see that

$$|A|^2 = (A, A) = -\text{Tr}\,(A^2). \tag{1.5.9}$$

In complete analogy with the electromagnetic field in the Abelian case, we can examine the non-commutativity of the gauge-covariant derivatives defined in (1.5.4) to get

$$D_\mu D_\nu \phi - D_\nu D_\mu \phi = (\partial_\mu A_\nu - \partial_\nu A_\mu + [A_\mu, A_\nu])\phi,$$

where $[\cdot, \cdot]$ is the Lie bracket (or commutator) of $\mathcal{G}$. Hence we are motivated to define the skew-symmetric Yang–Mills field (curvature) 2-tensor $F_{\mu\nu}$ as

$$F_{\mu\nu} = \partial_\mu A_\nu - \partial_\nu A_\mu + [A_\mu, A_\nu]. \tag{1.5.10}$$

Using (1.5.8) and (1.5.10), we see that $F_{\mu\nu}$ transforms itself according to

$$F_{\mu\nu} \mapsto F'_{\mu\nu} = \partial_\mu A'_\nu - \partial_\nu A'_\mu + [A'_\mu, A'_\nu] = \Omega F_{\mu\nu} \Omega^{-1}. \tag{1.5.11}$$

Hence we obtain the analogous invariant term,

$$\frac{1}{4}\text{Tr}\,(F_{\mu\nu} F^{\mu\nu}). \tag{1.5.12}$$

With (1.5.5) and (1.5.12), we arrive at the final form of our locally gauge-invariant Lagrangian action density

$$\mathcal{L} = \frac{1}{4}\text{Tr}\,(F_{\mu\nu} F^{\mu\nu}) + \frac{1}{2}(D_\mu \phi)(D^\mu \phi)^\dagger - V(|\phi|^2). \tag{1.5.13}$$

The action density (1.5.13) couples the matter scalar field ϕ with a Yang–Mills gauge field A_μ. In the case where the potential function gives rise to spontaneous symmetry breaking, ϕ governs a system of Higgs particles and (1.5.13) is called the *Yang–Mills–Higgs model*. The Euler–Lagrange equations of (1.5.13) are called the *Yang–Mills–Higgs equations*, which are non-Abelian extension of the system of equations (1.4.11) and (1.4.12).

In the case where the matter component is neglected, (1.5.13) becomes

$$\mathcal{L} = \frac{1}{4}\text{Tr}\,(F_{\mu\nu} F^{\mu\nu}), \tag{1.5.14}$$

which is simply called the (pure) Yang–Mills theory. The Euler–Lagrange equations of (1.5.14) are called the *Yang–Mills equations*, which are non-Abelian extension of the Maxwell equations in vacuum,

$$\partial_\nu F^{\mu\nu} = 0. \tag{1.5.15}$$

Since $F_{\mu\nu}$ contains nonvanishing commutators, the system (1.5.15) is non-linear. This nonlinearity may serve as 'current' sources to generate self-induced nontrivial solutions.

Like the Maxwell electromagnetic field, the Yang–Mills fields are also *mediating* fields. As mentioned above, in the non-Abelian case the commutators in (1.5.10) introduce nonlinearity and new physics appears: these non-Abelian gauge fields are in fact *nuclear force* fields which become significant only in short distances. More precisely, like the $U(1)$ group giving rise to electromagnetic interactions, the $SU(2)$ group gives rise to weak, $SU(3)$ strong, and $SU(5)$ grand unified interactions.

1.6   General Relativity and Cosmology

We start from a quick introduction to Riemannian geometry and a derivation of the metric energy-momentum tensor. We next derive the Einstein equations for gravitation. We then discuss some direct cosmological consequences from the Einstein equations and the origin of the cosmological constant and its interpretation as vacuum mass-energy density.

1.6.1   Einstein field equations

Let $(g_{\mu\nu})$ be the metric tensor of spacetime. The spacetime line element or the first fundamental form is defined by

$$ds^2 = g_{\mu\nu}dx^\mu dx^\nu,$$

which is also a measurement of the proper time (see (1.2.3)). A freely moving particle in spacetime follows a curve that stationarizes the action

$$\int ds.$$

We now derive the equations of motion from the above action principle.

We use the notation $x^\mu(s)$ to denote the desired curve (trajectory of the particle) and $\delta x^\mu(s)$ a small variation, both parametrized by s. Then, to the first order of variation, we have

$$\begin{aligned}
\delta(ds^2) &= 2ds\delta(ds) = (\delta g_{\mu\nu})dx^\mu dx^\nu + 2g_{\mu\nu}dx^\mu \delta(dx^\nu) \\
&= (g_{\mu\nu,\alpha}\delta x^\alpha)dx^\mu dx^\nu + 2g_{\mu\nu}dx^\mu d(\delta x^\nu),
\end{aligned}$$

where and in the sequel we use the notation

$$f_{,\alpha}, \quad A_{\mu,\alpha}, \quad F_{\mu\nu,\alpha}, \quad T^{\mu\nu}{}_{,\alpha},$$

etc., to denote the conventional partial derivative with respect to the variable x^α of various quantities. Using v^μ to denote the components of the 4-velocity,

$$v^\mu(s) = \frac{dx^\mu(s)}{ds},$$

we then obtain

$$
\begin{aligned}
\delta(ds) &= \left(\frac{1}{2} g_{\mu\nu,\alpha} v^\mu v^\nu \delta x^\alpha + g_{\mu\nu} v^\mu \frac{d}{ds}(\delta x^\nu) \right) ds \\
&= (\delta x^\alpha)\left(\frac{1}{2} g_{\mu\nu,\alpha} v^\mu v^\nu - \frac{d}{ds}(g_{\mu\alpha} v^\mu) \right) ds + d(g_{\mu\nu} v^\mu \delta x^\nu).
\end{aligned}
$$

Inserting the above into the stationary condition

$$\delta \int ds = 0$$

and using the fact that δx^μ vanishes at the two end points of the curve, we arrive at the equations of motion

$$\frac{d}{ds}(g_{\mu\alpha} v^\mu) - \frac{1}{2} g_{\mu\nu,\alpha} v^\mu v^\nu = 0.$$

Again, since $g_{\mu\nu}$ is symmetric, we have

$$
\begin{aligned}
\frac{d}{ds}(g_{\mu\alpha} v^\mu) &= g_{\mu\alpha} \frac{dv^\mu}{ds} + g_{\mu\alpha,\nu} v^\mu v^\nu \\
&= g_{\mu\alpha} \frac{dv^\mu}{ds} + \frac{1}{2}(g_{\alpha\mu,\nu} + g_{\alpha\nu,\mu}) v^\mu v^\nu.
\end{aligned}
$$

Consequently the equations of motion become

$$g_{\mu\alpha} \frac{dv^\mu}{ds} + \Gamma_{\alpha\mu\nu} v^\mu v^\nu = 0 \quad \text{or} \quad \frac{dv^\alpha}{ds} + \Gamma^\alpha_{\mu\nu} v^\mu v^\nu = 0, \tag{1.6.1}$$

where $\Gamma_{\mu\nu\alpha}$ and $\Gamma^\alpha_{\mu\nu}$ are called the *Christoffel symbols*, which are defined by

$$\Gamma_{\mu\nu\alpha} = \frac{1}{2}(g_{\mu\nu,\alpha} + g_{\mu\alpha,\nu} - g_{\nu\alpha,\mu}), \quad \Gamma^\alpha_{\mu\nu} = g^{\alpha\beta} \Gamma_{\mu\nu\beta}.$$

The curves that are solutions of (1.6.1) are called *geodesics*.

We see immediately that $\Gamma_{\mu\nu\alpha} = \Gamma_{\mu\alpha\nu}$. Besides, it is also useful to note that the definition of $\Gamma_{\mu\nu\alpha}$ gives us the identity

$$\Gamma_{\mu\nu\alpha} + \Gamma_{\nu\mu\alpha} = g_{\mu\nu,\alpha}. \tag{1.6.2}$$

One of the most important applications of the Christoffel symbols is their role in the definition of covariant derivatives for covariant and contravariant quantities,

$$A_{\mu;\alpha} = A_{\mu,\alpha} - \Gamma^{\beta}_{\mu\alpha} A_{\beta},$$

$$T_{\mu\nu;\alpha} = T_{\mu\nu,\alpha} - \Gamma^{\beta}_{\mu\alpha} T_{\beta\nu} - \Gamma^{\beta}_{\nu\alpha} T_{\mu\beta},$$

$$A^{\mu}_{\ ;\alpha} = A^{\mu}_{\ ,\alpha} + \Gamma^{\mu}_{\beta\alpha} A^{\beta},$$

$$T^{\mu\nu}_{\ \ ;\alpha} = T^{\mu\nu}_{\ \ ,\alpha} + \Gamma^{\mu}_{\beta\alpha} T^{\beta\nu} + \Gamma^{\nu}_{\beta\alpha} T^{\mu\beta}. \tag{1.6.3}$$

We will sometimes use ∇_{α} to denote covariant derivative. A direct consequence of the above definition and the identity (1.6.2) is that

$$g_{\mu\nu;\alpha} = g_{\mu\nu,\alpha} - \Gamma^{\beta}_{\mu\alpha} g_{\beta\nu} - \Gamma^{\beta}_{\nu\alpha} g_{\mu\beta}$$

$$= g_{\mu\nu,\alpha} - \Gamma_{\mu\alpha\nu} - \Gamma_{\nu\alpha\mu} = 0. \tag{1.6.4}$$

Similarly, $g^{\mu\nu}_{\ \ ;\alpha} = 0$. Therefore we have seen that the covariant and contravariant metric tensors, $g_{\mu\nu}$ and $g^{\mu\nu}$, behave like constants under covariant differentiation.

Let A_{μ} be a test covariant vector. Following (1.6.3), we obtain through an easy calculation the commutator

$$A_{\mu;\alpha;\beta} - A_{\mu;\beta;\alpha} = [\nabla_{\alpha}, \nabla_{\beta}] A_{\mu} = R^{\nu}_{\mu\alpha\beta} A_{\nu}, \tag{1.6.5}$$

where

$$R^{\nu}_{\mu\alpha\beta} = \Gamma^{\nu}_{\mu\beta,\alpha} - \Gamma^{\nu}_{\mu\alpha,\beta} + \Gamma^{\gamma}_{\mu\beta} \Gamma^{\nu}_{\gamma\alpha} - \Gamma^{\gamma}_{\mu\alpha} \Gamma^{\nu}_{\gamma\beta} \tag{1.6.6}$$

is a mixed 4-tensor called the *Riemann* curvature tensor. There hold the simple properties

$$R^{\nu}_{\mu\alpha\beta} = -R^{\nu}_{\mu\beta\alpha}, \tag{1.6.7}$$

$$R^{\nu}_{\mu\alpha\beta} + R^{\nu}_{\alpha\beta\mu} + R^{\nu}_{\beta\mu\alpha} = 0. \tag{1.6.8}$$

Furthermore, similar to (1.6.5), for covariant 2-tensors, we have

$$T_{\mu\nu;\alpha;\beta} - T_{\mu\nu;\beta;\alpha} = R^{\gamma}_{\mu\alpha\beta} T_{\gamma\nu} + R^{\gamma}_{\nu\alpha\beta} T_{\mu\gamma}. \tag{1.6.9}$$

Therefore, in particular, for a covariant vector field A_{μ}, we have

$$A_{\mu;\nu;\alpha;\beta} - A_{\mu;\nu;\beta;\alpha} = R^{\gamma}_{\mu\alpha\beta} A_{\gamma;\nu} + R^{\gamma}_{\nu\alpha\beta} A_{\mu;\gamma}.$$

We now make permutations of the indices ν, α, β and add the three resulting equations. In view of (1.6.5), the left-hand side is

$$(A_{\mu;\alpha;\beta;\nu} - A_{\mu;\beta;\alpha;\nu}) + \text{permutations}$$

$$= (R^{\gamma}_{\mu\alpha\beta} A_{\gamma})_{;\nu} + \text{permutations}$$

$$= (R^{\gamma}_{\mu\alpha\beta} A_{\gamma;\nu} + R^{\gamma}_{\mu\alpha\beta;\nu} A_{\gamma}) + \text{permutations}. \tag{1.6.10}$$

In view of (1.6.8), the right-hand side is

$$R^{\gamma}_{\mu\alpha\beta} A_{\gamma;\nu} + \text{permutations}. \tag{1.6.11}$$

Equating (1.6.10) and (1.6.11), we arrive at

$$R^\gamma_{\mu\alpha\beta;\nu} A_\gamma + \text{permutations} = 0.$$

Since A_μ is arbitrary, we find that

$$R^\gamma_{\mu\nu\alpha;\beta} + R^\gamma_{\mu\alpha\beta;\nu} + R^\gamma_{\mu\beta\nu;\alpha} = 0. \tag{1.6.12}$$

This result is also known as the Bianchi identity.

The *Ricci* tensor $R_{\mu\nu}$ is defined from $R^\nu_{\mu\alpha\beta}$ through contraction,

$$R_{\mu\nu} = R^\alpha_{\mu\nu\alpha}. \tag{1.6.13}$$

It is clear that $R_{\mu\nu}$ is symmetric. The *scalar curvature* R is then defined by

$$R = g^{\mu\nu} R_{\mu\nu}. \tag{1.6.14}$$

In the Bianchi identity (1.6.12), set $\gamma = \nu$ and multiply by $g^{\mu\alpha}$. We obtain

$$(g^{\mu\alpha} R^\nu_{\mu\nu\alpha})_{;\beta} + (g^{\mu\alpha} R^\nu_{\mu\alpha\beta})_{;\nu} + (g^{\mu\alpha} R^\nu_{\mu\beta\nu})_{;\alpha} = 0,$$

which is simply

$$2R^\beta_{\alpha;\beta} - R_{;\alpha} = 0.$$

Multiplying the above by $g^{\mu\alpha}$, we have the following very important result,

$$G^{\mu\nu}_{\ ;\nu} = 0, \tag{1.6.15}$$

where

$$G^{\mu\nu} = R^{\mu\nu} - \frac{1}{2} g^{\mu\nu} R, \tag{1.6.16}$$

or its covariant partner, $G_{\mu\nu}$, is called the *Einstein tensor*.

We next consider physics over the curved spacetime of metric $(g_{\mu\nu})$ governed by a matter field u which is either a scalar field or a vector field. Thus we need to replace the action (1.3.1) by a geometric one,

$$S = \int \mathcal{L}(x, Du, g) \sqrt{-|g|}\, dx, \tag{1.6.17}$$

where we have emphasized the influence of the metric tensor $g = (g_{\mu\nu})$ and used the canonical volume element $\sqrt{-|g|}\, dx$. Here $|g|$ is the determinant of the metric g which is negative due to its signature. Since physics is independent of coordinates, $\mathcal{L}$ must be a scalar. For example, the real Klein–Gordon action density now reads

$$\mathcal{L}(x, u, Du, g) = \frac{1}{2} g^{\mu\nu} \partial_\mu u \partial_\nu u - V(u), \tag{1.6.18}$$

which is g-dependent. In other words, physics can no longer be purely material.

It is easily seen that the Euler–Lagrange equations, or the equations of motion, of (1.6.17) are now

$$\frac{1}{\sqrt{-|g|}}\partial_\mu\left(\sqrt{-|g|}\frac{\partial\mathcal{L}}{\partial(\partial_\mu u^a)}\right)=\frac{\partial\mathcal{L}}{\partial u^a},$$

$$a=1,2,\cdots,m;\quad u=(u^a). \tag{1.6.19}$$

We are prepared to find conservation laws in view of Noether's theorem. For convenience, we assume u is a real scalar field. We fix a small domain Ω and consider the coordinates transformation

$$x'^\mu=x^\mu+\xi^\mu(x),\quad\mu=0,1,2,3$$

over it, where $\xi^\mu(x)$ vanishes near the boundary of Ω. Then, with the correspondence

$$u'(x')=u(x),\quad g_{\mu\nu}(x)=\frac{\partial x'^\alpha}{\partial x_\mu}\frac{\partial x'^\beta}{\partial x_\nu}g'^{\alpha\beta}(x'),$$

we have by renaming dummy integration variables the relation

$$
\begin{aligned}
S_\Omega &\equiv \int_\Omega \mathcal{L}(u(x),Du(x),g(x))\sqrt{-|g(x)|}\,\mathrm{d}x\\
&= \int_\Omega \mathcal{L}(u'(x'),D'u'(x'),g'(x'))\sqrt{-|g'(x')|}\,\mathrm{d}x'\\
&= \int_\Omega \mathcal{L}(u'(x),Du'(x),g'(x))\sqrt{-|g'(x)|}\,\mathrm{d}x\equiv S'_\Omega.
\end{aligned}
$$

Thus, up to the first-order terms of

$$(\delta u)(x)=u'(x)-u(x),\quad(\delta g)(x)=g'(x)-g(x),$$

we have

$$
\begin{aligned}
0 &= \delta S_\Omega = S'_\Omega - S_\Omega\\
&= \int_\Omega\left\{\frac{\partial\mathcal{L}}{\partial u}+\frac{\partial\mathcal{L}}{\partial(\partial_\mu u)}\partial_\mu(\delta u)\right\}\sqrt{-|g|}\,\mathrm{d}x-\frac{1}{2}\int_\Omega T^{\mu\nu}(\delta g_{\mu\nu})\sqrt{-|g|}\mathrm{d}x,
\end{aligned}
$$

$$\tag{1.6.20}$$

where $T^{\mu\nu}$ is a symmetric tensor given by

$$T^{\mu\nu}=-\frac{2}{\sqrt{-|g|}}\frac{\partial}{\partial g_{\mu\nu}}\left(\mathcal{L}(x,u,Du,g)\sqrt{-|g|}\right). \tag{1.6.21}$$

Using (1.6.19) and integrating by parts, we see that the first integral on the right-hand side of (1.6.20) vanishes. In order to recognize the second integral, we observe by neglecting higher-order terms in ξ^α that

$$
\begin{aligned}
g_{\mu\nu}(x) &= g'_{\alpha\beta}(x')(\delta^\alpha_\mu + \xi^\alpha{}_{,\mu})(\delta^\beta_\nu + \xi^\beta{}_{,\nu}) \\
&= g'_{\mu\nu}(x') + g'_{\mu\alpha}(x')\xi^\alpha{}_{,\nu} + g'_{\nu\alpha}(x)\xi^\alpha{}_{,\mu} \\
&= g'_{\mu\nu}(x) + g_{\mu\nu,\alpha}(x)\xi^\alpha + g_{\mu\alpha}(x)\xi^\alpha{}_{,\nu} + g_{\nu\alpha}(x)\xi^\alpha{}_{,\mu}.
\end{aligned}
$$

In view of the above, we obtain

$$
\begin{aligned}
\delta g_{\mu\nu} &= -g_{\mu\nu,\alpha}\xi^\alpha - g_{\mu\alpha}\xi^\alpha{}_{,\nu} - g_{\nu\alpha}\xi^\alpha{}_{,\mu} \\
&= (g_{\mu\alpha,\nu} + g_{\nu\alpha,\mu} - g_{\mu\nu,\alpha})\xi^\alpha - \xi_{\mu,\nu} - \xi_{\nu,\mu} \\
&= 2\xi_\alpha\Gamma^\alpha_{\mu\nu} - \xi_{\mu,\nu} - \xi_{\nu,\mu}.
\end{aligned} \tag{1.6.22}
$$

Therefore, integrating by parts again, we have

$$
\begin{aligned}
0 &= \int_\Omega T^{\mu\nu}(\delta g_{\mu\nu})\sqrt{-|g|}\,dx \\
&= \int_\Omega T^{\mu\nu}(-\xi_{\mu,\nu} - \xi_{\nu,\mu} + 2\xi_\alpha\Gamma^\alpha_{\mu\nu})\sqrt{-|g|}\,dx \\
&= \int_\Omega T^{\mu\nu}(-\xi_{\mu;\nu} - \xi_{\nu;\mu})\sqrt{-|g|}\,dx \\
&= -2\int_\Omega (T^{\mu\nu}\xi_\mu)_{;\nu}\sqrt{-|g|}\,dx + 2\int_\Omega T^{\mu\nu}{}_{;\nu}\xi_\mu\sqrt{-|g|}\,dx.
\end{aligned} \tag{1.6.23}
$$

Since the divergence formula,

$$
A^\mu{}_{;\mu} = \frac{1}{\sqrt{-|g|}}\frac{\partial(\sqrt{-|g|}A^\mu)}{\partial x^\mu}, \tag{1.6.24}
$$

holds, the first term on the right-hand side of (1.6.23) vanishes. Hence we arrive at

$$
\int_\Omega T^{\mu\nu}{}_{;\nu}\xi_\mu\sqrt{-|g|}\,dx = 0.
$$

The arbitrariness of ξ_μ then implies the following conservation laws in curved spacetime,

$$
T^{\mu\nu}{}_{;\nu} = 0. \tag{1.6.25}
$$

In order to see the meaning of the tensor $T^{\mu\nu}$, we need to express (1.6.21) more explicitly. It is readily seen that

$$
\frac{\partial|g|}{\partial g_{\mu\nu}} = |g|g^{\mu\nu}. \tag{1.6.26}
$$

Thus, applying (1.6.26) in (1.6.21), we find that

$$
T^{\mu\nu} = -2\frac{\partial\mathcal{L}}{\partial g_{\mu\nu}} - g^{\mu\nu}\mathcal{L}. \tag{1.6.27}
$$

On the other hand, from $g^{\alpha\beta}g_{\alpha\beta} = 4$, we have

$$\frac{\partial g^{\alpha\beta}}{\partial g_{\mu\nu}}g_{\alpha\beta} = -g^{\alpha\beta}\frac{\partial g_{\alpha\beta}}{\partial g_{\mu\nu}} = -g^{\alpha\beta}\delta_\alpha^\mu\delta_\beta^\nu = -g^{\mu\nu}.$$

Consequently, we obtain the useful formula

$$\frac{\partial g^{\alpha\beta}}{\partial g_{\mu\nu}} = -g^{\alpha\mu}g^{\beta\nu}. \tag{1.6.28}$$

In view of (1.6.28), the tensor (1.6.27) becomes

$$T^{\mu\nu} = 2g^{\alpha\mu}g^{\beta\nu}\frac{\partial\mathcal{L}}{\partial g^{\alpha\beta}} - g^{\mu\nu}\mathcal{L}. \tag{1.6.29}$$

In the case of a scalar field defined by the Lagrangian action density (1.3.28), it is easy to verify that

$$2g^{\alpha\mu}g^{\beta\nu}\frac{\partial\mathcal{L}}{\partial g^{\alpha\beta}} = \frac{\partial\mathcal{L}}{\partial(\partial_\mu u)}\cdot\partial^\nu u.$$

Therefore (1.6.29) coincides with the canonical energy-momentum tensor derived in §1.3.2 in flat spacetime limit. Thus (1.6.29) may well be recognized as the curved spacetime version of the energy-momentum tensor. A great advantage of this choice is that $T^{\mu\nu}$ is symmetric and obeys the covariant conservation law (1.6.25) automatically.

The basic principle which led Einstein to write down his fundamental equations for gravitation states that the geometry of a spacetime is determined by the matter it contains. Mathematically, Einstein's idea was to consider the equation

$$Q^{\mu\nu} = -\kappa T^{\mu\nu}, \tag{1.6.30}$$

where $Q^{\mu\nu}$ is a 2-tensor generated from the spacetime metric $(g_{\mu\nu})$ which is purely geometric, $T^{\mu\nu}$ is the energy-momentum tensor which is purely material, κ is a constant called the Einstein gravitational constant, and the negative sign in front of κ is inserted for convenience. This equation imposes severe restriction to the possible form of the 2-tensor $Q^{\mu\nu}$. For example, $Q^{\mu\nu}$ should also satisfy the same conservation law (or the divergence-free condition),

$$Q^{\mu\nu}{}_{;\nu} = 0, \tag{1.6.31}$$

as $T^{\mu\nu}$ (see (1.6.25)). The simplest candidate for $Q^{\mu\nu}$ could be $g^{\mu\nu}$. However, since $g^{\mu\nu}$ is nondegenerate, this choice is seen to be incorrect because it makes $T^{\mu\nu}$ nondegenerate, which is absurd in general. The next candidate could be the Ricci curvature $R^{\mu\nu}$. Since $R^{\mu\nu}$ does not satisfy the required identity (1.6.31), it has to be abandoned. Consequently, based on both the compatibility condition (1.6.31) and simplicity consideration, we are naturally led to the choice of the Einstein tensor, $G^{\mu\nu}$, defined in (1.6.16). Therefore we obtain the *Einstein equations*,

$$G^{\mu\nu} = -\kappa T^{\mu\nu} \quad \text{or} \quad G_{\mu\nu} = -\kappa T_{\mu\nu}. \tag{1.6.32}$$

It can be shown that the equation (1.6.32) recovers Newton's law of gravitation,

$$F = -G\frac{m_1 m_2}{r^2}$$

which gives the magnitude of an attractive force between two point particles of masses m_1 and m_2 with a distance r apart, in the static spacetime and slow motion limit, if and only if $\kappa = 8\pi G$. Recall that the constant G is called the Newton universal gravitational constant, which is extremely small compared to other quantities.

In summary, we have just derived the Einstein gravitational field equations,

$$G_{\mu\nu} = -8\pi G T_{\mu\nu}. \tag{1.6.33}$$

1.6.2 Cosmological consequences

In modern cosmology, the universe is believed to be *homogeneous* (the number of stars per unit volume is uniform throughout large regions of space) and *isotropic* (the number of stars per unit solid angle is the same in all directions). This basic property is known as the *Cosmological Principle* and has been evidenced by astronomical observations. A direct implication of such a principle is that synchronized clocks may be placed throughout the universe to give a uniform measurement of time (*Cosmic time*). Another is that the space curvature, K, is constant at any fixed cosmic time t. Hence we have the following simple mathematical descriptions for the space.

(a) If $K = K(t) > 0$, the space is closed and may be defined as a 3-sphere embedded in the flat Euclidean space of the form

$$x^2 + y^2 + z^2 + w^2 = a^2, \quad a = a(t) = \frac{1}{\sqrt{K(t)}}. \tag{1.6.34}$$

(b) If $K = K(t) < 0$, the space is open and may be defined similarly by the equation

$$-x^2 - y^2 - z^2 + w^2 = a^2, \quad a = a(t) = \frac{1}{\sqrt{|K(t)|}}, \tag{1.6.35}$$

which is embedded in the flat Minkowski space with the line element

$$dx^2 + dy^2 + dz^2 - dw^2. \tag{1.6.36}$$

(c) If $K = K(t) = 0$, the space is the Euclidean space $\mathbb{R}^3$. In particular the space is open and its line element is given by

$$d\ell^2 = dx^2 + dy^2 + dz^2. \tag{1.6.37}$$

Use the conventional spherical coordinates (r, θ, χ) to replace the Cartesian coordinates (x, y, z). We have

$$x = r \cos \theta \sin \chi, \quad y = r \sin \theta \sin \chi, \quad z = r \cos \chi. \tag{1.6.38}$$

Thus, in the cases (a) and (b), we have

$$\pm r^2 + w^2 = a^2, \quad \pm r dr = w dw. \tag{1.6.39}$$

Substituting (1.6.38) and (1.6.39) into the line elements of the Euclidean space and of the Minkowski space given by (1.6.36), respectively, we obtain the induced line element $d\ell^2$ of the space,

$$d\ell^2 = \frac{a^2}{(a^2 \mp r^2)} dr^2 + r^2 d\theta^2 + r^2 \sin^2 \theta d\chi^2. \tag{1.6.40}$$

Finally, inserting (1.6.40) into the spacetime line element and making the rescaling $r \mapsto ar$, we have

$$ds^2 = dt^2 - a^2(t) \left(\frac{1}{(1 - kr^2)} dr^2 + r^2 d\theta^2 + r^2 \sin^2 \theta d\chi^2 \right), \tag{1.6.41}$$

where $k = \pm 1$ or $k = 0$ according to $K > 0, K < 0$ or $K = 0$. This is the most general line element of a homogeneous and isotropic spacetime and is known as the *Robertson–Walker metric*.

In cosmology, the large-scale viewpoint allows us to treat stars or galaxies as particles of a perfect 'gas' that fills the universe and is characterized by its mass-energy density ρ, counting both rest mass and kinetic energy per unit volume, and pressure p, so that the associated energy-momentum tensor $T_{\mu\nu}$ is given by

$$T_{\mu\nu} = (\rho + p)v_\mu v_\nu + p g_{\mu\nu}, \tag{1.6.42}$$

where v_μ is the 4-velocity of the gas particles and $g_{\mu\nu}$ is the spacetime metric. The cosmological principle requires that ρ and p depend on time t only.

We now consider some possible consequences of a homogeneous and isotropic universe in view of the Einstein theory. From (1.6.6) and (1.6.13), we can represent the Ricci tensor in terms of the Christoffel symbols by

$$R_{\mu\nu} = \Gamma^\alpha_{\mu\alpha,\nu} - \Gamma^\alpha_{\mu\nu,\alpha} + \Gamma^\alpha_{\mu\beta}\Gamma^\beta_{\alpha\nu} - \Gamma^\alpha_{\mu\nu}\Gamma^\beta_{\alpha\beta}. \tag{1.6.43}$$

Naturally, we label our coordinates according to $x^0 = t, x^1 = r, x^2 = \theta, x^3 = \chi$. Then the nonzero Christoffel symbols induced from the Robertson–Walker line element are

$$\Gamma^0_{11} = \frac{a(t)a'(t)}{(1 - kr^2)}, \quad \Gamma^0_{22} = a(t)a'(t)r^2, \quad \Gamma^0_{33} = a(t)a'(t)r^2 \sin^2 \theta,$$

$$\Gamma^1_{01} = \frac{a'(t)}{a(t)}, \quad \Gamma^1_{11} = \frac{kr}{(1-kr^2)},$$

$$\Gamma^1_{22} = -r(1-kr^2), \quad \Gamma^1_{33} = -r(1-kr^2)\sin^2\theta,$$

$$\Gamma^2_{02} = \frac{a'(t)}{a(t)}, \quad \Gamma^2_{12} = \frac{1}{r}, \quad \Gamma^2_{33} = -\sin\theta\cos\theta,$$

$$\Gamma^3_{03} = \frac{a'(t)}{a(t)}, \quad \Gamma^3_{13} = \frac{1}{r}, \quad \Gamma^3_{23} = \cot\theta, \tag{1.6.44}$$

where $a'(t) = da(t)/dt$. Inserting (1.6.44) into (1.6.43), we see that the Ricci tensor $R_{\mu\nu}$ becomes diagonal with

$$R_{00} = \frac{3a''}{a}, \quad R_{11} = -\frac{aa'' + 2(a')^2 + 2k}{1-kr^2},$$

$$R_{22} = -(aa'' + 2(a')^2 + 2k)r^2,$$

$$R_{33} = -(aa'' + 2(a')^2 + 2k)r^2\sin^2\theta. \tag{1.6.45}$$

Hence, in view of (1.6.14), the scalar curvature (1.6.14) becomes

$$R = \frac{6}{a^2}(aa'' + (a')^2 + k). \tag{1.6.46}$$

On the other hand, from (1.6.1) and (1.6.44), we see that the geodesics of the metric (1.6.41), which are the trajectories of moving stars and galaxies when net local interactions are neglected, are given by $r, \theta, \chi = $ constant. Thus in (1.6.42) we have $v_0 = 1$ and $v_i = 0$, $i = 1, 2, 3$. Therefore $T_{\mu\nu}$ is also diagonal with

$$T_{00} = \rho, \quad T_{11} = \frac{pa^2}{1-kr^2}, \quad T_{22} = pa^2r^2, \quad T_{33} = pa^2r^2\sin^2\theta. \tag{1.6.47}$$

Substituting (1.6.45), (1.6.46), and (1.6.47) into the Einstein equations (1.6.33), we arrive at the following two equations,

$$\frac{3a''}{a} = -4\pi G(\rho + 3p),$$

$$aa'' + 2(a')^2 + 2k = 4\pi G(\rho - p)a^2. \tag{1.6.48}$$

Eliminating a'' from these equations, we obtain the well-known *Friedmann equation*

$$(a')^2 + k = \frac{8\pi}{3}G\rho a^2. \tag{1.6.49}$$

We can show that, in the category of time-dependent solutions, the Einstein cosmological equations, (1.6.48), are in fact equivalent to the single Friedmann equation (1.6.49). To this end, recall that both systems are to be subject to the conservation law for the energy-momentum tensor, namely, $T^{\mu\nu}_{;\nu} = 0$ or

$$\rho' + 3(\rho + p)\frac{a'}{a} = 0. \tag{1.6.50}$$

Differentiating (1.6.49) and using (1.6.50), we get the first equation in (1.6.48). Inserting (1.6.49) into the first equation in (1.6.48), we get the second equation in (1.6.48).

The relative rate of change of the radius of the universe is recognized as Hubble's 'constant', $H(t)$, which is given by

$$H(t) = \frac{a'(t)}{a(t)}. \tag{1.6.51}$$

Recent estimates for Hubble's constant put it at about $(18 \times 10^9 \text{ years})^{-1}$. In particular, $a' > 0$ at present. However, since the first equation in (1.6.48) indicates that $a'' < 0$ everywhere, we can conclude that $a' > 0$ for all time in the past. In other words, the universe has undergone a process of expansion in the past.

We now investigate whether the universe has a beginning time. For this purpose, let t_0 denote the present time and t denote any past time, $t < t_0$. The property $a'' < 0$ again gives us $a'(t) > a'(t_0)$, which implies that

$$a(t_0) - a(t) > a'(t_0)(t_0 - t).$$

Thus there must be a finite time t in the past, $t < t_0$, when a vanishes. Such a time may be defined as the time when the universe begins. For convenience, we may assume that the universe begins at $t = 0$, namely, $a(0) = 0$. Hence we arrive at the general picture of the *Big Bang cosmology* that the universe started at a finite time in the past from a singular point and has been expanding in all its history of evolution.

It is easy to see that the equations (1.6.48) do not allow *static* (time-independent) solutions. When Einstein applied his gravitational equations to cosmology, he hoped to obtain a homogeneous, isotropic, static, and compact universe. Therefore he was led to his modified equations

$$G_{\mu\nu} + \Lambda g_{\mu\nu} = -8\pi G\, T_{\mu\nu}, \tag{1.6.52}$$

where Λ is a constant called the *cosmological constant*. Of course the added cosmological term, $\Lambda g_{\mu\nu}$, does not violate the required divergence-free condition. Although static models of the universe have long been discarded since Hubble's discovery in 1929 that the universe is expanding, a nonvanishing cosmological constant gives important implications in the theoretical studies of the early-universe cosmology. In fact, the equations (1.6.52) may also be rewritten

$$G_{\mu\nu} = -8\pi G \left(T_{\mu\nu} + T_{\mu\nu}^{(\text{vac})} \right),$$

$$T_{\mu\nu}^{(\text{vac})} = \frac{\Lambda}{8\pi G} g_{\mu\nu}, \tag{1.6.53}$$

where $T_{\mu\nu}^{(\text{vac})}$ is interpreted as the energy-momentum tensor associated with the vacuum: the vacuum polarization of quantum field theory endows the vacuum with a nonzero energy-momentum tensor, which is completely unobservable except by its gravitational effects. In particular,

$$\rho^{(\text{vac})} = T_{00}^{(\text{vac})} = \frac{\Lambda}{8\pi G} \tag{1.6.54}$$

may be viewed as the mass-energy density of the vacuum. This viewpoint imposes a natural restriction on the sign of the cosmological constant, $\Lambda \geq 0$.

Multiplying (1.6.52) by the metric $g^{\mu\nu}$ and summing over repeated indices, we find

$$R = 8\pi G T + 4\Lambda, \quad T = g^{\mu\nu}T_{\mu\nu}. \tag{1.6.55}$$

Inserting (1.6.55) into (1.6.52), we obtain the more elegant equations

$$R_{\mu\nu} - \Lambda g_{\mu\nu} = -8\pi G \left(T_{\mu\nu} - \frac{1}{2}g_{\mu\nu}T \right). \tag{1.6.56}$$

In particular, in the absence of matter, we have the vacuum Einstein equations

$$R_{\mu\nu} = \Lambda g_{\mu\nu}. \tag{1.6.57}$$

Any spacetime satisfying (1.6.57) is called an *Einstein space* and its metric $g_{\mu\nu}$ is called an *Einstein metric*.

1.7 Remarks

We have presented in this chapter a minimal exposure of some basic concepts in field theory which will be frequently encountered in subsequent chapters. To pursue an in-depth study of these and other related topics, we can recommend several excellent textbooks and articles. We recommend, for electromagnetism and special relativity, the books by Barut [28], Becker [32], Landau and Lifshitz [178]; for quantum mechanics and symmetry groups, the book by Greiner and Müller [135]; for gauge field theory and quantum field theory, the books by Chaichian and Nelipa [70], Greiner and Reinhardt [136], Ryder [269], Sokolov, Ternov, Zhukovskii, and Borisov [289]; for general relativity and cosmology, the books by Dirac [93], Foster and Nightingale [110], Kenyon [167], Landau and Lifshitz [178], Misner, Thorne, and Wheeler [217], Peebles [244], Rindler [265], Weinberg [328]; for mathematical structure, the survey articles by Actor [2] and Eguchi, Gilkey, and Hanson [102], the books by Felsager [108], Monastyrsky [218], Nash and Sen [223], and von Westenholz [323]; for the quantum meaning of classical solutions, the article by Jackiw [149] and the monograph by Rajaraman [259].

2

Sigma Models

In this chapter we study the classical sigma model coupled with a gauge field. Using techniques in nonlinear functional analysis, we are able to establish a precise description of the static solutions in various topological classes. In §2.1 we review the Belavin–Polyakov solution of the classical sigma model. In §2.2 we describe the gauged sigma model and state our existence theorems. In §2.3 we reduce the problem into a semilinear elliptic equation. In §2.4 we present a complete mathematical analysis and prove all the results stated.

2.1   Sigma Model and Belavin–Polyakov Solution

We start with a review of the classical sigma model governing a Heisenberg ferromagnet. We illustrate the topological characteristics of static solutions and derive their lower energy bounds. We then present the Belavin–Polyakov solution [33] and show that these lower energy bounds can all be saturated.

2.1.1   Sigma model for Heisenberg ferromagnet

The field configuration describing a Heisenberg ferromagnet is a spin vector ϕ defined over the $(2+1)$-dimensional Minkowski spacetime $\mathbb{R}^{2,1}$ and taking range in the unit sphere, S^2, of $\mathbb{R}^3$, namely,

$$\phi = (\phi_1, \phi_2, \phi_3), \quad \phi_1^2 + \phi_2^2 + \phi_3^2 = |\phi|^2 = 1. \tag{2.1.1}$$

The dynamics of the field ϕ is governed by the Lagrangian action density

$$\mathcal{L} = \frac{1}{2}\partial_\mu\phi \cdot \partial^\mu\phi. \tag{2.1.2}$$

Since the components ϕ_a ($a = 1, 2, 3$) are not independent variables due to (2.1.1), we need extra care in deriving the equations of motion of (2.1.2). A standard method is to incorporate the constraint (2.1.1) by a Lagrange multiplier $\lambda = \lambda(x)$, rewrite the action density as

$$\mathcal{L} = \frac{1}{2}\partial_\mu\phi \cdot \partial^\mu\phi + \lambda(|\phi|^2 - 1), \tag{2.1.3}$$

and treat the components ϕ_a ($a = 1, 2, 3$) as independent variables. Thus the Euler–Lagrange equations of (2.1.3) are, in vector form,

$$\partial_\mu\partial^\mu\phi + \lambda\phi = \left(\frac{\partial^2}{\partial t^2} - \Delta\right)\phi + \lambda\phi \equiv \Box\phi + \lambda\phi = 0. \tag{2.1.4}$$

However, from (2.1.1), we have

$$\lambda = \lambda(x) = -\phi \cdot \Box\phi. \tag{2.1.5}$$

Substituting (2.1.5) into (2.1.4), we obtain the equations of motion of the action density (2.1.2) subject to the constraint (2.1.1),

$$\Box\phi - (\phi \cdot \Box\phi)\phi = 0. \tag{2.1.6}$$

The global existence problem of the equation (2.1.6) has been studied extensively. Here we focus on static solutions which are independent of time, $\partial_t\phi = 0$. Hence (2.1.6) becomes

$$\Delta\phi - (\phi \cdot [\Delta\phi])\phi = 0. \tag{2.1.7}$$

It is easily seen that the Hamiltonian density of (2.1.2) is

$$\mathcal{H} = \frac{1}{2}\{(\partial_1\phi)^2 + (\partial_2\phi)^2\}, \tag{2.1.8}$$

which gives us the total energy

$$E(\phi) = \frac{1}{2}\int_{\mathbb{R}^2}\{(\partial_1\phi)^2 + (\partial_2\phi)^2\}\,\mathrm{d}x. \tag{2.1.9}$$

It is clear that (2.1.7) are the Euler–Lagrange equations of (2.1.9). Finite energy condition implies that ϕ goes to a constant unit vector, say, ϕ_∞, at infinity, which makes ϕ a continuous map from S^2 to S^2. Hence ϕ represents a homotopy class in the homotopy group

$$\pi_2(S^2) = \mathbb{Z} \tag{2.1.10}$$

and is thus characterized by an integer N. This integer N is also the Brouwer degree [57, 139, 216], $\deg(\phi)$, of ϕ which measures the number of times S^2 being covered by itself under the map ϕ. Our fundamental problem is to find an energy-minimizing configuration among each topological class $\mathcal{C}_N$ where

$$\mathcal{C}_N = \left\{ \phi : \mathbb{R}^2 \to S^2 \; \middle| \; E(\phi) < \infty, \deg(\phi) = N \right\}, \quad N \in \mathbb{Z}. \quad (2.1.11)$$

The classical work of Belavin and Polyakov concludes that such a problem may be solved completely and explicitly. In fact their result is known to be the earliest and simplest complete integration of a physically interesting field theory model and immediately triggered a broad spectrum of activities in other related areas in gauge theory. The most important mathematical feature here, which is still actively pursued in various other models in field theory, is the possible existence of a first integral known as self-duality or a Bogomol'nyi type structure. In this situation the energy is a sum of some quadratic terms and a topological term, the latter serving as a lower energy bound. Of course, the necessary and sufficient condition that such a lower bound can be attained is that all the quadratic terms vanish which give rise to a first integral of the problem.

It is important to recall that $\deg(\phi)$ has the following integral representation,

$$\deg(\phi) = \frac{1}{4\pi} \int_{\mathbb{R}^2} \phi \cdot (\partial_1 \phi \times \partial_2 \phi) \, dx. \quad (2.1.12)$$

Intuitively, it is easily seen that the above integral indeed defines the number of times the unit sphere S^2 of a total surface area 4π is being covered under its parametrization $\phi : \mathbb{R}^2 \to S^2$.

On the other hand, since $\phi \cdot \phi = 1$ and $\phi \cdot (\partial_j \phi) = 0$, we have after some manipulation of vector products the following reduction,

$$\begin{aligned} (\phi \times \partial_j \phi) \cdot (\phi \times \partial_j \phi) &= ([\phi \times \partial_j \phi] \times \phi) \cdot (\partial_j \phi) \\ &= (\partial_j \phi - \phi[\partial_j \phi \cdot \phi]) \cdot (\partial_j \phi) \\ &= (\partial_j \phi)^2, \quad j = 1, 2. \end{aligned}$$

Therefore we obtain the expansion

$$|\partial_1 \phi \pm \phi \times \partial_2 \phi|^2 + |\partial_2 \phi \mp \phi \times \partial_1 \phi|^2 = 2(\partial_1 \phi)^2 + 2(\partial_2 \phi)^2 \mp 4\phi \cdot (\partial_1 \phi \times \partial_2 \phi).$$

Integrating the above equality over $\mathbb{R}^2$ and using the energy formula (2.1.9) and the degree formula (2.1.12), we have the, now classical, topological lower energy bound,

$$E(\phi) \geq 4\pi |\deg(\phi)|. \quad (2.1.13)$$

The lower bound (2.1.13) is saturated if and only if ϕ satisfies the following self-dual or anti-self-dual equation,

$$\partial_j \phi = \mp \epsilon_{jk}(\phi \times [\partial_k \phi]), \quad j, k = 1, 2 \quad (2.1.14)$$

for $\deg(\phi) = \pm|\deg(\phi)|$, where the skew-symmetric tensor ϵ_{jk} is defined by $\epsilon_{12} = 1$. It is straightforward to examine that (2.1.14) implies (2.1.7). In this sense (2.1.14) may be viewed as a first integral of (2.1.7), although the most interesting fact about (2.1.14) is that the solutions of (2.1.14) are global minimizers of the energy over the designated topological class, $\mathcal{C}_N$.

2.1.2   Solution by rational functions

We shall now see how the system (2.1.14) may be solved explicitly.

In order to simplify (2.1.14), consider the stereographic projection from the south pole $\mathbf{s} = (0, 0, -1)$ of the unit sphere S^2. We can then represent ϕ, at least partially, as a map

$$\mathbf{u} = (u_1, u_2) : \mathbb{R}^2 \to \mathbb{R}^2,$$
$$u_1 = \frac{\phi_1}{1 + \phi_3}, \quad u_2 = \frac{\phi_2}{1 + \phi_3}, \tag{2.1.15}$$

where, of course,

$$\phi_3 = \begin{cases} \sqrt{1 - \phi_1^2 - \phi_2^2} & \text{when } \phi \text{ lies on the upper sphere } S_+^2, \\ -\sqrt{1 - \phi_1^2 - \phi_2^2} & \text{when } \phi \text{ lies on the lower sphere } S_-^2. \end{cases} \tag{2.1.16}$$

Thus the transformation (2.1.15) is not well defined at the south pole

$$\phi_1 = 0, \quad \phi_2 = 0, \quad \phi_3 = -1.$$

In fact, let

$$p \in P = \phi^{-1}(\mathbf{s}). \tag{2.1.17}$$

Then by L'Hôpital's rule we easily see that the continuity of ϕ at $x = p$ implies the interesting property

$$\lim_{x \to p} |\mathbf{u}(x)|^2 = \lim_{x \to p} (u_1^2 + u_2^2) = \lim_{x \to p} \frac{\phi_1^2 + \phi_2^2}{(1 - \sqrt{1 - [\phi_1^2 + \phi_2^2]})^2} = \infty. \tag{2.1.18}$$

Furthermore, away from P, the system (2.1.14) becomes the following linear system of equations,

$$\partial_1 u_1 = \pm \partial_2 u_2, \quad \partial_1 u_2 = \mp \partial_2 u_1,$$

which are the Cauchy–Riemann equations for the complex holomorphic function $u(z) = u_1(z) + i u_2(z)$ with $z = x^1 \pm i x^2$. To see what kind of solutions are interesting to us, we fix a positive integer N and look for those solutions that correspond to $\deg(\phi) = N$. In order to construct a degree N solution, we may require that $P = \phi^{-1}(\mathbf{s})$ contain N points. Similarly, we denote the north pole $(0, 0, 1) \in S^2 \subset \mathbb{R}^3$ by $\mathbf{n}$ and set

$$Q = \phi^{-1}(\mathbf{n}). \tag{2.1.19}$$

Then (2.1.15) and (2.1.16) say that u vanishes on Q.

For greater generality, we assume that Q contains M points where the integer M is to be determined later. Thus

$$P = \Big\{ p_1, p_2, \cdots, p_N \Big\}, \quad Q = \Big\{ q_1, q_2, \cdots, q_M \Big\}, \tag{2.1.20}$$

where possible multiplicities of the points p's and q's are allowed, and, we can immediately write down the corresponding solution with regard to (2.1.20) as

$$u(z) = c\Big(\prod_{j=1}^{N} (z - p_j)^{-1} \Big)\Big(\prod_{k=1}^{M} (z - q_k) \Big), \tag{2.1.21}$$

where $c \neq 0$ is a free complex parameter.

It will be necessary to examine whether there is any restriction from the finite-energy requirement. To this end, we invert the transformation (2.1.15) and represent the map ϕ as

$$\phi_1 = \frac{2u_1}{1 + |u|^2}, \quad \phi_2 = \frac{2u_2}{1 + |u|^2}, \quad \phi_3 = \frac{1 - |u|^2}{1 + |u|^2} \tag{2.1.22}$$

in terms of the complex-valued function $u = u_1 + iu_2$. Thus the energy (2.1.9) takes the form

$$E(\phi) = E(u) = 2\int_{\mathbb{R}^2} \frac{|\partial_1 u|^2 + |\partial_2 u|^2}{(1 + |u|^2)^2}, \tag{2.1.23}$$

where and in the sequel we often suppress the Lebesgue measure dx in the integral when there is no risk of confusion. Therefore, for a meromorphic function such as the one expressed in (2.1.21), we have

$$E(\phi) = E(u) = 4\int_{\mathbb{R}^2} \frac{|u'(z)|^2}{(1 + |u(z)|^2)^2}.$$

It is clear that the integrand is well-behaved at the poles p's for any M and N, as it should be. So the only concern is at infinity. It can be seen that the integrand also behaves well enough at infinity to ensure a finite energy if and only if $M \neq N$.

Moreover, we have, after some algebra, the relation

$$\phi \cdot (\partial_1 \phi \times \partial_2 \phi) = \frac{2i}{(1 + |u|^2)^2}(\partial_1 u \partial_2 \overline{u} - \partial_1 \overline{u} \partial_2 u). \tag{2.1.24}$$

Consequently, in view of (2.1.12), (2.1.24), and the following easily verified identity,

$$|\partial_1 u \pm i\partial_2 u|^2 = |\partial_1 u|^2 + |\partial_2 u|^2 \mp i\partial_1 u \partial_2 \overline{u} \pm i\partial_1 \overline{u} \partial_2 u, \tag{2.1.25}$$

the energy (2.1.23) becomes

$$E(u) = 2 \int_{\mathbb{R}^2} \frac{|\partial_1 u \pm i \partial_2 u|^2}{(1+|u|^2)^2} \pm 2i \int_{\mathbb{R}^2} \frac{\partial_1 u \partial_2 \bar{u} - \partial_1 \bar{u} \partial_2 u}{(1+|u|^2)^2}$$

$$= 2 \int_{\mathbb{R}^2} \frac{|\partial_1 u \pm i \partial_2 u|^2}{(1+|u|^2)^2} + 4\pi |\deg(\phi)|. \tag{2.1.26}$$

Note that the validity of the above is independent of any specific asymptotic behavior of a field configuration at infinity. Hence, again, $E \geq 4\pi |\deg(\phi)|$ and such a topological lower bound is saturated if and only if u satisfies the linear equation

$$\partial_1 u \pm i \partial_2 u = 0, \tag{2.1.27}$$

which is the Cauchy–Riemann system we arrived at earlier. Thus the general solution is as described in the formula (2.1.21).

2.1.3 Topology

We now examine that the solution (2.1.21) indeed gives rise to a map $\phi : \mathbb{R}^2 \to S^2$ of degree $\pm N$. Without loss of generality, we work out the case $\deg(\phi) \geq 0$ only. The other case can be seen through a change of orientation.

It will be convenient to introduce the current density

$$J_k = \frac{i}{1+|u|^2} (u \partial_k \bar{u} - \bar{u} \partial_k u), \quad k = 1, 2. \tag{2.1.28}$$

We also observe that there holds the elegant relation

$$J_{12} = \partial_1 J_2 - \partial_2 J_1 = \frac{2i}{(1+|u|^2)^2} (\partial_1 u \partial_2 \bar{u} - \partial_1 \bar{u} \partial_2 u). \tag{2.1.29}$$

As a consequence of (2.1.12), (2.1.24), (2.1.29), and the divergence theorem, we have

$$4\pi \deg(\phi) = \int_{\mathbb{R}^2} J_{12}$$

$$= \lim_{r \to \infty} \oint_{|x|=r} J_k \, dx^k - \lim_{r \to 0} \sum_{j=1}^{N} \oint_{|x-p_j|=r} J_k \, dx^k. \tag{2.1.30}$$

On the other hand, using (2.1.21) and (2.1.28), we have for $|x| \to \infty$ the asymptotic behavior

$$J_k = O(|x|^{-\delta}), \quad k = 1, 2,$$

$$\delta = 1 + 2 \max\{0, N - M\}, \tag{2.1.31}$$

which implies that the first term on the right-hand side of (2.1.30), which is an integral along a circle near infinity, is zero if $N > M$.

In order to calculate the rest of the terms on the right-hand side of (2.1.30), it is convenient to use the complex derivatives

$$\partial = \frac{1}{2}(\partial_1 - i\partial_2), \quad \bar{\partial} = \frac{1}{2}(\partial_1 + i\partial_2). \tag{2.1.32}$$

In view of (2.1.27) and (2.1.32), we have

$$\partial_1 u = \partial u, \quad \partial_2 u = i\partial u, \quad \partial_1 \bar{u} = \bar{\partial}\bar{u}, \quad \partial_2 \bar{u} = -i\bar{\partial}\bar{u}. \tag{2.1.33}$$

From (2.1.21), we see that, near $z = p_j$, there holds

$$\partial u = u\left(-\frac{1}{z - p_j} + O(1)\right), \quad j = 1, 2, \cdots, N. \tag{2.1.34}$$

With

$$\begin{aligned} y^1 &= \operatorname{Re}\{z - p_j\} \quad \text{(real part)}, \\ y^2 &= \operatorname{Im}\{z - p_j\} \quad \text{(imaginary part)}, \end{aligned}$$

it is seen from (2.1.28), (2.1.33), and (2.1.34) that

$$\begin{aligned} J_1 &= \frac{2|u|^2}{1 + |u|^2}\left(\frac{y^2}{|z - p_j|^2} + O(1)\right), \\ J_2 &= -\frac{2|u|^2}{1 + |u|^2}\left(\frac{y^1}{|z - p_j|^2} + O(1)\right) \end{aligned} \tag{2.1.35}$$

near p_j.

Using either the polar coordinates $y^1 = r\cos\theta$ and $y^2 = r\sin\theta$ or Green's theorem, it follows from (2.1.35) that

$$\begin{aligned} \lim_{r \to 0} \oint_{|z - p_j| = r} J_k \, dx^k &= \lim_{r \to 0}\left(\frac{2}{r^2} \oint_{|z - p_j| = r} y^2 \, dy^1 - y^1 \, dy^2\right) \\ &= -4\pi. \end{aligned} \tag{2.1.36}$$

Thus, inserting (2.1.36) into (2.1.30), we obtain the anticipated result, $\deg(\phi) = N$.

Since the largest value of M is $M = N - 1$, we see that these general solutions are labelled by $2N + 2(N-1) + 2 = 4N$ continuous real parameters for which the $2N$ (or $2M = 2(N-1)$) parameters characterize the locations of the poles (or zeros), $p_1, p_2, \cdots, p_N$ (or $q_1, q_2, \cdots, q_M$), and the additional two parameters come from the complex conformal factor c in (2.1.21).

2.2   Gauged Sigma Model

A minimal extension of the sigma model just discussed is to include a gauge field. In this section we present such an extension due to Schroers [274]. We first develop the gauged sigma model and derive the same topological characteristics for the solutions as in the classical sigma model. We then state our existence theorems. It will be seen that the gauged sigma model offers a much richer structure.

2.2.1   Field theory and self-dual equations

Consider static fields. Since the functional (2.1.9) is invariant under the group of rotations, $O(3)$, applied on ϕ, we may always assume that the asymptotic state ϕ_∞ of a finite-energy field configuration ϕ is simply $\mathbf{n}$ (the north pole). This choice breaks the $O(3)$ symmetry down to $SO(2) = U(1)$, which is clearly represented by the elements of $O(3)$ that preserve the direction $\mathbf{n}$. Consequently one is tempted to include a minimal potential term like $(1-\mathbf{n}\cdot\phi)^2$ in the energy integral to automatically respect such a choice of the asymptotics. However, although the new energy is still clearly bounded from below and satisfies (2.1.13), a simple scaling argument as that in the Derrick theorem indicates that there can be no nontrivial solutions. In fact, let us consider the modified energy

$$E(\phi) = \frac{1}{2} \int_{\mathbb{R}^2} \{(\partial_1\phi)^2 + (\partial_2\phi)^2 + (1 - \mathbf{n} \cdot \phi)^2\}. \qquad (2.2.1)$$

If ϕ is a critical point, then $\phi_t(x) = \phi(tx)$ $(t > 0)$ satisfies

$$\left.\frac{\mathrm{d}E(\phi_t)}{\mathrm{d}t}\right|_{t=1} = 0,$$

which leads to

$$\int_{\mathbb{R}^2} (1 - \mathbf{n} \cdot \phi)^2 = 0.$$

Namely, ϕ must be trivial, $\phi = \mathbf{n}$. To overcome this difficulty, we recall that a standard device in gauge theory is to add a gauge vector field. Thus we need to explore the internal symmetry of (2.2.1).

We first consider our problem over the Minkowski spacetime $\mathbb{R}^{2,1}$. Thus the Lagrangian action density is

$$\mathcal{L} = \frac{1}{2}\partial_\mu\phi \cdot \partial^\mu\phi - \frac{1}{2}(1 - \mathbf{n} \cdot \phi)^2. \qquad (2.2.2)$$

As mentioned earlier, the subgroup of $O(3)$ that preserves the vacuum state

$$\phi_{\mathbf{v}} = \mathbf{n} \qquad (2.2.3)$$

is the Abelian rotational group $SO(2)$ represented by

$$\phi = (\phi_1, \phi_2, \phi_3)$$
$$\mapsto (\phi_1 \cos\theta - \phi_2 \sin\theta, \phi_1 \sin\theta + \phi_2 \cos\theta, \phi_3), \quad \theta \in \mathbb{R}. \quad (2.2.4)$$

It is clear that (2.2.2) is invariant under the global symmetry (2.2.4). Of course, we may also introduce the complex field $\psi = \phi_1 + i\phi_2$ to compress the $SO(2)$ symmetry (2.2.4) into a $U(1)$ symmetry,

$$\psi \mapsto e^{i\theta}\psi, \quad \phi_3 \mapsto \phi_3, \quad (2.2.5)$$

and rewrite the action density (2.2.2) as

$$\mathcal{L} = \frac{1}{2}\partial_\mu\psi\partial^\mu\overline{\psi} + \frac{1}{2}\partial_\mu\phi_3\partial^\mu\phi_3 - \frac{1}{2}(1 - \phi_3)^2. \quad (2.2.6)$$

In order to enlarge the global internal symmetry of (2.2.6) into a local one, we have to replace the partial derivatives on ψ by gauge-covariant derivatives,

$$D_\mu\psi = \partial_\mu\psi - iA_\mu\psi, \quad \mu = 0, 1, 2, \quad (2.2.7)$$

and require that the gauge vector field $A = (A_\mu)$ transforms itself accordingly with (2.2.5) when θ is a local quantity (i.e., a function over spacetime),

$$A_\mu \mapsto A_\mu + \partial_\mu\theta. \quad (2.2.8)$$

Introduce the Maxwell field

$$F_{\mu\nu} = \partial_\mu A_\nu - \partial_\nu A_\mu. \quad (2.2.9)$$

Then we obtain from (2.2.6) the following complete local $U(1)$-invariant action density,

$$\mathcal{L} = -\frac{1}{4}F_{\mu\nu}F^{\mu\nu} + \frac{1}{2}D_\mu\psi\overline{D^\mu\psi} + \frac{1}{2}\partial_\mu\phi_3\partial^\mu\phi_3 - \frac{1}{2}(1 - \phi_3)^2. \quad (2.2.10)$$

Finally, a simple calculation shows that we can replace A_μ by $-A_\mu$ and rewrite (2.2.10) more concisely as

$$\mathcal{L} = -\frac{1}{4}F_{\mu\nu}F^{\mu\nu} + \frac{1}{2}D_\mu\phi \cdot D^\mu\phi - \frac{1}{2}(1 - \mathbf{n} \cdot \phi)^2. \quad (2.2.11)$$

where the new gauge-covariant derivatives on ϕ are defined by

$$D_\mu\phi = \partial_\mu\phi + A_\mu(\mathbf{n} \times \phi), \quad \mu = 0, 1, 2. \quad (2.2.12)$$

The action density (2.2.11) defines the gauged sigma model. From now on we focus on static solutions. We assume the temporal gauge, $A_0 = 0$, and use the notation $\mathbf{A} = (A_1, A_2)$. Hence the total energy derived from (2.2.11) is

$$E(\phi, \mathbf{A}) = \frac{1}{2} \int_{\mathbb{R}^2} \{(D_1\phi)^2 + (D_2\phi)^2 + (1 - \mathbf{n} \cdot \phi)^2 + F_{12}^2\}, \qquad (2.2.13)$$

which implies the boundary condition

$$\lim_{|x| \to \infty} \mathbf{n} \cdot \phi(x) = 1. \qquad (2.2.14)$$

This condition will be observed throughout the rest of our study.

We identify the curvature F_{12} as the magnetic field and consider the magnetic flux

$$\Phi = \Phi(\phi, \mathbf{A}) = \int_{\mathbb{R}^2} F_{12}. \qquad (2.2.15)$$

In order to find a similar topological lower energy bound for (2.2.13), we shall write down an expression that represents $\deg(\phi)$ in terms of both ϕ and $\mathbf{A}$. To motivate such a construction, we recall the following vector identity for arbitrary 3-vectors $\mathbf{A}, \mathbf{B}, \mathbf{C}$,

$$\mathbf{A} \times (\mathbf{B} \times \mathbf{C}) = \mathbf{B}(\mathbf{A} \cdot \mathbf{C}) - \mathbf{C}(\mathbf{A} \cdot \mathbf{B}). \qquad (2.2.16)$$

Thus we can use (2.2.16) and the fact that $\phi \cdot \phi = 1$ to get

$$
\begin{aligned}
&\phi \cdot (\partial_1\phi \times \partial_2\phi) = \phi \cdot ([D_1\phi - A_1(\mathbf{n} \times \phi)] \times [D_2\phi - A_2(\mathbf{n} \times \phi)]) \\
&= \phi \cdot (D_1\phi \times D_2\phi) - \phi \cdot (A_2 D_1\phi \times [\mathbf{n} \times \phi] - A_1 D_2\phi \times [\mathbf{n} \times \phi]) \\
&= \phi \cdot (D_1\phi \times D_2\phi) - \phi \cdot (A_2 \partial_1\phi \times [\mathbf{n} \times \phi] - A_1 \partial_2\phi \times [\mathbf{n} \times \phi]) \\
&= \phi \cdot (D_1\phi \times D_2\phi) + (A_2\mathbf{n} \cdot \partial_1\phi - A_1\mathbf{n} \cdot \partial_2\phi) \\
&= \phi \cdot (D_1\phi \times D_2\phi) + F_{12}(1 - \mathbf{n} \cdot \phi) + H_{12}, \qquad (2.2.17)
\end{aligned}
$$

where

$$H_{12} = \partial_1 H_2 - \partial_2 H_1, \quad H_j = A_j(\mathbf{n} \cdot \phi - 1), \quad j = 1, 2, \qquad (2.2.18)$$

and it will become clear later that H_{12} is a 'tail' term. In fact, if the 'current density' H_j fulfills the asymptotic estimate

$$|H_j| = |A_j(\mathbf{n} \cdot \phi - 1)| = O(|x|^{-(1+\varepsilon)}) \quad (\varepsilon > 0), \quad j = 1, 2, \qquad (2.2.19)$$

near infinity, then we obtain from the divergence theorem that

$$\int_{\mathbb{R}^2} H_{12} = \lim_{r \to \infty} \oint_{|x|=r} H_k \, dx^k = 0. \qquad (2.2.20)$$

As a consequence of (2.2.19) and (2.2.20), we conclude that, within the category of the solutions satisfying (2.2.19), the degree formula (2.1.12) takes the form

$$\deg(\phi) = \frac{1}{4\pi} \int_{\mathbb{R}^2} \{\phi \cdot (D_1\phi \times D_2\phi) + F_{12}(1 - \mathbf{n} \cdot \phi)\}. \qquad (2.2.21)$$

Thus, in view of (2.2.21), the energy (2.2.13) can be recast into the form

$$E(\phi, \mathbf{A}) = 4\pi |\deg(\phi)|$$
$$+\frac{1}{2}\int_{\mathbb{R}^2}\{(D_1\phi \pm \phi \times D_2\phi)^2 + (F_{12} \mp [1 - \mathbf{n}\cdot\phi])^2\}, \quad (2.2.22)$$

for $\deg(\phi) = \pm|\deg(\phi)|$. Consequently, as (2.1.13) for the $O(3)$ sigma model, the energy (2.2.13) has the *same* topological lower bound and that this bound, (2.1.13), is attained if and only if $(\phi, \mathbf{A})$ solves the self-dual or anti-self-dual equations

$$D_1\phi = \mp\phi \times D_2\phi, \quad (2.2.23)$$
$$F_{12} = \pm(1 - \mathbf{n}\cdot\phi). \quad (2.2.24)$$

For definiteness, we concentrate on the self-dual case, $\deg(\phi) = N \geq 0$. The other case may be obtained by a suitable transformation.

Note that, unlike the classical sigma model, the system, (2.2.23) and (2.2.24), is *nonlinear* and cannot be solved explicitly. In fact, several new features are present in the model and a understanding of the solutions of (2.2.23), (2.2.24) relies on nonlinear functional analysis.

2.2.2   Multisolitons: existence theorems

Since $\phi \in S^2$, we have $-1 \leq \mathbf{n}\cdot\phi \leq 1$. Inserting this property in the equation (2.2.24) we see that the magnetic field $B = F_{12}$ satisfies $0 \leq B \leq 2$. In other words the flux-lines are penetrating the plane along the same direction and the locations of the spots where the magnetic field achieves the greatest possible penetration is given by the those points $p \in \mathbb{R}^2$ at which

$$B(p) = B_{\max} = 2. \quad (2.2.25)$$

Of course, due to the structure of the equation (2.2.24), the locations of the peaks of the magnetic field given in (2.2.25) are equivalently represented by the valleys of the projection of the map ϕ along $\mathbf{n}$, at which

$$\mathbf{n} \cdot \phi(p) = (\mathbf{n} \cdot \phi)_{\min} = -1. \quad (2.2.26)$$

Such a simple identification leads to the important observation that the number of points at which the peaks or valleys take place is actually the degree N of the map ϕ. To see this fact we still use $\mathbf{s}$ to denote the south pole of S^2 and we can express the set of the locations of valleys defined in (2.2.26) as P given in (2.1.17). It will be seen as before that $\#P$ (the number of points)$= N = \deg(\phi)$, which is the old topological invariant. We now state and comment on our main results.

Theorem 2.2.1. *The system of equations (2.2.23) and (2.2.24) does not have any solution of unit degree satisfying the finite-energy boundary condition (2.2.14).*

This nonexistence property is a new feature which does not occur in the classical sigma model. Hence, to obtain the existence of solutions, we need to consider the case that the degree of the map ϕ is greater than one. We have

Theorem 2.2.2. *For any $N \geq 2$ the system (2.2.23), (2.2.24) has a family of finite-energy smooth solutions containing at least $(4N - 3)$ independent continuous parameters to fulfill the topological condition $\deg(\phi) = N$.*

In fact this theorem may be stated in the following precise form.

Theorem 2.2.3. *For any integers M, N satisfying $N \geq 2$, $M < N - 1$, the parameter α in the range*

$$1 - \frac{M + 2}{N} < \alpha < 1 - \frac{M + 1}{N}, \tag{2.2.27}$$

and the points

$$p_1, p_2, \cdots, p_N, q_1, q_2, \cdots, q_M \in \mathbb{R}^2$$

which may appear repeatedly in the collection, the system (2.2.23), (2.2.24) has a solution $(\phi_{(\alpha)}, \mathbf{A}_{(\alpha)})$ so that

$$\left\{ p_1, p_2, \cdots, p_N \right\} = \phi_{(\alpha)}^{-1}(\mathbf{s}), \quad \left\{ q_1, q_2, \cdots, q_M \right\} = \phi_{(\alpha)}^{-1}(\mathbf{n}),$$

the energy lower bound (2.1.13) is saturated at the quantized level,

$$E(\phi_{(\alpha)}, \mathbf{A}_{(\alpha)}) = 4\pi N \tag{2.2.28}$$

with $\deg(\phi_{(\alpha)}) = N$, and the total magnetic flux (2.2.15) is given by

$$\Phi(\phi_{(\alpha)}, \mathbf{A}_{(\alpha)}) = 2\pi N \alpha. \tag{2.2.29}$$

Furthermore, set

$$\beta = 2(N - M) - 2N\alpha. \tag{2.2.30}$$

Then the solution decays at infinity according to

$$\begin{aligned}
\phi_1^2, \phi_2^2, 1 - \phi_3^2 &= O(|x|^{-\beta}), \\
B = 1 - \mathbf{n} \cdot \phi &= O(|x|^{-\beta}), \\
|D_1\phi|^2 + |D_2\phi|^2 &= o(|x|^{-\beta}).
\end{aligned} \tag{2.2.31}$$

In other words, for given integer $N \geq 2$ and the points $p_1, p_2, \cdots, p_N \in \mathbb{R}^2$, the system (2.2.23), (2.2.24) has a solution so that these points p's are realized as the locations of the maximal peaks of the magnetic field and the total energy is proportional to the number N, which may well be viewed as the number of particles in the system. Besides, the magnetic flux (2.2.29) can be designated in a continuous interval by using the parameter α in (2.2.27) and the decay exponent β in (2.2.31) can also be specified in the continuous range $2 < \beta < 2(N - M)$ so that α and β are related to each other through (2.2.30).

The count of the number of parameters stated in Theorem 2.2.2 now follows easily from Theorem 2.2.3: the points $p_1, p_2, \cdots, p_N$ give $2N$ parameters, which are simply their coordinates; the points $q_1, q_2, \cdots, q_M$ can give up to $2(N - 2)$ parameters because we may have $M = N - 2$; the parameter α or β gives one more degree of freedom.

In the Ginzburg–Landau theory of superconductivity the greatest penetration of the magnetic field occurs at the zeros of the order parameter where the normal phase is restored. Such a phenomenon is known as the Meissner effect. In our gauged sigma model, however, the greatest penetration of the magnetic field B occurs at the point set $\phi^{-1}(s)$ in $\mathbb{R}^2$. If B is viewed as a vorticity field the system clearly represents multivortices. If B is viewed as a physical force such as the magnetic field the system represents multisolitons, which may be thought as a collection of identical particles because the maximal peaks of B indicate well-defined force concentration at those points. The surprising feature of the solution is that if the points $p_1, p_2, \cdots, p_N$ are viewed as magnetic monopoles, or vortices, then the points $q_1, q_2, \cdots, q_M$ behave like antimonopoles, or antivortices. This phenomenon may be seen from (2.2.27) and (2.2.29): although, according to the equation (2.2.24) the magnetic field at p's and q's has the same sign (in fact the magnetic field at the points q's vanishes), the formula (2.2.27) says that the presence of the points q's reduces the total flux. Besides, by (2.2.28), the total energy only depends on the number N of the points p's but not the number M of the points q's. When $M = 0$ but N is large, the expression (2.2.29) gives rise to a flux which is close to what is to be expected in the Ginzburg–Landau theory.

To conclude this section, we also notice that Theorem 2.2.1 is in fact a consequence of the following more general statement.

Theorem 2.2.4. *Let N be a positive integer. Then the system of equations (2.2.23) and (2.2.24) has a solution $(\phi, \mathbf{A})$ satisfying the finite-energy boundary condition (2.2.14) along with $\deg(\phi) = N$ and $\#\phi^{-1}(\mathbf{n}) = M$ if and only if $N - 2 \geq M$.*

Since $M \geq 0$ it is seen that there is a degree N solution if and only if $N \geq 2$ and Theorem 2.2.1 follows immediately.

In the next section, we rewrite the system (2.2.23), (2.2.24) into an elliptic scalar equation with some Dirac function type source terms representing vortices and antivortices. In the subsequent section, we present a complete mathematical analysis of this equation using the weighted Sobolev spaces and establish all the results stated here.

2.3 Governing Equations and Characterization

Using (2.1.15), we reduce (2.2.23) and (2.2.24) into the system

$$D_1 u = -iD_2 u, \tag{2.3.1}$$

$$B = \frac{2|u|^2}{1 + |u|^2}, \tag{2.3.2}$$

where $D_j u = \partial_j u + iA_j u$, $j = 1, 2$, and $x \in \mathbb{R}^2$, $x \notin P = \phi^{-1}(\mathbf{s})$.

We need to recall the $\bar{\partial}$-Poincaré lemma argument in [157] to reduce this system. Indeed, we use the complex derivative (2.1.32) to rewrite (2.3.1) as

$$\bar{\partial} u = -i\omega u, \quad \omega = \frac{1}{2}(A_1 + iA_2). \tag{2.3.3}$$

In view of the $\bar{\partial}$-Poincaré lemma, the equation $\bar{\partial} f = i\omega$ can be solved locally [137]. Inserting this result into the above equation, we see that $\bar{\partial}(ue^f) = 0$. Hence, locally, we can represent u as $u(z) = h(z)e^{-f(z)}$ where h is holomorphic and e^{-f} never vanishes. Thus, any zero of u will have an integer multiplicity.

We assume that $\phi^{-1}(\mathbf{n})$ is finite and set as before

$$Q = \phi^{-1}(\mathbf{n}) = \left\{ q_1, q_2, \cdots, q_\ell \right\}. \tag{2.3.4}$$

Then $\phi_1(q_j) = 0, \phi_2(q_j) = 0, \phi_3(q_j) = 1$, and, so, $u(q_j) = 0, j = 1, 2, \cdots, \ell$. Consequently there are positive integers $m_1, m_2, \cdots, m_\ell$ so that, according to the discussion just made, up to a positive factor,

$$|u(x)| = |x - q_j|^{m_j}, \quad j = 1, 2, \cdots, \ell. \tag{2.3.5}$$

Similarly, we shall see that near a point $p \in P = \phi^{-1}(\mathbf{s})$, the function u may be written as the product of a nonvanishing function and a meromorphic function with p as the only singular point. In fact p is a pole. To see this, we consider $U = 1/u$ near p. Then (2.1.18) says that U is continuous at p and $U(p) = 0$. Furthermore, simple calculation gives us

$$\overline{D_j} U = -U^2 D_j u, \quad j = 1, 2, \quad x \neq p.$$

Here $\overline{D_j} U = (\partial_j - iA_j)U$. Using (2.3.1), we have $\overline{D_1} U = -i\overline{D_2} U$. Therefore the above argument and the removable singularity theorem tell us that, near p, U is the product of a nonvanishing function and a holomorphic function. Hence p must be a zero of U with an integer multiplicity. Such a property is crucial for the rest of the work here.

In particular, assume as before

$$P = \phi^{-1}(\mathbf{s}) = \left\{ p_1, p_2, \cdots, p_k \right\}. \tag{2.3.6}$$

Then there are positive integers $n_1, n_2, \cdots, n_k$ so that, up to a positive factor,

$$|u(x)| = \frac{1}{|x - p_j|^{n_j}}, \quad j = 1, 2, \cdots, k. \tag{2.3.7}$$

We are now ready to derive from (2.3.1) and (2.3.2) the governing non-linear elliptic equation. First, away from the set $P \cup Q$, we have in view of (2.3.3) that

$$
\begin{aligned}
B & = \partial_1 A_2 - \partial_2 A_1 = \mathrm{Im}\{4\partial\omega\} \\
& = -2\mathrm{i}(\partial\omega - \overline{\partial\omega}) = 2\partial\overline{\partial}\ln u + 2\overline{\partial}\partial\ln\overline{u} \\
& = \frac{1}{2}(\partial_1^2 + \partial_2^2)\ln|u|^2 = \frac{1}{2}\Delta\ln|u|^2.
\end{aligned}
$$

Therefore, using the information stated in (2.3.5) and (2.3.7) and converting to the new variable $v = \ln|u|^2$, we see that the system of equations (2.3.1) and (2.3.2) is reduced into the following scalar equation with the Dirac function type point source terms

$$\Delta v = \frac{4e^v}{1 + e^v} - 4\pi \sum_{j=1}^{k} n_j \delta_{p_j} + 4\pi \sum_{j=1}^{\ell} m_j \delta_{q_j}, \quad x \in \mathbb{R}^2. \tag{2.3.8}$$

Equation (2.3.8) is a typical planar 'vortex' equation. Its most interesting feature is that vortices of opposite charges (at p's and q's, respectively) coexist. Although the vorticity or magnetic field B does not change its sign the vortices at p's may still be called antivortices, whereas, at q's, vortices, due to their opposite magnetic properties.

The next section is a study of (2.3.8) based on functional analysis. Since finite-energy condition $\phi(x) \to \mathbf{n}$ as $|x| \to \infty$ requires $u(x) \to 0$ as $|x| \to \infty$, we need to look for solutions of (2.3.8) satisfying $v(x) \to -\infty$ as $|x| \to \infty$.

2.4   Mathematical Analysis

We shall use a variational method. We first introduce a background function to regularize the problem and derive the range of an important parameter. We then apply the weighted Sobolev space techniques and the calculus of variations to find a subsolution-supersolution pair. We next prove the existence of a bounded solution by a monotone iteration method and establish some important asymptotic estimates. As a further step we show how to recover a solution configuration of the equations, examine the values of the quantized energy and fractional magnetic flux, and calculate the topological degree of the obtained solutions. Finally we present a proof that there does not exist a solution of unit degree.

2.4.1 Regularized equation and range of parameter

To save space, we allow repetition of p's and q's to accommodate multiplicities. Thus, in place of (2.3.8), we have the following governing equation,

$$\Delta v = \frac{4e^v}{1+e^v} - 4\pi \sum_{j=1}^{N} \delta_{p_j} + 4\pi \sum_{j=1}^{M} \delta_{q_j}, \quad x \in \mathbb{R}^2. \qquad (2.4.1)$$

To proceed, we need to define $\rho(t)$ to be a smooth monotone increasing function over $t > 0$ so that

$$\rho(t) = \begin{cases} \ln t, & t \leq \frac{1}{2}, \\ 0, & t \geq 1, \\ \leq 0, & \text{for all } t > 0. \end{cases}$$

Let $\delta > 0$ be such that

$$B_\delta(p_j) = \{x \,|\, |x - p_j| < \delta\}, \quad j = 1, 2, \cdots, N$$

are disjoint, namely, $B_\delta(p_j) \cap B_\delta(p_{j'}) = \emptyset$, $p_j \neq p_{j'}$. Consider the functions

$$u_j(x) = 2\rho\left(\frac{|x - p_j|}{\delta}\right), \quad j = 1, 2, \cdots, N.$$

Then $u_j \leq 0$, $u_j = 0$ for $|x - p_j| \geq \delta$, and $\Delta u_j = 4\pi \delta_{p_j}(x) - u_{0,j}$, where, of course,

$$\text{supp}(u_{0,j}) \subset \left\{x \,\Big|\, \frac{\delta}{2} \leq |x - p_j| \leq \delta\right\}$$

and $u_{0,j}$ is smooth. Besides, we have

$$\int_{\mathbb{R}^2} u_{0,j}\, dx = \int_{|x-p_j|\geq \delta/3} (-\Delta u_j)\, dx$$

$$= \oint_{|x-p_j|=\delta/3} \frac{\partial u_j}{\partial n} ds = 4\pi.$$

Define the background function $v_1 = \sum_{j=1}^{N} u_j$. Then

$$\Delta v_1 = 4\pi \sum_{j=1}^{N} \delta_{p_j} - g_1,$$

where

$$g_1 = \sum_{j=1}^{N} u_{0,j}, \quad \int_{\mathbb{R}^2} g_1 = 4\pi N.$$

The function v_1 has the useful property that

$$v_1 \leq 0, \quad v_1 = 0 \quad \text{on } \mathbb{R}^2 - \Omega_\delta,$$

where $\Omega_\delta = \cup_{j=1}^N B_\delta(p_j)$. Similarly, we define the function $v_2 = \sum_{j=1}^M u'_j$, where

$$u'_j(x) = 2\rho\left(\frac{|x - q_j|}{\delta}\right), \quad j = 1, 2, \cdots, M,$$

and we assume that $\delta > 0$ is small to make $B_\delta(q_j)$ $(j = 1, 2, \cdots, M)$ and $B_\delta(p_j)$ $(j = 1, 2, \cdots, N)$ disjoint. Then we also have $v_2 \leq 0$ and

$$\Delta v_2 = 4\pi \sum_{j=1}^M \delta_{q_j} - g_2, \quad g_2 \in C_c^\infty(\mathbb{R}^2), \quad \int_{\mathbb{R}^2} g_2 = 4\pi M.$$

With the above preparation, we introduce the substitution $v = -v_1 + v_2 + V$ in (2.4.1). Then V satisfies

$$\Delta V = \frac{4e^{-v_1+v_2+V}}{1 + e^{-v_1+v_2+V}} - (g_1 - g_2). \tag{2.4.2}$$

Note that e^{-v_1} has finite poles. Choose $v_3 \in C^\infty(\mathbb{R}^2)$ so that

$$v_3(x) = -\ln|x|, \quad |x| \geq 1, \quad x \in \mathbb{R}^2.$$

As before, we have

$$\int_{\mathbb{R}^2}(-\Delta v_3) = 2\pi.$$

Let $\beta > 2$ and put $K = e^{\beta v_3}$. Then $V = \beta v_3 + w$ transforms (2.4.2) into

$$\Delta w = \frac{4Ke^{v_2+w}}{e^{v_1} + Ke^{v_2+w}} - g, \tag{2.4.3}$$

where $g = g_1 - g_2 + \beta \Delta v_3$ is of compact support and satisfies

$$\int_{\mathbb{R}^2} g = 2\pi(2[N - M] - \beta) > 0 \tag{2.4.4}$$

if the parameter β obeys the requirement

$$2 < \beta < 2(N - M). \tag{2.4.5}$$

We will solve (2.4.3) by constructing a suitable sub- and supersolution pair.

2.4.2 Subsolution and variational method

Choose a function $h \in C_c^\infty(\mathbb{R}^2)$ that has the same total integral over $\mathbb{R}^2$ as g (see (2.4.4)),

$$\int_{\mathbb{R}^2} h = 2\pi(2[N - M] - \beta), \tag{2.4.6}$$

and $h \geq 0$. We first study an easier version of (2.4.3), namely,

$$\Delta w = \frac{4Ke^w}{e^{v_1} + Ke^w} - h. \tag{2.4.7}$$

Our equation here shares many common features with that in the prescribed Gauss curvature problem in $\mathbb{R}^2$. Thus we can also rely on some tools from the weighted Sobolev space theory [213, 214]. The weight function that we will use is K,

$$K = e^{\beta v_3} > 0, \quad K = O(|x|^{-\beta}), \quad \beta > 2. \tag{2.4.8}$$

We denote the usual L^p space over $\mathbb{R}^2$ simply by L^p and its norm by $\| \ \|_p$. Define the weighted measure $d\mu = K \, dx$ and use $L^p(d\mu)$ to denote the induced L^p space. Let $\mathcal{X}$ be the space of L^2 functions w such that

$$\|w\|_{\mathcal{X}}^2 = \|\nabla w\|_2^2 + \|w\|_{L^2(d\mu)}^2 < \infty.$$

Then $\mathcal{X}$ contains all constant functions and, thus,

$$\mathcal{X}' = \left\{ w \in \mathcal{X} \ \middle| \ \int_{\mathbb{R}^2} w \, d\mu = 0 \right\}$$

is a closed subspace of $\mathcal{X}$. Recall that there are positive constants

$$\gamma < \min(4\pi, 2\pi[\beta - 2]) \quad \text{and} \quad C = C(\gamma),$$

where β is as given in (2.4.8) so that the following Trudinger–Moser type inequality holds [214],

$$\int_{\mathbb{R}^2} e^{|w|} \, d\mu \leq C e^{\|\nabla w\|_2^2/4\gamma}, \quad w \in \mathcal{X}'. \tag{2.4.9}$$

For our problem here the range of γ will not be important.

Furthermore, the embedding $\mathcal{X} \to L^2(d\mu)$ is completely continuous [214].

We now prove the existence of a solution of (2.4.7) by obtaining a critical point of the functional

$$I(w) = \int_{\mathbb{R}^2} \left\{ \frac{1}{2}|\nabla w|^2 + 4\ln(e^{v_1} + Ke^w) - hw \right\}$$

in the admissible space

$$A = \left\{ w \in \mathcal{X} \ \middle| \ \int_{\mathbb{R}^2} \frac{4Ke^w}{e^{v_1} + Ke^w} = \int_{\mathbb{R}^2} h \right\}.$$

Lemma 2.4.1. *The functional I is bounded from below over A, or more precisely,*

$$I(w) \geq C_1 \|\nabla w\|_2^2 - C_2, \tag{2.4.10}$$

where $C_1, C_2 > 0$ are constants.

Proof. To prove (2.4.10), we first use the fact that $\mathrm{supp}(v_1) \subset \Omega_\delta$ to rewrite $I(w)$ in the form

$$I(w) = \int_{\mathbb{R}^2} \left\{ \frac{1}{2} |\nabla w|^2 + 4\ln(1 + Ke^w) - hw \right\} + 4 \int_{\Omega_\delta} \ln \left(\frac{e^{v_1} + Ke^w}{1 + Ke^w} \right). \tag{2.4.11}$$

Since the function $\sigma(t) = (a + t)/(1 + t)$ $(a \leq 1)$ increases, we have $\sigma(t) \geq \sigma(0)$ for $t \geq 0$. Therefore $(e^{v_1} + Ke^w)/(1 + Ke^w) > e^{v_1}$ and

$$4 \int_{\Omega_\delta} \ln \left(\frac{e^{v_1} + Ke^w}{1 + Ke^w} \right) \geq 4 \int_{\Omega_\delta} v_1 > -C_\delta \quad \text{(finite)}.$$

Inserting the above into (2.4.11), we have

$$I(w) \geq \int_{\mathbb{R}^2} \left\{ \frac{1}{2} |\nabla w|^2 + 4\ln(1 + Ke^w) - hw \right\} - C_\delta. \tag{2.4.12}$$

On the other hand, from the simple relation

$$\ln(1 + Ke^w) = \beta v_3 + w + \ln(1 + e^{-(\beta v_3 + w)})$$

and substitution $W = \ln(1 + Ke^w)$, we have the representation

$$w = W - \beta v_3 - \ln(1 + e^{-(\beta v_3 + w)}).$$

Thus, in terms of W, we derive from (2.4.12) that

$$\begin{aligned} I(w) \geq\ & \frac{1}{2} \int_{\mathbb{R}^2} |\nabla w|^2 + 4 \int_{\mathbb{R}^2} W - \int_{\mathbb{R}^2} hW \\ & + \int_{\mathbb{R}^2} h(\beta v_3) + \int_{\mathbb{R}^2} h\ln(1 + e^{-(\beta v_3 + w)}) - C_\delta. \end{aligned} \tag{2.4.13}$$

Using

$$\nabla W = \frac{Ke^w}{1 + Ke^w}(\beta \nabla v_3 + \nabla w)$$

and the fact that

$$\begin{aligned} \int_{\mathbb{R}^2} \left(\frac{Ke^w}{1 + Ke^w} \right)^2 &\leq \int_{\mathbb{R}^2} \frac{Ke^w}{1 + Ke^w} \\ &\leq \int_{\mathbb{R}^2} \frac{Ke^w}{e^{v_1} + Ke^w} = \pi \left([N - M] - \frac{\beta}{2} \right) \end{aligned}$$

(see the definition of the admissible space $\mathcal{A}$ and (2.4.6)), we have

$$\|\nabla W\|_2 \leq \|\nabla w\|_2 + C.$$

Applying this inequality to (2.4.13), we get

$$I(w) \geq \frac{1}{4}\|\nabla w\|_2^2 + \frac{1}{4}\|\nabla W\|_2^2 + 4\int_{\mathbb{R}^2} W - \int_{\mathbb{R}^2} hW - C_1. \qquad (2.4.14)$$

Since $W \geq 0$, we have $W \geq W^2/(1+W)$. So the above gives us

$$I(w) \geq \frac{1}{4}\|\nabla w\|_2^2 + \frac{1}{4}\|\nabla W\|_2^2 + 4\int_{\mathbb{R}^2} \frac{W^2}{1+W} - \int_{\mathbb{R}^2} hW - C_1. \qquad (2.4.15)$$

To proceed further, we need to invoke the following standard interpolation inequality over $\mathbb{R}^2$,

$$\int_{\mathbb{R}^2} W^4 \leq 2\int_{\mathbb{R}^2} W^2 \int_{\mathbb{R}^2} |\nabla W|^2. \qquad (2.4.16)$$

Since h is of compact support, we may use (2.4.16) to obtain the bound

$$\begin{aligned}
\int_{\mathbb{R}^2} hW &\leq \|h\|_{\frac{4}{3}}\|W\|_4 \\
&\leq \varepsilon\|W\|_2 + C(\varepsilon)\|\nabla W\|_2 + C \\
&\leq \varepsilon\|W\|_2 + \frac{1}{8}\|\nabla W\|_2^2 + C(\varepsilon), \qquad (2.4.17)
\end{aligned}$$

where $\varepsilon > 0$ is small. Now, using (2.4.16) again, we have

$$\begin{aligned}
\|W\|_2^4 &= \left(\int_{\mathbb{R}^2} \frac{W}{1+W}(1+W)W\right)^2 \\
&\leq \int_{\mathbb{R}^2} \frac{W^2}{1+W} \int_{\mathbb{R}^2} (W+W^2)^2 \\
&\leq 2\int_{\mathbb{R}^2} \frac{W^2}{1+W}\left(\int_{\mathbb{R}^2} W^2 + 2\int_{\mathbb{R}^2} W^2 \int_{\mathbb{R}^2} |\nabla W|^2\right).
\end{aligned}$$

Consequently,

$$\|W\|_2 \leq 1 + \int_{\mathbb{R}^2} \frac{W^2}{1+W} + \|\nabla W\|_2^2. \qquad (2.4.18)$$

Inserting (2.4.16) and (2.4.18) into (2.4.15), we obtain the following important lower bound estimate,

$$I(w) \geq \frac{1}{4}\|\nabla w\|_2^2 + C_1\left(\|\nabla W\|_2^2 + \int_{\mathbb{R}^2} \frac{W^2}{1+W}\right) - C_2, \qquad (2.4.19)$$

where $C_1, C_2 > 0$ are constants.

Lemma 2.4.2. *The optimization problem*

$$\min\left\{I(w)\,\middle|\,w \in \mathcal{A}\right\} \equiv \gamma_0 \qquad (2.4.20)$$

has a solution.

Proof. Let $\{w_j\}$ be a minimizing sequence of (2.4.20) and set
$$W_j = \ln(1 + Ke^{w_j}), \quad j = 1, 2, \cdots.$$
Then (2.4.18) and (2.4.19) imply that $\{W_j\}$ is a bounded sequence in L^2. Besides, from (2.4.14) and the boundedness of $\{W_j\}$ in L^2, we deduce that $\{W_j\}$ is also bounded in L^1. Recall that (2.4.19) implies already that $\{\|\nabla w_j\|_2\}$ is bounded.

We decompose w_j into the form
$$w_j = \underline{w}_j + w'_j, \quad \underline{w}_j \in \mathbb{R}, \quad w'_j \in \mathcal{X}'. \tag{2.4.21}$$

We need to show that $\{\underline{w}_j\}$ is a bounded sequence in $\mathbb{R}$. We will apply the following Poincaré type inequality [214],
$$\int_{\mathbb{R}^2} w^2 \, \mathrm{d}\mu \leq C \int_{\mathbb{R}^2} |\nabla w|^2, \quad w \in \mathcal{X}'. \tag{2.4.22}$$

We first have in view of $v_1 \leq 0$ and $\mathrm{supp}(v_1) \subset \Omega_\delta$ that
$$\frac{1}{2}\|\nabla w_j\|_2^2 + 4\|W_j\|_1 - \underline{w}_j \int_{\mathbb{R}^2} h - \int_{\mathbb{R}^2} hw'_j$$
$$\geq \frac{1}{2}\|\nabla w_j\|_2^2 + 4\|W_j\|_1 + 4\int_{\Omega_\delta} \ln\left(\frac{e^{v_1} + Ke^{w_j}}{1 + Ke^{w_j}}\right) - \int_{\mathbb{R}^2} hw_j$$
$$= I(w_j) \geq \gamma_0. \tag{2.4.23}$$

Therefore, using (2.4.22) in (2.4.23), we obtain
$$\underline{w}_j \int_{\mathbb{R}^2} h \leq |\gamma_0| + \frac{1}{2}\|\nabla w_j\|_2^2 + 4\|W_j\|_1 - \int_{\mathbb{R}^2} hw'_j$$
$$\leq |\gamma_0| + \|\nabla w_j\|_2^2 + 4\|W_j\|_1 + C_2, \tag{2.4.24}$$

which means $\{\underline{w}_j\}$ is bounded from above.

We now use the constraint in the definition of the admissible space to show that $\{\underline{w}_j\}$ is also bounded from below. We have
$$\frac{1}{4}\int_{\mathbb{R}^2} h = \int_{\Omega_\delta} \frac{Ke^{w_j}}{e^{v_1} + Ke^{w_j}} + \int_{\mathbb{R}^2 - \Omega_\delta} \frac{Ke^{w_j}}{1 + Ke^{w_j}}$$
$$< \int_{\Omega_\delta} \frac{Ke^{w_j}}{e^{v_1} + Ke^{w_j}} + \int_{\mathbb{R}^2} \frac{Ke^{w_j}}{1 + Ke^{w_j}}$$
$$< |\Omega_\delta| + e^{\underline{w}_j}\int_{\mathbb{R}^2} e^{w'_j} \, \mathrm{d}\mu.$$

Of course we may assume that $|\Omega_\delta| < \frac{1}{4}\int_{\mathbb{R}^2} h$. Thus the inequality (2.4.9) leads us to
$$\underline{w}_j > \ln\left(\frac{1}{4}\int_{\mathbb{R}^2} h - |\Omega_\delta|\right) - \ln\left(\int_{\mathbb{R}^2} e^{w'_j} \, \mathrm{d}\mu\right)$$
$$> -C_1 - C_2\|\nabla w_j\|_2^2. \tag{2.4.25}$$

Using the boundedness of $\{\|\nabla w_j\|\}$ in (2.4.25), we see that $\{\underline{w}_j\}$ is bounded from below.

In summary we have shown that for the minimizing sequence $\{w_j\}$ of (2.4.20) the corresponding sequence $\{\underline{w}_j\}$ is bounded.

From the boundedness of $\{\|\nabla w_j\|\}$ and (2.4.21) we see that $\{w'_j\}$ is bounded in $\mathcal{X}$. By going to a suitable subsequence if necessary, we may assume that there is an element $w \in \mathcal{X}$ so that

$$w_j = \underline{w}_j + w'_j \to w \quad \text{as } j \to \infty \text{ weakly in } \mathcal{X}.$$

Recall that the embedding $\mathcal{X} \to L^2(\mathrm{d}\mu)$ is completely continuous. So we may assume without loss of generality that $w_j \to w$ strongly in $L^2(\mathrm{d}\mu)$.

To see that the constraint in the admissible space $\mathcal{A}$ is preserved, we note that, for any given small number $\varepsilon > 0$ and an open neighborhood Ω of the points $p_1, p_2, \cdots, p_N$ with $|\Omega| < \varepsilon/2$, we have

$$
\begin{aligned}
\omega_j &= \int_{\mathbb{R}^2} \left| \frac{Ke^{w_j}}{e^{v_1} + Ke^{w_j}} - \frac{Ke^w}{e^{v_1} + Ke^w} \right| \\
&= \int_\Omega \left| \frac{Ke^{w_j}}{e^{v_1} + Ke^{w_j}} - \frac{Ke^w}{e^{v_1} + Ke^w} \right| + \int_{\mathbb{R}^2 - \Omega} \left| \frac{Ke^{w_j}}{e^{v_1} + Ke^{w_j}} - \frac{Ke^w}{e^{v_1} + Ke^w} \right| \\
&\le 2|\Omega| + \int_{\mathbb{R}^2 - \Omega} \frac{Ke^{|w| + |w_j|}}{e^{v_1}} |w_j - w| \\
&< \varepsilon + C_\varepsilon \|w_j - w\|_{L^2(\mathrm{d}\mu)} \sup_j \left(e^{|\underline{w}_j| + |\underline{w}| + C(\|\nabla w_j\|_2^2 + \|\nabla w\|_2^2)} \right).
\end{aligned}
$$

$$(2.4.26)$$

Letting $j \to \infty$ in (2.4.26), we see that

$$\omega_0 = \limsup_{j \to \infty} \omega_j \le \varepsilon.$$

In other words, $\omega_0 = 0$ and the constraint is indeed preserved.

Besides we also have, for ξ lying between w_j and w,

$$
\begin{aligned}
\tau_j &= \int_{\mathbb{R}^2} \left| \ln(e^{v_1} + Ke^{w_j}) - \ln(e^{v_1} + Ke^w) \right| \\
&\le \int_\Omega |w_j - w| + \int_{\mathbb{R}^2 - \Omega} \frac{Ke^\xi}{e^{v_1} + Ke^\xi} |w_j - w| \\
&\le C_1 \|w_j - w\|_{L^2(\mathrm{d}\mu)} + \int_{\mathbb{R}^2 - \Omega} \frac{Ke^{|w_j| + |w|}}{e^{v_1}} |w_j - w|.
\end{aligned}
$$

The above estimate is similar to that obtained in (2.4.26). Hence $\tau_j \to 0$ as $j \to \infty$.

Thus we see that the functional $I(\cdot)$ is weakly lower semicontinuous on $\mathcal{X}$ and $I(w) \le \liminf I(w_j) = \gamma_0$ and $w \in \mathcal{A}$. In other words w is a solution of (2.4.20).

Lemma 2.4.3. *The solution w of the optimization problem (2.4.20) obtained in Lemma 2.4.2 is a solution of the modified equation (2.4.7).*

Proof. In fact, by the rule of the Lagrange multipliers, we have

$$\int_{\mathbb{R}^2} \left\{ \nabla w \cdot \nabla \xi + \frac{4K e^w}{e^{v_1} + K e^w} \xi - h\xi \right\} = \lambda \int_{\mathbb{R}^2} \frac{K e^{v_1 + w}}{(e^{v_1} + K e^w)^2} \xi, \quad (2.4.27)$$

where $\lambda \in \mathbb{R}$ is a constant and $\xi \in \mathcal{X}$ is an arbitrary trial function. Let $\xi \equiv 1$ in (2.4.27). Using the definition of $\mathcal{A}$ and $w \in \mathcal{A}$ we immediately obtain $\lambda = 0$. Returning to (2.4.27) with $\lambda = 0$ we see that w is a weak solution of (2.4.7). By the standard elliptic theory we find that w is a classical solution of (2.4.7).

To achieve a tighter control of the solutions near infinity, we now begin to work in the parameter range $\beta < 4$ in the following.

Lemma 2.4.4. *For any constant $c < 0$, the equation (2.4.3) has a subsolution w_- satisfying $w_- < c$ in $\mathbb{R}^2$.*

Proof. In fact, we can first consider the linear equation

$$\Delta w = h - g. \quad (2.4.28)$$

It is easily seen through minimizing the functional

$$J(w) = \int_{\mathbb{R}^2} \left\{ \frac{1}{2} |\nabla w|^2 + (h - g)w \right\}$$

over the admissible space $\mathcal{X}'$ that (2.4.28) has a unique solution, say, w_1, in $\mathcal{X}'$. We show now that w_1 approaches a constant at infinity.

For $\delta \in \mathbb{R}$ and $s \in \mathbb{N}$ (the set of nonnegative integers) define $W_{s,\delta}^2$ to be the closure of the set of C^∞ functions over $\mathbb{R}^2$ with compact supports under the norm

$$\|\xi\|_{W_{s,\delta}^2}^2 = \sum_{|\alpha| \le s} \|(1 + |x|)^{\delta + |\alpha|} D^\gamma \xi\|_2^2.$$

Let $C_0(\mathbb{R}^2)$ be the set of continuous functions on $\mathbb{R}^2$ vanishing at infinity.

We recall the following three facts from the well-developed theory of the weighted Sobolev spaces [214].

Lemma 2.4.5. *(i) If $s > 1$ and $\delta > -1$ then $W_{s,\delta}^2 \subset C_0(\mathbb{R}^2)$.*

(ii) For $-1 < \delta < 0$, the Laplace operator $\Delta : W_{2,\delta}^2 \to W_{0,\delta+2}^2$ is 1-1 and the range of Δ has the characterization

$$\Delta(W_{2,\delta}^2) = \left\{ F \in W_{0,\delta+2}^2 \,\middle|\, \int_{\mathbb{R}^2} F = 0 \right\}.$$

(iii) If $\xi \in \mathcal{X}$ and $\Delta \xi = 0$ then $\xi =$ constant.

We now continue the proof of Lemma 2.4.4.

Choose any $0 > \delta > -1$. Then $h - g \in L(\mathbb{R}^2) \cap W^2_{0,\delta+2}$ and $\int_{\mathbb{R}^2}(h-g) = 0$. Thus the statements (i) and (ii) in Lemma 2.5 imply that there is an element $\xi \in W^2_{2,\delta}$ such that $\Delta\xi = h - g$ and $\xi = 0$ at infinity. In particular, $\xi \in L^2(d\mu)$. However, from $\nabla\xi \in W^2_{0,\delta+1}$ and $\delta > -1$, we deduce that $\nabla\xi \in L^2(\mathbb{R}^2)$. Hence $\xi \in \mathcal{X}$. Since $\xi - w_1 \in \mathcal{X}$ and $\Delta(\xi - w_1) = 0$, the fact (iii) implies that $\xi - w_1 =$ constant. So w_1 approaches a constant at infinity.

Let w_2 be the solution of the equation (2.4.7) obtained earlier. We show that w_2 has a similar asymptotic behavior at infinity. In fact, if we use F to denote the right-hand side of (2.4.7), then $\int_{\mathbb{R}^2} F = 0$. Besides, we also have, for a bounded domain Ω satisfying $\mathrm{supp}(v_1) \subset \Omega$,

$$\int_{\mathbb{R}^2}(1+|x|^{\delta+2})^2\left(\frac{Ke^{w_2}}{e^{v_1}+Ke^{w_2}}\right)^2 \leq \int_\Omega(1+|x|^{\delta+2})^2 + C_0\int_{\mathbb{R}^2}e^{2w_2}\,d\mu,$$

where

$$C_0 = \sup_{x\in\mathbb{R}^2}\left\{K(x)(1+|x|^{\delta+2})^2\right\}$$

is finite if $2(\delta+2) \leq \beta$. A convenient choice of β is $\beta = 2(\delta+2)$ or

$$\delta = -2 + \frac{\beta}{2} = -1 + \frac{\beta-2}{2}.$$

Therefore $F \in W^2_{0,\delta+2}$ and the condition $2 < \beta < 4$ is equivalent to $-1 < \delta < 0$, which makes the statements (i) and (ii) in Lemma 2.4.5 applicable. In particular we may find a solution of $\Delta\zeta = F$ in the space $W^2_{2,\delta}$. We now repeat the argument for w_1 to show that $\zeta - w_2 =$ constant. Consequently, w_2 approaches a constant at infinity as well.

Since w_1 and w_2 are both bounded, we can choose a constant $C > 0$ so that $w_- \equiv w_1 - C + w_2 < c < 0$ and $w_1 - C < 0$. Moreover, since the function

$$F(t) = \frac{Ke^t}{e^{v_1}+Ke^t}, \quad t \in \mathbb{R} \tag{2.4.29}$$

is nondecreasing, we have

$$F(w_-) = F(w_1 - C + w_2) \leq F(w_2).$$

Consequently, from (2.4.7) and (2.4.28), we have

$$
\begin{aligned}
\Delta w_- &= 4F(w_2) - g \\
&\geq 4F(w_-) - g \\
&\geq \frac{4Ke^{v_2+w_-}}{e^{v_1}+Ke^{v_2+w_-}} - g,
\end{aligned}
$$

because we also have $v_2 \leq 0$. In other words, w_- is a subsolution of the original governing equation (2.4.3).

The proof of Lemma 2.4.4 is complete.

2.4.3   Existence of supersolution

We now construct a suitable supersolution for (2.4.3).

We consider the modified equation

$$\Delta w = \frac{4Ke^w}{1 + Ke^w} - \tilde{g}, \qquad (2.4.30)$$

where $\tilde{g} = g_1 + \beta \Delta v_3$. We then introduce a function $\tilde{h} \in C_c^\infty(\mathbb{R}^2)$ such that $\tilde{h} \geq 0$ and

$$\int_{\mathbb{R}^2} \tilde{h} = \int_{\mathbb{R}^2} \tilde{g} = 2\pi(2N - \beta).$$

Let $\tilde{w}_1$ be a bounded solution of

$$\Delta w = \tilde{h} - \tilde{g}. \qquad (2.4.31)$$

We may assume that $\tilde{w}_1 \geq 0$.

As before, we study the following easier version of (2.4.30),

$$\Delta w = \frac{4Ke^w}{1 + Ke^w} - \tilde{h}. \qquad (2.4.32)$$

The function $\tilde{h}$ enjoys all the crucial properties we used for h in the equation (2.4.7). Hence we may simplify the argument in Lemmas 2.4.1–2.4.3 to get a solution, say, $\tilde{w}_2$, for (2.4.32) which goes to a constant limiting value at infinity.

Then it follows from (2.4.31) and (2.4.32) that the function $\tilde{w} = \tilde{w}_1 + \tilde{w}_2$ satisfies

$$\begin{aligned}
\Delta \tilde{w} &= \frac{4Ke^{\tilde{w}_2}}{1 + Ke^{\tilde{w}_2}} - \tilde{g} \\
&\leq \frac{4Ke^{\tilde{w}_1 + \tilde{w}_2}}{1 + Ke^{\tilde{w}_1 + \tilde{w}_2}} - \tilde{g} \\
&\leq \frac{4Ke^{\tilde{w}}}{1 + Ke^{\tilde{w}}} - \tilde{g} + 4\pi \sum_{j=1}^{M} \delta_{q_j}
\end{aligned}$$

in the sense of distributions, where we have again used the property that the function $F(\cdot)$ defined in (2.4.29) with e^{v_1} set to 1 is increasing. Set $w_+ = \tilde{w} - v_2$. The above inequality implies that w_+ fulfills

$$\Delta w_+ \leq \frac{Ke^{v_2 + w_+}}{e^{v_1} + Ke^{v_2 + w_+}} - g, \qquad (2.4.33)$$

which means that w_+ is a supersolution of (2.4.3) in the sense of distributions. Since w_+ is bounded from below, we may assume that $w_- < w_+$ everywhere in $\mathbb{R}^2$.

2.4.4 Existence of bounded solution

We show that (2.4.3) has a solution lying between w_- and w_+.

For any $r > 0$, use B_r to denote the circular region $\{x \in \mathbb{R}^2 \,|\, |x| < r\}$. We first show that the boundary value problem

$$\Delta w = \frac{4Ke^{v_2+w}}{e^{v_1} + Ke^{v_2+w}} - g, \quad x \in B_r, \tag{2.4.34}$$

$$w = w_+, \quad x \in \partial B_r \tag{2.4.35}$$

has a unique solution satisfying $w_- \leq w \leq w_+$ when r is so large that $r > |p_j| \ (1 \leq j \leq N)$ and $r > |q_j| \ (1 \leq j \leq M)$. To this end, we invoke the standard iterative scheme

$$\Delta w_n - C_0 w_n = \frac{4Ke^{v_2+w_{n-1}}}{e^{v_1} + Ke^{v_2+w_{n-1}}} - C_0 w_{n-1} - g, \quad x \in B_r, \tag{2.4.36}$$

$$w_n = w_+, \quad x \in \partial B_r \tag{2.4.37}$$

$$w_0 = w_+, \quad n = 1, 2, \cdots, \tag{2.4.38}$$

where $C_0 \geq 2$. We show that the sequence $\{w_n\}$ defined by (2.4.36)–(2.4.38) obeys the following monotonicity property

$$w_- < \cdots < w_n < \cdots < w_2 < w_1 < w_+. \tag{2.4.39}$$

We prove (2.4.39) by induction.

Since w_1 satisfies an equation with L^s ($s > 2$) right-hand side,

$$\Delta w_1 - C_0 w_1 = \frac{4Ke^{\tilde{w}}}{e^{v_1} + Ke^{\tilde{w}}} - C_0(\tilde{w} - v_2) - g,$$

$w_1 \in C^{1,\alpha}(\overline{B_r})$ $(0 < \alpha < 1)$. In particular, $w_1 < w_+$ near the set $Q = \{q_1, q_2, \cdots, q_M\}$. In $B_r - Q$, we have $\Delta(w_1 - w_+) \geq C_0(w_1 - w_+)$. Hence the maximum principle implies $w_1 < w_+$ everywhere. On the other hand, since $w_- < w_+$, we have

$$(\Delta - C_0)(w_- - w_1) \geq \left(\frac{4Ke^{v_1+v_2+\xi}}{[e^{v_1} + Ke^{v_2+\xi}]^2} - C_0 \right)(w_- - w_+)$$

$$\geq (2 - C_0)(w_- - w_+) \geq 0.$$

Here and in the sequel, ξ denotes an intermediate quantity from the mean value theorem. Therefore the maximum principle again gives us $w_- < w_1$.

Suppose we have already established the property that $w_- < w_k$ and $w_k < w_{k-1}$ for some $k \geq 1$. Then (2.4.36) gives us

$$(\Delta - C_0)(w_{k+1} - w_k) = \left(\frac{4Ke^{v_1+v_2+\xi}}{[e^{v_1} + Ke^{v_2+\xi}]^2} - C_0 \right)(w_k - w_{k-1})$$

$$\geq (2 - C_0)(w_k - w_{k-1}) \geq 0.$$

Hence $w_{k+1} < w_k$. Furthermore,

$$(\Delta - C_0)(w_- - w_{k+1}) \geq \left(\frac{4Ke^{v_1+v_2+\xi}}{[e^{v_1} + Ke^{v_2+\xi}]^2} - C_0 \right)(w_- - w_k)$$
$$\geq (2 - C_0)(w_- - w_k) \geq 0.$$

So, again, $w_- < w_{k+1}$. Consequently (2.4.39) is proven.

Since w_- is a bounded function, we see that the pointwise limit

$$w = \lim_{n \to \infty} w_n \tag{2.4.40}$$

exists. Letting $n \to \infty$ in (2.4.36) and using elliptic embedding theorems, we know that the limit (2.4.40) may be achieved in any strong sense and w is a smooth solution of (2.4.34) and (2.4.35). Of course such a solution is unique and satisfies $w_- \leq w \leq w_+$ as claimed.

We now use w_n to denote the solution of (2.4.34) and (2.4.35) with $r = n$ (n is large enough so that $n > |p_j|$ and $n > |q_k|$ for $j = 1, 2, \cdots, N$ and $k = 1, 2, \cdots, M$, respectively). Since on ∂B_n, $w_n = w_+ > w_{n+1}$, we see from $\Delta(w_n - w_{n+1}) = C(x)(w_n - w_{n+1})$ $(C(x) \geq 0)$ that $w_n > w_{n+1}$ in B_n as well. Thus, for each fixed $n_0 \geq 1$, we have the monotone sequence $w_{n_0} > w_{n_0+1} > \cdots > w_n > \cdots > w_-$ on B_{n_0}. This result indicates that the sequence $\{w_n\}$ converges to a solution, say w, of the equation (2.4.3) over the entire $\mathbb{R}^2$ and there holds $w_- \leq w \leq w_+$.

2.4.5   Asymptotic limit

We first show that the solution of (2.4.3) obtained in the previous paragraph also approaches a finite limit at infinity.

We first show that the solution w obtained above lies in $\mathcal{X}$. For this purpose let the cut-off function $\xi \in C^\infty(\mathbb{R}^2)$ be such that

$$\xi = 1 \text{ in } \left\{ x \,\middle|\, |x| \leq 1 \right\}, \quad \xi = 0 \text{ in } \left\{ x \,\middle|\, |x| \geq 2 \right\}, \quad 0 \leq \xi \leq 1 \text{ everywhere.}$$

Set $\xi_\rho(x) = \xi(x/\rho)$ $(\rho > 0)$. Multiplying (2.4.3) by $\xi_\rho^2 w$ and integrating by parts, we have

$$\int_{\mathbb{R}^2} \xi_\rho^2 |\nabla w|^2 = -2 \int_{\mathbb{R}^2} \xi_\rho w(\nabla \xi_\rho \cdot \nabla w) - \int_{\mathbb{R}^2} \xi_\rho^2 w \left(\frac{4Ke^{v_2+w}}{e^{v_1} + Ke^{v_2+w}} - g \right).$$
$$\tag{2.4.41}$$

However, the definition of ξ_ρ gives us

$$\int_{\mathbb{R}^2} |\nabla \xi_\rho|^2 \leq \frac{1}{\rho^2} \sup_{\mathbb{R}^2} |\nabla \xi|^2 \int_{|x| \leq 2\rho} = 4\pi \sup_{\mathbb{R}^2} |\nabla \xi|^2.$$

(Another way of getting the above bound is to use the scaling invariance to arrive at $\int_{\mathbb{R}^2} |\nabla \xi_\rho|^2 = \int_{\mathbb{R}^2} |\nabla \xi|^2 \leq 4\pi \sup |\nabla \xi|^2$.) Inserting this result into (2.4.41), we find that

$$\int_{\mathbb{R}^2} \xi_\rho^2 |\nabla w|^2 \leq C_1 + C_2 \int_{\mathbb{R}^2} \xi_\rho^2 \left| \frac{4Ke^{v_2+w}}{e^{v_1} + Ke^{v_2+w}} - g \right|, \qquad (2.4.42)$$

where $C_1, C_2 > 0$ depend on w and we have used the fact that w is a bounded function and the Schwartz inequality. Letting $\rho \to \infty$ in (2.4.42) we have $|\nabla w| \in L^2$. Consequently, we obtain the relation $w \in \mathcal{X}$.

Lemma 2.4.6. *For the solution w above, we have the vanishing integral,*

$$\int_{\mathbb{R}^2} \Delta w = 0.$$

Proof. With the cut-off function ξ_ρ defined earlier, we have the bound

$$\left| \int_{\mathbb{R}^2} \xi_\rho \Delta w \right| \leq \int_{\mathbb{R}^2} |\nabla \xi_\rho \cdot \nabla w|$$

$$\leq \left(\int_{\mathbb{R}^2} |\nabla \xi_\rho|^2 \right)^{1/2} \left(\int_{\rho \leq |x| \leq 2\rho} |\nabla w|^2 \right)^{1/2}$$

$$\leq C \left(\int_{\rho \leq |x| \leq 2\rho} |\nabla w|^2 \right)^{1/2}.$$

Letting $\rho \to \infty$, we arrive at $\int_{\mathbb{R}^2} \Delta w = 0$ as expected.

Using Lemma 2.4.6, we can then mimic the argument in proof of Lemma 2.4.4 to show that w approaches a constant at infinity.

We next show that $\partial_j w \to 0$ ($j = 1, 2$) at infinity. From (2.4.3), we have

$$\Delta(\partial_j w) = \frac{4Ke^{v_1+v_2+w}}{(e^{v_1} + Ke^{v_2+w})^2}(-\partial_j v_1 + \partial_j v_2 + \beta \partial_j v_3 + \partial_j w) - (\partial_j g)$$

$$= \frac{4Ke^{v_1+v_2+w}}{(e^{v_1} + Ke^{v_2+w})^2}(\partial_j w) + g_c, \qquad (2.4.43)$$

where g_c is of compact support. Since we have shown that $\partial_j w \in L^2$, the L^2-estimates applied in (2.4.43) give us $\partial_j w \in W^{2,2}$. In particular $\partial_j w \to 0$ as $|x| \to \infty$ as expected since we are in two dimensions.

Another application of Lemma 2.4.6 is that we may integrate (2.4.3) and use (2.4.4) to get

$$\int_{\mathbb{R}^2} \frac{Ke^{v_2+w}}{e^{v_1} + Ke^{v_2+w}} = \frac{\pi}{2}(2[N - M] - \beta). \qquad (2.4.44)$$

2.4.6 Recovery of original field configurations

Let w be the solution of (2.4.3) just obtained. Then $v = -v_1 + v_2 + \beta v_3 + w$ solves the governing equation (2.4.1). We have by the properties of w the following sharp estimate near infinity,

$$e^v = e^{-v_1 + v_2 + \beta v_3 + w} = O(|x|^{-\beta}). \tag{2.4.45}$$

We now use the standard prescription

$$
\begin{aligned}
u(z) &= \exp\left(\frac{1}{2}v(z) + i\theta(z)\right), \\
\theta(z) &= -\sum_{j=1}^{N} \arg(z - p_j) + \sum_{j=1}^{M} \arg(z - q_j), \\
A_1(z) &= \operatorname{Re}\{2i\bar{\partial}\ln u(z)\}, \quad A_2(z) = \operatorname{Im}\{2i\bar{\partial}\ln u(z)\} \tag{2.4.46}
\end{aligned}
$$

to get a solution $(u, \mathbf{A})$ of the system (2.3.1), (2.3.2). Applying (2.1.22), we obtain a solution $(\phi, \mathbf{A})$ for the self-dual system (2.2.23), (2.2.24). Thus, from (2.4.45), we can establish all the decay estimates stated in Theorem 2.2.3.

2.4.7 Magnetic flux and minimum energy value

It is straightforward to get the magnetic flux from (2.3.2) and (2.4.44) as follows,

$$\Phi = \int_{\mathbb{R}^2} B = \int \frac{2|u|^2}{1 + |u|^2} = \int \frac{2Ke^{v_2 + w}}{e^{v_1} + Ke^{v_2}} = \pi(2[N - M] - \beta).$$

As in Theorem 2.2.3, we rewrite the *fractional* total flux or 'total curvature' Φ as $\Phi = 2\pi N\alpha$. Thus $\beta = 2(N - M) - 2N\alpha$ and the condition $2 < \beta < 4$ is translated into the expected interval for α,

$$1 - \frac{M + 2}{N} < \alpha < 1 - \frac{M + 1}{N}.$$

Note that α is allowed to assume any value in the above interval.

To compute the energy, we note that the current density H_k $(k = 1, 2)$ defined in (2.2.18) satisfies the decay estimate (2.2.19). Hence, by (2.2.22), (2.2.23), and (2.2.24), we have the minimum energy,

$$E(\phi, \mathbf{A}) = 4\pi |\deg(\phi)|. \tag{2.4.47}$$

2.4.8 Brouwer degree of map

Let $(u, \mathbf{A})$ be a solution of (2.3.1), (2.3.2) just constructed so that u has poles at $p_1, p_2, \cdots, p_N$ and (ϕ, A) be the corresponding solution of (2.2.23),

(2.2.24). Then $\phi^{-1}(\mathbf{s}) = \{p_1, p_2, \cdots, p_N\}$. In the following, we will show that $N = \deg(\phi)$, when ϕ is viewed as a map from S^2 to S^2.

We will present two different proofs. The first one is directly from the definition of $\deg(\phi)$, under the assumption that the poles are all simple. The second one is for the general case when the poles have arbitrary multiplicities.

For convenience we use the reciprocal of u,

$$U = \frac{1}{u} = \frac{u_1}{|u|^2} - \mathrm{i}\frac{u_2}{|u|^2} \equiv U_1 + \mathrm{i}U_2,$$

to study ϕ_1, ϕ_2 near the preimages (the poles of u) of the south pole under ϕ. By (2.1.22), we have

$$\phi_1 = \frac{2U_1}{|U|^2 + 1}, \quad \phi_2 = -\frac{2U_2}{|U|^2 + 1}. \tag{2.4.48}$$

It is clear that, at a pole p of u, we have $U(p) = 0$ and $\phi_1(p) = \phi_2(p) = 0$. Thus, by virtue of (2.4.48), we have the expressions

$$\frac{\partial \phi_1}{\partial x^j}(p) = 2\frac{\partial U_1}{\partial x^j}(p), \quad \frac{\partial \phi_2}{\partial x^j}(p) = -2\frac{\partial U_2}{\partial x^j}(p), \quad j = 1, 2.$$

Since U satisfies the equation $\overline{D}_1 U = -\mathrm{i}\overline{D}_2 U$, therefore we obtain the Jacobian of (2.4.48) at p in the form

$$\left.\frac{\partial(\phi_1, \phi_2)}{\partial(x^1, x^2)}\right|_{x=p} = -2(|\nabla U_1|^2 + |\nabla U_2|^2)(p), \tag{2.4.49}$$

which is nonvanishing if p is a simple zero of U or a simple pole of u. In this situation p is a regular point of ϕ.

First we assume that U has only simple zeros or u has only simple poles. Then, at each pole p of u,

$$\partial_1\phi \times \partial_2\phi = \left(0, 0, \left.\frac{\partial(\phi_1, \phi_2)}{\partial(x^1, x^2)}\right|_{x=p}\right).$$

Hence, by (2.4.49), we see that ϕ preserves orientation at p. This observation enables us to conclude that

$$\deg(\phi) = \#\phi^{-1}(\mathbf{s}) = N. \tag{2.4.50}$$

Next, we consider the general case that the poles have arbitrary multiplicities.

From (2.1.24), we have

$$\phi \cdot (\partial_1\phi \times \partial_2\phi) = \frac{2\mathrm{i}}{(1 + |u|^2)^2}(\partial_1 u \partial_2 \overline{u} - \partial_1 \overline{u} \partial_2 u)$$

$$= \frac{2i}{(1+|u|^2)^2}([D_1u\overline{D_2u} - \overline{D_1u}D_2u] + iA_2\partial_1|u|^2 - iA_1\partial_2|u|^2)$$

$$= 2\left(F_{12}\frac{|u|^2}{1+|u|^2} + i\frac{D_1u\overline{D_2u} - \overline{D_1u}D_2u}{(1+|u|^2)^2}\right)$$

$$+2\left(\frac{-1}{(1+|u|^2)^2}(A_2\partial_1|u|^2 - A_1\partial_2|u|^2) - F_{12}\frac{|u|^2}{1+|u|^2}\right)$$

$$= J_{12} + H_{12}, \tag{2.4.51}$$

where J_k and H_k are defined by

$$J_k = \frac{i}{1+|u|^2}(u\overline{D_ku} - \overline{u}D_ku), \quad k = 1, 2, \tag{2.4.52}$$

$$H_k = -2A_k\frac{|u|^2}{1+|u|^2} = A_k(\mathbf{n} \cdot \phi - 1), \quad k = 1, 2. \tag{2.4.53}$$

Here we have used the useful commutator identities

$$(D_jD_k - D_kD_j)u = iF_{jk}u, \quad j = 1, 2. \tag{2.4.54}$$

Again, using (2.2.20) in (2.4.51), we obtain

$$\int_{\mathbb{R}^2} \phi \cdot (\partial_1\phi \times \partial_2\phi) = \int_{\mathbb{R}^2} J_{12}$$

$$= \lim_{r\to\infty} \oint_{|x|=r} J_k \, dx^k - \sum_{j=1}^{\ell} \lim_{r\to 0} \oint_{|x-p_j|=r} J_k \, dx^k, \tag{2.4.55}$$

where all path integrals are counterclockwise and each pole, p_j, has a multiplicity n_j, with the total anticharge

$$\sum_{j=1}^{s} n_j = N. \tag{2.4.56}$$

On the other hand, in view of (2.4.46), we have

$$D_1u = (\partial + \overline{\partial})u + i\left(i\frac{\overline{\partial}u}{u} - i\frac{\partial\overline{u}}{\overline{u}}\right)u = u\partial v, \tag{2.4.57}$$

$$D_2u = i(\partial - \overline{\partial})u + i\left(\frac{\overline{\partial}u}{u} + \frac{\partial\overline{u}}{\overline{u}}\right)u = iu\partial v. \tag{2.4.58}$$

Inserting (2.4.57) and (2.4.58) into (2.4.52) and using the decay rate $|u|^2 = e^v = O(|x|^{-\beta})$ near infinity obtained earlier (see (2.4.45)), we have

$$J_k = O(|x|^{-\beta}) \quad \text{as } |x| \to \infty, \quad k = 1, 2. \tag{2.4.59}$$

In particular,

$$\lim_{r \to \infty} \oint_{|x|=r} J_k \, \mathrm{d}x^k = 0. \tag{2.4.60}$$

Moreover, using (2.4.52), (2.4.57), and (2.4.58) again, we have

$$\oint_{|x-p_j|=r} J_k \, \mathrm{d}x^k = \mathrm{i} \oint_{|x-p_j|=r} \frac{|u|^2}{1+|u|^2} ([\overline{\partial} - \partial]v \, \mathrm{d}x^1 - \mathrm{i}[\overline{\partial} + \partial]v \, \mathrm{d}x^2)$$

$$= \oint_{|x-p_j|=r} \frac{e^v}{1+e^v} (-\partial_2 v \, \mathrm{d}x^1 + \partial_1 v \, \mathrm{d}x^2). \tag{2.4.61}$$

Recall from (2.3.7) that, near p_j $(j = 1, 2, \cdots, s)$, v has the local representation

$$v(x) = -n_j \ln |x - p_j|^2 + w_j(x) \tag{2.4.62}$$

where w_j is a smooth function. Therefore, near $x = p_j$, we may write

$$\nabla v = -2n_j \frac{x - p_j}{|x - p_j|^2} + \nabla w_j, \quad j = 1, 2, \cdots, s. \tag{2.4.63}$$

Inserting (2.4.62) and (2.4.63) into (2.4.61) and taking the $\rho \to 0$ limit, we obtain immediately

$$\lim_{r \to 0} \int_{|x-p_j|=\rho} J_k \, \mathrm{d}x^k = -4\pi n_j, \quad j = 1, 2, \cdots, s. \tag{2.4.64}$$

In view of (2.4.56), (2.4.60), and (2.4.64), we obtain from (2.4.55) the general conclusion,

$$\deg(\phi) = \frac{1}{4\pi} \int_{\mathbb{R}^2} \phi \cdot (\partial_1 \phi \times \partial_2 \phi)$$

$$= \frac{1}{4\pi} \int_{\mathbb{R}^2} J_{12} = N. \tag{2.4.65}$$

2.4.9 Nonexistence of solution of unit degree

The condition (2.4.5) says that we need $N \geq M + 2$ to ensure the existence of a solution. In this subsection we show that this condition is also *necessary*. In other words we prove that whenever $N < M + 2$ (or $N \leq M + 1$) there can be no solution over $\mathbb{R}^2$ satisfying (2.2.14). Such a result implies the important conclusion that there is no solution with $\deg(\phi) = \pm 1$.

Suppose that the system of equations, (2.2.23) and (2.2.24), has a multi-soliton solution $(\phi, \mathbf{A})$ over $\mathbb{R}^2$ so that $\#\phi^{-1}(\mathbf{s}) = N$, $\#\phi^{-1}(\mathbf{n}) = M$. Then the stereographic projection (2.1.15) and the variable $v = \ln |u|^2$ leads us to (2.3.8) with $N = \sum n_j \geq 0$ and $M = \sum m_j \geq 0$. In (2.3.8), set

$$v(x) = -\sum_{j=1}^{N} \ln |x - p_j|^2 + \sum_{j=1}^{M} \ln |x - q_j|^2 + w(x).$$

Then it becomes

$$\Delta w = \frac{4K_0 e^w}{1 + K_0 e^w} \equiv h(x), \quad K_0(x) = \prod_{j=1}^{N} |x - p_j|^{-2} \prod_{j=1}^{M} |x - q_j|^2. \quad (2.4.66)$$

Recall that

$$1 - \mathbf{n} \cdot \phi(x) = 1 - \phi_3(x) = 1 - \frac{1 - |u(x)|^2}{1 + |u(x)|^2} = \frac{2|u(x)|^2}{1 + |u(x)|^2}$$

vanishes at infinity in view of (2.1.22) and (2.2.14). Hence there is an $r_0 > 0$ such that

$$K_0(x) e^{w(x)} = e^{v(x)} = |u(x)|^2 \leq 1, \quad x \in \mathbb{R}^2, \quad |x| \geq r_0. \quad (2.4.67)$$

Consider the polar coordinates (ρ, θ) or (r, θ) for $\mathbb{R}^2$. Integrating (2.4.66) in the region $0 \leq \rho \leq r, 0 \leq \theta \leq 2\pi$, we have

$$\int_0^r \int_0^{2\pi} (\Delta w) \, \rho \, d\theta d\rho = \int_0^r \int_0^{2\pi} h \rho \, d\theta d\rho. \quad (2.4.68)$$

Set now a new function of the radial variable r as follows,

$$W(r) = \frac{1}{2\pi} \int_0^{2\pi} w(r, \theta) \, d\theta. \quad (2.4.69)$$

With the notation (2.4.69), we are led from (2.4.68) by the 2π-periodicity of w the following equation,

$$rW_r(r) = \frac{1}{2\pi} \int_0^r \int_0^{2\pi} \left([\rho w_\rho]_\rho + \frac{1}{\rho} \frac{\partial^2 w}{\partial^2 \theta} \right) d\theta d\rho$$

$$= \frac{1}{2\pi} \int_0^r \int_0^{2\pi} h\rho \, d\theta d\rho, \quad (2.4.70)$$

where $W_r = dW/dr$. Differentiating (2.4.70) with respect to $r > 0$ we find

$$\frac{1}{r}(rW_r)_r = \frac{1}{2\pi} \int_0^{2\pi} h \, d\theta. \quad (2.4.71)$$

On the other hand, choose $r_0 > 0$ large enough such that

$$K_0(x) = 2 \prod_{j=1}^{N} |x - p_j|^{-2} \prod_{j=1}^{M} |x - q_j|^2 \geq |x|^{-2(N-M)}, \quad |x| \geq r_0.$$

Therefore, in view of (2.4.67) and the above, we have

$$h(x) \geq |x|^{-2(N-M)} e^w, \quad |x| \geq r_0. \quad (2.4.72)$$

Consequently, we derive from (2.4.71) and (2.4.72) the inequality

$$\frac{1}{r}(rW_r)_r \geq \frac{e^W}{r^{2(N-M)}}, \quad r \geq r_0 \tag{2.4.73}$$

where we have used the Jensen inequality

$$\frac{1}{2\pi}\int_0^{2\pi} e^w \, d\theta \geq \exp\left(\frac{1}{2\pi}\int_0^{2\pi} w \, d\theta\right) = e^W.$$

To study (2.4.73) it is convenient to introduce the new variables $t = \ln r, \sigma(t) = W(e^t) = W(r)$, and the initial point $t_0 = \ln r_0$.
Assume now $1 - [N - M] \geq 0$. Then (2.4.73) becomes

$$\sigma_{tt} \geq e^{2(1-[N-M])t+\sigma} \geq C_0 e^\sigma, \quad t \geq t_0, \quad C_0 \equiv e^{2(1-[N-M])t_0} \tag{2.4.74}$$

Here $\sigma_t = d\sigma/dt$.

Recall that in (2.4.71) the function h is positive except at the points $q_1, q_2, \cdots, q_M$. Hence $rW_r(r) > 0$ for all $r > 0$. Returning to the variables t, σ we have $\sigma_t(t) > 0$ for all t.

Multiplying (2.4.74) by $\sigma_t > 0$ and integrating, we have

$$\sigma_t^2(t) - \sigma_t^2(t_0) \geq C_0(e^{\sigma(t)} - e^{\sigma(t_0)}).$$

Using the fact $\sigma_t > 0$ again and setting $\eta_0 = \sigma_t^2(t_0) - C_0 e^{\sigma(t_0)}$, we obtain then

$$\sigma_t(t) \geq \sqrt{\eta_0 + C_0 e^{\sigma(t)}}, \quad t \geq t_0,$$

or

$$\int_{\sigma(t_0)}^{\sigma(t)} \frac{d\sigma}{\sqrt{\eta_0 + C_0 e^\sigma}} \geq t - t_0, \quad t \geq t_0. \tag{2.4.75}$$

However the left-hand side of (2.4.75) has an upper bound since

$$\int_{\sigma(t_0)}^{\infty} \frac{d\sigma}{\sqrt{\eta_0 + C_0 e^\sigma}} < \infty.$$

Thus (2.4.75) cannot be valid for all $t \geq t_0$. This proves that the system of the equations (2.2.23) and (2.2.24) does not have a solution satisfying the finite-energy boundary condition (2.2.14) and $N \leq M + 1$.

For a solution $(\phi, \mathbf{A})$ of the system of equations (2.2.23) and (2.2.24), we have $N = \deg(\phi)$. Since $M \geq 0$, we see that there is no degree one ($N = 1$) solution.

2.5   Remarks

The work of Belavin and Polyakov [33] on the solution of the classical sigma model belongs to a broad category of beautiful results in field theory.

We recommend the monograph of Rajaraman [259] for a comprehensive account of related development. A basic principle behind is the structure of self-duality systematically explored by Bogomol'nyi [42], which reduces the original second-order governing equations into a first-order system of equations. For example, we have seen that the second-order harmonic maps equations are reduced into the Cauchy–Riemann equations, which are of first order, in the sigma model case just considered. It was Schroers [274] who first extended the classical sigma model to allow the coupling of a gauge field and his work has generated numerous activities [11, 119, 127, 171, 275] in field theory. This gauged sigma model gives a rich spectrum of new mathematical features. First, in each topological class, the magnetic flux takes value in an explicitly determined open interval. Second, there is no solution of unit topological charge. Third, when the topological charge is not unit, the solutions in each topological class exist and fulfill the same minimum topological energy levels. Hence we see that quantization of flux is not a characteristic condition for topological solutions. The proofs of the results here for the gauged sigma model are the modified versions of those presented in [346]. Note that a similar existence problem [54] arises in the Thomas–Fermi model.

The nonexistence theorem for a degree one solution is quite interesting. Note that such a result is only proven for the self-dual system (2.2.23), (2.2.24) and it is not known whether the same conclusion holds for the Euler–Lagrange equations of the model. Indeed it is not clear whether the energy functional (2.2.13) permits the existence of any critical points whose energies are greater than those given by the formula (2.2.28). The absence of degree one solutions of (2.2.23), (2.2.24) already indicates that (2.1.13) is never attained at $\deg(\phi) = \pm 1$. Thus if one could prove that the energy (2.2.13) has a minimizer, or a critical point, among all degree one field configurations, it would mean that the system allows non-self-dual solutions as $SU(2)$ instantons [43, 242, 270, 286]. At this moment the question in either direction is open. In particular, we propose the following.

Open Problem 2.1. *Does the energy (2.2.13) have a finite energy, degree one, critical point?*

It is not clear whether the result on the number of parameters of the self-dual solutions for the case $N \geq 2$ stated in Theorem 2.2.2 is optimal, although it seems to be exhaustive. As pointed out by Schroers in [274], one could hope to obtain more information by using index theorems to calculate the dimensions of fluctuations around a degree N solution, as was done, for example, for the Abelian Higgs vortices [327].

Open Problem 2.2. *What is the maximum number of free parameters a solution of (2.2.23), (2.2.24) should have?*

The result on fractional values of the magnetic flux stated in Theorem 2.2.3 says that the parameter α is confined in the range (2.2.27). The interesting fact is that a larger N (or M) leads immediately to a larger (or smaller) range of the flux Φ given in (2.2.29). Thus, magnetically, we could imagine that the points q's in $\phi^{-1}(\mathbf{n})$ counter-balance the points p's in $\phi^{-1}(\mathbf{s})$.

It is well known that the solutions of the time-dependent harmonic map equations develop singularities. In other words, there are some smooth initial data so that the solutions starting from these data become non-smooth in finite time [279, 298]. The introduction of a gauge field sets a more realistic stage for physics to perform and may eliminate such a singularity-generation phenomenon. Thus we propose

Open Problem 2.3. *Prove that the solution of the Euler–Lagrange equations of the action density (2.2.11) initiated from a smooth configuration stays smooth for all time.*

Note that the global existence problem of the time-dependent Yang–Mills equations has been thoroughly studied [84, 100, 124, 128, 278, 339].

Another important application of the gauged sigma model is to generate magnetically opposite vortex-lines and cosmic strings. We shall return to this subject later.

A similar but more difficult problem is the existence of energy minimizers in the Skyrme model for heavy particles where the topological degree that characterizes the solutions is identified as the baryon number [104, 105, 106, 202, 210, 211, 337]. For a brief survey on this subject, see Lieb [194].

3
Multiple Instantons and Characteristic Classes

The Hodge theorem states that, on a compact oriented manifold, each de Rham cohomology class can be represented by a harmonic form. Such a result has an important parallel in the Yang–Mills theory: each second Chern–Pontryagin class on S^4 can be represented by a family of self-dual instantons. The purpose of this chapter is to establish the general theorem that, for each $m = 1, 2, \cdots$, a similarly defined $2m$-th cohomology class on S^{4m} generalizing the Chern–Pontryagin class on S^4 can be represented by a family of self-dual instantons. In §3.1 we review some basic facts in 4 dimensions. In §3.2 we solve the Liouville equation. In §3.3 we present the explicit solutions of Witten in 4 dimensions based in the solution of the Liouville equation. In §3.4 we discuss the problem in all $4m$ dimensions and state our general representation theorem. In §3.5–§3.7 we prove the theorem.

3.1 Classical Yang–Mills Fields

We shall first discuss the variational formulation of the Yang–Mills theory in the Euclidean space and derive the self-dual equations. We next relate the energy levels of self-dual solutions to topology. We then present the solution of 't Hooft and its extensions.

3.1.1 Action principle and self-dual equations

In general, we have seen that the Yang–Mills fields are Lie algebra valued vector fields. In the classical model of Yang and Mills [336], the gauge group is $SU(2)$ with the associated Lie algebra $su(2)$ generated by the 2×2 matrices t_1, t_2, t_3 satisfying the commutation relation

$$[t_a, t_b] \equiv t_a t_b - t_b t_a = \varepsilon_{abc} t_c, \quad a, b, c = 1, 2, 3, \tag{3.1.1}$$

where the symbol ε_{abc} is skewsymmetric with respect to permutation of subscripts and $\varepsilon_{123} = 1$. In fact, in terms of the Pauli matrices σ_a ($a = 1, 2, 3$),

$$\sigma_1 = \begin{pmatrix} 0 & 1 \\ 1 & 0 \end{pmatrix}, \quad \sigma_2 = \begin{pmatrix} 0 & -i \\ i & 0 \end{pmatrix}, \quad \sigma_3 = \begin{pmatrix} 1 & 0 \\ 0 & -1 \end{pmatrix}, \tag{3.1.2}$$

these generators are realized by the relation $t_a = \sigma_a/2i$ ($a = 1, 2, 3$).

Let $A = (A_\mu)$ ($\mu = 1, 2, 3, 4$) be an $su(2)$-valued gauge field over the Euclidean space $\mathbb{R}^4$. Then A_μ may be represented by

$$A_\mu = A_\mu^a t_a. \tag{3.1.3}$$

In analogy to the Maxwell (electromagnetic) field, the field strength tensor (or curvature) $F_{\mu\nu}$ induced from A_μ is defined by

$$F_{\mu\nu} = \partial_\mu A_\nu - \partial_\nu A_\mu + [A_\mu, A_\nu]. \tag{3.1.4}$$

Since the metric tensor is now $(\delta_{\mu\nu})$, there is no difference between tensors with lower or upper indices: $A_\mu = A^\mu, F_{\mu\nu} = F^{\mu\nu}$, etc. Hence our action density may be written as

$$\mathcal{L} = \frac{1}{2} \text{Tr} \, (F_{\mu\nu})^2. \tag{3.1.5}$$

From (3.1.5) we obtain the total energy

$$E(A) = -\frac{1}{2} \int_{\mathbb{R}^4} \text{Tr} \, (F_{\mu\nu}^2) \, dx. \tag{3.1.6}$$

One may argue that the name 'total energy' for (3.1.6) is not quite proper physically unless the system is viewed as a static system. Indeed, this is exactly the view we adopt here since the time variable appears in the problem equally (mathematically) as the spatial variables and the spacetime is simply the Euclidean space $\mathbb{R}^4$. In particular, the time variable is denoted in this section by x^4 instead of x^0.

We can express the energy (3.1.6) more concretely. It is straightforward to examine that $2\text{Tr} \, (t_a t_b) = -\delta_{ab}$. Consequently, we have

$$E(A) = \frac{1}{4} \int_{\mathbb{R}^4} F_{\mu\nu}^a F_{\mu\nu}^a \, dx. \tag{3.1.7}$$

A critical point of (3.1.7) satisfies the Euler–Lagrange equations of (3.1.7) or (3.1.5),

$$\partial_\mu F_{\mu\nu} + [A_\mu, F_{\mu\nu}] = 0, \tag{3.1.8}$$

which are the vacuum Yang–Mills equations, generalizing the electromagnetic Maxwell equations. The solutions of (3.1.8) are called the Yang–Mills fields.

There are also other widely adopted convenient representations of the Yang–Mills fields. If $A = (A_\mu)$ is viewed as an $su(2)$-valued differential form, $A = A_\mu dx^\mu$, then the curvature F can be expressed as

$$
\begin{aligned}
F &= dA + A \wedge A \\
&= \frac{1}{2}(\partial_\mu A_\nu - \partial_\nu A_\mu + [A_\mu, A_\nu]) \, dx^\mu \wedge dx^\nu \\
&= \frac{1}{2} F_{\mu\nu} \, dx^\mu \wedge dx^\nu \\
&= dA + \frac{1}{2}[A, A] \\
&\equiv DA, \tag{3.1.9}
\end{aligned}
$$

where the wedge-product of two matrix-valued differential forms is simply the matrix product of the two matrices with differential forms as entries where the entries are multiplied via wedge and D is the connection associated with A. If for each fixed μ, $A_\mu = A_\mu^a t^a$ is identified with the 3-vector $\mathbf{A}_\mu = (A_\mu^1, A_\mu^2, A_\mu^3)$ in $\mathbb{R}^3$, then $\mathbf{F}_{\mu\nu} = F_{\mu\nu}$ (the field strength) is also a 3-vector for fixed μ, ν, and, because of the relation (3.1.1), the formula (3.1.4) is simply

$$\mathbf{F}_{\mu\nu} = \partial_\mu \mathbf{A}_\nu - \partial_\nu \mathbf{A}_\mu + \mathbf{A}_\mu \times \mathbf{A}_\nu \tag{3.1.10}$$

and the total energy (3.1.7) becomes

$$E(\mathbf{A}) = \frac{1}{4} \int_{\mathbb{R}^4} |\mathbf{F}_{\mu\nu}|^2 \, dx, \tag{3.1.11}$$

where $|\cdot|$ denotes the standard norm of $\mathbb{R}^3$ and the summation convention is adopted for repeated indices, μ, ν, as always.

It will be instructive to rederive the governing equations from (3.1.11). Let $\mathbf{B}$ be a trial configuration vector. Then the critical condition

$$\frac{d}{dt} E(\mathbf{A} + t\mathbf{B}) \bigg|_{t=0} = 0$$

and the divergence theorem gives us

$$
\begin{aligned}
0 &= \frac{1}{2} \int_{\mathbb{R}^4} \left\{ \partial_\nu \mathbf{F}_{\mu\nu} \cdot \mathbf{B}_\mu + \mathbf{F}_{\mu\nu} \cdot (\mathbf{B}_\mu \times \mathbf{A}_\nu) \right\} dx \\
&= \frac{1}{2} \int_{\mathbb{R}^4} (\partial_\nu \mathbf{F}_{\mu\nu} + \mathbf{A}_\nu \times \mathbf{F}_{\mu\nu}) \cdot \mathbf{B}_\mu \, dx.
\end{aligned}
$$

Since **B** is arbitrary, we have

$$\partial_\mu \mathbf{F}_{\mu\nu} + \mathbf{A}_\mu \times \mathbf{F}_{\mu\nu} = 0, \tag{3.1.12}$$

which are exactly the Yang–Mills equations (3.1.8) stated earlier.

In general, the second-order nonlinear field equations (3.1.12) or (3.1.8) are hard to solve. Here we show that, as in the sigma model case, there is a self-dual structure to explore so that the problem may be reduced significantly. We first illustrate this idea with a simple example, the Maxwell equations.

Let a_μ be a real-valued vector field and $f_{\mu\nu} = \partial_\mu a_\nu - \partial_\nu a_\mu$ the induced field strength. The vacuum Maxwell equations are

$$\partial_\mu f_{\mu\nu} = 0, \tag{3.1.13}$$

which are of second-order in the partial derivatives of a_μ. Define now the Hodge dual $*f$ of $f = (f_{\mu\nu})$ by

$$*f_{\mu\nu} = \frac{1}{2}\epsilon_{\mu\nu\alpha\beta}f_{\alpha\beta}. \tag{3.1.14}$$

Then it is easy to examine the validity of the following Bianchi identity

$$\partial_\mu(*f_{\mu\nu}) = 0, \tag{3.1.15}$$

which is simply the commutativity of partial derivatives (see §1.4.1). Hence, if $f_{\mu\nu}$ is self-dual or anti-self-dual with respect to the Hodge $*$, namely,

$$f_{\mu\nu} = \pm * f_{\mu\nu}, \tag{3.1.16}$$

the Maxwell equations, (3.1.13), are automatically satisfied because of the identity (3.1.15). Consequently, the second-order system (3.1.13) is reduced to a first-order system, (3.1.16).

We next consider the original Yang–Mills equations, (3.1.12) or (3.1.8). With the Hodge dual $*$ like that defined by (3.1.14),

$$*F_{\mu\nu} = \frac{1}{2}\epsilon_{\mu\nu\alpha\beta}F_{\alpha\beta}, \tag{3.1.17}$$

and the Bianchi identity,

$$D_\mu(*F_{\mu\nu}) \equiv \partial_\mu(*F_{\mu\nu}) + [A_\mu, *F_{\mu\nu}] = 0, \tag{3.1.18}$$

we see that the following self-dual or anti-self-dual equations are the corresponding reduction of (3.1.8),

$$F_{\mu\nu} = \pm * F_{\mu\nu}. \tag{3.1.19}$$

Again, note that the system (3.1.8) is of the second order but (3.1.19) the first order. Although solutions of (3.1.19) satisfy (3.1.8), the result of

Sibner, Sibner, and Uhlenbeck [286] (see also [242, 270]) shows that there are solutions of (3.1.8) which do not satisfy either self-dual or anti-self-dual equations, (3.1.19). The solutions of (3.1.19) are called self-dual or anti-self-dual solutions or instantons.

Since $F_{\mu\nu}$ is covariant under the gauge transformation

$$
\begin{aligned}
\tilde{A}_\mu(x) &= g^{-1}(x)A_\mu(x)g(x) + g^{-1}(x)\partial_\mu g(x), \\
\tilde{F}_{\mu\nu}(x) &= g^{-1}(x)F_{\mu\nu}(x)g(x), \\
g(x) &\in SU(2), \quad x \in \mathbb{R}^4,
\end{aligned}
$$

the energy, the Yang–Mills equations, and the self-dual equations are all invariant.

3.1.2 Energetic and topological characterizations

In this subsection, we study some basic properties of the solutions of the self-dual (or anti-self-dual) equations, (3.1.19).

First, since the finite-energy condition requires that $F_{\mu\nu}$ vanish rapidly at infinity of $\mathbb{R}^4$, the gauge field A_μ should be asymptotically trivial. This makes it natural to impose a boundary condition of the form

$$
A_\mu(x) \to g^{-1}(x)\partial_\mu g(x) \quad \text{as } |x| \to \infty. \tag{3.1.20}
$$

In fact, for the specific solutions considered here, precise decay estimates can be established so that the field configurations smoothly extend to $S^4 = \mathbb{R}^4 \cup \{\infty\}$.

Concerning the smooth extension of the Yang–Mills fields, a well-known general result is the Uhlenbeck removable singularity theorem [317, 318], which states that a Yang–Mills field with finite energy over a 4-manifold cannot have isolated singularities, that apparent point singularities can be removed by gauge transformations, and that, in particular, a Yang–Mills field for a bundle over $\mathbb{R}^4$ which has finite energy may be extended to a smooth field over a smooth bundle over S^4. See also [284] and references therein for other related developments.

The $SU(2)$-valued map g over $\mathbb{R}^4$ given in (3.1.20) captures the topology of the problem because asymptotically g defines a continuous map from a sphere near infinity of $\mathbb{R}^4$, which is S^3, to $SU(2)$ which, topologically, is also S^3. In other words, g maps S^3 into itself, which is characterized by its topological degree, N, which classifies the induced principal bundle over S^4 and is often identified as the second Chern class c_2.

In fact, this topological invariant has a more natural realization. To see this, we represent the field strength or curvature F as an $su(2)$-valued differential 2-form given by (3.1.9). Recall that F may also be viewed as a 2×2 matrix with entries of differential 2-forms,

$$
\Omega^{jk} = \Omega^{jk}_{\mu\nu}\,\mathrm{d}x^\mu \wedge \mathrm{d}x^\nu, \quad j, k = 1, 2.
$$

The fact that the elements of $su(2)$ are traceless skew-Hermitian matrices implies the properties

$$\Omega_{\mu\nu}^{11} = -\Omega_{\mu\nu}^{22}, \quad \Omega_{\mu\nu}^{12} = -\overline{\Omega}_{\mu\nu}^{21}.$$

Hence the second Chern or first Pontryagin form $c_2(F)$ is formally written

$$-4\pi^2 c_2(F) = \det(F) = \det \begin{pmatrix} \Omega^{11} & \Omega^{12} \\ \Omega^{21} & \Omega^{22} \end{pmatrix}$$

$$= \Omega^{11} \wedge \Omega^{22} - \Omega^{12} \wedge \Omega^{21}$$

$$= \Omega_{\mu\nu}^{11} dx^{\mu} \wedge dx^{\nu} \wedge \Omega_{\alpha\beta}^{22} dx^{\alpha} \wedge dx^{\beta} - \Omega_{\mu\nu}^{12} dx^{\mu} \wedge dx^{\nu} \wedge \Omega_{\alpha\beta}^{21} dx^{\alpha} \wedge dx^{\beta}$$

$$= (\Omega_{\mu\nu}^{11} \Omega_{\alpha\beta}^{22} - \Omega_{\mu\nu}^{12} \Omega_{\alpha\beta}^{21}) dx^{\mu} \wedge dx^{\nu} \wedge dx^{\alpha} \wedge dx^{\beta}$$

$$= -\frac{1}{2} \Big\{ (\Omega_{\mu\nu}^{11} \Omega_{\alpha\beta}^{11} + \Omega_{\mu\nu}^{22} \Omega_{\alpha\beta}^{22})$$

$$-(\Omega_{\mu\nu}^{12} \overline{\Omega}_{\alpha\beta}^{12} + \Omega_{\mu\nu}^{21} \overline{\Omega}_{\alpha\beta}^{21}) \Big\} dx^{\mu} \wedge dx^{\nu} \wedge dx^{\alpha} \wedge dx^{\beta}.$$

In other words, we have obtained the classical expression

$$c_2(F) = -\frac{1}{8\pi^2} \text{Tr}\,(F \wedge F). \tag{3.1.21}$$

Therefore, we have the following integral representation of the second Chern class,

$$c_2 = \int_{\mathbb{R}^4} c_2(F) = -\frac{1}{8\pi^2} \int_{\mathbb{R}^4} \text{Tr}\,(F \wedge F) = N. \tag{3.1.22}$$

Using (3.1.9) and (3.1.17), it is straightforward to show that

$$F \wedge F = \frac{1}{2}(F_{\mu\nu} * F_{\mu\nu}) dx^1 \wedge dx^2 \wedge dx^3 \wedge dx^4,$$

$$F \wedge *F = \frac{1}{2}(F_{\mu\nu} F_{\mu\nu}) dx^1 \wedge dx^2 \wedge dx^3 \wedge dx^4. \tag{3.1.23}$$

Thus, we may represent the Chern class (3.1.22) concretely as

$$c_2 = -\frac{1}{16\pi^2} \int_{\mathbb{R}^4} \text{Tr}(F_{\mu\nu} * F_{\mu\nu}) dx = N. \tag{3.1.24}$$

We can now pursue an understanding of the self-dual or anti-self-dual equations (3.1.19). It can be verified by (3.1.17) that

$$F_{\mu\nu} F_{\mu\nu} = *F_{\mu\nu} * F_{\mu\nu}. \tag{3.1.25}$$

Thus the energy (3.1.6) becomes

$$E(A) = -\frac{1}{4} \int_{\mathbb{R}^4} \text{Tr}\,\{(F_{\mu\nu})^2 + (*F_{\mu\nu})^2\} dx$$

$$= -\frac{1}{4} \int_{\mathbb{R}^4} \text{Tr}\,\{(F_{\mu\nu} \mp *F_{\mu\nu})^2\} dx \mp \frac{1}{2} \int_{\mathbb{R}^4} \text{Tr}\,(F_{\mu\nu} * F_{\mu\nu}) dx. \tag{3.1.26}$$

Combining (3.1.22) and (3.1.26), we arrive at the topological energy lower bound

$$E(A) \geq 8\pi^2|c_2| = 8\pi^2|N|. \tag{3.1.27}$$

It is clear that the above lower bound is attained if and only if the gauge field A satisfies the self-dual or anti-self-dual equations (3.1.19).

It will be useful to note that, in terms of differential forms, the energy (3.1.6) is

$$E = -\int_{\mathbb{R}^4} \mathrm{Tr}\,(F \wedge *F). \tag{3.1.28}$$

3.1.3 't Hooft instantons

It is convenient to use the Pauli matrices (3.1.2). Note that $x^\mu = x_\mu$ ($\mu = 1,2,3,4$). The boundary condition (3.1.20) gives us a hint to choose the gauge field A_μ to be

$$A_\mu(x) = \frac{x^2}{x^2 + \lambda^2} g^{-1}(x)\partial_\mu g(x), \quad x^2 = x_1^2 + x_2^2 + x_3^2 + x_4^2, \tag{3.1.29}$$

where $\lambda > 0$ is a parameter and the group element $g \in SU(2)$ is defined by

$$g(x) = \frac{x_\mu \omega_\mu}{\sqrt{x^2}} \tag{3.1.30}$$

with the 2×2 ω-matrices defined by

$$\omega_a = i\sigma_a, \quad a = 1,2,3; \quad \omega_4 = \begin{pmatrix} 1 & 0 \\ 0 & 1 \end{pmatrix}. \tag{3.1.31}$$

Introduce the 't Hooft tensors

$$\eta_{\mu\nu} = -\frac{1}{4}(\omega_\mu^\dagger \omega_\nu - \omega_\nu^\dagger \omega_\mu), \quad \bar{\eta}_{\mu\nu} = -\frac{1}{4}(\omega_\mu \omega_\nu^\dagger - \omega_\nu \omega_\mu^\dagger). \tag{3.1.32}$$

It is straightforward to examine that these tensors are either self-dual or anti-self-dual,

$$\eta_{\mu\nu} = *\eta_{\mu\nu}, \quad \bar{\eta}_{\mu\nu} = - *\bar{\eta}_{\mu\nu}. \tag{3.1.33}$$

We need to represent the gauge field (3.1.29) in terms of the 't Hooft tensors so that self-duality becomes apparent to achieve. For this purpose, we first note by using (3.1.30) and $g^{-1} = g^\dagger$ that

$$g^{-1}\partial_\mu g = \frac{x_\nu \omega_\nu^\dagger \omega_\mu}{x^2} - \frac{x_\mu}{x^2}\omega_4.$$

Similarly,

$$(\partial_\mu g^{-1})g = \frac{x_\nu \omega_\mu^\dagger \omega_\nu}{x^2} - \frac{x_\mu}{x^2}\omega_4.$$

However, in view of $(\partial_\mu g^{-1})g = -g^{-1}\partial_\mu g$, we have

$$g^{-1}\partial_\mu g = \frac{1}{2}(g^{-1}\partial_\mu g - [\partial_\mu g^{-1}]g) = \frac{x_\nu}{2x^2}(\omega_\nu^\dagger \omega_\mu - \omega_\mu^\dagger \omega_\nu). \qquad (3.1.34)$$

Inserting (3.1.34) into (3.1.29) and using the definition (3.1.32), we have

$$A_\mu(x) = \frac{2x_\nu}{x^2 + \lambda^2}\eta_{\mu\nu}. \qquad (3.1.35)$$

We next compute the field strength tensor, $F_{\mu\nu}$. From (3.1.35), we get

$$\partial_\mu A_\nu - \partial_\nu A_\mu = \frac{4}{x^2 + \lambda^2}\eta_{\mu\nu} + \frac{4}{(x^2 + \lambda^2)^2}(x_\mu x_\alpha \eta_{\nu\alpha} - x_\nu x_\alpha \eta_{\mu\alpha}).$$

Besides, from (3.1.29) and (3.1.30), we have

$$\begin{aligned}
[A_\mu, A_\nu] &= \frac{x^4}{(x^2 + \lambda^2)^2}([\partial_\mu g^{-1}][\partial_\nu g] - [\partial_\nu g^{-1}][\partial_\mu g]) \\
&= -\frac{4x^2}{(x^2 + \lambda^2)^2}\eta_{\mu\nu} + \frac{4}{(x^2 + \lambda^2)^2}(x_\nu x_\alpha \eta_{\mu\alpha} - x_\mu x_\alpha \eta_{\nu\alpha}).
\end{aligned}$$

Consequently, by the property (3.1.33), we obtain the self-dual tensor field,

$$F_{\mu\nu}(x) = \frac{4\lambda^2}{(x^2 + \lambda^2)^2}\eta_{\mu\nu}. \qquad (3.1.36)$$

One of the interesting features of this solution is that its energy density peaks at the origin $x = 0$ with a level determined by λ. In other words, this solution looks like a particle, or an instanton, located at $x = 0$ with a size specified by a parameter.

For convenience, we may represent $\eta_{\mu\nu}$ in terms of the standard basis, $\{t_a\}_{a=1,2,3}$, of the Lie algebra $su(2)$, in the form

$$\eta_{\mu\nu} = \eta_{\mu\nu}^a t_a.$$

Various properties of the real-valued tensors $\eta_{\mu\nu}^a$ are stated in [310], of which, the most useful one for our purpose here is

$$\eta_{\mu\nu}^a \eta_{\mu\nu}^b = 4\delta_{ab}. \qquad (3.1.37)$$

Inserting (3.1.36) and (3.1.37) into (3.1.24) and using Tr $(t_a t_b) = -\delta_{ab}/2$, we have

$$c_2 = \frac{6}{\pi^2}\int_{\mathbb{R}^4}\frac{\lambda^4}{(x^2 + \lambda^2)^4}\,dx = 1. \qquad (3.1.38)$$

Hence we have constructed an instanton of unit topological charge, $c_2 = 1$. This one-instanton solution was discovered by Belavin, Polyakov, Schwartz, and Tyupkin [34] and is known as the BPST solution.

We then show that the above method may be generalized to obtain instantons of an arbitrary topological charge, $c_2 = N$. To this end, we recall that (3.1.34) may be rewritten as

$$g^{-1}\partial_\mu g = \frac{2x_\nu}{x^2}\eta_{\mu\nu}.$$ (3.1.39)

On the other hand, define

$$\tilde{A}_\mu(x) = \left(\partial_\nu \ln\left[1 + \frac{\lambda^2}{x^2}\right]\right)\eta_{\mu\nu}.$$ (3.1.40)

We derive from (3.1.39) the relation

$$\begin{aligned}
\tilde{A}_\mu &= \left(\frac{2x_\nu}{x^2 + \lambda^2} - \frac{2x_\nu}{x^2}\right)\eta_{\mu\nu} = A_\mu - g^{-1}\partial_\mu g \\
&= A_\mu + (\partial_\mu g^{-1})g = gA_\mu g^{-1} + g\partial_\mu g^{-1}.
\end{aligned}$$

In other words, the gauge fields A_μ and $\tilde{A}_\mu$ defined in (3.1.29), or (3.1.35), and (3.1.40), respectively, are equivalent. Consequently, the field strength tensor induced from $\tilde{A}_\mu$ is also self-dual and we get a gauge-equivalent self-dual instanton. Hence we may write the obtained solution in the form

$$A_\mu = (\partial_\nu \ln f)\eta_{\mu\nu},$$ (3.1.41)

where $f = 1 + \lambda^2/x^2$. At first glance, this procedure does not lead to any new solutions. However, it suggests that we may obtain more solutions if we simply use (3.1.41) as an ansatz for which f is a positive-valued function to be determined by our self-duality requirement. It turns out that a general choice of f is

$$f(x) = 1 + \sum_{j=1}^{N} \frac{\lambda_j^2}{(x - p_j)^2}, \quad \lambda_j > 0, \quad p_j \in \mathbb{R}^4, \quad j = 1, 2, \cdots, N,$$ (3.1.42)

which contains $5N$ continuous parameters and is called the 't Hooft solution [310]. In fact this solution describes N instantons located at the points $p_1, p_2, \cdots, p_N$ with sizes determined by the parameters $\lambda_1, \lambda_2, \cdots, \lambda_N$. It can be examined that $c_2 = N$ (we omit the details). The 't Hooft instantons have been extended by Jackiw and Rebbi [151] and Ansourian and Ore [10] into a form containing $5N + 4$ parameters which is the most general explicit self-dual solution known, although, according to a result [14, 276] based on the Atiyah–Singer index theorem [16], the number of free parameters of a general self-dual instanton in the class $c_2 = N$ is $8N - 3$. This conclusion was first arrived at by physicists [58, 155] using plausible arguments: $4N$ parameters determine the positions and N parameters the sizes of the instantons as in the 't Hooft solution case, $3N$ parameters determine the asymptotic orientations of the instantons in the internal space $SU(2) = S^3$ from which the 3 parameters originated from the global $SU(2)$ gauge equivalence must be subtracted. For a general construction of 4-dimensional Yang–Mills instantons, see [12, 15].

3.2 Liouville Equation and Solution

Our solutions in $4m$ dimensions are analogous to Witten's self-dual instantons in 4 dimensions. Witten's method relies on the explicit solution of the Liouville equation. Hence, in this section, we prepare ourselves to consider the integration of the Liouville equation. There are four major methods of independent interest: namely, Liouville's original approach [200], the Bäcklund transformation [212], the inverse scattering method [9], and the method of separation of variables [188]. The purpose of this section is to present the first two independent methods because of their simplicity and instructiveness. In §6.2.2, we shall present the fourth method in connection with the solution of the Toda systems.

3.2.1 Liouville method

The elliptic Liouville equation that concerns us is of the form

$$\Delta u = \mp \frac{8}{a^2} e^u, \quad x \in \mathbb{R}^2, \tag{3.2.1}$$

where $a > 0$ is a constant. For greater generality, we shall first study (3.2.1) under its 'hyperbolic' disguise,

$$\frac{\partial^2}{\partial s \partial t} \ln f \pm \frac{2}{a^2} f = 0. \tag{3.2.2}$$

In essence, Liouville's method may be viewed as a separation of variable technique. To proceed, assume that f has the representation

$$f = \frac{\partial g}{\partial s} \equiv g_s. \tag{3.2.3}$$

Then (3.2.2) becomes

$$\left(\frac{\partial}{\partial t} \ln f \pm \frac{2}{a^2} g \right)_s = 0.$$

Consequently, we obtain the equation

$$g_{st} = \mp \frac{2}{a^2} g g_s + P(t) g_s,$$

where P is an arbitrary function depending only on t. Integrating the above with respect to the variable s, we find

$$g_t = \mp \frac{1}{a^2} g^2 + P(t) g + Q(t), \tag{3.2.4}$$

where, as before, Q is an arbitrary function of t.

Assume that $h(t)$ is a particular solution of (3.2.4), namely, $h(t)$ satisfies

$$h'(t) = \mp \frac{1}{a^2} h^2(t) + P(t)h(t) + Q(t). \tag{3.2.5}$$

We seek for the solution of (3.2.4) that takes the form

$$g = h(t) - \frac{1}{\xi} \tag{3.2.6}$$

Inserting (3.2.6) into (3.2.4) and using (3.2.5), we arrive at the first-order linear equation

$$\frac{\partial \xi}{\partial t} + \left(P(t) \mp \frac{2}{a^2} h(t) \right) \xi = \mp \frac{1}{a^2} \tag{3.2.7}$$

with the integrating factor

$$\mu(t) = \exp\left(\int \left[P(t) \mp \frac{2}{a^2} h(t) \right] dt \right).$$

Therefore the general solution of (3.2.7) is given by

$$\xi = \frac{1}{\mu(t)} \left(\mp \frac{1}{a^2} \int \mu(t)\, dt + G(s) \right)$$

in which G is an arbitrary function of s.

With the notation

$$F(t) = \frac{1}{a^2} \int \mu(t)\, dt,$$

the general solution of (3.2.7) is represented by the expression

$$\xi(s,t) = \frac{1}{a^2 F'(t)} (\mp F(t) + G(s)). \tag{3.2.8}$$

From (3.2.3), (3.2.6), and (3.2.8), we obtain the general solution of (3.2.2) given by the following formula involving two arbitrary functions,

$$f(s,t) = \frac{a^2 F'(t) G'(s)}{(F(t) \mp G(s))^2}. \tag{3.2.9}$$

Sometimes people replace $F(t)$ by $-1/F(t)$ and rewrite (3.2.9) as

$$f(s,t) = \frac{a^2 F'(t) G'(s)}{(1 \pm F(t) G(s))^2}. \tag{3.2.10}$$

Now return to the elliptic Liouville equation (3.2.1). Using the complex variables

$$s \mapsto z = x_1 + i x_2, \quad t \mapsto \bar{z} = x_1 - i x_2$$

and observing that $4\partial^2_{z\bar{z}} = \partial^2_1 + \partial^2_2 = \Delta$, we recover (3.2.1) with $f = e^u$ so that the corresponding formula of general solution becomes

$$u(x_1, x_2) = u(z) = \ln\left(\frac{a^2 F'(z)G'(\bar{z})}{(1 \pm F(z)G(\bar{z}))^2}\right). \qquad (3.2.11)$$

Here F and G are differentiable functions of the complex variables z and $\bar{z}$, respectively. Consequently they are complex-valued functions and the solution u given in (3.2.11) is also complex in general. In order to obtain a real-valued solution, we impose $G(\bar{z}) = \overline{F(z)}$. Then $G'(\bar{z}) = \overline{F'(z)}$ and (3.2.11) becomes

$$u(z) = \ln\left(\frac{a^2 |F'(z)|^2}{(1 \pm |F(z)|^2)^2}\right), \qquad (3.2.12)$$

where $F(z)$ is an arbitrary holomorphic function of the complex variable z. This formula suggests that the equation $\Delta u = \lambda e^u$ ($\lambda > 0$) has no entire solution which is a well-known classical result [236, 240, 271, 335].

3.2.2   Bäcklund transformation method

Consider (3.2.2). Using the change of variables $s = x - y, t = x + y$, a suitable rescaling to absorb the parameter $a > 0$ and the $\pm$ sign, and the substitution $u = \frac{1}{2}\ln f$ in the equation which is half of the function u of the previous subsection and should not cause confusion, it suffices in view of the Bäcklund transformation to study (3.2.2) in its more convenient form

$$u_{xx} - u_{yy} = e^{2u}. \qquad (3.2.13)$$

It will be useful to use the hyperbolic 'complex' variables

$$w = x + jy, \quad h = u + jv,$$

where j satisfies $j^2 = 1$ and $\bar{w} = x - jy$ denotes the associated conjugate variable. In fact such a rule may be realized by a natural 2×2 matrix representation

$$1 \mapsto \begin{pmatrix} 1 & 0 \\ 0 & 1 \end{pmatrix}, \quad j \mapsto \begin{pmatrix} 0 & 1 \\ 1 & 0 \end{pmatrix},$$

and w, h may be viewed as 2×2 matrix-valued quantities with components x, y and u, v, respectively. Thus we have

$$\partial_w = \frac{1}{2}(\partial_x - j\partial_y), \quad \partial_{\bar{w}} = \overline{\partial_w} = \frac{1}{2}(\partial_x + j\partial_y), \quad 4\partial^2_{w\bar{w}} = \partial^2_x - \partial^2_y$$

and the corresponding Euler formula

$$e^{j\alpha} = \cosh\alpha + j\sinh\alpha.$$

Hence we have $e^{\overline{h}} = \overline{(e^h)}$, $\partial_w(e^h) = e^h \partial_w h$, and other useful properties.

With the above preparation, the Bäcklund transformation for the hyperbolic Liouville equation (3.2.13) is generated from the Dirac 'square root' of (3.2.13), namely,

$$2\partial_{\overline{w}} h = e^{\overline{h}}. \tag{3.2.14}$$

We can directly check that (3.2.14) implies (3.2.13). In fact, applying the operator $2\partial_w$ on (3.2.14), we have

$$4\partial^2_{w\overline{w}} h = 2\partial_w e^{\overline{h}} = \overline{(2\partial_{\overline{w}} e^h)} = \overline{(2e^h \partial_{\overline{w}} h)} = \overline{(e^{h+\overline{h}})} = e^{h+\overline{h}}. \tag{3.2.15}$$

Rewriting (3.2.15) in its components, we obtain the following two real scalar equations

$$\begin{aligned} u_{xx} - u_{yy} &= e^{2u}, \\ v_{xx} - v_{yy} &= 0. \end{aligned} \tag{3.2.16}$$

The first equation is exactly (3.2.13) and the second one is a free wave equation.

Therefore, to solve (3.2.13) or (3.2.16), it suffices to solve (3.2.14). For this purpose, we express (3.2.14) in its component form

$$\begin{aligned} u_x + v_y &= e^u \cosh v, \\ u_y + v_x &= -e^u \sinh v. \end{aligned} \tag{3.2.17}$$

The system (3.2.17) defines the desired Bäcklund transformation for (3.2.13) and may be viewed as a pair of nonlinear Cauchy–Riemann equations in the hyperbolic complex variable $w = x + \mathrm{j}y$. It is clear that the solution of the first equation in (3.2.16) may be obtained from solving (3.2.17) in which v satisfies the second equation in (3.2.16).

As a consequence, we write down for v the general solution

$$v = V_1(x+y) + V_2(x-y) = V_1(t) + V_2(s),$$

where $V_1(t)$ and $V_2(s)$ are arbitrary differentiable functions of variables t and s, respectively.

By adding the two equations in (3.2.17), we have

$$2u_t + 2V_1'(t) = e^{u - V_1(t) - V_2(s)}.$$

With the new variable $\xi = e^{-u}$, we arrive at the following linear equation,

$$\frac{\partial \xi}{\partial t} - V_1'(t)\xi = -\frac{1}{2} e^{-V_1(t) - V_2(s)}. \tag{3.2.18}$$

The integrating factor of (3.2.18) is $e^{-V_1(t)}$. Thus its solution can be written as

$$\xi = e^{V_1(t)}\left(-\frac{1}{2} e^{-V_2(s)} \int e^{-2V_1(t)}\, dt + Q(s) \right),$$

where $Q(s)$ depends only on s and is to be determined. Making the replacement

$$-\frac{1}{2}\int e^{-2V_1(t)}\,dt = F(t) \quad \text{so that} \quad e^{V_1(t)} = \frac{1}{\sqrt{-2F'(t)}},$$

we have

$$\xi = \frac{F(t)}{\sqrt{-2F'(t)}}e^{-V_2(s)} + \frac{Q(s)}{\sqrt{-2F'(t)}}. \qquad (3.2.19)$$

Similarly, by subtracting the two equations in (3.2.17), we obtain

$$\xi = \frac{G(s)}{\sqrt{-2G'(s)}}e^{V_1(t)} + \frac{P(t)}{\sqrt{-2G'(s)}}, \qquad (3.2.20)$$

where $G(s)$ satisfies

$$e^{-V_2(s)} = \frac{1}{\sqrt{-2G'(s)}}$$

and $P(t)$ depends only on t and is also to be determined.

Comparing (3.2.19) and (3.2.20), we obtain the relations

$$P(t) = \frac{F(t)}{\sqrt{-2F'(t)}}, \quad Q(s) = \frac{G(s)}{\sqrt{-2G'(s)}}.$$

Consequently, we can recall the assignment $\xi = e^{-u}$ to arrive at the following expected solution formula for (3.2.13),

$$u = \frac{1}{2}\ln\left(\frac{4F'(t)G'(s)}{(F(t)+G(s))^2}\right) = \frac{1}{2}\ln\left(\frac{4F'(x+y)G'(x-y)}{(F(x+y)+G(x-y))^2}\right). \qquad (3.2.21)$$

3.3   Witten's Instanton

In this section, we study the multiple instantons of Witten. Such solutions are of particular interest to us because they lead to a construction of higher-dimensional self-dual instantons representing a family of higher-order characteristic classes, extending the second Chern–Pontryagin class in 4 dimensions, to be presented in the next section. We shall first present Witten's dimensional reduction. We then show how to use the Liouville equation to find Witten's N-instantons explicitly.

3.3.1   Field configurations and equations

Witten's instanton is symmetric with respect to rotation of the spatial coordinates x_j $(j = 1, 2, 3)$ and is of the form

$$A_j^a = \epsilon_{jak}\frac{x_k}{r^2}(1 - \phi_2(r,t)) + \frac{1}{r^3}(\delta_{ja}r^2 - x_jx_a)\phi_1(r,t) + \frac{x_jx_a}{r^2}a_1(r,t),$$

$$A_4^a = \frac{x_a}{r}a_2(r,t), \quad a,j,k = 1,2,3, \qquad (3.3.1)$$

where $r^2 = x_1^2 + x_2^2 + x_3^2$, $t = x_4$ is the temporal coordinate, and a_1, a_2, ϕ_1, ϕ_2 are real-valued functions. Thus, in view of (3.3.1), the field strength tensor becomes

$$
F_{4j}^a = -\left(\frac{\partial\phi_2}{\partial t} + a_2\phi_1\right)\frac{\epsilon_{jak}x_k}{r^2} + \left(\frac{\partial\phi_1}{\partial t} - a_2\phi_2\right)\frac{(\delta_{aj}r^2 - x_a x_j)}{r^3}
$$
$$
+ \left(\frac{\partial a_1}{\partial t} - \frac{\partial a_2}{\partial r}\right)\frac{x_a x_j}{r^2},
$$

$$
\frac{1}{2}\epsilon_{jkk'}F_{kk'}^a = -\frac{\epsilon_{jak'}x_{k'}}{r^2}\left(\frac{\partial\phi_1}{\partial r} - a_1\phi_2\right) - \left(\frac{\partial\phi_2}{\partial r} + a_1\phi_1\right)\frac{(\delta_{aj}r^2 - x_a x_j)}{r^3}
$$
$$
+ (1 - \phi_1^2 - \phi_2^2)\frac{x_a x_j}{r^4}. \tag{3.3.2}
$$

Inserting (3.3.2) into (3.1.7) and (3.1.22), we obtain the reduced expressions for the total energy

$$
E = \frac{1}{4}\int_{\mathbb{R}^3}d x \int_{\mathbb{R}}dt\{F_{\mu\nu}^a F_{\mu\nu}^a\}
$$
$$
= 4\pi\int_{-\infty}^{\infty}dt\int_0^{\infty}dr\left\{|D_j\phi|^2 + \frac{1}{4}r^2 f_{jk}^2 + \frac{1}{2r^2}(1 - |\phi|^2)^2\right\} \tag{3.3.3}
$$

and the Chern class

$$
c_2 = -\frac{1}{16\pi^2}\int_{\mathbb{R}^4}\mathrm{Tr}\,(F_{\mu\nu} * F_{\mu\nu})
$$
$$
= -\frac{1}{2\pi}\int_{-\infty}^{\infty}dt\int_0^{\infty}dr\left\{(1 - |\phi|^2)f_{12} - i(D_1\phi\overline{D_2\phi} - \overline{D_1\phi}D_2\phi)\right\}, \tag{3.3.4}
$$

where now ϕ is a complex field defined by $\phi = \phi_1 + i\phi_2$, ∂_1 and ∂_2 denote $\partial/\partial r$ and $\partial/\partial t$, respectively, $f_{jk} = \partial_j a_k - \partial_k a_j$ ($j, k = 1, 2$), $D_j\phi = \partial_j\phi + ia_j\phi$ ($j = 1, 2$). In terms of these, the self-dual equations (3.1.19) become

$$
D_1\phi + iD_2\phi = 0,
$$
$$
r^2 f_{12} = |\phi|^2 - 1. \tag{3.3.5}
$$

Recall that we are looking for multi-instanton solutions with localized field concentrations. In view of the second equation in (3.3.5) and the energy density given in (3.3.3), we are led to the observation that the field concentration spots are where ϕ vanishes. Suppose that

$$
p_1, p_2, \cdots, p_N \in \mathbb{R}_+^2 = \{(t, r)\,|\,-\infty < t < \infty, 0 < r < \infty\}
$$

are zeros of ϕ. Then the substitution $u = \ln|\phi|$ transforms (3.3.5) into the following scalar nonlinear elliptic equation with sources,

$$
\Delta u = \frac{1}{r^2}(e^{2u} - 1) + 2\pi\sum_{j=1}^{N}\delta_{p_j}, \quad x \in \mathbb{R}_+^2. \tag{3.3.6}
$$

Conversely, if u is a solution of (3.3.6), we may use the method described in §2.4.6 to recover a solution of (3.3.5) by setting

$$\phi(z) = e^{u(z)+i\theta(z)}, \quad \theta(z) = \sum_{j=1}^{N} \arg(z - p_j),$$

$$a_1(z) = \text{Re}\{2i\overline{\partial}\ln\phi(z)\}, \quad a_2(z) = \text{Im}\{2i\overline{\partial}\ln u(z)\}. \quad (3.3.7)$$

3.3.2 Explicit instanton solutions

We now use the method of Witten [331] to construct the solution of (3.3.6) explicitly. We momentarily neglect the singular source term and consider

$$r^2 \Delta u = e^{2u} - 1. \quad (3.3.8)$$

It is seen that (3.3.8) reduces to the Liouville equation

$$\Delta v = e^{2v} \quad (3.3.9)$$

after the transformation

$$u = \ln r + v. \quad (3.3.10)$$

By (3.2.12), we have the representation

$$v(z) = \ln\left(\frac{2|F'(z)|}{1 - |F(z)|^2}\right), \quad z = r + it. \quad (3.3.11)$$

The solution is free of singularities if $F'(z) \neq 0$ and $|F(z)| < 1$.

We now use (3.3.10) to get the original solution u from (3.3.11), i.e.,

$$u(z) = \ln\left(\frac{2r|F'(z)|}{1 - |F(z)|^2}\right), \quad z = r + it. \quad (3.3.12)$$

Recall that the finite-energy condition imposed on (3.3.3) implies the boundary condition $|\phi| = 1$ or $u = 0$ for $r = 0$. So we must have $|F(z)| = 1$ at $r = 0$. Besides, to obtain an entire solution, we also need $|F(z)| < 1$ for all $r > 0$. These properties motivate the choice

$$F(z) = \prod_{n=0}^{N}\left(\frac{a_n - z}{\overline{a}_n + z}\right), \quad \text{Re}\{a_n\} > 0, \quad n = 0, 1, \cdots, N. \quad (3.3.13)$$

When $N = 0$, it is straightforward to check through (3.3.12) and (3.3.13) that $u(z) \equiv 0$. This corresponds to the trivial solution.

When $N \geq 1$, the zeros of $F'(z)$ will give rise to the desired singularities of $u(z)$. To see this, we rewrite $F'(z)$ as $F'(z) = f(z)Q(z)$ where

$$f(z) = \sum_{n=0}^{N}\left\{(a_n + \overline{a}_n) \cdot \prod_{n'=0,n'\neq n}^{N}(z^2 - [a_{n'} - \overline{a}_{n'}]z - |a_{n'}|^2)\right\},$$

$$Q(z) = \prod_{n=0}^{N}\frac{1}{(\overline{a}_n + z)^2}.$$

Since $\text{Re}\{a_n\} > 0$, $n = 0, 1, \cdots, N$, the zeros of $F'(z)$ on the Poincaré half-plane $\text{Re}\{z\} \geq 0$ are the zeros of $f(z)$. On the other hand, with the change of variables, $g(z) = f(iz)$, we see that $g(z)$ is a real-coefficient polynomial of degree $2N$. Hence $g(z)$ has N pairs of complex-conjugate zeros on the complex plane. This result implies that the zeros of $f(z)$ are of the form $z_j^{\pm} = \pm r_j + it_j$, $r_j \geq 0$, $j = 1, 2, \cdots, N$. Of course, $r_j \neq 0$, $j = 1, 2, \cdots, N$ because $f(it_j) < 0$ for all j. Thus we have obtained exactly N zeros, $z_j = r_j + it_j$, $j = 1, 2, \cdots, N$, in the right half-plane. The positions of these z_j's can be specified arbitrarily by adjusting the parameters a_n's. Therefore the function u defined in (3.3.12) solves (3.3.6) with $p_j = z_j$, $j = 1, 2, \cdots, N$. Returning to the original variables through (3.3.7), we obtain Witten's explicit multi-instanton of charge $c_2 = N$. This last topological conclusion will be verified systematically within our general formulation set in the subsequent sections.

3.4   Instantons and Characteristic Classes

With collaborators, the work of Tchrakian [306, 307, 308] shows that one can systematically develop the Yang–Mills theory in $4m$ dimensions so that higher-dimensional self-dual instantons are characterized by a family of $2m$-th order characteristic classes extending the Chern–Pontryagin class in 4 dimensions. Independently, Grossman, Kephart, and Stasheff [138] made the same formulation in 8 dimensions. In this section, we first present their elegant general formulation and state our main representation theorem. We then relate the problem to the solution of a quasilinear elliptic equation in the Poincaré half-plane.

3.4.1   Self-duality and Witten–Tchrakian equations

In this subsection we consider the Yang–Mills theory over S^{4m} and derive the self-dual equations for instantons in the general framework of Tchrakian [306, 307, 308]. The most natural principal bundle to house the gauge fields over S^{4m} is the frame bundle associated with the tangent bundle. Hence, we are led to the largest possible structure group, $SO(4m)$. In 4 dimensions, we have $SO(4)$, which contains two copies of $SO(3)$. Since the Lie algebra of $SO(3)$ is the same as that of $SU(2)$, the $SU(2)$-gauge theory considered in the previous sections is a special case of the $SO(4)$-gauge theory. Thus, we now formulate a general $SO(4m)$ pure Yang–Mills gauge theory over S^{4m}.

Recall that, over an n-dimensional Riemannian manifold M with a metric $g = (g_{\mu\nu})$, one defines the Hodge dual $*$, which maps the set of real-valued p-forms, Λ^p, to the set of real-valued $(n-p)$-forms, Λ^{n-p}, according to

$$*(dx^{\mu_1} \wedge \cdots \wedge dx^{\mu_p}) =$$

$$\frac{\sqrt{g}}{(n-p)!}g^{\mu_1\nu_1}\cdots g^{\mu_p\nu_p}\epsilon_{\nu_1\cdots\nu_p\nu_{p+1}\cdots\nu_n}\,dx^{\nu_{p+1}}\wedge\cdots\wedge dx^{\nu_n}, \quad (3.4.1)$$

where g under the square-root denotes the determinant of the metric and ϵ is the skew-symmetric Kronecker symbol satisfying $\epsilon_{12\cdots n}=1$. In particular, when n is even and $p=n/2$, (3.4.1) is invariant under conformal deformation of metric. Furthermore,

$$**\omega=(-1)^{p(n-p)}\omega, \quad \omega\in\Lambda^p \qquad (3.4.2)$$

and

$$(\alpha,\beta)=\int_M\alpha\wedge*\beta, \quad \alpha,\beta\in\Lambda^p, \qquad (3.4.3)$$

defines an inner product on Λ^p. It is clear from (3.4.2) that, when both p and n are even, $*:\Lambda^p\to\Lambda^{n-p}$ is an isometry with respect to the inner product (3.4.3), namely,

$$(\alpha,\beta)=(*\alpha,*\beta), \quad \alpha,\beta\in\Lambda^p. \qquad (3.4.4)$$

Let A be an $so(4m)$-valued connection 1-form over S^{4m} and F its induced curvature 2-form. Motivated from (3.1.28) in 4 dimensions, we introduce the energy functional

$$E=-\int_{S^{4m}}\mathrm{Tr}\,(F(m)\wedge*F(m)) \qquad (3.4.5)$$

in $4m$ dimensions, where

$$F(m)=\underbrace{F\wedge\cdots\wedge F}_{m} \qquad (3.4.6)$$

is a $2m$-form generalizing the 2-form F. For $so(4m)$-valued differential forms over S^{4m}, the inner product (3.4.3) needs to take the modified form

$$(\alpha,\beta)=-\int_{S^{4m}}\mathrm{Tr}\,(\alpha\wedge*\beta). \qquad (3.4.7)$$

In view of this, the energy (3.4.5) is nothing but the squared norm of the generalized curvature $F(m)$: $E=\|F(m)\|^2$. Of course, (3.4.4) again holds.

In place of the Chern class (3.1.21), we introduce the characteristic class

$$s_{2m}(F)=-\mathrm{Tr}\,(F(m)\wedge F(m))=-\mathrm{Tr}\,(\underbrace{F\wedge\cdots\wedge F}_{2m}). \qquad (3.4.8)$$

Of course, $s_2(F)=8\pi^2c_2(F)$. In general, $s_{2m}(F)$ is a polynomial of the Chern–Pontryagin forms. The associated topological charge is then defined as

$$s_{2m}=\int_{S^{4m}}s_{2m}(F)=-\int_{S^{4m}}\mathrm{Tr}\,(F(m)\wedge F(m)). \qquad (3.4.9)$$

We now decompose $F(m)$ into its self-dual and anti-self-dual parts,

$$F(m) = F^+(m) + F^-(m), \quad F^\pm(m) = \frac{1}{2}(F(m) \pm *F(m)). \qquad (3.4.10)$$

Using (3.4.4), we see that $F^+(m)$ and $F^-(m)$ are orthogonal,

$$(F^+(m), F^-(m)) = 0.$$

Therefore, inserting (3.4.10) into (3.4.5) and (3.4.9) and using the property $*F^\pm(m) = \pm F^\pm(m)$ and the orthogonality of $F^+(m)$ and $F^-(m)$, we obtain

$$
\begin{aligned}
E &= (F(m), F(m)) \\
&= (F^+(m) + F^-(m), F^+(m) + F^-(m)) \\
&= \|F^+(m)\|^2 + \|F^-(m)\|^2, \qquad (3.4.11) \\
s_{2m} &= (F(m), *F(m)) \\
&= (F^+(m) + F^-(m), F^+(m) - F^-(m)) \\
&= \|F^+(m)\|^2 - \|F^-(m)\|^2. \qquad (3.4.12)
\end{aligned}
$$

Consequently, we can combine (3.4.11) and (3.4.12) to arrive at

$$
\begin{aligned}
E &= 2\|F^\mp(m)\|^2 \pm (\|F^+(m)\|^2 - \|F^-(m)\|^2) \\
&= 2\|F^\mp(m)\|^2 + |s_{2m}| \\
&\geq |s_{2m}|. \qquad (3.4.13)
\end{aligned}
$$

The above topological lower bound is attained for $s_{2m} = \pm|s_{2m}|$ if and only if the curvature satisfies $F^\mp(m) = 0$ or the self-dual or anti-self-dual Yang–Mills equation

$$F(m) = \pm * F(m). \qquad (3.4.14)$$

It will be instructive to consider (3.4.14) in view of the Euler–Lagrange equation of the energy (3.4.5). For this purpose, recall that the connection 1-form A and the curvature 2-form F are related through

$$F = dA + A \wedge A, \qquad (3.4.15)$$

that F is transformed under a gauge group element g according to the rule

$$F \mapsto g^{-1}Fg, \qquad (3.4.16)$$

and that the connection or gauge-covariant derivative, D, operates on any Lie algebra valued p-form ω with the transformation property (3.4.16) in such a way that [223]

$$D\omega = d\omega + A \wedge \omega + (-1)^{p+1}\omega \wedge A. \qquad (3.4.17)$$

Hence, in particular,

$$
\begin{aligned}
DF &= dF + A \wedge F - F \wedge A \\
&= d(A \wedge A) + A \wedge F - F \wedge A \\
&= dA \wedge A - A \wedge dA + A \wedge F - F \wedge A \\
&= dA \wedge A - A \wedge dA + A \wedge (dA + A \wedge A) - (dA + A \wedge A) \wedge A \\
&= 0, \tag{3.4.18}
\end{aligned}
$$

which is the usual Bianchi identity.

In general, assume

$$
DF(k) = 0, \quad k \geq 1; \quad F(1) = F. \tag{3.4.19}
$$

Then $F(k+1) = F \wedge F(k)$ and

$$
\begin{aligned}
DF(k+1) &= D(F \wedge F(k)) \\
&= d(F \wedge F(k)) + A \wedge F \wedge F(k) - F \wedge F(k) \wedge A \\
&= dF \wedge F(k) + A \wedge F \wedge F(k) - F \wedge A \wedge F(k) \\
&\quad + F \wedge dF(k) + F \wedge A \wedge F(k) - F \wedge F(k) \wedge A \\
&= DF \wedge F(k) + F \wedge DF(k) \\
&= 0.
\end{aligned}
$$

Thus we have shown that (3.4.19) is true for all k. This generalized Bianchi identity will be useful for us to understand (3.4.14).

We now consider the variation of A,

$$
A \mapsto A_t = A + tB, \quad F \mapsto F_t = d(A + tB) + (A + tB) \wedge (A + tB).
$$

Set

$$
K = \left(\frac{dF_t}{dt} \right)_{t=0} = dB + B \wedge A + A \wedge B.
$$

Therefore

$$
\begin{aligned}
\left(\frac{d}{dt} E(A_t) \right)_{t=0} &= -2 \int_{S^{4m}} \mathrm{Tr} \left[\left(\frac{d}{dt} F_t(m) \right)_{t=0} \wedge *F(m) \right] \\
&= -2 \int_{S^{4m}} \mathrm{Tr} \left[(K \wedge F \wedge \cdots \wedge F + F \wedge K \wedge \cdots \wedge F \right. \\
&\quad \left. + \cdots + F \wedge \cdots \wedge F \wedge K) \wedge *F(m) \right]. \tag{3.4.20}
\end{aligned}
$$

However,

$$
\begin{aligned}
&\int_{S^{4m}} \mathrm{Tr} \left[A \wedge B \wedge F(m-1) \wedge *F(m) \right] \\
&= \int_{S^{4m}} \mathrm{Tr} \left[F(m-1) \wedge *F(m) \wedge A \wedge B \right] \\
&= -\int_{S^{4m}} \mathrm{Tr} \left[B \wedge F(m-1) \wedge *F(m) \wedge A \right].
\end{aligned}
$$

Consequently,

$$\int_{S^{4m}} \mathrm{Tr} \left[K \wedge \underbrace{F \wedge \cdots \wedge F}_{m-1} \wedge * F(m) \right]$$

$$= \int_{S^{4m}} \mathrm{Tr} \left[(dB + B \wedge A + A \wedge B) \wedge F(m-1) \wedge * F(m) \right]$$

$$= \int_{S^{4m}} \mathrm{Tr} \left\{ B \wedge [d(F(m-1) \wedge * F(m)) \right.$$

$$\left. + A \wedge (F(m-1) \wedge * F(m)) - (F(m-1) \wedge * F(m)) \wedge A] \right\}$$

$$= \int_{S^{4m}} \mathrm{Tr} \left[B \wedge D(F(m-1) \wedge * F(m)) \right].$$

Similarly,

$$\int_{S^{4m}} \mathrm{Tr} \left[F \wedge K \wedge \underbrace{F \wedge \cdots \wedge F}_{m-2} \wedge * F(m) \right]$$

$$= \int_{S^{4m}} \mathrm{Tr} \left[K \wedge \underbrace{F \wedge \cdots \wedge F}_{m-2} \wedge * F(m) \wedge F \right]$$

$$= \int_{S^{4m}} \mathrm{Tr} \left[B \wedge D(F(m-2) \wedge * F(m) \wedge F) \right],$$

etc., with the last one in the list of the form

$$\int_{S^{4m}} \mathrm{Tr} \left[\underbrace{F \wedge \cdots \wedge F}_{m-1} \wedge K \wedge * F(m) \right]$$

$$= \int_{S^{4m}} \mathrm{Tr} \left[K \wedge * F(m) \wedge F(m-1) \right]$$

$$= \int_{S^{4m}} \mathrm{Tr} \left[B \wedge D(* F(m) \wedge F(m-1)) \right].$$

Inserting all these results into (3.4.20), we obtain

$$\left(\frac{d}{dt} E(A_t) \right)_{t=0} = -2 \int_{S^{4m}} \mathrm{Tr} \left[B \wedge \{ D(F(m-1) \wedge * F(m)) \right.$$

$$+ D(F(m-2) \wedge * F(m) \wedge F)$$

$$+ D(F(m-3) \wedge * F(m) \wedge F(2))$$

$$\left. + \cdots + D(* F(m) \wedge F(m-1)) \} \right]$$

$$= 0.$$

Since B is an arbitrary test 1-form, we arrive at the following Euler–Lagrange equation of the energy (3.4.5),

$$D(F(m-1) \wedge * F(m) + F(m-2) \wedge * F(m) \wedge F$$

$$+F(m-3) \wedge *F(m) \wedge F(2) + \cdots + *F(m) \wedge F(m-1))$$
$$= 0, \tag{3.4.21}$$

which may be called the generalized Yang–Mills equation in $4m$ dimensions. When $m = 1$, it is the classical one,

$$D(*F) = 0. \tag{3.4.22}$$

If $F(m)$ is self-dual or anti-self-dual, i.e., $F(m)$ satisfies the equation (3.4.14), the generalized Yang–Mills equation (3.4.21) is reduced to

$$DF(2m - 1) = 0,$$

which is automatically fulfilled because of the generalized Bianchi identity (3.4.19).

As in the classical 4-dimensional Yang–Mills theory case, we shall concentrate on (3.4.14). Below is our main representation theorem.

Theorem 3.4.1. *For any integer N, the self-dual or anti-self-dual equation (3.4.14) on S^{4m} has a $2N$-parameter family of N-instanton solutions so that the topological charge s_{2m} is proportional to N and the minimum energy is $E = |s_{2m}|$, which is proportional to N.*

We shall establish this theorem. Without loss of generality, we may concentrate on the self-dual case, $s_{2m} \geq 0$. It will be convenient to work on the Euclidean space $\mathbb{R}^{4m}$ instead of the sphere S^{4m}. Such a reduction is possible because, through a stereographic projection, $\mathbb{R}^{4m}$ is conformal to a punctured sphere, say, $S^{4m} - \{P\}$. However, in view of (3.4.1), we know that the Hodge dual $*F(m)$ is conformally invariant. Hence the Yang–Mills theory on $\mathbb{R}^{4m}$ is identical to that on $S^{4m} - \{P\}$. Finally, in analogy to Uhlenbeck's removable singularity theorem, the finite-energy condition implies that the solutions on $\mathbb{R}^{4m}$ behave well at infinity so that, when viewed on S^{4m}, they extend smoothly to the point P. Therefore, in this way, we have actually obtained a family of solutions on S^{4m}. Thus, from now on, we consider the Yang–Mills theory on $\mathbb{R}^{4m}$. The energy and topological charge are

$$E = -\int_{\mathbb{R}^{4m}} \mathrm{Tr}\,(F(m) \wedge *F(m)), \tag{3.4.23}$$

$$s_{2m} = -\int_{\mathbb{R}^{4m}} \mathrm{Tr}\,(F(m) \wedge F(m)), \tag{3.4.24}$$

where the $*$-operator is induced from the standard metric $(\delta_{\mu\nu})$ on $\mathbb{R}^{4m}$. The self-dual equation (3.4.14) is now valid on $\mathbb{R}^{4m}$ as observed.

In order to obtain N-instanton solutions for (3.4.14), one may use the approach of Witten as described in the last section. First, one reformulates the original $SU(2)$ problem on $\mathbb{R}^4$ (or S^4) in the $SO(4)$ context as mentioned earlier. Next, one extends such a reduction into the general $SO(4m)$ setting over $\mathbb{R}^{4m}$ (or S^{4m}). The algebra is quite involved [306, 307, 308] and is skipped. Here we will only record the final form of the problem: a field configuration is represented by a complex scalar field ϕ and a real-valued vector field $\mathbf{a} = (a_1, a_2)$, both defined on the Poincaré half-plane

$$\mathbb{R}_+^2 = \{(r,t) \,|\, r > 0, \ -\infty < t < \infty\},$$

where $r = \sqrt{x_1^2 + x_2^2 + \cdots + x_{4m-1}^2}$ and $t = x_{4m}$; up to a positive numerical factor the energy functional (3.4.23) is

$$E^{(m)} = \int_{-\infty}^{\infty} dt \int_0^{\infty} dr \,(1 - |\phi|^2)^{2(m-2)}$$

$$\times \left\{ r^2([1 - |\phi|^2]f_{12} - i\,[m-1][D_1\phi\overline{D_2\phi} - \overline{D_1\phi}D_2\phi])^2 \right.$$

$$+2m(2m-1)(1 - |\phi|^2)^2(|D_1\phi|^2 + |D_2\phi|^2)$$

$$\left. +\frac{(2m-1)^2}{r^2}(1 - |\phi|^2)^4 \right\}, \tag{3.4.25}$$

the topological charge (3.4.24) is

$$s^{(m)} = -\int_{-\infty}^{\infty} dt \int_0^{\infty} dr \left\{ (1 - |\phi|^2)f_{12} \right.$$

$$\left. -i(2m-1)(D_1\phi\overline{D_2\phi} - \overline{D_1\phi}D_2\phi) \right\}(1 - |\phi|^2)^{2(m-1)}, \tag{3.4.26}$$

and the self-dual equation (3.4.14) becomes

$$D_1\phi = -iD_2\phi,$$

$$\frac{(2m-1)}{r^2}(1 - |\phi|^2)^2 = -(1 - |\phi|^2)f_{12}$$

$$+i(m-1)(D_1\phi\overline{D_2\phi} - \overline{D_1\phi}D_2\phi),$$

$$x_1 = r, \quad x_2 = t, \quad x = (x_1, x_2) \in \mathbb{R}_+^2, \tag{3.4.27}$$

where $f_{jk} = \partial_j a_k - \partial_k a_j$ and $D_j\phi = \partial_j\phi + ia_j\phi$ $(j, k = 1, 2)$. It is comforting to note that, when $m = 1$, we recover Witten's results (3.3.3), (3.3.4), and (3.3.5). The above general equations for arbitrary $m = 1, 2, \cdots$ arising from the Yang–Mills theory in $4m$ dimensions were derived by Tchrakian [71] and may be called the Witten–Tchrakian equations. A numerical study of (3.4.27) was carried out in [61].

It is direct to see the relation between the energy $E^{(m)}$ and the topological charge $s^{(m)}$ expressed in (3.4.25) and (3.4.26), respectively. In fact, the integrand of $E^{(m)}$ can be rewritten as

$$
\mathcal{H}^{(m)} =
$$

$$
(1 - |\phi|^2)^{2(m-2)} \left\{ \left(r([1 - |\phi|^2]f_{12} - \mathrm{i}\,[m-1][D_1\phi\overline{D_2\phi} - \overline{D_1\phi}D_2\phi]) \right. \right.
$$

$$
\left. + \frac{(2m-1)}{r}(1 - |\phi|^2)^2 \right)^2 + 2m(2m-1)(1 - |\phi|^2)^2|D_1\phi + \mathrm{i}D_2\phi|^2 \right\}
$$

$$
-2(2m-1)(1 - |\phi|^2)^{2(m-1)} \left\{ (1 - |\phi|^2)f_{12} \right.
$$

$$
\left. -\mathrm{i}\,(2m-1)(D_1\phi\overline{D_2\phi} - \overline{D_1\phi}D_2\phi) \right\}. \tag{3.4.28}
$$

Using (3.4.28) in (3.4.25) and (3.4.26), we obtain the following topological lower bound for the energy

$$
E^{(m)} \geq 2(2m-1)s^{(m)}. \tag{3.4.29}
$$

This lower bound is saturated if and only if the field configuration $(\phi, \mathbf{a})$ satisfies the Witten–Tchrakian equations (3.4.27).

As in the case of Witten, our N-instanton solutions of the self-dual Yang–Mills equations on S^{4m} or R^{4m} stated in Theorem 3.4.1 will be obtained through a family of N-soliton solutions of the equations (3.4.27) on the Poincaré half-plane $\mathbb{R}_+^2$ characterized by N zeros of the complex field ϕ.

Here is our main existence and uniqueness theorem for the N-soliton solutions of the elegant Witten–Tchrakian equations (3.4.27).

Theorem 3.4.2. *For any N points $p_1, p_2, \cdots, p_N$ in $\mathbb{R}_+^2$ and any integer $m = 1, 2, \cdots$, the Witten–Tchrakian equations (3.4.27) have a unique solution $(\phi, \mathbf{a})$ so that ϕ vanishes exactly at these points, $|\phi| = 1$ at the boundary and the infinity of the Poincaré half-plane, the topological charge is given by $s^{(m)} = 2\pi N$, and the solution carries the quantized minimum energy $E^{(m)} = 4\pi(2m-1)N$.*

The N prescribed zeros stated in Theorem 3.4.2 give rise to the $2N$ parameters stated in Theorem 3.4.1. Other delicate properties of the solutions will be presented in subsequent analysis.

3.4.2   Quasilinear elliptic equation

Let $p_1, p_2, \cdots, p_N \in \mathbb{R}_+^2$ (with possible multiplicities) be as given in Theorem 3.4.2. Then the substitution $u = \ln|\phi|$ transforms (3.4.27) into the following equivalent scalar equation,

$$
(e^{2u} - 1)\Delta u = \frac{(2m-1)}{r^2}(e^{2u} - 1)^2 - 2(m-1)e^{2u}|\nabla u|^2 - 2\pi \sum_{j=1}^{N} \delta_{p_j},
$$

$$
x \in \mathbb{R}_+^2, \tag{3.4.30}
$$

where δ_p is the Dirac measure concentrated at p. We are to look for a solution u of (3.4.30) so that $u(x) \to 0$ (hence $|\phi(x)| \to 1$) as $x \to \partial \mathbb{R}^2_+$ or as $|x| \to \infty$.

It is clear that (3.4.30) is quasilinear for $m \neq 1$. In the subsequent sections, we present a complete study of this equation which establishes Theorem 3.4.2.

3.5 Existence of Weak Solution

We now consider (3.4.30) with $m \geq 2$. The analysis for the case $m = 1$ is straightforward. Since the maximum principle implies that $u(x) \leq 0$ everywhere, it will be more convenient to use the new variable

$$v = f(u) = 2(-1)^m \int_0^u (e^{2s} - 1)^{m-1} ds, \quad u \leq 0. \tag{3.5.1}$$

It is easily seen that

$$f : (-\infty, 0] \to [0, \infty)$$

is strictly decreasing and convex. For later use, we note that

$$\begin{aligned} f'(u) &= 2(-1)^m (e^{2u} - 1)^{m-1}, \\ f''(u) &= 4(-1)^m (m-1) e^{2u} (e^{2u} - 1)^{m-2}. \end{aligned}$$

Set

$$u = F(v) = f^{-1}(v), \quad v \geq 0.$$

Then the equation (3.4.30) is simplified to semilinear one,

$$\Delta v = \frac{2(-1)^m (2m-1)}{r^2} (e^{2F(v)} - 1)^m - 4\pi \sum_{j=1}^N \delta_{p_j} \quad \text{in } \mathbb{R}^2_+. \tag{3.5.2}$$

To approach (3.5.2), we introduce its modification of the form

$$\Delta v = \frac{2(2m-1)}{r^2} R(v) - 4\pi \sum_{j=1}^N \delta_{p_j} \tag{3.5.3}$$

where the right-hand-side function $R(v)$ is defined by

$$R(v) = \begin{cases} (-1)^m (e^{2F(v)} - 1)^m, & v \geq 0, \\ mv, & v < 0. \end{cases}$$

Then it is straightforward to check that $R(\cdot) \in C^1$. In order to obtain a solution of the original equation (3.5.2), it suffices to get a solution of (3.5.3) satisfying $v(x) \geq 0$ in $\mathbb{R}^2_+$ and $v(x) \to 0$ as $x \to \partial \mathbb{R}^2_+$ or as $|x| \to \infty$.

The main technical difficulty in (3.5.2) or (3.5.3) is the singular boundary of $\mathbb{R}_+^2$. We will employ a limiting argument to overcome this difficulty. We first solve (3.5.3) on a given bounded domain away from $r = 0$ under the homogeneous Dirichlet boundary condition. It will be seen that the obtained solution is indeed nonnegative and thus (3.5.2) is recovered. Such a property also allows us to control its energy and pointwise bounds conveniently. We then choose a sequence of bounded domains to approximate the full $\mathbb{R}_+^2$. The corresponding sequence of solutions is shown to converge to a weak solution of (3.5.2). This weak solution is actually a positive classical solution of (3.5.2) which necessarily vanishes asymptotically as desired. Then suitable decay rates are established by using certain comparison functions.

To proceed, choose a function, say, v_0, satisfying the requirement that it is compactly supported in $\mathbb{R}_+^2$ and smooth everywhere except at $p_1, p_2, \cdots, p_N$ so that

$$\Delta v_0 + 4\pi \sum_{j=1}^{N} \delta_{p_j} = g(x) \in C_0^\infty(\mathbb{R}_+^2).$$

Let Ω be any given bounded domain containing the support of v_0 and $\overline{\Omega} \subset \mathbb{R}_+^2$ (where and in the sequel, all bounded domains have smooth boundaries). Then $v = v_0 + w$ changes (3.5.3) into a regular form without the Dirac measure right-hand-side source terms, which is the equation in the following boundary value problem,

$$\Delta w = \frac{2(2m-1)}{r^2} R(v_0 + w) - g \quad \text{in } \Omega,$$
$$w = 0 \quad \text{on } \partial\Omega. \tag{3.5.4}$$

We first apply a variational method to prove the existence of a solution to (3.5.4). As usual, we use $W_0^{1,2}(\Omega)$ to denote the Hilbert space obtained through taking completion of the set of compactly supported smooth functions in Ω under the norm

$$\|w\|^2 = \int_\Omega \{|w|^2 + |\nabla w|^2\} \, \mathrm{d}r\mathrm{d}t.$$

Lemma 3.5.1. *The problem (3.5.4) has a unique solution.*

Proof. It is seen that (3.5.4) is the variational equation of the functional

$$I(w) = \int_\Omega \left\{ \frac{1}{2}|\nabla w|^2 + \frac{2(2m-1)}{r^2} Q(v_0 + w) - gw \right\} \mathrm{d}x,$$
$$w \in W_0^{1,2}(\Omega), \tag{3.5.5}$$

where $\mathrm{d}x = \mathrm{d}r\mathrm{d}t$ and the function $Q(s)$ is defined by

$$Q(s) = \int_0^s R(s') \, \mathrm{d}s' = \begin{cases} (-1)^m \int_0^s (e^{2F(s')} - 1)^m \, \mathrm{d}s', & s \geq 0, \\ \int_0^s ms' \, \mathrm{d}s' = \frac{m}{2}s^2, & s < 0, \end{cases} \tag{3.5.6}$$

which is positive except at $s = 0$. This property and the Poincaré inequality indicate that the functional (3.5.5) is coercive and bounded from below on $W_0^{1,2}(\Omega)$. On the other hand, since $F(s) \leq 0$ for $s \geq 0$, we have

$$\left|\frac{d}{ds}Q(s)\right| = |R(s)| \leq \max\{1, m|s|\}.$$

This feature says that the functional (3.5.5) is continuous on $W_0^{1,2}(\Omega)$ because $\overline{\Omega}$ is away from the boundary of $\mathbb{R}_+^2$ and, so, the weight $2(2m-1)/r^2$ is bounded. Besides, the definition of $F(s)$ gives us the result

$$\frac{d^2}{ds^2}Q(s) = \begin{cases} me^{2F(s)}, & s \geq 0, \\ m, & s < 0, \end{cases}$$

which says that the functional (3.5.5) is also convex. Thus, by convex analysis, the functional is weakly lower semicontinuous on $W_0^{1,2}(\Omega)$ and the existence and uniqueness of a critical point is ensured. The standard elliptic theory then implies that such a critical point is a classical solution of (3.5.4).

Lemma 3.5.2. *Let w be the solution of (3.5.4) obtained in Lemma 3.5.1. Then w satisfies $v_0 + w > 0$ in Ω.*

Proof. The function $v = v_0 + w$ satisfies (3.5.3) and assumes arbitrarily large values near p_j ($j = 1, 2, \cdots, N$). Since $\text{supp}(v_0) \subset \Omega$, we have $v = 0$ on $\partial\Omega$. The fact that $R(v) < 0$ for $v < 0$ and the maximum principle then lead us to the conclusion that $v > 0$ in Ω as stated.

We now choose a sequence of bounded domains $\{\Omega_n\}$ satisfying

$$\text{supp}(v_0) \subset \Omega_1, \ \overline{\Omega_n} \subset \Omega_{n+1}, \ \overline{\Omega_n} \subset \mathbb{R}_+^2, \ n = 1, 2, \cdots, \ \lim_{n\to\infty} \Omega_n = \mathbb{R}_+^2.$$

Lemma 3.5.3. *Let w_n be the solution of (3.5.4) for $\Omega = \Omega_n$ obtained in Lemma 3.5.1 and $I(\cdot; \Omega_n)$ be the functional (3.5.5) with $\Omega = \Omega_n$. Then we have the monotonicity*

$$I(w_n; \Omega_n) \geq I(w_{n+1}; \Omega_{n+1}), \quad n = 1, 2, \cdots.$$

Proof. In fact, for given n, the function w_n is the unique minimizer of the functional $I(\cdot; \Omega_n)$ on $W_0^{1,2}(\Omega_n)$. Now set $w_n = 0$ on $\Omega_{n+1} - \Omega_n$. Then $w_n \in W_0^{1,2}(\Omega_{n+1})$ and $I(w_n; \Omega_n) = I(w_n; \Omega_{n+1})$. However w_{n+1} is the global minimizer of $I(\cdot; \Omega_{n+1})$ on $W_0^{1,2}(\Omega_{n+1})$. Therefore the stated monotonicity follows.

To see that the energies are bounded from below, we need

Lemma 3.5.4. *For any $W_0^{1,2}(\mathbb{R}_+^2)$ function w, there holds the Poincaré inequality*

$$\int_{\mathbb{R}_+^2} \frac{1}{r^2} w^2(x)\, dx \le 4 \int_{\mathbb{R}_+^2} |\nabla w(x)|^2\, dx. \tag{3.5.7}$$

Proof. For $w \in C_0^1(\mathbb{R}_+^2)$ we have after integration by parts

$$\int_0^\infty \frac{1}{r^2} w^2(r,t)\, dr = 2 \int_0^\infty \frac{1}{r} w(r,t) \frac{d}{dr} w(r,t)\, dr.$$

Thus the Schwartz inequality gives us

$$\int_{\mathbb{R}_+^2} \frac{1}{r^2} w^2(x)\, dx \le 4 \int_{\mathbb{R}_+^2} \left| \frac{dw}{dr}(x) \right|^2 dx,$$

which is actually stronger than (3.5.7). Therefore the lemma follows.

Lemma 3.5.5. *Let $\{w_n\}$ be the solution sequence stated in Lemma 3.5.3. Then $w_n < w_{n+1}$ on Ω_n, $n = 1, 2, \cdots$.*

Proof. Set $v_n = v_0 + w_n$. Then Lemma 3.5.2 says that $v_n > 0$ in Ω_n. In particular $v_{n+1} > 0$ on $\overline{\Omega}_n$. Thus the equation

$$\Delta(v_{n+1} - v_n) = \frac{2m(2m-1)}{r^2} e^{2F(\xi_n)} (v_{n+1} - v_n),$$

where ξ_n lies between v_n and v_{n+1} and the boundary property $v_{n+1} - v_n > 0$ on $\partial\Omega_n$ imply that $v_{n+1} - v_n > 0$ in Ω_n as expected.

Lemma 3.5.6. *Let $\{w_n\}$ be the sequence stated in Lemma 3.5.3. There are positive constants C_1, C_2 independent of $n = 1, 2, \cdots$ so that*

$$I(w_n; \Omega_n) \ge C_1 \|\nabla w_n\|_{L^2(\mathbb{R}_+^2)}^2 - C_2, \quad n = 1, 2, \cdots.$$

Proof. The expression (3.5.6) says that $Q \ge 0$. Since g is of compact support in $\mathbb{R}_+^2$, the Schwartz inequality and Lemma 3.5.4 give us

$$I(w_n; \Omega_n) \ge \frac{1}{4} \int_{\mathbb{R}_+^2} |\nabla w_n|^2\, dx - 4 \int_{\mathbb{R}_+^2} r^2 g^2\, dx.$$

Lemma 3.5.7. *For a given bounded subdomain Ω_0 with $\overline{\Omega}_0 \subset \mathbb{R}_+^2$, the sequence $\{w_n\}$ is weakly convergent in $W^{1,2}(\Omega_0)$. The weak limit, say, w_{Ω_0}, is a solution of the equation (3.5.4) with $\Omega = \Omega_0$ (neglecting the boundary condition) which satisfies $w_{\Omega_0}(x) > 0$.*

Proof. Using Lemmas 3.5.3 and 3.5.6, we see that there is a constant $C > 0$ such that

$$\sup_n \|\nabla w_n\|^2_{L^2(\mathbb{R}^2_+)} \le C. \tag{3.5.8}$$

From (3.5.7) and (3.5.8) we obtain the boundedness of $\{w_n\}$ in $W^{1,2}(\Omega_0)$. Combining this with the monotonicity property stated in Lemma 3.5.5 we conclude that $\{w_n\}$ in weakly convergent in $W^{1,2}(\Omega_0)$. It then follows from the compact embedding $W^{1,2}(\Omega_0) \to L^2(\Omega_0)$ that $R(v_0 + w_n)$ is convergent in $L^2(\Omega_0)$. On the other hand, since for sufficiently large n, we have $\Omega_0 \subset \Omega_n$, consequently

$$\int_{\mathbb{R}^2_+} \left\{ \nabla w_n \cdot \nabla \xi + \frac{2(2m-1)}{r^2} R(v_0 + w_n)\xi - g\xi \right\} dx = 0,$$

$$\forall \xi \in C_0^1(\Omega_0). \tag{3.5.9}$$

Letting $n \to \infty$ in (3.5.9) we see that w_{Ω_0} is a weak solution of (3.5.4) (without considering the boundary condition). The standard elliptic regularity theory then implies that it is also a classical (hence, smooth) solution. Since $w_n > 0$, we have $w_{\Omega_0} \ge 0$. The maximum principle then yields $w_{\Omega_0} > 0$ in Ω_0. Thus our lemma follows.

Set $w(x) = w_{\Omega_0}(x)$ for $x \in \Omega_0$ for any given Ω_0 stated in Lemma 3.5.7. In this way we obtain a global solution of the equation in (3.5.4) over the full $\mathbb{R}^2_+$. Lemmas 3.5.3 and 3.5.6 imply that there is a constant $C > 0$ to make

$$I(w) \le C, \quad \|\nabla w\|_{L^2(\mathbb{R}^2_+)} \le C. \tag{3.5.10}$$

In the next section we establish the desired asymptotic behavior of the obtained solution w. The boundedness result (3.5.10) is not sufficient to ensure the decay of w at $r = 0$ and at infinity. We need also to show that w is pointwise bounded as a preparation.

3.6 Asymptotic Estimates

For technical reasons which will become clear later, we need to show first that the solution w is pointwise bounded. This will be accomplished by the following lemma.

Lemma 3.6.1. *Let $\{w_n\}$ be the sequence of local solutions stated in Lemma 3.5.3 and the domain Ω satisfy*

$$\text{supp}\,(v_0) \subset \Omega \subset \overline{\Omega} \subset \Omega_1.$$

There exists a constant $C > 0$ independent of n so that

$$\sup_{x \in \Omega_n} w_n(x) \le \sup_{x \in \partial\Omega} \{w_n(x)\} + C \sup_{x \in \Omega} |g(x)|, \quad n = 1, 2, \cdots. \tag{3.6.1}$$

Proof. Set $D_n = \Omega_n - \overline{\Omega}$. We consider w_n on D_n and Ω separately.

Note that w_n satisfies $\Delta w_n \geq -g$ and $v_0 + w_n > 0$ in Ω. Hence the inequality (3.6.1) is standard if on the left-hand side of (3.6.1) the domain Ω_n is replaced by its subdomain Ω because $v_0 = 0$ on $\partial\Omega$ implies $w_n > 0$ on $\partial\Omega$ in view of Lemma 3.5.2 applied to w_n. In this situation the constant C only depends on the size of Ω (cf. [123]).

Now consider the other case, $x \in D_n$. Set $\eta_n = \sup\{w_n(x) \,|\, x \in \partial\Omega\}$. Then the property $v_0 = 0, g = 0$ in D_n gives us

$$\Delta(w_n - \eta_n) \geq \frac{2(2m-1)}{r^2}(-1)^m([e^{2F(w_n)} - 1]^m - [e^{2F(\eta_n)} - 1]^m) \text{ in } D_n. \tag{3.6.2}$$

Since the function $(-1)^m(e^{2F(s)} - 1)^m$ is strictly increasing for $s \geq 0$ and $w_n - \eta_n \leq 0$ on ∂D_n, we obtain by the maximum principle the result $w_n \leq \eta_n$ in D_n.

Therefore (3.6.1) follows immediately.

Lemma 3.6.2. *Let w be the solution of the governing equation*

$$\Delta w = \frac{2(2m-1)}{r^2}R(v_0 + w) - g \tag{3.6.3}$$

over the full $\mathbb{R}_+^2$ obtained in the last section. Then w is bounded.

Proof. Since $w_n < w$ in Ω_n, we have in particular

$$\sup_{x \in \partial\Omega}\left\{w_n(x)\right\} < \sup_{x \in \partial\Omega}\left\{w(x)\right\}, \quad n = 1, 2, \cdots.$$

Hence Lemma 3.6.1 says that there is a constant $C > 0$ independent of n so that

$$\sup_{x \in \Omega_n}\left\{w_n(x)\right\} \leq C, \quad n = 1, 2, \cdots. \tag{3.6.4}$$

A simple application of the embedding theory gives us the pointwise convergence $w_n \to w$ as $n \to \infty$. Thus (3.6.4) yields the boundedness of w from above. However, $v_0 + w > 0$ (see Lemma 3.5.2) implies already the boundedness of w from below. The lemma is consequently proven.

Lemma 3.6.2 enables us to establish the asymptotic behavior of w near infinity and at the boundary $r = 0$ as was done for the multi-meron solutions [62, 129, 160, 263]. The proof of the following lemma is adapted from [160].

Lemma 3.6.3. *Let w be the solution stated in Lemma 3.6.2. Then for $x = (r, t) \in \mathbb{R}_+^2$ we have the uniform limits*

$$\lim_{r \to 0} w(x) = \lim_{|x| \to \infty} w(x) = 0. \tag{3.6.5}$$

Proof. Given $x = (r, t)$, let D be the disk centered at x with radius $r/2$. The Dirichlet Green's function $G(x', x'')$ of the Laplacian Δ on D (satisfying $G(x', x'') = 0$ for $|x'' - x| = r/2$) is defined by the expression

$$G(x', x'') = \frac{1}{2\pi} \ln \sqrt{|x' - x|^2 + |x'' - x|^2 - 2(x' - x) \cdot (x'' - x)}$$

$$- \frac{1}{2\pi} \ln \sqrt{\left(\frac{2|x' - x||x'' - x|}{r}\right)^2 + \left(\frac{r}{2}\right)^2 - 2(x' - x) \cdot (x'' - x)}$$

where $x', x'' \in D$ but $x' \neq x''$.

Hence w at $x' \in D$ can be represented as

$$w(x') =$$
$$\int_D dx'' \left\{ (-1)^m \frac{2(2m-1)}{r''^2} (e^{2F(v_0+w)} - 1)^m - gw \right\} (x'') G(x', x'')$$
$$+ \int_{\partial D} dS'' \left\{ \frac{\partial G}{\partial n''}(x', x'') \right\} w(x''), \tag{3.6.6}$$

where $x'' = (r'', t'')$ and $\partial/\partial n''$ denotes the outer normal derivative on D with respect to the variable x''. We need to first evaluate $|r(\nabla_x w)(x)|$. This can be done by differentiating (3.6.6) and then setting $x = x'$. Note that

$$(\nabla_{x'} G(x', x''))_{x'=x} = \frac{1}{2\pi}\left(\frac{4}{r^2} - \frac{1}{|x'' - x|^2}\right)(x'' - x),$$

$$\left(\nabla_{x'} \frac{\partial G}{\partial n''}(x', x'')\right)_{x'=x} = \nabla_{x'}\left(\frac{x'' - x}{|x'' - x|} \cdot \nabla_{x''} G(x', x'')\right)\Big|_{x'=x}$$

$$= \frac{8}{\pi r^3}(x'' - x), \quad x'' \in \partial D.$$

Now let

$$C_1 = \sup_{\mathbb{R}_+^2}\left\{ |2(2m-1)(e^{2F(v_0+w)} - 1)^m(x) - r^2 g(x) w(x)| \right\},$$

$$C_2 = \sup_{\mathbb{R}_+^2}\left\{ |w(x)| \right\}.$$

Differentiate (3.6.6) with respect to x', set $x' = x$, apply the above results, and use $r'' \geq r/2$. We have

$$|\nabla w(x)| \leq \frac{2C_1}{\pi r^2} \int_D \frac{1}{|x'' - x|} dx'' + \frac{8C_2}{\pi r^3} \int_{\partial D} |x'' - x| dS''$$

$$\leq \frac{C}{r}, \tag{3.6.7}$$

where C is a constant independent of $r > 0$. Thus the claimed bound for $|r\nabla w(x)|$ over $\mathbb{R}_+^2$ is established.

To show that (3.6.5) holds for w, we argue by contradiction. Let $x_n = (r_n, t_n)$ be a sequence in $\mathbb{R}_+^2$ satisfying either $r_n \to 0$ or $|x_n| \to \infty$ but $|w(x_n)| \geq$ some $\varepsilon > 0$. Without loss of generality we may also assume that the sequence is so chosen that the disks centered at x_n with radius $r_n/2$ are non-overlapping. Then set

$$D_n = \left\{ x \in \mathbb{R}_+^2 \ \Big| \ |x - x_n| < \varepsilon_0 r_n \right\}, \quad \varepsilon_0 = \min\left\{ \frac{1}{2}, \frac{\varepsilon}{4C} \right\},$$

where $C > 0$ is the constant given in (3.6.7). For $x = (r, t) \in D_n$ we have $3r_n/2 \geq r \geq r_n/2$. Thus, integrating ∇w over the straight line L from x_n to $x \in D_n$ and using $|\nabla w(x')| < 2C/r_n$ ($\forall x' \in D_n$), we obtain the estimate

$$
\begin{aligned}
|w(x)| &= \left| w(x_n) + \int_L \nabla w(x') \cdot dl' \right| \\
&\geq \varepsilon - \frac{2C}{r_n} \frac{\varepsilon}{4C} r_n \\
&= \frac{\varepsilon}{2}, \quad x \in D_n.
\end{aligned}
$$

Therefore we arrive at the contradiction

$$
\begin{aligned}
\int_{\mathbb{R}_+^2} \frac{w^2}{r^2} \, dx &\geq \sum_{n=1}^{\infty} \int_{D_n} \frac{w^2}{r^2} dx \\
&\geq \sum_{n=1}^{\infty} \left(\frac{2}{3r_n} \right)^2 \left(\frac{\varepsilon}{2} \right)^2 \pi(\varepsilon_0 r_n)^2 \\
&= \infty
\end{aligned}
$$

because in view of Lemma 3.5.4 and (3.5.10) we have $w/r \in L^2(\mathbb{R}_+^2)$.

So (3.6.5) must hold and thus the proof of the lemma is complete.

We now strengthen the above result and prove

Lemma 3.6.4. *Let w be the solution stated in Lemma 3.6.3. There are constants $r_0 > 0$ (small) and $\rho_0 > 0$ (large) so that for any $0 < \varepsilon < 1$ there is a constant $C(\varepsilon) > 0$ to make the following asymptotic bounds valid,*

$$
\begin{aligned}
0 < w(x) &< C(\varepsilon) r^{2m - \varepsilon}, \quad 0 < r < r_0; & (3.6.8) \\
0 < w(x) &< C(\varepsilon) r^{2m - \varepsilon} |x|^{-2(2m - \varepsilon)}, \quad |x| > \rho_0, & (3.6.9)
\end{aligned}
$$

where $x = (r, t) \in \mathbb{R}_+^2$. In other words, roughly speaking, there hold asymptotically $w(x) = O(r^{2m})$ as $r \to 0$ and $w(x) = O(|x|^{-2m})$ as $|x| \to \infty$.

Proof. First let $r_0 > 0$ be small so that

$$\text{supp}(v_0) \subset \left\{ x = (r, t) \in \mathbb{R}_+^2 \ \Big| \ r > r_0 \right\}.$$

Consider the infinite strip $R_0 = \{x = (r,t) \in \mathbb{R}_+^2 \,|\, 0 < r < r_0\}$ and set

$$\sigma(x) = Cr^\beta. \tag{3.6.10}$$

Then $r^2 \Delta \sigma = \beta(\beta - 1)\sigma$. On the other hand, the solution w satisfies

$$
\begin{aligned}
r^2 \Delta w &= 2(2m-1)(-1)^m (e^{2F(w)} - 1)^m \\
&= 2m(2m-1)(-1)^m (e^{2F(\xi)} - 1)^{m-1} e^{2F(\xi)} 2F'(\xi) w \\
&= 2m(2m-1) e^{2F(\xi)} w, \qquad \xi \in (0, w). \tag{3.6.11}
\end{aligned}
$$

Now take $\beta = 2m - \varepsilon$. Since $w \to 0$ uniformly as $r \to 0$, we may choose r_0 small enough so that $2m(2m-1)e^{2F(w)} > \beta(\beta-1)$ for $x \in R_0$. Consequently

$$r^2 \Delta(w - \sigma) > \beta(\beta - 1)(w - \sigma), \quad x \in R_0. \tag{3.6.12}$$

Let C in (3.6.10) be so large that $(w - \sigma)_{r=r_0} < 0$. Using this and the property $w - \sigma \to 0$ as $r \to 0$ and $w \to 0$ as $|x| \to \infty$ in (3.6.12) we obtain (3.6.8), namely,

$$0 < w(x) < Cr^\beta, \quad 0 < r < r_0. \tag{3.6.13}$$

Next, put $S_0 = \{x \in \mathbb{R}_+^2 \,|\, |x| > \rho_0\}$, where $\rho_0 > 0$ is so large that $\mathrm{supp}(v_0) \subset \mathbb{R}_+^2 - \overline{S_0}$. Define the comparison function

$$\sigma(x) = C_1 r^\beta (1 + r^2 + t^2)^{-\beta}, \quad x \in S_0, \tag{3.6.14}$$

where $\beta = 2m - \varepsilon$. Then

$$r^2 \Delta \sigma = \beta\left((\beta - 1) - \frac{4(\beta + 1)r^2}{(1 + |x|^2)^2}\right)\sigma. \tag{3.6.15}$$

Using $w \to 0$ as $|x| \to \infty$ we obtain (3.6.12) for $x \in S_0$ where ρ_0 is sufficiently large. From (3.6.13) and (3.6.14) we see that the constant $C_1 > 0$ may be chosen so that $(w - \sigma)_{|x|=\rho_0} < 0$. Using this property and Lemma 3.6.3 in (3.6.12) with R_0 being replaced by S_0 we have $w < \sigma$ throughout S_0. This is (3.6.9) and the proof is complete.

In order to show that our solutions give rise to desired topology, we still need to prove that $|\nabla w|$ decays sufficiently fast near the boundary and infinity of $\mathbb{R}_+^2$.

Lemma 3.6.5. *For the solution w obtained earlier, its derivatives decay near $r = 0$ according to the rates*

$$-Cr^{2m-\varepsilon} < \frac{\partial w}{\partial r} < Cr^{2m-(1+\varepsilon)}, \quad 0 < r < r_0,$$

$$\left|\frac{\partial w}{\partial t}\right| < Cr^{2m-\varepsilon}, \quad 0 < r < r_0,$$

where $C = C(\varepsilon)$ is a constant independent of r, t and $\varepsilon \in (0, 1)$ is arbitrary.

Proof. Since v_0 is compactly supported, v behaves like w asymptotically. Besides, let $\delta > 0$ be small such that $r_0 - \delta > 0$ and

$$\text{supp}(v_0) \subset \left\{ (r, t) \in \mathbb{R}_+^2 \,\Big|\, r > r_0 + \delta \right\}.$$

Define $S_\delta = \{(r, t) \in \mathbb{R}_+^2 \,|\, 0 < r_0 - \delta < r < r_0 + \delta\}$. Then, over S_δ, w satisfies

$$\Delta w = \frac{2(2m-1)}{r^2} R(w), \quad R(w) = (-1)^m (e^{2F(w)} - 1)^m, \quad w \geq 0. \quad (3.6.16)$$

The fact that $R'(0) = m$ and Lemma 3.6.4 imply the existence of some $C > 0$ such that

$$|R(w)| \leq C|x|^{-2(2m-\varepsilon)}, \quad x \in S_\delta. \quad (3.6.17)$$

Hence we have Δw (as well as w) $\in L^p(S_\delta)$ $(p > 2)$. Consequently, $w \in W^{2,p}(S_\delta)$ and $|\nabla w| \to 0$ as $|x| \to \infty$. In particular, $|\nabla w|$ is bounded over the infinite strip S_δ. Let $h > 0$ be a small number and set

$$w_h(r, t) = \frac{w(r, t+h) - w(r, t)}{h}.$$

Then w_h is also bounded over S_δ, $w_h(r, t) \to 0$ as $r \to 0$, and satisfies

$$\begin{aligned} r^2 \Delta w_h &= 2(2m-1)R'(\tilde{w})w_h \\ &= 2m(2m-1)e^{2F(\tilde{w})}w_h, \quad 0 < r < r_0, \quad (3.6.18) \end{aligned}$$

where $\tilde{w}(r, t)$ is between $w(r, t+h)$ and $w(r, t)$. Let σ be given in (3.6.10). Since $w(r, t) \to 0$ uniformly as $r \to 0$, we may assume r_0 (independent of h) to be small so that

$$r^2 \Delta(w_h - \sigma) \geq 2m(2m-1)e^{2F(\tilde{w})}(w_h - \sigma), \quad 0 < r < r_0.$$

Since $|w_h|$ has an upper bound on $r = r_0$ independent of h, we can choose the constant $C > 0$ in (3.6.10) large so that

$$|w_h(r_0, t)| \leq \sigma(r_0, t), \quad \forall h > 0, t \in \mathbb{R}.$$

Using the maximum principle over the strip R_0, we obtain $w_h \leq \sigma$.
Similarly, we also have

$$r^2 \Delta(w_h + \sigma) \leq 2m(2m-1)e^{2F(\tilde{w})}(w_h + \sigma), \quad 0 < r < r_0.$$

By the same argument we have $-\sigma \leq w_h$ over R_0. Hence we arrive at the uniform bound

$$|w_h(x)| \leq Cr^\beta, \quad \forall h > 0, x \in R_0.$$

Consequently, in the limit $h \to 0$, we have

$$\left|\frac{\partial w}{\partial t}\right| \le Cr^\beta, \quad 0 < r < r_0.$$

On the other hand, using the fact that $\varepsilon > 0$ in (3.6.8) may be made arbitrarily small, we see that $\Delta w \in L^p(\Omega_{t_0})$ for a suitable $p > 2$ for any $t_0 > 0$ where

$$\Omega_{t_0} = \left\{(r,t)\,\middle|\,0 < r < r_0,\ -t_0 < t < t_0\right\}.$$

Therefore $w \in W^{2,p}(\Omega_{t_0})$. In particular, the partial derivatives of w, as well as w, are continuous over $\overline{\Omega_{t_0}}$. As a consequence, from the fact that $w(0,t) = 0$ and (3.6.8), we have

$$\frac{\partial w}{\partial r}(0,t) = \lim_{r \to 0} \frac{w(r,t)}{r} = 0. \tag{3.6.19}$$

In order to get the decay estimate for

$$w_1 = \frac{\partial w}{\partial r},$$

we differentiate (3.6.16) and use (3.6.8) to arrive at

$$2m(2m-1)e^{2F(w)}w_1 - C_1 r^{2m-(1+\varepsilon)} < r^2 \Delta w_1$$
$$< 2m(2m-1)e^{2F(w)}w_1. \tag{3.6.20}$$

With the function σ defined in (3.6.10), we obtain from the right half of (3.6.20) that

$$r^2\Delta(w_1 + \sigma) < 2m(2m-1)e^{2F(w)}(w_1 + \sigma), \quad 0 < r < r_0,$$

where $r_0 > 0$ is small. Hence $-\sigma \le w_1$ $(0 < r < r_0)$ when the number $C > 0$ in (3.6.10) is sufficiently large.

On the other hand, set $\beta = 2m - (1+\varepsilon)$ in (3.6.10). With a large enough C in (3.6.10), we have

$$\begin{aligned}
-r^2\Delta\sigma &= -2m(2m-1)e^{2F(w)}\sigma + (2m(2m-1)e^{2F(w)} - \beta[\beta-1])\sigma \\
&> -2m(2m-1)e^{2F(w)}\sigma + C_1 r^{2m-(1+\varepsilon)}. \tag{3.6.21}
\end{aligned}$$

Inserting (3.6.21) into the left half of (3.6.20), we have

$$r^2\Delta(w_1 - \sigma) > 2m(2m-1)e^{2F(w)}(w_1 - \sigma), \quad 0 < r < r_0.$$

By the maximum principle, $w_1 - \sigma = 0$ at $r = 0$ (see (3.6.19)), and $w_1 < \sigma$ at $r = r_0$ again, we have $w_1 < \sigma$ for $0 < r < r_0$ as expected.

This completes the proof of the lemma.

We next study the decay estimate of $|\nabla w|$ for $|x| \to \infty$.

Lemma 3.6.6. *We have the asymptotic estimates for $|x| > \rho_0$,*

$$-Cr^{2m-\varepsilon}|x|^{-2(2m-\varepsilon)} < \frac{\partial w}{\partial r} < Cr^{2m-(1+\varepsilon)}|x|^{-2(\gamma(m)-\varepsilon)},$$

$$\gamma(m) = m - \frac{1}{2} + \sqrt{m^2 - \frac{1}{4}},$$

$$\left|\frac{\partial w}{\partial t}\right| < Cr^{2m-\varepsilon}|x|^{-2(2m-\varepsilon)},$$

where $C = C(\varepsilon)$ is a constant depending on the parameter $\varepsilon \in (0,1)$ which may be made arbitrarily small and $\rho_0 > 0$ is sufficiently large.

Proof. Let S_0 denote the set used in the proof of Lemma 3.6.4 where $S_0 = \{x \in \mathbb{R}^2_+ \,|\, |x| > \rho_0\}$ satisfies $\text{supp}(v_0) \subset \mathbb{R}^2_+ - \overline{S_0}$. Put $w_1 = \partial w/\partial r$. Then we have

$$r^2 \Delta w_1 = 2m(2m-1)e^{2F(w)}w_1 - \frac{4(2m-1)}{r}R(w), \quad x \in \mathbb{R}^2_+ - \overline{S_0}. \quad (3.6.22)$$

Since $R'(0) = m$ and $w \to 0$ as $|x| \to \infty$, we have, in view of Lemma 3.6.4, that

$$|R(w)| \le C(\varepsilon)r^{2m-\varepsilon}|x|^{-2(2m-\varepsilon)}, \quad |x| > \rho_0, \quad (3.6.23)$$

for any given $\varepsilon \in (0,1)$. In particular, we see from (3.6.22) that $\Delta w_1 \in L^2(\mathbb{R}^2_+ - \overline{S_0})$. On the other hand, we already had the fact that $w_1 \in L^2(\mathbb{R}^2_+)$. Hence, in view of the L^2-estimates, we have $w_1 \in W^{2,2}(\mathbb{R}^2_+ - \overline{S_0})$. Consequently, $w_1 \to 0$ as $|x| \to \infty$.

We now apply the fact $R(w) \ge 0$ and (3.6.23) in (3.6.22) to get

$$2m(2m-1)e^{2F(w)}w_1 - Cr^{2m-(1+\varepsilon)}|x|^{-2(2m-\varepsilon)}$$
$$< r^2 \Delta w_1$$
$$< 2m(2m-1)e^{2F(w)}w_1, \quad |x| > \rho_0. \quad (3.6.24)$$

Again we have two different situations. First, if $\beta = 2m - \varepsilon$ in (3.6.14), then we may obtain from (3.6.15) and the right-hand half of (3.6.24) that

$$r^2 \Delta(w_1 + \sigma) < 2m(2m-1)e^{2F(w)}(w_1 + \sigma), \quad |x| > \rho_0,$$

where ρ_0 is sufficiently large. Let C_1 in (3.6.14) be large so that $w_1 + \sigma > 0$ at $|x| = \rho_0$. By the above inequality and the property that $w_1 + \sigma \to 0$ as $|x| \to \infty$, we have $w_1 + \sigma > 0$ for all $|x| > \rho_0$.

In order to derive the other half of the decay estimate, we need a comparison function more general than (3.6.14),

$$\sigma = C_1 r^\beta (1 + r^2 + t^2)^{-\gamma}. \quad (3.6.25)$$

There holds

$$r^2 \Delta \sigma = \left(\beta(\beta - 1) - \frac{4(\beta + 1)\gamma r^2}{(1 + r^2 + t^2)^2} - \frac{4(\beta - \gamma)\gamma r^2}{(1 + r^2 + t^2)^2}(r^2 + t^2) \right) \sigma. \quad (3.6.26)$$

It is seen that (3.6.26) reduces to (3.6.15) when $\gamma = \beta$. In fact, any estimate with $\gamma \leq \beta$ is easy to obtain. For our purposes, we need to achieve $\gamma > \beta$ instead. We have as before

$$-r^2 \Delta \sigma > -2m(2m - 1)e^{2F(w)}\sigma$$
$$+ \left(2m(2m - 1)e^{2F(w)} - \beta(\beta - 1) + \frac{4(\beta - \gamma)\gamma r^2}{(1 + r^2 + t^2)^2}(r^2 + t^2) \right) \sigma$$
$$> -2m(2m - 1)e^{2F(w)}\sigma$$
$$+ (2m(2m - 1)e^{2F(w)} - \beta[\beta - 1] - 4\gamma[\gamma - \beta])\sigma. \quad (3.6.27)$$

The second term on the right-hand side of the above inequality is crucial. We observe that the quadratic function of γ,

$$2m(2m - 1) - (2m - 1)(2m - 2) - 4\gamma(\gamma - [2m - 1])$$

is positive if $\gamma > 0$ is slightly below the critical number, $\gamma(m)$, stated in the lemma. Hence, for $\beta = 2m - 1 - \varepsilon$ and $\gamma = \gamma - \varepsilon$ with $\varepsilon > 0$ sufficiently small, we have by (3.6.27) that

$$-r^2 \Delta \sigma > -2m(2m - 1)e^{2F(w)}\sigma + c\sigma, \quad |x| > \rho_0,$$

where ρ_0 is sufficiently large and $c > 0$ is a constant. Recall that $\gamma(m) < 2m$. Hence the above result and the definition of σ (see (3.6.25)) lead us to

$$-r^2 \Delta \sigma > -2m(2m - 1)e^{2F(w)}\sigma + Cr^{2m-1-\varepsilon}|x|^{-2(2m-\varepsilon)}, \quad |x| > \rho_0,$$
$$(3.6.28)$$

where, again, ρ_0 is large. Combining the right-hand half of (3.6.24) and (3.6.28), we have

$$r^2 \Delta(w_1 - \sigma) > 2m(2m - 1)e^{2F(w)}(w_1 - \sigma), \quad |x| > \rho_0.$$

For the above fixed ρ_0, let C_1 in (3.6.25) be large to ensure both $w_1 - \sigma < 0$ at $|x| = \rho_0$ and (3.6.28). Since $w_1 - \sigma \to 0$ as $|x| \to 0$ or $r \to 0$, we obtain the result $w_1 < \sigma$ for all $|x| > \rho_0$ by the maximum principle.

The asymptotic estimate for $w_2 = \partial w / \partial t$ may be derived similarly from the equation

$$r^2 \Delta w_2 = 2m(2m - 1)e^{2F(w)}w_2, \quad |x| > \rho_0,$$

the property that $w_2 \in L^2(\mathbb{R}_+^2)$, and the method of the proof of Lemma 3.6.4.

The proof is complete.

3.7 Topological Charge

Let u be the solution of (3.4.30) obtained in the previous sections. We can use the assignment (3.3.7) to construct an N-instanton solution. We need to compute the topological charge s^m of such a solution. For definiteness, we rewrite (3.4.30) as

$$(e^{2u} - 1)\Delta u = \frac{(2m-1)}{r^2}(e^{2u} - 1)^2 - 2(m-1)e^{2u}|\nabla u|^2 - 2\pi \sum_{j=1}^{\ell} n_j \delta_{p_j},$$

$$x \in \mathbb{R}_+^2, \tag{3.7.1}$$

where the points $p_1, p_2, \cdots, p_\ell$ are distinct, $n_1, n_2, \cdots, n_\ell \geq 1$ are integers, and $N = \sum_{j=1}^{\ell} n_j$. In view of (3.3.7), we have, away from the set $P = \{p_1, p_2, \cdots, p_\ell\}$ and with $a = a_1 + ia_2$, that

$$\begin{aligned} f_{12} &= \partial_1 a_2 - \partial_2 a_1 = -i(\partial a - \bar{\partial}\bar{a}) \\ &= 2\partial\bar{\partial}\ln|\phi|^2 = \Delta\ln|\phi| = \Delta u. \end{aligned} \tag{3.7.2}$$

On the other hand, using (3.3.7) again, we have the representations

$$D_1\phi = (\partial + \bar{\partial})\phi - \left(\frac{\bar{\partial}\phi}{\phi} - \frac{\partial\bar{\phi}}{\bar{\phi}}\right)\phi = 2(\partial u)\phi, \tag{3.7.3}$$

$$D_2\phi = i(\partial - \bar{\partial})\phi + i\left(\frac{\bar{\partial}\phi}{\phi} + \frac{\partial\bar{\phi}}{\bar{\phi}}\right)\phi = 2i(\partial u)\phi. \tag{3.7.4}$$

Consequently, we obtain

$$|D_1\phi|^2 + |D_2\phi|^2 = 8|\phi|^2|\partial u|^2 = 2e^{2u}|\nabla u|^2. \tag{3.7.5}$$

We rewrite the topological charge (3.4.26) as

$$s^{(m)} = \int_{\mathbb{R}_+^2} \rho^{(m)} \, \mathrm{d}x \tag{3.7.6}$$

where, applying the first equation in (3.4.27), (3.7.2), and (3.7.5), the charge density $\rho^{(m)}$ away from the set P is given by

$$\begin{aligned} \rho^{(m)} &= -(1 - |\phi|^2)^{2(m-1)}\left\{(1 - |\phi|^2)f_{12} - (2m-1)(|D_1\phi|^2 + |D_2\phi|^2)\right\} \\ &= -(1 - e^{2u})^{2(m-1)}\left\{(1 - e^{2u})\Delta u - 2(2m-1)e^{2u}|\nabla u|^2\right\} \\ &= -\Delta U, \end{aligned} \tag{3.7.7}$$

where $U = U(r, t)$ is defined by

$$U(r, t) = \int_0^{u(r,t)} (1 - e^{2s})^{2m-1} \, \mathrm{d}s. \tag{3.7.8}$$

With the notation of the previous section, the decomposition $u = v_0 + w$, and the fact that v_0 is compactly supported, we know that u and its derivatives vanish like w and its derivatives, respectively, near the boundary and infinity of $\mathbb{R}_+^2$. Therefore we can use the divergence theorem to get

$$\int_{\mathbb{R}_+^2} \rho^{(m)} \, dx$$

$$= \lim_{r_0 \to 0, t_0 \to \infty} \left\{ \int_{r=r_0, -t_0 \le t \le t_0} \frac{\partial U}{\partial r} \, dt - \int_{r \ge r_0, |x| = t_0} \frac{\partial U}{\partial n} \, dS \right\}$$

$$+ \lim_{\delta \to 0} \sum_{j=1}^{\ell} \oint_{|x - p_j| = \delta} \frac{\partial U}{\partial n} \, dS$$

$$= \lim_{\delta \to 0} \sum_{j=1}^{\ell} \oint_{|x - p_j| = \delta} (1 - e^{2u})^{2m-1} \frac{\partial u}{\partial n} \, dS, \tag{3.7.9}$$

where on any circular path $\partial/\partial n$ denote the outward normal derivative in the positive radial direction along the circle $|x - p_j| = \delta$, $j = 1, 2, \cdots, \ell$. Since u satisfies (3.7.1), it has the form

$$u(x) = \ln |x - p_j|^{n_j} + W_j(x) \tag{3.7.10}$$

near the point p_j where W_j is smooth ($j = 1, 2, \cdots, \ell$). Inserting (3.7.10) into (3.7.9), we finally obtain the expected result,

$$s^{(m)} = \int_{\mathbb{R}_+^2} \rho^{(m)} \, dx = 2\pi \sum_{j=1}^{\ell} n_j = 2\pi N. \tag{3.7.11}$$

The proof of Theorem 3.4.2, and hence Theorem 3.4.1, is complete.

3.8 Remarks

In this chapter we have made a uniform treatment of the Yang–Mills theory in $4m$ dimensions via self-dual instantons and their topological representations which extend the Chern–Pontryagin class in 4 dimensions. It has been seen, in general, that the problem does not allow an explicit solution as in the 4-dimensional case and one has to rely on a delicate analysis of the governing nonlinear elliptic equation. The solution of the general problem presented here has appeared in [291]. See [71, 182, 204, 205, 280, 309] for other related studies in higher dimensions. Below we mention some open problems which may deserve further attention.

Open Problem 3.8.1. *Determine the maximal number of free parameters in the solutions or the dimensions of the moduli space of solutions of the $4m$-dimensional self-dual Yang–Mills equations (3.4.14).*

Our study has shown that, when the instanton number is N, the solutions contain at least $2N$ parameters. In view of [58, 155], we may attempt an intuitive counting of the number of free parameters in the general solution of (3.4.14) in $4m$ dimensions as follows: for an N-instanton solution, we need $4mN$ parameters and N parameters to determine the positions and sizes of the N localized instanton lumps. Besides, using the dimension formula $\dim(SO(n)) = n(n-1)/2$ and the fact that our generalized Yang–Mills theory has an $SO(4m)$ internal symmetry which is of the dimension $\dim(SO(4m)) = 2m(4m-1)$, we need $2m(4m-1)N$ extra parameters to determine the asymptotic orientations of these N instantons at infinity, from which the $2m(4m-1)$ parameters originated from the global $SO(4m)$ gauge equivalence is to be subtracted. Hence it appears that a plausible number-of-parameter count for the general N-instanton solution in $4m$ dimensions is given by

$$4mN + N + 2m(4m-1)N - 2m(4m-1)$$
$$= (8m^2 + 2m + 1)N - 2m(4m-1). \tag{3.8.1}$$

For $m = 1$ (4 dimensions), this is $10N - 6$. Furthermore, if we make restriction of the Yang–Mills theory to one of the chiral representations, $SO(4m)_\pm$, of $SO(4m)$, we have only half of the dimension of $SO(4m)$: $\dim(SO(4m)_\pm) = m(4m-1)$. For example, the Witten–Tchrakian equations (3.4.27) arise from such a restriction. Now the parameter count is instead

$$4mN + N + m(4m-1)N - m(4m-1)$$
$$= (4m^2 + 3m + 1)N - m(4m-1). \tag{3.8.2}$$

For $m = 1$ (4 dimensions), the chiral representations of $SO(4)$ are simply two copies of $SO(3)$ which has the same Lie algebra as $SU(2)$ and the number count is $8N - 3$, which is the magic number obtained earlier for the classical $SU(2)$ Yang–Mills theory in 4 dimensions [14, 58, 155, 276]. Recall that the proof of Atiyah, Hitchin, and Singer [14] in 4 dimensions is a study of the fluctuation modes around an N-instanton. Our theorem for the existence of N-instanton solutions in all $4m$ dimensions lays a foundational step for a general analysis of this type.

Open Problem 3.8.2. *Prove in all $4m$ dimensions that there exist critical points of the energy (3.4.23) which are not the solutions of the self-dual or anti-self-dual equations (3.4.14).*

This problem has an interesting history in the mathematical study of gauge theory. First, the work of Burzlaff [59, 60] showed that there exists a static critical point of an $SU(3)$ Yang–Mills–Higgs theory which is not the solution of its self-dual (or anti-self-dual) equations. Based on a min-max analysis, the work of Taubes [305] then showed the existence of an infinite family of non-self-dual static solutions of the $SU(2)$ Yang–Mills–Higgs theory. For a long time, the question whether there exist non-self-dual solutions in the pure $SU(2)$ Yang–Mills theory on S^4 remained open and the only available result was the stability theorem established by Bourguignon and Lawson [50], which states that any stable critical point of the Yang–Mills energy in 4 dimensions must be either self-dual or anti-self-dual. Fairly recently, Sibner, Sibner, and Uhlenbeck [286] proved that there exists among the trivial topological class $c_2 = 0$ a family of non-self-dual critical points for the $SU(2)$ Yang–Mills theory on S^4. Moreover, Sadun and Segert [270] obtained the same conclusion for $c_2 \neq \pm 1$. Parker [242], Bor [43], and Bor and Montgomery [44] have established similar results, again on S^4. It will be desirable to solve the problem in all $4m$ dimensions.

4

Generalized Abelian Higgs Equations

In this chapter we present a thorough study of a most natural generalization of the Abelian Higgs theory in $(2+1)$ dimensions containing m Higgs scalar fields. We are led to an $m \times m$ system of nonlinear elliptic equations which is not integrable. The main tool here is to use the Cholesky decomposition theorem to reveal a variational structure of the problem. In §4.1 we formulate our problem and state an existence and uniqueness theorem. In §4.2 we treat the problem as a pure differential equation problem and state a series of general results for the system defined on a compact surface and the full plane. In §4.3–§4.5, we establish the existence part of our results stated in §4.1 and §4.2. In §4.6 we establish the nonexistence results stated in §4.2. In §4.7 we extend our study to the situation when the coefficient matrix of the nonlinear elliptic system is arbitrary.

4.1 Field Theory Structure

We first introduce the generalized Abelian Higgs theory with $(U(1)^m)$ as its gauge group and derive the system of equations governing global minimizers of the energy among static fields. We then state our main existence and uniqueness theorem for multiple soliton solutions and convert the problem into an $m \times m$ system of nonlinear elliptic equations.

4.1.1   Formulation and main existence theorem

In the classical Abelian Higgs model, one starts from a complex scalar field ϕ that lies in the fundamental or defining representation of $U(1)$. Local gauge-invariance requires the presence of a real-valued gauge field lying in the Lie algebra of $U(1)$ which is the origin of the Maxwellian electromagnetism. Schroers [275] studies the extension of this theory into the situation that the gauge group is framed by

$$(U(1))^m = \underbrace{U(1) \times \cdots \times U(1)}_{m},$$

which is realized as the maximal torus of the unitary group $U(m)$. Thus, naturally the corresponding Higgs scalar field, say, w, should lie in the fundamental or defining representation of $U(m)$, which is of course $\mathbb{C}^m$. Thus we may choose the generators, t^a $(a = 1, 2, \cdots, m)$ of the maximal torus to be $m \times m$ diagonal matrices with integer entries. Unless specified, no summation convention is assumed over repeated group indices $a, b, c = 1, 2, \cdots, m$ in this chapter.

Associated with each of the generators there is a real-valued gauge field A_μ^a so that the $(U(1))^m$ Lie algebra valued gauge field takes the form

$$A_\mu = \sum_{a=1}^{m} A_\mu^a t_a,$$

where $\mu = 0, 1, 2$ is the $(2 + 1)$-dimensional Minkowski spacetime variable index, which gives rise to m electromagnetic fields or curvatures

$$F_{\mu\nu}^a = \partial_\mu A_\nu^a - \partial_\nu A_\mu^a.$$

The gauge-covariant derivatives or connections are defined by

$$D_\mu w = \partial_\mu w + i \sum_{a=1}^{m} t_a A_\mu^a w.$$

Thus, using † to denote the Hermitian conjugate, namely, the matrix transpose and complex conjugate, we recall that the Lagrangian action density of the generalized Abelian Higgs theory is written as [275]

$$\mathcal{L} = -\sum_{a=1}^{m} \frac{1}{4e_a^2} F_{\mu\nu}^a F^{a\mu\nu} + \frac{1}{2}(D_\mu w)^\dagger (D^\mu w) - \sum_{a=1}^{m} \frac{e_a^2}{8}(R_a - w^\dagger t_a w)^2,$$

where the signature of the $(2+1)$-dimensional Minkowski space is $(+--)$, $e_a > 0$ is a coupling constant that resembles the positron charge e $(a = 1, 2, \cdots, m)$, and $R_a > 0$ is a constant that defines the vacuum level of the quantity $w^\dagger t_a w$ $(a = 1, 2, \cdots, m)$. The potential density terms, $(R_a -$

$w^\dagger t_a w)^2$, are also called the Fayet–Iliopoulos D-terms [107], for which $H_a = R_a w^\dagger t_a w$ is an Hamiltonian for the $U(1)$-action on the symplectic space $(\mathbb{C}^m, dw^\dagger \wedge dw)$ generated by t_a [19], as remarked by Schroers [275].

We are interested in static solutions. We impose the temporal gauge $A_0^a = 0$ $(a = 1, 2, \cdots, m)$ and assume that all the field configurations depend on the spatial variables x_1, x_2 only. Therefore, from the action density $\mathcal{L}$, we can calculate the energy as follows,

$$E = \frac{1}{2} \int_{\mathbb{R}^2} dx \left\{ \sum_{a=1}^m \frac{1}{e_a^2} (F_{12}^a)^2 \right.$$
$$\left. + (D_1 w)^\dagger (D_1 w) + (D_2 w)^\dagger (D_2 w) + \sum_{a=1}^m \frac{e_a^2}{4} (R_a - w^\dagger t_a w)^2 \right\}.$$

If we study the Euler–Lagrange equations of the energy E, we can conclude from some elliptic *a priori* estimates that the finite energy condition requires the exponential decay

$$F_{12}^a \to 0, \quad D_1 w \to 0, \quad D_2 w \to 0, \quad w^\dagger t_a w \to R_a, \quad a = 1, 2, \cdots, m.$$

As in the Abelian Higgs model [157] or the $O(3)$ sigma model studied in detail in Chapter 2, we rewrite the energy E in the form of quadratures,

$$E = \frac{1}{2} \int_{\mathbb{R}^2} dx \left\{ |D_1 w \pm i D_2 w|^2 + \sum_{a=1}^m \left| \frac{1}{e_a} F_{12}^a \pm \frac{e_a}{2} (R_a - w^\dagger t_a w) \right|^2 \right\}$$
$$\mp \frac{1}{2} \int_{\mathbb{R}^2} dx \left\{ i(D_1 w)^\dagger (D_2 w) - i(D_2 w)^\dagger (D_1 w) \right.$$
$$\left. + \sum_{a=1}^m F_{12}^a (R_a - w^\dagger t_a w) \right\}.$$

It is straightforward to check that there holds the useful identity

$$w^\dagger (D_1 D_2 - D_2 D_1) w = i \sum_{a=1}^m w^\dagger t_a w F_{12}^a.$$

From the above result, we can express E as

$$E = \frac{1}{2} \int_{\mathbb{R}^2} dx \left\{ |D_1 w \pm i D_2 w|^2 + \sum_{a=1}^m \left| \frac{1}{e_a} F_{12}^a \pm \frac{e_a}{2} (R_a - w^\dagger t_a w) \right|^2 \right\}$$
$$\mp \frac{1}{2} \sum_{a=1}^m R_a \int_{\mathbb{R}^2} F_{12}^a dx,$$

where we have dropped some vanishing boundary terms resulting from integration by parts.

As in the Abelian Higgs model, there are m topological invariants which are physically identified as the total normalized magnetic fluxes resembling the first Chern classes,

$$-\frac{1}{2\pi}\int_{\mathbb{R}^2} F_{12}^a\, dx, \quad a = 1, 2, \cdots, m.$$

Consequently we are led to the lower energy bound

$$E \geq \pi|T|, \quad T = \sum_{a=1}^{m} R_a \Phi_a.$$

This lower bound is attained if and only if the following system of equations, which comes from the quadratic terms in E, is satisfied,

$$D_1 w \pm i D_2 w = 0, \tag{4.1.1}$$

$$F_{12}^a \pm \frac{e_a^2}{2}(R_a - w^\dagger t_a w) = 0, \quad a = 1, 2, \cdots, m. \tag{4.1.2}$$

Note that the signs of the above equations may be mixed. In order to avoid unnecessary complication, we make a suitable arrangement so that either the plus (self-dual) or minus (anti-self-dual) sign is assumed throughout all the equations.

We now consider the components of the $\mathbb{C}^m$-valued Higgs field $w = (w_a)_{1 \leq a \leq m}$.

Let the integer-valued diagonal matrix t_a be given by

$$t_a = \text{diag}\Big\{Q_{a1}, Q_{a2}, \cdots, Q_{am}\Big\}. \tag{4.1.3}$$

Since $\{t_a \mid a = 1, 2, \cdots, m\}$ is a basis for the Lie algebra of $(U(1))^m$, which is simply $\mathbb{R}^m$ as a vector space, the matrices $t_1, t_2, \cdots, t_m$ are linearly independent over the field $\mathbb{R}$. Hence, the $m \times m$ matrix

$$Q = (Q_{ab}), \quad a, b = 1, 2, \cdots, m \tag{4.1.4}$$

is invertible. Thus there are uniquely determined real numbers $r_1, r_2, \cdots, r_m$ such that

$$R_a = \sum_{b=1}^{m} Q_{ab} r_b, \quad a = 1, 2, \cdots, m. \tag{4.1.5}$$

Consequently, since t^a is diagonal $(a = 1, 2, \cdots, m)$ and the matrix Q given in (4.1.4) is invertible, the boundary condition

$$w^\dagger t_a w = R_a, \quad a = 1, 2, \cdots, m \tag{4.1.6}$$

at infinity is transformed into the equivalent one,

$$|w_a|^2 = r_a, \quad a = 1, 2, \cdots, m. \tag{4.1.7}$$

In particular, it is necessary to have the restriction

$$r_a \geq 0, \quad a = 1, 2, \cdots, m. \tag{4.1.8}$$

If we require that each copy of the $U(1)$ symmetry be spontaneously broken, we shall strengthen (4.1.8) into

$$r_a > 0, \quad a = 1, 2, \cdots, m. \tag{4.1.9}$$

As in the classical Abelian Higgs theory [157], multiple soliton solutions of (4.1.1) and (4.1.2) are characterized by a prescribed distribution of zeros of each of the complex components of w. Our main existence theorem is stated as follows.

Theorem 4.1.1. *Suppose that (4.1.9) is valid. For any given positive integers $N_1, N_2, \cdots, N_m$ and points*

$$p_{11}, p_{12}, \cdots, p_{1N_1}, p_{21}, p_{22}, \cdots, p_{2N_2}, \cdots, p_{m1}, p_{m2}, \cdots, p_{mN_m} \in \mathbb{R}^2,$$

the self-dual or anti-self-dual system (4.1.1), (4.1.2) has a unique solution (w, A_j) so that each of the complex component w_a of w vanishes exactly at $p_{a1}, p_{a2}, \cdots, p_{aN_a}$, $a = 1, 2, \cdots, m$, and the quantities

$$|w_a|^2 - r_a, \quad F_{12}^a \ (a = 1, 2, \cdots, m), \quad |D_j w|^2 \ (j = 1, 2)$$

all decay to zero exponentially fast at infinity. Moreover, the solution has the following quantized total fluxes (charges) and energy,

$$\begin{aligned}
\Phi_a &= -\int_{\mathbb{R}^2} F_{12}^a \, dx \\
&= \pm 2\pi \sum_{b=1}^{m} Q^{ab} N_b, \quad a = 1, 2, \cdots, m, \\
E &= \pi \left| \sum_{a=1}^{m} R_a \Phi_a \right|,
\end{aligned}$$

where (Q^{ab}) is the inverse matrix of Q.

This theorem may also be interpreted as giving the existence of m different varieties of magnetic vortices, with N_a vortices in the ath variety, $a = 1, 2, \cdots, m$.

4.1.2 Nonlinear elliptic system

We introduce the new vector potentials or connection 1-forms and field strength tensors or curvatures according to

$$B_j^a = \sum_{b=1}^{m} Q_{ab} A_j^b, \quad j = 1, 2, \quad a = 1, 2, \cdots, m, \tag{4.1.10}$$

$$G^a_{jk} = \sum_{b=1}^{m} Q_{ab} F^b_{jk}, \quad j,k = 1, 2, \quad a = 1, 2, \cdots, m. \quad (4.1.11)$$

Again, we complexify the variables and derivatives following the rules

$$z = x^1 + ix^2, \quad \partial_\pm = \partial_1 \pm i\partial_2, \quad B^a_\pm = B^a_1 \pm iB^a_2, \quad a = 1, 2, \cdots, m.$$

Recall that $\partial_\pm$ is related to the standard complex derivatives ∂ and $\bar{\partial}$ by

$$\partial_+ = 2\bar{\partial}, \quad \partial_- = 2\partial.$$

Like the pair ∂ and $\bar{\partial}$, the operators ∂_+ and ∂_- are also commutative. With the above notation, we can rewrite (4.1.1) and (4.1.2) as

$$\partial_\pm w_a + iB^a_\pm w_a = 0, \quad (4.1.12)$$

$$G^a_{12} \pm \frac{1}{2} \sum_{b=1}^{m} M_{ab}(r_a - |w_b|^2) = 0, \quad (4.1.13)$$

$$G^a_{12} = \partial_1 B^a_2 - \partial_2 B^a_1, \quad a = 1, 2, \cdots, m,$$

where the matrix $M = (M_{ab})$ is defined by

$$M_{ab} = \sum_{c=1}^{m} e^2_c Q_{ac} Q_{bc}, \quad a, b = 1, 2, \cdots, m,$$

which is of course symmetric and positive definite.

We can put (4.1.12) into the form

$$\partial_\pm w_a = -iB^a_\pm w_a \quad \text{or} \quad B^a_\pm = i\partial_\pm \ln w_a, \quad a = 1, 2, \cdots, m.$$

Consequently, away from the zeros of w_a, the curvature G^a_{12} becomes

$$G^a_{12} = \frac{1}{2} \mp i(\partial_\mp B^a_\pm - \partial_\pm \overline{B^a}_\pm) = \pm \frac{1}{2}(\partial_\mp \partial_\pm \ln w_a + \partial_\pm \partial_\mp \ln \overline{w_a})$$

$$= \pm \frac{1}{2} \partial_- \partial_+ \ln |w_a|^2 = \pm 2\partial\bar{\partial} \ln |w_a|^2 = \pm \frac{1}{2} \Delta \ln |w_a|^2.$$

On the other hand, the equation (4.1.12) implies that, locally, up to a non-vanishing factor, w_a is analytic in the variable $z = x^1 + ix^2$ (see §2.3). Hence there are finitely many zeros of w_a, say, $p_{a1}, p_{a2}, \cdots, p_{aN_a}$ in $\mathbb{R}^2 = \mathbb{C}$, all of integer multiplicities. With the substitution $u_a = \ln |w_a|^2$ $(a = 1, 2, \cdots, m)$, we finally convert the system (4.1.12), (4.1.13) into the following system of nonlinear elliptic equations with sources over $\mathbb{R}^2$,

$$\Delta u_a = \sum_{b=1}^{m} M_{ab}(e^{u_b} - r_b) + 4\pi \sum_{s=1}^{N_a} \delta_{p_{as}}, \quad a = 1, 2, \cdots, m, \quad (4.1.14)$$

supplemented with the boundary condition, translated from (4.1.7),

$$\lim_{|x|\to\infty} u_a(x) = \ln r_a, \quad a = 1, 2, \cdots, m. \tag{4.1.15}$$

Conversely, if $(u_a(x))$ solves (4.1.14), (4.1.15), then the prescription

$$w_a(z) = e^{\frac{1}{2}u_a(z)\pm i\theta_a(z)}, \quad \theta_a(z) = \sum_{s=1}^{N_a} \arg(z - p_{as}),$$

$$B_1^a(z) = \mathrm{Re}\{i\partial_\pm \ln w_a(z)\}, \quad B_2^a(z) = \pm\mathrm{Im}\{i\partial_\pm \ln w_a(z)\},$$

$$a = 1, 2, \cdots, m$$

gives us a solution of (4.1.12), (4.1.13), with corresponding boundary behavior. One of the results stated in the next section will give us the quantized fluxes

$$\int_{\mathbb{R}^2} G_{12}^a \, dx = -2\pi N_a, \quad a = 1, 2, \cdots, m. \tag{4.1.16}$$

Using (4.1.16) in (4.1.11), we obtain the value of each of the Φ_a ($a = 1, 2, \cdots, m$) stated in Theorem 4.1.1. Thus, we can now concentrate on the nonlinear system (4.1.14) subject to the boundary condition (4.1.15).

4.2 General Problems and Solutions

Motivated by its field-theoretical origin discussed in the last section, we present here a complete understanding [354] of the existence and uniqueness problem of the system of nonlinear elliptic equations

$$\Delta u_j = \sum_{k=1}^{n} a_{jk}(e^{u_k} - r_k) + 4\pi \sum_{k=1}^{N_j} \delta_{p_{jk}} \quad \text{on } S, \quad j = 1, 2, \cdots, n, \tag{4.2.1}$$

where Δ is the Laplace–Beltrami operator induced from the Riemannian metric of the 2-surface S with the convention that $\Delta = \partial_1^2 + \partial_2^2$ in the flat space limit, $A = (a_{jk})$ is an $n \times n$ positive definite real symmetric matrix, $r_j > 0$ ($j = 1, 2, \cdots, n$) are constants, δ_p is the Dirac measure concentrated at $p \in S$, p_{jk} ($j = 1, 2, \cdots, n, k = 1, 2, \cdots, N_j$) are points in S, and the surface S is either closed (compact without boundary) or the full plane $\mathbb{R}^2$. In the latter case, we need to supplement (4.2.1) with the 'physical' boundary conditions $e^{u_j} = r_j$ ($j = 1, 2, \cdots, n$) at infinity.

For convenience, we shall first fix our notation.

(a) For the data given in (4.2.1), we set $\mathbf{r} = (r_1, r_2, \cdots, r_n)^\tau$, $\mathbf{N} = (N_1, N_2, \cdots, N_n)^\tau$ and $\mathbf{U} = (e^{u_1}, e^{u_2}, \cdots, e^{u_n})^\tau$, unless otherwise stated.

(b) For vectors $\mathbf{a} = (a_1, a_2, \cdots, a_n)^\tau$ and $\mathbf{b} = (b_1, b_2, \cdots, b_n)^\tau$, we write $\mathbf{a} > (\geq)\,\mathbf{b}$ if $a_j > (\geq)\,b_j$ for all $j = 1, 2, \cdots, n$.

(c) Suppose that the eigenvalues of the positive definite matrix $A = (a_{jk})$ (see (4.2.1)) are $\lambda_1, \lambda_2, \cdots, \lambda_n$. Set

$$\lambda_0 = 2 \min \left\{ \lambda_1, \lambda_2, \cdots, \lambda_n \right\}. \tag{4.2.2}$$

(d) The space of square integrable functions on S with square integrable distributional derivatives, equipped with the usual inner product, is denoted by H. The n-fold product of H is denoted by $H(n)$.

(e) The usual L^p norm for functions defined over S is written $\| \cdot \|_p$ and the inner product on L^2 is written $(\cdot, \cdot)_2$.

(f) Let J be a functional. Then dJ stands for the corresponding Fréchet derivative and we adopt the notation

$$(dJ(u))(v) = \lim_{t \to 0} \frac{J(u + tv) - J(u)}{t}.$$

Sometimes we also rewrite the above as

$$\frac{dJ}{du}(v) \equiv (dJ(u))(v).$$

If we view (4.2.1) as a vortex system, then the total number of vortices of all types is

$$N = |\mathbf{N}| \equiv \sum_{j=1}^{n} N_j.$$

Our main results are stated as three theorems.

Theorem 4.2.1. *Let S be a closed 2-surface and $|S|$ its volume. Then the elliptic system (4.2.1) has a solution if and only if there holds*

$$\frac{4\pi}{|S|} A^{-1} \mathbf{N} < \mathbf{r}. \tag{4.2.3}$$

Furthermore, if there is a solution, the solution must be unique.

Note that in (4.2.3) we do not actually require $\mathbf{r} > 0$. However, when $\mathbf{r} > 0$, the condition (4.2.3) may always be satisfied with sufficiently large $|S|$. Roughly speaking, larger spaces allow the existence of more vortices. This observation suggests that, for $S = \mathbb{R}^2$, the vortex numbers defined by the vector $\mathbf{N}$ may be arbitrary. The following result confirms such a speculation.

Theorem 4.2.2. *Let $S = \mathbb{R}^2$ and the data in (4.2.1) be arbitrarily given. Then the system (4.2.1) has a unique solution $\mathbf{u} = (u_1, u_2, \cdots, u_n)^\tau$ that satisfies $e^{u_j} = r_j$ ($j = 1, 2, \cdots, n$) at infinity. Furthermore this solution* *fulfills the following sharp decay estimates at infinity,*

$$\sum_{j=1}^{n} (u_j(x) - \ln r_j)^2 \leq C(\varepsilon) e^{-(1-\varepsilon)\sqrt{\lambda_0}|x|}, \tag{4.2.4}$$

$$\sum_{j=1}^{n} |\nabla u_j(x)|^2 \leq C(\varepsilon) e^{-(1-\varepsilon)\sqrt{\lambda_0 r_0}|x|}, \tag{4.2.5}$$

where $\varepsilon : 0 < \varepsilon < 1$, *is an arbitrary number and* $C(\varepsilon) > 0$ *is a constant,* λ_0 *is as defined by (4.2.2), and* $r_0 = \min\{r_1, r_2, \cdots, r_n\}$. *Furthermore there hold the quantized integrals in the full plane,*

$$\int_{\mathbb{R}^2} \sum_{k=1}^{n} a_{jk}(e^{u_k} - r_k)\, dx = -4\pi N_j, \quad j = 1, 2, \cdots, n. \tag{4.2.6}$$

Note that the expressions stated in (4.2.6) correspond to the quantized fluxes in the field theory model studied in the last section which are also valid for the solution obtained in the compact setting in which the space $\mathbb{R}^2$ is replaced by a closed 2-surface. On the other hand, however, unlike the problem in the compact case, the condition $\mathbf{r} > \mathbf{0}$ in the problem over $\mathbb{R}^2$ now becomes a crucial assumption. Physically it corresponds to positive vacuum expectation values of the Higgs fields characterized by the fact that the temperature is strictly below a critical temperature for which the gauge symmetry is completely broken. The following nonexistence theorem says that, when some components of $\mathbf{r}$ vanish, the system (4.2.1) may fail to possess a solution.

Theorem 4.2.3. *Consider (4.2.1) over the full plane* $\mathbb{R}^2$. *If the vector* $\mathbf{r} = \mathbf{0}$, *then (4.2.1) has no solution. If some components of* $\mathbf{r}$ *are positive but the rest of it vanish, there are situations under which the system has no solution. More precisely,*

(i) the general $n \times n$ *system has no solution when* $r_1 > 0, \cdots, r_{n-1} > 0$ *but* $r_n = 0$ *if the lower triangular matrix* L *in the Cholesky decomposition (see (4.3.3) below) of the coefficient matrix* $A = (a_{jk})$ *satisfies the property that the off-diagonal entries of the nth row of* L *are all nonpositive;*

(ii) for $n = 2$ *the system has no solution when* $r_1 > 0, r_2 = 0$ *or* $r_1 = 0, r_2 > 0$ *if* $a_{12} = a_{21} \leq 0$;

(iii) for $n = 2$, $a_{12} = a_{21} > 0$, *and* $r_1 > 0, r_2 = 0$, *the system has no solution with finite 'potential energy',* $\int e^{u_2} < \infty$, *if the two vortex numbers* N_1 *and* N_2 *satisfy*

$$\frac{a_{12}}{a_{11}} N_1 \leq 1 + N_2.$$

Note that although the derivations of (ii) and (iii) above are of independent interest, it is comforting to see that the inequality stated in (iii) contains the condition $a_{12} \leq 0$ in (ii). Technically none of the above except

(iii) requires any finite-energy condition to be observed for the nonexistence of a solution.

4.3 Compact Surface Case

In this section, we prove Theorem 4.2.1. The proof splits into a few steps. First we derive (4.2.3) as necessary condition. Next we formulate a constrained variational principle in which we minimize an objective functional under n functionally independent constraints. The crucial part is to show that the Lagrangian multipliers will give rise to the correct values of the coefficients in the equations to be solved. This situation is similar to that in the prescribed Gaussian curvature problem [17, 21, 35, 73, 76, 140, 164, 165, 191, 220]. Then we prove the existence of a solution by showing that the constrained minimization problem indeed has a solution. The key tool is the Trudinger–Moser inequality. The structure of the problem implies that the value of the optimal constant in the Trudinger–Moser inequality is irrelevant, which ensures our existence proof without any additional condition except the necessary condition (4.2.3). Finally the uniqueness follows from the convexity of a suitable energy functional.

4.3.1 Necessary condition

Consider (4.2.1). Let u_j^0 be such that

$$\Delta u_j^0 = 4\pi \sum_{k=1}^{N_j} \delta_{p_{jk}} - \frac{4\pi}{|S|} N_j, \qquad j = 1, 2, \cdots, n \qquad (4.3.1)$$

(of course, such a background function is unique up to an additive constant [17]). With $u_j = u_j^0 + v_j$, $\mathbf{v} = (v_1, v_2, \cdots, v_n)^\tau$ and the notation set in the last section, we rewrite (4.2.1) in the matrix form

$$\Delta \mathbf{v} = A\mathbf{U} - A\mathbf{r} + \frac{4\pi}{|S|}\mathbf{N}. \qquad (4.3.2)$$

Integrating (4.3.2) and using the properties

$$\int_S \Delta(A^{-1}\mathbf{v}) \, d\Omega = \mathbf{0} \quad \text{and} \quad \int_S \mathbf{U} \, d\Omega > \mathbf{0},$$

we immediately arrive at (4.2.3). In the sequel, we often omit the volume element $d\Omega$ of S in all integrals when no risk of confusion may arise.

4.3.2 Variational principle

At first glance, it is not clear whether (4.3.2) has a variational principle. The crucial step in our study is to find a powerful variational structure of the problem: it turns out that when the matrix A is factored properly, one is able to see that the problem allows a variational treatment. The next subsection is a detailed study of this problem.

Since the matrix A is positive definite, we know that there is a unique lower triangular $n \times n$ matrix $L = (L_{jk})$ for which all the diagonal entries are positive, i.e., $L_{jj} > 0$, $j = 1, 2, \cdots, n$, so that

$$A = LL^{\tau}. \tag{4.3.3}$$

This is the well-known Cholesky decomposition theorem. Note that the condition stated in part (i) of Theorem 4.2.3 involves the matrix L. Thus it may be useful to record here the relation between the entries of A and L as follows [131],

$$L_{11} = \sqrt{a_{11}}, \quad L_{j1} = \frac{a_{j1}}{L_{11}}, \quad j = 2, \cdots, n,$$

$$L_{jj} = \sqrt{a_{jj} - \sum_{k=1}^{j-1} L_{jk}^2}, \quad j = 2, \cdots, n,$$

$$L_{jk} = \frac{a_{jk} - \sum_{k'=1}^{k-1} L_{jk'} L_{kk'}}{L_{kk}}, \quad j = k+1, \cdots, n, \ k = 2, \cdots, n,$$

whose derivation is based on an inductive argument. Introduce the new variable vector

$$\mathbf{w} = L^{-1}\mathbf{v} \quad \text{or} \quad \mathbf{v} = L\mathbf{w}. \tag{4.3.4}$$

Then (4.3.2) takes the form

$$\Delta\mathbf{w} = L^{\tau}\mathbf{U} - L^{\tau}\mathbf{r} + \frac{4\pi}{|S|}L^{-1}\mathbf{N}. \tag{4.3.5}$$

Use the notation

$$\mathbf{b} = (b_1, b_2, \cdots, b_n)^{\tau} = L^{\tau}\mathbf{r} - \frac{4\pi}{|S|}L^{-1}\mathbf{N}. \tag{4.3.6}$$

It is then more transparent to work on the component form of (4.3.5), which is

$$\Delta w_j = \sum_{k=j}^{n} L_{kj} \exp\left(u_k^0 + \sum_{k'=1}^{k} L_{kk'} w_{k'}\right) - b_j \quad \text{on } S,$$

$$j = 1, 2, \cdots, n. \tag{4.3.7}$$

In order to prove the existence of a solution to the system (4.3.7), we introduce a constrained variational principle. For this purpose, we see by a direct integration of (4.3.5) that

$$I_j(\mathbf{w}) = \int_S \exp\left(u_j^0 + \sum_{k=1}^{j} L_{jk} w_k\right) = K_j, \quad j = 1, 2, \cdots, n, \tag{4.3.8}$$

where $\mathbf{K} = (K_1, K_2, \cdots, K_n)^\tau = |S|(L^\tau)^{-1}\mathbf{b}$. As a consequence of (4.3.7) and (4.3.8), we are led to the formulation of the following optimization problem,

$$\sigma \equiv \min\left\{ I_0(\mathbf{w}) \,\middle|\, \mathbf{w} \in \mathcal{A} \right\}, \tag{4.3.9}$$

where the objective functional I_0 and the admissible space $\mathcal{A}$ are defined by

$$I_0(\mathbf{w}) = \int_S \frac{1}{2} \sum_{j=1}^n |\nabla w_j|^2 - \sum_{j=1}^n b_j w_j,$$

$$\mathcal{A} = \left\{ \mathbf{w} \in H(n) \,\middle|\, \mathbf{w} \text{ satisfies the condition (4.3.8)} \right\},$$

respectively. We now establish our variational principle by showing that a solution of (4.3.9) must satisfy the system (4.3.7). The crucial part is to prove that the constraints (4.3.8) do not give rise to undesired terms in the variational equations.

In fact, let $\mathbf{w}$ be a solution of (4.3.9), then by the theory of Lagrange multipliers, there are real numbers $\mu_1, \mu_2, \cdots, \mu_n$ so that

$$dI_0(\mathbf{w}) - \sum_{j=1}^n \mu_j dI_j(\mathbf{w}) = 0,$$

because the derivatives $dI_j(\mathbf{w})$ $(j = 1, 2, \cdots, n)$ are linearly independent. In other words, for any test function $\mathbf{f} = (f_1, f_2, \cdots, f_n)^\tau \in H(n)$, we have

$$\int_S \sum_{j=1}^n \left(\nabla w_j \cdot \nabla f_j - b_j f_j \right)$$

$$= \sum_{j=1}^n \mu_j \int_S \exp\left(u_j^0 + \sum_{k=1}^j L_{jk} w_k \right) \sum_{k'=1}^j L_{jk'} f_{k'}. \tag{4.3.10}$$

Since f_j $(j = 1, 2, \cdots, n)$ are arbitrary, we may fix j and let $f_k \equiv 0$ for $k \neq j$. Therefore (4.3.10) can be simplified to

$$\int_S \nabla w_j \cdot \nabla f_j - b_j f_j = \sum_{k=j}^n \mu_k L_{kj} \int_S \exp\left(u_k^0 + \sum_{k'=1}^k L_{kk'} w_{k'} \right) f_j,$$

$$j = 1, 2, \cdots, n. \tag{4.3.11}$$

It is easily seen that (4.3.11) is an upper triangular system. Consequently, we can determine the values of $\mu_1, \mu_2, \cdots, \mu_n$ by backward search. In fact, set $f_j \equiv 1$ for $j = 1, 2, \cdots, n$. The last equation, $j = n$, in (4.3.11) reads

$$-b_n|S| = \mu_n L_{nn} \int_S \exp\left(u_n^0 + \sum_{k'=1}^n L_{nk'} w_{k'} \right).$$

Using (4.3.8) in the above expression and applying the relation between $\mathbf{K}$ and $\mathbf{b}$, namely,

$$\sum_{j'=j}^{n} L_{j'j} K_{j'} = |S| b_j, \qquad j = 1, 2, \cdots, n,$$

we obtain $\mu_n = -1$. For $j = n - 1$, we have

$$
\begin{aligned}
-b_{n-1}|S| &= \sum_{k=n-1}^{n} \mu_k L_{kn-1} \int_S \exp\left(u_k^0 + \sum_{k'=1}^{k} L_{kk'} w_{k'} \right) \\
&= \mu_{n-1} L_{n-1n-1} K_{n-1} - L_{nn-1} K_n.
\end{aligned}
$$

Thus we again have $\mu_{n-1} = -1$. In fact, applying this argument subsequently, we find by (4.3.8) that $\mu_j = -1$ for all $j = 1, 2, \cdots, n$. Inserting such a conclusion into (4.3.11) we see that $\mathbf{w}$ is a weak solution of the equations $\Delta \mathbf{w} = L^\tau \mathbf{U} - \mathbf{b}$. The standard elliptic regularity theory then implies that $\mathbf{w}$ is also a classical solution. Thus the system (4.3.5) or (4.3.7) is solved and the desired variational principle follows.

4.3.3 Existence of solution

Since S is a closed surface, any function $f \in H$ may be decomposed uniquely into the sum $f = \underline{f} + f'$ where $\underline{f}$ is a constant and f' satisfies $\int_S f' = 0$. This decomposition is useful because it enables us to employ the well-known Trudinger–Moser inequality [17, 18, 221, 313] of the form

$$
\int_S e^f \, d\Omega \leq C(\varepsilon) \exp\left(\left[\frac{1}{16\pi} + \varepsilon \right] \int_S |\nabla f|^2 \, d\Omega \right),
$$
$$
f \in H, \qquad \int_S f \, d\Omega = 0, \tag{4.3.12}
$$

where $d\Omega$ is the area element of S and $C(\varepsilon) > 0$ is a constant depending only on $\varepsilon > 0$. An immediate consequence of this inequality is that the map $H \to L(S)$, $f \mapsto e^f$, is completely continuous.

It is interesting to note that (4.3.12) has a sharp form, where $\varepsilon = 0$, due to the recent work of Fontana [109], which will not be needed here for our purposes though.

We now assume (4.2.3). It is seen that such a condition simply implies that K_j's on the right-hand sides of (4.3.8) are all positive and, consequently, the admissible space $\mathcal{A}$ is nonempty. For any

$$\mathbf{w} = (w_1, w_2, \cdots, w_n)^\tau \in \mathcal{A},$$

with the decompositions $w_j = \underline{w}_j + w'_j$ $(j = 1, 2, \cdots, n)$, we convert (4.3.8) into the form

$$\sum_{k=1}^{j} L_{jk}\underline{w}_k = \ln K_j - \ln\left(\int_S \exp\left[u_j^0 + \sum_{k=1}^{j} L_{jk}w'_k\right]\right), \quad j = 1, 2, \cdots, n.$$

$$(4.3.13)$$

Since by Jensen's inequality we have the lower bounds

$$\ln\left(\int_S \exp\left[u_j^0 + \sum_{k=1}^{j} L_{jk}w'_k\right]\right) \geq \frac{1}{|S|}\int_S u_j^0 + \ln|S|, \quad j = 1, 2, \cdots, n,$$

(4.3.13) says that there is a constant vector $\mathbf{C}_0$ so that $L\underline{w} \leq \mathbf{C}_0$. Using the fact that L is a lower triangular matrix with positive diagonal entries, and hence, so is L^{-1}, we conclude that the entries of $\mathbf{C}_0$ can be made suitably large to achieve $L^{-1}(\mathbf{C}_0 - L\underline{w}) > 0$. Namely, $\underline{w} \leq L^{-1}\mathbf{C}_0$. Consequently, using $\mathbf{b} > 0$, we can bind the functional I_0 from below on $\mathcal{A}$,

$$I_0(\mathbf{w}) \geq \left(\int_S \frac{1}{2}\sum_{j=1}^{n} |\nabla w'_j|^2\right) - |S|\,\mathbf{b} \cdot L^{-1}\mathbf{C}_0. \qquad (4.3.14)$$

In particular, for a minimizing sequence $\{\mathbf{w}\}$ of the problem (4.3.9), the corresponding sequence $\{\mathbf{w}'\}$ is bounded in $H(n)$ due to (4.3.14) and the Poincaré inequality

$$\int_S f^2 \leq C\int_S |\nabla f|^2, \quad f \in H, \quad \int_S f = 0,$$

where $C > 0$ is a suitable constant. From (4.3.12) and (4.3.13) we see that there is a constant vector of positive entries, say, $\mathbf{C}_0$, so that $L\underline{w} \geq -\mathbf{C}_0$. A similar argument allows us to assume then $\underline{w} > -L^{-1}\mathbf{C}_0$. Therefore the sequence $\{\underline{w}\}$ is bounded, which leads to the boundedness of $\{\mathbf{w}\}$ in $H(n)$. Without loss of generality, we may assume that the sequence $\{\mathbf{w}\}$ itself is weakly convergent in $H(n)$. By the well-established compact embedding theorem and the structure of the functionals I_j $(j = 0, 1, 2, \cdots, n)$, which says that I_0 is weakly lower semicontinuous and I_j's are weakly continuous, we see that the weak limit in $H(n)$ is a solution of (4.3.9). This furnishes a proof of existence.

4.3.4 Uniqueness

The uniqueness of the solution may be seen by a convexity argument. In fact it is straightforward to verify that a solution of (4.3.7) is a critical point of the functional

$$J(\mathbf{w}) = \int_S \frac{1}{2}\sum_{j=1}^{n} |\nabla w_j|^2 - \sum_{j=1}^{n} b_j w_j + F(x, w_1, w_2, \cdots, w_n), \qquad (4.3.15)$$

where the nonlinear potential term F is defined as

$$F(x, w_1, w_2, \cdots, w_n) = \sum_{j=1}^{n} \exp\left(u_j^0(x) + \sum_{k=1}^{j} L_{jk} w_k\right). \qquad (4.3.16)$$

It may be examined that the Hessian of F has the representation given in the form

$$\left(\frac{\partial^2 F}{\partial w_j \partial w_k}\right) = L^\tau \text{diag}\left\{\exp\left(u_j^0 + \sum_{k=1}^{j} L_{jk} w_k\right)\right\}_{1 \le j \le n} L, \qquad (4.3.17)$$

which is of course positive definite except at the points $p_{jk} \in S$, $j = 1, 2, \cdots, n, k = 1, 2, \cdots, N_j$. Hence $J(\mathbf{w})$ is strictly convex. In particular, J can at most have one critical point in $H(n)$. Thus the uniqueness follows.

4.4 Solution on Plane: Existence

We now turn our attention to the problem on the full plane, $\mathbb{R}^2$. When $n = 1$ (the scalar case), the result is obtained in Jaffe and Taubes [157]. Our proof in the general case follows their ideas through solving an absolute minimization problem. The general framework comes from realizing some crucial facts in the optimization problem for a convex functional which is based on the following well-set steps: Formulate a variational principle in which a convex C^1 energy functional I is defined over a suitable Hilbert space X (for our problem here $X = H(n)$). Therefore I must be weakly lower semicontinuous. Show that I is coercive in the sense that there are constants $C_1, C_2 > 0$ so that $(\mathrm{d}I(s))(s) \ge C_1 \|s\|_X - C_2$, $s \in X$. This step requires a careful rearrangement of the various terms in $(\mathrm{d}I(s))(s)$ and the use of suitable embedding inequalities. Combining the above steps, the existence and uniqueness of a critical point of I in X is obtained. Then the standard elliptic regularity theory implies that the critical point is a smooth solution of the original equations.

4.4.1 Variational problem

As in the last section, we need to introduce some background functions:

$$u_j^0 = -\sum_{k=1}^{N_j} \ln(1 + \mu|x - p_{jk}|^{-2}), \quad \Delta u_j^0 = 4\pi \sum_{k=1}^{N_j} \delta_{p_{jk}} - g_j,$$

$$g_j = 4\sum_{k=1}^{N_j} \frac{\mu}{(\mu + |x - p_{jk}|^2)^2}, \quad j = 1, 2, \cdots, n, \quad x \in \mathbb{R}^2, \qquad (4.4.1)$$

where $\mu > 0$ is a parameter to be specified later. It is useful to note that the vector $\mathbf{g} = (g_1, g_2, \cdots, g_n)^\tau$ satisfies

$$\int_{\mathbb{R}^2} g_j = 4\pi N_j, \quad j = 1, 2, \cdots, n. \qquad (4.4.2)$$

With $S = \mathbb{R}^2$ and $u_j = \ln r_j + u_j^0 + v_j$ $(j = 1, 2, \cdots, n)$, the system (4.2.1) becomes

$$\Delta v_j = \sum_{k=1}^{n} a_{jk} r_k (e^{u_k^0 + v_k} - 1) + g_j, \quad j = 1, 2, \cdots, n. \qquad (4.4.3)$$

Again we will need the new variable vector $\mathbf{w}$ defined as in (4.3.4). So (4.4.3) reads

$$\Delta w_j = \sum_{k=j}^{n} L_{kj} r_k \left(\exp\left[u_k^0 + \sum_{k'=1}^{k} L_{kk'} w_{k'} \right] - 1 \right) + h_j \quad \text{in } \mathbb{R}^2,$$

$$j = 1, 2, \cdots, n, \qquad (4.4.4)$$

where $\mathbf{h} = (h_1, h_2, \cdots, h_n)^\tau = L^{-1}\mathbf{g}$. It is direct to check that (4.4.4) are the variational equations of the energy functional

$$I(\mathbf{w}) = \int_{\mathbb{R}^2} \frac{1}{2} \sum_{j=1}^{n} |\nabla w_j|^2 + \sum_{j=1}^{n} h_j w_j$$

$$+ \int_{\mathbb{R}^2} \sum_{j=1}^{n} r_j \left(\exp\left[u_j^0 + \sum_{k=1}^{j} L_{jk} w_k \right] - e^{u_j^0} - \sum_{k=1}^{j} L_{jk} w_k \right). (4.4.5)$$

4.4.2   Coercivity

It is convenient to rewrite (4.4.5) as

$$I(\mathbf{w})$$

$$= \frac{1}{2} \sum_{j=1}^{n} \|\nabla w_j\|_2^2 + \sum_{j=1}^{n} r_j \left(e^{u_j^0}, \exp\left[\sum_{k=1}^{j} L_{jk} w_k \right] - 1 - \sum_{k=1}^{j} L_{jk} w_k \right)_2$$

$$+ \sum_{j=1}^{n} (h_j, w_j)_2 + \sum_{j=1}^{n} r_j \left(e^{u_j^0}, \sum_{k=1}^{j} L_{jk} w_j \right)_2 - \int_{\mathbb{R}^2} \sum_{j=1}^{n} r_j \sum_{k=1}^{j} L_{jk} w_k$$

$$= \frac{1}{2} \sum_{j=1}^{n} \|\nabla w_j\|_2^2 + \sum_{j=1}^{n} r_j \left(e^{u_j^0}, \exp\left[\sum_{k=1}^{j} L_{jk} w_k \right] - 1 - \sum_{k=1}^{j} L_{jk} w_k \right)_2$$

$$+ \sum_{j=1}^{n} \left(w_j, h_j + \sum_{k=j}^{n} L_{kj} r_k \left[e^{u_k^0} - 1 \right] \right)_2. \qquad (4.4.6)$$

The form of (4.4.6) allows us to obtain

$$(d I(\mathbf{w}))(\mathbf{w}) - \sum_{j=1}^{n} \|\nabla w_j\|_2^2$$

$$= \sum_{j=1}^{n} \left(w_j, \sum_{k=j}^{n} L_{kj} r_k e^{u_k^0} \left[\exp\left\{ \sum_{k'=1}^{k} L_{kk'} w_{k'} \right\} - 1 \right] \right)_2$$

$$+ \sum_{j=1}^{n} \left(w_j, h_j + \sum_{k=j}^{n} L_{kj} r_k (e^{u_k^0} - 1) \right)_2$$

$$= \sum_{j=1}^{n} \left(w_j, h_j + \sum_{k=j}^{n} L_{kj} r_k \left[\exp\left\{ u_k^0 + \sum_{k'=1}^{k} L_{kk'} w_{k'} \right\} - 1 \right] \right)_2$$

$$= \sum_{j=1}^{n} r_j \left(\sum_{k=1}^{j} L_{jk} w_k, \left[\exp\left\{ u_j^0 + \sum_{k'=1}^{j} L_{jk'} w_{k'} \right\} - 1 \right] + H_j \right)_2,$$

where H_j depends linearly on $\mathbf{h}$. To estimate the right-hand side of the above, we consider the quantity

$$M(v) = (v, e^{u^0 + v} - 1 + h)_2,$$

where v, u^0, h stand for one of the functions $\sum_{k=1}^{j} L_{jk} w_k, u_j^0, H_j$, respectively.

To proceed, we start also from the decomposition $v = v_+ - v_-$ with $v_+ = \max\{0, v\}$ and $v_- = \max\{0, -v\}$. Then $M(v) = M(v_+) + M(-v_-)$. The first term, $M(v_+)$, is of no harm whatever h is because the fact that

$$e^{u^0 + v_+} - 1 + h = e^{u^0 + v_+} - 1 - (u^0 + v_+) + (v_+ + u^0 + h)$$
$$\geq v_+ + u^0 + h$$

and that $u^0, h \in L^2$ yield the lower bound

$$M(v_+) \geq \int_{\mathbb{R}^2} v_+^2 + \int_{\mathbb{R}^2} v_+(u^0 + h)$$
$$\geq \frac{1}{2} \|v_+\|_2^2 - C_1.$$

This simple result shows that $M(v_+)$ is well behaved. On the other hand, using the inequality

$$1 - e^{-x} \geq \frac{x}{1+x}, \qquad x \geq 0,$$

we can estimate $M(-v_-)$ from below as follows:

$$M(-v_-) = (v_-, 1 - h - e^{u^0})_2 + (v_-, e^{u^0}[1 - e^{-v_-}])_2$$

$$\geq \left(v_-, \left\{1 - h - e^{u^0} + \frac{v_-}{1 + v_-}e^{u^0}\right\}\right)_2$$

$$= \int_{\mathbb{R}^2} \frac{v_-}{1 + v_-}([1 + v_-][1 - h - e^{u^0}] + v_-e^{u^0})$$

$$= \int_{\mathbb{R}^2} \frac{v_-}{1 + v_-}([1 - h]v_- + [1 - h - e^{u^0}]).$$

By the definition of g_j $(j = 1, 2, \cdots, n)$, we can make μ sufficiently large so that $h < 1/2$ everywhere. It is easily checked that both h and $1 - e^{u^0}$ belong to L^2. So

$$\int_{\mathbb{R}^2} \frac{v_-}{1 + v_-}([1 - h]v_- + [1 - h - e^{u^0}]) \leq C_2\left(\int_{\mathbb{R}^2} \frac{v_-^2}{1 + v_-}\right)^{\frac{1}{2}}.$$

Thus there is a constant $C_3 > 0$ to make the lower estimate

$$M(-v_-) \geq \frac{1}{4}\int_{\mathbb{R}^2} \frac{v_-^2}{1 + v_-} - C_3$$

valid. Recall the lower estimate for $M(v_+)$ obtained earlier. We now conclude that

$$M(v) \geq \frac{1}{4}\int_{\mathbb{R}^2} \frac{v^2}{1 + |v|} - C$$

holds for some constant $C > 0$. Using this result, we arrive at

$$(dI(\mathbf{w}))(\mathbf{w}) - \sum_{j=1}^{n}\|\nabla w_j\|_2^2 \geq \sum_{j=1}^{n}\mu_j \int_{\mathbb{R}^2} \frac{\left(\sum_{k=1}^{j} L_{jk}w_k\right)^2}{1 + \left|\sum_{k=1}^{j} L_{jk}w_k\right|} - C, \quad (4.4.7)$$

where $\mu_1, \mu_2, \cdots, \mu_n$ and C are some positive irrelevant constants.

Moreover, since the matrix L is invertible, we have a positive constant C_0 to make

$$\sum_{j=1}^{n}\left(\sum_{k=1}^{j} L_{jk}w_k\right)^2 \geq C_0\sum_{j=1}^{n} w_j^2$$

valid. Substitute this into (4.4.7). We obtain

$$(dI(\mathbf{w}))(\mathbf{w}) \geq \sum_{j=1}^{n}\|\nabla w_j\|_2^2 + c_1\int_{\mathbb{R}^2} \frac{\sum_{j=1}^{n} w_j^2}{(1 + \sum_{j=1}^{n}|w_j|)^2} - c_2, \quad (4.4.8)$$

where $c_1, c_2 > 0$ are constants.

To proceed further, we recall the standard embedding inequality [36, 123, 175, 176]

$$\int_{\mathbb{R}^2} f^4 \leq 2\int_{\mathbb{R}^2} f^2 \int_{\mathbb{R}^2} |\nabla f|^2, \quad f \in H. \quad (4.4.9)$$

We will use (4.4.9) to show that the two first terms on the right-hand side of (4.4.8) are strong enough to achieve the desired coercivity inequality.

In fact, by virtue of (4.4.9), we see that

$$
\left(\sum_{j=1}^{n} \|w_j\|_2^2 \right)^2
$$

$$
\leq \left(\int_{\mathbb{R}^2} \frac{\sum_{j=1}^{n} |w_j|}{1 + \sum_{j=1}^{n} |w_j|} \left[1 + \sum_{j=1}^{n} |w_j| \right] \sum_{j=1}^{n} |w_j| \right)^2
$$

$$
\leq C \int_{\mathbb{R}^2} \frac{\sum_{j=1}^{n} w_j^2}{(1 + \sum_{j=1}^{n} |w_j|)^2} \int_{\mathbb{R}^2} \sum_{j=1}^{n} (w_j^2 + w_j^4)
$$

$$
\leq C \int_{\mathbb{R}^2} \frac{\sum_{j=1}^{n} w_j^2}{(1 + \sum_{j=1}^{n} |w_j|)^2} \left(\int_{\mathbb{R}^2} \sum_{j=1}^{n} w_j^2 \right) \left(1 + \int_{\mathbb{R}^2} \sum_{j=1}^{n} |\nabla w_j|^2 \right)
$$

$$
\leq \frac{1}{2} \left(\int_{\mathbb{R}^2} \sum_{j=1}^{n} w_j^2 \right)^2 + C \left(\left[\int_{\mathbb{R}^2} \frac{\sum_{j=1}^{n} w_j^2}{(1 + \sum_{j=1}^{n} |w_j|)^2} \right]^4 + \right.
$$

$$
\left. + \left[\int_{\mathbb{R}^2} \sum_{j=1}^{n} |\nabla w_j|^2 \right]^4 + 1 \right), \tag{4.4.10}
$$

where, and in the sequel, the symbol C in the above denotes an absolute constant which may vary its value from place to place. Hence a simple interpolation inequality applied to (4.4.10) yields

$$
\sum_{j=1}^{n} \|w_j\|_2 \leq C \left(1 + \sum_{j=1}^{n} \|\nabla w_j\|_2^2 + \int_{\mathbb{R}^2} \frac{\sum_{j=1}^{n} w_j^2}{(1 + \sum_{j=1}^{n} |w_j|)^2} \right). \tag{4.4.11}
$$

Consequently, inserting (4.4.11) into (4.4.8), we arrive at the expected coercivity inequality

$$
(\mathrm{d}I(\mathbf{w}))(\mathbf{w}) \geq C_1 \left(\sum_{j=1}^{n} \|w_j\|_2 + \sum_{j=1}^{n} \|\nabla w_j\|_2 \right) - C_2 \tag{4.4.12}
$$

for suitable constants $C_1, C_2 > 0$.

Our next step is to use (4.4.12) to prove that the system (4.4.3) has a solution by showing that (4.4.5) has a critical point.

4.4.3 Existence and uniqueness of critical point

It is not hard to show that $I : H(n) \to \mathbb{R}$ is a C^1-functional. Using (4.3.16) and (4.3.17) we see that I is also convex. Recall that any convex C^1-functional over a Hilbert space must be weakly lower semicontinuous.

Moreover, from (4.4.12), we can find a large enough number $R > 0$ so that

$$\inf \left\{ (dI(\mathbf{w}))(\mathbf{w}) \ \bigg| \ \mathbf{w} \in H(n), \ \|\mathbf{w}\|_{H(n)} = R \right\} \geq 1. \qquad (4.4.13)$$

Consider the optimization problem

$$\sigma = \min \left\{ I(\mathbf{w}) \ \bigg| \ \|\mathbf{w}\|_{H(n)} \leq R \right\}. \qquad (4.4.14)$$

Let $\{\mathbf{w}\}$ be a minimizing sequence of (4.4.14). Without loss of generality, we may assume that this sequence is also weakly convergent. Let $\tilde{\mathbf{w}}$ be its weak limit. Thus, using the fact that I is weakly lower semicontinuous, we have $I(\tilde{\mathbf{w}}) \leq \sigma$. Of course $\|\tilde{\mathbf{w}}\|_{H(n)} \leq R$ because norm is also weakly lower semicontinuous. Hence $I(\tilde{\mathbf{w}}) = \sigma$ and $\tilde{\mathbf{w}}$ solves (4.4.14). We show next that $\tilde{\mathbf{w}}$ is a critical point of the functional I. In fact, we only need to show that $\tilde{\mathbf{w}}$ is an interior point, namely,

$$\|\tilde{\mathbf{w}}\|_{H(n)} < R.$$

Suppose otherwise that $\|\tilde{\mathbf{w}}\|_{H(n)} = R$. Then, in view of (4.4.13), we have

$$
\begin{aligned}
\lim_{t \to 0} \frac{I(\tilde{\mathbf{w}} - t\tilde{\mathbf{w}}) - I(\tilde{\mathbf{w}})}{t} &= \frac{d}{dt} I(\tilde{\mathbf{w}} - t\tilde{\mathbf{w}}) \bigg|_{t=0} \\
&= -(dI(\tilde{\mathbf{w}}))(\tilde{\mathbf{w}}) \leq -1. \qquad (4.4.15)
\end{aligned}
$$

Therefore, when $t > 0$ is sufficiently small, we see by virtue of (4.4.15) that

$$I(\tilde{\mathbf{w}} - t\tilde{\mathbf{w}}) < I(\tilde{\mathbf{w}}) = \sigma.$$

However, since $\|\tilde{\mathbf{w}} - t\tilde{\mathbf{w}}\|_{H(n)} = (1-t)R < R$, we arrive at a contradiction to the definition of $\tilde{\mathbf{w}}$ or (4.4.14).

Finally the strict convexity of I says that I can only have at most one critical point, so we have the conclusion that I has exactly one critical point in $H(n)$. Of course this critical point is a solution of (4.4.3) which must be smooth by virtue of the elliptic regularity theory and also unique in the space $H(n)$.

4.5 Solution on Plane: Asymptotic Behavior

For the solution obtained in the last section, we first establish some pointwise decay properties. We then find the desired asymptotic estimates near infinity. The tools are suitable L^p estimates and some elliptic comparison inequalities.

4.5.1 Pointwise decay near infinity

Since we are in two dimensions, there holds the embedding inequality [35, 123, 157, 176]

$$\|f\|_p \le \left(\pi\left[\frac{p-2}{2}\right]\right)^{\frac{p-2}{2p}} \|f\|_H, \qquad f \in H, \qquad p > 2, \qquad (4.5.1)$$

which implies $e^f - 1 \in L^2$ for $f \in H$. In fact, the MacLaurin series leads to

$$(e^f - 1)^2 = f^2 + \sum_{k=3}^{\infty} \frac{2^k - 2}{k!} f^k.$$

Combining the above with (4.5.1), we have, formally,

$$\|e^f - 1\|_2^2 \le \|f\|_2^2 + \sum_{k=3}^{\infty} \frac{2^k - 2}{k!} \left(\pi\frac{k-2}{2}\right)^{\frac{k-2}{2}} \|f\|_H^k. \qquad (4.5.2)$$

It is readily shown that (4.5.2) is a convergent series, which verifies our claim.

Let us now examine (4.4.4) again. First, by virtue of $\mathbf{w} \in H(n)$ and

$$\exp\left(u_k^0 + \sum_{k'=1}^{k} L_{kk'}w_{k'}\right) - 1 = e^{u_k^0}\left(\exp\left[\sum_{k'=1}^{k} L_{kk'}w_{k'}\right] - 1\right) + (e^{u_k^0} - 1),$$

it is seen that the right-hand sides of all the equations in (4.4.4) belong to L^2. Next the well-known L^2-estimates for elliptic equations indicate that $w_j \in W^{2,2}$ ($j = 1, 2, \cdots, n$). Such a result implies that $\mathbf{w}(x) \to \mathbf{0}$ as $|x| \to \infty$ because of some standard Sobolev embeddings and the fact that we are in two dimensions.

By a slight extension of the argument, we can see that the same conclusion also holds for $|\nabla w_j|$. In fact we first recognize that the crucial terms on the right-hand sides of (4.4.4) may be rewritten as

$$\exp\left(u_j^0 + \sum_{k'=1}^{k} L_{kk'}w_{k'}\right) - 1$$

$$= (e^{u_j^0} - 1)\exp\left(\sum_{k'=1}^{k} L_{kk'}w_{k'}\right) + \left(\exp\left[\sum_{k'=1}^{k} L_{kk'}w_{k'}\right] - 1\right),$$

$$(4.5.3)$$

which lies in L^p for any $p > 2$ due to the embedding $H \to L^p$ and the definition of u_j^0. Consequently all the terms on the right-hand sides of the equations in (4.4.4) belong to L^p. Besides, we have seen that $w_j \in W^{2,2} \subset W^{1,p}$ ($p > 2$). Thus the elliptic L^p-estimates imply that $w_j \in W^{2,p}$ ($\forall p > 2$). As a consequence, we must have $|\nabla w_j|(x) \to 0$ as $|x| \to \infty$, $j = 1, 2, \cdots, n$, as expected.

4.5.2 Exponential decay estimates

Let $\mathbf{u} = (u_1, u_2, \cdots, u_n)^\tau$ be the solution of (4.2.1) found in §4.4. We have obtained the behavior $u_j \to \ln r_j$ as $|x| \to \infty$ $(j = 1, 2, \cdots, n)$ in the last subsection in terms of the configuration field $\mathbf{w}$. Our purpose now is to derive the promised exponential rate for these asymptotics.

Consider (4.2.1) outside the disk $D_R = \{x \in \mathbb{R}^2 \mid |x| < R\}$ where

$$R > \max\left\{ |p_{jk}| \,\bigg|\, j = 1, 2, \cdots, n, \ k = 1, 2, \cdots, N_j \right\}.$$

We rewrite (4.2.1) in $\mathbb{R}^2 - D_R$ as

$$\begin{aligned}
\Delta u_j &= \sum_{k=1}^{n} a_{jk}(u_k - \ln r_k) + \sum_{k=1}^{n} a_{jk}(e^{u_k} - r_k - [u_k - \ln r_k]), \\
j &= 1, 2, \cdots, n.
\end{aligned} \tag{4.5.4}$$

Let O be an $n \times n$ orthogonal matrix so that

$$O^\tau A O = \operatorname{diag}\left\{ \lambda_1, \lambda_2, \cdots, \lambda_n \right\}. \tag{4.5.5}$$

Introduce a new variable vector, $\mathbf{U} = (U_1, U_2, \cdots, U_n)^\tau$, defined by

$$\mathbf{U} = O^\tau (u_1 - \ln r_1, u_2 - \ln r_2, \cdots, u_n - \ln r_n)^\tau. \tag{4.5.6}$$

Substitute (4.5.6) into (4.5.4). By (4.5.5) and the behavior $\mathbf{U}(x) \to \mathbf{0}$ as $|x| \to \infty$, we have

$$\Delta U_j = \lambda_j U_j + \sum_{k=1}^{n} b_{jk}(x) U_k, \qquad j = 1, 2, \cdots, n, \tag{4.5.7}$$

where $b_{jk}(x)$ $(j, k = 1, 2, \cdots, n)$ depend on $\mathbf{U}(x)$ and $b_{jk}(x) \to 0$ as $|x| \to \infty$ $(j, k = 1, 2, \cdots, n)$. Set $U^2 = U_1^2 + U_2^2 + \cdots U_n^2$. Then (4.5.7) gives us

$$\Delta U^2 \geq \lambda_0 U^2 - b(x)U^2, \qquad x \in \mathbb{R}^2 - D_R,$$

where $b(x) \to 0$ as $|x| \to \infty$. Consequently, for any $\varepsilon : 0 < \varepsilon < 1$, we can find a suitably large $R_\varepsilon > R$ so that

$$\Delta U^2 \geq \left(1 - \frac{\varepsilon}{2}\right) \lambda_0 U^2, \qquad x \in \mathbb{R}^2 - D_{R_\varepsilon}. \tag{4.5.8}$$

Thus, using a comparison function argument and the property $U^2 = 0$ at infinity, we can obtain a constant $C(\varepsilon) > 0$ to make

$$U^2(x) \leq C(\varepsilon) e^{-(1-\varepsilon)\sqrt{\lambda_0}|x|}$$

valid, which leads to (4.2.4) stated in Theorem 4.2.2.

Let ∂ temporarily denote any of the two partial derivatives, ∂_1 and ∂_2. Then (4.5.4) yields

$$\Delta(\partial u_j) = \sum_{k=1}^{n} a_{jk} e^{u_k} (\partial u_k), \qquad j = 1, 2, \cdots, n. \qquad (4.5.9)$$

It will be convenient to look at the matrix form of (4.5.9). With the notation

$$D = \mathrm{diag}\Big\{ r_1, r_2, \cdots, r_n \Big\}, \quad E(x) = \mathrm{diag}\Big\{ e^{u_1(x)}, e^{u_2(x)}, \cdots, e^{u_n(x)} \Big\},$$

(4.5.9) reads, after substituting $\mathbf{v} = (\partial u_1, \partial u_2, \cdots, \partial u_n)^\tau$,

$$\Delta \mathbf{v} = AD\mathbf{v} + A(E(x) - D)\mathbf{v}. \qquad (4.5.10)$$

We will use the orthogonal matrix O defined in (4.5.5) again, but with the new variable vector $\mathbf{V}$ determined through

$$D\mathbf{v} = O\mathbf{V}. \qquad (4.5.11)$$

Then $f \equiv \mathbf{v}^\tau D\mathbf{v} = \sum r_j v_j^2$ satisfies

$$\begin{aligned}
\Delta f &\geq 2\mathbf{v}^\tau \Delta D\mathbf{v} \\
&= 2\mathbf{v}^\tau DAD\mathbf{v} + 2\mathbf{v}^\tau DA(E(x) - D)\mathbf{v} \\
&\geq \lambda_0 \mathbf{V}^\tau \mathbf{V} + 2\mathbf{v}^\tau DA(E(x) - D)\mathbf{v} \\
&\geq \lambda_0 \mathbf{v}^\tau D^2 \mathbf{v} + 2\mathbf{v}^\tau DA(E(x) - D)\mathbf{v} \\
&\geq \lambda_0 r_0 f - b(x)f, \qquad x \in \mathbb{R}^2 - D_R, \qquad (4.5.12)
\end{aligned}$$

where, recall that $r_0 = \min\{r_1, r_2, \cdots, r_n\}$ and that $b(x)$ is a function with the behavior $b(x) \to 0$ as $|x| \to \infty$. Therefore, as before, we conclude that for any $\varepsilon : 0 < \varepsilon < 1$, there is a constant $C(\varepsilon) > 0$ so that

$$f \leq C(\varepsilon) e^{(1-\varepsilon)\sqrt{\lambda_0 r_0}|x|}, \qquad |x| > R. \qquad (4.5.13)$$

Consequently the estimate (4.2.5) in Theorem 4.2.2 is also proven.

4.5.3   Uniqueness and quantized integrals

Let $\mathbf{u} = (u_1, u_2, \cdots, u_n)^\tau$ be a solution of (4.2.1) satisfying

$$\mathbf{u} \to (\ln r_1, \ln r_2, \cdots, \ln r_n)^\tau \qquad (4.5.2)$$

as $|x| \to \infty$. We have seen that the convergence is actually exponentially fast. Consider the variable vector $\mathbf{w}$ as in §4.4. The background functions defined in (4.4.1) then imply that $\mathbf{w}$ vanishes at infinity at least as fast as $|x|^{-2}$. So $w_j \in L^2$ $(j = 1, 2, \cdots, n)$. However, all the right-hand sides of the equations in (4.4.4) also belong to L^2. Thus, the L^2-estimates yield $\mathbf{w} \in H(n)$, in particular. In other words, $\mathbf{w}$ is the unique critical point of the functional (4.4.5) in the space $H(n)$.

Again, from (4.4.1) and the exponential decay property of $|\nabla u_j|$'s stated in (4.2.5) or (4.5.13), we see that $|\nabla w_j| = O(|x|^{-3})$ at infinity. Therefore, in view of the divergence theorem, we have

$$\int_{\mathbb{R}^2} \Delta v_j = \int_{\mathbb{R}^2} \Delta w_j = 0, \quad j = 1, 2, \cdots, n.$$

These results and (4.4.2) immediately imply that

$$\begin{aligned}
\int_{\mathbb{R}^2} \Delta u_j &= \int_{\mathbb{R}^2} \Delta u_j^0 \\
&= 4\pi N_j - \int_{\mathbb{R}^2} g_j = 0, \quad j = 1, 2, \cdots, n,
\end{aligned}$$

which lead by integrating (4.2.1) the promised expressions given in (4.2.6).

4.6 Nonexistence Results

We are yet to prove the statements made in Theorem 4.2.3 concerning nonexistence for the system (4.2.1) when S is $\mathbb{R}^2$. It will be seen from the study here that, indeed, the condition $\mathbf{r} > \mathbf{0}$ is important for existence.

We shall concentrate on the case $\mathbf{r} = \mathbf{0}$ first. The system (4.2.1) takes the form

$$\Delta \mathbf{u} = A\mathbf{U} + \mathbf{s}, \quad \mathbf{s} = 4\pi \left(\sum_{k=1}^{N_1} \delta_{p_{1k}}, \sum_{k=1}^{N_2} \delta_{p_{2k}}, \cdots, \sum_{k=1}^{N_n} \delta_{p_{nk}} \right)^\tau. \quad (4.6.1)$$

where, as before, $\mathbf{U} = (e^{u_1}, e^{u_2}, \cdots, e^{u_n})^\tau$. With $U = \sum e^{u_j}$ and using the arithmetic mean-geometric mean inequality, we see that the equation (4.6.1) gives us

$$\begin{aligned}
\Delta U &= \mathbf{U}^\tau \Delta \mathbf{u} + \sum_{j=1}^{n} e^{u_j} |\nabla u_j|^2 \\
&\geq \mathbf{U}^\tau \Delta \mathbf{u} = \mathbf{U}^\tau A \mathbf{U} \\
&\geq c|\mathbf{U}|^2 \geq \frac{c}{n} U^2. \quad (4.6.2)
\end{aligned}$$

We now show that there is no globally defined function $U \neq 0$ on D_R which satisfies (4.6.2). For this goal, we use (r, θ) to denote the polar coordinates and set

$$W(r) = \frac{1}{2\pi} \int_0^{2\pi} U(r, \theta) \, d\theta.$$

Using the 2π-periodicity of U with respect to the angular coordinate θ, we obtain

$$rW_r = \frac{1}{2\pi} \int_0^r \int_0^{2\pi} \left([\rho U_\rho]_\rho + \frac{1}{\rho} \frac{\partial^2 U}{\partial \theta^2} \right) d\theta d\rho$$

$$= \frac{1}{2\pi} \int_0^r \int_0^{2\pi} h(\rho, \theta) \rho d\theta d\rho, \qquad (4.6.3)$$

where we have set $h = \Delta U$ in $\mathbb{R}^2$. Since h is a positive-valued smooth function in view of (4.6.2), except at $x = p_{jk}$ $(j = 1, 2, \cdots, n, k = 1, 2, \cdots, N_j)$, the formula (4.6.3) says that $W_r > 0$. Furthermore, differentiating (4.6.3), we have

$$\frac{1}{r}(rW_r)_r = \frac{1}{2\pi} \int_0^{2\pi} h \, d\theta, \qquad r > 0. \qquad (4.6.4)$$

We next restrict our attention to the interval $r > R$ where $R > 0$ is so large that the disk of radius R and centered at the origin contains all the vortex points p_{jk}'s. Thus inserting (4.6.2) into (4.6.4) and then using the Jensen inequality, we arrive at

$$\frac{1}{r}(rW_r)_r \geq \frac{c}{2n\pi} \int_0^{2\pi} U^2 \, d\theta$$

$$\geq \frac{c}{n} \left(\frac{1}{2\pi} \int_0^{2\pi} U(r, \theta) \, d\theta \right)^2$$

$$= \frac{c}{n} W^2, \quad r > R. \qquad (4.6.5)$$

In (4.6.5), we use the new variable $t = \ln r$. Thus we have $W_{tt} \geq CW^2$, $t > t_0 = \ln R$, $C = cR^2/n$. Multiplying this differential inequality by $W_t > 0$ and integrating, we obtain

$$W_t^2(t) - W_t^2(t_0) > \frac{C}{3}(W^3(t) - W^3(t_0)), \qquad t > t_0.$$

Integrating the above inequality, we obtain

$$\infty > \int_{W(t_0)}^\infty \frac{dW}{\sqrt{W^3 - c_1}} > \int_{W(t_0)}^{W(t)} \frac{dW}{\sqrt{W^3 - c_1}} \geq c_2 \int_{t_0}^t ds, \qquad (4.6.6)$$

where $c_1, c_2 > 0$ are constants with $W^3(t_0) > c_1$. From (4.6.6) we immediately see that W must blow up in finite 'time', $t > t_0$. Hence U cannot be a global function as claimed.

Naturally one may wonder whether (4.2.1) may allow a solution when only some members among the components of $\mathbf{r}$ are zero but the rest of them remain positive. The rest of this section focuses on this problem. The statements made in Theorem 4.2.3 are some first-step results under the simplest conditions one can think of beyond the case $r_j = 0$ $(\forall j)$ just studied.

As mentioned, we next concentrate on the special case that $r_n = 0$ but $r_j > 0$ for $j = 1, \cdots, n-1$ $(n \geq 2)$ (part (i) in Theorem 4.2.3). We shall see that there may be no globally defined solution either, whatsoever, when the specific condition stated in part (i) of Theorem 4.2.3 holds. Of course, it will also be interesting to investigate the nonexistence problem under more general situations.

To proceed, we use the transformation defined as in (4.3.4), namely, $\mathbf{w} = L^{-1}(\mathbf{u} - \mathbf{u}^0)$. With the notation $L^{-1} = (L'_{jk})$ $(j, k = 1, 2, \cdots, n)$, we know that

$$w_j = \sum_{k=1}^{j} L'_{jk}(u_k - u_k^0), \quad j = 1, 2, \cdots, n, \tag{4.6.7}$$

because L^{-1} is also a lower triangular matrix. The system (4.2.1) becomes $\Delta \mathbf{w} = L^{\tau}(\mathbf{U} - \mathbf{r}) + L^{-1}\mathbf{g}$ or

$$\Delta w_j = \sum_{k=j}^{n} L_{kj}\left(\exp\left[u_k^0 + \sum_{k'=1}^{k} L_{kk'}w_{k'}\right] - r_k \right) + \sum_{k=1}^{j} L'_{jk}g_k, \quad j = 1, 2, \cdots, n$$

whose last equation is the simplest one,

$$\Delta w_n = L_{nn} \exp\left(u_n^0 + \sum_{j=1}^{n} L_{nj}w_j \right) + \sum_{k=1}^{n} L'_{nk}g_k. \tag{4.6.8}$$

We now recall the condition that all the off-diagonal entries of L are non-positive. Therefore we have

$$L'_{nk} \geq 0, \quad k = 1, \cdots, n-1. \tag{4.6.9}$$

Since $g_j > 0$ for all j and $L'_{nn} > 0$, we see that $\Delta w_n \equiv h > 0$. As before, define

$$W(r) = \frac{1}{2\pi} \int_0^{2\pi} w_n(r, \theta) \, d\theta.$$

Then, integrating (4.6.8) as in (4.6.3) and using $h > 0$, we find that $W_r > 0$ for all $r > 0$.

On the other hand, recall the boundary condition $e^{u_j} = r_j > 0$ $(j = 1, \cdots, n-1)$ at infinity. In particular the relations (4.6.7) say that the functions $w_1, \cdots, w_{n-1}$ are bounded in $\mathbb{R}^2$, which implies that the function

$$\exp\left(\sum_{j=1}^{n-1} L_{nj}w_j \right)$$

is bounded from below by a positive constant. Substituting this result into (4.6.8) and using the property that $L_{nn} > 0$, we see that there are constants $c_1, c_2 > 0$ to make

$$\Delta w_n \geq c_1 e^{c_2 w_n} \tag{4.6.10}$$

valid in $D_R = \{x \in \mathbb{R}^2 \,|\, |x| > R\}$. The right-hand side of (4.6.10) is again a convex function in the variable w_n. Hence a similar argument as before shows that W satisfies

$$W_{tt} > C_1 e^{C_2 W}, \quad t > t_0 = \ln R, \quad t = \ln r.$$

Multiplying the above inequality by $W_t > 0$ and integrating, we arrive at

$$W_t^2(t) - W_t^2(t_0) > \frac{C_1}{C_2}(e^{W(t)} - e^{W(t_0)}), \quad t > t_0,$$

which leads to the following inequality, which is similar to (4.6.6),

$$\infty > \int_{W(t_0)}^{\infty} \frac{dW}{\sqrt{e^W - c_3}} > \int_{W(t_0)}^{W(t)} \frac{dW}{\sqrt{e^W - c_3}} > c_4 \int_{t_0}^{t} ds,$$

where $c_3, c_4 > 0$ are constants with $c_3 < e^{W(t_0)}$. Hence W blows up in finite $t > t_0$, which implies that w_n cannot be globally defined and the expected nonexistence thus again follows.

We now show that part (ii) of Theorem 4.2.3 follows immediately as an example. Indeed, when $n = 2$, the system (4.2.1) becomes

$$\Delta u_1 = a_{11}(e^{u_1} - r_1) + a_{12}(e^{u_2} - r_2) + 4\pi \sum_{j=1}^{N_1} \delta_{p_j}, \quad (4.6.11)$$

$$\Delta u_2 = a_{21}(e^{u_1} - r_1) + a_{22}(e^{u_2} - r_2) + 4\pi \sum_{j=1}^{N_2} \delta_{q_j}. \quad (4.6.12)$$

Here we assume that $a_{12} = a_{21} \le 0$. Since the coefficient matrix $A = (a_{jk})$ is positive definite, we can define $a > 0, b \ge 0, c > 0$ by

$$a = \sqrt{a_{11}}, \quad b = -\frac{a_{12}}{\sqrt{a_{11}}}, \quad c = \sqrt{a_{22} - \frac{a_{12}^2}{a_{11}}}. \quad (4.6.13)$$

It is easily checked that the Cholesky decomposition (4.3.3) is determined by the matrix

$$L = \begin{pmatrix} a & 0 \\ -b & c \end{pmatrix}. \quad (4.6.14)$$

From (4.6.14), we see that the condition (4.6.9) is fulfilled. Hence we may conclude that the system (4.6.11), (4.6.12) has no solution for $r_1 > 0, r_2 = 0$. By symmetry, we see that there is no solution either if $r_1 = 0, r_2 > 0$.

One may ask whether there are solutions when $a_{12} = a_{21} > 0$. Despite some effort, an existence result has not been obtained. The statement made in part (iii) of Theorem 4.2.3 presents a partial nonexistence result. Here we provide its proof.

Recall that we have assumed in (4.6.11), (4.6.12) the condition $r_1 > 0, r_2 = 0$. Without loss of generality, we may also assume $r_1 = 1$ because otherwise we can make the shift $u_1 \mapsto u_1 + \ln r_1$, $u_2 \mapsto u_2 + \ln r_1$ to put the system into the desired simplified form. Similar to (4.6.13), we define

$$a = \sqrt{a_{11}}, \quad b = \frac{a_{12}}{\sqrt{a_{11}}}, \quad c = \sqrt{a_{22} - \frac{a_{12}^2}{a_{11}}}. \qquad (4.6.15)$$

Then the decomposition (4.3.3) gives us the matrices

$$L = \begin{pmatrix} a & 0 \\ b & c \end{pmatrix}, \quad L^{-1} = \begin{pmatrix} 1/a & 0 \\ -b/ac & 1/c \end{pmatrix}. \qquad (4.6.16)$$

It suffices to put $\mu = 1$ in (4.4.1) and simply set

$$u_1^0 = -\sum_{j=1}^{N_1} \ln(1 + |x - p_j|^{-2}), \quad u_2^0 = -\sum_{j=1}^{N_2} \ln(1 + |x - q_j|^{-2}),$$

$$g_1 = 4 \sum_{j=1}^{N_1} \frac{1}{(1 + |x - p_j|^2)^2}, \quad g_2 = 4 \sum_{j=1}^{N_2} \frac{1}{(1 + |x - q_j|^2)^2}. \qquad (4.6.17)$$

With these preparations, (4.6.11) and (4.6.12) become under the substitution $u_j = u_j^0 + v_j$ $(j = 1, 2)$ the form

$$\Delta v_1 = a_{11}(e^{u_1^0 + v_1} - 1) + a_{12}e^{u_2^0 + v_2} + g_1, \qquad (4.6.18)$$

$$\Delta v_2 = a_{21}(e^{u_1^0 + v_1} - 1) + a_{22}e^{u_2^0 + v_2} + g_2. \qquad (4.6.19)$$

As before we introduce the new variable vector $\mathbf{w} = (w_1, w_2)^\tau$ by setting $\mathbf{w} = L^{-1}\mathbf{v}$. Therefore (4.6.16) gives us the transformed system from (4.6.18), (4.6.19) as follows,

$$\Delta w_1 = a(e^{u_1^0 + aw_1} - 1) + be^{u_2^0 + bw_1 + cw_2} + \frac{1}{a}g_1, \qquad (4.6.20)$$

$$\Delta w_2 = ce^{u_2^0 + bw_1 + cw_2} - \frac{b}{ac}g_1 + \frac{1}{c}g_2. \qquad (4.6.21)$$

Remember that the numbers a, b, c are all positive. We shall only be interested in solutions satisfying the finite (potential) energy condition $\int e^{u_2} < \infty$. Therefore the convergent integral

$$\int_{\mathbb{R}^2} e^{u_2^0 + bw_1 + cw_2} \, dx \equiv 2\pi\beta \qquad (4.6.22)$$

well defines a positive number β. We concentrate on the equation (4.6.21).

Consider the Newton potential

$$v(x) = \frac{1}{2\pi} \int_{\mathbb{R}^2} (\ln|x - y| - \ln|y|)h(y) \, dy,$$

where h is the right-hand side of the equation (4.6.21). By (4.6.22) there holds

$$\int_{\mathbb{R}^2} h(x)\,dx = 2\pi\left(c\beta - 2N_1\frac{b}{ac} + 2N_2\frac{1}{c}\right).$$

Since v satisfies $\Delta v = h$ in $\mathbb{R}^2$, we see that $w_2 - v$ is an entire harmonic function. On the other hand, because the boundary condition $e^{u_1} \to 1$, $e^{u_2} \to 0$ as $|x| \to \infty$ implies that $w_2 \to -\infty$ as $|x| \to \infty$, we see that w_2 is bounded from above. Hence we may find a constant $C > 0$ so that $w_2(x) - v(x) \le C(\ln|x| + 1)$ for $|x| \ge 1$ (say). Consequently, $w_2 = v + c_0$ for some suitable constant c_0 (see Lemma 4.6.1 below). Such a result enables us to derive the relation

$$\lim_{|x|\to\infty} \frac{w_2(x)}{\ln|x|} = \lim_{|x|\to\infty} \frac{v(x)}{\ln|x|} = \frac{1}{2\pi}\int_{\mathbb{R}^2} h(x)\,dx$$
$$= c\beta - 2N_1\frac{b}{ac} + 2N_2\frac{1}{c} \equiv -\sigma.$$

This expression tells us that, in order to ensure the convergence of the integral (4.6.22), we must have the necessary condition $c\sigma \ge 2$, namely,

$$\frac{b}{a}N_1 - N_2 \ge 1 + \frac{1}{2}c^2\beta > 1,$$

which is indeed the inequality stated in part (iii) of Theorem 4.2.3.

The proof of Theorem 4.2.3 is thus complete.

Part of the preceding argument is based on the following slight extension of the Liouville theorem on entire harmonic functions on $\mathbb{R}^2$.

Lemma 4.6.1. *If U is an entire harmonic function in $\mathbb{R}^2$, $\Delta U = 0$, satisfying*

$$U(x) \le C(\ln|x| + 1), \qquad |x| \ge 1$$

for some constant $C > 0$, then U must be a constant.

Proof. Set $V(x) = \ln|x| + 1, |x| \ge 1$. Then it is easily checked that

$$\Delta\left(\frac{U}{V}\right) = -\frac{2}{V}\nabla V \cdot \nabla\left(\frac{U}{V}\right), \qquad |x| \ge 1.$$

Using the maximum principle in the above, we obtain

$$\max_{|x|\ge 1}\left|\frac{U(x)}{V(x)}\right| \le \max\left\{C, \max_{|x|=1}|U(x)|\right\} \equiv C_1.$$

In particular, there holds the bound

$$|U(x)| \le C_1(\ln|x| + 1), \qquad |x| \ge 1. \tag{4.6.23}$$

To show that U must be a constant, it suffices to prove that $\nabla U = \mathbf{0}$ everywhere in $\mathbb{R}^2$. Without loss of generality, we show that $(\nabla U)(0) = \mathbf{0}$.

Let B_R be the disk centered at the origin $x = 0$ with radius $R > 0$ and set

$$\Gamma(\rho) = \frac{1}{2\pi} \ln \rho, \qquad \rho > 0.$$

Then Green's function over B_R that vanishes on ∂B_R is defined as [123]

$$G(x,y) = \Gamma(\sqrt{|x|^2 + |y|^2 - 2x \cdot y})$$
$$-\Gamma\left(\sqrt{\left(\frac{|x||y|}{R}\right)^2 + R^2 - 2x \cdot y}\right), \quad x, y \in B_R, \quad x \neq y,$$

and

$$U(x) = \oint_{\partial B_R} U(y) \frac{\partial G}{\partial n_y}(x,y) \, ds_y, \tag{4.6.24}$$

where a direct calculation gives us the expression

$$\frac{\partial G}{\partial n_y} = \frac{y}{|y|} \cdot \nabla_y G(x,y) = \frac{1}{2\pi}\left(\frac{R^2 - |x|^2}{R(|x|^2 + R^2 - 2x \cdot y)}\right), \quad y \in \partial B_R.$$

Consequently, there holds

$$\nabla_x\left(\frac{\partial G}{\partial n_y}(x,y)\right) = -\frac{1}{\pi R}\left(\frac{x}{|x|^2 + R^2 - 2x \cdot y} + \frac{(R^2 - |x|^2)(x - y)}{(|x|^2 + R^2 - 2x \cdot y)^2}\right).$$

Setting $x = 0$, we obtain

$$\nabla_x\left(\frac{\partial G}{\partial n_y}(x,y)\right)\bigg|_{x=0} = \frac{y}{\pi R^3}, \quad y \in \partial B_R. \tag{4.6.25}$$

Differentiating (4.6.24) and using (4.6.25), we have in view of (4.6.23) the estimate

$$|(\nabla U)(0)| \leq \oint_{\partial B_R} \left|\frac{y U(y)}{\pi R^3}\right| ds_y$$

$$\leq \frac{2}{R} C_1 (\ln R + 1), \quad R > 1.$$

Letting $R \to \infty$ in the above, we see that $(\nabla U)(0) = \mathbf{0}$ as expected.

We now pursue a possible physical interpretation for the situation that some of the components of the vacuum expectation vector

$$\mathbf{r} = (r_1, r_2, \cdots, r_n)$$

may vanish. For this purpose, it is most direct, perhaps, to relate r_j's to the temperature dependence of the model within the following standard framework.

First let T denote the temperature and $T_c > 0$ some critical temperature. Then the r_j's are T-dependent numbers given by

$$r_j = R_j \left(1 - \left[\frac{T}{T_c} \right]^2 \right), \qquad j = 1, 2, \cdots, n, \qquad (4.6.26)$$

where R_j's are some positive constants. Finite-energy condition $r_j \geq 0$ ($j = 1, 2, \cdots, n$) says that the system makes sense only when the temperature is subcritical: $T \leq T_c$. Our results (Theorems 4.2.2 and 4.2.3) then imply that there are vortex solutions in the full plane if and only if T is strictly below T_c because there can be no solution at the critical temperature $T = T_c$ in view of the beginning statement in Theorem 4.2.3.

Next we assume that the critical temperature T_c is replaced by n not necessarily all distinct critical temperatures T_c^j's, ordered in such a way that

$$T_c^1 \geq T_c^2 \geq \cdots \geq T_c^n > 0.$$

Thus (4.6.26) becomes

$$r_j = R_j \left(1 - \left[\frac{T}{T_c^j} \right]^2 \right), \qquad j = 1, 2, \cdots, n.$$

When the temperature is sufficiently low to make $T < \min\{T_c^j\}$ valid, $r_j > 0$ ($j = 1, 2, \cdots, n$) again and the existence of multivortex solutions is ensured as in the previous case of a unique critical temperature. However, in the present situation, we can find a $j_0 > 1$ so that $T_c^j = T_c^n$ ($j = j_0, \cdots, n$) but $T_c^{j_0-1} > T_c^{j_0}$. Thus, when the temperature T of the system reaches the first critical temperature, $T = T_c^{j_0} = T_c^n$, we have $r_j > 0$ ($j = 1, \cdots, j_0 - 1$) but $r_j = 0$ ($j = j_0, \cdots, n$). In other words, we are in a phase where the gauge symmetry is only partially broken.

4.7 Arbitrary Coefficient Matrix Case

In this section we comment that some existence results may be established for the more extended situation that the coefficient matrix A in the system (4.2.1) is not positive definite or not even symmetric. For greater generality, we study the system

$$\Delta \mathbf{v} = A \mathbf{V} - \mathbf{g}, \qquad x \in S, \qquad (4.7.1)$$

where the Dirac measure type source terms may be viewed as absorbed into a well-defined background as in §4.2, the matrix A is simply nonsingular,

$$\mathbf{V} = (H_1 e^{v_1}, H_2 e^{v_2}, \cdots, H_n e^{v_n})^\tau, \qquad \mathbf{g} = (g_1, g_2, \cdots, g_n)^\tau$$

with H_j, g_j $(j = 1, 2, \cdots, n)$ sufficiently regular (say C^α), and S is a closed 2-surface. The scalar form $(n = 1)$ is exactly the classical 2-dimensional conformal deformation equation for the prescribed Gaussian curvature problem [17]. When A fails to be a symmetric positive definite matrix, there is a lack of physical motivation at this moment and our study of the system (4.7.1) is of only mathematical interest.

To proceed, we will look for an analogous variational principle as in §4.2. Recall that, when A is nonsingular, the more general Crout decomposition theorem ensures the existence of two $n \times n$ matrices, $L = (L_{jk})$, which is again lower triangular with $L_{jj} = 1$ $(j = 1, 2, \cdots, n)$, and, $R = (R_{jk})$, which is upper triangular and has nonvanishing diagonal entries, so that

$$A = LR, \tag{4.7.2}$$

provided a certain sufficient condition for the matrix A holds (for example, a general condition [144] is to assume that all principal minors of A are nonvanishing, which contains the condition of the Cholesky decomposition as a special case). Furthermore L and R can be explicitly constructed from $A = (a_{jk})$ according to the scheme [297]

$$
\begin{aligned}
R_{1k} &= a_{1k}, \quad k = 1, 2, \cdots, n, \\
L_{j1} &= \frac{a_{j1}}{R_{11}}, \quad j = 2, \cdots, n, \\
R_{jk} &= a_{jk} - \sum_{k'=1}^{j-1} L_{jk'} R_{k'k}, \quad k = j, j+1, \cdots, n, \quad j \geq 2, \\
L_{kj} &= \frac{a_{kj} - \sum_{k'=1}^{j-1} L_{kk'} R_{k'j}}{R_{jj}}, \quad k = j+1, \cdots, n, \quad j \geq 1,
\end{aligned}
\tag{4.7.3}
$$

which will be useful in the sufficient conditions derived later for the existence of a solution of the system (4.7.1).

Again we apply the transformation (4.3.4) to (4.7.1). Setting $\mathbf{b} = L^{-1}\mathbf{g}$, we have $\Delta \mathbf{w} = R\mathbf{V} - \mathbf{b}$. Note that $\mathbf{b} = (b_1, b_2, \cdots, b_n)$ is now a vector function. Thus the transformed system may be written

$$\Delta w_j = \sum_{k=j}^{n} R_{jk} H_k \exp\left(\sum_{k'=1}^{k} L_{kk'} w_{k'}\right) - b_j, \quad j = 1, 2, \cdots, n. \tag{4.7.4}$$

For simplicity we now impose the condition that

$$H_j \not\equiv 0, \quad H_j(x) > 0 \text{ for some } x \in S, \quad j = 1, 2, \cdots, n. \tag{4.7.5}$$

Integrating (4.7.4), we obtain the following n constraints, which resemble the Euler characteristics constraints in the conformal deformation equation case or the conditions (4.3.8) in the vortex equation case,

$$\int_S H_j \exp\left(\sum_{k=1}^{j} L_{jk} w_k\right) = B_j \text{ where } \mathbf{B} = R^{-1} \int_S \mathbf{b} = A^{-1} \int_S \mathbf{g}. \tag{4.7.6}$$

It is seen that the condition (4.7.5) and the matrix L being lower triangular ensure the nonemptyness of the admissible space

$$\mathcal{S} = \left\{ \mathbf{w} \,\middle|\, \mathbf{w} \in H(n) \text{ and satisfies } (4.7.6) \right\} \tag{4.7.7}$$

provided that $\mathbf{B} > 0$. In the section, we always assume this to hold.

Consider the optimization problem (4.3.9) with $\mathcal{A}$ being replaced by $\mathcal{S}$ defined in (4.7.7). We shall first verify the same variational principle proved in §4.2 for our general situation here. For this purpose, assume that $\mathbf{w}$ is a solution of the revised (4.3.9). Then the method of Lagrangian multipliers again implies the existence of n real numbers, $\lambda_1, \lambda_2, \cdots, \lambda_n$, so that

$$\int_S (\nabla w_j \cdot \nabla f_j - b_j f_j) = \sum_{k=j}^{n} L_{kj} \lambda_k \int_S H_k \exp\left(\sum_{k'=1}^{k} L_{kk'} w_{k'} \right) f_j,$$
$$\forall f_j \in H. \tag{4.7.8}$$

Setting $f_1 \equiv 1$ $(j = 1, 2, \cdots, n)$ in (4.7.8) and using (4.7.6), we have

$$-\underline{b}_j \equiv -\int_S b_j = \sum_{k=j}^{n} L_{kj} \lambda_k B_k.$$

It is more transparent now to write the above into its matrix form $-\underline{\mathbf{b}} = L^\tau \Lambda \mathbf{B}$, or

$$-R\mathbf{B} = L^\tau \Lambda \mathbf{B}, \qquad \Lambda = \mathrm{diag}\left\{ \lambda_1, \lambda_2, \cdots, \lambda_n \right\}. \tag{4.7.9}$$

Since both R and $L^\tau \Lambda$ are upper triangular matrices and $\mathbf{B} > 0$, so $L^\tau \Lambda = -R$. Hence, (4.7.8) become

$$\int_S (\nabla w_j \cdot \nabla f_j - b_j f_j) = -\sum_{k=j}^{n} R_{jk} \int_S H_k \exp\left(\sum_{k'=1}^{k} L_{kk'} w_{k'} \right) f_j,$$
$$\forall f_j \in H,$$

and the standard elliptic regularity theory implies that $\mathbf{w}$ is a classical $(C^{2,\alpha})$ solution of the original system (4.7.1). Consequently, the Lagrangian multipliers take the desired values and the constraints (4.7.6) are also natural as before.

We need now to examine when the revised problem (4.3.9) (with $\mathcal{A}$ being replaced by $\mathcal{S}$) allows a solution. To this end, we rewrite (4.7.6) as

$$\sum_{k=1}^{j} L_{jk} \underline{w}_k = \ln B_j - \ln \left(\int_S H_j \exp\left[\sum_{k=1}^{j} L_{jk} w'_k \right] \right) \equiv q_j,$$

where the decomposition $w_j = \underline{w}_j + w'_j$ satisfies $\underline{w}_j \in \mathbb{R}$ and $\int_S w'_j = 0$ ($j = 1, 2, \cdots, n$). Thus, with the notation $L^{-1} = (L'_{jk})$, the fact that $\underline{w} = L^{-1}\mathbf{q}$ where $\underline{w} = (\underline{w}_j)$ and $\mathbf{q} = (q_j)$, and the Poincaré inequality, we obtain

$$
\begin{aligned}
I_0(\mathbf{w}) &= \sum_{j=1}^{n} \left(\frac{1}{2} \|\nabla w'_j\|_2^2 - \int_S b_j w'_j - \underline{b}_j \underline{w}_j \right) \\
&\geq \frac{1}{2}(1-\varepsilon) \sum_{j=1}^{n} \|\nabla w'_j\|_2^2 \\
&\quad + \sum_{j=1}^{n} \underline{b}_j \sum_{j'=1}^{j} L'_{jj'} \ln \left(\int_S H_{j'} \exp \left[\sum_{k'=1}^{j'} L_{j'k'} w'_{k'} \right] \right) - C_1(\varepsilon), \\
&\equiv \frac{1}{2}(1-\varepsilon) \sum_{j=1}^{n} \|\nabla w'_j\|_2^2 + \sum_{j=1}^{n} \underline{b}_j \sum_{j'=1}^{j} L'_{jj'} \beta_{j'} - C_1(\varepsilon), \quad (4.7.10)
\end{aligned}
$$

where $\varepsilon > 0$ is an arbitrarily small number and $C_1(\varepsilon) > 0$ depends on ε. On the other hand, the Moser–Trudinger inequality (4.3.12) gives us

$$
\int_S H_j \exp \left(\sum_{k=1}^{j} L_{jk} w'_k \right) \leq C_2(\varepsilon) \exp \left(\left[\frac{1}{16\pi} + \varepsilon \right] |L|^2 \|\nabla \mathbf{w}\|_2^2 \right). \quad (4.7.11)
$$

where $|L|$ is the norm of the mapping $L : \mathbb{R}^n \to \mathbb{R}^n$. In the following, we also use $|\cdot|$ to denote the norm of a vector in $\mathbb{R}^n$ induced from the standard inner product. The second term on the right-hand side of (4.7.10) satisfies

$$
\left| \sum_{j=1}^{n} \underline{b}_j \sum_{j'=1}^{j} L'_{jj'} \beta_{j'} \right| = |(\underline{\mathbf{b}}, L^{-1}\beta)_2| \leq |\underline{\mathbf{b}}| \, |L^{-1}| \, |\beta|
$$

$$
\leq |\underline{\mathbf{b}}| \, |L^{-1}| \left(C_3(\varepsilon) + n \left[\frac{1}{16\pi} + \varepsilon \right] |L|^2 \|\nabla \mathbf{w}'\|_2^2 \right). \quad (4.7.12)
$$

Inserting (4.7.11) and (4.7.12) into (4.7.10) and assuming

$$
n|\underline{\mathbf{b}}| \, |L^{-1}| \, |L|^2 < 8\pi, \quad (4.7.13)
$$

then the number $\varepsilon > 0$ may be chosen to be sufficiently small to obtain

$$
I_0(\mathbf{w}) \geq \varepsilon \|\nabla \mathbf{w}'\|_2^2 - C(\varepsilon). \quad (4.7.14)
$$

Therefore I_0 is bounded from below on S. Besides, (4.7.14) also says that if $\{\mathbf{w}\}$ is a minimizing sequence for the problem (4.3.9) (after revision $A \to S$), then $\{\mathbf{w}'\}$ is bounded in $H(n)$. The Trudinger–Moser inequality thus implies that the corresponding sequence $\{\mathbf{q}\}$ is also bounded, which leads us to the boundedness of $\{\underline{\mathbf{w}}\}$ via the relation $\underline{\mathbf{w}} = L^{-1}\mathbf{q}$. Using these properties and a weak compactness argument, we easily see that the existence of a minimizer follows.

In summary, we state the existence result obtained for the system (4.7.1) as follows.

Theorem 4.7.1. *Consider the $n \times n$ system (4.7.1) and let the matrix L be defined by (4.7.2) and (4.7.3). Set $\mathbf{b} = L^{-1}\mathbf{g}$. Suppose that the condition (4.7.5) holds and*

$$A^{-1} \int_S \mathbf{g} > 0.$$

Then, under the assumption (4.7.13), the system (4.7.1) has a classical solution.

4.8   Remarks

In this chapter, an existence theory has been developed for the general system of nonlinear elliptic equations (4.2.1) via a variational approach. For the case when the system is over the full plane $\mathbb{R}^2$ and some of the numbers r_j's vanish, there are still many interesting issues to be resolved.

Open Problem 4.8.1. *In addition to the nonexistence results stated in Theorem 4.2.3, it would be interesting to know whether the system allows any solution to exist when some of the r_j's are zero.*

It will also be interesting to establish some more general nonexistence results. As mentioned earlier, so far, this is an open problem even in the case of a 2×2 system. In particular, it would be interesting to know whether the inequality obstruction stated in (iii) of Theorem 4.2.3 is sharp.

The existence problem may also be viewed from the point of gauge symmetry breaking. The case where $\mathbf{r} = \mathbf{0}$ corresponds to the completely restored vacuum symmetry and allows no solution. The question as to whether there exists any solution when only some components of $\mathbf{r}$ vanish but the rest of its components remain positive is based on the idea that a partially broken symmetry may already be enough to onset mixed states.

Open Problem 4.8.2. *It is not clear what is the most general condition under which (4.7.1) defined over a closed surface has a solution.*

This is known to be a difficult question even in the scalar equation case arising from differential geometry [17, 35, 21, 73, 76, 140, 165, 191, 220].

Open Problem 4.8.3. *Develop an existence and nonexistence theory for the system (4.7.1) defined over the full plane when the coefficient matrix A is general.*

Some specific cases are: the matrix A is negative definite, A is not symmetric, A is singular, or A is nonsingular but A has no Crout decomposition. Again, in the scalar case, the equation arises from the prescribed Gaussian curvature problem on $\mathbb{R}^2$ in differential geometry [20, 79, 80, 166, 213, 214, 225, 271].

5

Chern–Simons Systems: Abelian Case

In this chapter we present a study of an important planar Abelian gauge field theory arising in condensed matter physics in which electromagnetism is governed by a Chern–Simons dynamics. In §5.1 we consider the gauged Schrödinger equation, which is nonrelativistic, and we present its explicit solution. In §5.2 we introduce the relativistic Chern–Simons model and state our main results concerning its topological solutions. In §5.3 we make a systematic analysis of the problem and prove the existence theorem stated in §5.2. In §5.4 we obtain all possible symmetric non-topological solutions. In §5.5 and §5.6 we construct spatially periodic solutions modelling a condensed lattice structure.

5.1 Schrödinger Equation

In this section, we present a brief discussion of the Schrödinger equation coupled with a Chern–Simons gauge field, which is an integrable nonrelativistic soliton model. First we introduce the field equation. Then we write down its multisoliton solutions explicitly by virtue of the Liouville equation.

5.1.1 Schrödinger fields and Chern–Simons dynamics

Various Chern–Simons models arise in the study of anyon physics [112, 113, 114]. Consider first the widely used nonlinear Schrödinger equation

$$\mathrm{i}\frac{\partial\psi}{\partial t} = -\frac{1}{2m}\Delta\psi - g|\psi|^2\psi \tag{5.1.1}$$

in $(2+1)$ dimensions, where the wave function ψ is a complex field and $m, g > 0$ are physical parameters. The work of Zakharov and Shabat [356] shows that (5.1.1) may be integrated explicitly by the method of inverse scattering as that for the Korteveg–de Vries equation [117, 181]. We maintain the standard statistical interpretation of the wave function that $\rho = |\psi(t,x)|^2$ is the probability density for finding the described particle at time t and spot $x \in \mathbb{R}^2$. Using (5.1.1), we have

$$
\begin{aligned}
\frac{\partial\rho}{\partial t} &= \frac{\partial}{\partial t}|\psi(t,x)|^2 = \psi\frac{\partial}{\partial t}\overline{\psi} + \overline{\psi}\frac{\partial}{\partial t}\psi \\
&= -\partial_k\left\{\frac{\mathrm{i}}{2m}(\psi\partial_k\overline{\psi} - \overline{\psi}\partial_k\psi)\right\}.
\end{aligned}
\tag{5.1.2}
$$

Thus, if we use the Lorentz covariant notation, $J = (J^\mu) = (\rho, J^k) = (\rho, \mathbf{J})$, where

$$
\begin{aligned}
\mathbf{J} &= (J^k)_{k=1,2}, \\
J^k &= -\frac{\mathrm{i}}{2m}(\psi\partial^k\overline{\psi} - \overline{\psi}\partial^k\psi) \\
&= \frac{\mathrm{i}}{2m}(\psi\partial_k\overline{\psi} - \overline{\psi}\partial_k\psi), \quad k = 1, 2,
\end{aligned}
\tag{5.1.3}
$$

is often called the probability current density, the relation (5.1.2) is simply a conservation law,

$$\partial_\mu J^\mu = 0 \quad\text{or}\quad \frac{\partial\rho}{\partial t} + \nabla\cdot\mathbf{J} = 0, \tag{5.1.4}$$

which can also be derived in view of Noether's theorem.

It is clear that the Schrödinger equation (5.1.1) is the Euler–Lagrange equation of the action density

$$\mathcal{L} = \mathrm{i}\overline{\psi}\partial_0\psi - \frac{1}{2m}|\partial_j\psi|^2 + \frac{g}{2}|\psi|^4. \tag{5.1.5}$$

In order to incorporate electromagnetism, we need to introduce a real-valued gauge vector field, $A = (A_\mu)$ ($A_\mu \in \mathbb{R}, \mu = 0, 1, 2$). We follow a standard procedure to replace the derivatives ∂_μ in the above by the gauge-covariant derivatives

$$D_\mu = \partial_\mu - \mathrm{i}A_\mu, \quad \mu = 0, 1, 2.$$

Hence we arrive at the gauged Schrödinger equation

$$i D_0 \psi = -\frac{1}{2m} D_j^2 \psi - g|\psi|^2 \psi \qquad (5.1.6)$$

and the associated current density

$$J = (\rho, J^k), \quad \rho = |\psi|^2, \quad J^k = \frac{i}{2m}(\psi \overline{D_k \psi} - \overline{\psi} D_k \psi), \quad k = 1, 2. \quad (5.1.7)$$

Note that, for any complex-valued functions ψ_1 and ψ_2, we have the identity

$$\partial_\mu(\psi_1 \overline{\psi_2}) = \psi_1 \overline{D_\mu \psi_2} + (D_\mu \psi_1)\overline{\psi_2}, \quad \mu = 0, 1, 2. \qquad (5.1.8)$$

Thus, in view of (5.1.6) and (5.1.8), we see that the current (5.1.7) obeys the *same* conservation law (5.1.4).

The equation (5.1.6) governs the wave function ψ. If A_μ is interpreted as a gauge potential for electromagnetism, then it is naturally governed by the Maxwell equation subject to the current (5.1.7), namely,

$$\partial_\nu F^{\mu\nu} = -J^\mu,$$

which roughly says that the rate of change of electromagnetic field,

$$F_{\mu\nu} = \partial_\mu A_\nu - \partial_\nu A_\mu, \quad \mu, \nu = 0, 1, 2,$$

is proportional to the matter current. In the Chern–Simons theory, however, one replaces such a classical relation by the concept that the electromagnetic field is directly proportional to the matter current. Hence, because $F_{\mu\nu}$ is skew-symmetric, we are led to the simple relation

$$F_{\mu\nu} = \frac{1}{\kappa}\epsilon_{\mu\nu\alpha}J^\alpha, \quad \mu, \nu, \alpha = 0, 1, 2, \qquad (5.1.9)$$

where $\kappa > 0$ is proportionality constant or a coupling parameter and the field strength tensor $F_{\mu\nu}$ gives rise to the induced electric and magnetic fields, $\mathbf{E} = (E_1, E_2, 0)$ (horizontal) and $\mathbf{B} = (0, 0, B)$ (vertical), by the standard prescription

$$E_j = F_{j0}, \quad j = 1, 2, \quad B = F_{12}.$$

Therefore, (5.1.9) becomes

$$B = \frac{1}{\kappa}\rho, \qquad (5.1.10)$$

$$E_j = -\frac{1}{\kappa}\epsilon_{jk}J^k, \quad j, k = 1, 2. \qquad (5.1.11)$$

The equation (5.1.9) or the system (5.1.10), (5.1.11) is the simplest Chern–Simons equation.

In $(J^\mu) = (\rho, \mathbf{J})$, if ρ is viewed as the density of electric charge as in the Maxwell electromagnetism, the equation (5.1.10) suggests that the total magnetic charge (flux) Φ and electric charge Q are related by

$$\Phi = \int_{\mathbb{R}^2} B \, dx = \frac{1}{\kappa} \int_{\mathbb{R}^2} \rho \, dx = \frac{Q}{\kappa}, \tag{5.1.12}$$

which says that electricity and magnetism must exist simultaneously and is one of the most important features of the Chern–Simons type field-theoretical models.

The coupled system of equations, (5.1.6) and (5.1.9), are the gauged Schrödinger–Chern–Simons equations. It is clear that this system is the Euler–Lagrange equations of the action density

$$\mathcal{L} = -\frac{\kappa}{2} \epsilon^{\mu\nu\alpha} A_\mu \partial_\nu A_\alpha + i\overline{\psi} D_0 \psi - \frac{1}{2m} |D_j \psi|^2 + \frac{g}{2} |\psi|^4. \tag{5.1.13}$$

Note that $\epsilon^{\mu\nu\alpha} A_\mu \partial_\nu A_\alpha = \frac{1}{2} \epsilon^{\mu\nu\alpha} A_\mu F_{\nu\alpha}$ is the Chern–Simons form whose topological meaning will not be discussed here.

5.1.2 Explicit static solution

We now describe the explicit solution of the nonrelativistic Chern–Simons theory (5.1.13) due to Jackiw and Pi [152]. We shall look for static solutions which are independent of the time variable $x_0 = t$. Hence the equations (5.1.6), (5.1.10), and (5.1.11) become

$$A_0 \psi = -\frac{1}{2m} D_j^2 \psi - g|\psi|^2 \psi, \tag{5.1.14}$$

$$F_{12} = \frac{1}{\kappa} |\psi|^2, \tag{5.1.15}$$

$$\partial_j A_0 = -\frac{i}{2m\kappa} \epsilon_{jk} (\psi \overline{D_k \psi} - \overline{\psi} D_k \psi). \tag{5.1.16}$$

We introduce the operators $D_\pm$ as follows,

$$D_\pm = D_1 \pm i D_2.$$

Then we have the identities

$$D_+ D_- \psi = D_j^2 \psi - i[D_1, D_2]\psi = D_j^2 \psi - F_{12}\psi, \tag{5.1.17}$$
$$D_- D_+ \psi = D_j^2 \psi + i[D_1, D_2]\psi = D_j^2 \psi + F_{12}\psi. \tag{5.1.18}$$

We first note in view of (5.1.8) that the equation (5.1.16) may take the form

$$\partial_j A_0 = \mp \frac{1}{2m\kappa} \partial_j |\psi|^2 \pm \frac{1}{2m\kappa} (\psi \overline{D_j \psi} + \overline{\psi} D_j \psi)$$

$$-\frac{i}{2m\kappa}\epsilon_{jk}(\psi\overline{D_k\psi} - \overline{\psi}D_k\psi)$$

$$= \mp\frac{1}{2m\kappa}\partial_j|\psi|^2 \pm \frac{\psi}{2m\kappa}\overline{(D_j\psi \pm i\epsilon_{jk}D_k\psi)}$$

$$\pm\frac{\overline{\psi}}{2m\kappa}(D_j\psi \pm i\epsilon_{jk}D_k\psi). \tag{5.1.19}$$

Next, using (5.1.17) and (5.1.18) in (5.1.14), we have

$$A_0\psi = -\frac{1}{2m}D_\mp D_\pm\psi \pm \frac{1}{2m}F_{12}\psi - g|\psi|^2\psi. \tag{5.1.20}$$

Thus, from (5.1.15), (5.1.19), and (5.1.20), we see that the system (5.1.14)–(5.1.16) may be reduced into

$$D_\pm\psi = 0, \tag{5.1.21}$$

$$F_{12} = \frac{1}{\kappa}|\psi|^2, \tag{5.1.22}$$

$$A_0 = \mp\frac{1}{2m\kappa}|\psi|^2, \tag{5.1.23}$$

if the coupling parameter g for the matter self-potential strength satisfies the critical condition

$$g = \pm\frac{1}{m\kappa}. \tag{5.1.24}$$

Hence, under the condition (5.1.24), the reduced system (5.1.21)–(5.1.23) may be viewed as a first integral of (5.1.14)–(5.1.16).

Use the notation

$$\partial_\pm = \partial_1 \pm i\partial_2, \quad A_\pm = A_1 \pm iA_2.$$

Then (5.1.21) becomes $A_\pm\psi = -i\partial_\pm\psi$. Away from the zeros of ψ, we have $A_\pm = -i\partial_\pm \ln\psi$. On the other hand, recall that F_{12} can be represented as

$$F_{12} = \mp\frac{1}{2}i(\partial_\mp A_\pm - \partial_\pm\overline{A}_\pm) = \mp\frac{1}{2}(\partial_\mp\partial_\pm \ln\psi + \partial_\pm\partial_\mp \ln\overline{\psi})$$

$$= \mp\frac{1}{2}\partial_-\partial_+ \ln|\psi|^2 = \mp2\partial\overline{\partial}\ln|\psi|^2 = \mp\frac{1}{2}\Delta\ln|\psi|^2.$$

It can be shown as before that (5.1.21) implies that the zeros of ψ are discrete and finite. Let the zeros of ψ be $p_1, p_2, \cdots, p_N \in \mathbb{R}^2$. Then in view of the above and (5.1.22), the substitution $u = \ln|\psi|^2$ transforms (5.1.21) and (5.1.22) into the equivalent Liouville equation

$$\Delta u = \pm\frac{2}{\kappa}e^u + 4\pi\sum_{j=1}^N \delta_{p_j}, \tag{5.1.25}$$

which has an entire solution over the full plane $\mathbb{R}^2$ only if

$$\kappa = \mp|\kappa|. \tag{5.1.26}$$

Hence (5.1.25) becomes

$$\Delta u = -\frac{2}{|\kappa|}e^u + 4\pi \sum_{j=1}^{N} \delta_{p_j}, \tag{5.1.27}$$

Using the complex variable $z = x_1 + ix_2$ and the formula (3.2.12), we can write u as

$$u(z) = \ln\left(\frac{4|\kappa||F'(z)|^2}{(1+|F(z)|^2)^2}\right), \tag{5.1.28}$$

where $F(z)$ is any holomorphic function of z so that $p_1, p_2, \cdots, p_N$ are the zeros of $F'(z)$. In particular, we may choose $F(z)$ to be a polynomial in z of degree $N+1$ and $p_1, p_2, \cdots, p_N$ are as prescribed. The solution pair (ϕ, A) of the self-dual equations are given by the scheme

$$\psi(z) \;=\; e^{\frac{1}{2}u(z)+i\theta(z)}, \quad \theta(z) = \sum_{j=1}^{N} \arg(z - p_j),$$

$$A_1(z) \;=\; -\mathrm{Re}\{i\partial_\pm \ln\psi(z)\}, \quad A_2(z) = -\mathrm{Im}\{i\partial_\pm \ln\psi(z)\}. \tag{5.1.29}$$

Consequently, we obtain the asymptotic decay rate

$$|\psi(x)|^2 = e^{u(x)} = O(|x|^{-2N-4}), \quad x = (x_1, x_2) \in \mathbb{R}^2, \quad |x| \to \infty. \tag{5.1.30}$$

We now calculate the magnetic flux Φ and electric charge Q.

From the expression (5.1.28) where $F(z)$ is a polynomial, we see that the decay estimate (5.1.30) may be achieved uniformly with respect to the locations of the zeros of $F'(z)$ distributed in a bounded region. Hence the integral

$$I = \int_{\mathbb{R}^2} e^u \, dx$$

depends continuously on $p_1, p_2, \cdots, p_N$. We consider the special case $F(z) = c_0 z^{N+1}$ ($c_0 \neq 0$). We have, by virtue of (5.1.27) with all p_j's coincide at the origin,

$$\int_{|x|<r} \frac{2}{|\kappa|}e^u \, dx = 4\pi N - \oint_{|x|=r} \frac{\partial u}{\partial n} \, ds$$

$$= 4\pi N - 4\pi\left(N - 2(N+1)\frac{|c_0|^2 r^{2N+2}}{1 + |c_0|^2 r^{2N+2}}\right). \tag{5.1.31}$$

Letting $r \to \infty$ in (5.1.31), we obtain

$$\int_{\mathbb{R}^2} e^u \, dx = 4\pi|\kappa|(N+1). \tag{5.1.32}$$

Since (5.1.32) is quantized, it is true for the general situation that $F'(z)$ is a polynomial with N arbitrarily distributed zeros. Such a calculation indicates that there is a contribution from infinity as well.

In view of (5.1.12), (5.1.22), and (5.1.32), we find the flux Φ and charge Q as follows,

$$\Phi = 4\pi \operatorname{sgn}(\kappa)\,(N+1), \quad Q = 4\pi|\kappa|(N+1). \tag{5.1.33}$$

We finally look for the energy of the obtained N-soliton solutions. We see that, neglecting unharmful exact divergence terms which do not affect the total energy after integration, the Hamiltonian density of (5.1.13) reads

$$
\begin{aligned}
\mathcal{H} &= \frac{\partial \mathcal{L}}{\partial(\partial_0 A_\mu)}\partial_0 A_\mu + \frac{\partial \mathcal{L}}{\partial(\partial_0\psi)}\partial_0\psi + \frac{\partial \mathcal{L}}{\partial(\partial_0\overline{\psi})}\partial_0\overline{\psi} - \mathcal{L} \\
&= \kappa A_0 F_{12} - A_0|\psi|^2 + \frac{1}{2m}|D_j\psi|^2 - \frac{g}{2}|\psi|^4 \\
&= \frac{1}{2m}|D_j\psi|^2 - \frac{g}{2}|\psi|^4,
\end{aligned}
$$

where we have applied one of the governing equations, (5.1.15), and the fact that the fields are static. Hence the total energy is given by

$$E = \int_{\mathbb{R}^2}\mathcal{H}\,\mathrm{d}x = \int_{\mathbb{R}^2}\left\{\frac{1}{2m}|D_j\psi|^2 - \frac{g}{2}|\psi|^4\right\}\mathrm{d}x, \tag{5.1.34}$$

which is *indefinite*. We note that, in view of (5.1.21) and (5.1.22), there holds the decomposition

$$
\begin{aligned}
|D_j\psi|^2 &= |D_1\psi \pm iD_2\psi|^2 \pm i(D_1\psi\overline{D_2\psi} - D_2\psi\overline{D_1\psi}) \\
&= \pm i(\partial_1[\psi\overline{D_2\psi}] - \partial_2[\psi\overline{D_1\psi}]) \mp i\psi(\overline{[D_1,D_2]\psi}) \\
&= \pm i(\partial_1[\psi\overline{D_2\psi}] - \partial_2[\psi\overline{D_1\psi}]) \pm F_{12}|\psi|^2 \\
&= \pm i(\partial_1[\psi\overline{D_2\psi}] - \partial_2[\psi\overline{D_1\psi}]) + \frac{1}{\kappa}|\psi|^4.
\end{aligned}
\tag{5.1.35}
$$

In view of (5.1.29), we have the relation

$$|D_1\psi|^2 + |D_2\psi|^2 = \frac{1}{2}e^u|\nabla u|^2.$$

Consequently, using (5.1.28) and (5.1.30), we have the estimate

$$\psi\overline{D_j\psi} = O(|x|^{-2N-5}) \quad \text{as } |x| \to \infty, \quad j = 1, 2. \tag{5.1.36}$$

Inserting (5.1.35) into (5.1.34), using (5.1.24) and (5.1.36), we obtain immediately the zero energy, $E = 0$.

In summary, we have seen that, when the coupling constants satisfy the critical condition stated in (5.1.24), the N-soliton solutions of the nonlinear Schrödinger–Chern–Simons equations can be constructed explicitly to yield quantized electric charge, magnetic flux, precise decay rate at infinity, and zero energy.

5.2 Relativistic Chern–Simons Model on Plane

In this section we start our study on the relativistic Abelian Chern–Simons theory developed independently in the seminal work of Hong, Kim, and Pac [143] and Jackiw and Weinberg [156]. We first introduce the field equations and state our main existence theorem for static topological solutions. We then formulate the problem in the form of a self-dual system.

5.2.1 Field equations and existence results

The Minkowski spacetime metric tensor $g_{\mu\nu}$ is diag$(1, -1, -1)$. In normalized units and assuming the critical coupling, the Lagrangian density of the relativistic Abelian Chern–Simons theory of Hong, Kim, and Pac [143] and Jackiw and Weinberg [156] is written

$$\mathcal{L} = -\frac{1}{4}\kappa\epsilon^{\mu\nu\alpha}A_\mu F_{\nu\alpha} + D_\mu\phi\overline{D^\mu\phi} - \frac{1}{\kappa^2}|\phi|^2(1-|\phi|^2)^2, \qquad (5.2.1)$$

where $\kappa \in \mathbb{R}$ is nonzero and ϕ is a complex scalar field which can be viewed as a Higgs field. For this reason, (5.2.1) is also called a Chern–Simons–Higgs model. The Euler–Lagrange equations of (5.2.1) are

$$\frac{1}{2}\kappa\epsilon^{\mu\nu\alpha}F_{\nu\alpha} \;=\; j^\mu = -\mathrm{i}(\phi\overline{D^\mu\phi} - \overline{\phi}D^\mu\phi), \qquad (5.2.2)$$

$$D_\mu D^\mu\phi \;=\; -\frac{1}{\kappa^2}(2|\phi|^2[|\phi|^2 - 1] + [|\phi|^2 - 1]^2)\phi. \qquad (5.2.3)$$

Again, it is easy to see that

$$(\rho, \mathbf{j}) = (j^\mu), \quad j^\mu = \mathrm{i}(\phi\overline{D^\mu\phi} - \overline{\phi}D^\mu\phi), \quad \mu = 0, 1, 2 \qquad (5.2.4)$$

is a conserved matter current density.

We are only interested in static solutions of the Chern–Simons–Higgs equations (5.2.2) and (5.2.3). The $\mu = 0$ component of (5.2.2) is a Chern–Simons Gauss law and relates the induced magnetic field to the electric charge density,

$$\kappa F_{12} = j^0 = \rho = 2A_0|\phi|^2, \qquad (5.2.5)$$

which implies the following flux-charge relation as before,

$$\kappa\Phi = \kappa\int_{\mathbb{R}^2} F_{12}\,\mathrm{d}x = \int_{\mathbb{R}^2} \rho\,\mathrm{d}x = Q. \qquad (5.2.6)$$

We next consider the energy of a static solution. It is straightforward to see as in the last section that the Hamiltonian density of (5.2.1) has the form

$$\mathcal{H} \;=\; \frac{\partial\mathcal{L}}{\partial(\partial_0 A_\mu)}\partial_0 A_\mu + \frac{\partial\mathcal{L}}{\partial(\partial_0\phi)}\partial_0\phi + \frac{\partial\mathcal{L}}{\partial(\partial_0\overline{\phi})}\partial_0\overline{\phi} - \mathcal{L}$$

$$= \kappa A_0 F_{12} - A_0^2|\phi|^2 + |D_j\phi|^2 + \frac{1}{\kappa^2}|\phi|^2(1 - |\phi|^2)^2$$

$$= A_0^2|\phi|^2 + |D_j\phi|^2 + \frac{1}{\kappa^2}|\phi|^2(1 - |\phi|^2)^2$$

$$= \frac{\kappa^2 F_{12}^2}{4|\phi|^2} + |D_j\phi|^2 + \frac{1}{\kappa^2}|\phi|^2(1 - |\phi|^2)^2, \tag{5.2.7}$$

where we have repeatedly used the Gauss law (5.2.5).

If (ϕ, A) is a solution so that the energy

$$E = \int_{\mathbb{R}^2} \mathcal{H}\, dx$$

is finite where $\mathcal{H}$ is as defined in (5.2.7), then it is necessary that one of the following boundary conditions

$$\phi \to 0 \quad \text{as } |x| \to \infty, \tag{5.2.8}$$

$$|\phi|^2 \to 1 \quad \text{as } |x| \to \infty, \tag{5.2.9}$$

holds. The boundary condition (5.2.9) is called topological. In this situation, ϕ maps a circle at infinity of $\mathbb{R}^2$ into S^1 and is identified with a homotopy class in the fundamental group

$$\pi_1(S^1) = \mathbb{Z}.$$

This homotopy class corresponds to an integer, N, which measures the number of times S^1 is covered by a circle near infinity of $\mathbb{R}^2$ under the map ϕ and is called the winding number or topological charge of the solution.

We will first consider topological solutions. Non-topological solutions, which satisfy the boundary condition (5.2.8), are more subtle and will be considered later.

Theorem 5.2.1. *Suppose that (ϕ, A) is a static finite-energy topological solution of the Chern–Simons equations (5.2.2) and (5.2.3) of winding number N. Then there holds the energy lower bound*

$$E = \int_{\mathbb{R}^2} \mathcal{H}\, dx \geq 2\pi|N|. \tag{5.2.10}$$

The energy lower bound (5.2.10) can always be saturated by a certain solution and this solution satisfies the following decay estimates at infinity,

$$|D_j\phi|^2 = O(e^{-m_1(1-\varepsilon)|x|}), \quad 1 - |\phi|^2, \; F_{12} = O(e^{-m_2(1-\varepsilon)|x|}), \tag{5.2.11}$$

where $m_1 = 2\sqrt{2}/|\kappa|$, $m_2 = 2/|\kappa|$, and $\varepsilon \in (0,1)$ is arbitrary. Moreover, in addition to the energy $E = \int \mathcal{H}$, the magnetic flux Φ and the electric charge Q defined in (5.2.6) are all quantized,

$$E = 2\pi|N|, \quad \Phi = 2\pi N, \quad Q = 2\pi N\kappa. \tag{5.2.12}$$

The integer $|N|$ is actually the algebraic number of zeros of the Higgs field ϕ in $\mathbb{R}^2$. Conversely, let $p_1, ..., p_m \in \mathbb{R}^2$ and $n_1, ..., n_m \in \mathbb{Z}_+$ (the set of positive integers). The equations (5.2.2) and (5.2.3) have a topological solution (ϕ, A) so that the zeros of ϕ are exactly $p_1, ..., p_m$ with the corresponding multiplicities $n_1, ..., n_m$ and the conditions in (5.2.12) are fulfilled with $N = \sum n_\ell$. The solution is maximal in the sense that the Higgs field ϕ has the largest possible magnitude among all the solutions realizing the same zero distribution and local vortex charges in the plane. Furthermore, the maximal solution may be approximated by a monotone iterative scheme defined over bounded domains in such a way that the truncation errors away from local regions are exponentially small as a function of the distance. Finally, if (ϕ, A) is a solution of winding number N, the pair $(\overline{\phi}, -A)$ is a solution of winding number $-N$.

This theorem was first obtained in [294]. An earlier weaker version was obtained in [324] through a variational approach as that for the Abelian Higgs theory [157].

5.2.2   Topological lower energy bound

We will first make some completion-of-square argument in the energy $\int \mathcal{H}$. We have

$$\frac{\kappa^2}{4} \frac{F_{12}^2}{|\phi|^2} + \frac{|\phi|^2}{\kappa^2}(1 - |\phi|^2)^2$$
$$= \left(\frac{\kappa}{2} \frac{F_{12}}{|\phi|} \mp \frac{|\phi|}{\kappa}(1 - |\phi|^2) \right)^2 \pm F_{12}(1 - |\phi|^2). \qquad (5.2.13)$$

Moreover, as in (5.1.35), we have

$$|D_j\phi|^2 = |D_1\phi \pm iD_2\phi|^2 \pm i(\partial_1[\phi\overline{D_2\phi}] - \partial_2[\phi\overline{D_1\phi}]) \pm F_{12}|\phi|^2. \qquad (5.2.14)$$

It can be shown for a finite-energy static solution of (5.2.2) and (5.2.3) that $|D_j\phi|^2$ decays exponentially fast at infinity. Here we skip these details.

If (ϕ, A) is a finite-energy solution of winding number N, then for $r > 0$ sufficiently large so that $|\phi| > 1/2$ whenever $|x| > r$, we have

$$2\pi N = \int_{|x|=r} d\arg\phi = -i \int_{|x|=r} d\ln\phi. \qquad (5.2.15)$$

Consequently, (5.2.15), the divergence theorem, and the exponential decay property of $|D_j\phi|^2$ give us the estimate

$$\left| \int_{|x|<r} F_{12} \, dx - 2\pi N \right| = \left| \int_{|x|=r} A_j \, dx_j + i \int_{|x|=r} d\ln\phi \right|$$
$$= \left| \int_{|x|=r} \phi^{-1}D_j\phi \, dx_j \right| \le Cre^{-\alpha r} \qquad (5.2.16)$$

for some small $\alpha > 0$. Letting $r \to \infty$ in (5.2.16), we have the flux formula

$$\int_{\mathbb{R}^2} F_{12} \, dx = 2\pi N. \tag{5.2.17}$$

Now, inserting (5.2.13) and (5.2.14) into (5.2.7), integrating over $\mathbb{R}^2$, and applying (5.2.17), we obtain

$$\int_{\mathbb{R}^2} \mathcal{H} \, dx = \pm 2\pi N$$

$$+ \int_{\mathbb{R}^2} \left\{ \left(\frac{\kappa}{2} \frac{F_{12}}{|\phi|} \mp \frac{|\phi|}{\kappa} (1 - |\phi|^2) \right)^2 + |D_1\phi \pm iD_2\phi|^2 \right\} dx. \tag{5.2.18}$$

Therefore we conclude that there holds the energy lower bound $E \geq 2\pi |N|$ and such a lower bound is attained if and only if (ϕ, A) is a solution of the system

$$D_1\phi \pm iD_2\phi = 0, \tag{5.2.19}$$

$$F_{12} = \pm \frac{2}{\kappa^2} |\phi|^2 (1 - |\phi|^2) \tag{5.2.20}$$

according to $N = \pm |N|$. In (5.2.19) and (5.2.20), it is straightforward to check that solutions of the system of the signs $\pm$ are indeed related through the correspondence $(\phi, A) \to (\overline{\phi}, -A)$. Besides, supplemented with (5.2.5), any solution of (5.2.19) and (5.2.20) is also a solution of (5.2.2) and (5.2.3).

The structure of (5.2.19) and the property (5.2.9) indicate that ϕ can only have finitely many zeros, say $p_1, p_2, \cdots, p_m$, each with an integer multiplicity, say, $n_1, n_2, \cdots, n_m$, and $\sum n_\ell = N$ gives us the winding number of the solution.

Conversely, with these data as prescribed, we pursue a solution that realizes the properties just stated, hence, proving Theorem 5.2.1. The zeros are often called vortices due to the mechanical and magnetic features they possess, their relevance to superconductivity [125, 203, 330], and their natural physical interpretation as defining mixed states [1, 227]. Local multiplicities $n_1, n_2, \cdots, n_m$ are identified as local vortex charges or local winding numbers of the vortices at $p_1, p_2, \cdots, p_m$.

5.3   Construction of Solution

In this section we present an iterative method for the proof of existence of the solution described in Theorem 5.2.1. A by-product of the method is that it provides us an effective numerical construction of the solution.

5.3.1 Iterative method and control of sequence

We aim at obtaining an N-vortex solution with vortices at $p_1, \cdots, p_m \in \mathbb{R}^2$ and local winding numbers $n_1, \cdots, n_m \in \mathbb{Z}_+$ so that $\sum n_\ell = N$. With the substitution $u = \ln |\phi|^2$, we therefore arrive at the governing equation

$$\Delta u = \frac{4}{\kappa^2} e^u (e^u - 1) + 4\pi \sum_{\ell=1}^{m} n_\ell \delta(x - p_\ell), \quad x \in \mathbb{R}^2. \tag{5.3.1}$$

We define a background function u_0,

$$u_0(z) = -\sum_{\ell=1}^{m} n_\ell \ln(1 + |x - p_\ell|^{-2}).$$

Then the substitution $v = u - u_0$ changes (5.3.1) into the form

$$\begin{aligned}
\Delta v &= \frac{4}{\kappa^2} e^{u_0 + v} (e^{u_0 + v} - 1) + 4 \sum_{\ell=1}^{m} n_\ell (1 + |x - p_\ell|^2)^{-2} \\
&= \frac{4}{\kappa^2} e^{u_0 + v} (e^{u_0 + v} - 1) + g, \quad \text{in } \mathbb{R}^2.
\end{aligned} \tag{5.3.2}$$

Our monotone iterative scheme can be described as follows.

Let $\Omega_0 \subset \mathbb{R}^2$ be a fixed bounded domain containing the prescribed zero set

$$Z(\phi) = \{p_1, \cdots, p_m\}$$

of the Higgs field ϕ and let $\Omega \supset \overline{\Omega}_0$ be a bounded domain with sufficiently regular (Lipschitzian, say) boundary. Let $K > 0$ be a constant verifying $K \geq 8/\kappa^2$.

We first introduce an iteration sequence on Ω,

$$\begin{aligned}
(\Delta - K) v_n &= \frac{4}{\kappa^2} e^{u_0 + v_{n-1}} (e^{u_0 + v_{n-1}} - 1) - K v_{n-1} + g \quad \text{in } \Omega, \\
v_n &= -u_0 \quad \text{on } \partial\Omega, \quad n = 1, 2, \cdots, \\
v_0 &= -u_0.
\end{aligned} \tag{5.3.3}$$

Lemma 5.3.1. *Let $\{v_n\}$ be the sequence defined by the iteration scheme (5.3.3). Then*

$$v_0 \geq v_1 \geq v_2 \geq \cdots \geq v_n \geq \cdots \tag{5.3.4}$$

Proof. We prove (5.3.4) by induction. It is easy to verify that $(\Delta - K)(v_1 - v_0) = 0$ in $\Omega - \{p_1, \cdots, p_m\}$ and $v_1 \in C^\infty(\Omega - \{p_1, \cdots, p_m\})$. For $\varepsilon > 0$ small, set $\Omega_\varepsilon = \Omega - \cup_{\ell=1}^{m} \{x \mid |x - p_\ell| \leq \varepsilon\}$. If $\varepsilon > 0$ is sufficiently small, we have $v_1 - v_0 \leq 0$ on $\partial\Omega_\varepsilon$. Hence the maximum principle implies $v_1 \leq v_0$ in Ω_ε. Therefore $v_1 \leq v_0$ in Ω.

In general, suppose there holds $v_0 \geq v_1 \geq \cdots \geq v_k$. We obtain from (5.3.3)

$$(\Delta - K)(v_{k+1} - v_k)$$

$$= \frac{4}{\kappa^2}e^{2u_0}(e^{2v_k} - e^{2v_{k-1}}) - K(v_k - v_{k-1}) - \frac{4}{\kappa^2}e^{u_0}(e^{v_k} - e^{v_{k-1}})$$

$$\geq K(e^{2u_0+2W} - 1)(v_k - v_{k-1}) - \frac{4}{\kappa^2}e^{u_0}(e^{v_k} - e^{v_{k-1}})$$

$$(v_k \leq W \leq v_{k-1} \leq v_0)$$

$$\geq K(e^{2u_0+2v_0} - 1)(v_k - v_{k-1}) = 0. \tag{5.3.5}$$

Since $v_{k+1} - v_k = 0$ on $\partial\Omega$, the maximum principle applied to (5.3.5) gives $v_{k+1} \leq v_k$ in Ω. This proves the lemma.

Now let

$$F(v) \equiv \int_\Omega dx \left\{ \frac{1}{2}|\nabla v|^2 + \frac{2}{\kappa^2}(e^{u_0+v} - 1)^2 + gv \right\}$$

be the natural functional associated to the Euler–Lagrange equation (5.3.2). Then the iterates $\{v_n\}$ enjoy the following monotonicity property.

Lemma 5.3.2. *There holds* $F(v_n) \leq F(v_{n-1}) \leq \cdots \leq F(v_1) \leq C$, *where* C *depends only on* Ω_0.

Proof. Multiplying (5.3.3) by $v_n - v_{n-1}$ and integrating by parts gives

$$\int_\Omega dx \left\{ |\nabla v_n|^2 - \nabla v_n \cdot \nabla v_{n-1} + K(v_n - v_{n-1})^2 \right\}$$

$$= -\frac{4}{\kappa^2}\int_\Omega dx \left\{ (v_n - v_{n-1})e^{u_0+v_{n-1}}(e^{u_0+v_{n-1}} - 1) + g(v_n - v_{n-1}) \right\}. \tag{5.3.6}$$

Now observe that for $u_0 + v \leq 0$ and $K \geq 4/\kappa^2$, the function

$$\varphi(v) \equiv \frac{2}{\kappa^2}(e^{u_0+v} - 1)^2 - \frac{K}{2}v^2$$

is concave in v. Hence,

$$\frac{2}{\kappa^2}(e^{u_0+v_n} - 1)^2 \leq \frac{2}{\kappa^2}(e^{u_0+v_{n-1}} - 1)^2 + \frac{K}{2}(v_n - v_{n-1})^2$$

$$+ \frac{4}{\kappa^2}(v_n - v_{n-1})e^{u_0+v_{n-1}}(e^{u_0+v_{n-1}} - 1). \tag{5.3.7}$$

Using (5.3.6) and (5.3.7) and $|\nabla v_n \cdot \nabla v_{n-1}| \leq 1/2(|\nabla v_n|^2 + |\nabla v_{n-1}|^2)$, we finally obtain

$$F(v_n) + \frac{K}{2}\|v_n - v_{n-1}\|^2_{L^2(\Omega)} \leq F(v_{n-1}), \tag{5.3.8}$$

which is a slightly stronger form of the required monotonicity.

Next we show that $F(v_1)$ can be bounded from above by a constant depending only on Ω_0. In fact, since $u_0 + v_1 = -v_0 + v_1 \leq 0$, we have $(e^{u_0+v_1} - 1)^2 \leq (u_0 + v_1)^2$. Therefore

$$F(v_1) \leq \frac{1}{2}\|\nabla v_1\|^2_{L^2(\Omega)} + \frac{4}{\kappa^2}\|v_1\|^2_{L^2(\Omega)} + \frac{4}{\kappa^2}\|u_0\|^2_{L^2(\mathbb{R}^2)} + \|g\|_{L^2(\mathbb{R}^2)}\|v_1\|_{L^2(\Omega)},$$

and it suffices to prove that $\|v_1\|_{W^{1,2}(\Omega_1)} \leq C$, where $C > 0$ depends only on Ω_0. To see this, we assume $\tilde{u}_0 \in C^\infty(\mathbb{R}^2)$ be such that $\tilde{u}_0 = u_0$ outside Ω_0. Then $\Delta\tilde{u}_0 = -g + f$ where f is smooth and of compact support. Hence $v_1 + \tilde{u}_0 = 0$ on $\partial\Omega$ and

$$(\Delta - K)(v_1 + \tilde{u}_0) = f + K(u_0 - \tilde{u}_0) \quad \text{in } \Omega.$$

Multiplying the above equation by $v_1 + \tilde{u}_0$, integrating by parts, and using the Schwartz inequality, we obtain $\|v_1\|_{W^{1,2}(\Omega)} \leq C$, where $C > 0$ depends only on K, $\|f\|_{L^2(\mathbb{R}^2)}$, $\|u_0\|_{L^2(\mathbb{R}^2)}$, and $\|\tilde{u}_0\|_{W^{1,2}(\mathbb{R}^2)}$.

Proposition 5.3.3. *There holds* $\|v_n\|_{W^{1,2}(\Omega)} \leq C$, $n = 1, 2, \cdots$, *where* C *depends only on* Ω_0.

Proof. We show that $F(v)$ controls the $W^{1,2}$ norm of v. Given $v \in W^{1,2}(\Omega)$ with $v = -u_0$ on $\partial\Omega$, define

$$\tilde{v} = \begin{cases} v & \text{in} & \Omega, \\ -u_0 & \text{in} & \mathbb{R}^2 - \Omega. \end{cases}$$

Then $\tilde{v} \in W^{1,2}(\mathbb{R}^2)$ and we have the standard interpolation inequality

$$\int_{\mathbb{R}^2} \tilde{v}^4 \, dx \leq 2 \int_{\mathbb{R}^2} \tilde{v}^2 \, dx \int_{\mathbb{R}^2} |\nabla\tilde{v}|^2 \, dx.$$

This implies

$$\int_\Omega v^4 \, dx \leq 2 \int_\Omega v^2 \, dx \int_\Omega |\nabla v|^2 \, dx + C\left(\int_\Omega dx\left\{|\nabla v|^2 + v^2\right\}\right) + C \quad (5.3.9)$$

with uniform constant C approaching zero as Ω tends to $\mathbb{R}^2$.

To estimate $F(v)$ from below we use (5.3.9) to get

$$\left|\int_\Omega gv \, dx\right| \leq \|g\|_{L^{\frac{4}{3}}(\Omega)}\|v\|_{L^4(\Omega)} \leq C\|v\|_{L^4(\Omega)}$$

$$\leq \varepsilon\|v\|_{L^2(\Omega)} + \frac{C}{\varepsilon}\|\nabla v\|_{L^2(\Omega)} + C$$

$$\leq \varepsilon\|v\|_{L^2(\Omega)} + \frac{1}{4}\int_\Omega |\nabla v|^2 \, dx + \frac{C}{\varepsilon^2}, \quad (5.3.10)$$

where (and in the sequel) $C > 0$ is a uniform constant which may change its value at different places and $\varepsilon > 0$ will be chosen below. Now

$$(e^{u_0 + v} - 1)^2 \geq \frac{|u_0 + v|^2}{(1 + |u_0 + v|)^2} \geq \frac{\frac{1}{2}|v|^2}{(1 + |v| + |u_0|)^2} - \frac{|u_0|^2}{(1 + |u_0|)^2}. \quad (5.3.11)$$

From (5.3.10) and (5.3.11), we obtain the lower bound

$$F(v) \geq \frac{1}{4} \int_\Omega |\nabla v|^2 \, dx + \frac{1}{\kappa^2} \int_\Omega \frac{|v|^2}{(1 + |v| + |u_0|)^2} \, dx$$
$$- \varepsilon \| v \|_{L^2(\Omega)} - \frac{C}{\varepsilon^2} - C. \quad (5.3.12)$$

Again, using (5.3.9) and neglecting the Lesbegue measure dx to save space, we can estimate

$$\left(\int_\Omega v^2 \right)^2 = \left(\int_\Omega \frac{|v|}{(1 + |v| + |u_0|)} (1 + |v| + |u_0|)|v| \right)^2$$
$$\leq C \int_\Omega \frac{|v|^2}{(1 + |v| + |u_0|)^2} \int_\Omega (|v|^2 + |v|^4 + |u_0|^4)$$
$$\leq C \int_\Omega \frac{|v|^2}{(1 + |v| + |u_0|)^2} \left(\int_\Omega v^2 \left[1 + \int_\Omega |\nabla v|^2 \right] + \int_\Omega |\nabla v|^2 + 1 \right)$$
$$\leq \frac{1}{2} \left(\int_\Omega v^2 \right)^2 + C \left(\left[\int_\Omega \frac{|v|^2}{(1 + |v| + |u_0|)^2} \right]^4 + \left[\int_\Omega |\nabla v|^2 \right]^4 + 1 \right).$$

Hence,

$$\| v \|_{L^2(\Omega)} \leq C \left\{ \int_\Omega dx \left(|\nabla v|^2 + \frac{v^2}{[1 + |v| + |u_0|]^2} \right) + 1 \right\}. \quad (5.3.13)$$

Finally, we obtain from (5.3.12) and (5.3.13)

$$\| v \|_{L^2(\Omega)} \leq C \left\{ F(v) + \varepsilon \| v \|_{L^2(\Omega)} + \frac{1}{\varepsilon^2} + 1 \right\}. \quad (5.3.14)$$

Let ε be so small that $\varepsilon(C + 1) < 1$. Thus (5.3.12) and (5.3.14) imply the desired bound

$$\| v \|_{W^{1,2}(\Omega)} \leq C \{ F(v) + 1 \}. \quad (5.3.15)$$

The proposition now follows from (5.3.15) and Lemma 5.3.2.

An immediate corollary of Proposition 5.3.3, Lemma 5.3.1, and standard elliptic regularity is the uniform convergence of the iteration scheme (5.3.3) to a smooth solution in any topology. We summarize this basic result as

Theorem 5.3.4. *The sequence (5.3.3) converges to a smooth solution v of the boundary value problem*

$$\Delta v = \frac{4}{\kappa^2} e^{u_0+v}(e^{u_0+v} - 1) + g \quad in \ \Omega, \tag{5.3.16}$$

$$v = -u_0 \quad on \ \partial\Omega.$$

The convergence may be taken in the $C^{k+\alpha}(\Omega) \cap W^{1,2}(\Omega)$ topology.

It is worth mentioning that all the results above are valid without change for the limiting case $\Omega = \mathbb{R}^2$. To clarify this point, we note that in such a situation the problem (5.3.16) becomes

$$\Delta v = \frac{4}{\kappa^2} e^{u_0+v}(e^{u_0+v} - 1) + g \quad in \ \mathbb{R}^2,$$

$$v \to 0 \quad as \ |x| \to \infty. \tag{5.3.17}$$

Therefore, (5.3.3) must formally be replaced by the following iterative scheme in $\mathbb{R}^2$,

$$(\Delta - K)v_n = \frac{4}{\kappa^2} e^{u_0+v_{n-1}}(e^{u_0+v_{n-1}} - 1) - Kv_{n-1} + g \quad in \ \mathbb{R}^2,$$

$$v_n \to 0 \quad as \ |x| \to \infty, \quad n = 1, 2, \cdots,$$

$$v_0 = -u_0. \tag{5.3.18}$$

In analogy to Theorem 5.3.4, we have

Theorem 5.3.5. *The scheme (5.3.18) defines a sequence $\{v_n\}$ in $W^{2,2}(\mathbb{R}^2)$ so that (5.3.4) is fulfilled in $\mathbb{R}^2$. As $n \to \infty$, v_n converges weakly in the space $W^{k,2}(\mathbb{R}^2)$ for any $k \geq 1$ to a smooth solution of (5.3.17). In fact this solution is maximal among all possible solutions of (5.3.17).*

Proof. We proceed by induction. When $n = 1$, (5.3.18) takes the form

$$(\Delta - K)v_1 = Ku_0 + g. \tag{5.3.19}$$

Since $u_0, g \in L^2(\mathbb{R}^2)$ and $\Delta - K : W^{2,2}(\mathbb{R}^2) \to L^2(\mathbb{R}^2)$ is a bijection, (5.3.19) defines a unique $v_1 \in W^{2,2}(\mathbb{R}^2)$. Thus we see in particular that v_1 vanishes at infinity as desired. On the other hand, there holds $(\Delta - K)(v_1 - v_0) = 0$ in the complement of $\{p_1, \cdots, p_m\}$. Hence the argument of Lemma 5.3.1 proves that $v_0 \geq v_1$.

We now assume for some $k \geq 1$ that the scheme (5.3.18) defines on $\mathbb{R}^2$ the functions $v_1, \cdots, v_k$ so that

$$v_1, \cdots, v_k \in W^{2,2}(\mathbb{R}^2) \quad and \quad v_0 \geq v_1 \geq \cdots \geq v_k. \tag{5.3.20}$$

We have, in view of (5.3.20), $u_0 + v_k \leq 0$. Thus $e^{u_0+v_k} \leq 1$ and

$$|e^{u_0+v_k} - 1| \leq |u_0 + v_k|. \tag{5.3.21}$$

As a consequence, for $n = k + 1$, the right-hand side of the first equation in
(5.3.18) lies in $L^2(\mathbb{R}^2)$ and thus the equation determines a unique $v_{k+1} \in$
$W^{2,2}(\mathbb{R}^2)$. From the fact that $v_{k+1} - v_k$ verifies (5.3.5) and vanishes at
infinity, we arrive at $v_{k+1} \le v_k$. Therefore (5.3.20) is true for any k.

By virtue of (5.3.21), the functional $F(v)$ is finite for $v = v_k$, $k = 1, 2, \cdots$.
Thus applying Lemma 5.3.2 and Proposition 5.3.3 to the sequence $\{v_n\}$ here
yields the bound $\|v_n\|_{W^{1,2}(\mathbb{R}^2)} \le C$, $n = 1, 2, \cdots$, where $C > 0$ is a constant.
Combining this result with (5.3.21) and using the L^2-estimates in (5.3.18),
we get $\|v_n\|_{W^{2,2}(\mathbb{R}^2)} \le C$. In fact a standard bootstrap argument shows that
in general one has $\|v_n\|_{W^{k,2}(\mathbb{R}^2)} \le C$, $n \ge$ some $n(k) \ge 1$, where $C > 0$ is a
constant depending only on $k \ge 1$. Therefore we see that there is a function
v so that v_n converges weakly in $W^{k,2}(\mathbb{R}^2)$ for any $k \ge 1$ to v and v is a
solution of (5.3.17).

Finally we show that v is maximal. Let w be another solution of (5.3.17).
Since $-v_0 + w = 0$ at infinity,

$$\Delta(-v_0 + w) = \frac{4}{\kappa^2} e^{-v_0 + w}(e^{-v_0 + w} - 1) \quad \text{in } \mathbb{R}^2 - \{p_1, \cdots, p_m\}, \quad (5.3.22)$$

and $-v_0 + w < 0$ in a small neighborhood of $\{p_1, \cdots, p_m\}$, applying the
maximum principle in (5.3.22) leads to $v_0 \ge w$. From this fact we can use
induction as in the proof of Lemma 5.3.6 in the next subsection to establish
the general inequality $v_n \ge w$, $n = 0, 1, 2, \cdots$. Hence, $v = \lim v_n \ge w$ and
the theorem follows.

The above theorem says that a solution of (5.3.2) on the full plane may
be constructed via our iterative scheme (5.3.18). However, from the point
of view of computation it is preferable to give a global convergence result
so that a full plane solution can be approximated by the solutions of the
system restricted to bounded domains. This will be accomplished in the
next section.

5.3.2 Global convergence theorems

We now consider convergence in the full plane. We continue to use the
notation in the previous subsection.

Lemma 5.3.6. *Let $V \in C^2(\Omega) \cap C^0(\bar{\Omega})$ be such that*

$$\Delta V \ge \frac{4}{\kappa^2} e^{u_0 + V}(e^{u_0 + V} - 1) + g \quad \text{in } \Omega, \quad V \le -u_0 \quad \text{on } \partial\Omega, \quad (5.3.23)$$

and $\{v_n\}$ be the sequence defined in (5.3.3). Then

$$v_0 \ge v_1 \ge v_2 \ge \cdots \ge v_n \ge \cdots \ge V. \quad (5.3.24)$$

Proof. We prove (5.3.24) by induction. Note that $v_0 \ge V$ in $\bar{\Omega}$. For such
an inequality already holds on $\partial\Omega$ by the definition of v_0 and for small

$\varepsilon > 0$, $u_0 + V \leq 0$ on $\partial\Omega_\varepsilon$. Hence, the result follows from the maximum principle applied to the inequalities

$$\Delta(u_0 + V) \geq \frac{4}{\kappa^2} e^{u_0 + V}(e^{u_0 + V} - 1) \quad \text{in } \Omega_\varepsilon,$$

$$u_0 + V \leq 0 \quad \text{on } \partial\Omega_\varepsilon.$$

Suppose there holds $v_k \geq V$ ($k = 0, 1, 2, \cdots$). We need to show that $v_{k+1} \geq V$. In fact, from (5.3.3) and (5.3.23), we get

$$(\Delta - K)(v_{k+1} - V)$$
$$\leq \frac{4}{\kappa^2} e^{2u_0}(e^{2v_k} - e^{2V}) - K(v_k - V) - \frac{4}{\kappa^2} e^{u_0}(e^{v_k} - e^V)$$
$$\leq K(e^{2u_0 + 2W} - 1)(v_k - V) \qquad (V \leq W \leq v_k)$$
$$\leq K(e^{2u_0 + 2v_0} - 1)(v_k - V) = 0. \tag{5.3.25}$$

Since for $k + 1 = n = 1, 2, \cdots$, the right-hand side of (5.3.3) always lies in $L^p(\Omega)$ for any $p \geq 2$, we see that $v_{k+1} \in W^{2,p}(\Omega)$. In particular $v_{k+1} \in C^{1+\alpha}(\overline{\Omega})$ ($0 < \alpha < 1$). On the other hand, we have $v_{k+1} - V \geq 0$ on $\partial\Omega$. Thus (5.3.25) and the weak maximum principle (see Gilbarg and Trudinger [123]) imply that $v_{k+1} \geq V$ in Ω. The lemma is proven.

Next, let $\{\Omega_n\}$ be a monotone sequence of bounded convex domains in $\mathbb{R}^2$ satisfying the same properties as those for Ω in defining the iterative scheme (5.3.3),

$$\Omega_1 \subset \Omega_2 \subset \cdots \subset \Omega_n \subset \cdots, \quad \cup_{n=1}^{\infty}\Omega_n = \mathbb{R}^2.$$

Lemma 5.3.7. *Let $v^{(j)}$ and $v^{(k)}$ be the solutions of (5.3.16) obtained from (5.3.3) by setting $\Omega = \Omega_j$ and $\Omega = \Omega_k$ respectively, $j, k = 1, 2, \cdots$. If $\Omega_j \subset \Omega_k$, then*

$$v^{(j)} \geq v^{(k)} \quad \text{in } \Omega_j. \tag{5.3.26}$$

Proof. By the construction of $v^{(k)}$, we have in particular that $v^{(k)} \leq -u_0$ in Ω_k. Thus $v^{(k)}$ is a subsolution of (5.3.16) for $\Omega = \Omega_j$. Thus by by Lemma 5.3.6, we get $v^{(j)} \geq v^{(k)}$ in Ω_j.

For convenience, from now on we extend the domain of definition of each $v^{(j)}$ to the entire $\mathbb{R}^2$ by setting $v^{(j)} = -u_0$ in $\mathbb{R}^2 - \Omega_j$. Thus $\{v^{(j)}\}$ is a sequence in $W^{1,2}(\mathbb{R}^2)$.

From Proposition 5.3.3, we can obtain a constant $C > 0$ independent of $j = 1, 2, \cdots$, so that $\|v^{(j)}\|_{W^{1,2}(\mathbb{R}^2)} \leq C$. As in the previous subsection, this leads to

Theorem 5.3.8. *The sequence of solutions $\{v^{(j)}\}$ defined in Lemma 5.3.7 converges weakly in $W^{1,2}(\mathbb{R}^2)$ to the maximal solution of (5.3.17) obtained in Theorem 5.3.5.*

Proof. Let w be the weak limit of the sequence $\{v^{(j)}\}$ in $W^{1,2}(\mathbb{R}^2)$. Then w is a solution of (5.3.2) satisfying $u_0 + w \leq 0$. Hence, as in the proof of Theorem 5.3.5, the right-hand side of (5.3.2) now lies in $L^2(\mathbb{R}^2)$. But $w \in W^{1,2}(\mathbb{R}^2)$, so the L^2-estimates applied in (5.3.2) give the result $w \in W^{2,2}(\mathbb{R}^2)$. Thus we see that $w = 0$ at infinity. In particular w is a solution of the problem (5.3.17). On the other hand, let the maximal solution of (5.3.17) be v. Then $v \geq w$. Recall that the proof of Theorem 5.3.5 has given us the comparison $u_0 + v = -v_0 + v \leq 0$ in $\mathbb{R}^2$. So v verifies (5.3.23) on each $\Omega = \Omega_j$. As a consequence, Lemma 5.3.6 implies that $v^{(j)} \geq v$ in Ω_j, $j = 1, 2, \cdots$. Therefore $w = \lim v^{(j)} \geq v$. This proves the desired result $v = w$.

In the sequel we shall denote by $\tilde{v}$ the maximal solution of (5.3.17) obtained in Theorem 5.3.5 or 5.3.8 and set $\tilde{u} = u_0 + \tilde{v}$. Therefore we can construct a finite energy solution pair $(\tilde{\phi}, \tilde{A})$ of (5.2.19) and (5.2.20), supplemented with (5.2.5), so that $|\tilde{\phi}|^2 = e^{\tilde{u}}$. In fact we can state

Proposition 5.3.9. *Let $u = u_0 + v$ where v is a solution of (5.3.2) which lies in $W^{2,2}(\mathbb{R}^2)$. Denote by (ϕ, A) the solution pair of (5.2.5), (5.2.19), and (5.2.20) constructed by the scheme (5.1.29) so that $|\phi|^2 = e^u$. Then (ϕ, A) is of finite energy.*

Proof. By $v \in W^{2,2}(\mathbb{R}^2)$, we see that $v \to 0$ at infinity. In particular, $\lim_{|x|\to\infty} u = 0$. Thus using the fact that $u < 0$ in a neighborhood of $\{p_1, \cdots, p_m\}$ and the maximum principle in (5.3.1) we have $u \leq 0$ in $\mathbb{R}^2$. This implies $|\phi|^2 = e^u \leq 1$.

Given $0 < \varepsilon < 1$, choose $t > 0$ sufficiently large to make

$$\frac{4}{\kappa^2} e^{2u} \geq \frac{4}{\kappa^2}(1-\varepsilon)^2, \quad |x| \geq t.$$

Set $m_2 = 2/|\kappa|$. Then from (5.3.1) we arrive at

$$\Delta u = m_2^2 e^u e^w u \quad (u \leq w \leq 0) \quad \leq m_2^2 (1-\varepsilon)^2 u, \quad |x| \geq t. \qquad (5.3.27)$$

From (5.3.27) we can show by the maximum principle that there is $C(\varepsilon) > 0$ so that

$$0 \geq u \geq -C(\varepsilon)e^{-m_2(1-\varepsilon)|x|}.$$

Hence $|\phi|^2 - 1 = e^u - 1 \in L(\mathbb{R}^2)$.

Since (ϕ, A) is a solution of (5.2.19) and (5.2.20), ϕ verifies the equation

$$D_j D_j \phi = \frac{2}{\kappa^2} |\phi|^2 (|\phi|^2 - 1)\phi.$$

For any $\psi \in W_0^{1,2}(\Omega)$ where $\Omega \subset \mathbb{R}^2$ is a bounded domain, we get by multiplying both sides of the above by $\overline{\psi}$ and integrating the equation

$$\text{Re} \int_{\mathbb{R}^2} dx \left\{ D_j \phi \overline{D_j \psi} + \frac{2}{\kappa^2} |\phi|^2 (|\phi|^2 - 1) \phi \overline{\psi} \right\} = 0.$$

Therefore, replacing ψ above by $\eta_t^2\phi$, we arrive at

$$
\begin{aligned}
0 &= \mathrm{Re}\int_{\Omega_{2t}} dx \left\{ D_j\phi\overline{D_j(\eta_t^2\phi)} + \frac{2}{\kappa^2}|\phi|^4(|\phi|^2-1)\eta_t^2 \right\} \\
&= 2\mathrm{Re}\int_{t\le|x|\le 2t} \left\{ (D_j\phi)\eta_t\overline{\phi}\eta'\left(\frac{|x|}{t}\right)\frac{x_j}{t|x|} \right\} dx \\
&\quad + \int_{\Omega_{2t}} \left\{ \eta_t^2|D_j\phi|^2 + \frac{2}{\kappa^2}|\phi|^4(|\phi|^2-1)\eta_t^2 \right\} dx. \qquad (5.3.28)
\end{aligned}
$$

Here η_t is defined by

$$
\eta_t(x) = \eta\left(\frac{|x|}{t}\right), \quad x \in \mathbb{R}^2, \quad t > 0,
$$

where $\eta \in C_0^\infty(\mathbb{R})$ is such that

$$
0 \le \eta \le 1, \quad \eta(s) = 1 \quad \text{for } s \le 1, \quad \eta(s) = 0 \quad \text{for } s \ge 2.
$$

Using $|\phi| \le 1$ and a simple interpolation inequality, (5.3.28) leads us to

$$
\int_{\Omega_{2t}} \eta_t^2|D_j\phi|^2\, dx \le C_1 + C_2\int_{\mathbb{R}^2} \big||\phi|^2 - 1\big|\, dx, \qquad (5.3.29)
$$

where $C_1, C_2 > 0$ are independent of $t > 0$. Letting $t \to \infty$ in (5.3.29) we see that $D_j\phi \in L^2(\mathbb{R}^2)$.

Moreover, from (5.2.19) and (5.2.20), we have

$$
A_0^2|\phi|^2, \quad \frac{1}{\kappa^2}|\phi|^2(|\phi|^2-1)^2 \le \frac{1}{\kappa^2}(e^u-1)^2 \in L(\mathbb{R}^2).
$$

Consequently (ϕ, A) is indeed of finite energy (see (5.2.7)).

Now let $(\phi^{(j)}, A^{(j)})$ be the solution pair of the truncated equations over Ω_j,

$$
\begin{aligned}
D_1\phi \pm iD_2\phi &= 0, \\
F_{12} \pm \frac{2}{\kappa^2}|\phi|^2(|\phi|^2-1) &= 0 \quad \text{in } \Omega_j, \\
|\phi| &= 1, \quad F_{12} = 0 \quad \text{on } \partial\Omega_j
\end{aligned}
$$

obtained from the function $v^{(j)}$ described in Lemma 5.3.7. For convenience, we understand that $|\phi^{(j)}| = 1, F_{12}^{(j)} = 0$ in $\mathbb{R}^2 - \Omega_j$. Such an assumption corresponds to the earlier extension of $v^{(j)}$ with setting $v^{(j)} = -u_0$ in $\mathbb{R}^2 - \Omega_j$. In the sequel, this convention is always implied unless otherwise stated.

Define the norm $|\ |_\delta$ where $\delta \in (0, 2/|\kappa|)$ by

$$
|\eta|_\delta = \sup_{x\in\mathbb{R}^2} \left| e^{\delta|x|}\eta(x) \right|.
$$

This expression says functions with finite $||\;|_\delta$ norms decay exponentially fast at infinity. Our global convergence theorem for the computation of a topological solution of (5.2.19) and (5.2.20) may be stated as follows.

Theorem 5.3.10. *Let the field configuration pair (ϕ, A) be an arbitrary topological solution of the self-dual Chern–Simons equations (5.2.19) and (5.2.20) with $Z(\phi) = \{p_1, \cdots, p_m\}$ and the algebraic multiplicities of the zeros $p_1, \cdots, p_m$ are $n_1, \cdots, n_m \in \mathbb{Z}_+$, respectively, and $\{(\phi^{(j)}, A^{(j)})\}$ be the solution sequence over Ω_j described above. Then*

$$(\tilde{\phi}, \tilde{A}) = \lim_{j \to \infty} (\phi^{(j)}, A^{(j)})$$

is a topological solution of (5.2.19) and (5.2.20) characterized by the same vortex distribution as (ϕ, A) and verifying $|\tilde{\phi}| \geq |\phi|$ in $\mathbb{R}^2$. Furthermore, the physical fields have the following convergence rate for any $\delta \in (0, 2/|\kappa|)$,

$$||\phi^{(j)}|^2 - |\tilde{\phi}|^2|_\delta \to 0, \quad |F_{12}^{(j)} - \tilde{F}_{12}|_\delta \to 0 \quad as\ j \to \infty.$$

In particular,

$$\lim_{j \to \infty} \int_{\Omega_j} F_{12}^{(j)}\, dx = \int_{\mathbb{R}^2} \tilde{F}_{12}\, dx = 2\pi N,$$

where $N = n_1 + \cdots + n_m$.

Proof. We have already seen in Theorem 5.3.8 and Proposition 5.3.9 that the $\tilde{v} = \lim_{j \to \infty} v^{(j)}$ is the maximal solution of (5.3.17) which generates a finite energy solution pair $(\tilde{\phi}, \tilde{A}) = \lim_{j \to \infty}(\phi_j, A_j)$ of (5.2.19) and (5.2.20). We observe that if (ϕ, A) is any finite energy topological solution of (5.2.19) and (5.2.20), then $|\phi| = 1$ at infinity. Thus $v = \ln|\phi|^2 - u_0$ verifies (5.3.17). Therefore $\tilde{v} \geq v$ in $\mathbb{R}^2$. Consequently $|\tilde{\phi}| \geq |\phi|$.

For $\delta \in (0, 2/|\kappa|)$, choose $\varepsilon \in (0, 1)$ to make $(2/|\kappa|)(1 - \varepsilon) > \delta$. Then the fact $||\phi^{(j)}|^2 - |\tilde{\phi}|^2|_\mu \to 0$ as $j \to \infty$ follows immediately from the decay estimates (5.2.11) since

$$\left| e^{u^{(j)}} - e^{\tilde{u}} \right| \leq 1 - e^{\tilde{u}} = 1 - |\tilde{\phi}|^2.$$

By virtue of the equation in (5.2.20), it is straightforward that $|F_{12}^{(j)} - \tilde{F}_{12}|_\delta \to 0$ as $j \to \infty$.

The proof of Theorem 5.3.10 is complete.

5.4   Symmetric Non-topological Solutions

Unlike topological solutions, the energy and charge of a non-topological solution [150, 153, 154] depend on its accurate decay rate at infinity which is rather difficult to obtain. Here we present a complete study of radially symmetric solutions. We first state the results. We next formulate the problem. We then solve the problem through a shooting analysis.

5.4.1   Existence theorem

One of the most interesting features of the non-topological solutions is that, given any integer N, there is a family of N-vortex solutions whose energy, electric, and magnetic charges can take arbitrarily prescribed values from explicitly determined open intervals depending on N.

Concerning such N-vortex solutions, we can state

Theorem 5.4.1. *For any $\tilde{x} \in \mathbb{R}^2$ and a given integer $N \geq 1$ and any $\alpha \geq \ln 2$, the self-dual equations (5.2.19) and (5.2.20) have a finite-energy solution $(\phi^{(\alpha)}, A^{(\alpha)})$ so that the only zero of $\phi^{(\alpha)}$ is $x = \tilde{x}$ and the multiplicity of the zero is N. Moreover, $(\phi^{(\alpha)}, A^{(\alpha)})$ is radially symmetric about the point $x = \tilde{x}$ and*

$$\max_{x \in \mathbb{R}^2} |\phi^{(\alpha)}(x)|^2 = e^{-\alpha}, \tag{5.4.1}$$

$$|\phi^{(\alpha)}(x)|^2 = O(r^{-\beta}), \quad |D_j \phi^{(\alpha)}|^2 = O(r^{-(2+\beta)}), \quad F_{12} = O(r^{-\beta}) \tag{5.4.2}$$

for large $r = |x|$, and there hold the following energy and charge formulas,

$$E = \int_{\mathbb{R}^2} \mathcal{H} \, dx = 2\pi N + \pi \beta, \quad \Phi = \int_{\mathbb{R}^2} F_{12} \, dx = 2\pi N + \pi \beta = \frac{Q}{\kappa}, \tag{5.4.3}$$

where $\beta > 2N + 4$ is a constant which may depend on α and $\tilde{x}$. In fact, β may be made to assume any value in the open interval $(2N + 4, \infty)$.

This result, first obtained in [295], indicates that non-topological solutions of the relativistic Chern–Simons theory (5.2.1) are rather different from those of the nonrelativistic theory (5.1.13). In fact, it is interesting to note that these nonrelativistic solutions, constructed from the Liouville equation, correspond to the borderline case of the non-topological solutions in the relativistic theory, $\beta = 2N + 4$, and hence, carry quantized energy and charges.

Since (5.4.1) is gauge-invariant, we see that different values of α give rise to gauge-distinct solutions. In particular, there is non-uniqueness.

Denote by E_1 the energy of a topological N-vortex solution and E_2 the energy of a non-topological solution stated in Theorem 5.4.1. Since $E_1 = 2\pi N$, we see from (5.4.3) that

$$E_2 > 2E_1 + 4\pi.$$

Similar comparisons may be made for magnetic charges (fluxes) and electric charges.

5.4.2   Two-point boundary value problem

Since (5.2.19) and (5.2.20) are invariant under space translations, we may assume that $\tilde{x}$ is the origin. Therefore we shall look for a solution (ϕ, A) of (5.2.19) and (5.2.20) so that ϕ has the local property

$$|\phi(x)| = |x|^N \eta(x) \quad \text{near } x = 0, \tag{5.4.4}$$

where $N \geq 1$ is a given integer and $\eta(x)$ is a nonvanishing function. In view of (5.4.4), it is standard that the substitution $u = \ln |\phi|^2$ reduces (5.2.19) and (5.2.20) to the elliptic equation

$$\Delta u = \frac{4}{\kappa^2} e^u (e^u - 1) + 4\pi N \delta(x), \quad x \in \mathbb{R}^2, \tag{5.4.5}$$

where $\delta(x)$ is the Dirac distribution. Let $v = u - N \ln |x|^2$. Then (3.2) becomes

$$\Delta v = \frac{4}{\kappa^2} |x|^{2N} e^v (|x|^{2N} e^v - 1), \quad x \in \mathbb{R}^2. \tag{5.4.6}$$

We will restrict our attention to radially symmetric solutions of (5.4.6), $v = v(r), r = |x| > 0$. Hence (5.4.6) takes on the form

$$v_{rr} + \frac{1}{r} v_r = \frac{4}{\kappa^2} r^{2N} e^v (r^{2N} e^v - 1), \quad r > 0. \tag{5.4.7}$$

Eventually we want to extend a solution of (5.4.7) defined in $\mathbb{R}^2 - \{0\}$ to recover a smooth solution of (5.4.6) in full $\mathbb{R}^2$ (so that u is asymptotic to $2N \ln |x|$ as x tends to zero). For this purpose, we need the following special form of the well-known removable singularity theorem [255].

Lemma 5.4.2. *Let Ω be a domain in $\mathbb{R}^2$ containing the origin $x = 0$ and f a harmonic function defined in the punctured domain $\Omega - \{0\}$. Then f can be extended to a harmonic function in Ω if and only if*

$$\lim_{|x| \to 0} \frac{f(x)}{\ln |x|} = 0. \tag{5.4.8}$$

Using Lemma 5.4.2 in our problem, we have

Lemma 5.4.3. *The solution v of (5.4.7) can be extended to a smooth solution of (5.4.6) in $\mathbb{R}^2$ if and only if*

$$\lim_{r \to 0} \frac{v(r)}{\ln r} = 0. \tag{5.4.9}$$

Proof. Let $v(r)$ be a solution of (5.4.7) verifying (5.4.9). Then for any $a > 0$, we have

$$\lim_{r \to 0} r^a e^{v(r)} = \lim_{r \to 0} e^{\ln r (a + \frac{v(r)}{\ln r})} = 0.$$

Thus it is easily seen that the right-hand side of (5.4.7) can be viewed as a Hölder continuous function over the full $\mathbb{R}^2$.

Let w be a solution of

$$\Delta w = \frac{4}{\kappa^2} r^{2N} e^v (r^{2N} e^v - 1)$$

in a small neighborhood of the origin, say, Ω. Then w is C^2-Hölder continuous and $f = v - w$ is harmonic in $\Omega - \{0\}$. However, (5.4.9) says that f fulfills the condition (5.4.8). Therefore, using Lemma 5.4.2, we see that v is C^2-Hölder continuous in Ω. A bootstrap argument then shows that v is C^∞ in $\mathbb{R}^2$.

Thus from now on we shall look for solutions of (5.4.7) under the condition (5.4.9). It will be most convenient to study the equation in the original variable $u(r) = N \ln r^2 + v(r)$ (see (5.4.5)). Hence (5.4.7) is changed into the simpler form

$$u_{rr} + \frac{1}{r} u_r = \frac{4}{\kappa^2} e^u (e^u - 1), \quad r > 0. \tag{5.4.10}$$

The boundary condition (5.4.9) now becomes

$$\lim_{r \to 0} \frac{u(r)}{\ln r} = 2N. \tag{5.4.11}$$

Recall that we need to find non-topological solutions of the self-dual system (5.2.19) and (5.2.20). Therefore the relation $|\phi|^2 = e^u$ implies that u is subject to the following boundary condition at $r = \infty$,

$$\lim_{r \to \infty} u(r) = -\infty. \tag{5.4.12}$$

Our goal now is to find solutions of (5.4.10) under the boundary conditions (5.4.11) and (5.4.12), which is a two-point boundary value problem over the infinite interval $(0, \infty)$.

5.4.3 Shooting analysis

We shall show that for suitable $r_0 > 0$, we can obtain global solutions of (5.4.10) coupled with some adequate initial data at $r = r_0$ to fulfill (5.4.11) and (5.4.12). In other words, we are going to solve (5.4.10)–(5.4.12) by a two-side shooting technique.

To motivate our shooting data, we first make a simple observation.

Lemma 5.4.4. *If $u(r)$ is a solution of (5.4.10) satisfying*

$$\lim_{r \to 0} u(r) = -\infty, \quad \lim_{r \to \infty} u(r) = -\infty,$$

then $u(r) < 0$ for all $r > 0$.

Proof. The conclusion can be seen directly from a maximum principle argument.

From Lemma 5.4.4 we see that a desired solution of (5.4.10)–(5.4.12) must have a global maximum $u_0 = -\alpha < 0$ at some $r = r_0 > 0$. Therefore we should look for solutions of (5.4.10) under the initial condition

$$u(r_0) = -\alpha, \quad u_r(r_0) = 0. \tag{5.4.13}$$

We expect that, when $r_0 > 0, \alpha > 0$ are suitably chosen, the unique solution of (5.4.10) under the condition (5.4.13) will verify both (5.4.11) and (5.4.12). Our study in this subsection shows that such a goal can be achieved.

To simplify the discussion, we introduce a change of independent variable

$$t = \ln r, \quad t_0 = \ln r_0. \tag{5.4.14}$$

Then (5.4.10) and (5.4.13) become

$$u'' = \frac{4}{\kappa^2}e^{2t}e^u(e^u - 1), \quad -\infty < t < \infty, \tag{5.4.15}$$

$$u(t_0) = -\alpha, \quad u'(t_0) = 0, \tag{5.4.16}$$

where, and in the sequel, $u' = du/dt$ and $u(t)$ denotes the dependence of the solution u of (5.4.10) on the new variable t (or vice versa, for simplicity).

Lemma 5.4.5. *For any $t_0 \in \mathbb{R}$ and $\alpha > 0$, (5.4.15) and (5.4.16) have a unique global solution $u(t)$. This solution satisfies $u(t) < 0$ and*

$$\lim_{t \to -\infty} u(t) = -\infty, \quad \lim_{t \to \infty} u(t) = -\infty. \tag{5.4.17}$$

Proof. Let $u(t)$ be a local solution of (5.4.15) and (5.4.16). Then in the interval of existence,

$$u'(t) = \frac{4}{\kappa^2} \int_{t_0}^{t} e^{2s}e^{u(s)}(e^{u(s)} - 1)ds. \tag{5.4.18}$$

We can show that, for all t, where $u(t)$ exists, there holds $u(t) < 0$. In fact, if there is a $\tilde{t} > t_0$ so that $u(\tilde{t}) \geq 0$, we may assume $\tilde{t}$ is such that

$$\tilde{t} = \inf\left\{t \geq t_0 \,\middle|\, u(t) \text{ exists and } u(t) \geq 0\right\}.$$

Then $\tilde{t} > t_0$ and $u(\tilde{t}) = 0$. Obviously $u(t) < 0$ for all $t_0 \leq t < \tilde{t}$. However, from (5.4.18), we see that $u'(t) < 0$ for $t_0 < t \leq \tilde{t}$. So $u(\tilde{t}) < 0$. This reaches a contradiction.

Similarly, if there is a $\tilde{t} < t_0$ so that $u(\tilde{t}) = 0$ and $u(t) < 0$ for $\tilde{t} < t \leq t_0$, then $u'(t) > 0$ for $\tilde{t} \leq t < t_0$. So $u(\tilde{t}) < 0$. This is again a contradiction.

From the property $u(t) < 0$ and (5.4.18), it is seen that $u'(t)$ cannot blow up in finite time. As a consequence the solution of (5.4.15) and (5.4.16) exists globally in $t \in (-\infty, \infty)$.

The behavior $u(t) \to -\infty$ as $t \to -\infty$ is easy to verify because $u(t) < 0$ and (5.4.18) imply that

$$\lim_{t \to -\infty} u'(t) = \frac{4}{\kappa^2} \int_{t_0}^{-\infty} e^{2s} e^{u(s)}(e^{u(s)} - 1)ds = C > 0.$$

Finally we show that $u(t) \to -\infty$ as $t \to \infty$. By virtue of $u(t) < 0$ and (5.4.18), we have $u'(t) < 0$ for $t > t_0$. Therefore, either $u(t) \to -\infty$ or $u(t) \to$ a finite number $a < -\alpha < 0$ as $t \to \infty$. However, the latter situation cannot happen. To see this, we assume otherwise. Thus by $a < u(t) \le -\alpha$ and (5.4.18) we find the estimate

$$u'(t) \ge \frac{4}{\kappa^2} \int_{t_0}^{t} e^{2s} \left[\min_{a \le u \le -\alpha} \left\{ e^u(e^u - 1) \right\} \right] ds$$

$$= -C \int_{t_0}^{t} e^{2s} ds = -\frac{C}{2}(e^{2t} - e^{2t_0}), \quad t > t_0, \qquad (5.4.19)$$

where $C > 0$ is a constant. A simple consequence of (5.4.19) is that $u(t) \to -\infty$ as $t \to \infty$. This contradicts our assumption.

In terms of the new variable t defined in (5.4.14), the boundary condition (5.4.11) reads

$$\lim_{t \to -\infty} \frac{u(t)}{t} = 2N. \qquad (5.4.20)$$

Lemma 5.4.6. *For any given $\alpha \ge \ln 2$, there is a $t_0 = t_0(\alpha)$ such that the unique solution of (5.4.15) and (5.4.16) verifies the condition (5.4.20).*

Proof. For $t_0 \in (-\infty, \infty)$ and $\alpha > 0$, let $u = u(t; t_0, \alpha)$ be the unique global solution of (5.4.15) and (5.4.16). Then $u < 0$ and $u \to -\infty$ as $t \to -\infty$ by Lemma 5.4.5. Therefore, using the L'Hôpital rule, (5.4.20) reads

$$\eta(t_0, \alpha) \equiv \lim_{t \to -\infty} u'(t; t_0, \alpha) = 2N, \qquad (5.4.21)$$

where, in view of (5.4.18), the function $\eta(t_0, \alpha)$ has the representation

$$\eta(t_0, \alpha) = -\frac{4}{\kappa^2} \int_{-\infty}^{t_0} e^{2s} e^{u(s; t_0, \alpha)}(e^{u(s; t_0, \alpha)} - 1)ds. \qquad (5.4.22)$$

Since $u < 0$ and u depends continuously on t_0, α, (5.4.22) says that η is a continuous function of t_0, α. In the following, we shall show that there are t_0, α to make η fulfill the condition (5.4.21).

Step 1. From (5.4.15) we get $u'' > -(4/\kappa^2)e^{2t}e^u$. Set $w = 2t + u$. Then

$$w'' > -\frac{4}{\kappa^2}e^w. \qquad (5.4.23)$$

However, since $u' \geq 0$ for $t \leq t_0$, we have $w' > 0$ when $t \leq t_0$. Multiplying (5.4.23) by w' and integrating on (t, t_0), we find

$$4 - (w'(t))^2 > \frac{8}{\kappa^2}(e^{w(t)} - e^{2t_0 - \alpha}), \quad t < t_0,$$

i.e.,

$$0 < u'(t; t_0, \alpha) < 2\sqrt{1 + \frac{2}{\kappa^2}e^{2t_0 - \alpha} - 2} \equiv K, \quad t < t_0. \tag{5.4.24}$$

From (5.4.24), we obtain another useful inequality,

$$-\alpha > u(t; t_0, \alpha) > -\alpha - K(t_0 - t), \quad t < t_0. \tag{5.4.25}$$

Step 2. It is straightforward to examine that $e^u(e^u - 1)$ is a decreasing function in $u \in (-\infty, -\ln 2]$. Therefore the condition $\alpha \geq \ln 2$, (5.4.25), and (5.4.15) imply

$$u'' < \frac{4}{\kappa^2}e^{2t}e^{-\alpha - K(t_0 - t)}(e^{-\alpha - K(t_0 - t)} - 1), \quad t < t_0. \tag{5.4.26}$$

Integrating (5.4.26) over $(-\infty, t_0)$ gives

$$-\eta(t_0, \alpha) = -\lim_{t \to -\infty} u'(t; t_0, \alpha)$$

$$\leq \frac{4}{\kappa^2}\int_{-\infty}^{t_0} e^{2s}e^{-\alpha - K(t_0 - s)}(e^{-\alpha - K(t_0 - s)} - 1)ds,$$

or

$$\eta(t_0, \alpha) \geq \frac{4}{\kappa^2}e^{2t_0 - \alpha}\left(\frac{1}{K + 2} - \frac{e^{-\alpha}}{2K + 2}\right)$$

$$\geq \frac{4}{\kappa^2}e^{2t_0 - \alpha}\left(\frac{1}{K + 2} - \frac{1}{2(2K + 2)}\right)$$

$$\geq \frac{2e^{2t_0 - \alpha}}{\kappa^2(K + 2)} = e^{2t_0 - \alpha}\Big/\kappa\sqrt{\kappa^2 + 2e^{2t_0 - \alpha}}. \tag{5.4.27}$$

Step 3. In view of (5.4.24), we have

$$0 < \eta(t_0, \alpha) \leq 2\sqrt{1 + \frac{2}{\kappa^2}e^{2t_0 - \alpha} - 2}.$$

Therefore, for any $\alpha \geq \ln 2 > 0$, we can find a suitable $t_0 = t_0'$ so that $\eta(t_0', \alpha) < 2N$. On the other hand, (5.4.27) says that for fixed $\alpha \geq \ln 2$, there is some $t_0 = t_0''$ to make $\eta(t_0'', \alpha) > 2N$. Consequently, there is a point $t_0 = t_0(\alpha)$ between t_0' and t_0'' so that $\eta(t_0, \alpha) = 2N$.

We next study the asymptotic behavior of the solution $u(t)$ of (5.4.15) and (5.4.16) (as $t \to \infty$) produced in Lemma 5.4.6. We have

Lemma 5.4.7. *There is a constant $\beta > 2N + 4$ so that*

$$\lim_{t \to \infty} u'(t) = -\beta. \tag{5.4.28}$$

Proof. Since $u'' < 0$ (see (5.4.15) and Lemma 5.4.5), we see that either $u'(t) \to -\infty$ or a finite number as $t \to \infty$. First suppose that $u'(t) \to -\infty$ as $t \to \infty$. Then there is a $\bar{t}$ so that $u'(t) < -3$ (say) for $t > \bar{t}$. Hence, $u(t) < -3t + C$ ($t > t_0$) for some constant C and

$$
\begin{aligned}
\lim_{t \to \infty} u'(t) &= \frac{4}{\kappa^2} \int_{t_0}^{\infty} e^{2s} e^{u(s)} (e^{u(s)} - 1) ds \\
&> -\frac{4}{\kappa^2} \int_{t_0}^{\infty} e^{2s} e^{-3s+C} ds = -\frac{4}{\kappa^2} e^{C-t_0} > -\infty,
\end{aligned}
$$

which is a contradiction.

Thus, in the sequel, we assume there is a $\beta > 0$ to make (5.4.28) hold. It remains to show that $\beta > 2N + 4$.

First of all, since $u'(t)$ is decreasing for $t \geq t_0$. Therefore $u'(t) > -\beta$, $t \geq t_0$ and $u(t) > -\beta t + C$, $t \geq t_0$, where C is a constant. It is obvious that $u(t) \to -\infty$ and $u'(t) \to -\beta$ as $t \to \infty$ and the above relation imply the convergence of the integral

$$\int_{t_0}^{\infty} e^{2s} e^{u(s)} ds.$$

As a consequence, we must have $\beta > 2$. Such a property in turn implies that

$$\lim_{t \to \infty} e^{2t} e^{u(t)} = \lim_{t \to \infty} e^{t(2 + \frac{u(t)}{t})} = 0. \tag{5.4.29}$$

Next, multiplying the equation (5.4.15) by u', integrating over $(-\infty, \infty)$, and using (5.4.29), we obtain

$$\beta^2 - 4N^2 = -\frac{16}{\kappa^2} \int_{-\infty}^{\infty} e^{2s} e^{u(s)} (e^{u(s)} - 1) \, ds + \frac{8}{\kappa^2} \int_{-\infty}^{\infty} e^{2s} e^{2u(s)} \, ds.$$

In other words,

$$\beta^2 - 4\beta - 2N(2N + 4) = \frac{8}{\kappa^2} \int_{-\infty}^{\infty} e^{2s} e^{2u(s)} ds > 0. \tag{5.4.30}$$

Therefore $\beta > 2N + 4$ as desired.

Let u be the solution of (5.4.15) and (5.4.16) satisfying the properties stated in Lemmas 5.4.6 and 5.4.7. Thus, in terms of the original variable $r = e^t$, the function u is a solution of (5.4.10)–(5.4.12). Thus, from the earlier discussion, we can see that u is in fact a radially symmetric classical solution of the N-vortex equation (5.4.5). Consequently,

$$\phi(z) \;=\; \exp\left(\frac{1}{2}u(z) + iN\arg(z)\right),$$

$$A_1(z) \;=\; -\mathrm{Re}\{2i\bar\partial \ln\phi(z)\}, \quad A_2(z) = -\mathrm{Im}\{2i\bar\partial \ln\phi(z)\}$$

is a solution of the self-dual equations (5.2.19) and (5.2.20) so that $\phi(0) = 0$, the multiplicity of this zero is N, and ϕ is nonvanishing elsewhere.

Since u is global strictly concave in $r > 0$ and $\max u(r) = u(r_0)$, we have

$$\max_{x\in\mathbb{R}^2} |\phi(x)|^2 = e^{u(r_0)} = e^{-\alpha},$$

which is (5.4.1).

Next, to see the asymptotic behavior of ϕ, we proceed as follows.

Let $\varepsilon > 0$ be arbitrarily small such that $2N + 4 < \beta - \varepsilon < \beta$. Then

$$\frac{|\phi(x)|^2}{r^{-(\beta-\varepsilon)}} = r^\sigma e^u = e^{t(\frac{u(t)}{t} + [\beta-\varepsilon])}, \tag{5.4.31}$$

where $r = |x|$ and $t = \ln r$. However, since

$$\lim_{t\to\infty}\frac{u(t)}{t} = \lim_{t\to\infty} u'(t) = -\beta,$$

the right-hand side of (5.4.31) goes to zero as $t \to \infty$. This proves that ϕ satisfies a slightly weaker decay estimate than that stated (5.4.2),

$$|\phi(x)|^2 = O(|x|^{-(\beta-\varepsilon)}) \quad \text{as } |x| \to \infty. \tag{5.4.32}$$

On the other hand, we can write down the relation

$$|D_j\phi|^2 = \frac{1}{2}e^u|\nabla u|^2 = \frac{1}{2}e^u u_r^2, \quad r = |x|.$$

However, from $ru_r = u'(t)$ and Lemma 5.4.7, we have $u_r = O(r^{-1})$ for large $r > 0$. As a consequence,

$$|D_j\phi|^2 = O(r^{-(\beta-\varepsilon+2)}) \quad \text{for large } r = |x| > 0.$$

Of course, by (5.2.20), F_{12} also fulfills (5.4.32).

Let $u = u(r)$ be the radially symmetric solution of (5.4.5) just obtained. Then we have

$$\lim_{r\to 0} ru_r(r) = \lim_{r\to 0}\frac{u(r)}{\ln r} = 2N, \quad \lim_{r\to\infty} ru_r(r) = -\beta. \tag{5.4.33}$$

Therefore, by (5.2.20) and (5.4.33), we obtain the electric charge and magnetic flux,

$$\frac{Q}{\kappa} = \Phi = \int_{\mathbb{R}^2} F_{12}\,dx = \frac{2}{\kappa^2}e^u(1 - e^u)\,dx = \frac{4\pi}{\kappa^2}\int_0^\infty re^u(1 - e^u)\,dr$$

$$= \pi\left(\lim_{r\to 0} ru_r(r) - \lim_{r\to\infty} ru_r(r)\right) = 2\pi N + \pi\beta. \qquad (5.4.34)$$

Similarly, we can obtain the value for the energy E as stated in (5.4.3).

The conclusion that β may assume arbitrary value in the interval $(2N + 4, \infty)$ and that ε in (5.4.32) actually vanishes will be established in Chapter 9 when we study a more general two-point boundary value problem.

Thus the proof of Theorem 5.4.1 is complete.

5.5   Solutions on Doubly Periodic Domains

The appearance of lattice structure in the form of spatially periodic vortex-lines is an important phenomenon in a condensed matter system. The purpose of this section is to establish an existence theorem for such solutions in the context of the Chern–Simons theory. We first reformulate the problem over a periodic lattice domain. We next state our existence results. We then present two different analytic approaches to the problem.

5.5.1   Boundary condition modulo gauge symmetry

We are to look for stationary solutions of the Chern–Simons equations over a periodic cell with a *gauge-periodic* boundary condition. That is to say, our field configurations are periodic over a cell domain modulo gauge transformations. Such an important concept was first investigated systematically by 't Hooft [311] and applied to the Ginzburg–Landau theory in [325] to prove the existence of Abrikosov's vortices in a superconducting slab. See also [72, 145, 148] for some related work.

We recall the gauge-symmetry of $\mathcal{L}$ given in (5.2.1) by the general expressions

$$\phi \mapsto \phi e^{i\omega}, \qquad A_\mu \mapsto A_\mu + \partial_\mu\omega.$$

Since we are interested in stationary field configurations, the above gauge-symmetry becomes

$$\phi \mapsto \phi e^{i\omega}, \quad A_0 \mapsto A_0, \quad A_j \mapsto A_j + \partial_j\omega, \qquad (5.5.1)$$

where ω is a real-valued function of the spatial coordinates x_j ($j = 1, 2$) only.

We are now ready to examine the 't Hooft boundary condition.

Consider a basic lattice cell Ω in $\mathbb{R}^2$ generated by independent vectors a^1 and a^2,

$$\Omega = \{x = (x_1, x_2) \in \mathbb{R}^2 \,|\, x = s_1 a^1 + s_2 a^2, \; 0 < s_1, s_2 < 1\}.$$

Define

$$\Gamma^k = \{x \in \mathbb{R}^2 \,|\, x = s_k a^k, \; 0 < s_k < 1\}, \quad k = 1, 2.$$

Then the boundary of Ω is given by

$$\partial \Omega = \Gamma^1 \cup \Gamma^2 \cup \{a^1 + \Gamma^2\} \cup \{a^2 + \Gamma^1\} \cup \{0, a^1, a^2, a^1 + a^2\}.$$

In view of the gauge transformation (5.5.1), we impose the following 't Hooft boundary condition on Ω,

$$
\begin{aligned}
\exp(i\xi_k(x + a^k))\phi(x + a^k) &= \exp(i\xi_k(x))\phi(x), \\
A_0(x + a^k) &= A_0(x), \\
(A_j + \partial_j \xi_k)(x + a^k) &= (A_j + \partial_j \xi_k)(x), \\
x &\in \Gamma^1 \cup \Gamma^2 - \Gamma^k, \\
k &= 1, 2, \quad\quad\quad\quad (5.5.2)
\end{aligned}
$$

where ξ_1, ξ_2 are real-valued smooth functions defined in a neighborhood of $\Gamma^2 \cup \{a^1 + \Gamma^2\}$, $\Gamma^1 \cup \{a^2 + \Gamma^1\}$, respectively.

It will be more convenient to denote the value of a function ξ at a point $x = s_1 a^1 + s_2 a^2 \in \bar{\Omega}$ by $\xi(s_1, s_2)$. Since ϕ is a single-valued complex function, its phase change around Ω can only be a multiple of 2π. Thus, the boundary condition (5.5.2) leads to the equation

$$
\begin{aligned}
&\xi_1(1, 1^-) - \xi_1(1, 0^+) + \xi_1(0, 0^+) - \xi_1(0, 1^-) \\
&+ \xi_2(0^+, 1) - \xi_2(1^-, 1) + \xi_2(1^-, 0) - \xi_2(0^+, 0) + 2\pi N \\
&= 0, \quad\quad\quad\quad\quad\quad\quad\quad\quad\quad\quad\quad\quad\quad (5.5.3)
\end{aligned}
$$

for some integer N. As a consequence of (5.5.2) and (5.5.3), we obtain

$$\Phi = \int_\Omega F_{12}\, dx = \int_{\partial\Omega} A_j\, dx_j = 2\pi N, \quad\quad (5.5.4)$$

which says that the magnetic flux and electric charge (see (5.2.5) are both quantized in a cell domain. Without loss of generality, we assume $N \geq 0$ in the sequel.

As before, we rewrite the Hamiltonian energy density (5.2.7) over the periodic domain Ω as

$$\mathcal{H} = \frac{1}{4}\left[\frac{\kappa}{|\phi|}F_{12} + \frac{2}{\kappa}|\phi|(|\phi|^2 - 1)\right]^2 + |D_1\phi + iD_2\phi|^2 + F_{12} + \Lambda, \quad (5.5.5)$$

where

$$\Lambda = \text{Im}\left\{\partial_j(\epsilon_{jk}\overline{\phi}D_k\phi)\right\}$$

is a total divergence whose integral over the cell Ω vanishes by virtue of (5.5.2) and(5.5.3). Thus, applying (5.5.4) in the decomposition (5.5.5), we find

$$E = E(\phi, A) = \int_\Omega \mathcal{H}\,dx \geq \Phi = 2\pi N,$$

with equality fulfilled if and only if the pair (ϕ, A) satisfies the self-dual equations

$$D_1\phi + iD_2\phi \;=\; 0, \tag{5.5.6}$$

$$F_{12} + \frac{2}{\kappa^2}|\phi|^2(|\phi|^2 - 1) \;=\; 0, \tag{5.5.7}$$

$$\kappa F_{12} - 2A_0|\phi|^2 \;=\; 0 \tag{5.5.8}$$

subject to the periodic boundary condition (5.5.2) and (5.5.3). It is straightforward to examine that the solutions of (5.5.6)–(5.5.8) satisfy the full Chern–Simons equations (5.2.2) and (5.2.3) in Ω. The problem is now defined on a general periodic lattice cell. The solutions of (5.5.6)–(5.5.8) subject to the boundary condition (5.5.2) and (5.5.3) are condensates which saturate the designated energy level labelled by the integer N. Such an energy level is actually determined by the number of vortices confined in the cell. One of the interesting results below is the conclusion that there are only finitely many possible energy levels for each given Chern–Simons coupling parameter κ.

5.5.2   Existence versus coupling parameter

Let the zeros of ϕ be $p_1, \cdots, p_m$ with multiplicities $n_1, \cdots, n_m$, respectively. Then $N = n_1 + \cdots + n_m$ is the total vortex number which leads to the phase condition (5.5.3), i.e., the vortex number gives the winding number of ϕ around the boundary of a lattice cell and thus determines the quantized magnetic and electric charges. Counting algebraic multiplicities, an N-vortex solution is represented by a solution so that ϕ has N zeros. Our basic existence problem for N-vortices is, under what conditions, the system (5.5.6)–(5.5.8) permits a solution satisfying the periodic boundary condition (5.5.2) and (5.5.3) and realizing a prescribed N-zero set for the Higgs field ϕ? Our main existence theorem is stated as follows.

Theorem 5.5.1. *Let $p_1, \cdots, p_m \in \Omega$, $n_1, \cdots, n_m$ be some positive integers, and $N = n_1 + \cdots + n_m$. There is a critical value of the coupling parameter, say, κ_c, satisfying the upper bound*

$$\kappa_c \leq \frac{1}{2}\sqrt{\frac{|\Omega|}{\pi N}}, \tag{5.5.9}$$

so that, for $0 < \kappa < \kappa_c$, the self-dual Chern–Simons equations (5.5.6)–(5.5.8) subject to the periodic boundary condition (5.5.2) and (5.5.3) have an N-vortex solution (ϕ, A), for which, $p_1, \cdots, p_m$ are the zeros of ϕ with multiplicities $n_1, \cdots, n_m$, respectively, and the energy, magnetic flux, and electric charge are given by the formulas

$$E = 2\pi N, \quad \Phi = 2\pi N, \quad Q = 2\pi\kappa N.$$

Moreover, the solution can also be so chosen that the magnitude of ϕ, $|\phi|$, has the largest possible values. Such a solution is called a maximal solution which represents a state that is most superconducting. If $\kappa > \kappa_c$, the equations (5.5.6)–(5.5.8) subject to (5.5.2) and (5.5.3) have no solution realizing the zeros $p_1, \cdots, p_m$ with respective multiplicities $n_1, \cdots, n_m$.

Furthermore, let the prescribed data be denoted by

$$P = \{p_1, \cdots, p_m; n_1, \cdots, n_m\},$$

where the $n's$ may also take zero value, and, write the dependence of κ_c on P by $\kappa_c(P)$. For $P' = \{p_1, \cdots, p_m; n'_1, \cdots, n'_m\}$, we write $P \le P'$ if $n_1 \le n'_1, \cdots, n_m \le n'_m$. Then κ_c is a decreasing function of P in the sense that

$$\kappa_c(P') \le \kappa_c(P), \qquad whenever \; P \le P'. \tag{5.5.10}$$

The inequality (5.5.9) says that, for any given coupling parameter κ, the periodic Chern–Simons system over Ω can only have finitely many saturated energy levels of the form $E = 2\pi N$.

In the following subsections, we present proofs of the these results. The construction employing sub/supersolutions should be useful again in creating numerical simulations of the multivortex solutions. We will show that the iterations can always start from a largest supersolution (with some point singularities) so that the desired solution obtained in the limit is the maximal solution. Although we do not have accurate estimates for the critical number κ_c, the analysis suggests that it seems to depend on the locations of vortices as well as the total vortex number. Another interesting approach is through a variational principle subject to an inequality type constraint. There is a Lagrange multiplier problem if the minimizer occurs at the boundary of the admissible set. A crucial part in the proof is to show that, as long as the parameter κ is not too large, the minimizers must be interior. These proofs are adapted from the original work [63].

5.5.3   Construction via sub- and supersolutions

For convenience, we introduce the new parameter $\lambda = 4/\kappa^2$. In this section, the prescribed zero set of ϕ is written $Z(\phi) = \{p_1, \cdots, p_N\}$ containing all possible multiplicities. Then the new variable $u = \ln|\phi|^2$ reduces the system

(5.5.6)–(5.5.8) to the equation

$$\Delta u = \lambda e^u (e^u - 1) + 4\pi \sum_{j=1}^{N} \delta_{p_j} \quad \text{in } \Omega, \qquad (5.5.11)$$

where δ_p is the Dirac distribution concentrated at $p \in \Omega$. The boundary condition consisting of (5.5.2) and (5.5.3) implies that we are now looking for a solution of (5.5.11) defined on the doubly periodic region Ω or the 2-torus $\Omega = \mathbb{R}^2/\Omega$. In the rest of the study, we always observe this assumption. Conversely, if u is a solution of (5.5.11), a periodic solution pair (A, ϕ) may be constructed by the same formulas we used earlier for the problem over the full plane.

Let u_0 be a solution of the equation (see [17])

$$\Delta u_0 = -\frac{4\pi N}{|\Omega|} + 4\pi \sum_{j=1}^{N} \delta_{p_j}. \qquad (5.5.12)$$

Inserting $u = u_0 + v$ into (5.5.11), we obtain

$$\Delta v = \lambda e^{u_0+v}(e^{u_0+v} - 1) + \frac{4\pi N}{|\Omega|}. \qquad (5.5.13)$$

Integrating this equation on Ω yields the constraint

$$\frac{\lambda}{|\Omega|} \int_\Omega \left(e^{u_0+v} - \frac{1}{2} \right)^2 = \frac{\lambda}{4} - \frac{4\pi N}{|\Omega|}. \qquad (5.5.14)$$

Thus we are led to the following necessary condition for existence,

$$\frac{\lambda}{4} > \frac{4\pi N}{|\Omega|}, \qquad (5.5.15)$$

as expected. We introduce a monotone iterative scheme to solve (5.5.13),

$$
\begin{aligned}
(\Delta - K)v_n &= \lambda e^{u_0+v_{n-1}}(e^{u_0+v_{n-1}} - 1) - Kv_{n-1} + \frac{4\pi N}{|\Omega|}, \\
n &= 1, 2, \cdots, \\
v_0 &= -u_0,
\end{aligned}
\qquad (5.5.16)
$$

where $K > 0$ is a constant to be determined.

Lemma 5.5.2. *Let $\{v_n\}$ be the sequence defined by the scheme (5.5.16) with $K \geq 2\lambda$. Then*

$$v_0 > v_1 > v_2 > \cdots > v_n > \cdots > v_- \qquad (5.5.17)$$

for any subsolution v_- of (5.5.13). Thus, if there exists a subsolution, the sequence $\{v_n\}$ converges to a solution of (5.5.13) in the space $C^k(\Omega)$ for any $k \geq 0$ and such a solution is the maximal solution of the equation.

Proof. We prove (5.5.17) by induction.

First, it is standard that $v_1 \in C^\infty(\Omega - \{p_1, \cdots, p_N\}) \cap C^\alpha(\Omega)$ $(0 < \alpha < 1)$. Since $(\Delta - K)(v_1 - v_0) = 0$ in $\Omega - \{p_1, \cdots, p_N\}$ and $v_1 - v_0 < 0$ on $\partial\Omega_\varepsilon$ where Ω_ε is the complement of $\cup_{j=1}^N \{x \mid |x - p_j| < \varepsilon\}$ in Ω with ε sufficiently small, the maximal principle implies $v_1 - v_0 < 0$ in Ω_ε. Hence $v_1 - v_0 < 0$ throughout.

Suppose that $v_0 > v_1 > \cdots > v_k$. We obtain from (5.5.16) and $K \geq 2\lambda$ that

$$
\begin{aligned}
&(\Delta - K)(v_{k+1} - v_k) \\
&= \lambda e^{2u_0}(e^{2v_k} - e^{2v_{k-1}}) - K(v_k - v_{k-1}) - \lambda e^{u_0}(e^{v_k} - e^{v_{k-1}}) \\
&\geq 2\lambda e^{2u_0 + 2w}(v_k - v_{k-1}) - K(v_k - v_{k-1}) \\
&\geq K(e^{2u_0 + 2v_0} - 1)(v_k - v_{k-1}) = 0,
\end{aligned}
$$

where $v_k \leq w \leq v_{k-1} \leq v_0$. The maximum principle implies $v_{k+1} - v_k < 0$ in Ω.

Next, we establish the lower bound in (5.5.17) in terms of the subsolution v_- of (5.5.13), namely, $v_- \in C^2(\Omega)$ and

$$
\Delta v_- \geq \lambda e^{u_0 + v_-}(e^{u_0 + v_-} - 1) + \frac{4\pi N}{|\Omega|}. \tag{5.5.18}
$$

Initially, we have in view of the definition of v_0 and (5.5.18) that

$$
\begin{aligned}
\Delta(v_- - v_0) &= \Delta(v_- + u_0) \\
&\geq \lambda e^{u_0 + v_-}(e^{u_0 + v_-} - 1) = \lambda e^{v_- - v_0}(e^{v_- - v_0} - 1)
\end{aligned}
$$

in $\Omega - \{p_1, \cdots, p_N\}$. So if $\varepsilon > 0$ is small, then $v_- - v_0 < 0$ on $\partial\Omega_\varepsilon$, and by the maximum principle, we have $v_- - v_0 < 0$ in Ω_ε. Hence $v_- - v_0 < 0$ throughout Ω.

Now suppose that $v_- < v_k$ for some $k \geq 0$. Then (5.5.17) and (5.5.18) give us

$$
\begin{aligned}
&(\Delta - K)(v_- - v_{k+1}) \\
&= \lambda e^{2u_0}(e^{2v_-} - e^{2v_k}) - K(v_- - v_k) - \lambda e^{u_0}(e^{v_-} - e^{v_k}) \\
&\geq 2\lambda e^{2u_0 + w}(v_- - v_k) - K(v_- - v_k) \\
&\geq K(e^{2u_0 + 2v_0} - 1)(v_- - v_k) = 0,
\end{aligned}
$$

where $v_- \leq w \leq v_k \leq v_0$. So the maximum principle again implies that $v_- < v_{k+1}$.

The statement of convergence follows from (5.5.17) and a standard bootstrap argument.

Lemma 5.5.3. *If $\lambda > 0$ is sufficiently large, the equation (5.5.13) has a subsolution v_- satisfying (5.5.18).*

Proof. Choose small $\varepsilon > 0$ so that the balls

$$B(p_j; 2\varepsilon) = \left\{ x \in \Omega \ \middle| \ |x - p_j| < 2\varepsilon \right\}, \qquad j = 1, 2, \cdots, N$$

satisfy $B(p_j; 2\varepsilon) \cap B(p_{j'}; 2\varepsilon) = \emptyset$ for $p_j \neq p_{j'}$. Let f_ε be a smooth function so that $0 \leq f_\varepsilon \leq 1$ and

$$
\begin{aligned}
f_\varepsilon(x) &= 1, \quad x \in B(p_j; \varepsilon), \quad j = 1, 2, \cdots, N, \\
f_\varepsilon(x) &= 0, \quad x \notin \cup_{j=1}^N B(p_j; 2\varepsilon).
\end{aligned}
$$

Then we have

$$C(\varepsilon) = \frac{1}{|\Omega|} \int_\Omega \frac{8\pi N}{|\Omega|} f_\varepsilon \, dx \leq \frac{32\pi^2 N^2}{|\Omega|^2} \varepsilon^2. \tag{5.5.19}$$

Define

$$g_\varepsilon = \frac{8\pi N}{|\Omega|} f_\varepsilon - C(\varepsilon).$$

Since $\int_\Omega g_\varepsilon \, dx = 0$, we know that the equation

$$\Delta w = g_\varepsilon \tag{5.5.20}$$

has a solution which is unique up to an additive constant.

First, from (5.5.19), we see that, for $x \in B(p_j; \varepsilon)$,

$$g_\varepsilon \geq \frac{4\pi N}{|\Omega|} \left(2 - \frac{8\pi N}{|\Omega|} \varepsilon^2 \right) > \frac{4\pi N}{|\Omega|} \tag{5.5.21}$$

if ε is small enough. In the following, we fix ε so that (5.5.21) is valid.

Next, we choose a solution of (5.5.20), say, w_0, to fulfill

$$e^{u_0 + w_0} < 1, \qquad x \in \Omega. \tag{5.5.22}$$

Therefore, for any $\lambda > 0$, we have the inequality

$$
\begin{aligned}
\Delta w_0 \ = \ g_\varepsilon &> \frac{4\pi N}{|\Omega|} \\
&\geq \ \lambda e^{u_0 + w_0} (e^{u_0 + w_0} - 1) + \frac{4\pi N}{|\Omega|}, \quad x \in B(p_j; \varepsilon), \\
j &= \ 1, 2, \cdots, N.
\end{aligned} \tag{5.5.23}
$$

Finally, set

$$
\begin{aligned}
\mu_0 &= \ \inf \left\{ e^{u_0 + w_0} \ \middle| \ x \in \Omega - \cup_{j=1}^N B(p_j; \varepsilon) \right\}, \\
\mu_1 &= \ \sup \left\{ e^{u_0 + w_0} \ \middle| \ x \in \Omega - \cup_{j=1}^N B(p_j; \varepsilon) \right\}.
\end{aligned}
$$

Then $0 < \mu_0 < \mu_1 < 1$ and $e^{u_0 + w_0}(e^{u_0 + w_0} - 1) \leq \mu_0(\mu_1 - 1) = -C_0 < 0$ for $x \in \Omega - \cup_{j=1}^{N} B(p_j; \varepsilon)$. As a consequence, we can choose $\lambda > 0$ sufficiently large to fulfill (5.5.23) in entire Ω. Thus, w_0 is a subsolution of (5.5.13).

Lemma 5.5.4. *There is a critical value of λ, say, λ_c, satisfying*

$$\lambda_c \geq \frac{16\pi N}{|\Omega|}, \tag{5.5.24}$$

so that, for $\lambda > \lambda_c$, the equation (5.5.13) has a solution, while for $\lambda < \lambda_c$, the equation has no solution.

Proof. Suppose that v is a solution of (5.5.13). Then $u = u_0 + v$ verifies (5.5.11) and is negative near the point $x = p_j$, $j = 1, 2, \cdots, N$. Using the maximum principle away from the points p_j's, we find that $u < 0$ in Ω.
Define

$$\Lambda = \left\{ \lambda > 0 \,\Big|\, \lambda \text{ is such that (5.5.13) has a solution} \right\}.$$

Then Λ is an interval. To show this fact, we prove that, if $\lambda' \in \Lambda$, then $[\lambda', \infty) \subset \Lambda$. In fact, denote by v' a solution of (5.5.13) at $\lambda = \lambda'$. Since $u_0 + v' < 0$, we see that v' is a subsolution of (5.5.13) for any $\lambda \geq \lambda'$. By virtue of Lemma 5.5.2, we obtain $\lambda \subset \Lambda$ as desired.

Set $\lambda_c = \inf \Lambda$. Then $\lambda > 16\pi N/|\Omega|$ for any $\lambda > \lambda_c$ by (5.5.15). Taking the limit $\lambda \to \lambda_c$, we arrive at (5.5.24). Thus the proof is concluded.

Recall the notation in Theorem 5.5.1 for the data of the prescribed zeros of the Higgs field,

$$P = \{p_1, \cdots, p_m; n_1, \cdots, n_m\}, \quad P' = \{p_1, \cdots, p_m; n_1', \cdots, n_m'\},$$

and the order $P \leq P'$. Then the corresponding statement in Theorem 5.5.1 is related to the solvability of the following form of the equation (5.5.11),

$$\Delta u = \lambda e^u(e^u - 1) + \sum_{j=1}^{N} n_j \delta_{p_j}, \tag{5.5.25}$$

in view of the parameter λ. We denote the dependence of λ_c on P by $\lambda_c(P)$ (see Lemma 5.5.4).

Lemma 5.5.5. $\lambda_c(P) \leq \lambda_c(P')$ *for $P \leq P'$. Hence (5.5.10) holds.*

Proof. We need only to show that, if $\lambda > \lambda_c(P')$, then $\lambda \geq \lambda_c(P)$.
Let u' be a solution of (5.5.25) with $n_j = n_j'$, $j = 1, 2, \cdots, m$ and u_0 satisfy

$$\Delta u_0 = -\frac{4\pi N}{|\Omega|} + 4\pi \sum_{j=1}^{m} n_j \delta_{p_j},$$

where $N = n_1 + \cdots + n_m$. Then the substitution $u' = u_0 + v_-$ leads to

$$\Delta v_- = \lambda e^{u_0 + v_-} (e^{u_0 + v_-} - 1) + \frac{4\pi N}{|\Omega|} + 4\pi \sum_{j=1}^{m} (n_j' - n_j)\delta_{p_j},$$

which implies in particular that v_- is a subsolution of (5.5.13) in the sense of distributions and (5.5.17) holds pointwise. However, since the singularity of v_- at $x = p_j$ is at most of the type $\ln |x - p_1|$, the inequality (5.5.17) still results in the convergence of the sequence $\{v_n\}$ to a solution of (5.5.13) in any C^k-norm. In fact, using (5.5.17), we see that $\{v_n\}$ converges almost everywhere and is bounded in the L^2-norm. Hence the sequence converges in L^2. Similarly, the right-hand side of (5.5.16) also converges in L^2. Applying the standard L^2-estimates we see that the sequence converges in $W^{2,2}$ to a strong solution of (5.5.13). Thus a classical solution is obtained. The convergence in C^k follows again from a bootstrap argument. This proves $\lambda \geq \lambda_c(P)$. Thus $\lambda_c(P) \leq \lambda_c(P')$ as expected and the lemma is proven.

It is clear that the lemmas of this section furnish the proofs of all the statements made in Theorem 5.5.1.

5.5.4   Alternative variational treatment

In this subsection, we shall formulate a variational solution of the equation (5.5.13) by using an inequality-type constraint. This problem is of independent interest due to the two exponential nonlinear terms in (5.5.13). Recall that a similar equation of the form $\Delta v = K_0 - K e^v$ arises in the prescribed curvature problem for a 2-surface, compact or noncompact, which has been studied extensively [20, 63, 164, 165, 166, 213, 226]. A basic structure of this latter problem is that it permits a constrained variational principle so that the Lagrange multiplier arising from the constraint naturally recovers the original coefficient in the equation. In our equation (5.5.13), the two exponential terms ruin such an approach because the Fréchet derivative of the constraint functional cannot assume a suitable form allowing the recovery of the original equation. Our variational treatment of (5.5.13) can be briefly sketched as follows. We first replace the equality constraint (5.5.14) by an inequality constraint which is equivalent to the solvability of the equality constraint and defines the admissible set, $\mathcal{A}$, for a suitable objective functional, I. We then show that when λ is large, the minimizer of I will stay in the interior of $\mathcal{A}$, hence we are able to avoid the Lagrange multiplier problem arising from the equality constraint. Finally we prove that the minimizer obtained in a smaller space is actually a critical point of I in the usual Sobolev space. Thus a solution of (5.5.13) is found.

We use the notation $U = e^{u_0}$. Then (5.5.13) takes the form

$$\Delta v = \lambda U e^v (U e^v - 1) + \frac{4\pi N}{|\Omega|}. \tag{5.5.26}$$

The function $U \geq 0$ is smooth since u_0 behaves like $\ln |x - p|^2$ near the prescribed vortex point p.

We shall work on the standard space $\mathcal{S} = W^{1,2}(\Omega)$ (the standard Sobolev space over the 2-torus Ω). Then

$$X = \left\{ v \in \mathcal{S} \ \middle| \ \int_\Omega v = 0 \right\}$$

is a closed subspace of $\mathcal{S}$ and $\mathcal{S} = \mathbb{R} \oplus X$. That is, for any $v \in \mathcal{S}$, there is a unique number $c \in \mathbb{R}$ and $v' \in X$ so that

$$v = c + v'. \tag{5.5.27}$$

Suppose that $v \in \mathcal{S}$ given in (5.5.27) satisfies (5.5.14). Then

$$e^{2c} \int_\Omega U^2 e^{2v'} - e^c \int_\Omega U e^{v'} + \frac{4\pi N}{\lambda} = 0. \tag{5.5.28}$$

Of course (5.5.28) is a quadratic equation in $t = e^c$ which has a solution if and only if

$$\left(\int_\Omega U e^{v'} \right)^2 - \frac{16\pi N}{\lambda} \int_\Omega U^2 e^{2v'} \geq 0. \tag{5.5.29}$$

In this case we may choose $c = c(v')$ in (5.5.28) to satisfy

$$e^c = \frac{\int_\Omega U e^{v'} + \sqrt{\left(\int_\Omega U e^{v'} \right)^2 - \frac{16\pi N}{\lambda} \int_\Omega U^2 e^{2v'}}}{2 \int_\Omega U^2 e^{2v'}}. \tag{5.5.30}$$

With v' satisfying (5.5.29) and c given by (5.5.30), we define a functional I on X by the expression

$$I(v') = \int_\Omega \left\{ \frac{1}{2} |\nabla v'|^2 + \frac{\lambda}{2} U^2 e^{2c+2v'} - \lambda U e^{c+v'} \right\} + 4\pi N c. \tag{5.5.31}$$

Set $\mathcal{A} = \{ v' \in X \mid v' \text{ satisfies (5.5.29)} \}$. Consider the optimization problem

$$\min \left\{ I(v') \ \middle| \ v' \in \mathcal{A} \right\}. \tag{5.5.32}$$

We shall find some condition under which the problem (5.5.32) has only interior minimizers.

Lemma 5.5.6. *For $v' \in X$ on the boundary of $\mathcal{A}$, namely,*

$$\left(\int_\Omega U e^{v'} \right)^2 - \frac{16\pi N}{\lambda} \int_\Omega U^2 e^{2v'} = 0, \tag{5.5.33}$$

we have $I(v') \geq -4\pi \ln \lambda - C$ for some constant $C > 0$ independent of λ.

Proof. From (5.5.30) and (5.5.33), we obtain

$$e^c = \frac{\int_\Omega U e^{v'}}{2 \int_\Omega U^2 e^{2v'}} = \frac{8\pi N}{\lambda \int_\Omega U e^{v'}}. \tag{5.5.34}$$

Therefore a simple calculation shows that

$$I(v') = -6\pi N + \frac{1}{2}\|\nabla v'\|_2^2 + 4\pi N c, \tag{5.5.35}$$

where and in the sequel we use $\|\ \|_2$ to denote the usual L^2-norm on Ω.
We rewrite (5.5.34) as

$$c = \ln 8\pi N - \ln \lambda - \ln\left(\int_\Omega U e^{v'}\right). \tag{5.5.36}$$

Let $p, q > 1$ be conjugate exponents to be determined so that $1/p + 1/q = 1$. In view of the Schwartz inequality and the Trudinger–Moser inequality (4.3.12), we have the following upper bound for $\ln(\int U e^{v'})$,

$$\ln\left(\int_\Omega U e^{v'}\right) \leq \frac{1}{p}\ln\left(\int_\Omega U^p\right) + \frac{1}{q}\ln\left(\int_\Omega e^{qv'}\right)$$
$$\leq \frac{1}{p}\ln\left(\int_\Omega U^p\right) + q\left(\frac{1}{16\pi} + \varepsilon\right)\|\nabla v'\|_2^2 + \frac{1}{q}\ln C(\varepsilon). \tag{5.5.37}$$

Using (5.5.36) and (5.5.37) in (5.5.35), we arrive at

$$I(v') \geq \left(\frac{1}{2} - 4\pi N q\left[\frac{1}{16\pi} + \varepsilon\right]\right)\|\nabla v'\|_2^2 - 4\pi N \ln \lambda - C(\varepsilon, q). \tag{5.5.38}$$

If $N = 1$, we can choose suitable $\varepsilon > 0$ and $q > 1$ above to make the coefficient of the first term on the right-hand side of (5.5.38) positive. If $N > 1$, we need an inequality derived by Nolasco and Tarantello [233, 234] for functions in $\mathcal{A}$.

In fact, we may rewrite the constraint (5.5.29) as

$$\int_\Omega U^2 e^{2v'} \leq \frac{\lambda}{16\pi N}\left(\int_\Omega U e^{v'}\right)^2. \tag{5.5.39}$$

Let $s \in (0,1)$ and $a = 1/(2-s)$. Then $sa + 2(1-a) = 1$. Since $p = 1/a, q = 1/(1-a) > 1$ are conjugate exponents, namely, $1/p + 1/q = 1$, we have in view of (5.5.39) that

$$\int_\Omega U e^{v'} = \int_\Omega (U^{sa} e^{sav'})(U^{2(1-a)} e^{2(1-a)v'})$$
$$\leq \left(\int_\Omega U^s e^{sv'}\right)^a \left(\int_\Omega U^2 e^{2v'}\right)^{1-a}$$
$$\leq \left(\frac{\lambda}{16\pi N}\right)^{1-a}\left(\int_\Omega U^s e^{sv'}\right)^a \left(\int_\Omega U e^{v'}\right)^{2(1-a)},$$

which reads,

$$\left(\int_\Omega U e^{v'}\right)^{2a-1} \le \left(\frac{\lambda}{16\pi N}\right)^{1-a} \left(\int_\Omega U^s e^{sv'}\right)^a.$$ (5.5.40)

Consequently, there holds

$$\int_\Omega U e^{v'} \le \left(\frac{\lambda}{16\pi N}\right)^{\frac{1-s}{s}} \left(\int_\Omega U^s e^{sv'}\right)^{\frac{1}{s}}$$

$$\le \left(\frac{\lambda}{16\pi N}\right)^{\frac{1-s}{s}} \left(\max_{x\in\Omega}\left\{U(x)\right\}\right)\left(\int_\Omega e^{sv'}\right)^{\frac{1}{s}}.$$ (5.5.41)

Applying (4.3.12) in (5.5.41), we obtain the following upper bound instead of (5.5.37),

$$\ln\left(\int_\Omega U e^{v'}\right) \le C_1(\varepsilon,s) + \left(\frac{1}{16\pi N}+\varepsilon\right)s\|\nabla v'\|_2^2 \quad v'\in\mathcal{A},$$ (5.5.42)

where $C(\varepsilon,s) > 0$ is a constant depending on ε and s. Thus, in place of (5.5.38), we have

$$I(v') \ge \left(\frac{1}{2} - 4\pi N s\left[\frac{1}{16\pi}+\varepsilon\right]\right)\|\nabla v'\|_2^2 - 4\pi N \ln\lambda - C_2(\varepsilon,s).$$ (5.5.43)

Of course, given N, we can choose s to make the coefficient of the first term on the right-hand side of (5.5.43) positive. Therefore the lemma follows.

We now evaluate I at an interior trial point in the admissible set $\mathcal{A}$. For convenience, we choose $v' = 0$ as a trial element.

Lemma 5.5.7. *Suppose that $\lambda > 0$ is sufficiently large so that*

$$\left(\int_\Omega U\right)^2 - \frac{16\pi N}{\lambda}\int_\Omega U^2 > 0,$$

i.e., $v' = 0$ lies in the interior of $\mathcal{A}$. Then there are constants $C_1, C_2 > 0$ independent of λ so that $I(0) \le -C_1\lambda + C_2$.

Proof. Assume that $c_0 = c$ is given by the expression (5.5.30) with $v' = 0$. The expression (5.5.31) with $c = c_0$ and $v' = 0$ enable us to obtain

$$I(0) = -\frac{\lambda}{2}e^{c_0}\int_\Omega U + 4\pi N\left(c_0 - \frac{1}{2}\right).$$ (5.5.44)

However, the equation (5.5.30) says that

$$\frac{\int_\Omega U}{2\int_\Omega U^2} < e^{c_0} < \frac{\int_\Omega U}{\int_\Omega U^2}.$$

Inserting this into (5.5.44), we obtain

$$I(0) < -\lambda \frac{\left(\int_\Omega U\right)^2}{4\int_\Omega U^2} + 4\pi N \ln\left(\frac{\int_\Omega U}{\int_\Omega U^2}\right).$$

Recall that U is independent of λ. Therefore the lemma follows.

From the above two lemmas, we see that there is a $\lambda_0 > 0$ so that

$$I(0) < -1 + I(w'), \qquad w' \in \partial\mathcal{A}, \quad \lambda > \lambda_0. \tag{5.5.45}$$

So it is hopeful to get an interior minimizer for (5.5.32). From now on we always assume that λ is such that (5.5.45) holds.

Lemma 5.5.8. *There are constants $C_1, C_2 > 0$ so that*

$$I(v') \geq C_1\|\nabla v'\|_2^2 - C_2, \qquad v' \in \mathcal{A}.$$

Proof. Using (5.5.29) and (5.5.30), we have

$$e^c \geq \frac{8\pi N}{\lambda}\left(\int_\Omega U e^{v'}\right)^{-1}.$$

As a consequence,

$$c \geq -\ln\left(\int_\Omega U e^{v'}\right) + \ln\left(\frac{8\pi N}{\lambda}\right). \tag{5.5.46}$$

On the other hand, the two exponential terms in $I(v')$ (see (5.5.31)) are easily controlled. In fact, using the Schwarz inequality, we have

$$\int_\Omega \left\{\frac{1}{2}U^2 e^{2c+2v'} - U e^{c+v'}\right\} = \frac{1}{2}\int_\Omega (U e^{c+v'} - 1)^2 - \frac{1}{2}|\Omega|$$
$$\geq -2|\Omega|. \tag{5.5.47}$$

Finally, inserting (5.5.46) and (5.5.47) into (5.5.31) and applying the Trudinger–Moser inequality (4.3.12) again as in the proof of Lemma 5.5.6, we arrive at the conclusion of the lemma.

Lemma 5.5.9. *The problem (5.5.32) has a minimizer v' which lies in the interior of the admissible set $\mathcal{A}$.*

Proof. Let $\{v'_n\}$ be a minimizing sequence of (5.5.32). From Lemma 5.5.8 and the Poincaré inequality, we see that $\{v'_n\}$ is bounded in X. Therefore we may assume without loss of generality that $\{v'_n\}$ weakly converges to an element of X, say, v'. Since the mapping $X \to L(\Omega)$ given by $f \mapsto e^f$ is well-defined and compact (see [17]), we know that $v' \in \mathcal{A}$ and $c(v'_n) \to c(v')$ as

$n \to \infty$. Applying this observation in (5.5.31) we see that v' is a minimizer of (5.5.32). Moreover, (5.5.45) implies

$$I(v') \le -1 + \inf\left\{ I(w') \;\middle|\; w' \in \partial A \right\}.$$

In other words, v' belongs to the interior of A.

Since our optimization problem is defined on the subspace X of S, it is not obvious whether a critical point of I in X gives rise to a solution of the equation (5.5.26). In the following, we will examine that the composition $c + v'$ with c defined by (5.5.30) indeed is a critical point of I in the full S and thus solves the equation (5.5.26).

Lemma 5.5.10. *Let v' be the minimizer produced in Lemma 5.5.9 and the number c defined by (5.5.30). Then $v = c + v'$ is a solution of (5.5.13).*

Proof. In fact, since v' is an interior minimizer, the Fréchet derivative of I at v' vanishes:

$$[dI(v')](w') = 0 \quad \text{for any } w' \in X.$$

It is more convenient to rewrite the above equation in the functional form

$$0 = \int_\Omega \left\{ \nabla v' \cdot \nabla w' + \lambda[U^2 e^{2c(v')+2v'} - U e^{c(v')+v'}]w' \right\}$$
$$+ \; [D_{w'}c(v')] \int_\Omega \left\{ \lambda[U^2 e^{2c(v')+2v'} - U e^{c(v')+v'}] + \frac{4\pi N}{|\Omega|} \right\}, \quad (5.5.48)$$

where the numerical factor in front of the second integral above, i.e.,

$$D_{w'}c(v') = \frac{d}{dt}c(v' + tw')\bigg|_{t=0},$$

is the directional derivative of c at v' along w'. On the other hand, in view of the equation (5.5.28), the second integral above actually vanishes. Thus (5.5.48) takes the simplified form

$$\int_\Omega \left\{ \nabla v' \cdot \nabla w' + \lambda[U^2 e^{2c+2v'} - U e^{c+v'}]w' \right\} = 0. \qquad (5.5.49)$$

Consider the decomposition $L^2(\Omega) = \mathbb{R} + Y$ where

$$Y = \left\{ f \in L^2(\Omega) \;\middle|\; \int_\Omega f = 0 \right\}.$$

Choose a suitable $\sigma \in \mathbb{R}$ such that

$$\lambda(U^2 e^{2c+2v'} - U e^{c+v'}) + \sigma \in Y.$$

Then the relation $X \subset Y$ and (5.5.49) imply that

$$
\begin{aligned}
0 &= \int_\Omega \left\{ \nabla v' \cdot \nabla w' + (\lambda[U^2 e^{2c(v')+2v'} - U e^{c(v')+v'}] + \sigma)w' \right\} \\
&= \int_\Omega \left\{ \nabla v' \cdot \nabla(a + w') + (\lambda[U^2 e^{2c(v')+2v'} - U e^{c(v')+v'}] + \sigma)(a + w') \right\}
\end{aligned}
$$

for any $a \in \mathbb{R}$. Consequently,

$$
\int_\Omega \left\{ \nabla v' \cdot \nabla w + (\lambda[U^2 e^{2c+2v'} - U e^{c+v'}] + \sigma)w \right\} = 0, \quad \forall w \in \mathcal{S}.
$$

This equation implies that v' is a smooth solution of

$$
\Delta v' = \lambda U e^{c+v'}(U e^{c+v'} - 1) + \sigma. \tag{5.5.50}
$$

Integrating (5.5.50) yields

$$
\lambda \int_\Omega (U e^{c+v'} - U^2 e^{2c+2c'}) = \sigma|\Omega|.
$$

Comparing the above equation with (5.5.28), we obtain immediately the relation $\sigma|\Omega| = 4\pi N$. Thus, by (5.5.50), we see that $v = c + v'$ solves (5.5.26) and the existence proof is complete.

5.6   Tarantello's Secondary Solution

The work of Tarantello [302] shows that for $\lambda > \lambda_c$ the equation (5.5.13) has at least two solutions. Her idea was first to establish the existence of a solution at $\lambda = \lambda_c$, hence necessarily that $\lambda_c > 16\pi N/|\Omega|$, and then use the solution at λ_c as a lower barrier to prove the existence of a secondary solution for any given $\lambda > \lambda_c$ through a min-max approach. In this section, we present Tarantello's solutions.

5.6.1   Critical coupling parameter

First, we observe that the maximum solutions of (5.5.13), $\{v_\lambda \mid \lambda > \lambda_c\}$, are a monotone family in the sense that $v_{\lambda_1} > v_{\lambda_2}$ whenever $\lambda_1 > \lambda_2 > \lambda_c$ because $u_0 + v_\lambda < 0$ ($\lambda > \lambda_c$). In fact, from (5.5.13) with $\lambda = \lambda_2$ and $v = v_{\lambda_2}$, we have

$$
\begin{aligned}
\Delta v_{\lambda_2} &= \lambda_1 e^{u_0 + v_{\lambda_2}}(e^{u_0 + v_{\lambda_2}} - 1) \\
&\quad + (\lambda_2 - \lambda_1)e^{u_0 + v_{\lambda_2}}(e^{u_0 + v_{\lambda_2}} - 1) + \frac{4\pi N}{|\Omega|} \\
&\geq \lambda_1 e^{u_0 + v_{\lambda_2}}(e^{u_0 + v_{\lambda_2}} - 1) + \frac{4\pi N}{|\Omega|}.
\end{aligned}
$$

Hence v_{λ_2} is a subsolution of (5.5.13) at $\lambda = \lambda_1$. Therefore $v_{\lambda_1} > v_{\lambda_2}$.

Lemma 5.6.1. *For $v_\lambda = c_\lambda + v'_\lambda$ where $c_\lambda \in \mathbb{R}$ and $v'_\lambda \in X$, we have the bound*

$$\|\nabla v'_\lambda\|_2 \leq C\lambda, \qquad (5.6.1)$$

where the constant C depends only on the size of the torus Ω. Moreover, $\{c_\lambda\}$ satisfies a similar estimate, $|c_\lambda| \leq C(1 + \lambda + \lambda^2)$. In particular, v_λ satisfies the bound $\|v_\lambda\|_S \leq C(1 + \lambda + \lambda^2)$.

Proof. Multiplying (5.5.13) by the function v'_λ, integrating over Ω, and using the Poincaré inequality, we have

$$\|\nabla v'_\lambda\|_2^2 = \lambda \int_\Omega e^{u_0 + v_\lambda}(1 - e^{u_0 + v_\lambda})v'_\lambda < 2\lambda \int_\Omega |v'_\lambda| \leq C\lambda \|\nabla v'_\lambda\|_2,$$

which proves (5.6.1). On the other hand, the property $u_0 + v_\lambda < 0$ gives us the upper bound

$$c_\lambda = \frac{1}{|\Omega|} \int_\Omega v_\lambda < -\frac{1}{|\Omega|} \int_\Omega u_0.$$

Furthermore, from (5.5.28) and the Trudinger–Moser inequality, we have

$$e^{c_\lambda} \geq \frac{4\pi N}{\lambda} \left(\int_\Omega e^{u_0 + v'_\lambda} \right)^{-1} > C_1 \lambda^{-1} e^{-C_2 \|\nabla v'_\lambda\|_2^2}.$$

Inserting (5.6.1) into the above, we see that a lower bound for c_λ of the form $-C(1 + \lambda + \lambda^2)$ is valid.

Theorem 5.6.2. *The set of λ for which the equation (5.5.13) has a solution is a closed interval. In other words, (5.5.13) has a solution for $\lambda = \lambda_c$ as well.*

Proof. For $\lambda_c < \lambda < \lambda_c + 1$ (say), the set $\{v_\lambda\}$ is bounded in S by Lemma 5.6.1. Since $\{v_\lambda\}$ is monotone with respect to λ, we see that there holds the following weak convergence in S,

$$\lim_{\lambda \to \lambda_c} v_\lambda = v_*, \quad v_* \in S. \qquad (5.6.2)$$

Hence $v_\lambda \to v_*$ strongly in $L^p(\Omega)$ for any $p \geq 1$. Consequently, $e^{v_\lambda} \to e^{v_*}$ strongly in $L^p(\Omega)$ for any $p \geq 1$ in view of (5.6.2) and the Trudinger–Moser inequality. Using this result in (5.5.13) and the L^2-estimates for elliptic equations, we obtain $v_* \in W^{2,2}(\Omega)$ and $v_\lambda \to v_*$ in $W^{2,2}(\Omega)$ as $\lambda \to \lambda_c$. In particular, letting $\lambda \to \lambda_c$ in (5.5.13), we see that v_* is a solution of (5.5.13) for $\lambda = \lambda_c$.

5.6.2   Local minimum

Suppose that

$$I(v) = \int_\Omega \left\{ \frac{1}{2}|\nabla v|^2 + F(x,v) \right\} dx$$

is a C^1-functional over S. We use the notation

$$f(x,v) = \frac{\partial F(x,v)}{\partial v}, \quad g(x,v) = \frac{\partial f(x,v)}{\partial v},$$

assume that $|g(x,v)| \le h(x,|v|)$ where $h(x,s) \ge 0$ increases in s, and that for some $p > 1$

$$\int_\Omega (h(x,|v|))^p \, dx$$

converges for any $v \in S$ in this subsection. As a preparation, we establish

Lemma 5.6.3. *If v_* is a weak subsolution of the Euler–Lagrange equation*

$$\Delta v = f(x,v)$$

in S in the sense that

$$\int_\Omega \{\nabla v_* \cdot \nabla w + f(x,v_*)w\} \, dx \le 0, \quad \forall w \in S, \quad w \ge 0 \ a.e. \ in \ \Omega, \tag{5.6.3}$$

then the solution of the constrained optimization problem

$$\min \left\{ I(v) \,\Big|\, v \in S_* \right\}, \quad S_* = \left\{ v \in S \,\Big|\, v \ge v_* \ a.e. \ in \ \Omega \right\} \tag{5.6.4}$$

is a critical point of the functional I in S.

Proof. Let v be a solution of (5.6.4). For any given $\varphi \in S$ and $t > 0$, we have

$$v_t \equiv \max\{v + t\varphi, v_*\} \in S_*.$$

Using h_+ and h_- to denote the positive and negative parts of a function h, respectively, we can examine that

$$v_t = v + t\varphi + w_t, \quad w_t = (v + t\varphi - v_*)_-.$$

Consequently, we have

$$\begin{aligned}
0 \ \le \ & \frac{1}{t}(I(v_t) - I(v)) \\
= \ & \frac{1}{2t}(\|\nabla(v + t\varphi + w_t)\|_2^2 - \|\nabla v\|_2^2) \\
& + \frac{1}{t}\int_\Omega \{F(x, v + t\varphi + w_t) - F(x,v)\} \, dx \\
= \ & \frac{1}{2t}\|\nabla(t\varphi + w_t)\|_2^2 + \int_\Omega \frac{1}{t}\nabla v \cdot \nabla(t\varphi + w_t) \, dx + \frac{1}{t}\int f(x,v)t\varphi \, dx \\
& + \frac{1}{t}\int_\Omega (F(x, v + t\varphi + w_t) - F(x,v) - f(x,v)t\varphi) \, dx.
\end{aligned}$$

From the above result, we obtain

$$(dI(v))(\varphi) = \int_\Omega (\nabla v \cdot \nabla \varphi + f(x,v)\varphi)\, dx$$

$$\geq -\frac{t}{2}\|\nabla\varphi\|_2^2 - \int_\Omega \nabla\varphi \cdot \nabla w_t \, dx - \frac{1}{2t}\|\nabla w_t\|_2^2 - \frac{1}{t}\int_\Omega \nabla v \cdot \nabla w_t \, dx$$

$$-\frac{1}{t}\int_\Omega \{F(x, v + t\varphi + w_t) - F(x,v) - f(x,v)t\varphi\}\, dx$$

$$= O(t) + \frac{1}{t}\int_\Omega \nabla(-t\varphi - v + v_*) \cdot \nabla w_t \, dx$$

$$-\frac{1}{2t}\|\nabla w_t\|_2^2 - \frac{1}{t}\int_\Omega \nabla v_* \cdot \nabla w_t \, dx - \frac{1}{t}\int_\Omega f(x,v_*)w_t \, dx$$

$$-\frac{1}{t}\int_\Omega \{F(x, v + t\varphi + w_t) - F(x,v) - f(x,v)(t\varphi + w_t)\}\, dx$$

$$+\frac{1}{t}\int_\Omega (f(x,v_*) - f(x,v))w_t \, dx$$

$$\geq O(t) - \frac{1}{t}\int_\Omega \left(\int_0^1 \int_0^1 sg(x, v + rs(t\varphi + w_t))\, drds\right)(t\varphi + w_t)^2 \, dx$$

$$+\frac{1}{t}\int_\Omega (f(x,v_*) - f(x,v))w_t \, dx, \tag{5.6.5}$$

where we have used the inequality (5.6.3) with the test function w_t and

$$\int_\Omega \nabla(-t\varphi - v + v_*) \cdot \nabla w_t \, dx = \|\nabla w_t\|_2^2.$$

On the other hand, since for $w_t > 0$ we have $\varphi < 0$ and $t\varphi < v_* - v$, hence,

$$|t\varphi + w_t| = |v_* - v| = v - v_* < -t\varphi = t|\varphi|.$$

Thus, in general, we have

$$(t\varphi + w_t)^2 \leq t^2\varphi^2, \quad x \in \Omega, \quad 0 \leq t \leq 1.$$

Besides, let

$$\Omega_t = \left\{x \in \Omega \,\middle|\, w_t \neq 0\right\} = \left\{x \in \Omega \,\middle|\, v(x) + t\varphi(x) - v_*(x) < 0\right\},$$

$$\Omega_0 = \left\{x \in \Omega \,\middle|\, v(x) = v_*(x)\right\}.$$

Then $|\Omega_t - \Omega_0| \to 0$ as $t \to 0$. Thus, using $w_t \leq |w_t + t\varphi| + t|\varphi| < 2t|\varphi|$, we have

$$\frac{1}{t}\int_\Omega (f(x,v_*) - f(x,v))w_t \, dx = \frac{1}{t}\int_{\Omega_t - \Omega_0} (f(x,v_*) - f(x,v))w_t \, dx$$

$$\geq -\int_{\Omega_t - \Omega_0} 2|f(x,v_*) - f(x,v)||\varphi|\, dx.$$

Inserting these estimates into the right-hand side of (5.6.5) and taking $t \to 0$, we obtain $(dI(v))(\varphi) \geq 0$. Replacing φ with $-\varphi$, we get the reversed inequality. Consequently, $(dI(v))(\varphi) = 0$ and the lemma follows.

We now turn to the Chern–Simons equation.

It is easily checked that (5.5.13) is the Euler–Lagrange equation of the functional

$$I_\lambda(v) = \int_\Omega \left\{ \frac{1}{2} |\nabla v|^2 + \frac{\lambda}{2} (e^{u_0 + v} - 1)^2 + \frac{4\pi N}{|\Omega|} v \right\} \qquad (5.6.6)$$

which satisfies all the conditions made on the functional I in Lemma 5.6.3.

Lemma 5.6.4. *For each $\lambda > \lambda_c$, the equation (5.5.13) has a solution which is a local minimum of the functional (5.6.6) in S.*

Proof. It may be examined that I_λ is weakly lower semicontinuous in S.

We use the notation of Theorem 5.6.2. It is clear that v_* is a subsolution of (5.5.13) for any $\lambda > \lambda_c$ because $u_0 + v_* < 0$.

Define

$$S_* = \left\{ v \in S \,\middle|\, v \geq v_* \quad \text{a.e. in } \Omega \right\}.$$

Then I_λ is bounded from below on S_*. We consider the optimization problem

$$\min \left\{ I_\lambda(v) \,\middle|\, v \in S_* \right\}. \qquad (5.6.7)$$

We prove that (5.6.7) has a solution.

Let $\{v_n\}$ be a minimizing sequence of (5.6.7). Then, with the decomposition

$$v_n = c_n + v_n', \quad c_n \in \mathbb{R}, \quad v_n' \in X, \quad n = 1, 2, \cdots,$$

we see that $\{\|\nabla v_n'\|_2\}$ is bounded because the definition of S_* already gives us lower-boundedness of $\{c_n\}$. The upper-boundedness of $\{c_n\}$ follows from

$$I_\lambda(v_n) \geq 4\pi N c_n.$$

Therefore, $\{v_n\}$ is a bounded sequence in S. Without loss of generality, we may assume that $\{v_n\}$ converges to an element $v \in S$ weakly as $n \to \infty$. Hence v is a solution of (5.6.7). Using Lemma 5.6.3, we know that v solves (5.5.13) and $v \geq v_*$ everywhere. The maximum principle then implies the strict inequality $v > v_*$.

It remains to show that v is a local minimum of the functional (5.6.6) in S. One can approach this problem following an argument of Brezis and Nirenberg [56].

Suppose otherwise that v is not a local minimum of I_λ in S. Then, for any integer $n \geq 1$, we have

$$\inf \left\{ I_\lambda(w) \,\middle|\, w \in S, \ \|w - v\|_S \leq \frac{1}{n} \right\} = \varepsilon_n < I_\lambda(v). \qquad (5.6.8)$$

It is easy to show that, for each n, there is an element $v_n \in S$ such that $\|v_n - v\|_S \leq 1/n$ and $I_\lambda(v_n) = \varepsilon_n$. Using the principle of Lagrangian multipliers, we see that there exists a number $\mu_n \leq 0$ such that

$$-\Delta v_n + \lambda e^{u_0 + v_n}(e^{u_0 + v_n} - 1) + \frac{4\pi N}{|\Omega|} = \mu_n(-\Delta[v_n - v] + v_n - v).$$

It will be more transparent to rewrite the above equation as

$$\Delta(v_n - v) = \frac{|\mu_n|}{1 + |\mu_n|}(v_n - v)$$

$$+ \frac{\lambda}{1 + |\mu_n|}(e^{u_0 + v_n}[e^{u_0 + v_n} - 1] - e^{u_0 + v}[e^{u_0 + v} - 1]). \quad (5.6.9)$$

Since $\|v_n - v\|_S \to 0$ as $n \to \infty$, the Trudinger–Moser inequality implies that the right-hand side of (5.6.9) converges to 0 as $n \to \infty$ in $L^2(\Omega)$. The L^2-estimates then indicates that $v_n \to v$ as $n \to \infty$ in the space $W^{2,2}(\Omega)$. In particular, $v_n \to v$ in the C^α-norm for any $0 < \alpha < 1$. The compactness of the torus Ω and the strict inequality $v > v_*$ imply that when n is sufficiently large, we have $v_n > v_*$ as well. This conclusion contradicts the definition of v as a solution of the problem (5.6.7) and the lemma follows.

5.6.3 Nonminimum via mountain-pass lemma

In order to apply the mountain-pass lemma, we need to establish a compactness property called the Palais–Smale (P.S.) condition as stated in the following lemma.

Lemma 5.6.5. *Any sequence* $\{v_n\}$ *in* S *satisfying*

$$I_\lambda(v_n) \to \alpha, \quad \|dI_\lambda(v_n)\|_S \to 0 \quad as\ n \to \infty$$

has a convergent subsequence, where we use the same notation $\|\cdot\|_S$ *to denote the norm of the dual space of* S.

Proof. From the condition stated, we have a number $\alpha \in \mathbb{R}$ and a sequence $\{\varepsilon_n\}$ $(\varepsilon_n > 0)$ such that $\varepsilon_n \to 0$ as $n \to \infty$ and

$$\frac{1}{2}\|\nabla v_n\|_2^2 + \frac{\lambda}{2}\|e^{u_0 + v_n} - 1\|_2^2 + \frac{4\pi N}{|\Omega|}\int_\Omega v_n \to \alpha \quad as\ n \to \infty, \quad (5.6.10)$$

$$\left|\int_\Omega \nabla v_n \cdot \nabla\varphi + \lambda\int_\Omega e^{u_0 + v_n}(e^{u_0 + v_n} - 1)\varphi + \frac{4\pi N}{|\Omega|}\int_\Omega \varphi\right| \leq \varepsilon_n\|\varphi\|_S, \quad (5.6.11)$$

where $\varphi \in S$ is arbitrary.

First, taking $\varphi = 1$ in (5.6.11), we have

$$
\begin{aligned}
\delta_n &\equiv \int_\Omega e^{u_0+v_n}(e^{u_0+v_n} - 1) + \frac{4\pi N}{\lambda} \\
&= \int_\Omega (e^{u_0+v_n} - 1)^2 + \int_\Omega e^{u_0+v_n} - |\Omega| + \frac{4\pi N}{\lambda}, \quad (5.6.12)
\end{aligned}
$$

and $\delta_n \to 0$ as $n \to \infty$. Therefore

$$
\begin{aligned}
\int_\Omega e^{2(u_0+v_n)} &\leq 2\int_\Omega ((e^{u_0+v_n} - 1)^2 + e^{u_0+v_n}) - |\Omega| \\
&= 2\left(|\Omega| - \frac{4\pi N}{\lambda}\right) - |\Omega| + 2\delta_n \\
&= |\Omega| - \frac{8\pi N}{\lambda} + 2\delta_n. \quad (5.6.13)
\end{aligned}
$$

We next make the decomposition $v_n = c_n + v_n'$, $c_n \in \mathbb{R}, v_n' \in X$ with $\int_\Omega v_n' \, dx = 0$, $n = 1, 2, \cdots$. Thus the assumption (5.6.10) implies that $\{c_n\}$ is bounded from above.

In view of (5.6.10), we may assume that $\alpha - 1 < I_\lambda(v_n) < \alpha + 1$ for all n. Hence, by (5.6.12), there holds

$$
\alpha - 1 + \frac{1}{2}(4\pi N - \lambda|\Omega|) - \frac{1}{2}\lambda\delta_n \leq \frac{1}{2}\|\nabla v_n'\|_2^2 + 4\pi N c_n \leq \alpha + 1. \quad (5.6.14)
$$

On the other hand, setting $\varphi = v_n'$ in (5.6.11), we obtain

$$
\|\nabla v_n'\|_2^2 + \lambda \int_\Omega e^{2(u_0+v_n)}v_n' - \lambda \int_\Omega e^{u_0+v_n}v_n' \leq \varepsilon_n\|v_n'\|_{\mathcal{S}}.
$$

However, since $\{c_n\}$ is bounded from above, we may use (5.6.13) and apply the Poincaré inequality to get

$$
\begin{aligned}
\|\nabla v_n'\|_2^2 &+ \lambda\int_\Omega e^{2(u_0+c_n)}(e^{2v_n'} - 1)v_n' \\
&\leq \lambda\int_\Omega e^{2(u_0+c_n)}|v_n'| + \lambda\|e^{u_0+v_n}\|_2\|v_n'\|_2 + \varepsilon_n\|v_n'\|_{\mathcal{S}} \\
&\leq C\|\nabla v_n'\|_2, \quad (5.6.15)
\end{aligned}
$$

where $C > 0$ is a constant. Because $(e^{2v_n'} - 1)v_n' \geq 0$ in Ω, we are led by (5.6.15) to the boundedness of $\{\|\nabla v_n'\|_2\}$. Inserting this result into (5.6.14), we see that $\{c_n\}$ is also bounded from below. Hence, $\{v_n\}$ is a bounded sequence in $\mathcal{S}$. By passing to a subsequence if necessary, we may assume for convenience that there is an element $v \in \mathcal{S}$ such that

$$
v_n \to v \quad \text{weakly in } \mathcal{S} \quad \text{and strongly in } L^p(\Omega), \quad \forall p \geq 1.
$$

Letting $n \to \infty$ in (5.6.11), we see that v is a critical point of I_λ since it satisfies the equation

$$\int_\Omega \left\{ \nabla v \cdot \nabla \varphi + \lambda e^{u_0 + v}(e^{u_0 + v} - 1)\varphi + \frac{4\pi N}{|\Omega|}\varphi \right\} = 0, \quad \forall \varphi \in \mathcal{S}. \quad (5.6.16)$$

Finally, in order to show that $v_n \to v$ strongly in $\mathcal{S}$ as $n \to \infty$, we choose $\varphi = v_n - v$ in (5.6.11) and (5.6.16) and subtract the resulting expressions to arrive at

$$\|\nabla(v_n' - v')\|_2^2 \leq \lambda \int_\Omega (|f(x,v)| + |f(x,v_n)|)|v_n - v| + \varepsilon_n \|v_n - v\|_\mathcal{S}$$

$$\leq C\|v_n - v\|_2 + \varepsilon_n(\|v_n\|_\mathcal{S} + \|v\|_\mathcal{S}),$$

where $f(x,v) = e^{u_0(x)+v}(e^{u_0(x)+v} - 1)$. Thus we have proved the convergence $\|\nabla(v_n' - v')\|_2 \to 0$, which leads to $\|v_n - v\|_\mathcal{S} \to 0$, as $n \to \infty$ as expected.

We are now ready to prove the existence of secondary solution of the equation (5.5.13).

Let v_λ denote the local minimum of I_λ produced in Lemma 5.6.4. Then there is a number $\delta > 0$ such that

$$I_\lambda(v_\lambda) < I_\lambda(w), \quad w \in \mathcal{S}, \quad \|w - v_\lambda\|_\mathcal{S} \leq \delta. \quad (5.6.17)$$

Here we assume that v_λ is a strict local minimum because otherwise we would already have additional solutions. Therefore we can assert that there exists a number $\delta_0 > 0$ such that

$$\inf\left\{ I_\lambda(w) \,\Big|\, w \in \mathcal{S}, \ \|w - v_\lambda\|_\mathcal{S} = \delta_0 \right\} > I_\lambda(v_\lambda). \quad (5.6.18)$$

Besides, we observe that the functional I_λ possesses a "mountain pass" structure. In fact, for any number $c > 0$, we have by $u_0 + v_\lambda < 0$ that

$$I_\lambda(v_\lambda - c) - I_\lambda(v_\lambda) = \frac{\lambda}{2}\int_\Omega ([e^{u_0+v_\lambda-c} - 1]^2 - [e^{u_0+v_\lambda} - 1]^2) - 4\pi Nc$$

$$< -4\pi Nc. \quad (5.6.19)$$

Hence, in view of (5.6.19), we can choose $c_0 > \delta_0$ sufficiently large to make

$$I_\lambda(v_\lambda - c_0) < I_\lambda(v_\lambda) - 1 \quad \text{and} \quad |\Omega|^{1/2}c_0 > \delta_0$$

valid.

Denote by $\mathcal{P}$ the set of all continuous paths in $\mathcal{S}$, $\gamma : [0,1] \to \mathcal{S}$, connecting the points v_λ and $v_\lambda - c_0$ with $\gamma(0) = v_\lambda$ and $\gamma(1) = v_\lambda - c_0$ and define

$$\alpha_0 = \inf\left\{ \alpha_\gamma \,\Big|\, \gamma \in \mathcal{P} \right\},$$

$$\alpha_\gamma = \sup\left\{ I_\lambda(\gamma(t)) \,\Big|\, 0 \leq t \leq 1 \right\} \quad (\gamma \in \mathcal{P}). \quad (5.6.20)$$

Then (5.6.18) yields $\alpha_0 > I_\lambda(v_\lambda)$. Following the well-known theory of Ambrosetti and Rabinowitz [8, 228], α_0 in (5.6.20) is a critical value of the functional I_λ in S. Since $\alpha_0 > I_\lambda(v_\lambda)$, this critical value gives rise to an additional solution of the Chern–Simons equation (5.5.13).

The significance of Tarantello's secondary solution of (5.5.13) is that it gives rise to a gauge-distinct solution from the maximal solution obtained in Theorem 5.5.1 of the system (5.5.6)–(5.5.8), although these two solutions lie on the same energy level, $E = 2\pi|N|$, and carry identical vortices. Thus degeneracy takes place in the Chern–Simons theory (5.2.1) even in a compact setting.

5.7 Remarks

In this chapter, we have presented a systematic study of the relativistic self-dual Chern–Simons equation when the gauge group is the Abelian circle, $U(1)$. We have seen that for the problem over a full plane, there exist topological and non-topological N-vortex solutions realizing any given winding number N, and, for the problem over a finite periodic domain Ω, the vortex number N is confined by the size of Ω and for each prescribed distribution of vortex locations, there are at least two gauge-distinct solutions realizing such vortex locations whenever existence is ensured subcritically through the Chern–Simons coupling parameter,

$$0 < \kappa < \kappa_c < \frac{1}{2}\sqrt{\frac{|\Omega|}{\pi N}}. \tag{5.7.1}$$

There are still some interesting open problems.

Open Problem 5.7.1. *In the category of static finite-energy solutions over* $\mathbb{R}^2$, *are the original system equations of motion (5.2.2) and (5.2.3) equivalent to the self-dual or anti-self-dual system of equations (5.2.19) and (5.2.20)?*

It is well known that for the classical Abelian Higgs model, also known in its static limit as the Ginzburg–Landau theory, with the Lagrangian action density

$$\mathcal{L} = -\frac{1}{4}F_{\mu\nu}F^{\mu\nu} + \frac{1}{2}(D_\mu\phi)\overline{(D^\mu\phi)} - \frac{1}{4}(1 - |\phi|^2)^2 \tag{5.7.2}$$

instead of (5.2.1), the answer to the same problem is affirmative [157, 304], although such an equivalence result does not hold for general Yang–Mills equations [43, 44, 59, 60, 242, 270, 286, 305].

For the Chern–Simons equations, topological solutions characterized by the boundary condition $|\phi| \to 1$ as $|x| \to \infty$ resemble solutions of the Abelian Higgs model for which the locations and multiplicities of vortices uniquely determine solution configurations [157, 303]. However, the same problem for the Chern–Simons case is not solved yet.

Open Problem 5.7.2. *Given the locations and multiplicities of N vortices, is there a unique topological N-vortex solution realizing such locations and multiplicities?*

In Chapter 9, we shall obtain a partial affirmative answer to the above problem: it will be shown that in the category of radially symmetric solutions, uniqueness of topological solutions holds.

We have seen in this chapter that non-topological solutions are more subtle and harder to obtain in general. In particular, even for radially symmetric solutions, there is non-uniqueness and the solutions possess arbitrary fractional energies and charges. In fact, a successful partial differential equation approach to obtaining arbitrarily distributed non-topological vortices has not been available yet.

Open Problem 5.7.3. *Obtain an existence theory for the non-topological Chern–Simons vortex solutions to the system of equations (5.2.19) and (5.2.20), given an arbitrarily prescribed N vortex distribution.*

Recently, Chae and Imanuvilov [68] obtained an existence theorem for an arbitrary distribution of non-topological vortices using an implicit function theorem argument confined in a neighborhood of the Liouville solution, which is a progress toward the solution of the problem. However, we have seen, in general, that non-topological solutions are far from the Liouville solution. Thus, in order to get a more complete description of solutions, a non-perturbative approach should be pursued.

We now turn to doubly periodic solutions.

We have seen that for any prescribed N vortices, there is a critical value κ_c satisfying

$$\kappa_c < \frac{1}{2}\sqrt{\frac{|\Omega|}{\pi N}} \tag{5.7.3}$$

for the Chern–Simons coupling parameter $\kappa > 0$, such that there is a solution realizing these vortices if and only if $0 < \kappa \leq \kappa_c$. It will be interesting to estimate κ_c. Hence we propose

Open Problem 5.7.4. *Is κ_c independent of the locations of the N prescribed N vortices? How large is the gap*

$$\frac{1}{2}\sqrt{\frac{|\Omega|}{\pi N}} - \kappa_c \tag{5.7.4}$$

in terms of $|\Omega|$, N, and other possible parameters?

In §5.6, we have established the degeneracy result that for each prescribed N vortex distribution there are at least two distinct solutions when $\kappa < \kappa_c$. It will be interesting to know whether there exist more solutions.

Open Problem 5.7.5. *How many distinct solutions does the equation (5.5.13) possess when*

$$\lambda > \lambda_c > \frac{16\pi N}{|\Omega|}?$$

(5.7.5)

The variational approach described in §5.5 may be useful for some technical reasons. For example, it has been used in the study the self-dual Chern–Simons solutions in the large λ or small κ limit [91, 302].

Self-duality may also be achieved when one adds a Maxwell term in the Chern–Simons Higgs theory, with the sacrifice of introducing an additional, neutral, scalar field into the theory [183]. For some mathematical analysis of such an extended system, see [69, 264].

The existence of static solutions of the Chern–Simons theory in the noncritical case is a completely open area even for radially symmetric solutions [87, 174, 243] and is of ultimate importance. The main difficulty is that the governing equations of motion consist of a nonlinear system. Although there is a variational principle, the Minkowski spacetime signature makes the variational functional indefinite.

Recently, there have been a lot of activities in the study of the Ginzburg–Landau equations [125] concerning the existence and behavior of solutions [29, 31, 37, 39, 142, 196, 258, 266, 267, 325, 338, 340, 343], vortex dynamics [30, 99, 195, 224, 268, 300], and computation [88, 95, 197, 326]. In light of these studies, the Chern–Simons equations offer new challenges.

6

Chern–Simons Systems: Non-Abelian Case

In this chapter we present static multisoliton solutions of the non-Abelian Chern–Simons equations. In §6.1 we review some basic facts about a complex semi-simple Lie algebra such as the Cartan–Weyl bases and Cartan matrices to be used in the development to follow. In §6.2 we consider the solution of the non-Abelian gauged Schrödinger equations coupled with a Chern–Simons dynamics via the Toda equations, which is a non-relativistic Chern–Simons theory. In §6.3 we introduce the relativistic Chern–Simons equations and state a general existence theorem. In §6.4 we reduce the governing equations into a nonlinear elliptic system and formulate a variational principle. In §6.5 we present an analysis of the elliptic system and prove all the results stated. In §6.6 we apply our existence theory to some concrete examples.

6.1   Lie Algebras and Cartan–Weyl Bases

In this section we review some basic notions from the theory of complex Lie algebras. In particular, we will need to express a semi-simple Lie algebra in terms of its Cartan–Weyl basis. We shall start from some illustrative examples. We then present a general discussion.

6.1.1 Simple examples

First we consider the Lie algebra of $SU(2)$. With the Pauli spin-matrices

$$\sigma_1 = \begin{pmatrix} 0 & 1 \\ 1 & 0 \end{pmatrix}, \quad \sigma_2 = \begin{pmatrix} 0 & -i \\ i & 0 \end{pmatrix}, \quad \sigma_3 = \begin{pmatrix} 1 & 0 \\ 0 & -1 \end{pmatrix},$$

we obtain a convenient basis $t_a = \sigma_a/2, a = 1, 2, 3$ satisfying the commutation relation

$$[t_a, t_b] = i\varepsilon_{abc} t_c, \quad a, b, c = 1, 2, 3.$$

Since these generators of $SU(2)$ do not contain a commutative pair, the maximum number of commuting generators is one. Hence the rank of $SU(2)$ or the dimension of the Cartan subalgebra (the largest Abelian subalgebra) is one.

We now make a transformation to switch to a special basis, J_-, J_0, J_+, called the 'momentum operators', defined by

$$J_0 = t_3 = \frac{1}{2}\sigma_3, \quad J_- = t_1 - it_2 = \frac{1}{2}(\sigma_1 - i\sigma_2), \quad J_+ = t_1 + it_2 = \frac{1}{2}(\sigma_1 + i\sigma_2).$$

Then it is easily examined that there holds the commutation relation

$$[J_+, J_-] = 2J_0, \quad [J_0, J_\pm] = \pm J_\pm. \tag{6.1.1}$$

The basis $\{J_-, J_0, J_+\}$ is a simplest example of the Cartan–Weyl basis for which J_-, J_+ are ladder generators and J_0 is a weight generator which spans the Cartan subalgebra. The weight of J_+ (J_-) is defined to be $\alpha = +1$ ($-\alpha = -1$) and we use the notation $E_\alpha = J_+$, $E_{-\alpha} = J_-$, and $H_1 = J_0$. Then we have

$$[E_\alpha, E_{-\alpha}] = 2H_1, \quad [H_1, E_{\pm\alpha}] = \pm\alpha E_{\pm\alpha}.$$

Next we consider $SU(3)$. A basis of eight special generators called F-spin matrices introduced by Gell-Mann are

$$F_1 = \frac{1}{2}\begin{pmatrix} 0 & 1 & 0 \\ 1 & 0 & 0 \\ 0 & 0 & 0 \end{pmatrix}, \quad F_2 = \frac{1}{2}\begin{pmatrix} 0 & -i & 0 \\ i & 0 & 0 \\ 0 & 0 & 0 \end{pmatrix},$$

$$F_3 = \frac{1}{2}\begin{pmatrix} 1 & 0 & 0 \\ 0 & -1 & 0 \\ 0 & 0 & 0 \end{pmatrix}, \quad F_4 = \frac{1}{2}\begin{pmatrix} 0 & 0 & 1 \\ 0 & 0 & 0 \\ 1 & 0 & 0 \end{pmatrix},$$

$$F_5 = \frac{1}{2}\begin{pmatrix} 0 & 0 & -i \\ 0 & 0 & 0 \\ i & 0 & 0 \end{pmatrix}, \quad F_6 = \frac{1}{2}\begin{pmatrix} 0 & 0 & 0 \\ 0 & 0 & 1 \\ 0 & 1 & 0 \end{pmatrix},$$

$$F_7 = \frac{1}{2}\begin{pmatrix} 0 & 0 & 0 \\ 0 & 0 & -i \\ 0 & i & 0 \end{pmatrix}, \quad F_8 = \frac{1}{2\sqrt{3}}\begin{pmatrix} 1 & 0 & 0 \\ 0 & 1 & 0 \\ 0 & 0 & -2 \end{pmatrix}.$$

These generators obey the following commutation relation,

$$[F_i, F_j] = ic_{ijk}F_k, \quad i, j, k = 1, 2, \cdots, 8,$$

where the independent nonvanishing structure constants are

$$c_{123} = 1, \quad c_{147} = \frac{1}{2}, \quad c_{156} = -\frac{1}{2}, \quad c_{246} = \frac{1}{2}, \quad c_{257} = \frac{1}{2},$$

$$c_{345} = \frac{1}{2}, \quad c_{367} = -\frac{1}{2}, \quad c_{458} = \frac{\sqrt{3}}{2}, \quad c_{678} = \frac{\sqrt{3}}{2}.$$

As for the case of $SU(2)$ where we introduce the momentum operators, here we use the 'spherical representation' of the F-spin generators by assigning

$$T_{\pm} = F_1 \pm iF_2, \quad T_3 = F_3, \quad V_{\pm} = F_4 \pm iF_5,$$

$$U_{\pm} = F_6 \pm iF_7, \quad Y = \frac{2}{\sqrt{3}}F_8.$$

The commutation relation is given by the following expressions,

$$[T_3, T_{\pm}] = \pm T_{\pm}, \quad [T_+, T_-] = 2T_3, \quad [T_3, U_{\pm}] = \mp\frac{1}{2}U_{\pm},$$

$$[U_+, U_-] = \frac{3}{2}Y - T_3, \quad [T_3, V_{\pm}] = \pm\frac{1}{2}V_{\pm}, \quad [V_+, V_-] = \frac{3}{2}Y + T_3,$$

$$[Y, T_{\pm}] = 0, \quad [Y, U_{\pm}] = \pm U_{\pm}, \quad [Y, V_{\pm}] = \pm V_{\pm},$$

$$[T_+, V_+] = [T_+, U_-] = [U_+, V_+] = 0, \quad [T_+, V_-] = -U_-,$$

$$[T_+, U_+] = V_+, \quad [U_+, V_-] = T_-, \quad [T_3, Y] = 0. \tag{6.1.2}$$

From (6.1.2), the maximum number of commuting generators of $SU(3)$ Lie algebra is two (e.g., $[T_3, Y] = 0$ or $[Y, T_{\pm}] = 0$). Thus the rank of $SU(3)$ or its Lie algebra $su(3)$ is two, which is a special case of a general conclusion that the rank of $SU(n)$ or $su(n)$ is $n - 1$.

The commutators stated in (6.1.2) give us a clear display of $su(3)$. In fact, we see that the T-, U-, V-spin matrices,

$$\left\{T_-, T_3, T_+\right\}, \quad \left\{U_-, U_3, U_+\right\}, \quad \left\{V_-, V_3, V_+\right\}$$

form three subalgebras, each identical to $su(2)$, called, respectively, the T-, U-, V-spin algebras. Here U_3 and V_3 are defined by

$$U_3 = \frac{3}{4}Y - \frac{1}{2}T_3, \quad V_3 = \frac{3}{4}Y + \frac{1}{2}T_3.$$

As before, there are again ladder generators, T_+, U_+, V_+ and T_-, U_-, V_-. Besides, there are two weight generators, T_3, Y, which commute with each other and form the Cartan subalgebra.

Although the Lie algebra $su(3)$ seems to be more complicated, we can show that it has the same structure as $su(2)$ just discussed. To see this, we use $H_1 = T_3$ and $H_2 = Y$ to denote the weight generators. From

$$[H_1, T_\pm] = \pm T_\pm, \quad [H_2, T_\pm] = 0,$$

we see that the weight of $T_\pm$ is given by the weight vector $\alpha = (\pm 1, 0)$ and we can use E_α to denote $T_\pm$. Thus, with $\alpha = (\alpha_1, \alpha_2)$, we have

$$[H_j, E_\alpha] = \alpha_j E_\alpha, \quad j = 1, 2. \tag{6.1.3}$$

Similarly, the generators $V_\pm$ and $U_\pm$ have the weight vectors

$$\alpha = \left(\pm \frac{1}{2}, \pm 1 \right) \quad \text{and} \quad \alpha = \left(\mp \frac{1}{2}, \pm 1 \right),$$

respectively. If we denote these generators again by E_α with α defined above, the relation (6.1.3) still holds. In fact, using α and β as weight vectors, we arrive at the commutators

$$\begin{aligned}
[E_\alpha, E_\beta] &= N_{\alpha\beta} E_{\alpha+\beta}, \quad \alpha \neq \beta, \\
[E_\alpha, E_{-\alpha}] &= \alpha^j H_j, \\
[H_j, E_\alpha] &= \alpha_j E_\alpha, \\
[H_j, H_k] &= 0,
\end{aligned} \tag{6.1.4}$$

where $N_{\alpha\beta}$ and α^j are constants and $(\alpha_j) = \alpha$ is the weight vector. For example, we can check that

$$[E_{(+1,0)}, E_{(-\frac{1}{2}, -1)}] = -E_{(\frac{1}{2}, -1)} \quad \text{because} \quad [T_+, V_-] = -U_-,$$

which indicates the important property that weights are additive. Besides,

$$[E_{(-\frac{1}{2}, +1)}, E_{(+\frac{1}{2}, -1)}] = -H_1 + \frac{3}{2} H_2 \quad \text{because} \quad [U_+, U_-] = \frac{3}{2} Y - T_3.$$

The above examples show that various semi-simple Lie algebras may be put into a standard form in terms of weight and ladder generators, or the Cartan–Weyl basis, and weight vectors. We shall study such a general formulation in the next subsection.

6.1.2 Classification theorem

In this subsection, we show how to classify general complex semi-simple Lie algebras in terms of the Cartan–Weyl bases.

We use $\{X_\ell\}$ to denote a given basis of an N-dimensional semi-simple Lie algebra $(L, [\,,\,])$ and $\{C_{mn}^\ell\}$ to denote the set of the structure constants,

$$[X_m, X_n] = C_{mn}^\ell X_\ell, \quad \ell, m, n = 1, 2, \cdots, N. \tag{6.1.5}$$

We consider the eigenvalue problem in L,

$$[A, X] = \lambda X, \quad A, X \in L. \tag{6.1.6}$$

It is easy to check that the $N \times N$ matrix of the linear transformation

$$\mathrm{ad}(A) : X \to [A, X], \quad X \in L, \tag{6.1.7}$$

with respect to the basis $\{X_m\}$ is

$$\Omega = (\Omega_{mn}) = (a^\ell C_{\ell n}^m),$$

where $A = a^\ell X_\ell$. Of course, for any nonzero A, the matrix Ω has N eigenvalues, some of which may be degenerate. It is clear that $\lambda = 0$ is always an eigenvalue because one can choose $X = A$. Cartan showed that the element A may be so chosen that only the eigenvalue $\lambda = 0$ is degenerate and the number of different eigenvalues of Ω is maximum if L is semi-simple. The multiplicity r of the eigenvalue $\lambda = 0$ is obviously independent of the choice of A and is called the rank of L. This definition actually coincides with an earlier one where the rank r is defined to be the dimensions of the largest Abelian subalgebra, or the Cartan subalgebra, of L (see below). Thus the Cartan subalgebra is the eigenspace of the transformation (6.1.7) corresponding to the eigenvalue $\lambda = 0$. Let $\{H_i \,|\, i = 1, 2, \cdots, r\}$ be a basis of this eigenspace. We record the result

$$[A, H_i] = 0, \quad i = 1, 2, \cdots, r. \tag{6.1.8}$$

Denote by λ now a nonzero eigenvalue of the transformation (6.1.7), called a root of L, and E_λ a chosen corresponding eigenvector. Then there are exactly $N - r$ such eigenpairs. We also record

$$[A, E_\lambda] = \lambda E_\lambda. \tag{6.1.9}$$

We first claim that $\{H_i\}$ is a commuting set,

$$[H_i, H_j] = 0, \quad i, j = 1, 2, \cdots, r. \tag{6.1.10}$$

To see that (6.1.10) is true, we assume otherwise that H_1 does not commute with some of the H_i's. It is easy to examine that $\{H_i\}$ generate a subalgebra because, in view of the Jacobian identity, we have

$$[A, [H_i, H_j]] = 0, \quad i, j = 1, 2, \cdots, r.$$

Hence, we may use the notation

$$[H_1, H_i] = c_{ij} H_j, \quad i, j = 1, 2, \cdots, r$$

and assert that the matrix (c_{ij}) is not a zero matrix.

Let $\delta_0 > 0$ be sufficiently small so that it is below the absolute values of all nonzero eigenvalues of the transformation (6.1.7). We replace the element A by a perturbation of it,

$$A' = A + \varepsilon H_1,$$

where $\varepsilon > 0$ is so small that the transformation

$$X \to [A', X], \quad X \in L \tag{6.1.11}$$

has at least $N - r$ different nonzero eigenvalues with absolute values above $\delta_0/2$.

However, restricted to the subspace generated by the elements $\{H_i\}$, the transformation (6.1.11) has nonzero matrix $\varepsilon(c_{ij})$ with respect to the basis $\{H_i\}$ which has at least a nonzero eigenvalue of the magnitude $O(\varepsilon)$ which can be made below $\delta_0/2$. This conclusion contradicts the assumption that A has been chosen to yield the largest number of different eigenvalues for the transformation (6.1.7).

Thus $\{H_i\}$ generate an Abelian subalgebra of L which is clearly a maximal Abelian subalgebra or a Cartan subalgebra, and contains A.

We represent A as

$$A = a^i H_i. \tag{6.1.12}$$

Then we have by (6.1.8) and (6.1.9),

$$
\begin{aligned}
[A, [H_i, E_\lambda]] &= -[H_i, [E_\lambda, A]] - [E_\lambda, [A, H_i]] \\
&= \lambda[H_i, E_\lambda], \quad i = 1, 2, \cdots, r.
\end{aligned} \tag{6.1.13}
$$

However, since λ is not degenerate, all the $[H_i, E_\lambda]$ $(i = 1, 2, \cdots, r)$ must be proportional to E_λ. Therefore there are numbers $\alpha_1, \alpha_2, \cdots, \alpha_r$ such that

$$[H_i, E_\lambda] = \alpha_i E_\lambda, \quad i = 1, 2, \cdots, r. \tag{6.1.14}$$

We show that the vector $\alpha = (\alpha_1, \alpha_2, \cdots, \alpha_r)$, now called the root vector, is uniquely determined by the eigenvalue λ. To see this, we use λ' to denote another nonzero eigenvalue and $\alpha' = (\alpha_1', \alpha_2', \cdots, \alpha_r')$ the corresponding root vector. If $\alpha = \alpha'$, we obtain from (6.1.14) that

$$[H_i, E_\lambda - E_{\lambda'}] = \alpha_i(E_\lambda - E_{\lambda'}), \quad i = 1, 2, \cdots, r. \tag{6.1.15}$$

Combining (6.1.12) into (6.1.15), we have

$$[A, E_\lambda - E_{\lambda'}] = s(E_\lambda - E_{\lambda'}), \quad s = \alpha_i a^i.$$

On the other hand, (6.1.9) gives us $[A, E_\lambda - E_{\lambda'}] = \lambda E_\lambda - \lambda' E_{\lambda'}$. Inserting this result into the above we find the E_λ and $E_{\lambda'}$ are linearly dependent, which is false.

Using this uniqueness, we write E_λ as E_α where α is the root vector associated with the eigenvalue λ. Hence, we rewrite (6.1.14) as

$$[H_i, E_\alpha] = \alpha_i E_\alpha, \quad i = 1, 2, \cdots, r. \tag{6.1.16}$$

By (6.1.9) and (6.1.14), we get the relation

$$\lambda = \alpha_i a^i. \tag{6.1.17}$$

This result implies that $\lambda \mapsto \alpha$ is a 1-1 correspondence. Thus, from now on, we replace the root vector α by the eigenvalue λ and use α to denote the eigenvalue as well. For example, (6.1.17) becomes

$$\alpha = a^i \alpha_i. \tag{6.1.18}$$

To get other commutators, we consider $[E_\alpha, E_\beta]$ where α and β are two nonzero eigenvalues of (6.1.7). Then, it is straightforward from the Jacobi identity that

$$[A, [E_\alpha, E_\beta]] = (\alpha + \beta)[E_\alpha, E_\beta]. \tag{6.1.19}$$

That is, $[E_\alpha, E_\beta] = 0$ or otherwise is an eigenvector associated with the nonzero eigenvalue $(\alpha + \beta)$ if $\alpha \neq -\beta$. Hence, there is a number $N_{\alpha\beta}$ such that

$$[E_\alpha, E_\beta] = N_{\alpha\beta} E_{\alpha+\beta}. \tag{6.1.20}$$

When $\beta = -\alpha$ is a nonzero eigenvalue, we have

$$[E_\alpha, E_{-\alpha}] = \alpha^i H_i \tag{6.1.21}$$

for some numbers $\alpha^1, \alpha^2, \cdots, \alpha^r$.

Let $i, j = 1, 2, \cdots, r$ and α, β run through all the nonzero eigenvalues of L.

We use μ, ν, σ, τ to denote the Lie algebra indices corresponding to a particular basis and $C^\sigma_{\mu\nu}$ the structure constants. We recall the classical result that a Lie algebra is semi-simple if and only if the symmetric tensor

$$g_{\mu\nu} = C^\tau_{\mu\sigma} C^\sigma_{\nu\tau} \tag{6.1.22}$$

is not degenerate, i.e.,

$$\det(g_{\mu\nu}) \neq 0.$$

With the tensor (6.1.22), we can define a bilinear form, called the Killing form, over the Lie algebra L by

$$\kappa(X, Y) = x^\mu g_{\mu\nu} y^\nu, \quad X = x^\mu X_\mu, \, Y = y^\mu X_\mu \in L.$$

In fact, it is easily checked that the Killing form κ may be obtained by taking trace of the product of the matrices of $\mathrm{ad}(X)$ and $\mathrm{ad}(Y)$, namely,

$$\kappa(X, Y) = \mathrm{Tr}\,(\mathrm{ad}(X)\,\mathrm{ad}(Y)), \quad X, Y \in L.$$

We need to study what the Killing form looks like in the Cartan–Weyl basis $\{H_i, E_\alpha\}$. In particular, we show that nonzero eigenvalues appear in $\pm\alpha$ pairs.

First, it is clear that

$$C_{ij}^\mu = 0. \tag{6.1.23}$$

Next, in view of (6.1.16), we have

$$C_{i\alpha}^\mu = \alpha_i \delta_\alpha^\mu. \tag{6.1.24}$$

Furthermore, (6.1.20) says that

$$C_{\alpha\beta}^\mu = N_{\alpha\beta} \delta_{(\alpha+\beta)}^\mu. \tag{6.1.25}$$

Finally, (6.1.21) implies that

$$C_{\alpha(-\alpha)}^\mu = \alpha^i \delta_i^\mu \tag{6.1.26}$$

if $-\alpha$ is a nonzero eigenvalue.

By (6.1.22)–(6.1.25) and

$$g_{\alpha\mu} = C_{\alpha j}^i C_{\mu i}^j + C_{\alpha i}^\beta C_{\mu\beta}^i + C_{\alpha\beta}^\nu C_{\mu\nu}^\beta,$$

we easily see that, if α is a nonzero eigenvalue, the only possible nonvanishing $g_{\alpha\mu}$ is for $\mu = -\alpha$, which proves that $-\alpha$ must also be an eigenvalue because otherwise it will violates the fact that $\det(g_{\mu\nu}) \neq 0$. This fact can also be seen from observing a simple property of the Killing form,

$$\kappa([X,Y],Z) = \kappa(X,[Y,Z]), \quad X,Y,Z \in L. \tag{6.1.27}$$

In fact, for nonzero eigenvalues α, β, we can show that E_α and E_β are orthogonal with respect to the Killing form κ if $\alpha + \beta \neq 0$. To see this, we fix i so that $\alpha_i + \beta_i \neq 0$. For definiteness, assume $\alpha_i \neq 0$. Hence,

$$E_\alpha = \alpha_i^{-1}[H_i, E_\alpha]. \tag{6.1.28}$$

Using (6.1.27), we have

$$
\begin{aligned}
\kappa(E_\alpha, E_\beta) &= \alpha_i^{-1}\kappa([H_i, E_\alpha], E_\beta) \\
&= -\alpha_i^{-1}\kappa(E_\alpha, [H_i, E_\beta]) \\
&= -\alpha_i^{-1}\beta_i\,\kappa(E_\alpha, E_\beta).
\end{aligned}
$$

Therefore $(\alpha_i+\beta_i)\kappa(E_\alpha, E_\beta) = 0$ or $\kappa(E_\alpha, E_\beta) = 0$. However, E_α is already orthogonal to H_i for any $i = 1, 2, \cdots, r$ as may be seen from using (6.1.27) and (6.1.28) to arrive at (since there always exists an i such that $\alpha_i \neq 0$)

$$
\begin{aligned}
\kappa(H_i, E_\alpha) &= \alpha_i^{-1}\kappa(H_i, [H_i, E_\alpha]) \\
&= \alpha_i^{-1}\kappa([H_i, H_i], E_\alpha) = 0.
\end{aligned}
$$

Thus, we have seen that if $-\alpha$ is not an eigenvalue, then E_α is orthogonal to the full L, which contradicts the fact that the Killing form κ is not degenerate because L is semi-simple. Hence $-\alpha$ must be an eigenvalue as well.

Restricted to the Cartan subalgebra, the Killing form κ is determined by the tensor

$$g_{ij} = C_{i\nu}^\mu C_{j\mu}^\nu. \tag{6.1.29}$$

Using (6.1.23) and (6.1.24) in (6.1.29), we find that

$$g_{ij} = \sum_\alpha C_{i\alpha}^\alpha C_{j\alpha}^\alpha = \sum_\alpha \alpha_i \alpha_j. \tag{6.1.30}$$

With the above preparation, we can now represent the coefficients α^i in (6.1.21) in terms of the components of the root vector α stated in (6.1.16). In fact, by (6.1.16), (6.1.21), and (6.1.27), we have

$$\begin{aligned}
\alpha_i \kappa(E_\alpha, E_{-\alpha}) &= \kappa([H_i, E_\alpha], E_{-\alpha}) \\
&= \kappa(H_i, [E_\alpha, E_{-\alpha}]) \\
&= \alpha^j \kappa(H_i, H_j) = \alpha^j g_{ij}.
\end{aligned}$$

Renormalizing the generators E_α and $E_{-\alpha}$, we can make $\kappa(E_\alpha, E_{-\alpha}) = 1$ ($\forall \alpha$). Consequently, we obtain the relation

$$\alpha^i = \alpha_j g^{ij},$$

from which we can define the inner product of two root vectors $\alpha = (\alpha_1, \alpha_2, \cdots, \alpha_r)$ and $(\beta_1, \beta_2, \cdots, \beta_r)$ by

$$(\alpha, \beta) = \alpha^i \beta_i. \tag{6.1.31}$$

6.1.3   Root vectors and Cartan matrices

Information about the set of root vectors, Φ, of a semi-simple Lie algebra of rank r is in fact contained in a smaller part, Δ, called the system of simple root vectors. The Cartan matrix is then formed from Δ. Since Φ spans the entire space of r-vectors, so does Δ. Formally, Δ is defined to satisfy the axiom that Δ is a linearly independent system so that each root vector β in Φ may be represented by the vectors in Δ in the form

$$\beta = \sum_{\alpha \in \Delta} k_\alpha \alpha,$$

where the coefficients k_α are integers which are either all nonnegative or nonpositive.

The existence proof for a simple root system Δ is omitted.

We use $\alpha_1, \alpha_2, \cdots, \alpha_r$ to denote the r simple root vectors of a semi-simple Lie algebra of rank r which should not be confused with the same notation for the coordinates of a root vector in the previous subsection, namely, $\Delta = \{\alpha_1, \alpha_2, \cdots, \alpha_r\}$. It can be shown that $(\alpha_i, \alpha_i) \neq 0$ for $i = 1, 2, \cdots, r$. The Cartan matrix $K = (K_{ij})$ is defined by

$$K_{ij} = 2\frac{(\alpha_i, \alpha_j)}{(\alpha_i, \alpha_i)}, \quad i, j = 1, 2, \cdots, r. \tag{6.1.32}$$

Thus the diagonal entries of K are all 2. It is a standard fact that, for any given semi-simple Lie algebra, all other entries are either $-3, -2, -1$, or 0. In the study of the Chern–Simons theory, the Lie algebra is typically assumed to be simply laced. In this case a renormalization can be made so that all simple root vectors are of the same length. In particular, the Cartan matrix is symmetric.

We now examine the example $L = su(3)$ discussed earlier in this section. For $su(3)$, $r = 2$, and the system of root vectors is

$$\Phi = \left\{\alpha_1 = \left(\frac{1}{2}, 1\right), \beta_1 = -\alpha_1, \alpha_2 = \left(\frac{1}{2}, -1\right),\right.$$
$$\left.\beta_2 = -\alpha_2, \alpha_3 = (1, 0), \beta_3 = -\alpha_3\right\}.$$

Thus we may choose

$$\Delta = \left\{\alpha_1 = \left(\frac{1}{2}, 1\right), \alpha_2 = \left(\frac{1}{2}, -1\right)\right\} \tag{6.1.33}$$

to be a simple root system. Besides, in view of (6.1.30), we can determine the restricted Killing tensor of $su(3)$,

$$g_{11} = 3, \quad g_{12} = g_{21} = 0, \quad g_{22} = 4.$$

Therefore

$$(g^{ij}) = (g_{ij})^{-1} = \begin{pmatrix} \frac{1}{3} & 0 \\ 0 & \frac{1}{4} \end{pmatrix}.$$

Using the choice of the simple root system Δ given in (6.1.33), we immediately obtain

$$(\alpha_1, \alpha_1) = \frac{1}{3}, \quad (\alpha_1, \alpha_2) = -\frac{1}{6}, \quad (\alpha_2, \alpha_2) = \frac{1}{3}.$$

Consequently, the Cartan matrix of $su(3)$ is

$$K = \begin{pmatrix} 2 & -1 \\ -1 & 2 \end{pmatrix}.$$

For the general Lie algebra $su(N)$, $N \geq 3$, $r = N - 1$, and the $(N - 1) \times (N - 1)$ Cartan matrix is

$$
K = \begin{pmatrix}
2 & -1 & 0 & \cdots & \cdots & 0 \\
-1 & 2 & -1 & 0 & \cdots & 0 \\
0 & -1 & 2 & -1 & \cdots & 0 \\
\vdots & & \ddots & \ddots & \ddots & \vdots \\
0 & & & \ddots & -1 & 2 & -1 \\
0 & \cdots & & 0 & -1 & 2
\end{pmatrix}.
\tag{6.1.34}
$$

Note that there are two conventions in choosing a basis for a matrix Lie algebra in the physics literature: one may have $[X_i, X_j] = c_{ij}^k X_k$ where c_{ij}^k's are structure constants, or, with the new basis $Y_j = iX_j$, one may have $[Y_i, Y_j] = ic_{ij}^k Y_k$. Similarly, a group element may either be represented as $\exp(\omega^k X_k)$ or $\exp(i\omega^k Y_k)$. For a unitary group, these conventions correspond to using anti-Hermitian or Hermitian matrices to represent its associated Lie algebra. We have maintained both conventions in this book in the discussion of various problems in order to keep up with their respective, different, field-theoretical origins, which should be clear from the context and should not cause confusion.

6.2 Non-Abelian Gauged Schrödinger Equations

In this section, we discuss self-dual non-Abelian Chern–Simons gauge equations coupled with a Schrödinger matter field in the adjoint representation of the gauge group. We show how these non-relativistic equations can be solved explicitly using the well-known solution techniques developed for the Toda systems.

6.2.1 Adjoint representation and elliptic problems

Let $\mathcal{G}$ be the semi-simple Lie algebra represented by Hermitian matrices of a given compact gauge group and A_μ be a $\mathcal{G}$-valued gauge field,

$$
A_\mu = A_\mu^a T_a, \quad \mu = 0, 1, 2,
$$

where $\{T_a\}$ is a basis of $\mathcal{G}$. The gauge-covariant derivatives of a matter field ϕ in the adjoint representation of G are

$$
D_\mu \phi = \partial_\mu \phi + [A_\mu, \phi].
$$

To ease the discussion, we specialize to $\mathcal{G} = su(N)$ as a prototype so that the inner product on $\mathcal{G}$ may be taken to be the matrix trace,

$$
(A, B) = \operatorname{Tr}(AB^\dagger), \quad A, B \in \mathcal{G}.
$$

As a consequence, the generators T_a may be normalized so that

$$(T_a, T_b) = \text{Tr } (T_a T_b^\dagger) = -\text{Tr}(T_a T_b) = \delta_{ab}.$$

With the non-Abelian Chern–Simons action density

$$\mathcal{L}_{CS} = \epsilon^{\mu\nu\alpha}\text{Tr } \left(A_\mu \partial_\nu A_\alpha + \frac{2}{3} A_\mu A_\nu A_\alpha \right)$$

and a positive coupling constant κ, we write down the following non-relativistic full Lagrangian density

$$\mathcal{L} = -\frac{\kappa}{2}\mathcal{L}_{CS} + \text{i Tr } (\phi^\dagger D_0\phi) - \frac{1}{2m}\text{Tr } ([D_i\phi]^\dagger[D_i\phi]) + \frac{1}{2m\kappa}\text{Tr } ([\phi, \phi^\dagger]^2),$$
$$(6.2.1)$$

where the last term determines the potential density of the matter field ϕ. Note that, since the identity

$$[\phi, \phi^\dagger]^\dagger = [\phi, \phi^\dagger]$$

holds, the potential density function is in fact of the form

$$\frac{1}{2m\kappa}\text{Tr } ([\phi, \phi^\dagger][\phi, \phi^\dagger]^\dagger),$$

which is of course positive definite.

The equations of motion, or the Euler–Lagrange equations, of (6.2.1) consist of a gauged Schödinger equation and a Chern–Simons equation,

$$iD_0\phi = -\frac{1}{2m}D_i^2\phi - \frac{1}{m\kappa}[[\phi, \phi^\dagger], \phi], \qquad (6.2.2)$$

$$\kappa F_{\mu\nu} = i\epsilon_{\mu\nu\alpha}J^\alpha, \qquad (6.2.3)$$

where the strength tensor $F_{\mu\nu}$ and the gauge-covariantly conserved current density J^μ satisfying $D_\mu J^\mu = 0$ are defined by

$$F_{\mu\nu} = \partial_\mu A_\nu - \partial_\nu A_\mu + [A_\mu, A_\nu],$$
$$J^0 = [\phi, \phi^\dagger],$$
$$J^i = -\frac{i}{2m}([\phi^\dagger, D_i\phi] - [(D_i\phi)^\dagger, \phi]).$$

Note that the corresponding ordinarily conserved current density j^μ satisfying $\partial_\mu j^\mu = 0$ is defined by

$$j^0 = \text{Tr } (\phi\phi^\dagger),$$
$$j^i = -\frac{i}{2m}\text{Tr } (\phi^\dagger[D_i\phi] - [D_i\phi]^\dagger\phi).$$

As before, the Hamiltonian density of the action density (6.2.1) is

$$\mathcal{H} = \frac{1}{2m}\text{Tr } ([D_i\phi]^\dagger[D_i\phi]) - \frac{1}{2m\kappa}\text{Tr } ([\phi, \phi^\dagger]^2). \qquad (6.2.4)$$

We are interested in static solutions. Using the Jacobi identity for the commutator $[\cdot, \cdot]$, it is straightforward to examine that the curvature identity

$$[D_1, D_2]\phi = [F_{12}, \phi] \tag{6.2.5}$$

holds. Besides, we also have

$$\text{Tr}\,\{\partial_\mu(\phi\psi^\dagger)\} = \text{Tr}\,\{(D_\mu\phi)\psi^\dagger\} + \text{Tr}\,\{\phi(D_\mu\psi)^\dagger\}. \tag{6.2.6}$$

Using (6.2.5) and (6.2.6), we obtain

$$
\begin{aligned}
j_{12} &= \partial_1 j_2 - \partial_2 j_1 \\
&= \frac{i}{m}\text{Tr}\,\{(D_1\phi)^\dagger(D_2\phi) - (D_1\phi)(D_2\phi)^\dagger\} \\
&\quad + \frac{i}{2m}\text{Tr}\,(\phi^\dagger[F_{12}, \phi] - \phi[F_{12}, \phi]^\dagger).
\end{aligned}
\tag{6.2.7}
$$

On the other hand, using the property of trace operation, we can examine that

$$\text{Tr}(\phi^\dagger[F_{12}, \phi]) = -\text{Tr}\,(\phi[F_{12}, \phi]^\dagger) = \text{Tr}(F_{12}[\phi, \phi^\dagger]). \tag{6.2.8}$$

Thus, with the notation $D_\pm = D_1 \pm iD_2$, we have by virtue of (6.2.7) that

$$
\begin{aligned}
&\text{Tr}\,([D_i\phi]^\dagger[D_i\phi]) \\
&= \text{Tr}\,([D_\pm\phi]^\dagger[D_\pm\phi]) \pm i\text{Tr}\,\{(D_1\phi)(D_2\phi)^\dagger - (D_1\phi)^\dagger(D_2\phi)\} \\
&= \text{Tr}\,([D_\pm\phi]^\dagger[D_\pm\phi]) \pm i\,\text{Tr}(F_{12}[\phi, \phi^\dagger]) \mp mj_{12}.
\end{aligned}
\tag{6.2.9}
$$

Recall that the Gauss law of the model is obtained by fixing $\mu = 1, \nu = 2$, which is

$$-i\kappa F_{12} = [\phi, \phi^\dagger]. \tag{6.2.10}$$

An equation of this type belongs to a general family of equations studied much earlier by Hitchin [141]. Inserting (6.2.10) into (6.2.9), we have

$$\text{Tr}\,([D_i\phi]^\dagger[D_i\phi]) = \text{Tr}\,([D_\pm\phi]^\dagger[D_\pm\phi]) \mp \frac{1}{\kappa}\text{Tr}\,([\phi, \phi^\dagger]^2) \mp mj_{12}. \tag{6.2.11}$$

Applying (6.2.11) in (6.2.4), we obtain

$$\mathcal{H} = \frac{1}{2m}\text{Tr}\,([D_-\phi]^\dagger[D_-\phi]) + \frac{1}{2}j_{12}.$$

However, since j_1, j_2 decay vanish rapidly at infinity, the divergence theorem leads to

$$\int_{\mathbb{R}^2} j_{12} = 0.$$

Consequently, we obtain the energy lower bound

$$E = \int_{\mathbb{R}^2} \text{Tr}\,([D_-\phi]^\dagger[D_-\phi]) \geq 0. \tag{6.2.12}$$

This lower bound is saturated if there holds

$$D_-\phi = 0 \quad \text{or} \quad D_1\phi = iD_2\phi. \tag{6.2.13}$$

The equations (6.2.10) and (6.2.13) are the non-relativistic self-dual Chern–Simons equations of the model (6.2.1). It can be shown that (6.2.10) and (6.2.13) imply (6.2.2) and (6.2.3). Thus we may focus on (6.2.10) and (6.2.13) from now on.

The governing system of equations, (6.2.10) and (6.2.13), are still too general to approach and a further simplification is to be considered. For this purpose, we use the standard Cartan–Weyl decomposition in terms of a Chevalley basis [66, 67, 82, 96] satisfying the following commutator and trace relations,

$$
\begin{aligned}
[H_a, H_b] &= 0, \\
[E_a, E_{-b}] &= \delta_{ab} H_a, \\
[H_a, E_{\pm b}] &= \pm K_{ab} E_{\pm b}, \\
\text{Tr}\,(E_a E_{-b}) &= \delta_{ab}, \\
\text{Tr}\,(H_a H_b) &= K_{ab}, \\
\text{Tr}\,(H_a E_{\pm b}) &= 0, \quad a, b = 1, 2, \cdots, r, \tag{6.2.14}
\end{aligned}
$$

where r is the rank of $\mathcal{G}$, $\{H_a\}$ are the Cartan subalgebra generators, and $\{E_a\}$ are the simple root ladder generators satisfying $E_a^\dagger = E_{-a}$. Within this framework, the gauge and adjoint matter fields A_μ and ϕ are chosen to be

$$A_\mu = i\sum_{a=1}^{r} A_\mu^a H_a, \tag{6.2.15}$$

$$\phi = \sum_{a=1}^{r} \phi^a E_a, \tag{6.2.16}$$

where A_μ^a are real-valued vector fields and ϕ^a are complex-valued scalar fields ($a = 1, 2, \cdots, r$). Since the generators H_a's are commutative, the gauge field A_μ is Abelian,

$$
\begin{aligned}
F_{\mu\nu} &= \partial_\mu A_\nu - \partial_\nu A_\mu + [A_\mu, A_\nu] \\
&= i\sum_{a=1}^{r}(\partial_\mu A_\nu - \partial_\nu A_\mu) = i\sum_{a=1}^{r} F_{\mu\nu}^a H_a. \tag{6.2.17}
\end{aligned}
$$

On the other hand, using $E_a^\dagger = E_{-a}$, we obtain from (6.2.14) that

$$[\phi, \phi^\dagger] = \sum_{a=1}^{r} |\phi^a|^2 H_a. \tag{6.2.18}$$

Thus, (6.2.10) becomes

$$\kappa F_{12}^a = |\phi^a|^2, \quad a = 1, 2, \cdots, r. \tag{6.2.19}$$

Besides, inserting (6.2.14), we have

$$[A_\mu, \phi] = i \sum_{a,b=1}^{r} A_\mu^a \phi^b K_{ab} E_b. \tag{6.2.20}$$

Using (6.2.20) in (6.2.13), we arrive at

$$(\partial_1 - i\partial_2)\phi^a + i\phi^a \sum_{b=1}^{r} (A_1^b - iA_2^b) K_{ba} = 0, \quad a = 1, 2, \cdots, r. \tag{6.2.21}$$

With the notation

$$\partial = \frac{1}{2}(\partial_1 - i\partial_2), \quad \overline{\partial} = \frac{1}{2}(\partial_1 + i\partial_2), \quad A^a = A_1^a + iA_2^a, \tag{6.2.22}$$

we can rewrite (6.2.21) away from the zeros of ϕ^a's as

$$2\partial \ln \phi^a = -i \sum_{b=1}^{r} \overline{A}^b K_{ba}, \quad a = 1, 2, \cdots, r. \tag{6.2.23}$$

However, since

$$F_{12}^a = \partial_1 A_2^a - \partial_2 A_1^a = -i(\partial A^a - \overline{\partial A}^a), \tag{6.2.24}$$

we have in view of (6.2.23) that

$$\frac{1}{2}\Delta \ln |\phi^a|^2 = 2\partial\overline{\partial} \ln |\phi^a|^2 = -\sum_{b=1}^{r} F_{12}^b K_{ba}. \tag{6.2.25}$$

Inserting (6.2.19) into (6.2.25), we see that, away from the zeros of ϕ^a's,

$$\Delta \ln |\phi^a|^2 = -\frac{2}{\kappa} \sum_{b=1}^{r} |\phi^b|^2 K_{ba}, \quad a = 1, 2, \cdots, r. \tag{6.2.26}$$

Applying the $\overline{\partial}$-Poincaré lemma to (6.2.21), we see easily that each of the functions $\overline{\phi}^a$'s may be decomposed locally as the product of a holomorphic function and a nonvanishing function. In particular the zeros of each of the functions ϕ^a's is discrete and of integer multiplicity. Therefore, setting $u_a = \ln |\phi^a|^2$, we arrive at the elliptic Toda system with sources,

$$\Delta u_a = -\frac{2}{\kappa} \sum_{b=1}^{r} K_{ba} e^{u_b} + 4\pi \sum_{j=1}^{N_a} \delta_{p_{aj}}, \quad a = 1, 2, \cdots, r, \tag{6.2.27}$$

where $p_{a1}, p_{a2}, \cdots, p_{aN_a}$ are possible zeros of the complex scalar field ϕ^a, $a = 1, 2, \cdots, r$.

6.2.2 Toda systems

Neglecting the singular source terms in (6.2.27), rescaling the variables, observing the assumption that the Cartan matrix is symmetric, and using the complex derivatives $\partial_z = \partial/\partial z = \partial$ and $\partial_{\bar{z}} = \partial/\partial\bar{z} = \bar{\partial}$, we obtain the classical Toda system,

$$\frac{\partial^2 u_a}{\partial z \partial \bar{z}} = -\sum_{b=1}^{r} K_{ab}\, e^{u_b}, \quad a = 1, 2, \cdots, r, \qquad (6.2.28)$$

which is known to be integrable in general [115, 172, 188, 189, 190, 208, 215, 237]. We shall follow the procedure developed by Leznov [188] to present a solution for the case when the Cartan matrix K is that of $SU(N)$ expressed in (6.1.34) with $N - 1 = r$.

Using the new dependent variables $v_1, v_2, \cdots, v_r$ with

$$u = Kv,$$

where $u = (u_a)$ and $v = (v_a)$ are column vectors, we see that (6.2.28) becomes

$$\frac{\partial^2 v_a}{\partial z \partial \bar{z}} = -e^{u_a}, \quad a = 1, 2, \cdots, r, \qquad (6.2.29)$$

or

$$\frac{\partial^2 v_1}{\partial z \partial \bar{z}} = -e^{2v_1 - v_2},$$

$$\frac{\partial^2 v_2}{\partial z \partial \bar{z}} = -e^{-v_1 + 2v_2 - v_3},$$

$$\cdots \qquad \cdots$$

$$\frac{\partial^2 v_a}{\partial z \partial \bar{z}} = -e^{-v_{a-1} + 2v_a - v_{a+1}},$$

$$\cdots \qquad \cdots$$

$$\frac{\partial^2 v_r}{\partial z \partial \bar{z}} = -e^{-v_{r-1} + 2v_r}. \qquad (6.2.30)$$

Introduce a new variable,

$$e^{-v_1} = X \equiv \det{}_1(X),$$

and use subscript to denote partial derivatives. The first equation in the system (6.2.30) is simply

$$X_z X_{\bar{z}} - X X_{z\bar{z}} = -e^{-v_2},$$

or

$$e^{-v_2} = \det \begin{pmatrix} X & X_z \\ X_{\bar{z}} & X_{z\bar{z}} \end{pmatrix} \equiv \det{}_2(X).$$

Using the above in the second equation in (6.2.30), we find that

$$\partial\bar\partial v_2 = \frac{\partial^2 v_2}{\partial z \partial \bar z} = -\frac{\det_2(X)(\partial\bar\partial\det_2(X)) - (\partial\det_2(X))(\bar\partial\det_2(X))}{(\det_2(X))^2}$$

$$= -\frac{\det_1(X)}{(\det_2(X))^2} e^{-v_3}. \tag{6.2.31}$$

However, there hold

$$\partial\det_2(X) = \det\begin{pmatrix} X & X_{zz} \\ X_{\bar z} & X_{zz\bar z} \end{pmatrix},$$

$$\bar\partial\det_2(X) = \det\begin{pmatrix} X & X_z \\ X_{\bar z\bar z} & X_{z\bar z\bar z} \end{pmatrix},$$

$$\partial\bar\partial\det_2(X) = \det\begin{pmatrix} X & X_{zz} \\ X_{\bar z\bar z} & X_{zz\bar z\bar z} \end{pmatrix}.$$

Inserting these relations, we obtain the identity

$$\frac{\det_2(X)(\partial\bar\partial\det_2(X)) - (\partial\det_2(X))(\bar\partial\det_2(X))}{(\det_1(X))}$$

$$= \det\begin{pmatrix} X & X_z & X_{zz} \\ X_{\bar z} & X_{z\bar z} & X_{zz\bar z} \\ X_{\bar z\bar z} & X_{z\bar z\bar z} & X_{zz\bar z\bar z} \end{pmatrix}$$

$$\equiv \det_3(X). \tag{6.2.32}$$

Note that for any natural number a, we can define $\det_a(X)$ similarly according to the above rule. Substituting (6.2.32) into (6.2.31), we have

$$e^{-v_3} = \det_3(X).$$

Since the identity (6.2.32) can be generalized into the form

$$\frac{\det_a(X)(\partial\bar\partial\det_a(X)) - (\partial\det_a(X))(\bar\partial\det_a(X))}{(\det_{a-1}(X))}$$

$$= \det_{a+1}(X), \quad a \geq 2, \tag{6.2.33}$$

we can use (6.2.33) to prove by induction that the first $r - 1$ equations are

$$e^{-v_a} = \det_a(X), \quad a = 2, \cdots, r, \tag{6.2.34}$$

which do not impose any restriction to the function X. Using (6.2.33) again, we can show that the last equation in (6.2.30) reads

$$\det_{r+1}(X) = 1, \tag{6.2.35}$$

which fixes the freedom of X. In other words, the solution of (6.2.30) is reduced into the solution of (6.2.35), which is a nonlinear scalar equation of order $2r$.

We look for X in the two independent variables, z and $\bar{z}$, of the form

$$X(z,\bar{z}) = \sum_{a=1}^{r+1} F_a(z) G_a(\bar{z}). \tag{6.2.36}$$

Then, it may be checked that (6.2.36) factors (6.2.35),

$$\det(F)\det(G) = 1, \tag{6.2.37}$$

where $F = (F_{ab})$ and $G = (G_{ab})$ are $(r+1) \times (r+1)$ matrices defined by

$$F_{ab} = \frac{\partial^{b-1} F_a}{\partial z^{b-1}}, \quad G_{ab} = \frac{\partial^{b-1} G_a}{\partial \bar{z}^{b-1}}, \quad a, b = 1, 2, \cdots, r+1.$$

For $r = 1$, (6.2.37) reads, say,

$$F_1 \frac{\partial F_2}{\partial z} - \frac{\partial F_1}{\partial z} F_2 = -1, \tag{6.2.38}$$

$$G_1 \frac{\partial G_2}{\partial \bar{z}} - \frac{\partial G_1}{\partial \bar{z}} G_2 = -1, \tag{6.2.39}$$

which are two decoupled equations. To solve (6.2.38), we rewrite it as

$$\frac{1}{F_1 F_2} = \frac{\partial}{\partial z} \ln \left(\frac{F_1}{F_2} \right).$$

Set $F_1/F_2 = F$ (an arbitrary function). The above equation leads to $F_2 = F_z^{-1/2}$. Consequently, $F_1 = F F_z^{-1/2}$. Similarly, $G_1 = G G_{\bar{z}}^{-1/2}, G_2 = G_{\bar{z}}^{-1/2}$.

Another, more systematic and elegant, way of constructing general solutions of (6.2.38) and (6.2.39) is to explore symmetry of these equations. To see this, we rewrite (6.2.38) as

$$\det \begin{pmatrix} F_1 & F_{1,z} \\ F_2 & F_{2,z} \end{pmatrix} = -1 \tag{6.2.40}$$

and observe that (6.2.40) is invariant under the transformation

$$F_1(z) \mapsto F_1(F(z))(F'(z))^{-1/2}, \quad F_2(z) \mapsto F_2(F(z))(F'(z))^{-1/2}, \tag{6.2.41}$$

where $F(z)$ is an arbitrary analytic function of z. It is easy to check that (6.2.40) has a particular solution, $F_1(z) = z, F_2(z) = 1$, which, through (6.2.41), gives rise to the general solution obtained earlier,

$$F_1(z) = F(z) F'(z)^{-1/2}, \quad F_2(z) = F'(z)^{-1/2}.$$

Hence, in view of (6.2.36), we have reproduced the general solution for the Liouville equation,

$$v = v_1 = \frac{1}{2} \ln \left(\frac{F'(z) G'(\bar{z})}{(1 + F(z) G(\bar{z}))^2} \right).$$

Let $r = 2$. We now show how the general solution of (6.2.37) may be obtained by the same method. Consider, for example, $\det(F) = -1$ or

$$\det \begin{pmatrix} F_1 & F_{1,z} & F_{1,zz} \\ F_2 & F_{2,z} & F_{2,zz} \\ F_3 & F_{3,z} & F_{3,zz} \end{pmatrix} = -1. \tag{6.2.42}$$

We may choose (f_1, f_2) to be the general solution of (6.2.40). Then

$$F_1(z) = \int_0^z f_1(\xi)\,d\xi, \quad F_2(z) = \int_0^z f_2(\xi)\,d\xi, \quad F_3(z) = 1 \tag{6.2.43}$$

is a particular solution of (6.2.42). On the other hand, it is clear that (6.2.42) is invariant under the transformation

$$F_a(z) \mapsto F_a(F(z))(F'(z))^{-1}, \quad a = 1, 2, 3. \tag{6.2.44}$$

Combining (6.2.43) and (6.2.44), we see that the general solution of (6.2.42) is given by

$$F_a(z) = F'(z)^{-1} \int_0^{F(z)} f_a(\xi)\,d\xi, \quad a = 1, 2, \quad F_3(z) = F'(z)^{-1}, \tag{6.2.45}$$

where $F(z)$ is an arbitrary analytic function.

The above procedure hints an inductive method to get the general solution corresponding to any rank number r of the equation

$$\det(F) = (-1)^{r(r+1)/2}. \tag{6.2.46}$$

In fact, it can be examined that (6.2.46) is invariant under the transformation

$$F_a(z) \mapsto F_a(F(z))F'(z)^{-r/2}, \quad a = 1, 2, \cdots, r + 1. \tag{6.2.47}$$

As before, suppose $F = (f_1, f_2, \cdots, f_r)$ is the general solution of

$$\det(F) = (-1)^{r(r-1)/2} \tag{6.2.48}$$

when the rank number is $r - 1 \geq 1$. Then

$$F_a(z) = \int_0^z f_a(\xi)\,d\xi, \quad a = 1, 2, \cdots, r, \quad F_{r+1}(z) = 1 \tag{6.2.49}$$

solves the equation corresponding to the rank number r case, (6.2.46). Using (6.2.47), we obtain the general solution of (6.2.48) as follows,

$$\begin{aligned} F_a(z) &= F'(z)^{-r/2} \int_0^{F(z)} f_a(\xi)\,d\xi, \quad a = 1, 2, \cdots, r, \\ F_{r+1}(z) &= F'(z)^{-r/2}, \end{aligned} \tag{6.2.50}$$

where $F(z)$ is an arbitrary analytic function.

Identically, for $\det(G) = (-1)^{r(r+1)/2}$, let $(g_1, g_2, \cdots, g_r)$ be the general solution of the equation corresponding to the rank number $r - 1$. Then

$$
\begin{aligned}
G_a(z) &= G'(z)^{-r/2} \int_0^{G(z)} g_a(\xi)\, d\xi, \quad a = 1, 2, \cdots, r, \\
G_{r+1}(z) &= G'(z)^{-r/2}
\end{aligned}
\tag{6.2.51}
$$

is the general solution for the rank number r. Hence, inserting (6.2.50) and (6.2.51) into (6.2.36), we see that the equation (6.2.35) or the system (6.2.30) is completely solved.

It is also convenient to express the solutions more explicitly. For $r = 1$, let $\varphi_1(z) = F'(z)$ and $\psi_1(z) = G'(z)$. We have

$$
\begin{aligned}
F_1(z) &= \varphi_2(z) \int^z \varphi_1(z_1)\, dz_1, \quad F_2(z) = \varphi_2(z), \\
G_1(z) &= \psi_2(z) \int^z \psi_1(z_1)\, dz_1, \quad G_2(z) = \psi_2(z),
\end{aligned}
\tag{6.2.52}
$$

where $\varphi_2 = \varphi_1^{-1/2}$ and $\psi_2 = \psi_1^{-1/2}$. For $r = 2$, since

$$
\int^{F(z)} f_a(\xi)\, d\xi = \int^z F'(\xi) f_a(F(\xi))\, d\xi, \quad a = 1, 2,
$$

the formula (6.2.45) gives us

$$
\begin{aligned}
F_1(z) &= F'(z)^{-1} \int^z F'(z_2) \varphi_1^{-1/2}(F(z_2))\, dz_2 \int^{z_2} F'(z_1)\varphi_1(F(z_1))\, dz_1, \\
F_2(z) &= F'(z)^{-1} \int^z F'(z_2)\varphi_1^{-1/2}(F(z_2))\, dz_2, \\
F_3(z) &= F'(z)^{-1}.
\end{aligned}
\tag{6.2.53}
$$

In (6.2.53), we can rename the arbitrary functions to obtain

$$
\begin{aligned}
F_1(z) &= \varphi_3(z) \int^z \varphi_2(z_2)\, dz_2 \int^{z_2} \varphi_1(z_1)\, dz_1, \\
F_2(z) &= \varphi_3(z) \int^z \varphi_2(z_2)\, dz_2, \\
F_3(z) &= \varphi_3(z).
\end{aligned}
\tag{6.2.54}
$$

Note that the only constraint for the functions $\varphi_1, \varphi_2, \varphi_3$ is $\varphi_3^{-1} = \varphi_1^{1/3}\varphi_2^{2/3}$.

For a general rank number r, we find, by induction,

$$
F_a(z) = \varphi_{r+1}(z) \int^z \varphi_r(z_r)\, dz_r \int^{z_r} \cdots \int^{z_a} \varphi_a(z_a)\, dz_a,
$$

$$a = 1, 2, \cdots r,$$

$$F_{r+1}(z) = \varphi_{r+1}(z), \quad (\varphi_{r+1})^{-1} = \prod_{a=1}^{r} \varphi_a^{a/(r+1)},$$

$$G_a(\overline{z}) = \psi_{r+1}(\overline{z}) \int^{\overline{z}} \psi_r(z_r) \, dz_r \int^{z_r} \cdots \int^{z_a} \psi_a(z_a) \, dz_a,$$

$$a = 1, 2, \cdots r,$$

$$G_{r+1}(\overline{z}) = \psi_{r+1}(\overline{z}), \quad (\psi_{r+1})^{-1} = \prod_{a=1}^{r} \psi_a^{a/(r+1)}. \tag{6.2.55}$$

Finally, we use (6.2.34) to recover the general solution of (6.2.30).

6.2.3 Explicit non-Abelian solutions

Here we briefly discuss how to construct solutions of (6.2.27) explicitly when the gauge group is $SU(N)$. Since $4\partial\overline{\partial} = \Delta$, we may use the translation

$$u_a - \ln(2\kappa) \mapsto u_a \tag{6.2.56}$$

to modify the system (6.2.27) into the standard Toda system, (6.2.28), where we have neglected the source terms temporarily. The relation (6.2.34) gives us the formula

$$-v_a = \ln\det_a(X), \quad a = 1, 2, \cdots, r. \tag{6.2.57}$$

Using (6.2.57) in the Toda system (6.2.29) and returning to the original variables through (6.2.56), we have

$$c^{u_a} = \frac{\kappa}{2} \Delta \ln\det_a(X), \quad a = 1, 2, \cdots, r. \tag{6.2.58}$$

To get real-valued solutions $u_1, u_2, \cdots, u_r$ from (6.2.58), it suffices to choose $G_a(\overline{z}) = \overline{F_a(z)}$ for $a = 1, 2, \cdots, r + 1$ in (6.2.36). Thus, note that $r = N - 1$ and define the $a \times N$ matrix $M_a(z)$ by

$$M_a(z) = \begin{pmatrix} F_1(z) & F_2(z) & \cdots & F_N(z) \\ F_1'(z) & F_2'(z) & \cdots & F_N'(z) \\ \cdots & \cdots & \cdots & \cdots \\ F_1^{(a-1)}(z) & F_2^{(a-1)}(z) & \cdots & F_N^{(a-1)}(z) \end{pmatrix}. \tag{6.2.59}$$

Then we obtain

$$\det_a(X) = \det(M_a M_a^\dagger). \tag{6.2.60}$$

Note that, here, no summation is assumed over repeated indices. Inserting (6.2.60) into (6.2.58), we can represent our general solution of the non-Abelian, non-relativistic, Chern–Simons equations (6.2.27) as

$$e^{u_a} = \frac{\kappa}{2} \Delta \ln\det(M_a M_a^\dagger), \quad a = 1, 2, \cdots, r, \tag{6.2.61}$$

where the complex holomorphic functions $F_1, F_2, \cdots, F_N$ in the matrices M_a's are as defined in (6.2.55) and some care must be taken to account for the prescribed singularities stated in (6.2.27) and sufficient decay rate near infinity.

The discovery of the self-dual non-relativistic Chern–Simons equations and their solution are presented in the original work of Dunne, Jackiw, Pi, and Trugenberger [98]. It is seen that explicit solvability imposes rigid restrictions on the coefficients of the equations.

6.3   Relativistic Chern–Simons Systems

Following [96, 97, 98], the non-Abelian relativistic self-dual Chern–Simons model in $(2 + 1)$ dimensions is described by the Lagrangian action density

$$\mathcal{L} = -\frac{\kappa}{2}\epsilon^{\mu\nu\alpha}\mathrm{Tr}\left(\partial_\mu A_\nu A_\alpha + \frac{2}{3}A_\mu A_\nu A_\alpha\right) + \mathrm{Tr}\left([D_\mu\phi]^\dagger[D^\mu\phi]\right) - V(\phi, \phi^\dagger),$$

(6.3.1)

where the potential energy density of the Higgs field is given by the special formula

$$V(\phi, \phi^\dagger) = \frac{1}{\kappa^2}\mathrm{Tr}\left\{([[\phi, \phi^\dagger], \phi] - v^2\phi)^\dagger([[\phi, \phi^\dagger], \phi] - v^2\phi)\right\}$$

(6.3.2)

in order to write the energy density of the model as a sum of squares and a divergence term, and $v > 0$ is a constant which measures either the scale of the broken symmetry or the subcritical temperature of the system. The Euler–Lagrange equations of motion of (6.3.1) are

$$D_\mu D^\mu\phi = -\frac{\partial V}{\partial \phi^\dagger},$$

(6.3.3)

$$\kappa F_{\mu\nu} = \varepsilon_{\mu\nu\alpha}J^\alpha,$$

(6.3.4)

where the non-Abelian relativistic covariantly conserved current density J^μ is defined by

$$J^\mu = [\phi^\dagger, (D^\mu\phi)] - [(D^\mu\phi)^\dagger, \phi]$$

(6.3.5)

and satisfies $D_\mu J^\mu = 0$. There is an Abelian current Q^μ, corresponding to the global $U(1)$ invariance of (6.3.1), satisfying $\partial_\mu Q^\mu = 0$ and defined by

$$Q^\mu = -i\,\mathrm{Tr}\,(\phi^\dagger D^\mu\phi - [D^\mu\phi]^\dagger\phi).$$

(6.3.6)

The Hamiltonian density of (6.3.1) is

$$\mathcal{H} = \mathrm{Tr}\,([D_0\phi]^\dagger[D_0\phi]) + \mathrm{Tr}\,([D_i\phi]^\dagger[D_i\phi]) + V(\phi, \phi^\dagger).$$

(6.3.7)

The Gauss law, which is obtained by setting $\mu = 1, \nu = 2$ in (6.3.4), reads

$$\kappa F_{12} = J^0 = [\phi^\dagger, (D_0\phi)] - [(D_0\phi)^\dagger, \phi].$$

(6.3.8)

In order to derive a self-dual reduction in the static limit for the equations of motion, (6.3.3) and (6.3.4), we note that $Q_i = -Q^i, D_i = -D^i$ and use (6.2.5), (6.2.6), and (6.2.8) to calculate

$$
\begin{aligned}
Q_{12} &= \partial_1 Q_2 - \partial_2 Q_1 \\
&= 2\mathrm{i}\,\mathrm{Tr}\,\{(D_1\phi)^\dagger(D_2\phi) - (D_1\phi)(D_2\phi)^\dagger\} + 2\mathrm{i}\,\mathrm{Tr}\,(F_{12}[\phi,\phi^\dagger]).
\end{aligned}
$$
$$(6.3.9)$$

In view of (6.3.8), (6.3.9), and the easily verified identity

$$\mathrm{Tr}\,([A,B][C,D]) = \mathrm{Tr}\,([[C,D],A]B),\tag{6.3.10}$$

we have by the middle part of (6.2.9) that

$$
\begin{aligned}
&\mathrm{Tr}\,\{(D_i\phi)^\dagger(D_i\phi)\} \\
&= \mp\frac{1}{2}Q_{12} + \mathrm{Tr}\,\{(D_\pm\phi)^\dagger(D_\pm\phi)\} \pm \mathrm{i}\,\mathrm{Tr}\,(F_{12}[\phi,\phi^\dagger]) \\
&= \mp\frac{1}{2}Q_{12} + \mathrm{Tr}\,\{(D_\pm\phi)^\dagger(D_\pm\phi)\} \\
&\quad \pm\frac{\mathrm{i}}{\kappa}\,\mathrm{Tr}\,\left([[\phi,\phi^\dagger],\phi](D_0\phi)^\dagger - [[\phi,\phi^\dagger],\phi]^\dagger(D_0\phi)\right).
\end{aligned}
$$
$$(6.3.11)$$

Moreover, using the notation $|A|^2 = \mathrm{Tr}\,(AA^\dagger)$ to simplify writing at a few places, the term containing $D_0\phi$ in the Hamiltonian density can be rewritten,

$$
\begin{aligned}
\mathrm{Tr}([D_0\phi]^\dagger[D_0\phi]) &= \left|D_0\phi \pm \frac{\mathrm{i}}{\kappa}([[\phi,\phi^\dagger],\phi] - v^2\phi)\right|^2 \\
&\quad \pm\frac{\mathrm{i}}{\kappa}\mathrm{Tr}\,\left(([[\phi,\phi^\dagger],\phi] - v^2\phi)^\dagger(D_0\phi) - ([[\phi,\phi^\dagger],\phi] - v^2\phi)(D_0\phi)^\dagger\right) \\
&\quad -\frac{1}{\kappa^2}\left|[[\phi,\phi^\dagger],\phi] - v^2\phi\right|^2.
\end{aligned}
$$
$$(6.3.12)$$

Substituting (6.3.9) and (6.3.12) into (6.3.7), using (6.3.2), and dropping the integral

$$\int_{\mathbb{R}^2} Q_{12},\tag{6.3.13}$$

which vanishes due to the decay property of a finite energy solution, we obtain

$$
\begin{aligned}
E &= \int_{\mathbb{R}^2}\mathcal{H} \\
&= \int_{\mathbb{R}^2}\left|D_0\phi \pm \frac{\mathrm{i}}{\kappa}([[\phi,\phi^\dagger],\phi] - v^2\phi)\right|^2
\end{aligned}
$$

$$+ \int_{\mathbb{R}^2} \mathrm{Tr} \; ([D_\pm \phi]^\dagger [D_\pm \phi]) \mp \mathrm{i} \frac{v^2}{\kappa} \int_{\mathbb{R}^2} \mathrm{Tr} \; (\phi^\dagger [D_0 \phi] - [D_0 \phi]^\dagger \phi)$$

$$\geq \frac{v^2}{\kappa} \left| \int_{\mathbb{R}^2} Q_0 \right|, \tag{6.3.14}$$

and this energy lower bound is attained if the field configuration satisfies the equations

$$D_\pm \phi = 0, \tag{6.3.15}$$

$$D_0 \phi = \mp \frac{\mathrm{i}}{\kappa} ([[\phi, \phi^\dagger], \phi] - v^2 \phi). \tag{6.3.16}$$

Using the Gauss law (6.3.8) in (6.3.16), we have the following self-dual equations,

$$D_1 \phi = \mp \mathrm{i} \, D_2 \phi, \tag{6.3.17}$$

$$\mathrm{i} \, F_{12} = \pm \frac{2}{\kappa^2} [v^2 \phi - [[\phi, \phi^\dagger], \phi], \phi^\dagger]. \tag{6.3.18}$$

This system has more sophisticated nonlinearity than Hitchin's equations [141]. It can be checked that the solutions of (6.3.17) and (6.3.18) also solve the Euler–Lagrange equations (6.3.3) and (6.3.4).

Under the ansatz given in (6.2.15) and (6.2.16), the system describes r Abelian Chern–Simons gauge fields A_μ^a coupled to r complex scalar Higgs fields ϕ^a, with the couplings determined by the Cartan matrix $K = (K_{ab})$ of the Lie algebra $\mathcal{G}$. Besides, the trivial solutions satisfying

$$[[\phi, \phi^\dagger], \phi] = v^2 \phi, \tag{6.3.19}$$

also called the principal embedding vacuum [97], in the broken symmetry case, $\phi_{(0)}^a \neq 0 \; (\forall a)$, are characterized by

$$|\phi_{(0)}^a|^2 = v^2 \sum_{b=1}^r (K^{-1})_{ba}, \quad a = 1, 2, \cdots, r, \tag{6.3.20}$$

for which the gauge fields all vanish [97]. In fact, inserting (6.2.16) into (6.3.19) and using $E_a^\dagger = E_{-a}$, we get

$$\sum_{a=1}^r \left(\sum_{b=1}^r |\phi^b|^2 \phi^a K_{ba} - v^2 \phi^a \right) E_a = 0. \tag{6.3.21}$$

Since we are looking for solutions satisfying $\phi^a \neq 0$ for $a = 1, 2, \cdots, r$, (6.3.21) gives us

$$\sum_{b=1}^r |\phi^b|^2 K_{ba} = v^2, \quad a = 1, 2, \cdots, r,$$

leading to (6.3.20). These solutions define the symmetry-breaking vacuum states. On the other hand, nontrivial multivortex solutions of (6.3.17) and (6.3.18) under the ansatz (6.2.15), (6.2.16) are characterized by the zeros of the complex scalar fields, ϕ^a's, and their asymptotic behavior at infinity. These zeros may be viewed as electrically and magnetically charged particles residing in equilibrium or defects in a superconductor so that the solution configuration interpolates the point normal state achieved at the defects and the superconducting vacuum at infinity.

Unlike the gauged non-Abelian Schrödinger equation solutions discussed in the previous section, the governing equations in the current situation are not integrable and any understanding of them will have to be based upon nonlinear functional analysis again [349]. Our main existence theorem for such solutions is stated as follows.

Theorem 6.3.1. *For any given nonnegative integers N_a and the corresponding prescribed point sets*

$$Z_a = \left\{ p_{a1}, p_{a2}, \cdots, p_{aN_a} \right\}, \quad a = 1, 2, \cdots, r,$$

the relativistic self-dual Chern–Simons equations (6.3.17), (6.3.18) have a solution satisfying (6.2.15), (6.2.16) so that Z_a is the zero set of ψ^a, the magnetic charge induced from F_{12}^a takes the quantized value

$$\Phi_a = \int_{\mathbb{R}^2} F_{12}^a = \mp 2\pi \sum_{b=1}^{r} (K^{-1})_{ba} N_b, \quad a = 1, 2, \cdots, r,$$

and there hold the decay estimates at infinity,

$$
\begin{aligned}
|\phi^a|^2 - |\phi_{(0)}^a|^2 &= O\left(e^{-\lambda_0 v^2 (\sqrt{2\sigma_0}/2\kappa)(1-\varepsilon)|x|}\right), \\
F_{12}^a &= O\left(e^{-\lambda_0 v^2 (\sqrt{2\sigma_0}/2\kappa)(1-\varepsilon)|x|}\right), \\
|D_j \phi^a|^2 &= O\left(e^{-\lambda_0 v^2 (\sqrt{2\sigma_0}/\kappa)(1-\varepsilon)|x|}\right), \quad j = 1, 2, \\
a &= 1, 2, \cdots, r,
\end{aligned}
\tag{6.3.22}
$$

where $\varepsilon \in (0, 1)$ is arbitrary, λ_0^2 is the smallest positive eigenvalue of KRK, with the diagonal matrix R being defined by

$$R =: v^{-2} \, diag\left\{ |\phi_{(0)}^1|^2, |\phi_{(0)}^2|^2, \cdots, |\phi_{(0)}^r|^2 \right\},$$

and σ_0 by

$$\sigma_0 = v^{-2} \min_{1 \le a \le r} \left\{ |\phi_{(0)}^a|^2 \right\}.$$

6.4 Elliptic System and its Variational Principle

For convenience, we first stay away from the zeros of ϕ^a's to consider the governing equations (6.3.17) and (6.3.18). Using the ansatz (6.2.15), (6.2.16), and the notation

$$\partial_{\pm} = \partial_1 \pm i\partial_2, \quad A_{\pm}^a = A_1^a \pm iA_2^a,$$

these equations become

$$\partial_{\pm} \ln \phi^a = -i \sum_{b=1}^{r} A_{\pm}^b K_{ba}, \tag{6.4.1}$$

$$F_{12}^a = \pm \frac{2}{\kappa^2} \left(\sum_{b=1}^{r} |\phi^a|^2 |\phi^b|^2 K_{ba} - v^2 |\phi^a|^2 \right). \tag{6.4.2}$$

Note that $\overline{\partial}_{\pm} = \partial_{\mp}, \overline{A}_{\pm}^a = A_{\mp}^a$, and

$$F_{12}^a = \partial_1 A_2^a - \partial_2 A_1^a = \frac{1}{2i}(\partial_- A_+ - \partial_+ A_-). \tag{6.4.3}$$

Thus, from (6.4.1), we have

$$\partial_{\mp} \partial_{\pm} \ln \phi^a = -i \sum_{b=1}^{r} \partial_{\mp} A_{\pm}^b K_{ba},$$

$$\partial_{\pm} \partial_{\mp} \ln \overline{\phi}^a = i \sum_{b=1}^{r} \partial_{\pm} A_{\mp}^b K_{ba}. \tag{6.4.4}$$

Adding the two equations in (6.4.4), using (6.4.3), noting $\partial_- \partial_+ = \partial_+ \partial_- = \Delta$, and inserting (6.4.2), we have

$$\Delta \ln |\phi^a|^2 = -i \sum_{b=1}^{r} (\partial_{\mp} A_{\pm}^b - \partial_{\pm} A_{\mp}^b) K_{ba}$$

$$= \pm 2 \sum_{b=1}^{r} F_{12}^b K_{ba}$$

$$= \frac{4}{\kappa^2} \left(\sum_{b=1}^{r} \sum_{c=1}^{r} |\phi^b|^2 |\phi^c|^2 K_{cb} K_{ba} - v^2 \sum_{b=1}^{r} |\phi^b|^2 K_{ba} \right),$$

$$a = 1, 2, \cdots, r, \quad x \in \mathbb{R}^2. \tag{6.4.5}$$

With the prescribed data stated in Theorem 6.3.1 and the substitution

$$|\phi^a|^2 = v^2 e^{u_a}, \quad a = 1, 2, \cdots, r, \quad \lambda = \frac{4v^4}{\kappa^2},$$

the system (6.4.5) becomes

$$
\Delta u_a = \lambda\left(\sum_{b=1}^{r}\sum_{c=1}^{r} e^{u_b}\, e^{u_c} K_{cb} K_{ba} - \sum_{b=1}^{r} e^{u_b}\, K_{ba}\right) + 4\pi \sum_{j=1}^{N_a} \delta_{p_{aj}},
$$
$$
a = 1,2,\cdots,r, \quad x \in \mathbb{R}^2. \tag{6.4.6}
$$

In order to compare our system of equations with the Toda system (6.2.28), we write down its revised form as follows,

$$
\Delta u_a = -\lambda\sum_{b=1}^{r} K_{ab} e^{u_b} + \lambda\sum_{b=1}^{r}\sum_{c=1}^{r} e^{u_b} K_{ab} e^{u_c} K_{bc} + 4\pi \sum_{j=1}^{N_a} \delta_{p_{aj}},
$$
$$
a = 1,2,\cdots,r, \quad x \in \mathbb{R}^2. \tag{6.4.7}
$$

Of course, when the Cartan matrix K is symmetric, these two systems coincide.

The system (6.4.7) may be viewed as a non-integrable [273] deformation of the Toda equations arising in the study of the non-Abelian gauged Schrödinger equations treated in the last section. Besides, it may be worthwhile to point out that although the Cartan matrix K has the special property that its diagonal entries are all the same, $K_{aa} = 2$, $a = 1,2,\cdots,r$ and its off-diagonal entries can only assume values from the non-positive integers $-3,-2,-1,0$ (cf. [135]), our method for solving (6.4.6) and (6.4.7) is not restricted to such a property. In fact, we will establish our existence theorem under the general condition that the coefficient matrix K has the decomposition

$$
K = PS, \tag{6.4.8}
$$

where P is a diagonal matrix with positive diagonal entries and S is a symmetric positive definite matrix, both of size $r \times r$. Such a condition is broad enough to contain the case where K is a Cartan matrix because, due to the representation (6.1.32), one may choose

$$
P = 2\mathrm{diag}\left\{(\alpha_1,\alpha_1)^{-1}, (\alpha_2,\alpha_2)^{-1}, \cdots, (\alpha_r,\alpha_r)^{-1}\right\}, \quad S = ((\alpha_a,\alpha_b)).
$$

Throughout the rest of this paper, we always assume the validity of the hypothesis (6.4.8) on the matrix K in the systems (6.4.6) and (6.4.7).

Here is our main existence theorem for (6.4.6) and (6.4.7).

Theorem 6.4.1. *Suppose that the coefficient matrix K in the system of equations (6.4.7) satisfies*

$$
\sum_{b=1}^{r} (K^{-1})_{ab} > 0, \quad a = 1,2,\cdots,r. \tag{6.4.9}
$$

Then (6.4.7) has a classical solution verifying the asymptotic condition

$$\lim_{|x|\to\infty} u_a(x) = v_a^0 \equiv \ln\left(\sum_{b=1}^{r}(K^{-1})_{ab}\right),$$

$$\lim_{|x|\to\infty} |\nabla u_a| = 0, \quad a = 1, 2, \cdots, r. \qquad (6.4.10)$$

Furthermore, if K is symmetric and λ_0^2 is the smallest positive eigenvalue of the matrix KRK where R is the diagonal matrix defined by

$$R = diag\left\{R_1, R_2, \cdots, R_r\right\},$$

$$R_a \equiv \sum_{b=1}^{r}(K^{-1})_{ab}, \quad a = 1, 2, \cdots, r,$$

then (6.4.10) can be strengthened to the following exponential decay estimate

$$\max_{1\leq a\leq r} |u_a(x) - v_a^0|^2 + \max_{1\leq a\leq r} |\nabla u_a|^2 = O\left(e^{-\lambda_0\sqrt{2\lambda\sigma_0}(1-\varepsilon)|x|}\right) \qquad (6.4.11)$$

near infinity, where $0 < \varepsilon < 1$ is arbitrary and σ_0 is the smallest member among the numbers $R_1, R_2, \cdots, R_r$. Besides, in this case, there are quantized integrals

$$\int_{\mathbb{R}^2} \sum_{b=1}^{r} K_{ab}e^{u_b} - \int_{\mathbb{R}^2} \sum_{b=1}^{r}\sum_{c=1}^{r} e^{u_b}K_{ab}e^{u_c}K_{bc}$$

$$= 4\pi\frac{N_a}{\lambda}, \quad a = 1, 2, \cdots, r. \qquad (6.4.12)$$

However, if K is not symmetric but the symmetrization of KRK, namely,

$$M = \frac{1}{2}(KRK + K^\tau RK^\tau),$$

is positive definite, then the same decay estimate (6.4.11) and the quantization of integrals (6.4.12) hold for the solution, where λ_0^2 now denotes the smallest eigenvalue of M. The same results hold for the system (6.4.6) if the matrix K is replaced with K^τ.

If K is the Cartan matrix of an arbitrary finite-dimensional semi-simple Lie algebra, the assumption (6.4.9) is redundant. In fact, let $\lambda_1, \lambda_2, \cdots, \lambda_r$ be the fundamental dominant weights relative to the given simple root vectors $\alpha_1, \alpha_2, \cdots, \alpha_r$. Then K is the transition matrix from the basis $\lambda_1, \lambda_2, \cdots, \lambda_r$ to the basis $\alpha_1, \alpha_2, \cdots, \alpha_r$ for the root space,

$$\alpha_a = \sum_{b=1}^{r} K_{ab}\lambda_b, \quad a = 1, 2, \cdots, r,$$

and K^{-1} defines the inverse of the transition whose entries are all nonnegative [147]. However, since K^{-1} exists, so the condition (6.4.9) is always
valid as claimed.

As mentioned earlier, we need to find variational principles for (6.4.6)
and (6.4.7). As in Chapter 4, our method depends crucially on the well-
known Cholesky decomposition theorem for positive definite matrices. We
proceed as follows.

Recall that, since the $r \times r$ matrix S is symmetric and positive definite,
by the theorem of Cholesky [131, 297] there is a (unique) lower triangular
$r \times r$ matrix L with positive diagonal entries so that

$$S = LL^\tau. \tag{6.4.13}$$

Write $\mathbf{1} = (1, 1, \cdots, 1)^\tau$ and $\mathbf{U} = (e^{u_1}, e^{u_2}, \cdots, e^{u_r})^\tau$ and introduce the
background functions

$$u_a^0(x) = -\sum_{j=1}^{N_a} \ln(1 + |x - p_{aj}|^{-2}),$$

$$g_a(x) = 4\sum_{j=1}^{N_a} \frac{1}{(1 + |x - p_{aj}|^2)^2},$$

$$a = 1, 2, \cdots, r, \quad x \in \mathbb{R}^2. \tag{6.4.14}$$

We first consider (6.4.7). Using the substitution $u_a = u_a^0 + v_a$, the system
(6.4.7) takes the regular form

$$\Delta v_a = -\lambda \sum_{b=1}^r K_{ab} e^{u_b^0 + v_b} + \lambda \sum_{b=1}^r \sum_{c=1}^r e^{u_b^0 + v_b} K_{ab} e^{u_c^0 + v_c} K_{bc} + g_a,$$

$$a = 1, 2, \cdots, r, \quad x \in \mathbb{R}^2. \tag{6.4.15}$$

Now we use transposition to save space, apply the transformation

$$\mathbf{w} = (PL)^{-1}\mathbf{v}, \quad \mathbf{v} = (v_1, v_2, \cdots, v_r)^\tau,$$
$$\mathbf{h} = (PL)^{-1}\mathbf{g}, \quad \mathbf{g} = (g_1, g_2, \cdots, g_r)^\tau, \tag{6.4.16}$$

and consider the energy functional, which is a crucial step in the method,

$$I(\mathbf{w}) = \int_{\mathbb{R}^2} \left\{ \frac{1}{2} \sum_{a=1}^r |\nabla w_a|^2 + \frac{\lambda}{2} |L^\tau \mathbf{U} - (PL)^{-1}\mathbf{1}|^2 + \mathbf{h} \cdot \mathbf{w} \right\},$$

$$\mathbf{U} = \left(e^{u_1^0 + \ell_{11} w_1}, \cdots, e^{u_a^0 + \sum_{b=1}^a \ell_{ab} w_b}, \cdots, e^{u_r^0 + \sum_{b=1}^r \ell_{rb} w_b} \right)^\tau, \tag{6.4.17}$$

where we adopt the notation $PL = (\ell_{ab})$. Define then the diagonal matrix

$$U = \text{diag}\left\{ e^{u_1^0 + \ell_{11} w_1}, \cdots, e^{u_a^0 + \sum_{b=1}^a \ell_{ab} w_b}, \cdots, e^{u_r^0 + \sum_{b=1}^r \ell_{rb} w_b} \right\}. \tag{6.4.18}$$

It is straightforward (but somewhat tedious) to verify that a critical point of the functional I, say, $\mathbf{w}$, satisfies the nonlinear elliptic equations

$$\Delta \mathbf{w} = -\lambda L^\tau \mathbf{U} + \lambda L^\tau PULL^\tau \mathbf{U} + \mathbf{h}. \tag{6.4.19}$$

Here it may be useful in the derivation of (6.4.19) to apply the relation

$$\frac{d}{d\mathbf{w}}\left(\mathbf{c}^\tau \cdot \mathbf{U}\right) = (PL)^\tau \mathbf{U}\mathbf{c}, \tag{6.4.20}$$

where $\mathbf{c}$ is a constant r-vector.

Returning to the original variable $\mathbf{v}$ via (6.4.16) and noting that the product PU is commutative, we obtain from (6.4.19) the system

$$\Delta \mathbf{v} = -\lambda K\mathbf{U} + \lambda KUK\mathbf{U} + \mathbf{g}. \tag{6.4.21}$$

It can be seen that (6.4.21) is the matrix form of the equations (6.4.15) as desired. Thus the expected variational principle is established.

Note that the form of the variational energy functional (6.4.17) imposes the boundary condition

$$L^\tau \mathbf{U} - (PL)^{-1}\mathbf{1} = \mathbf{0} \tag{6.4.22}$$

at infinity. It is this condition that leads to our main assumption in Theorem 6.4.1 which ensures the existence of a unique positive constant solution, $\mathbf{U} > \mathbf{0}$, to the above equation and will be employed in the next section.

We next consider (6.4.6). The same substitution as before gives us the system

$$\Delta \mathbf{v} = -\lambda K^\tau \mathbf{U} + \lambda K^\tau UK^\tau \mathbf{U} + \mathbf{g}. \tag{6.4.23}$$

Introduce the new variables

$$\mathbf{w} = L^{-1}\mathbf{v}, \quad \mathbf{h} = L^{-1}\mathbf{g}. \tag{6.4.24}$$

Then (6.4.8) and (6.4.13) imply that (6.4.23) takes the modified form

$$\Delta \mathbf{w} = -\lambda L^\tau P\mathbf{U} + \lambda L^\tau PULL^\tau P\mathbf{U} + \mathbf{h}. \tag{6.4.25}$$

In place of (6.4.17), we write down our energy functional for (6.4.25) as

$$I(\mathbf{w}) = \int_{\mathbb{R}^2} \left\{ \frac{1}{2}\sum_{a=1}^r |\nabla w_a|^2 + \frac{\lambda}{2}|L^\tau P\mathbf{U} - L^{-1}\mathbf{1}|^2 + \mathbf{h}\cdot\mathbf{w} \right\}. \tag{6.4.26}$$

We verify that (6.4.25) is the Euler–Lagrange equation of (6.4.26). For this purpose, we note that the relation (6.4.20) becomes

$$\frac{d}{d\mathbf{w}}(\mathbf{c}^\tau \cdot \mathbf{U}) = L^\tau \mathbf{U}\mathbf{c}. \tag{6.4.27}$$

On the other hand, the nonlinear part in the functional (6.4.26) can be expressed as

$$
\begin{aligned}
F(\mathbf{w}) &= |L^\tau P\mathbf{U} - L^{-1}\mathbf{1}|^2 \\
&= (L^\tau P\mathbf{U} - L^{-1}\mathbf{1})^\tau (L^\tau P\mathbf{U} - L^{-1}\mathbf{1}) \\
&= (PLL^\tau P\mathbf{U})^\tau \mathbf{U} - 2(P\mathbf{1})^\tau \mathbf{U} + r.
\end{aligned}
\tag{6.4.28}
$$

Applying (6.4.27) to differentiate (6.4.28) and noting $PU = UP$ and $U\mathbf{1} = \mathbf{U}$, we obtain

$$
(\mathrm{d}F)(\mathbf{w}) = \frac{\mathrm{d}F}{\mathrm{d}\mathbf{w}} = 2L^\tau PULL^\tau P\mathbf{U} - 2L^\tau P\mathbf{U}.
\tag{6.4.29}
$$

Consequently, the fact that (6.4.25) is the variational equation of the functional (6.4.26) follows immediately.

The boundary condition at infinity now reads instead

$$
L^\tau P\mathbf{U} = L^{-1}\mathbf{1},
\tag{6.4.30}
$$

namely, $\mathbf{U} = (K^\tau)^{-1}\mathbf{1}$. This condition is identical to the one that defines the 'principal embedding vacuum' (6.3.20) as stated in §6.3.

Since the analysis for (6.4.6) and that for (6.4.7) are the same, we will only present our study on (6.4.7) in detail in the rest of this chapter.

6.5   Existence of Minimizer

We now prove the existence of a solution of the system (6.4.7) satisfying the topological boundary condition (6.4.22) by minimizing the functional (6.4.17). We then establish the stated asymptotic decay estimates.

6.5.1   Boundary condition

We first derive suitable asymptotics for the unknown solution $\mathbf{w}$ of (6.4.19) viewed as a critical point of the functional (6.4.17).

Since $\mathbf{u}^0 = (u_1^0, u_2^0, \cdots, u_r^0)^\tau$ vanishes at infinity, the finite-energy condition for the functional (6.4.17), namely, (6.4.22), implies that $\mathbf{v}$ approaches a constant vector $\mathbf{v}^0$ satisfying

$$
L^\tau \mathbf{V}^0 = (PL)^{-1}\mathbf{1}, \quad \mathbf{V}^0 = \left(\mathrm{e}^{v_1^0}, \mathrm{e}^{v_2^0}, \cdots, \mathrm{e}^{v_r^0} \right)^\tau
\tag{6.5.1}
$$

at infinity. As before, denote by $K^{-1} = ((K^{-1})_{ab})$ the inverse of the matrix K given in (6.4.13). The relation (6.5.1) leads to the following expressions for the asymptotic states of the unknown v_a's:

$$
v_a^0 = \ln \left(\sum_{b=1}^{r} (K^{-1})_{ab} \right), \quad a = 1, 2, \cdots, r.
\tag{6.5.2}
$$

With $\mathbf{v}^0 = (v_1^0, v_2^0, \cdots, v_r^0)^\tau$, we now shift $\mathbf{w}$ according to the rule

$$\mathbf{w} \mapsto \mathbf{w} + (PL)^{-1}\mathbf{v}^0.$$

Then the functional (6.4.17) becomes

$$I(\mathbf{w}) = \int_{\mathbb{R}^2} \left\{ \frac{1}{2} \sum_{a=1}^{r} |\nabla w_a|^2 + \frac{\lambda}{2} |L^\tau R \mathbf{U} - (PL)^{-1}\mathbf{1}|^2 + \mathbf{h} \cdot \mathbf{w} + \mathbf{h} \cdot (PL)^{-1}\mathbf{v}^0 \right\},$$

(6.5.3)

where $\mathbf{U}$ is as defined in (6.4.17) and the matrix R is given by

$$R = \mathrm{diag}\left\{ e^{v_1^0}, e^{v_2^0}, \cdots, e^{v_r^0} \right\}.$$

According to (6.5.1), the finite-energy condition implies that the shifted unknown vector $\mathbf{w}$ satisfies the boundary condition

$$\mathbf{U} = \mathbf{1} \tag{6.5.4}$$

at infinity because $R\mathbf{1} = \mathbf{V}^0$. Or equivalently, $w_a \to 0$ as $|x| \to \infty$. This is a convenient boundary condition to work with.

We claim that there is a constant $c_0 > 0$ so that

$$|L^\tau R \mathbf{U} - (PL)^{-1}\mathbf{1}|^2 \geq c_0 |\mathbf{U} - \mathbf{1}|^2, \tag{6.5.5}$$

which is a stronger version of (6.5.4). In fact, by (6.5.1), we can rewrite the left-hand side of (6.5.5) as $|L^\tau R(\mathbf{U} - \mathbf{1})|^2$. Thus (6.5.5) follows from the fact that $L^\tau R$ is nonsingular.

6.5.2   Minimization

We are ready to use functional analysis to prove the existence of a minimizer of the energy (6.5.3) in the standard Sobolev spaces. For this purpose, let $W^{k,p}$ denote the space of scalar or vector-valued functions with distributional derivatives up to order k which are all lying in $L^p(\mathbb{R}^2)$. The norm of L^p will be written $\| \cdot \|_p$.

Consider the optimization problem

$$\min \left\{ I(\mathbf{w}) \mid \mathbf{w} \in W^{1,2} \right\}. \tag{6.5.6}$$

Because the matrix PL is nonsingular, we may switch to the variable $\mathbf{v}$ from $\mathbf{w}$ via (6.4.16) back and forth according to convenience. Using (6.5.5) and the variable $\mathbf{v}$, we can find suitable constants $c_1, c_2, c_3 > 0$ such that

$$I(\mathbf{w}) \geq c_1 \sum_{a=1}^{r} (\|\nabla v_a\|_2^2 + \|e^{u_a^0 + v_a} - 1\|_2^2) - c_2 \sum_{a=1}^{r} \| |h| v_a \|_1 - c_3. \tag{6.5.7}$$

It is easily checked that

$$|e^s - 1| \geq \frac{|s|}{1 + |s|}, \quad \forall s \in \mathbb{R}.$$

Consequently, we have the lower bound

$$\begin{aligned}
\|e^{u_a^0 + v_a^0} - 1\|_2^2 &\geq \int_{\mathbb{R}^2} \frac{|u_a^0 + v_a^0|^2}{(1 + |u_a^0 + v_a^0|)^2} \\
&\geq \frac{1}{2} \int_{\mathbb{R}^2} \frac{|v_a|^2}{(1 + |u_a^0| + |v_a|)^2} - \int_{\mathbb{R}^2} \frac{|u_a^0|^2}{(1 + |u_a^0|)^2}. \quad (6.5.8)
\end{aligned}$$

Note that since (6.4.14) implies $u_a^0 = O(|x|^{-2})$ for large $|x|$, the last integral on the right-hand side of (6.5.8) is convergent.

We recall the following standard embedding inequality in two dimensions,

$$\|v\|_4^2 \leq \sqrt{2} \|v\|_2 \|\nabla v\|_2, \quad v \in W^{1,2}. \tag{6.5.9}$$

Thus, in view of (6.5.9), we have for any meaningful insertion u_0 the upper bound

$$\begin{aligned}
\|v\|_2^4 &= \left(\int_{\mathbb{R}^2} \frac{|v|}{(1 + |u_0| + |v|)} (1 + |u_0| + |v|) |v| \right)^2 \\
&\leq 2 \int_{\mathbb{R}^2} \frac{|v|^2}{(1 + |u_0| + |v|)^2} \int_{\mathbb{R}^2} (|u_0|^4 + |v|^2 + |v|^4) \\
&\leq C(1 + \|v\|_2^2 + \|v\|_2^2 \|\nabla v\|_2^2) \int_{\mathbb{R}^2} \frac{|v|^2}{(1 + |u_0| + |v|)^2} \\
&\leq \frac{1}{2} \|v\|_2^4 + C\left(\left[\int_{\mathbb{R}^2} \frac{|v|^2}{(1 + |u_0| + |v|)^2} \right]^4 + \|\nabla v\|_2^8 + 1 \right).
\end{aligned}$$

Here and in the sequel, C denotes a generic positive constant. Thus we get

$$\|v\|_2 \leq C \left(1 + \|\nabla v\|_2^2 + \int_{\mathbb{R}^2} \frac{|v|^2}{(1 + |u_0| + |v|)^2} \right). \tag{6.5.10}$$

On the other hand, for any $v \in W^{1,2}$, we have in view of (6.5.9) that

$$\begin{aligned}
\| |\mathbf{h}| v \|_1 &\leq \|\mathbf{h}\|_{4/3} \|v\|_4 \\
&\leq \varepsilon \|v\|_2 + \frac{C}{\varepsilon} \|\nabla v\|_2 + C \\
&\leq \varepsilon (\|v\|_2 + \|\nabla v\|_2^2) + \frac{C}{\varepsilon^3}. \tag{6.5.11}
\end{aligned}$$

Now insert (6.5.8), (6.5.10), and (6.5.11) into (6.5.7) with $u_a^0 = u^0, v_a = v$ ($a = 1, 2, \cdots, r$). We arrive at the following coercive inequality,

$$\begin{aligned}
I(\mathbf{w}) &\geq C_1 \|\mathbf{v}\|_{W^{1,2}} - C_2 \\
&\geq C_3 \|\mathbf{w}\|_{W^{1,2}} - C_4. \tag{6.5.12}
\end{aligned}$$

where the constants C's above are all positive because of the fact that (6.4.16) defines an invertible transformation $\mathbf{w} \mapsto \mathbf{v}$ from $W^{1,2}$ to itself.

It is seen that (6.5.3) is finite everywhere on $W^{1,2}$. In fact, similar to (6.5.5), we have

$$
\begin{aligned}
|L^r R U - (PL)^{-1} \mathbf{1}|^2 &\leq c_1 |\mathbf{U} - \mathbf{1}|^2 \\
&\leq 2c_1 \sum_{a=1}^{r} (|e^{u_a^0}(e^{v_a} - 1)|^2 + |e^{u_a^0} - 1|^2) \\
&\leq c_2 \sum_{a=1}^{r} (|e^{w_a} - 1|^2 + |e^{u_a^0} - 1|^2)
\end{aligned}
$$

for some constants $c_1, c_2 > 0$. Using the MacLaurin series

$$
(e^f - 1)^2 = f^2 + \sum_{k=3}^{\infty} \frac{2^k - 2}{k!} f^k
$$

and (6.5.16) in the next subsection, we may find a bound $C_\mathbf{w} > 0$ depending on $\mathbf{w} \in W^{1,2}$ so that $\int |L^r R U - (PL)^{-1} \mathbf{1}|^2 \leq C_\mathbf{w}$, which proves the finiteness of (6.5.3) on $W^{1,2}$.

In view of (6.5.12) we see that the functional (6.5.3) is bounded from below on $W^{1,2}$. Set

$$
\eta_0 = \inf \left\{ I(\mathbf{w}) \,|\, \mathbf{w} \in W^{1,2} \right\}.
$$

and let $\{\mathbf{w}^{(n)}\}$ be a sequence in $W^{1,2}$ satisfying $I(\mathbf{w}^{(n)}) \to \eta_0$ as $n \to \infty$. The inequality (6.5.12) says that $\{\mathbf{w}^{(n)}\}$ is bounded in $W^{1,2}$. Without loss of generality, we can assume that $\{\mathbf{w}^{(n)}\}$ weakly converges to an element $\mathbf{w} \in W^{1,2}$. We now show that $\mathbf{w}$ is a solution to the problem (6.5.6).

Of course, the finiteness of $I(\mathbf{w})$ implies that for any $\varepsilon > 0$, there is bounded domain Ω so that the truncated energy (6.5.3) over Ω (in other words the integral in (6.5.3) is now taken over Ω instead of $\mathbb{R}^2$), $I_\Omega(\mathbf{w})$, satisfies

$$
I_\Omega(\mathbf{w}) > I(\mathbf{w}) - \varepsilon.
$$

Recall the Trudinger–Moser inequality of the form

$$
\int_\Omega e^f \leq C_1 e^{C_2 \|f\|_{W^{1,2}(\Omega)}^2}, \quad f \in W^{1,2}(\Omega).
$$

and the compact embedding $W^{1,2}(\Omega) \to L^p(\Omega)$ $(p \geq 2)$. Thus the structure of I_Ω implies that I_Ω is weakly lower semi-continuous over $W^{1,2}(\Omega)$. Consequently,

$$
\lim_{n \to \infty} I_\Omega(\mathbf{w}^{(n)}) \geq I_\Omega(\mathbf{w}),
$$

where, without loss of generality, we have assumed the convergence of the sequence of numbers $\{I_\Omega(\mathbf{w}^{(n)})\}$ because otherwise we can always focus on a convergent subsequence.

Besides, we may also assume that the bounded domain Ω is so chosen that

$$-\varepsilon < \int_{\mathbb{R}^2-\Omega} \mathbf{h}\cdot\mathbf{w}^{(n)} + \mathbf{h}\cdot(PL)^{-1}\mathbf{v}^0 < \varepsilon, \quad \forall n.$$

Hence, we obtain

$$\begin{aligned} I(\mathbf{w}^{(n)}) &\geq I_\Omega(\mathbf{w}^{(n)}) + \int_{\mathbb{R}^2-\Omega} \mathbf{h}\cdot\mathbf{w}^{(n)} + \mathbf{h}\cdot(PL)^{-1}\mathbf{v}^0 \\ &\geq I_\Omega(\mathbf{w}^{(n)}) - \varepsilon. \end{aligned}$$

Letting $n \to \infty$ in the above, we have $\eta_0 \geq I_\Omega(\mathbf{w}) - \varepsilon > I(\mathbf{w}) - 2\varepsilon$. Since $\varepsilon > 0$ can be arbitrarily small, we conclude that $I(\mathbf{w}) \leq \eta_0$. This proves that $\mathbf{w}$ solves (6.5.6). Thus $\mathbf{w}$ is a weak solution of (6.4.19). Using the elliptic regularity theory, we see that $\mathbf{w}$ is a smooth solution. Therefore a solution of (6.4.21) is obtained.

6.5.3 Asymptotic behavior

We now turn to the study of the behavior of the solution at infinity.

We first rewrite (6.4.19), after taking the shift $\mathbf{v} \mapsto \mathbf{v}^0 + \mathbf{v}$, in the form

$$\Delta\mathbf{v} = \lambda(KRU - KR)(KRU - 1) - \lambda KR(KRU - 1) + \mathbf{g}. \qquad (6.5.13)$$

Since the matrix KR is invertible and $KR\mathbf{1} = \mathbf{1}$, we have

$$|KRU - KR| \leq c|U - 1|, \quad |KRU - 1| \leq c|U - 1|, \qquad (6.5.14)$$

where $c > 0$ is a suitable constant. Using (6.5.14) in (6.5.13), we see that there are constants $c_1, c_2 > 0$ so that

$$\left| \text{the right-hand side of (6.5.13)} \right|$$

$$\leq c_1 \sum_{a=1}^{r} (e^{u_a^0+v_a} - 1)^2 + c_2 \sum_{a=1}^{r} |e^{u_a^0+v_a} - 1| + |\mathbf{g}|. \qquad (6.5.15)$$

The existence proof carried out in the last section already showed $e^{u_a^0+v_a} - 1 \in L^2$ $(a = 1, 2, \cdots, r)$. We need to derive now $e^{u_a^0+v_a} - 1 \in L^4$ by using $v_a \in W^{1,2}$ $(a = 1, 2, \cdots, r)$ established earlier also in the existence proof part. Denote by u^0 and v any pair among $u_1^0, u_2^0, \cdots, u_r^0$ and $v_1, v_2, \cdots, v_r$, respectively. We proceed as follows.

Since $u^0 < 0$, we have

$$|e^{u^0+v} - 1| \leq |e^v - 1| + |e^{u^0} - 1|.$$

So it suffices to verify that $e^v - 1 \in L^4$ for $v \in W^{1,2}$. For this purpose, we recall the following embedding inequality in two dimensions,

$$\|f\|_p \leq \left(\pi\left[\frac{p-2}{2}\right]\right)^{\frac{p-2}{2p}} \|f\|_{W^{1,2}}, \quad p > 2. \qquad (6.5.16)$$

We then use the MacLaurin series

$$(e^v - 1)^4 = \sum_{k=4}^{\infty} \frac{4^k - 4(3^k + 1) + 6 \cdot 2^k}{k!} v^k$$

and (6.5.16) to obtain a formal upper estimate,

$$\|e^v - 1\|_4^4 \leq \sum_{k=4}^{\infty} \frac{4^k - 4(3^k + 1) + 6 \cdot 2^k}{k!} \left(\pi\frac{k-2}{2}\right)^{\frac{k-2}{2}} \|v\|_{W^{1,2}}^k. \qquad (6.5.17)$$

The series (6.5.17) is convergent. This proves the desired conclusion.

Return now to (6.5.15). We see that the right-hand side of (6.5.13) lies in L^2. Thus $\mathbf{v} \in W^{2,2}$. Because we are in two dimensions, $\mathbf{v}$ must vanish at infinity. Besides, the embedding $W^{2,2} \to W^{1,p}$ $(p > 2)$ implies $\mathbf{v} \in W^{1,p}$. The boundedness of $\mathbf{v}$ makes it clear that $e^v - 1 \in L^p$ for any $p > 2$. Therefore, by L^p-estimates and (6.5.13), (6.5.15), we have $\mathbf{v} \in W^{2,p}$ $(p > 2)$. Hence, $\partial_j \mathbf{v}$ also vanishes at infinity $(j = 1, 2)$.

We now consider the decay rates for the original vector variable $\mathbf{u}$. Away from the vortex points p_{aj}'s, the corresponding matrix form of the system (6.4.7) is

$$\Delta \mathbf{u} = -\lambda K\mathbf{U} + \lambda KUK\mathbf{U},$$

$$U = \mathrm{diag}\left\{e^{u_1}, e^{u_2}, \cdots, e^{u_r}\right\},$$

$$\mathbf{U} = \left(e^{u_1}, e^{u_2}, \cdots, e^{u_r}\right)^{\tau}. \qquad (6.5.18)$$

Note that, in (6.5.18), the definitions of U and $\mathbf{U}$ are updated to simplify notation.

Set $\mathbf{U}_0 = (e^{v_1^0}, e^{v_2^0}, \cdots, e^{v_r^0})^{\tau} = K^{-1}\mathbf{1}$. Then (6.5.18) may be rewritten as

$$\Delta \mathbf{u} = \lambda(KUK)(\mathbf{U} - \mathbf{U}_0). \qquad (6.5.19)$$

To proceed further, we rewrite (6.5.19) as

$$\Delta(\mathbf{u} - \mathbf{v}^0) = \lambda(KRKR)(\mathbf{u} - \mathbf{v}^0) + \lambda(KUKU_\xi - KRKR)(\mathbf{u} - \mathbf{v}^0), \qquad (6.5.20)$$

where the diagonal matrix U_ξ is defined by

$$U_\xi = \mathrm{diag}\left\{e^{u_1^\xi}, e^{u_2^\xi}, \cdots, e^{u_r^\xi}\right\}$$

with u_a^ξ lying between $u_a(x)$ and v_a^0 $(a = 1, 2, \cdots, r)$.

The eigenvalues $\sigma_1, \sigma_2, \cdots, \sigma_r$ of the matrix KRK are arranged in an increasing order so that $0 < \lambda_0^2 \equiv \sigma_1 \leq \sigma_2 \leq \cdots \leq \sigma_r$. Choose an $r \times r$ orthogonal matrix O to satisfy

$$O^\tau (KRK) O = \mathrm{diag}\left\{\sigma_1, \sigma_2, \cdots, \sigma_r\right\}.$$

Set $f = (\mathbf{u} - \mathbf{v}^0)^\tau R(\mathbf{u} - \mathbf{v}^0)$ and define $\mathbf{w}$ by $R(\mathbf{u} - \mathbf{v}^0) = O\mathbf{w}$. Then the equation (6.5.20) implies

$$
\begin{aligned}
\Delta f \;\geq\;& 2(\mathbf{u} - \mathbf{v}^0)^\tau R \Delta(\mathbf{u} - \mathbf{v}^0) \\
=\;& 2\lambda (R[\mathbf{u} - \mathbf{v}^0])^\tau (KRK)(R[\mathbf{u} - \mathbf{v}^0]) \\
& + 2\lambda(\mathbf{u} - \mathbf{v})^\tau R(KUKU_\xi - KRKR)(\mathbf{u} - \mathbf{v}) \\
\geq\;& 2\lambda\lambda_0^2 \mathbf{w}^\tau \mathbf{w} - b(x)f \\
\geq\;& 2\lambda\lambda_0^2 (\mathbf{u} - \mathbf{v}^0)^\tau R^2(\mathbf{u} - \mathbf{v}^0) - b(x)f \\
\geq\;& (2\lambda\lambda_0^2 \sigma_0 - b(x))f, \qquad\qquad\qquad (6.5.21)
\end{aligned}
$$

where σ_0 is the smallest among the numbers $e^{v_1^0}, e^{v_2^0}, \cdots, e^{v_r^0}$ and $b(x) \to 0$ as $|x| \to \infty$. Using a suitable comparison function, standard techniques in elliptic inequalities, and the fact that $f \to 0$ as $|x| \to \infty$, we conclude that for any $\varepsilon \in (0, 1)$ there is a constant $C(\varepsilon) > 0$ so that

$$
\begin{aligned}
f \;=\;& \sum_{a=1}^{r} e^{v_a^0}(u_a - v_a^0)^2 \\
\leq\;& C(\varepsilon) \exp\left(-\lambda_0 \sqrt{2\lambda\sigma_0}(1 - \varepsilon)|x|\right) \quad \text{for } |x| \text{ large.} \quad (6.5.22)
\end{aligned}
$$

Similarly we can get the decay rate for the derivatives. In fact, for fixed $j = 1, 2$, let $\mathbf{v} = \partial_j \mathbf{u}$. Here and in the sequel we only consider the problem away from the vortex points. Thus, differentiating (6.5.19), we have

$$
\begin{aligned}
\Delta \mathbf{v} \;=\;& \lambda(KUKU)\mathbf{v} + \lambda(KUVK)(\mathbf{U} - \mathbf{U}_0), \\
V \;=\;& \mathrm{diag}\left\{v_1, v_2, \cdots, v_r\right\}. \qquad\qquad\qquad (6.5.23)
\end{aligned}
$$

Set $h = \mathbf{v}^\tau R\mathbf{v}$ and define $R\mathbf{v} = O\mathbf{w}$ as before. Then (44) yields

$$
\begin{aligned}
\Delta h \;\geq\;& 2\mathbf{v}^\tau R \Delta \mathbf{v} \\
=\;& 2\lambda(R\mathbf{v}^\tau)(KRK)(R\mathbf{v}) + 2\lambda\mathbf{v}^\tau R(KUKU - KRKR)\mathbf{v} \\
& + 2\lambda\mathbf{v}^\tau R(KUVK)(\mathbf{U} - \mathbf{U}_0),
\end{aligned}
$$

which is analogous to the first half of (6.5.21). Thus we see that the estimate (6.5.22) is valid for h,

$$h = \sum_{a=1}^{r} e^{v_a^0}|\partial_j u_a|^2$$

$$\leq C(\varepsilon) \exp\left(-\lambda_0 \sqrt{2\lambda\sigma_0}(1-\varepsilon)|x|\right) \quad \text{for } |x| \text{ large.} \quad (6.5.24)$$

The desired decay estimates stated in (6.4.12) follow from (6.5.22) and (6.5.24).

We then consider the case where K is not symmetric. The equation (6.5.20) takes the form

$$\Delta(\mathbf{u}-\mathbf{v}^0) = \frac{1}{2}\lambda(KRK + K^\tau RK^\tau)(R[\mathbf{u}-\mathbf{v}^0])$$
$$+\frac{1}{2}\lambda(KRK - K^\tau RK^\tau)(R[\mathbf{u}-\mathbf{v}^0])$$
$$+\lambda(KUKU_\xi - KRKR)(\mathbf{u}-\mathbf{v}^0).$$

Use λ_0^2 to denote the smallest positive eigenvalue of the matrix $KRK + K^\tau RK^\tau$. Since $(R[\mathbf{u}-\mathbf{v}^0])^\tau(KRK - K^\tau RK^\tau)(R[\mathbf{u}-\mathbf{v}^0]) = 0$, we see that the scalar function f still satisfies the last part of (6.5.21). Hence (6.5.22) is valid.

The same symmetrization method may be applied to show that the derivatives of $\mathbf{u}$ also obey (6.5.24) as expected. This completes the verification of the exponential decay estimates.

6.5.4 Quantized integrals

Here, again, the matrix K is symmetric. We now turn back to the notation in §6.5.3. Since $|\nabla u_a|$ decays exponentially fast at infinity and u_a^0 is as defined in (6.4.14), the function v_a must satisfy $|\nabla v_a| = O(|x|^{-3})$ asymptotically at infinity. Consequently the divergence theorem and the smoothness of v_a over $\mathbb{R}^2$ enable us to arrive at the conclusion

$$\int_{\mathbb{R}^2} \Delta v_a = \lim_{\rho\to\infty} \int_{|x|=\rho} \frac{\partial v_a}{\partial n}\, ds = 0, \quad a = 1, 2, \cdots, r.$$

On the other hand, it is straightforward to check that the functions g_a ($a = 1, 2, \cdots, r$) given in (6.4.14) satisfy

$$\int_{\mathbb{R}^2} g_a = 4\pi N_a, \quad a = 1, 2, \cdots, r.$$

In view of the above two results, we immediately obtain the expected conclusion concerning the quantized integrals stated in Theorem 6.4.1.

6.5.5 Original field configuration

In order to recover a field configuration pair (ϕ^a, A_μ^a) from a solution (u_a) of (6.4.6), we define

$$\phi^a(z) = v e^{\frac{1}{2}u_a(z)+i\theta_a(z)}, \quad \theta_a(z) = \sum_{j=1}^{N_a} \arg(z - p_{aj}),$$

$$A_{\pm}^a(z) = i\sum_{b=1}^{r}(K^{-1})_{ba}\partial_{\pm}\ln\phi^b(z), \quad a=1,2,\cdots,r, \quad (6.5.25)$$

which gives us A_1^a, A_2^a in particular. As for A_0^a, we may use (6.3.16) subject to (6.2.15) and (6.2.16).

To get the fluxes, we observe in view of (6.4.5) and (6.4.12) with K^τ replacing K that

$$\int_{\mathbb{R}^2} F_{12}^{a'} = \pm\frac{\lambda}{2}\sum_{a=1}^{r}(K^{-1})_{aa'}\int_{\mathbb{R}^2}\left(\sum_{b=1}^{r}\sum_{c=1}^{r}e^{u_b}e^{u_c}K_{cb}K_{ba} - \sum_{b=1}^{r}e^{u_b}K_{ba}\right)$$

$$= \mp 2\pi\sum_{a=1}^{r}(K^{-1})_{aa'}N_a, \quad a'=1,2,\cdots,r, \quad (6.5.26)$$

as stated in Theorem 6.3.1. The decay estimates for (ϕ^a, A_μ^a) near infinity may be obtained from the relation (6.5.25) and the asymptotic estimates on the functions $u_a, a=1,2,\cdots,r$.

6.6 Some Examples

We first consider the most important case when the gauge group is $SU(N)$.

The first non-Abelian member, $N = 2$, as well as the general $SU(N)$ model in the semiunitary gauge, was studied by Lee in [184, 185]. The governing Chern–Simons vortex equation is identical to that in the Abelian theory [143, 156]. Hence, its solutions are well understood as presented in Chapter 5.

The next member is $N = 3$. In [163], Kao and Lee studied self-dual $SU(3)$ Chern–Simons vortices. These soliton-like solutions carry fractional spins and non-Abelian charges. In their model the Higgs particle lies in the adjoint representation of the gauge group. The vortices are shown to be governed by the 2×2 system of nonlinear elliptic equations

$$\Delta u_1 = -2e^{u_1} + e^{u_2} + 4e^{2u_1} - 2e^{2u_2} - e^{u_1+u_2} + 4\pi\sum_{j=1}^{N_1}\delta_{p_{1j}},$$

$$\Delta u_2 = e^{u_1} - 2e^{u_2} - 2e^{2u_1} + 4e^{2u_2} - e^{u_1+u_2} + 4\pi\sum_{j=1}^{N_2}\delta_{p_{2j}} \quad (6.6.1)$$

after a rescaling. The nonlinear terms in (6.6.1) may be rewritten in the matrix form,

$$-\begin{pmatrix} 2 & -1 \\ -1 & 2 \end{pmatrix}\begin{pmatrix} e^{u_1} \\ e^{u_2} \end{pmatrix}$$

$$+\begin{pmatrix} 2 & -1 \\ -1 & 2 \end{pmatrix}\begin{pmatrix} e^{u_1} & 0 \\ 0 & e^{u_2} \end{pmatrix}\begin{pmatrix} 2 & -1 \\ -1 & 2 \end{pmatrix}\begin{pmatrix} e^{u_1} \\ e^{u_2} \end{pmatrix}$$

which says that (6.6.1) is covered as a special case of the general system (6.4.7) for which the Cartan matrix is

$$K = \begin{pmatrix} 2 & -1 \\ -1 & 2 \end{pmatrix}$$

with

$$(K^{-1})_{11} + (K^{-1})_{12} = (K^{-1})_{21} + (K^{-1})_{22} = 1.$$

Consequently, $v_1^0 = v_2^0 = 0$ and the existence of a multivortex solution (u_1, u_2) of (6.6.1) so that

$$u_1(x), u_2(x) \to 0, \quad |\nabla u_1|, |\nabla u_2| \to 0 \quad \text{as } |x| \to \infty$$

exponentially fast according to the asymptotic rate

$$u_1^2(x) + u_2^2(x) + |\nabla u_1(x)|^2 + |\nabla u_2(x)|^2 = O(e^{-\sqrt{2}(1-\varepsilon)|x|})$$

near infinity is ensured in view of Theorem 6.4.1 because $R = \text{diag}\{1,1\}$ and the smaller eigenvalue of $KRK = K^2$ is 1. Besides, there hold the following quantized integrals

$$\int_{\mathbb{R}^2} 2e^{u_1} - e^{u_2} - 4e^{2u_1} + 2e^{2u_2} + e^{u_1+u_2} = 4\pi N_1,$$

$$\int_{\mathbb{R}^2} -e^{u_1} + 2e^{u_2} + 2e^{2u_1} - 4e^{2u_2} + e^{u_1+u_2} = 4\pi N_2.$$

For the general gauge group $SU(N)$, $N \geq 3$, the rank r is $N-1$ and the governing system has $N-1$ unknowns and $N-1$ nonlinear vortex equations [96]. The structure of the system is completely determined by the $(N-1) \times (N-1)$ Cartan matrix (6.1.34). Inserting (6.1.34) into (6.4.7), it is straightforward as for the case $N = 3$ of Kao and Lee [163] above to state a similar existence theorem. Here we omit the detail.

Next, we consider the Lie algebra $\mathbf{G}_2$. The Cartan matrix is [135, 147]

$$K = \begin{pmatrix} 2 & -1 \\ -3 & 2 \end{pmatrix}. \tag{6.6.2}$$

Thus the associated system (6.4.7) (with $\lambda = 1$) is of the form

$$\Delta u_1 = -2e^{u_1} + e^{u_2} + 4e^{2u_1} - 2e^{2u_2} + e^{u_1+u_2} + 4\pi \sum_{j=1}^{N_1} \delta_{1j},$$

$$\Delta u_2 = 3e^{u_1} - 2e^{u_2} - 6e^{2u_1} + 4e^{2u_2} - 3e^{u_1+u_2} + 4\pi \sum_{j=1}^{N_2} \delta_{2j}. \tag{6.6.3}$$

The property

$$(K^{-1})_{11} + (K^{-1})_{12} = 3, \quad (K^{-1})_{21} + (K^{-1})_{22} = 5$$

leads to the result $v_1^0 = \ln 3$ and $v_2^0 = \ln 5$. Consequently, the statement in Theorem 6.4.1 says that (6.6.3) has a classical solution (u_1, u_2) satisfying $u_1(x) \to 3$, $u_2(x) \to 5$, and $|\nabla u_1(x)|^2 + |\nabla u_2(x)|^2 \to 0$ as $|x| \to \infty$. Using $R = \text{diag}\{3, 5\}$, we have

$$M = \frac{1}{2}(KRK + K^\tau RK^\tau) = \begin{pmatrix} 27 & -32 \\ -32 & 29 \end{pmatrix},$$

which is not positive definite. So (6.4.11) and (6.4.12) stated in Theorem 6.4.1 are not directly applicable here.

6.7 Remarks

In conclusion, through a study of the nonlinear elliptic systems (6.4.6) and (6.4.7), we obtained in this chapter a class of solutions commonly referred to as topological solutions, which give rise to the multivortex solutions of the relativistic non-Abelian self-dual Chern–Simons equations (6.3.17), (6.3.18). These solutions approach the so-called principal embedding vacuum at infinity.

We mention some open problems.

Open Problem 6.7.1. *Establish the existence of non-topological solutions of (6.4.6) and (6.4.7) for $r \geq 2$ which are characterized by the boundary condition*

$$\lim_{|x| \to \infty} e^{u_a(x)} = 0, \quad a = 1, 2, \cdots, r. \tag{6.7.1}$$

Of course, a milder project may be a thorough study of the symmetric solutions of (6.4.6) and (6.4.7) when all the vortex points coincide as was done for the Abelian case in Chapter 5.

Open Problem 6.7.2. *Identify a characteristic condition under which any solution of the equations (6.4.6) and (6.4.7) must be radially symmetric if all the vortex points p_{aj}'s coincide.*

This problem is unsettled even in the scalar case, $r = 1$. In the Liouville equation limit, there is a study by Chen and Li [78] (see also [74]). More recently, Chanillo and Kiessling [75] developed a theory of symmetric solutions for the system

$$-\Delta u_a = \exp\left(\sum_{b=1}^{n} M_{ab} u_b\right), \quad a = 1, 2, \cdots, n, \tag{6.7.2}$$

where $M = (M_{ab})$ is a positive definite matrix. In the new variable $\mathbf{v} = M\mathbf{u}$, the system (6.7.2) becomes

$$-\Delta v_a = \sum_{b=1}^{n} M_{ab}\, e^{v_b}, \quad a = 1, 2, \cdots, n, \tag{6.7.3}$$

which is clearly a Toda system if M is a Cartan matrix.

Open Problem 6.7.3. *Prove the existence of doubly periodic solutions or solutions over a compact surface of (6.4.6) and (6.4.7) as in the Abelian theory.*

We have seen in Chapter 5 a complete resolution of the existence problem in the Abelian case. Besides, it will also be interesting to know whether the conclusion of the existence of a secondary solution holds in the non-Abelian case, $r \geq 2$, as well.

It is a curious thing that if one accidently writes the governing equations as

$$\Delta u_a = -\lambda \sum_{b=1}^{r} K_{ab} e^{u_b} + \lambda \sum_{b=1}^{r} \sum_{c=1}^{r} e^{u_b} K_{bc} e^{u_c} K_{ac} + 4\pi \sum_{j=1}^{N_a} \delta_{p_{aj}},$$

$$a = 1, 2, \cdots, r, \quad x \in \mathbb{R}^2, \tag{6.7.4}$$

then our method no longer works when the matrix K is not symmetric. Of course, when K is symmetric, (6.7.4) is the same as (6.4.6) or (6.4.7).

7
Electroweak Vortices

In this chapter we present multivortex solutions in the theory of Weinberg–Salam for unified electromagnetic and weak interactions. In §7.1 we present an illustrative study of a simplified theory describing the interactions of electromagnetism and the W-particles. In §7.2 we consider the electroweak theory of Weinberg–Salam and state an existence theorem for multivortex solutions. In §7.3 we prove the existence theorem through a multiconstrained variational approach. In §7.4 we study an extended electroweak theory containing two Higgs fields, a topic of much recent focus.

7.1 Massive non-Abelian Gauge Theory

In this section, we discuss a massive $SU(2)$ gauge field model as a beginning example for the existence of self-dual non-Abelian vortices related to the electroweak theory. Although the example is simple, it will serve as an illustrative prototype to yield many clues to the general problems in this chapter.

7.1.1 Governing equations

It was Abrikosov [1] who first predicted the appearance of spatially periodic vortex-lines in a superconducting slab in the context of the Ginzburg–Landau theory [125]. In a series of joint work of Ambjorn and Olesen [4, 5, 6, 7], it was found that spatially periodic vortices also exist in the electroweak theory of Weinberg–Salam [177]. We begin our study by considering the existence problem proposed in [4]. Note that the existence of electroweak vortices has been actively studied from various points of view [146, 158, 159, 206, 319, 320].

As before, we use $\{\sigma_a\}_{a=1,2,3}$ to denote the Pauli matrices,

$$\sigma_1 = \begin{pmatrix} 0 & 1 \\ 1 & 0 \end{pmatrix}, \; \sigma_2 = \begin{pmatrix} 0 & -i \\ i & 0 \end{pmatrix}, \; \sigma_3 = \begin{pmatrix} 1 & 0 \\ 0 & -1 \end{pmatrix}.$$

Then $t_a = \sigma_a/2$, $a = 1, 2, 3$ is a set of generators of $SU(2)$ satisfying the commutation relation

$$[t_a, t_b] = i\epsilon_{abc}t_c.$$

It is easy to check that Tr $(t_a t_b) = \delta_{ab}/2$. A group element of $SU(2)$ may be written

$$\omega = \exp(-i\lambda^a t_a); \quad \lambda^a \in \mathbb{R}, \quad a = 1, 2, 3,$$

and any $su(2)$-valued gauge potential A_μ may be represented in the matrix form

$$A_\mu = A_\mu^a t_a.$$

For each μ, we may also identify A_μ with an isovector, $\mathbf{A}_\mu = (A_\mu^a)_{a=1,2,3}$.

The massive, non-Abelian, gauge theory model under consideration governs the dynamics of a special particle mediating electroweak interactions, called the W-particle, which is represented by the complex field

$$W_\mu = \frac{1}{\sqrt{2}}(A_\mu^1 + iA_\mu^2),$$

and its equations of motions are derived from the Lagrangian density

$$\begin{aligned} \mathcal{L} &= -\frac{1}{2} \text{ Tr } (F_{\mu\nu}F^{\mu\nu}) + m_W^2 \overline{W}_\mu W^\mu \\ &= -\frac{1}{4} F_{\mu\nu}^a F^{a\mu\nu} + m_W^2 \overline{W}_\mu W^\mu \end{aligned} \tag{7.1.1}$$

over the $(3+1)$-dimensional Minkowski spacetime $\mathbb{R}^{3,1}$ of signature $(+ - - -)$, where the field strength tensor $F_{\mu\nu}$, which may again be viewed as an $\mathbb{R}^3$-valued field, is defined by

$$F_{\mu\nu} = F_{\mu\nu}^a t_a = \partial_\mu A_\nu - \partial_\nu A_\mu + ie[A_\mu, A_\nu], \tag{7.1.2}$$

and $m_W > 0$ is the W-particle mass. Note that in (7.1.2) $e > 0$ denotes the charge of positron, or $-e$ is the charge of electron.

We may regard A_μ^3 as an electromagnetic gauge potential, $A_\mu^3 = P_\mu$ (P stands for photon), with the associated field strength tensor

$$P_{\mu\nu} = \partial_\mu P_\nu - \partial_\nu P_\mu.$$

With the notation $D_\mu = \partial_\mu - ieP_\mu$, we have

$$F^1_{\mu\nu} = \frac{1}{\sqrt{2}}([D_\mu W_\nu - D_\nu W_\mu] + \overline{[D_\mu W_\nu - D_\nu W_\mu]}),$$

$$F^2_{\mu\nu} = -\frac{i}{\sqrt{2}}([D_\mu W_\nu - D_\nu W_\mu] - \overline{[D_\mu W_\nu - D_\nu W_\mu]}),$$

$$F^3_{\mu\nu} = P_{\mu\nu} + ie(\overline{W}_\mu W_\nu - W_\mu \overline{W}_\nu).$$

Inserting the above relations into (7.1.1) and using the identity

$$(f_{\mu\nu} + \overline{f}_{\mu\nu})(f^{\mu\nu} + \overline{f}^{\mu\nu}) - (f_{\mu\nu} - \overline{f}_{\mu\nu})(f^{\mu\nu} - \overline{f}^{\mu\nu}) = 4f_{\mu\nu}\overline{f}^{\mu\nu}$$

for a complex-valued skewsymmetric 2-tensor $f_{\mu\nu}$, we obtain

$$\mathcal{L} = -\frac{1}{4}P_{\mu\nu}P^{\mu\nu} - \frac{1}{2}(D_\mu W_\nu - D_\nu W_\mu)\overline{(D^\mu W^\nu - D^\nu W^\mu)} + m_W^2 \overline{W}_\mu W^\mu$$

$$+ieP_{\mu\nu}W^\mu\overline{W}^\nu + \frac{e^2}{2}([\overline{W}_\mu \overline{W}^\mu][W_\nu W^\nu] - [\overline{W}_\mu W^\mu]^2), \qquad (7.1.3)$$

which describes the interaction between the weak force, W-particles, and the electromagnetic force, P-photons.

Vortex-like solutions are characterized by the further ansatz

$$W_0 = W_3 = 0, \quad P_0 = P_3 = 0,$$

$$W_j, \quad P_j \qquad \text{depending only on } x^1, x^2, \quad j = 1, 2.$$

Then equations of motion of (7.1.1) are reduced to

$$D_j(D_j W_k - D_k W_j) = m_W^2 W_k - ieP_{jk}W_j - e^2(W_j^2 \overline{W}_k - [\overline{W}_j W_j]W_k),$$

$$\partial_j P_{jk} = ie(\overline{W}_j[D_k W_j] - W_j\overline{[D_k W_j]})$$

$$-ie(\overline{W}_j[D_j W_k] - W_j\overline{[D_j W_k]})$$

$$+ie\partial_j(\overline{W}_k W_j - W_k \overline{W}_j), \qquad (7.1.4)$$

which are the Euler–Lagrange equations of the reduced energy density

$$\mathcal{H} = \frac{1}{4}P_{jk}^2 + \frac{1}{2}|D_j W_k - D_k W_j|^2 + m_W^2|W_j|^2 + ieP_{jk}\overline{W}_j W_k$$

$$-\frac{e^2}{2}(\overline{W}_j^2 W_k^2 - |W_j|^4). \qquad (7.1.5)$$

We now assume that W_1, W_2 can be represented by a single complex scalar field W through

$$W_1 = W, \quad W_2 = iW. \qquad (7.1.6)$$

As a consequence, the equations of motions (7.1.4) are reduced to an over-determined system,

$$
\begin{aligned}
D_1(D_1W + iD_2W) &= m_W^2 W - eP_{12}W + 2e^2|W|^2W, \\
D_2(D_2W - iD_1W) &= m_W^2 W - eP_{12}W + 2e^2|W|^2W, \\
\partial_j P_{jk} &= ie(\overline{W}[D_kW] - W\overline{[D_kW]}) \\
&\quad + 3e\epsilon_{jk}(\overline{W}[D_jW] + W\overline{[D_jW]}).
\end{aligned} \quad (7.1.7)
$$

On the other hand, the representation (7.1.6) simplifies the energy (7.1.5) into the form

$$
\mathcal{H} = \frac{1}{2}P_{12}^2 + |D_1W + iD_2W|^2 + 2m_W^2|W|^2 - 2eP_{12}|W|^2 + 2e^2|W|^4. \quad (7.1.8)
$$

By virtue of the relation

$$
(D_jD_k - D_kD_j)W = -ieP_{jk}W,
$$

the associated Euler–Lagrange equations of the energy (7.1.8) may be written as

$$
\begin{aligned}
D_jD_jW &= 2m_W^2 W - 3eP_{12}W + 4e^2|W|^2W, \\
\partial_j P_{jk} &= ie(\overline{W}[D_kW] - W\overline{[D_kW]}) \\
&\quad + 3e\epsilon_{jk}(\overline{W}[D_jW] + W\overline{[D_jW]}).
\end{aligned} \quad (7.1.9)
$$

In general, a solution of (7.1.9) may not be a solution of (7.1.7). However, both (7.1.7) (hence (7.1.4)) and (7.1.9) can be solved by a first-order system.

7.1.2 Periodic boundary condition

Consider a rectangular cell domain $\Omega = (0, a^1) \times (0, a^2)$ in $\mathbb{R}^2$. The 't Hooft periodic condition is a condition in which the gauge potential is periodic up to gauge transformations. In our case, such a periodic boundary condition may conveniently be realized in the x^1-direction over Ω by

$$
\omega(0, x^2)A_j(0, x^2)\omega^{-1}(0, x^2) - \frac{i}{e}\omega(0, x^2)(\partial_j\omega^{-1})(0, x^2)
$$

$$
= \omega(a^1, x^2)A_j(a^1, x^2)\omega^{-1}(a^1, x^2) - \frac{i}{e}\omega(a^1, x^2)(\partial_j\omega^{-1})(a^1, x^2),
$$

$$
0 < x^2 < a^2, \quad (7.1.10)
$$

where we specify

$$
\omega(x^1, x^2) = \exp(-i\xi(x^1, x^2)t_3) \in SU(2) \quad (7.1.11)
$$

such that ξ is a real scalar function.

To see the interesting implications of the boundary condition (7.1.10), we need to invoke at this stage the well-known Campbell–Hausdorff formula,

$$\exp(-A)B\exp(A) =$$

$$B + \frac{1}{1!}[B, A] + \frac{1}{2!}[[B, A], A] + \frac{1}{3!}[[[B, A], A], A] + \cdots, \quad (7.1.12)$$

where A, B are $n \times n$ complex matrices.

By virtue of the commutation relations of the group generators $\{t_a\}$ and (7.1.10)–(7.1.12), we find by a straightforward calculation that

$$\omega A_j \omega^{-1} = \exp(-i\xi t_3) A_j^a t_a \exp(i\xi t_3)$$

$$= \left(A_j^1 - \frac{1}{1!}\xi A_j^2 - \frac{1}{2!}\xi^2 A_j^1 + \frac{1}{3!}\xi^3 A_j^2 + \cdots \right)t_1$$

$$+ \left(A_j^2 + \frac{1}{1!}\xi A_j^1 - \frac{1}{2!}\xi^2 A_j^2 - \frac{1}{3!}\xi^3 A_j^1 + \cdots \right)t_2 + A_j^3 t_3,$$

$$-\frac{i}{e}\omega\partial_j\omega^{-1} = \frac{1}{e}(\partial_j\xi)t_3.$$

Hence, under the W-ansatz, $W_j = (A_j^1 + iA_j^2)/\sqrt{2}, W_1 = W, W_2 = iW, P_j = A_j^3$ discussed earlier, the boundary condition (7.1.10) becomes

$$W(0, x^2)\exp(i\xi(0, x^2)) = W(a^1, x^2)\exp(i\xi(a^1, x^2)),$$

$$P_j(0, x^2) + \frac{1}{e}(\partial_j\xi)(0, x^2) = P_j(a^1, x^2) + \frac{1}{e}(\partial_j\xi)(a^1, x^2) \quad (7.1.13)$$

for $0 < x^2 < a^2$.

Similarly, the periodic condition imposed on the gauge field A_j in the x^2-direction,

$$\rho(x^1, 0)A_j(x^1, 0)\rho^{-1}(x^1, 0) - \frac{i}{e}\rho(x^1, 0)(\partial_j\rho^{-1})(x^1, 0)$$

$$= \rho(x^1, a^2)A_j(x^1, a^2)\rho^{-1}(x^1, a^2) - \frac{i}{e}\rho(x^1, a^2)(\partial_j\rho^{-1})(x^1, a^2),$$

for $0 < x^1 < a^1$ with

$$\rho(x^1, x^2) = \exp(-i\zeta(x^1, x^2)t_3) \in SU(2),$$

leads to

$$W(x^1, 0)\exp(i\zeta(x^1, 0)) = W(x^1, a^2)\exp(i\zeta(x^1, a^2)),$$

$$P_j(x^1, 0) + \frac{1}{e}(\partial_j\zeta)(x^1, 0) = P_j(x^1, a^2) + \frac{1}{e}(\partial_j\zeta)(x^1, a^2) \quad (7.1.14)$$

for $0 < x^1 < a^1$.

The relations (7.1.13) and (7.1.14) are exactly the 't Hooft periodic boundary condition for the reduced $U(1)$ gauge field theory (7.1.8) with the gauge symmetry

$$W \mapsto e^{i\omega}W, \quad P_j \mapsto P_j + \frac{1}{e}\partial_j\omega.$$

The requirement that W be single-valued implies in particular

$$\xi(a^1, a^{2^-}) - \xi(a^1, 0^+) + \xi(0, 0^+) - \xi(0, a^{2^-})$$
$$+\zeta(0^+, a^2) - \zeta(a^{1^-}, a^2) + \zeta(a^{1^-}, 0) - \zeta(0^+, 0) + 2\pi N$$
$$= 0 \tag{7.1.15}$$

where $N \in \mathbb{Z}$. By virtue of (7.1.13)–(7.1.15) the total flux through Ω is quantized and independent of the size of Ω,

$$\Phi = \int_\Omega P_{12}\mathrm{d}x = \int_{\partial\Omega} P_j\mathrm{d}x^j = \frac{2\pi N}{e}. \tag{7.1.16}$$

7.1.3 First-order system and existence theorem

Using (7.1.16), we see that the energy density (7.1.8) leads to the energy lower bound

$$E = \int_\Omega \mathcal{H}\mathrm{d}x$$
$$= \int_\Omega \left\{ \left(P_{12}\frac{m_W^2}{e} - \frac{1}{2}\frac{m_W^4}{e^2} \right) \right.$$
$$\left. + \frac{1}{2}\left(P_{12} - \left[\frac{m_W^2}{e} + 2e|W|^2 \right] \right)^2 + |D_1W + iD_2W|^2 \right\}$$
$$\geq 2\pi N\frac{m_W^2}{e^2} - \frac{1}{2}\frac{m_W^4}{e^2}|\Omega|.$$

Such a lower bound is saturated by the solutions of the following first-order system which was first discovered in the work of Ambjorn and Olesen [4],

$$D_1W + iD_2W = 0,$$
$$P_{12} - \left(\frac{m_W^2}{e} + 2e|W|^2 \right) = 0, \tag{7.1.17}$$

where the field configurations over Ω are subject to the $U(1)$-periodic boundary condition (7.1.13)–(7.1.15). It is important to see that a solution of (7.1.17) satisfies both (7.1.9) and (7.1.7) (hence (7.1.4)). Therefore it suffices to construct solutions of (7.1.17). Besides, the second equation in (7.1.17) and (7.1.16) imply that N in (7.1.15) must be a positive integer.

We next see how such an integer is realized by some local properties of a solution.

For simplicity, let us look for solutions of (7.1.17) which do not vanish on the boundary of Ω. It is convenient to view Ω as a subset of the complex plane $\mathbb{C}$ and use $z = x^1 + \mathrm{i}x^2$ to denote a point in Ω. The set of zeros of W will be denoted by $Z(W)$. With the notation

$$\partial = \frac{1}{2}(\partial_1 - \mathrm{i}\partial_2), \qquad P = P_1 + \mathrm{i}P_2,$$

the first equation in (7.1.17) becomes

$$\bar{\partial}W = \frac{1}{2}\mathrm{i}ePW. \qquad (7.1.18)$$

Suppose $p \in Z(W)$. W has the representation

$$W(z) = (z - p)^n h(x^1, x^2) \qquad (7.1.19)$$

in a neighborhood of $z = p$, where n is a positive integer (the multiplicity of the zero $z = p$) and h is a smooth nonvanishing scalar function. In particular, we see that $Z(W)$ is a finite set.

Suppose now $Z(W) = \{p_1, p_2 \cdots, p_m\}$ and the multiplicity of the zero $z = p_\ell$ is $n_\ell > 0, \ell = 1, 2, \cdots, m$. The first equation in (7.1.17) or (7.1.18) may be rewritten

$$P = -\frac{2\mathrm{i}}{e}\bar{\partial}\ln W, \qquad \text{away from } Z(W). \qquad (7.1.20)$$

Therefore, outside $Z(W)$,

$$P_{12} = -\mathrm{i}(\partial P - \overline{\partial P}) = -\frac{2}{e}\partial\bar{\partial}\ln|W|^2 - -\frac{1}{2e}\triangle\ln|W|^2.$$

By virtue of the above equation, the representation (7.1.19) of W near a zero, and the second equation in (7.1.17), we see that $u \equiv \ln|W|^2$ satisfies

$$\Delta u = -2m_W^2 - 4e^2\exp(u) + 4\pi\sum_{\ell=1}^m n_\ell\delta(z - p_\ell) \quad \text{in } \Omega,$$

$$u \quad \text{is periodic on } \partial\Omega. \qquad (7.1.21)$$

Conversely, if u is a solution of (7.1.21), then (W, P) is a smooth solution of the governing equations (7.1.17) subject to the 't Hooft periodic boundary condition (7.1.13), (7.1.14), where

$$W(z) = \exp\frac{1}{2}(u(z) + \mathrm{i}\theta(z)) \quad \text{with } \theta(z) = 2\sum_{\ell=1}^m n_\ell\arg(z - p_\ell),$$

and P is determined by the formula (7.1.20). The pair (W, P) is actually smoothly defined over entire Ω. Moreover, it is easily examined that the integer N in (7.1.15) is given by $N = n_1 + n_2 + \cdots + n_m$, which is the total vortex number of the solution.

Here is the main existence result.

Theorem 7.1.1. *For any $p_1, p_2, \cdots, p_m \in \Omega$ and $n_1, n_2, \cdots, n_m \in \mathbb{Z}_+$ with $n_1 + n_2 + \cdots + n_m = N$, if the system (7.1.17) subject to the periodic boundary condition (7.1.13)–(7.1.15) has a solution (W, P_j) so that*

$$Z(W) = \left\{ p_1, p_2, \cdots, p_m \right\},$$

the multiplicity of the zero p_ℓ of W is $n_\ell, \ell = 1, 2, \cdots, m$, then the mass of the W-particle satisfies

$$0 < m_W < \sqrt{\frac{2\pi N}{|\Omega|}}. \tag{7.1.22}$$

If, in addition to (7.1.22), there holds

$$\frac{2\pi(N - 2)}{|\Omega|} < m_W^2, \tag{7.1.23}$$

then for the any prescribed points $p_1, p_2, \cdots, p_m \in \Omega$ and $n_1, n_2, \cdots, n_m \in \mathbb{Z}_+$, the system (7.1.17) subject to (7.1.13)–(7.1.15) has a solution (W, P_j) so that p_ℓ is a zero of W, the multiplicity of the zero p_ℓ is n_ℓ, $\ell = 1, 2, \cdots, m$, and the total flux is

$$\Phi = \int_\Omega P_{12}\, dx = \frac{2\pi N}{e}.$$

If $N = 1, 2$, (7.1.23) is trivially satisfied. Therefore, in this case, (7.1.22) is a necessary and sufficient condition for a solution to exist.

7.1.4   *Variational proof*

We concentrate on (7.1.21) Ω where Ω is viewed as a torus.

Let u_0 solve

$$\Delta u_0 = -\frac{4\pi N}{|\Omega|} + 4\pi \sum_{\ell=1}^{m} n_\ell \delta(z - p_\ell). \tag{7.1.24}$$

Then, in a neighborhood of p_ℓ, $u_0 - \ln |z - p_\ell|^{2n_\ell}$ is smooth.

We now make the substitution $v = u - u_0$ in (7.1.21). Then we find

$$\Delta v = \frac{4\pi N}{|\Omega|} - 2m_W^2 - 4e^2 U_0 \exp(v) \quad \text{with } U_0 = \exp(u_0). \tag{7.1.25}$$

The function U_0 is smooth and nonnegative.

An integration of (7.1.25) yields

$$4\pi N - 2m_W^2 |\Omega| = 4e^2 \int U_0 \exp(v) > 0, \qquad (7.1.26)$$

which leads to the necessary condition (7.1.22). Note that we frequently suppress the Lebesgue measure dx and the domain Ω under an integration when there is no risk of confusion.

It is convenient to rewrite (7.1.26) in the form

$$J(v) \equiv \int U_0 \exp(v) = C_0 \quad \text{with } C_0 = \frac{1}{e^2}\left(\pi N - \frac{1}{2}m_W^2 |\Omega|\right). \qquad (7.1.27)$$

Then (7.1.25) becomes

$$\Delta v = \frac{4e^2}{|\Omega|} C_0 - 4e^2 U_0 \exp(v). \qquad (7.1.28)$$

Let $W^{1,2} = W^{1,2}(\Omega)$ (the Sobolev space of (a^1, a^2)-periodic L^2 functions whose first-order distributional derivatives are also in L^2). Here and in the sequel $L^p = L^p(\Omega)$ and the norm of L^p will be denoted by $\| \ \|_p$.

Lemma 7.1.2. *A solution to the following constrained minimization problem*

$$\min\left\{I(w) \ \Big| \ w \in W^{1,2}, J(w) = C_0\right\}, \qquad (7.1.29)$$

$$I(w) = \int \left\{\frac{1}{2}|\nabla w|^2 + \frac{4e^2}{|\Omega|}C_0 w\right\}, \qquad (7.1.30)$$

is a smooth solution of (7.1.28).

Proof. Suppose $v \in W^{1,2}$ is a solution to (7.1.29). The standard elliptic regularity theory implies that v is a smooth solution of the equation

$$\Delta v = \frac{4e^2}{|\Omega|}C_0 + \lambda U_0 \exp(v), \qquad (7.1.31)$$

where λ is a Lagrangian multiplier. An integration of (7.1.31) by parts yields $\lambda = -4e^2$ because $J(v) = C_0$. This proves the lemma.

Our existence result for (7.1.28) may be stated as follows.

Lemma 7.1.3. *If, in addition to (7.1.22), the condition (7.1.23) also holds, then (7.1.28) has a solution.*

Proof. By Lemma 7.1.2, it is equivalent to show that (7.1.29) has a solution. For this purpose, we make the following decomposition for each $w \in W^{1,2}$,

$$w = \underline{w} + w', \tag{7.1.32}$$

where $\underline{w}$ denotes the integral mean of w, $\underline{w} = (\int w)/|\Omega|$ and $\underline{w'} = 0$.

From the condition (7.1.22) it is seen that $C_0 > 0$ and the admissible set

$$S(C_0) = \left\{ w \in W^{1,2} \;\middle|\; J(w) = C_0 \right\}$$

is not empty.

For each $w \in S(C_0)$, we have by virtue of (7.1.32) and (7.1.27),

$$\exp(\underline{w}) \int U_0 \exp(w') = C_0.$$

Therefore,

$$\underline{w} = \ln C_0 - \ln \left(\int U_0 \exp(w') \right). \tag{7.1.33}$$

Let s, t be a pair of conjugate exponents satisfying $1 < s, t < \infty, 1/s + 1/t = 1$. Then, using the Schwartz inequality and Trudinger–Moser inequality (4.3.12), we obtain

$$\ln \left(\int U_0 \exp(w') \right) \le \frac{1}{s} \ln \left(\int U_0^s \right) + \frac{1}{t} \ln \left(\int \exp(tw') \right)$$

$$\le \frac{1}{s} \ln \left(\int U_0^s \right) + \frac{1}{t} \ln C(\epsilon) + \left(\frac{1}{16\pi} + \epsilon \right) t \, \|\nabla w'\|_2^2. \tag{7.1.34}$$

As a consequence of (7.1.33) and (7.1.34), there holds for $w \in S(C_0)$ the lower bound

$$I(w) = \frac{1}{2} \|\nabla w'\|_2^2 + 4e^2 C_0 \underline{w}$$

$$\ge \sigma(t, \epsilon) \|\nabla w'\|_2^2 + 4e^2 C_0 \left(\ln C_0 - \frac{1}{s} \ln \left[\int U_0^s \right] - \frac{1}{t} \ln C(\epsilon) \right), \tag{7.1.35}$$

where

$$\sigma(t, \epsilon) = \frac{1}{2} - 4e^2 C_0 \left(\frac{1}{16\pi} + \epsilon \right) t.$$

From the condition (7.1.23), we see that the constants t, ϵ can be suitably chosen to make $\sigma(t, \epsilon) > 0$.

Let $\{w_j\}$ be a minimizing sequence of the problem (7.1.29). The inequality (7.1.35) and the relation (7.1.33) imply that $\{w_j\}$ is a bounded sequence in $W^{1,2}$. For simplicity we assume that $w_j \to$ some $v \in W^{1,2}$ weakly as $j \to \infty$. Since the constraint functional J is weakly continuous on $W^{1,2}$, we see that $v \in \mathcal{S}(C_0)$. On the other hand, the weakly lower semicontinuity of the functional I over $W^{1,2}$ enables us to conclude that $I(v) \leq \liminf I(w_j)$. Therefore, we have shown that v is a solution of (7.1.29) and the lemma is proven.

7.2   Classical Electroweak Theory

In this section we study the existence of spatially periodic multivortex solutions in the classical electroweak theory of Weinberg–Salam. In order to isolate physics, it is necessary to focus on the unitary gauge. As in the massive non-Abelian theory just considered, we are led again to exploring 't Hooft periodic boundary condition and self-duality originally discovered in the work of Ambjorn and Olesen [5, 6, 7]. An existence theorem will be established by a multi-constrained variational principle.

7.2.1   Unitary gauge framework

Let ϕ be a complex doublet,

$$\phi = \left(\begin{array}{c} \phi_1 \\ \phi_2 \end{array} \right), \quad \phi_1, \phi_2 \text{ are complex-valued scalar fields.}$$

The electroweak gauge group is $SU(2) \times U(1)$ which transforms ϕ according to the rule

$$\begin{aligned} \phi &\mapsto \exp(-i\omega_a t_a)\phi, \quad \omega_a \in \mathbb{R}, \quad a = 1, 2, 3, \\ \phi &\mapsto \exp(-i\xi t_0)\phi, \quad \xi \in \mathbb{R}, \end{aligned} \tag{7.2.1}$$

where

$$t_0 = \frac{1}{2} \left(\begin{array}{cc} 1 & 0 \\ 0 & 1 \end{array} \right)$$

is a generator of $U(1)$ in the above matrix representation. The $SU(2)$ and $U(1)$ gauge fields are denoted, respectively, by $A_\mu = A_\mu^a t_a$ (or $\mathbf{A}_\mu = (A_\mu^a)$ as an isovector) and B_μ. Both A_μ^a and B_μ are real 4-vectors. The field strength tensors and the $SU(2) \times U(1)$ gauge-covariant derivative are defined by

$$\begin{aligned} F_{\mu\nu} &= \partial_\mu A_\nu - \partial_\nu A_\mu + ig[A_\mu, A_\nu], \\ H_{\mu\nu} &= \partial_\mu B_\nu - \partial_\nu B_\mu, \\ D_\mu \phi &= \partial_\mu \phi + ig A_\mu^a t_a \phi + ig' B_\mu t_0 \phi, \end{aligned} \tag{7.2.2}$$

where $g, g' > 0$ are coupling constants.

The Lagrangian density of the bosonic electroweak theory is

$$\mathcal{L} = -\frac{1}{4}(F^{a\mu\nu}F_{\mu\nu}^a + H^{\mu\nu}H_{\mu\nu}) + (D^\mu\phi)^\dagger \cdot (D_\mu\phi) - \lambda(\varphi_0^2 - \phi^\dagger\phi)^2, \tag{7.2.3}$$

where λ, φ_0 are positive parameters and $\dagger$ denotes the Hermitian conjugate. Physically λ gives rise to the Higgs particle mass and φ_0 determines the energy scale of symmetry-breaking.

A pair of new vector fields P_μ and Z_μ are obtained as a rotation of the pair A_μ^3 and B_μ,

$$
\begin{aligned}
P_\mu &= B_\mu \cos\theta + A_\mu^3 \sin\theta, \\
Z_\mu &= -B_\mu \sin\theta + A_\mu^3 \cos\theta.
\end{aligned}
$$

In terms of P_μ, Z_μ, D_μ is written

$$
\begin{aligned}
D_\mu &= \partial_\mu + ig(A_\mu^1 t_1 + A_\mu^2 t_2) + iP_\mu(g\sin\theta t_3 + g'\cos\theta t_0) \\
&\quad + iZ_\mu(g\cos\theta t_3 - g'\sin\theta t_0).
\end{aligned}
$$

Requiring that the coefficient of P_μ be the charge operator $eQ = e(t_3 + t_0)$ where $-e$ is the charge of the electron, we obtain the relations

$$
\begin{aligned}
e &= g\sin\theta = g'\cos\theta, \\
e &= \frac{gg'}{(g^2 + g'^2)^{1/2}}, \\
\cos\theta &= \frac{g}{(g^2 + g'^2)^{1/2}}.
\end{aligned}
\tag{7.2.4}
$$

Such a θ is called the Weinberg (mixing) angle. In the sequel, we will always assume that θ is fixed this way. The D_μ takes the form

$$
D_\mu = \partial_\mu + ig(A_\mu^1 t_1 + A_\mu^2 t_2) + iP_\mu eQ + iZ_\mu eQ',
$$

where $Q' = \cot\theta t_3 - \tan\theta t_0$ is the neutral charge operator.

From (7.2.4), when we go to the unitary gauge in which

$$
\phi = \begin{pmatrix} 0 \\ \varphi \end{pmatrix},
$$

where φ is a real scalar field, there holds

$$
D_\mu\phi = \begin{pmatrix} \frac{i}{2}g(A_\mu^1 - iA_\mu^2)\varphi \\ \partial_\mu\varphi - \frac{ig}{2\cos\theta}Z_\mu\varphi \end{pmatrix}.
$$

Define now the complex vector field

$$
W_\mu = \frac{1}{\sqrt{2}}(A_\mu^1 + iA_\mu^2)
$$

and $\mathcal{D}_\mu = \partial_\mu - igA_\mu^3$. With the notation

$$
P_{\mu\nu} = \partial_\mu P_\nu - \partial_\nu P_\mu, \quad Z_{\mu\nu} = \partial_\mu Z_\nu - \partial_\nu Z_\mu,
$$

the Lagrangian (7.2.3) takes the form

$$
\mathcal{L} = -\frac{1}{2}\overline{(\mathcal{D}^\mu W^\nu - \mathcal{D}^\nu W^\mu)}(\mathcal{D}_\mu W_\nu - \mathcal{D}_\nu W_\mu) - \frac{1}{4}Z^{\mu\nu}Z_{\mu\nu} - \frac{1}{4}P^{\mu\nu}P_{\mu\nu}
$$
$$
-\frac{1}{2}g^2([W^\mu \overline{W}_\mu]^2 - [W^\mu W_\mu]\overline{[W^\nu W_\nu]})
$$
$$
-ig(Z^{\mu\nu}\cos\theta + P^{\mu\nu}\sin\theta)\overline{W}_\mu W_\nu + \frac{1}{2}g^2\varphi^2 W^\mu \overline{W}_\mu + \partial^\mu\varphi\partial_\mu\varphi
$$
$$
+\frac{1}{4\cos^2\theta}g^2\varphi^2 Z^\mu Z_\mu - \lambda(\varphi_0^2 - \varphi^2)^2. \tag{7.2.5}
$$

Thus the theory is now reformulated in the unitary gauge. The W and Z fields represent two massive vector bosons which eliminate the curious massless Goldstone particles in the original setting (7.2.3). These fields mediate short-range (weak) interactions. The remaining massless gauge (photon) field P arising from the residual $U(1)$ symmetry mediates long-range (electromagnetic) interactions.

As in the last section, we assume that electroweak excitation is in the third direction. Thus, we arrive at the vortex ansatz

$$
\begin{aligned}
A_0^a &= A_3^a = B_0 = B_3 = 0, \\
A_j^a &= A_j^a(x^1, x^2), \quad B_j = B_j(x^1, x^2), \quad j = 1, 2, \\
\phi &= \phi(x^1, x^2).
\end{aligned} \tag{7.2.6}
$$

As a consequence, if the corresponding W_1 and W_2 are represented by a complex scalar field W according to $W_1 = W, W_2 = iW$ (this implies the relation $A_2^1 = -A_1^2$, $A_2^2 = A_1^1$), the energy density associated with (7.2.5) takes the form

$$
\mathcal{H} = |D_1 W + iD_2 W|^2 + \frac{1}{2}P_{12}^2 + \frac{1}{2}Z_{12}^2 - 2g(Z_{12}\cos\theta + P_{12}\sin\theta)|W|^2
$$
$$
+2g^2|W|^4 + (\partial_j\varphi)^2 + \frac{1}{4\cos^2\theta}g^2\varphi^2 Z_j^2 + g^2\varphi^2|W|^2 + \lambda(\varphi_0^2 - \varphi^2)^2. \tag{7.2.7}
$$

There is a residual $U(1)$ symmetry in the model which may clearly be seen from the invariance of (7.2.7) under the gauge transformation

$$
W \mapsto \exp(i\zeta)W, \quad P_j \mapsto P_j + \frac{1}{e}\partial_j\zeta, \quad Z_j \mapsto Z_j, \quad \varphi \mapsto \varphi, \tag{7.2.8}
$$

due to (7.2.4).

7.2.2 't Hooft periodic boundary conditions

In this section, we discuss the 't Hooft periodic boundary conditions. Since we are interested in vortex-like solutions, only the two-dimensional case will be examined. That is, we assume that the field configurations are in the form (7.2.6).

Consider, more generally, a fundamental domain Ω of a lattice in $\mathbb{R}^2$ generated by independent vectors $\mathbf{a}_1$ and $\mathbf{a}_2$,

$$\Omega = \{\mathbf{x} = (x^1, x^2) \in \mathbb{R}^2 \mid \mathbf{x} = s^1 \mathbf{a}_1 + s^2 \mathbf{a}_2, \, 0 < s^1, s^2 < 1\}.$$

Define

$$\Gamma_{\mathbf{a}_k} = \{\mathbf{x} \in \mathbb{R}^2 \mid \mathbf{x} = s^k \mathbf{a}_k, \, 0 < s^k < 1\}, \quad k = 1, 2.$$

Then $\partial\Omega = \Gamma_{\mathbf{a}_1} \cup \Gamma_{\mathbf{a}_2} \cup \{\mathbf{a}_1 + \Gamma_{\mathbf{a}_2}\} \cup \{\mathbf{a}_2 + \Gamma_{\mathbf{a}_1}\} \cup \{0, \mathbf{a}_1, \mathbf{a}_2, \mathbf{a}_1 + \mathbf{a}_2\}$.

Let $A_j = A_j^a t_a$, B_j, and ϕ be the gauge potentials and the Higgs boson fields respectively. The 't Hooft periodic boundary conditions are such that the triple (A_j, B_j, ϕ) are doubly periodic in $\mathbb{R}^2$ up to gauge transformations. For our purpose we impose this periodicity as follows,

$$(\exp(-i\xi_k(t_3 + t_0))\phi)(\mathbf{x} + \mathbf{a}_k) = (\exp(-i\xi_k(t_3 + t_0))\phi)(\mathbf{x}),$$

$$(\omega_k A_j \omega_k^{-1} - \frac{i}{g}\omega_k \partial_j \omega_k^{-1})(\mathbf{x} + \mathbf{a}_k) = (\omega_k A_j \omega_k^{-1} - \frac{i}{g}\omega_k \partial_j \omega_k^{-1})(\mathbf{x}),$$

$$(B_j + \frac{1}{g'}\partial_j \xi)(\mathbf{x} + \mathbf{a}_k) = (B_j + \frac{1}{g'}\partial_j \xi)(\mathbf{x}),$$

$$\mathbf{x} \in (\Gamma_{\mathbf{a}_1} \cup \Gamma_{\mathbf{a}_2}) - \Gamma_{\mathbf{a}_k},$$

$$k = 1, 2, \tag{7.2.9}$$

where ξ_1, ξ_2 are real-valued smooth functions defined in a neighborhood of $\Gamma_{\mathbf{a}_2} \cup \{\mathbf{a}_1 + \Gamma_{\mathbf{a}_2}\}, \Gamma_{\mathbf{a}_1} \cup \{\mathbf{a}_2 + \Gamma_{\mathbf{a}_1}\}$, respectively, and

$$\omega_k(\mathbf{x}) = \exp(-i\xi_k(\mathbf{x})t_3) \in SU(2), \quad \exp(-i\xi_k(\mathbf{x})t_0) \in U(1).$$

Let us see what these conditions imply for the field configurations in the unitary gauge. It is easy to verify that the first relation in (7.2.9) says that φ is periodic,

$$\varphi(\mathbf{x} + \mathbf{a}_k) = \varphi(\mathbf{x}), \quad \mathbf{x} \in (\Gamma_{\mathbf{a}_1} \cup \Gamma_{\mathbf{a}_2}) - \Gamma_{\mathbf{a}_k}, \quad k = 1, 2. \tag{7.2.10}$$

To proceed further, we apply again the Campbell–Hausdorff formula (7.1.12) to get

$$\omega A_j \omega^{-1} = \exp(-i\xi t_3) A_j^a t_a \exp(i\xi t_3)$$

$$= \left(A_j^1 - \frac{1}{1!}\xi A_j^2 - \frac{1}{2!}\xi^2 A_j^1 + \frac{1}{3!}\xi^3 A_j^2 + \cdots\right)t_1$$

$$+ \left(A_j^2 + \frac{1}{1!}\xi A_j^1 - \frac{1}{2!}\xi^2 A_j^2 - \frac{1}{3!}\xi^3 A_j^1 + \cdots\right)t_2 + A_j^3 t_3,$$

$$-\frac{i}{g}\omega \partial_j \omega^{-1} = \frac{1}{g}(\partial_j \xi)t_3.$$

Thus, after some calculation, we obtain

$$\exp(i\xi_k(\mathbf{x}+\mathbf{a}_k))W(\mathbf{x}+\mathbf{a}_k) = \exp(i\xi_k(\mathbf{x}))W(\mathbf{x}),$$
$$(A_j^3 + \frac{1}{g}\partial_j\xi_k)(\mathbf{x}+\mathbf{a}_k) = (A_j^3 + \frac{1}{g}\partial_j\xi_k)(\mathbf{x}),$$
$$\mathbf{x} \in (\Gamma_{\mathbf{a}_1} \cup \Gamma_{\mathbf{a}_2}) - \Gamma_{\mathbf{a}_k},$$
$$k = 1, 2. \tag{7.2.11}$$

Combining the above equation with the last relation in the boundary condition (7.2.9) and using (7.2.4), we have

$$(P_j + \frac{1}{e}\partial_j\xi_k)(\mathbf{x}+\mathbf{a}_k) = (P_j + \frac{1}{e}\partial_j\xi_k)(\mathbf{x}),$$
$$Z_j(\mathbf{x}+\mathbf{a}_k) = Z_j(\mathbf{x}),$$
$$\mathbf{x} \in (\Gamma_{\mathbf{a}_1} \cup \Gamma_{\mathbf{a}_2}) - \Gamma_{\mathbf{a}_k},$$
$$k = 1, 2. \tag{7.2.12}$$

We summarize the boundary conditions (7.2.10)–(7.2.12) we have obtained as follows,

$$\varphi(\mathbf{x}+\mathbf{a}_k) = \varphi(\mathbf{x}),$$
$$\exp(i\xi_k(\mathbf{x}+\mathbf{a}_k))W(\mathbf{x}+\mathbf{a}_k) = \exp(i\xi_k(\mathbf{x}))W(\mathbf{x}),$$
$$(P_j + \frac{1}{e}\partial_j\xi_k)(\mathbf{x}+\mathbf{a}_k) = (P_j + \frac{1}{e}\partial_j\xi_k)(\mathbf{x}),$$
$$Z_j(\mathbf{x}+\mathbf{a}_k) = Z_j(\mathbf{x}),$$
$$\mathbf{x} \in (\Gamma_{\mathbf{a}_1} \cup \Gamma_{\mathbf{a}_2}) - \Gamma_{\mathbf{a}_k},$$
$$k = 1, 2. \tag{7.2.13}$$

The relations (7.2.13) are exactly the 't Hooft periodic boundary conditions for the reduced $U(1)$ model (7.2.7) over the lattice with fundamental domain Ω, because in such a situation a gauge transformation is defined according to the formula (7.2.8). Hence we have shown that the 't Hooft periodic conditions for the full $SU(2) \times U(1)$ theory and the theory in the residual $U(1)$ symmetry are in fact equivalent.

For convenience, we momentarily denote the value of a function ξ at a point $\mathbf{x} = s^1\mathbf{a}_1 + s^2\mathbf{a}_2 \in \overline{\Omega}$ by $\xi(s^1, s^2)$. Since W is a single-valued complex scalar field, there must exist an integer $N \in \mathbb{Z}$ so that

$$\xi_1(1,1^-) - \xi_1(1,0^+) + \xi_1(0,0^+) - \xi_1(0,1^-)$$
$$+\xi_2(0^+,1) - \xi_2(1^-,1) + \xi_2(1^-,0) - \xi_2(0^+,0) + 2\pi N$$
$$= 0. \tag{7.2.14}$$

As a consequence of (7.2.13) and (7.2.14), there holds again

$$\Phi = \int_\Omega P_{12}\, d\mathbf{x} = \int_{\partial\Omega} P_j\, dx^j = \frac{2\pi N}{e}. \tag{7.2.15}$$

That is, the total magnetic flux through Ω is quantized and independent of the size of Ω. On the other hand, it is easily seen that the flux through Ω induced by the massive vector boson Z is zero.

7.2.3 Lower energy bound and its saturation

Using (7.2.15) and the boundary condition (7.2.13), we see that the energy density (7.2.7) leads to the following energy lower bound over Ω,

$$
\begin{aligned}
E &= \int \mathcal{H} \\
&= \int \left\{ |\mathcal{D}_1 W + i\mathcal{D}_2 W|^2 + \frac{1}{2}\left(P_{12} - \frac{g}{2\sin\theta}\varphi_0^2 - 2g\sin\theta|W|^2 \right)^2 \right\} \\
&\quad + \int \left\{ \frac{1}{2}\left(Z_{12} - \frac{g}{2\cos\theta}(\varphi^2 - \varphi_0^2) - 2g\cos\theta|W|^2 \right)^2 \right. \\
&\quad \left. + \left(\frac{g\varphi}{2\cos\theta}Z_j + \epsilon_{jk}\partial_k\varphi \right)^2 \right\} \\
&\quad + \int \left\{ \left(\lambda - \frac{g^2}{8\cos^2\theta} \right)(\varphi_0^2 - \varphi^2)^2 - \frac{g^2}{8\sin^2\theta}\varphi_0^4 \right. \\
&\quad \left. + \frac{g\varphi_0^2}{2\sin\theta}P_{12} - \frac{g\varphi_0^2}{2\cos\theta}Z_{12} - \frac{g}{2\cos\theta}\partial_k(\epsilon_{jk}Z_j\varphi^2) \right\} \\
&\geq \frac{g\varphi_0^2}{\sin\theta}\left(\frac{\pi N}{e} - \frac{g\varphi_0^2}{8\sin\theta}|\Omega| \right),
\end{aligned}
$$

if $\lambda \geq \frac{g^2}{8\cos^2\theta}$. In the critical case where

$$
\lambda = \frac{g^2}{8\cos^2\theta}, \tag{7.2.16}
$$

i.e., the Higgs mass is equal to the mass of Z vector boson, the above energy lower bound may be saturated by the solutions of the self-dual system

$$
\begin{aligned}
\mathcal{D}_1 W + i\mathcal{D}_2 W &= 0, \\
P_{12} &= \frac{g}{2\sin\theta}\varphi_0^2 + 2g\sin\theta|W|^2, \\
Z_{12} &= \frac{g}{2\cos\theta}(\varphi^2 - \varphi_0^2) + 2g\cos\theta|W|^2, \\
Z_j &= -\frac{2\cos\theta}{g}\epsilon_{jk}\partial_k\ln\varphi, \quad j,k = 1,2, \tag{7.2.17}
\end{aligned}
$$

subject to the 't Hooft periodic boundary condition (7.2.13).

It is straightforward to verify that solutions of (7.2.17) give rise to solutions of the original electroweak theory. From the second equation in

(7.2.17) and the flux formula (7.2.15), it is seen that the integer N in the relation (7.2.14) must be positive.

The zeros of W, which are the locations of electroweak vortices, give rise to an array of parallel vortex-lines along the x^3-axis. These vortices are called W-condensed vortices.

The main existence result for W-condensed vortices is stated as follows.

Theorem 7.2.1. *If the system of electroweak equations, (7.2.17), subject to the boundary condition (7.2.13) has an N-vortex solution (φ, W, P_j, Z_j) so that W has N zeros in Ω, then the vortex number N must satisfy*

$$g^2\varphi_0^2 < \frac{4\pi N}{|\Omega|} < \frac{g^2\varphi_0^2}{\cos^2\theta}. \tag{7.2.18}$$

If, in addition to (7.2.18), there holds the condition

$$\frac{4\pi N}{|\Omega|} < \frac{8\pi \sin^2\theta}{|\Omega|} + g^2\varphi_0^2, \tag{7.2.19}$$

then for any $p_1, p_2, \cdots, p_m \in \Omega$ and $n_1, n_2, \cdots, n_m \in \mathbf{Z}_+$ with $n_1 + n_2 + \cdots + n_m = N$, the system (7.2.17) subject to the 't Hooft periodic boundary condition (7.2.13) has a smooth vortex-line solution (φ, W, P_j, Z_j) so that $\varphi > 0$ and $Z(W) = \{p_1, p_2, \cdots, p_m\}$, the multiplicity of the zero p_ℓ of W is $n_\ell, \ell = 1, 2, \cdots, m$, and the total flux $\Phi = 2\pi N/e$.

We prove this theorem in the next section. Note that the condition (7.2.19) is analogous to (7.1.23).

7.3   Multi-constrained Variational Approach

In this section we establish Theorem 7.2.1. We first derive the governing system of nonlinear elliptic equations. We then prove the existence of solutions through a constrained variational approach.

7.3.1   Elliptic equations

We introduce the complex field,

$$\alpha = A_1^3 + iA_2^3.$$

We see that the first equation in (7.2.17) takes the form

$$\bar{\partial}W = \frac{1}{2}ig\alpha W, \tag{7.3.1}$$

which implies in particular that $Z(W)$ (zeros of W) is a finite set. The unitary gauge assumption makes it necessary to impose the condition that the real Higgs field φ has no zero. Let $Z(W) = \{p_1, p_2, \cdots, p_m\}$ and assume that the multiplicity of the zero p_ℓ is $n_\ell > 0, \ell = 1, 2, \cdots, m$. $N = n_1 + n_2 + \cdots + n_m$ is the total vortex number.

The first equation in (7.2.17) or (7.3.1) may be rewritten

$$\alpha = -\frac{2\mathrm{i}}{g} \bar{\partial} \ln W, \quad \text{away from } Z(W). \tag{7.3.2}$$

Therefore, outside $Z(W)$, the equations (7.2.17) may be reduced by virtue of (7.3.2) and $Z_{12} = (2 \cos \theta / g) \triangle \ln \varphi$ to

$$-\triangle \ln |W|^2 = g^2 \varphi^2 + 4g^2 |W|^2,$$

$$\triangle \ln \varphi = \frac{g^2}{4 \cos^2 \theta} (\varphi^2 - \varphi_0^2) + g^2 |W|^2. \tag{7.3.3}$$

Hence the substitution $|W|^2 = e^u, \varphi^2 = e^w$ allows us to rewrite (7.3.3) in the full domain Ω in the form

$$-\triangle u = g^2 e^w + 4g^2 e^u - 4\pi \sum_{\ell=1}^{m} n_\ell \delta(z - p_\ell),$$

$$\triangle w = \frac{g^2}{2 \cos^2 \theta} (e^w - \varphi_0^2) + 2g^2 e^u, \quad \text{in } \Omega;$$

$$u, w \quad \text{are periodic on } \partial \Omega. \tag{7.3.4}$$

Conversely, if (u, w) is a solution of (7.3.4), then we can define the quartet (φ, W, P_j, Z_j) according to

$$\varphi(z) = \exp\left(\frac{1}{2} w(z)\right),$$

$$W(z) = \exp\left(\frac{1}{2} u(z) + \mathrm{i} \sum_{\ell=1}^{m} n_\ell \arg(z - p_\ell)\right),$$

$$Z_j(z) = -\frac{2 \cos \theta}{g} \epsilon_{jk} \partial_k \ln \varphi(z),$$

$$P_j(z) = \csc \theta A_j^3(z) - \cot \theta Z_j(z), \tag{7.3.5}$$

where A_j^3 is determined through (7.3.2) (the definition of the function α extends smoothly to the full Ω). It is not hard to check that (φ, W, P_j, Z_j) is a solution of the governing system (7.2.17) satisfying the periodic boundary condition (7.2.13) so that the total vortex number in (7.2.14) is given by $N = n_1 + n_2 + \cdots + n_m$.

7.3.2 Existence via minimization

Let u_0 be as defined in (7.1.24). Now set $v = u - u_0$. Obviously the function $U_0 = e^{u_0}$ is smooth and nonnegative. Hence, the system (7.3.4) becomes

$$\triangle v = \frac{4\pi N}{|\Omega|} - g^2 e^w - 4g^2 U_0 e^v,$$

$$\Delta w \; = \; \frac{g^2}{2\cos^2\theta}(e^w - \varphi_0^2) + 2g^2 U_0 e^v. \tag{7.3.6}$$

It may be hard to treat the above system directly. To proceed further we introduce the following transformation of dependent variables,

$$\eta = v + 2w, \quad v = v. \tag{7.3.7}$$

Then (7.3.6) is equivalent to

$$\Delta\eta \; = \; -H + g^2\tan^2\theta\exp\left(\frac{1}{2}[\eta - v]\right),$$

$$\Delta v \; = \; \frac{4\pi N}{|\Omega|} - g^2\exp\left(\frac{1}{2}[\eta - v]\right) - 4g^2 U_0\exp(v), \tag{7.3.8}$$

where the constant H is defined by

$$H = \frac{g^2\varphi_0^2}{\cos^2\theta} - \frac{4\pi N}{|\Omega|}.$$

An integration by parts of the first equation in (7.3.8) yields a constraint for the solution,

$$\int\exp\left(\frac{1}{2}[\eta - v]\right) = C_1 \equiv \frac{|\Omega|}{g^2}(\cot^2\theta)H > 0. \tag{7.3.9}$$

On the other hand, using (7.3.9) and the second equation in (7.3.8), we obtain another constraint,

$$\int U_0\exp(v) = C_2 \equiv \frac{|\Omega|}{4g^2\sin^2\theta}\left(\frac{4\pi N}{|\Omega|} - g^2\varphi_0^2\right) > 0. \tag{7.3.10}$$

These are constraints for both the solutions and the ranges of physical parameters and are summarized as (7.2.18).

Let $W^{1,2}$ be the usual Sobolev space (the set of (a_1, a_2)-periodic L^2 functions whose distributional derivatives are also in L^2, equipped with the standard inner product).

We now show that the modified system (7.3.8) leads to a variational reformulation of the problem. Let us first define the functionals I_σ, J_1, J_2 on $W^{1,2}$ by the expressions

$$I_\sigma(\eta, v) \; = \; \int\left\{\frac{1}{2}|\nabla v|^2 + \frac{\sigma}{2}|\nabla\eta|^2 + \frac{4\pi N}{|\Omega|}v - \sigma H\eta\right\},$$

$$J_1(\eta, v) \; = \; \int\exp\left(\frac{1}{2}[\eta - v]\right),$$

$$J_2(\eta, v) \; = \; \int U_0\,e^v.$$

Lemma 7.3.1. *Consider the following constrained minimization problem*

$$\min\{I_\sigma(\eta, v) \mid (\eta, v) \in W^{1,2},\ J_k(\eta, v) = C_k,\ k = 1, 2\}. \qquad (7.3.11)$$

If $\sigma = \cot^2 \theta$, then a solution of (7.3.11) is a smooth solution of the system (7.3.8).

Proof. It is clear that the Fréchet derivatives dJ_1, dJ_2 of the constraint functionals are linearly independent.

Given $\sigma > 0$, let (η, v) be a solution of (7.3.11). Then by standard elliptic regularity theory (η, v) must be smooth and there exist Lagrange multipliers $\lambda_\sigma, \mu_\sigma$ depending of course on σ so that

$$\Delta \eta \;=\; -H + \frac{\lambda_\sigma}{2\sigma} \exp\left(\frac{1}{2}[\eta - v]\right),$$

$$\Delta v \;=\; \frac{4\pi N}{|\Omega|} - \frac{\lambda_\sigma}{2} \exp\left(\frac{1}{2}[\eta - v]\right) + \mu_\sigma U_0\, e^v. \qquad (7.3.12)$$

Integrating the first equation in (7.3.12) and using $J_1(\eta, v) = C_1$, we obtain

$$\lambda_\sigma = 2\sigma g^2 \tan^2 \theta,$$

which means that (η, v) verifies the first equation in (7.3.8) for any $\sigma > 0$.

To recover the second equation in (7.3.8), we choose $\sigma = \cot^2 \theta$. Therefore, by virtue of $\lambda_\sigma = 2g^2$ and integrating the second equation in (7.3.12), we have $\mu_\sigma = -4g^2$. In particular, (η, v) solves the second equation in (7.3.8) as well. The lemma is proved.

In the rest of this section, we fix $\sigma = \cot^2 \theta$ and suppress the subscript of I_σ for simplicity. The admissible set of the variational problem (7.3.11) will be denoted by

$$\mathcal{S} = \left\{ (\eta, v) \in W^{1,2} \;\middle|\; J_k(\eta, v) = C_k, k = 1, 2 \right\}.$$

When (7.2.18) is satisfied, $C_1, C_2 > 0$, and thus $\mathcal{S} \neq \emptyset$.

We now state our existence result for the system (7.3.4) as follows.

Lemma 7.3.2. *If, in addition to (7.2.18), the condition (7.2.19) also holds, then for any distribution $p_1, p_2, \cdots, p_m \in \Omega$ and $n_1, n_2, \cdots, n_m \in \mathbf{Z}_+$ with $n_1 + n_2 + \cdots + n_m = N$, the system (7.3.4) has a solution.*

Proof. It suffices to prove that (7.3.6) or (7.3.8) has a solution. However, by virtue of Lemma 7.3.1, it is sufficient to show the existence of a minimizer of the constrained optimization problem (7.3.11).

We first prove that, under the condition (7.2.19), the objective functional I is bounded from below on $\mathcal{S}$. For this purpose, we rewrite each $f \in W^{1,2}$ as follows,

$$f = \underline{f} + f',$$

where f denotes the integral mean of f, $\underline{f} = (\int f)/|\Omega|$ and $\underline{f'} = 0$. Hence, I may be put for $(\eta, v) \in \mathcal{S}$ in the form

$$I(\eta, v) = \int \left\{ \frac{1}{2}|\nabla v'|^2 + \frac{\sigma}{2}|\nabla \eta'|^2 \right\} + 4\pi N\underline{v} - \sigma H|\Omega|\underline{\eta}. \tag{7.3.13}$$

Let us now evaluate

$$\Lambda(\eta, v) \equiv 4\pi N\underline{v} - \sigma H|\Omega|\underline{\eta}$$

in (7.3.13) in terms of η', v', and the constraints.

From (7.3.10), we have

$$\exp(\underline{v}) \int U_0\, e^{v'} = C_2.$$

Thus

$$\underline{v} = \ln C_2 - \ln \left(\int U_0\, e^{v'} \right). \tag{7.3.14}$$

On the other hand, (7.3.9) implies in a similar manner

$$\underline{\eta} = \underline{v} + 2\ln C_1 - 2\ln \left(\int \exp \left(\frac{1}{2}[\eta' - v'] \right) \right). \tag{7.3.15}$$

As a consequence,

$$\Lambda(\eta, v) = (4\pi N - \sigma H|\Omega|)\underline{v} + 2\sigma H|\Omega| \ln \left(\int \exp \left(\frac{1}{2}[\eta' - v'] \right) \right) + C_3,$$

where $C_3 = -2\sigma H|\Omega| \ln C_1$.

The second term in the expression of $\Lambda(\eta, v)$ above is bounded from below as may be seen from the convexity of the exponential function and Jensen's inequality,

$$\ln \left(\int \exp \left(\frac{1}{2}[\eta' - v'] \right) \right) \geq \ln \left(|\Omega| \exp \left(\frac{1}{|\Omega|} \int \frac{1}{2}[\eta' - v'] \right) \right) = \ln |\Omega|.$$

Therefore, using (7.3.14),

$$\Lambda(\eta, v) \geq C_4\underline{v} + C_3 + 2\sigma H|\Omega| \ln |\Omega|$$
$$= C_4 \ln C_2 + C_3 + 2\sigma H|\Omega| \ln |\Omega| - C_4 \ln \left(\int U_0\, e^{v'} \right), \tag{7.3.16}$$

where

$$C_4 \equiv 4\pi N - \sigma H|\Omega| = \frac{|\Omega|}{\sin^2 \theta} \left(\frac{4\pi N}{|\Omega|} - g^2\varphi_0^2 \right) > 0$$

due to the condition (7.2.18).

We now estimate the last term on the right-hand side of (7.3.16). Let p, q be a pair of conjugate exponents, $1 < p, q < \infty, 1/p + 1/q = 1$. From the Schwartz inequality and Trudinger–Moser inequality (4.3.12), it follows that

$$\ln\left(\int U_0 \exp(v')\right) \leq \frac{1}{p}\ln\left(\int U_0^p\right) + \frac{1}{q}\ln\left(\int e^{qv'}\right)$$

$$\leq \frac{1}{p}\ln\left(\int U_0^p\right) + \frac{1}{q}\ln C(\epsilon) + \left(\frac{1}{16\pi} + \epsilon\right)q\|\nabla v'\|_2^2. \quad (7.3.17)$$

By virtue of (7.3.16) and (7.3.17), we obtain the lower bound

$$\begin{aligned}
I(\eta, v) &= \frac{1}{2}\|\nabla v'\|_2^2 + \frac{\sigma}{2}\|\nabla \eta'\|_2^2 + \Lambda(\eta, v) \\
&\geq \kappa(q, \epsilon)\|\nabla v'\|_2^2 + \frac{\sigma}{2}\|\nabla \eta'\| + C_3 + 2\sigma H|\Omega|\ln|\Omega| \\
&\quad + C_4\left(\ln C_2 - \frac{1}{p}\ln\left[\int U_0^p\right] - \frac{1}{q}\ln C(\epsilon)\right), \quad (7.3.18)
\end{aligned}$$

where

$$\begin{aligned}
\kappa(q, \epsilon) &= \frac{1}{2} - C_4\left(\frac{1}{16\pi} + \epsilon\right)q \\
&= \frac{1}{2}\left(1 - \frac{2|\Omega|}{\sin^2\theta}\left[\frac{4\pi N}{|\Omega|} - g^2\varphi_0^2\right]\left[\frac{1}{16\pi} + \epsilon\right]q\right).
\end{aligned}$$

Using (7.2.19), it is seen that the constants $q > 1$ and $\epsilon > 0$ can be suitably chosen to make $\kappa(q, \epsilon) > 0$. Hence, I has a lower bound on $\mathcal{S}$.

Finally, let $\{(\eta_j, v_j)\} \subset \mathcal{S}$ be a minimizing sequence of the variational problem (7.3.11). The inequality (7.3.18) implies that $\{(\eta_j', v_j')\}$ is bounded in $W^{1,2}$. On the other hand, the relations (7.3.14) and (7.3.15) say that $\{\underline{v}_j\}$ and $\{\underline{\eta}_j\}$ are bounded sequences as well. Hence, $\{(\eta_j, v_j)\}$ itself is bounded in $W^{1,2}$. For simplicity, we assume that $(\eta_j, v_j) \to$ some $(\eta, v) \in W^{1,2}$ weakly as $j \to \infty$. Since the functionals J_1, J_2 are weakly continuous on $W^{1,2}$, there holds $(\eta, v) \in \mathcal{S}$. However, the weak lower semicontinuity of the functional I over $W^{1,2}$ enables us to make the comparison $I(\eta, v) \leq \liminf I(\eta_j, v_j)$. Thus (η, v) solves (7.3.11) and the proof of the lemma is complete.

From Lemma 7.3.2 and the discussion of §7.2, we are immediately led to the existence result, Theorem 7.2.1, for multivortex solutions of the electroweak theory

7.3.3 Alternative formulation

Our existence theorem is obtained through the transformation (7.3.7) which reformulates the problem into a 'lower triangular' system so that a constrained variational principle may be formulated which leads to the sufficiency condition, (7.2.19). We do not know whether (7.2.19) may further be improved. At a first glance, this condition seems to depend on our special choice of the change of variables (7.3.7). For example, the transformation

$$\zeta = \frac{v}{2\cos^2\theta} + w, \quad w = w \tag{7.3.19}$$

also reduces the system (7.3.6) into a variational problem which makes one think that a new, probably weaker, sufficient condition for the existence of multivortices of the model might be worked out and a possible improvement upon (7.2.19) would result. The following brief discussion provides a negative answer to such a speculation.

In fact, substituting (7.3.19) into (7.3.6), we have

$$\begin{aligned}
\Delta\zeta &= H' - 2g^2\tan^2\theta U_0 \exp(2\cos^2\theta[\zeta - w]), \\
\Delta w &= \beta_0(e^w - \varphi_0^2) + 2g^2 U_0 \exp(2\cos^2\theta[\zeta - w]),
\end{aligned} \tag{7.3.20}$$

where the constant coefficients H' and β_0 are defined by

$$H' = \frac{1}{2\cos^2\theta}\left(\frac{4\pi N}{|\Omega|} - g^2\varphi_0^2\right), \quad \beta_0 = \frac{g^2}{2\cos^2\theta}.$$

Integrating (7.3.20), we find the constraints for a solution as follows,

$$\int U_0 \exp(2\cos^2\theta[\zeta - w]) = C_1' \equiv \cot^2\theta\frac{H'|\Omega|}{2g^2}, \tag{7.3.21}$$

$$\int e^w = C_2' \equiv \cot^2\theta\frac{|\Omega|}{g^2}\left(\frac{g^2\varphi_0^2}{\cos^2\theta} - \frac{4\pi N}{|\Omega|}\right). \tag{7.3.22}$$

It can be shown as before that, if $\sigma = \cot^2\theta$, a minimizer of the constrained optimization problem

$$\min\left\{I(\zeta, w)\,\middle|\,(\zeta, w) \in S\right\}, \tag{7.3.23}$$

$$I(\zeta, w) = \int\left\{\frac{\sigma}{2}|\nabla\zeta|^2 + \frac{1}{2}|\nabla w|^2 + \sigma H'\zeta - \beta w\right\}, \quad \beta = \beta_0\varphi_0^2,$$

$$S \equiv \left\{(\zeta, w) \in W^{1,2}\,\middle|\,(\zeta, w)\ \text{satisfies (7.3.21) and (7.3.22)}\right\}$$

is a smooth solution of the system (7.3.20).

On the other hand, we may rewrite $I(\zeta, w)$ in the form

$$\begin{aligned}
I(\zeta, w) &= \frac{\sigma}{2}\|\nabla\zeta'\|_2^2 + \frac{1}{2}\|\nabla w'\|_2^2 + |\Omega|(\sigma H'\underline{\zeta} - \beta\underline{w}) \\
&= \frac{\sigma}{2}\|\nabla\zeta'\|_2^2 + \frac{1}{2}\|\nabla w'\|_2^2 + \beta'\ln\left(\int e^{w'}\right) \\
&\quad - |\Omega|\sigma H'\frac{1}{2\cos^2\theta}\ln\left(\int U_0\exp(2\cos^2\theta[\zeta' - w'])\right) + C_3',
\end{aligned}$$

where the constants C_3' and β' are defined by

$$\beta' = |\Omega|(\beta - \sigma H') = \frac{|\Omega|}{2\sin^2\theta}\left(\frac{g^2\varphi_0^2}{\cos^2\theta} - \frac{4\pi N}{|\Omega|}\right) > 0,$$

$$C_3' = -\beta' \ln C_2' + \frac{1}{2\cos^2\theta}|\Omega|\sigma H' \ln C_1',$$

and $\int \zeta' = 0$, $\int w' = 0$.

Jensen's inequality again implies that $\ln\left(\int e^{w'}\right) \geq \ln|\Omega|$.

Let p, q be a pair of conjugate exponents. From the Schwartz inequality and Trudinger–Moser inequality (4.3.12), we obtain the following lower bound for $I(\zeta, w)$,

$$I(\zeta, w) \geq \kappa'\|\nabla\zeta'\|_2^2 + \kappa''\|\nabla w'\|_2^2 + C_4', \tag{7.3.24}$$

where the coefficients κ', κ'', and tail term C_4' are given by

$$\kappa' = \sigma\left(\frac{1}{2} - 2|\Omega|H'q\cos^2\theta\left[\frac{1}{16\pi} + \epsilon\right]\left[1 + \frac{1}{r}\right]\right),$$

$$\kappa'' = \left(\frac{1}{2} - 2|\Omega|H'q\cos^2\theta\left[\frac{1}{16\pi} + \epsilon\right]\sigma[1 + r]\right),$$

$$r > 0 \text{ is a constant,}$$

$$C_4' = C_3' - \frac{|\Omega|\sigma H'}{2\cos^2\theta}\left(\frac{1}{p}\ln\left[\int U_0^p\right] + \frac{1}{q}\ln C(\epsilon)\right) + \beta'\ln|\Omega|.$$

Suppose now there is a suitable $r > 0$ to make

$$1 > \frac{|\Omega|}{8\pi}\left(\frac{4\pi N}{|\Omega|} - g^2\varphi_0^2\right)\left(1 + \frac{1}{r}\right),$$

$$1 > \frac{|\Omega|}{8\pi}\left(\frac{4\pi N}{|\Omega|} - g^2\varphi_0^2\right)\cot^2\theta(1 + r). \tag{7.3.25}$$

As a consequence of this condition, it is immediately clear that we can choose suitable $q > 1$ and $\epsilon > 0$ so that $\kappa', \kappa'' > 0$. Thus (7.3.24) implies that (7.3.23) has a minimizer and the existence of multivortex solutions again follows.

However, the two conditions (7.2.19) and (7.3.25) are actually equivalent. To see this, we first assume that (7.2.19) is true. Let $r = \tan^2\theta$. It is seen that both requirements in (7.3.25) are verified. Hence (7.2.19) implies (7.3.25). Suppose now (7.3.25) holds for some $r > 0$. If $r \geq \tan^2\theta$, then the second inequality in (7.3.25) implies (7.2.19); while if $r < \tan^2\theta$, or $1/r > \cot^2\theta$, then (7.2.19) follows from the first inequality in (7.3.25). Thus (7.3.25) implies (7.2.19) as well.

7.4   Two-Higgs Model

In this section we present a complete study of the multivortex equations, arising in the two-Higgs electroweak model, discovered in the work of Bimonte and Lozano [41]. We start from a discussion of the physical relevance of such extended models and introduce the governing equations. We then establish our existence and uniqueness theorems for spatially periodic and full plane solutions using a variational method.

7.4.1   Physical background

Various generalized electroweak models have been the focus of many latest studies. The common feature of these extensions is that more than one Higgs doublets are introduced so that it may be possible to obtain truly non-Abelian vortices in flat spacetime. It has been argued by physicists that electroweak models with at least two Higgs multiplets may arise from supersymmetric or supergravity grand unified theories and stability may be reached in physically relevant parameter regimes. Such a scenario has already found experimental support in particle physics laboratories. For example, see the articles by Ellis, Kelley, and Nanopoulos [103] and Langacker and Luo [179] and references therein. Here we present a complete solution for the recently discovered self-dual equations by Bimonte and Lozano [41] governing two electroweak Higgs doublets, which may be viewed as a minimal extension of the standard model that allows a Bogomol'nyi phase [42]. These same equations also appeared in the study of Edelstein and Nunez on supersymmetric electroweak strings [101]. Perivolaropoulos [245] has conducted numerical and asymptotic analysis for some two-Higgs systems under radial symmetry ansätze.

7.4.2   Field theory model and equations

Let ϕ_1 and ϕ_2 be two complex doublets in the fundamental representation of $SU(2)$. Of course, we still let $U(1)$ act on ϕ_q $(q = 1, 2)$ trivially as before. The gauge group is now $SU(2) \times U(1)_Y \times U(1)_{Y'}$ with Y, Y' the two $U(1)$-hypercharge labels. The gauge fields associated to $SU(2), U(1)_Y$, and $U(1)_{Y'}$ are denoted by A_μ, B_μ, and $\tilde{B}_\mu$, respectively, with the corresponding field strengths $F_{\mu\nu}, H_{\mu\nu}$, and $\tilde{H}_{\mu\nu}$. The Lagrangian density is

$$\mathcal{L} = -\frac{1}{2} \text{Tr}\,(F_{\mu\nu} F^{\mu\nu}) - \frac{1}{4}(H_{\mu\nu} H^{\mu\nu} + \tilde{H}_{\mu\nu} \tilde{H}^{\mu\nu})$$

$$+ \sum_{q=1}^{2} (D^{(q)\mu} \phi_q)^\dagger \cdot (D^{(q)}{}_\mu \phi_q) - V(\phi_1, \phi_2),$$

where, with no summation convention assumed on the index $q = 1, 2$, the gauge-covariant derivatives are defined by

$$D_\mu^{(q)}\phi_q = \left(\partial_\mu + \frac{i}{2}g\tau_a A_\mu^a + \frac{i}{2}g'Y_q B_\mu + \frac{i}{2}g_1 Y_q'\tilde{B}_\mu\right)\phi_q, \qquad q = 1, 2,$$

which mixes the weak and electromagnetic interactions. Recall that τ_a are the Pauli matrices given earlier. The quantities g, g', g_1 and Y_q, Y_q' are positive physical constants. The Higgs potential density V is determined by the expressions

$$V(\phi_1, \phi_2) = \frac{1}{2}R^a R^a + \frac{1}{2}R^2 + \frac{1}{2}\tilde{R}^2,$$

$$R^a = \frac{g}{2}(\phi_1^\dagger \tau_a \phi_1 - \phi_2^\dagger \tau_a \phi_2),$$

$$R = \frac{g'}{2}(Y_1\phi_1^\dagger\phi_1 - Y_2\phi_2^\dagger\phi_2 - \rho),$$

$$\tilde{R} = \frac{g_1}{2}(Y_1'\phi_1^\dagger\phi_1 - Y_2'\phi_2^\dagger\phi_2 - \tilde{\rho}).$$

Assume that the field configurations are independent of the time variable x^0 and the vertical direction x^3 with $A_\mu^3, B_\mu, \tilde{B}_\mu = 0$ (for $\mu = 0, 3$). Then the following self-dual equations are satisfied in order to saturate a topological energy lower bound,

$$F_{12}^a = -R^a, \quad H_{12} = -R, \quad \tilde{H}_{12} = -\tilde{R},$$
$$D_1^{(1)}\phi_1 = iD_2^{(1)}\phi_1, \quad D_1^{(2)}\phi_2 = -iD_2^{(2)}\phi_2. \qquad (7.4.1)$$

The asymptotic behavior of ϕ_1, ϕ_2 of a finite-energy solution requires the existence of a positive number v_0 so that

$$\rho = \frac{1}{2}v_0^2(Y_1 - Y_2), \qquad \tilde{\rho} = \frac{1}{2}v_0^2(Y_1' - Y_2'). \qquad (7.4.2)$$

To simplify (7.4.1), we impose the ansatz

$$A_j^1 = A_j^2 = 0, \quad j = 1, 2; \quad \phi_q = \begin{pmatrix} 0 \\ \psi_q \end{pmatrix}, \quad q = 1, 2. \qquad (7.4.3)$$

It can be seen that the ansatz (7.4.3) is self-consistent and $F_{12}^3 = \partial_1 A_2^3 - \partial_2 A_1^3$ (the commutator vanishes). Consequently, we derive from (7.4.1) the following first-order equations,

$$-2i\partial\psi_1 = \overline{\alpha^{(1)}}\psi_1,$$
$$-2i\bar{\partial}\psi_2 = \alpha^{(2)}\psi_2,$$
$$F_{12}^3 = \frac{g}{2}(|\psi_1|^2 - |\psi_2|^2),$$
$$H_{12} = -\frac{g'}{2}(Y_1|\psi_1|^2 - Y_2|\psi_2|^2 - \rho),$$
$$\tilde{H}_{12} = -\frac{g_1}{2}(Y_1'|\psi_1|^2 - Y_2'|\psi_2|^2 - \tilde{\rho}), \qquad (7.4.4)$$

where the complex-valued vector fields $\alpha^{(q)}$ $(q = 1, 2)$ are defined by

$$\alpha_j^{(q)} = \frac{g}{2}A_j^3 - \frac{g'}{2}Y_q B_j - \frac{g_1}{2}Y_q'\tilde{B}_j, \qquad j = 1, 2,$$

$$\alpha^{(q)} = \alpha_1^{(q)} + i\alpha_2^{(q)}. \tag{7.4.5}$$

The first two equations in (7.4.4) say that ψ_1 (or ψ_2) is an anti-holomorphic (or holomorphic) section. Thus both anti-self-duality and self-duality are present. The last three equations imply that F_{12}^3, H_{12}, and $\tilde{H}_{12}$ are linearly dependent,

$$g'g_1(Y_1Y_2' - Y_2Y_1')F_{12}^3 + gg_1(Y_2' - Y_1')H_{12} - g'g(Y_2 - Y_1)\tilde{H}_{12} = 0.$$

Hence we can invoke another ansatz

$$g'g_1(Y_1Y_2' - Y_2Y_1')A_j^3 + gg_1(Y_2' - Y_1')B_j - g'g(Y_2 - Y_1)\tilde{B}_j = 0, \quad (7.4.6)$$

$j = 1, 2$. We will assume (7.4.6) throughout the rest of this section. We are to obtain multivortex solutions of (7.4.4) on a periodic lattice cell as well as on the full plane.

7.4.3 Periodic multivortices

In this subsection we establish an existence theorem for the equations (7.4.4) subject to a periodic boundary condition. The obtained condition for existence is necessary and sufficient.

From (7.4.4) and the formula

$$\alpha_{12}^{(q)} = -i(\partial\alpha^{(q)} - \overline{\partial\alpha^{(q)}}), \qquad q = 1, 2,$$

we obtain

$$\begin{aligned} 2\Delta\ln|\psi_1|^2 &= |\psi_1|^2(g^2 + g'^2Y_1^2 + g_1^2Y_1'^2) \\ &\quad - |\psi_2|^2(g^2 + g'^2Y_1Y_2 + g_1^2Y_1'Y_2') \\ &\quad - (g'^2Y_1\rho + g_1^2Y_1'\tilde{\rho}), \\ 2\Delta\ln|\psi_2|^2 &= -|\psi_1|^2(g^2 + g'^2Y_1Y_2 + g_1^2Y_1'Y_2') \\ &\quad + |\psi_2|^2(g^2 + g'^2Y_2^2 + g_1^2Y_2'^2) \\ &\quad + (g'^2Y_2\rho + g_1^2Y_2'\tilde{\rho}). \end{aligned} \tag{7.4.7}$$

We are to look for multivortex solutions so that ψ_1 and ψ_2 vanish at the prescribed vortex locations $p_1, p_2, \cdots, p_{N_1}$ and $q_1, q_2, \cdots, q_{N_2}$, respectively. For convenience, we now use the notation

$$\frac{v_0^2}{4}\begin{pmatrix} g^2 + g'^2Y_1^2 + g_1^2Y_1'^2 & -(g^2 + g'^2Y_1Y_2 + g_1^2Y_1'Y_2') \\ -(g^2 + g'^2Y_1Y_2 + g_1^2Y_1'Y_2') & g^2 + g'^2Y_2^2 + g_1^2Y_2'^2 \end{pmatrix}$$

$$\equiv \begin{pmatrix} a_{11} & a_{12} \\ a_{21} & a_{22} \end{pmatrix} \equiv M \tag{7.4.8}$$

and

$$u_1 = \ln\left(\frac{2|\psi_1|^2}{v_0^2}\right), \quad u_2 = \ln\left(\frac{2|\psi_2|^2}{v_0^2}\right), \quad |M| = \det(M) > 0.$$

Then (7.4.7) becomes

$$\Delta u_1 = a_{11}(e^{u_1} - 1) + a_{21}(e^{u_2} - 1) + 4\pi \sum_{j=1}^{N_1} \delta_{p_j},$$

$$\Delta u_2 = a_{21}(e^{u_1} - 1) + a_{22}(e^{u_2} - 1) + 4\pi \sum_{j=1}^{N_2} \delta_{q_j}. \quad (7.4.9)$$

We will find a necessary and sufficient condition to ensure the existence of a doubly periodic solution of these equations.

We are to look for vortex solutions of the dual system (7.4.4) over a fundamental periodic lattice cell, say, Ω, as before. Such a situation requires us to solve (7.4.9) on Ω so that Ω is treated as a 2-torus. As before, the measure of Ω is denoted by $|\Omega|$.

Let u_0' and u_0'' be source function satisfying

$$\Delta u_0' = -\frac{4\pi N_1}{|\Omega|} + 4\pi \sum_{j=1}^{N_1} \delta_{p_j},$$

$$\Delta u_0'' = -\frac{4\pi N_2}{|\Omega|} + 4\pi \sum_{j=1}^{N_2} \delta_{q_j}.$$

Then, on the 2-torus Ω, the functions $v_1 = u_1 - u_0'$ and $v_2 = u_2 - u_0''$ verify a slightly modified version of (7.4.9) of the form

$$\Delta v_1 = a_{11}(e^{u_0'+v_1} - 1) + a_{12}(e^{u_0''+v_2} - 1) + \frac{4\pi N_1}{|\Omega|},$$

$$\Delta v_2 = a_{21}(e^{u_0'+v_1} - 1) + a_{22}(e^{u_0''+v_2} - 1) + \frac{4\pi N_2}{|\Omega|}. \quad (7.4.10)$$

Integrating (7.4.10) over Ω and remembering that there is no longer any boundary term involved, we easily obtain by simple linear algebra the following two basic constraints,

$$\int_\Omega e^{u_0'+v_1} \, dx = |\Omega| - \frac{4\pi}{|M|}(a_{22}N_1 - a_{12}N_2), \quad (7.4.11)$$

$$\int_\Omega e^{u_0''+v_2} \, dx = |\Omega| - \frac{4\pi}{|M|}(a_{11}N_2 - a_{21}N_1). \quad (7.4.12)$$

Therefore, we see that a necessary condition for the existence of a solution is

$$\frac{4\pi}{|M|}(a_{22}N_1 - a_{12}N_2) < |\Omega|, \quad \frac{4\pi}{|M|}(a_{11}N_2 - a_{21}N_1) < |\Omega|. \quad (7.4.13)$$

We shall show that (7.4.13) is also sufficient. To this end, we introduce the transform

$$\begin{cases} w_1 &= \dfrac{1}{\sqrt{|M|}} v_1, \\[2mm] w_2 &= \dfrac{1}{|M|}(a_{11}v_2 - a_{21}v_1), \end{cases}$$

$$\begin{cases} v_1 &= \sqrt{|M|}\, w_1, \\[2mm] v_2 &= \dfrac{1}{a_{11}}(|M|w_2 + a_{21}\sqrt{|M|}\, w_1). \end{cases} \qquad (7.4.14)$$

In view of (7.4.14), the system (7.4.10) takes the form

$$\Delta w_1 = \frac{a_{11}}{\sqrt{|M|}} e^{u_0' + \sqrt{|M|}w_1} + \frac{a_{12}}{\sqrt{|M|}} e^{u_0'' + (a_{21}\sqrt{|M|}w_1 + |M|w_2)/a_{11}} - C_1,$$

$$\Delta w_2 = e^{u_0'' + (a_{21}\sqrt{|M|}w_1 + |M|w_2)/a_{11}} - C_2, \qquad (7.4.15)$$

where

$$C_1 = \frac{1}{\sqrt{|M|}}\left(a_{11} + a_{12} - \frac{4\pi N_1}{|\Omega|}\right), \quad C_2 = 1 - \frac{4\pi}{|\Omega||M|}(a_{11}N_2 - a_{21}N_1).$$

Thus constraints (7.4.11)–(7.4.13) become

$$\frac{a_{11}}{\sqrt{|M|}} \int_\Omega e^{u_0' + \sqrt{|M|}w_1}\, dx = C_1|\Omega| - \frac{a_{12}}{\sqrt{|M|}} C_2|\Omega| \equiv C_3|\Omega| > 0, \quad (7.4.16)$$

$$\int_\Omega e^{u_0'' + (a_{21}\sqrt{|M|}w_1 + |M|w_2)/a_{11}}\, dx = C_2|\Omega| > 0. \qquad (7.4.17)$$

Consider the functional

$$I(w_1, w_2) = \int_\Omega \left\{ \frac{1}{2}|\nabla w_1|^2 + \frac{1}{2}|\nabla w_2|^2 - C_1 w_1 - C_2 w_2 \right\} dx.$$

Let $W^{1,2}(\Omega)$ be the space of L^2-functions over the 2-torus Ω so that their distributional derivatives also lie in L^2. We show that the solution to (7.4.15) can be reduced to the following optimization problem:

$$\min \left\{ I(w_1, w_2) \,\middle|\, w_1, w_2 \in W^{1,2}(\Omega);\ w_1, w_2 \text{ satisfy (7.4.16) and (7.4.17)} \right\}.$$
$$(7.4.18)$$

In view of the Trudinger–Moser inequality (4.3.12), the two constraints, (7.4.16) and (7.4.17), are well defined.

Lemma 7.4.1. *Let (w_1, w_2) be a solution to (7.4.18). Then (w_1, w_2) also solves (7.4.15).*

Proof. Let (w_1, w_2) be a critical point of I in $W^{1,2}(\Omega)$ satisfying the constraints (7.4.16) and (7.4.17). Then the Lagrange multiplier rule says that there are numbers λ_1 and λ_2 so that

$$\Delta w_1 = \lambda_1 e^{u_0' + \sqrt{|M|} w_1} + \lambda_2 \frac{a_{21} \sqrt{|M|}}{a_{11}} e^{u_0'' + (a_{21}\sqrt{|M|} w_1 + |M| w_2)/a_{11}} - C_1,$$

$$\Delta w_2 = \lambda_2 \frac{|M|}{a_{11}} e^{u_0'' + (a_{21}\sqrt{|M|} w_1 + |M| w_2)/a_{11}} - C_2. \tag{7.4.19}$$

Integrating the second equation in (7.4.19) and using (7.4.17), we have $\lambda_2 = a_{11}/|M|$. Inserting this result into the first equation in (7.4.19) and taking integration, we obtain in view of the constraint (7.4.16) that $\lambda_1 = a_{11}/\sqrt{|M|}$. Therefore, the original system (7.4.15) is recovered and the lemma is proven.

To proceed, we write any pair w_1, w_2 in the admissible class

$$C = \left\{ (w_1, w_2) \,\middle|\, w_1, w_2 \in W^{1,2}(\Omega); \quad w_1, w_2 \text{ satisfy (7.4.16) and (7.4.17)} \right\}$$

in the form

$$w_q = \underline{w}_q + w_q', \qquad \underline{w}_q \in \mathbb{R}, \qquad \int_\Omega w_q' \, dx = 0, \qquad q = 1, 2.$$

Hence, from (7.4.16) and (7.4.17), we have

$$\sqrt{|M|}\underline{w}_1 = \ln \frac{C_3 |\Omega| \sqrt{|M|}}{a_{11}} - \ln \left(\int_\Omega e^{u_0' + \sqrt{|M|} w_1'} \, dx \right), \tag{7.4.20}$$

$$\frac{1}{a_{11}} (a_{21} \sqrt{|M|}\, \underline{w}_1 + |M| \underline{w}_2)$$

$$= \ln C_2 |\Omega| - \ln \left(\int_\Omega e^{u_0'' + (a_{21}\sqrt{|M|} w_1' + |M| w_2')/a_{11}} \, dx \right). \tag{7.4.21}$$

Using (7.4.21) in the functional $I(w_1, w_2)$, we arrive at

$$I(w_1, w_2) = \int_\Omega \left\{ \frac{1}{2} |\nabla w_1'|^2 + \frac{1}{2} |\nabla w_2'|^2 \right\} dx - |\Omega|(C_1 \underline{w}_1 + C_2 \underline{w}_2)$$

$$= \frac{1}{2} (\|\nabla w_1'\|_2^2 + \|\nabla w_2'\|_2^2) - C_3 |\Omega| \, \underline{w}_1$$

$$- \frac{a_{11} C_2 |\Omega|}{|M|} \left(\ln C_2 |\Omega| - \ln \left[\int_\Omega e^{u_0'' + (a_{21}\sqrt{|M|} w_1' + |M| w_2')/a_{11}} \, dx \right] \right). \tag{7.4.22}$$

Using $C_3 > 0$ in (7.4.22) and (7.4.20), along with the Jensen inequality to get

$$\ln \left(\int_\Omega e^{u_0' + \sqrt{|M|} w_1'} \, dx \right) \geq \frac{1}{|\Omega|} \int_\Omega u_0' \, dx,$$

$$\ln\left(\int_\Omega e^{u_0'' + (a_{21}\sqrt{|M|}w_1' + |M|w_2')/a_{11}}\,dx\right) \geq \frac{1}{|\Omega|}\int_\Omega u_0''\,dx,$$

we find the estimate

$$I(w_1, w_2) \geq \frac{1}{2}(\|\nabla w_1'\|_2^2 + \|\nabla w_2'\|_2^2) - C, \qquad (7.4.23)$$

where $C > 0$ is a constant independent of w_1, w_2. In particular I is bounded from below in the admissible space $\mathcal{C}$.

Lemma 7.4.2. *If the condition (7.4.13) holds, then (7.4.18) has a solution. In other words, the system (7.4.10) has a solution if and only if (7.4.13) is fulfilled.*

Proof. When (7.4.13) is satisfied, the constants C_2 and C_3 are positive in (7.4.16) and (7.4.17). Hence the admissible class $\mathcal{C}$ is not empty. Let $(w_1^{(k)}, w_2^{(k)})$ be a minimizing sequence of (7.4.18). The inequality (7.4.23) says that $\{w_1^{(k)'}\}$ and $\{w_1^{(k)'}\}$ are bounded sequences in $W^{1,2}(\Omega)$. From (7.4.20) and (7.4.21), it is seen that $\{\underline{w}_1^{(k)}\}$ and $\{\underline{w}_2^{(k)}\}$ are bounded sequences in $\mathbb{R}$. Then a weak compactness argument shows that there is a subsequence of $(w_1^{(k)}, w_2^{(k)})$ that goes to a minimizer of (7.4.18). The lemma is proven.

Lemma 7.4.3. *If (7.4.15) has a solution, then the solution must be unique.*

Proof. Consider the following functional,

$$J(w_1, w_2) = \frac{1}{2}\|\nabla w_1\|_2^2 + \frac{1}{2}\|\nabla w_2\|_2^2 - |\Omega|(C_1\underline{w}_1 + C_2\underline{w}_2)$$

$$+ \int_\Omega \left\{\frac{a_{11}}{|M|}e^{u_0'' + \sqrt{|M|}w_1} + \frac{a_{11}}{|M|}e^{u_0'' + (a_{21}\sqrt{|M|}w_1 + |M|w_2)/a_{11}}\right\}dx.$$

It is straightforward to check by calculating the Hessian that J is strictly convex. Thus J has at most one critical point. However, any solution of (7.4.15) must be a critical point of J. This proves the lemma.

In order to construct solutions for the dual system (7.4.4) from solutions of (7.4.9) obtained in the last subsection, we need to examine the residual gauge symmetry of (7.4.4) because any periodic boundary condition modulo gauge transformations may impose additional restrictions on the phase jumps of ψ_q ($q = 1, 2$) along the boundary of the cell region Ω as observed earlier.

We formally set the gauge symmetry

$$\psi_q \mapsto e^{i\xi_q}\psi_q, \qquad q = 1, 2,$$

$$A_j^3 \mapsto A_j^3 + \partial_j\chi_1,$$

$$B_j \mapsto B_j + \partial_j\chi_2, \quad \tilde{B}_j \mapsto \tilde{B}_j + \partial_j\chi_3, \quad j = 1, 2, \qquad (7.4.24)$$

where ξ_q $(q = 1, 2)$ are real-valued functions and χ_k $(k = 1, 2, 3)$ are to be determined accordingly. Clearly the only thing we need to achieve is to obtain the gauge symmetry in the first two equations of (7.4.4). This requirement can be fulfilled when χ_k $(k = 1, 2, 3)$ satisfy by virtue of (7.4.5) and (7.4.6) the relations

$$g\partial_j\chi_1 - g'Y_1\partial_j\chi_2 - g_1Y_1'\partial_j\chi_3 = 2\partial_j\xi_1,$$
$$g\partial_j\chi_1 - g'Y_2\partial_j\chi_2 - g_1Y_2'\partial_j\chi_3 = 2\partial_j\xi_2,$$
$$g'g_1(Y_1Y_2' - Y_2Y_1')\partial_j\chi_1 + gg_1(Y_2' - Y_1')\partial_j\chi_2 = g'g(Y_2 - Y_1)\partial_j\chi_3.$$

$$(7.4.25)$$

It is easily examined that the coefficient matrix of (7.4.25) is nonsingular if and only if

$$g^2g'^2(Y_1 - Y_2)^2 + g^2g_1^2(Y_1' - Y_2')^2 + g'^2g_1^2(Y_1Y_2' - Y_2Y_1')^2 \neq 0,$$

which is equivalent to the condition

$$Y_1 \neq Y_2 \quad \text{or} \quad Y_1' \neq Y_2'. \tag{7.4.26}$$

Whenever (7.4.26) is verified, the system (7.4.25) has a unique solution for any ξ_q $(q = 1, 2)$. On the other hand, if (7.4.26) is violated so that $Y_1 = Y_2$ and $Y_1' = Y_2'$, then $D^{(1)} = D^{(2)}$ and the two Higgs scalars are just a duplicate of one another. This is obviously a trivial case one should avoid. Thus (7.4.26) is a general condition we should observe for the two Higgs system under discussion. Consequently, there are no restrictions to ξ_q $(q = 1, 2)$ and the locations and numbers of the zeros of ψ_1, ψ_2 confined in a periodic cell domain may be arbitrary when the condition (7.4.13) is satisfied.

We now calculate the fluxes of the weak and magnetic fields. Using (7.4.11) and (7.4.12) and

$$e^{u_1} = \frac{2}{v_0^2}|\psi_1|^2, \quad e^{u_2} = \frac{2}{v_0^2}|\psi_2|^2,$$

we obtain the quantities

$$\int_\Omega |\psi_1|^2 \, dx = \frac{v_0^2}{2}\left(|\Omega| - \frac{4\pi}{|M|}[a_{22}N_1 + |a_{12}|N_2]\right),$$

$$\int_\Omega |\psi_2|^2 \, dx = \frac{v_0^2}{2}\left(|\Omega| - \frac{4\pi}{|M|}[|a_{12}|N_1 + a_{11}N_2]\right).$$

These results combined with the last three equations in (7.4.4) give us the quantized fluxes,

$$\Phi_F = \int_\Omega F_{12}^3 \, dx$$

$$= \frac{g\pi}{|M|} v_0^2 \Big([|a_{12}| - a_{22}] N_1 + [a_{11} - |a_{12}|] N_2 \Big)$$

$$= \frac{g\pi}{4|M|} v_0^4 (g'^2 [Y_2 N_1 + Y_1 N_2][Y_1 - Y_2] + g_1^2 [Y_2' N_1 + Y_1' N_2][Y_1' - Y_2']),$$

$$\Phi_G = \int_\Omega G_{12}\, dx$$

$$= \frac{g'\pi}{|M|} v_0^2 \Big([Y_1 a_{22} - Y_2 |a_{12}|] N_1 + [Y_1 |a_{12}| - Y_2 a_{11}] N_2 \Big)$$

$$= \frac{g'\pi}{4|M|} v_0^4 (g^2 [N_1 + N_2][Y_1 - Y_2] + g_1^2 [Y_2' N_1 + Y_1' N_2][Y_1 Y_2' - Y_1' Y_2]),$$

$$\Phi_{\tilde{G}} = \int_\Omega \tilde{G}_{12}\, dx$$

$$= \frac{g_1\pi}{|M|} v_0^2 \Big([Y_1' a_{22} - Y_2' |a_{12}|] N_1 + [Y_1' |a_{12}| - Y_2' a_{11}] N_2 \Big)$$

$$= \frac{g_1\pi}{4|M|} v_0^4 (g^2 [N_1 + N_2][Y_1' - Y_2'] + g'^2 [Y_2 N_1 + Y_1 N_2][Y_1' Y_2 - Y_1 Y_2']).$$

$$(7.4.27)$$

In summary, under the general nontriviality condition (7.4.26), we have the following existence results.

Theorem 7.4.4. *Let N_1, N_2 be two positive integers and*

$$p_1, p_2, \cdots, p_{N_1}, \quad q_1, q_2, \cdots, q_{N_2}$$

be points in the periodic cell domain Ω (periodicity up to the gauge symmetry transform (7.4.24) and (7.4.25)). We are to obtain multivortex solutions of (7.4.4) of the extended electroweak model with two Higgs scalars, ψ_1, ψ_2, so that ψ_1 and ψ_2 vanish precisely at prescribed vortex locations $p_1, p_2, \cdots, p_{N_1}$ and $q_1, q_2, \cdots, q_{N_2}$, respectively.

(i) Uniqueness: For any prescription of vortices there is at most one solution.

(ii) Existence: Given any prescription of vortices, there is a solution if and only if the topological charges N_1, N_2 satisfy (7.4.13), namely,

$$g^2(N_1 + N_2) + g'^2 Y_q (Y_2 N_1 + Y_1 N_2) + g_1^2 Y_q' (Y_2' N_1 + Y_1' N_2)$$

$$< \frac{|\Omega|\,|M|}{\pi v_0^2}, \quad q = 1, 2. \tag{7.4.28}$$

Besides, the solution carries the quantized fluxes of the weak and magnetic fields over Ω given by the formulas in (7.4.27).

The condition (7.4.28) says that small energy scales of the symmetry-breaking characterized by small values of v_0 allow the existence of large vortex numbers N_1, N_2. Thus one may expect to have arbitrary numbers of vortex charges N_1, N_2 when symmetry is restored by setting $v_0 = 0$. In fact, we can show, to the contrary, that there is no solution in such a situation.

7.4.4 Planar solutions

In the last subsection, we obtained an existence theorem for doubly periodic vortices resembling the Abrikosov vortices in superconductivity. The conditions we found indicate that the numbers of B- and $\tilde{B}$-fluxlines confined in a lattice cell Ω are confined by the size of Ω and larger cells can accommodate more vortices. Consequently it is natural to expect that when the problem is considered over the full plane, the obstructions to the vortex charges should disappear. The result in the present section confirms such an expectation. In the following we shall prove the existence and uniqueness of a multivortex solution of the equations (7.4.9) over full $\mathbb{R}^2$ for an arbitrarily given vortex prescription. Since there are no constraints of the form (7.4.11) and (7.4.12), we will use a direct variational method.

We need to introduce some background functions depending on a real parameter $\mu > 0$,

$$
u_0' = -\sum_{j=1}^{N_1} \ln(1 + \mu|x - p_j|^{-2}), \quad u_0'' = -\sum_{j=1}^{N_2} \ln(1 + \mu|x - q_j|^{-2}),
$$

$$
g_0' = 4\sum_{j=1}^{N_1} \frac{\mu}{(\mu + |x - p_j|^2)^2}, \quad g_0'' = 4\sum_{j=1}^{N_2} \frac{\mu}{(\mu + |x - q_j|^2)^2}. \quad (7.4.29)
$$

Set $v_1 = u_1 - u_0'$ and $v_2 = u_2 - u_0''$. The equations (7.4.9) on $\mathbb{R}^2$ take the form

$$
\Delta v_1 = a_{11}(e^{u_0' + v_1} - 1) + a_{12}(e^{u_0'' + v_2} - 1) + g_0',
$$

$$
\Delta v_2 = a_{21}(e^{u_0' + v_1} - 1) + a_{22}(e^{u_0'' + v_2} - 1) + g_0''. \quad (7.4.30)
$$

It is important to notice that the integrals of g_0', g_0'' over $\mathbb{R}^2$ are independent of the value of μ. In fact, we have

$$
\int_{\mathbb{R}^2} g_0' \, dx = 4\pi N_1, \quad \int_{\mathbb{R}^2} g_0'' \, dx = 4\pi N_2. \quad (7.4.31)
$$

Using the transformation (7.4.14), we rewrite (7.4.30) as

$$
\Delta w_1 = \frac{a_{11}}{\sqrt{|M|}} (e^{u_0' + \sqrt{|M|}w_1} - 1)
$$

$$
+ \frac{a_{12}}{\sqrt{|M|}} (e^{u_0'' + (a_{21}\sqrt{|M|}w_1 + |M|w_2)/a_{11}} - 1) + h_1,
$$

$$
\Delta w_2 = (e^{u_0'' + (a_{21}\sqrt{|M|}w_1 + |M|w_2)/a_{11}} - 1) + h_2. \quad (7.4.32)
$$

where

$$h_1 = \frac{1}{\sqrt{|M|}} g_0', \qquad h_2 = \frac{1}{|M|}(a_{11} g_0'' - a_{21} g_0'). \tag{7.4.33}$$

It is clear that system (7.4.32) is the variational equations of the functional

$$
\begin{aligned}
I(w_1, w_2) &= \int_{\mathbb{R}^2} \left\{ \frac{1}{2}|\nabla w_1|^2 + \frac{1}{2}|\nabla w_2|^2 + \frac{a_{11}}{|M|} e^{u_0'}(e^{\sqrt{|M|}\, w_1} - 1) + \right. \\
&\quad \left. + \frac{a_{11}}{|M|} e^{u_0''}(e^{(a_{21}\sqrt{|M|}\, w_1 + |M| w_2)/a_{11}} - 1) \right\} dx \\
&\quad + \int_{\mathbb{R}^2} \left\{ \left(h_1 - \frac{1}{\sqrt{|M|}}[a_{11} + a_{12}] \right) w_1 + (h_2 - 1) w_2 \right\} dx \\
&= \frac{1}{2}\|\nabla w_1\|_2^2 + \frac{1}{2}\|\nabla w_2\|_2^2 + \frac{a_{11}}{|M|}\left(e^{u_0'}, e^{\sqrt{|M|}\, w_1} - 1 - \sqrt{|M|}\, w_1 \right)_2 \\
&\quad + \frac{a_{11}}{|M|}\left(e^{u_0''}, e^{(a_{21}\sqrt{|M|}\, w_1 + |M| w_2)/a_{11}} - 1 - \frac{[a_{21}\sqrt{|M|}\, w_1 + |M| w_2]}{a_{11}} \right)_2 \\
&\quad + \left(w_1, h_1 + \frac{a_{11}}{\sqrt{|M|}}[e^{u_0'} - 1] + \frac{a_{12}}{\sqrt{|M|}}[e^{u_0''} - 1] \right)_2 \\
&\quad + (w_2, h_2 + [e^{u_0''} - 1])_2, \tag{7.4.34}
\end{aligned}
$$

where $(\cdot, \cdot)_2$ denotes the inner product of $L^2(\mathbb{R}^2)$. This functional is similar to the one for the self-dual vortex solutions in the Abelian Higgs model (see (3.1) in III. 3 in Jaffe and Taubes [157]). Thus we will borrow some techniques used there with suitable adaption.

First, it can be checked that I is differentiable and strictly convex for $(w_1, w_2) \in W^{1,2}(\mathbb{R}^2) \times W^{1,2}(\mathbb{R}^2)$.

Next, using $dI(w_1, w_2)$ to denote the the Fréchet derivative of the functional I at the point (w_1, w_2), we prove

Lemma 7.4.5. *There is a sufficiently large number $\mu > 0$ in (7.4.29) so that*

$$(dI(w_1, w_2))(w_1, w_2) \geq C_1(\|w_1\|_{W^{1,2}} + \|w_2\|_{W^{1,2}}) - C_2, \quad w_1, w_2 \in W^{1,2}(\mathbb{R}^2),$$

where $C_1, C_2 > 0$ are constants independent of w_1, w_2.

Proof. A simple calculation gives us the difference

$$
\begin{aligned}
&(dI(w_1, w_2))(w_1, w_2) - (\|\nabla w_1\|_2^2 + \|\nabla w_2\|_2^2) \\
&= \frac{a_{11}}{\sqrt{|M|}} \left(w_1, e^{u_0'}(e^{\sqrt{|M|}\, w_1} - 1) \right)_2 \\
&\quad + \frac{a_{21}}{\sqrt{|M|}} \left(w_1, e^{u_0''}[e^{(a_{21}\sqrt{|M|}\, w_1 + |M| w_2)/a_{11}} - 1] \right)_2
\end{aligned}
$$

$$+\left(w_2, e^{u_0''}[e^{(a_{21}\sqrt{|M|}w_1+|M|w_2)/a_{11}} - 1]\right)_2$$

$$+\left(w_1, h_1 + \frac{a_{11}}{\sqrt{|M|}}[e^{u_0'} - 1] + \frac{a_{12}}{\sqrt{|M|}}[e^{u_0''} - 1]\right)_2$$

$$+(w_2, h_2 + [e^{u_0''} - 1])_2$$

$$= \left(w_1, \frac{a_{11}}{\sqrt{|M|}}[e^{u_0'+\sqrt{|M|}w_1} - 1]\right.$$

$$+\frac{a_{21}}{\sqrt{|M|}}[e^{u_0''+(a_{21}\sqrt{|M|}w_1+|M|w_2)/a_{11}} - 1] + h_1\Big)_2$$

$$+\left(w_2, e^{u_0''+(a_{21}\sqrt{|M|}w_1+|M|w_2)/a_{11}} - 1 + h_2\right)_2$$

$$= \left(w_1, \frac{a_{11}}{\sqrt{|M|}}[e^{u_0'+\sqrt{|M|}w_1} - 1] + h_1 - \frac{a_{21}}{\sqrt{|M|}}h_2\right)_2$$

$$+\left(\frac{a_{21}}{\sqrt{|M|}}w_1 + w_2, e^{u_0''+(a_{21}\sqrt{|M|}w_1+|M|w_2)/a_{11}} - 1 + h_2\right)_2. \quad (7.4.35)$$

In order to estimate the right-hand side of (7.4.35), we temporarily use u_0 to denote u_0' or u_0'', w to denote

$$\sqrt{|M|}w_1 \quad \text{or} \quad \frac{1}{a_{11}}\left(a_{21}\sqrt{|M|}w_1 + |M|w_2\right),$$

and h to denote any linear combination of h_1 and h_2, and we consider the quantity

$$Q(w) = (w, e^{u_0+w} - 1 + h)_2.$$

It will be convenient to start from the decomposition $w = w_+ - w_-$ with $w_+ = \max\{0, w\}$ and $w_- = \max\{0, -w\}$. Then $Q(w) = Q(w_+) + Q(-w_-)$. The first term, $Q(w_+)$, is of no harm whatever $\mu > 0$ is because the fact that

$$e^{u_0+w_+} - 1 + h = e^{u_0+w_+} - 1 - (u_0 + w_+) + (w_+ + u_0 + h)$$

$$\geq w_+ + u_0 + h$$

and that $u_0, h \in L^2$ yield the lower bound

$$Q(w_+) \geq \int_{\mathbb{R}^2} w_+^2 + \int_{\mathbb{R}^2} w_+(u_0 + h)$$

$$\geq \frac{1}{2}\|w_+\|_2^2 - C_1.$$

This simple result shows that $Q(w_+)$ is well behaved. On the other hand, using the inequality

$$1 - e^{-x} \geq \frac{x}{1+x}, \qquad x \geq 0,$$

we can estimate $Q(-w_-)$ from below as follows:

$$
\begin{aligned}
Q(-w_-) &= (w_-, 1 - h - e^{u_0})_2 + (w_-, e^{u_0}[1 - e^{-w_-}])_2 \\
&\geq \left(w_-, \left\{1 - h - e^{u_0} + \frac{w_-}{1 + w_-} e^{u_0}\right\}\right)_2 \\
&= \int_{\mathbb{R}^2} \frac{w_-}{1 + w_-}([1 + w_-][1 - h - e^{u_0}] + w_- e^{u_0}) \\
&= \int_{\mathbb{R}^2} \frac{w_-}{1 + w_-}([1 - h]w_- + [1 - h - e^{u_0}]).
\end{aligned}
$$

By the definition of g_0' and g_0'' stated in (7.4.29) and the relation (7.4.33), we can make μ sufficiently large so that $h < 1/2$ everywhere. It is easily checked that both h and $1 - e^{u_0}$ belong to L^2. So

$$
\int_{\mathbb{R}^2} \frac{w_-}{1 + w_-}([1 - h]w_- + [1 - h - e^{u_0}]) \leq C_2 \left(\int_{\mathbb{R}^2} \frac{w_-^2}{1 + w_-}\right)^{\frac{1}{2}}.
$$

Thus there is a constant $C_3 > 0$ to make the lower estimate

$$
Q(-w_-) \geq \frac{1}{4} \int_{\mathbb{R}^2} \frac{w_-^2}{1 + w_-} - C_3
$$

valid. Recall the lower estimate for $Q(w_+)$ obtained earlier. We thus conclude that

$$
Q(w) \geq \frac{1}{4} \int_{\mathbb{R}^2} \frac{w^2}{1 + |w|} - C
$$

holds for some constant $C > 0$. Using this result, we see that the right-hand side of (7.4.35) is bounded from below by the quantity

$$
b_1 \int_{\mathbb{R}^2} \frac{w_1^2}{1 + |w_1|} \, dx + b_2 \int_{\mathbb{R}^2} \frac{(a_{21}\sqrt{|M|}w_1 + |M|w_2)^2}{1 + |a_{21}\sqrt{|M|}w_1 + |M|w_2|} \, dx - b_0, \quad (7.4.36)
$$

where $b_0, b_1, b_2 > 0$ are constants. Therefore using an elementary interpolation inequality, we obtain from (7.4.35) and (7.4.36) the lower bound

$$
\begin{aligned}
(dI(w_1, w_2))(w_1, w_2) &- (\|\nabla w_1\|_2^2 + \|\nabla w_2\|_2^2) \\
&\geq b_1' \int_{\mathbb{R}^2} \frac{w_1^2 + w_2^2}{(1 + |w_1| + |w_2|)^2} \, dx - b_0', \quad (7.4.37)
\end{aligned}
$$

where, as before, $b_0', b_1' > 0$ are suitable constants.

We need the embedding inequality

$$
\int_{\mathbb{R}^2} f^4 \, dx \leq 2 \int_{\mathbb{R}^2} f^2 \, dx \int_{\mathbb{R}^2} |\nabla f|^2 \, dx, \qquad f \in W^{1,2}(\mathbb{R}^2) \quad (7.4.38)
$$

to extract useful information from the inequality (7.4.37). We will show that (7.4.37) and (7.4.38) are just good enough to enable us to arrive at the conclusion of this lemma.

In view of (7.4.38) and omitting the Lebesgue measure dx in the integrals to save space, we have

$$\left(\|w_1\|_2^2 + \|w_2\|_2^2 \right)^2$$

$$\leq \left(\int_{\mathbb{R}^2} \frac{|w_1| + |w_2|}{1 + |w_1| + |w_2|} (1 + |w_1| + |w_2|)(|w_1| + |w_2|) \right)^2$$

$$\leq C \int_{\mathbb{R}^2} \frac{w_1^2 + w_2^2}{(1 + |w_1| + |w_2|)^2} \int_{\mathbb{R}^2} \left(\sum_{q=1}^{2} [w_q^2 + w_q^4] \right)$$

$$\leq C \int_{\mathbb{R}^2} \frac{w_1^2 + w_2^2}{(1 + |w_1| + |w_2|)^2} \left(\int_{\mathbb{R}^2} \sum_{q=1}^{2} w_q^2 \right) \left(1 + \int_{\mathbb{R}^2} \sum_{q=1}^{2} |\nabla w_q|^2 \right)$$

$$\leq \frac{1}{2} \left(\int_{\mathbb{R}^2} \sum_{q=1}^{2} w_q^2 \right)^2$$

$$+ C \left(\left[\int_{\mathbb{R}^2} \frac{w_1^2 + w_2^2}{(1 + |w_1| + |w_2|)^2} \right]^4 + \left[\int_{\mathbb{R}^2} \sum_{q=1}^{2} |\nabla w_q|^2 \right]^4 + 1 \right),$$

where $C > 0$ denotes a uniform constant which may vary its value at different places. Hence,

$$\sum_{q=1}^{2} \|w_q\|_2 \leq C \left(1 + \sum_{q=1}^{2} \|\nabla w_q\|_2^2 + \int_{\mathbb{R}^2} \frac{w_1^2 + w_2^2}{(1 + |w_1| + |w_2|)^2} \, dx \right). \tag{7.4.39}$$

Combing (7.4.37) and (7.4.39), we see that the proof of the lemma is complete.

Lemma 7.4.6. *The functional I defined in (7.4.34) has a unique critical point (w_1, w_2) in $W^{1,2}(\mathbb{R}^2) \times W^{1,2}(\mathbb{R}^2)$ which solves the nonlinear elliptic system (7.4.32).*

Proof. For convenience, we use $\mathbf{w}$ to denote the pair (w_1, w_2) and X the space $W^{1,2}(\mathbb{R}^2) \times W^{1,2}(\mathbb{R}^2)$. Since I is convex and differentiable in $\mathbf{w} \in X$, it is weakly lower semicontinuous. Lemma 7.4.5 says that we can find a large enough number $r > 0$ so that

$$\inf \left\{ (dI(\mathbf{w}))(\mathbf{w}) \,\middle|\, \mathbf{w} \in X, \|\mathbf{w}\|_X = r \right\} \geq 1. \tag{7.4.40}$$

Consider the optimization problem

$$\sigma = \min \left\{ I(\mathbf{w}) \,\middle|\, \|\mathbf{w}\|_X \leq r \right\}. \tag{7.4.41}$$

Let $\{\mathbf{w}\}$ be a minimizing sequence of (7.4.41). Without loss of generality, we may assume that this sequence is also weakly convergent. Let $\tilde{\mathbf{w}}$ be its weak limit. Thus, using the fact that I is weakly lower semicontinuous, we have $I(\tilde{\mathbf{w}}) \leq \sigma$. Of course, $\|\tilde{\mathbf{w}}\|_X \leq r$ because norm is also weakly lower semicontinuous. Hence $I(\tilde{\mathbf{w}}) = \sigma$ and $\tilde{\mathbf{w}}$ solves (7.4.41). We show next that $\tilde{\mathbf{w}}$ is a critical point of the functional I. In fact, we only need to show that $\tilde{\mathbf{w}}$ is an interior point, namely,

$$\|\tilde{\mathbf{w}}\|_X < r.$$

Suppose otherwise that $\|\tilde{\mathbf{w}}\|_X = r$. Then, in view of (7.4.40), we have

$$\lim_{t \to 0} \frac{I(\tilde{\mathbf{w}} - t\tilde{\mathbf{w}}) - I(\tilde{\mathbf{w}})}{t} = \frac{d}{dt} I(\tilde{\mathbf{w}} - t\tilde{\mathbf{w}})\Big|_{t=0}$$
$$= -(dI(\tilde{\mathbf{w}}))(\tilde{\mathbf{w}}) \leq -1. \qquad (7.4.42)$$

Therefore, when $t > 0$ is sufficiently small, we see by virtue of (7.4.42) that

$$I(\tilde{\mathbf{w}} - t\tilde{\mathbf{w}}) < I(\tilde{\mathbf{w}}) = \sigma.$$

However, since $\|\tilde{\mathbf{w}} - t\tilde{\mathbf{w}}\|_{H(n)} = (1-t)R < R$, we arrive at a contradiction to the definition of $\tilde{\mathbf{w}}$ or (7.4.41).

Finally the strict convexity of I says that I can only have at most one critical point, so we have the conclusion that I has exactly one critical point in X. Of course this critical point is a weak solution of (7.4.32), which must be smooth by virtue of the elliptic regularity theory and also unique in the space X.

In the above, we proved the existence of a classical solution of (7.4.32). The next step is to show that such a solution gives rise to a finite-energy vortex solution of the two-Higgs dual system (7.4.4) which is not automatically guaranteed by the study made earlier. We now elaborate on this problem.

Lemma 7.4.7. *Let (w_1, w_2) be obtained as in Lemma 7.4.6. Then $w_q \to 0$ as $|x| \to \infty$, $x \in \mathbb{R}^2$, $q = 1, 2$.*

Proof. We recall first the standard embedding inequality for $p > 2$,

$$\|f\|_p \leq \left(\pi \left[\frac{p-2}{2}\right]\right)^{\frac{p-2}{2p}} \|f\|_{W^{1,2}}, \qquad f \in W^{1,2}(\mathbb{R}^2). \qquad (7.4.43)$$

We show that $e^f - 1 \in L^2$ for $f \in W^{1,2}$. In fact, the Taylor expansion gives us

$$(e^f - 1)^2 = f^2 + \sum_{k=3}^{\infty} \frac{2^k - 2}{k!} f^k.$$

Using (7.4.43), we obtain formally the series

$$\|e^f - 1\|_2^2 \leq \|f\|_2^2 + \sum_{k=3}^{\infty} \frac{2^k - 2}{k!} \left(\pi \left[\frac{k-2}{2} \right] \right)^{\frac{k-2}{2}} \|f\|_{W^{1,2}}^k. \qquad (7.4.44)$$

Calculus methods may be used to show that (7.4.44) is a convergent power series in $\|f\|_{W^{1,2}}$. This verifies the claim we made.

Return to the system (7.4.32). Since by the above observation and the property that $w_1, w_2 \in W^{1,2}(\mathbb{R}^2)$ the terms

$$e^{u_0' + \sqrt{|M|} w_1} - 1 = e^{u_0'} (e^{\sqrt{|M|} w_1} - 1) + (e^{u_0'} - 1),$$

$$e^{u_0'' + (a_{21}\sqrt{|M|} w_1 + |M| w_2)/a_{11}} - 1 = e^{u_0''} (e^{(a_{21}\sqrt{|M|} w_1 + |M| w_2)/a_{11}} - 1)$$
$$+ (e^{u_0''} - 1) \qquad (7.4.45)$$

in (7.4.32) both lie in $L^2(\mathbb{R}^2)$, the definitions of h_1, h_2 (see (7.4.29) and (7.4.33)) imply that the right-hand sides of the two equations in (7.4.32) are all in $L^2(\mathbb{R}^2)$. The well-known L^2-estimates for elliptic equations show that w_1, w_2 are elements in $W^{2,2}(\mathbb{R}^2)$. Then an embedding theorem says that w_1, w_2 both approach zero as $|x| \to \infty$. The lemma is proven.

Lemma 7.4.8. *Let w_q $(q = 1, 2)$ be as stated in Lemma 7.4.7. We also have the property that $|\nabla w_q(x)| \to 0$ as $|x| \to \infty$, $q = 1, 2$.*

Proof. Since $w_q(x) \to 0$ as $|x| \to \infty$, $q = 1, 2$, the conclusion that the terms in (7.4.45) all lie in $L^2(\mathbb{R}^2)$ may be generalized to the conclusion that they all lie in $L^p(\mathbb{R}^2)$ for any $p > 2$. Thus the right-hand sides of (7.4.32) all in L^p $(p > 2)$. The proof of Lemma 7.4.7 already shows that $w_q \in W^{2,2}(\mathbb{R}^2)$, $q = 1, 2$. Thus the Sobolev embedding indicates that $w_q \in W^{1,p}(\mathbb{R}^2)$ for $p > 2, q = 1, 2$. Hence the elliptic L^p-estimates imply that $w_q \in W^{2,p}(\mathbb{R}^2)$ for any $p > 2$, $q = 1, 2$. Consequently, $|\nabla w_q(x)| \to 0$ as $|x| \to \infty$, $q = 1, 2$, as expected.

Let $\lambda_1, \lambda_2 > 0$ be the eigenvalues of the positive definite matrix M defined in (7.4.8) and set

$$\lambda_0 = 2 \min \left\{ \lambda_1, \lambda_2 \right\}.$$

From w_1, w_2 in Lemma 7.4.7, we get the pair v_1, v_2 by (7.4.14). Then a solution (u_1, u_2) is obtained as a solution of (7.4.9) on the full $\mathbb{R}^2$.

Lemma 7.4.9. *For the pair u_1, u_2 stated above, there holds the exponential decay estimate*

$$u_1^2(x) + u_2^2(x) \leq C(\varepsilon) e^{-(1-\varepsilon)\sqrt{\lambda_0}|x|}$$

when $|x|$ is sufficiently large; where $\varepsilon \in (0, 1)$ is arbitrary and $C(\varepsilon) > 0$ is a constant.

Proof. Let O be an orthogonal 2×2 matrix so that

$$O^\tau M O = \begin{pmatrix} \lambda_1 & 0 \\ 0 & \lambda_2 \end{pmatrix}. \tag{7.4.46}$$

Consider (7.4.9) over $\mathbb{R}^2$ outside the disk $D_{R_0} = \{x \in \mathbb{R}^2 \,|\, |x| \leq R_0\}$ where

$$R > \max_j \left\{ |p_j|, |q_j| \right\}.$$

Then (7.4.9) takes the truncated form

$$\begin{aligned} \Delta u_1 &= a_{11}u_1 + a_{12}u_2 + a_{11}(e^{u_1} - 1 - u_1) + a_{12}(e^{u_2} - 1 - u_2), \\ \Delta u_2 &= a_{21}u_1 + a_{22}u_2 + a_{21}(e^{u_1} - 1 - u_1) + a_{22}(e^{u_2} - 1 - u_2), \end{aligned} \tag{7.4.47}$$

Introduce a new set of variables U_1, U_2 so that

$$\begin{pmatrix} u_1 \\ u_2 \end{pmatrix} = O \begin{pmatrix} U_1 \\ U_2 \end{pmatrix}.$$

Then, using (7.4.46) and (7.4.47), we get

$$\begin{aligned} \Delta U_1 &= \lambda_1 U_1 + b_{11}(x)U_1 + b_{12}(x)U_2, \\ \Delta U_2 &= \lambda_2 U_2 + b_{21}(x)U_1 + b_{22}(x)U_2, \end{aligned} \tag{7.4.48}$$

where b_{jk} depend on U_1, U_2 and $b_{jk}(x) \to 0$ $(j, k = 1, 2)$ as $|x| \to \infty$ because U_1, U_2 enjoy this property by virtue of Lemma 7.4.7. Therefore we arrive from (7.4.48) at the elliptic inequality

$$\Delta(U_1^2 + U_2^2) \geq \lambda_0(U_1^2 + U_2^2) - b(x)(U_1^2 + U_2^2),$$

where $b(x) \to 0$ as $|x| \to \infty$. Consequently, for any $\varepsilon > 0$, there exists a suitable $R_0 > 0$ so that

$$\Delta(U_1^2 + U_2^2) \geq \left(1 - \frac{\varepsilon}{2}\right)\lambda_0(U_1^2 + U_2^2), \qquad |x| > R_0. \tag{7.4.49}$$

Thus it follows from using a suitable comparison function and applying the maximum principle in (7.4.49) that there is a constant $C(\varepsilon) > 0$ to make

$$U_1^2(x) + U_2^2(x) \leq C(\varepsilon)e^{-(1-\varepsilon)\sqrt{\lambda_0}|x|} \tag{7.4.50}$$

hold for $|x| > R_0$. Since O is orthogonal, we have $u_1^2 + u_2^2 = U_1^2 + U_2^2$. Using (7.4.50), we see that the lemma is proven.

Lemma 7.4.10. *For u_1, u_2 in Lemma 7.4.9, we have in addition*

$$|\nabla u_1|^2 + |\nabla u_2|^2 \leq C(\varepsilon)e^{-(1-\varepsilon)\sqrt{\lambda_0}|x|},$$

where $\varepsilon, C(\varepsilon)$ are as described there.

Proof. Differentiate (7.4.47) outside D_R. We have

$$
\begin{aligned}
\Delta(\partial_j u_1) &= a_{11}(\partial_j u_1) + a_{12}(\partial_j u_2) + a_{11}(e^{u_1} - 1)(\partial_j u_1) \\
&\quad + a_{12}(e^{u_2} - 1)(\partial_j u_2), \\
\Delta(\partial_j u_2) &= a_{21}(\partial_j u_1) + a_{22}(\partial_j u_2) + a_{21}(e^{u_1} - 1)(\partial_j u_1) \\
&\quad + a_{22}(e^{u_2} - 1)(\partial_j u_2).
\end{aligned}
\tag{7.4.51}
$$

Comparing (7.4.51) with (7.4.47) and using Lemma 7.4.8, we see that the stated decay estimate holds.

We can now construct the solutions of the dual system (7.4.4) over the full $\mathbb{R}^2$ using u_1, u_2. In fact, let $\theta_p(z)$ and $\theta_q(z)$ be defined by

$$
\theta_p(z) = \sum_{j=1}^{N_1} \arg(z - p_j), \quad \theta_q(z) = \sum_{j=1}^{N_2} \arg(z - q_j).
$$

Set $\psi_1(z) = \exp(\frac{1}{2}u_1(z) - i\theta_p(z))$ and $\psi_2(z) = \exp(\frac{1}{2}u_2(z) + i\theta_q(z))$. Then ψ_1, ψ_2 are smooth functions which vanish precisely at $p_1, p_2, \cdots, p_{N_1}$ and $q_1, q_2, \cdots, q_{N_2}$, respectively. With these functions, we obtain $A_j^3, B_j, \tilde{B}_j$ by solving the linear system

$$
\begin{aligned}
g A_j^3 - g' Y_1 B_j - g_1 Y_1' \tilde{B}_j &= \alpha_j^{(1)}, \\
g A_j^3 - g' Y_2 B_j - g_1 Y_2' \tilde{B}_j &= \alpha_j^{(2)}, \\
g' g_1 (Y_1 Y_2' - Y_2 Y_1') A_j^3 + g g_1 (Y_2' - Y_1') B_j &= g' g (Y_2 - Y_1) \tilde{B}_j,
\end{aligned}
\tag{7.4.52}
$$

$j = 1, 2$, where we define

$$
\begin{aligned}
\alpha_1^{(1)} &= \operatorname{Re}\{2 i \bar{\partial} \ln \bar{\psi}_1\}, \quad \alpha_2^{(1)} = \operatorname{Im}\{2 i \bar{\partial} \ln \bar{\psi}_1\}, \\
\alpha_1^{(2)} &= -\operatorname{Re}\{2 i \bar{\partial} \ln \psi_2\}, \quad \alpha_2^{(2)} = -\operatorname{Im}\{2 i \bar{\partial} \ln \psi_2\}.
\end{aligned}
$$

Recall that the condition (7.4.26) ensures that (7.4.52) has a unique solution. The quintuplet $(\psi_1, \psi_2, A_j^3, B_j, \tilde{B}_j)$ solves (7.4.4) on entire $\mathbb{R}^2$.

Besides, using Lemmas 7.4.9 and 7.4.10, the system (7.4.4), and the equations

$$
|D^{(q)} \phi_q|^2 = \frac{1}{2} e^{u_q} |\nabla u_q|^2, \quad q = 1, 2,
$$

we find the following exponential type decay estimates for the physical fields,

$$
\begin{aligned}
|\phi_q|^2 - \frac{1}{2} v_0^2, \quad F_{12}^3, \quad H_{12}, \quad \tilde{H}_{12} &= O(e^{-\sqrt{\lambda_0}|x|/2}); \\
|D^{(q)} \psi_q|^2 &= O(e^{-\sqrt{\lambda_0}|x|}), \quad q = 1, 2,
\end{aligned}
\tag{7.4.53}
$$

where the precise meaning of $h(x) = O(e^{-\sqrt{\lambda_0}|x|})$, for example, is that, for any $\varepsilon \in (0,1)$, there is a constant $C(\varepsilon) > 0$ to make

$$|h(x)| \leq C(\varepsilon)e^{-(1-\varepsilon)\sqrt{\lambda_0}|x|}$$

valid. These estimates imply that the solution just constructed carries a finite energy.

Next, we calculate the fluxes.

Lemma 7.4.11. *Let v_1, v_2 be the solution of (7.4.30) just obtained. Then*

$$\int_{\mathbb{R}^2} \Delta v_1 \, dx = \int_{\mathbb{R}^2} \Delta v_2 \, dx = 0.$$

Proof. For v_1, we have $v_1 = u_1 - u_0'$. Hence, in view of Lemma 7.4.9, we have $|\nabla v_1| = O(|x|^{-3})$ as $|x| \to \infty$ because $|\nabla u_0'|$ decays like that at infinity. Using the divergence theorem, we easily show that Δv_1 has zero integral on $\mathbb{R}^2$. The same is true for v_2.

Applying Lemma 7.4.11 and (7.4.31) in (7.4.30), we find

$$\int_{\mathbb{R}^2} \left(|\psi_1|^2 - \frac{v_0^2}{2} \right) dx = \frac{v_0^2}{2} \int_{\mathbb{R}^2} (e^{u_1} - 1) \, dx = \frac{2\pi v_0^2}{|M|} (a_{12}N_2 - a_{22}N_1),$$

$$\int_{\mathbb{R}^2} \left(|\psi_2|^2 - \frac{v_0^2}{2} \right) dx = \frac{v_0^2}{2} \int_{\mathbb{R}^2} (e^{u_2} - 1) \, dx = \frac{2\pi v_0^2}{|M|} (a_{12}N_1 - a_{11}N_2).$$

Integrating the last three equations in (7.4.4) over $\mathbb{R}^2$ and using the above results, we see that the fluxes $\Phi_F, \Phi_G, \Phi_{\tilde{G}}$ are given by the same formulas (7.4.27) as for the fluxes in a periodic cell domain. This is certainly an interesting property.

We summarize our study for the existence of multivortices in the two-Higgs electroweak theory as follows.

Theorem 7.4.12. *Consider the dual system (7.4.4) in the entire $\mathbb{R}^2$ arising in the extended Weinberg–Salam model with two Higgs doublets. Given any prescribed points*

$$p_1, p_2, \cdots, p_{N_1}, \quad q_1, q_2, \cdots, q_{N_2}, \tag{7.4.54}$$

the system (7.4.4) has a unique finite-energy multivortex solution configuration $(\psi_1, \psi_2, A_j^3, B_j, \tilde{B}_j)$ so that ψ_1 and ψ_2 vanish precisely at the points p's and q's given in (7.4.54) and their winding numbers or the topological vortex charges around a circle near infinity are N_1 and N_2, respectively. Furthermore the physical field strengths decay exponentially at infinity according to (7.4.53) and the weak and the two electromagnetic fluxes over $\mathbb{R}^2$ obey the same quantization property (7.4.27) as that for periodic vortices.

We note that the 2×2 system (7.4.9) is in fact covered as a special case of the general $n \times n$ system studied in Chapter 4. The reason that we have presented a self-contained proof here is to provide a concrete analysis of a physical model and avoid complexity created from abstraction.

7.5 Remarks

In this chapter, we have presented a series of results concerning the existence of W-condensed multivortex solutions of the system of nonlinear equations arising in the electroweak theory, originally published in [292, 341]. For the massive non-Abelian model and the Weinberg–Salam model, it is seen that the best constant in the Trudinger–Moser inequality (4.3.12) stands out as an obstacle to a general resolution of the existence problem. Hence, as in the prescribed curvature problem [17] in geometric analysis, we encounter the same difficulty here and our existence theory may be considered complete only when the vortex number N is small, $N = 1, 2$. This leads us to

Open Problem 7.5.1. For $N \geq 3$, find a necessary and sufficient condition for (7.1.21) to have a solution.

In particular, a milder open problem is whether an existence theorem may be established when (7.1.23) becomes an equality,

$$\frac{2\pi(N-2)}{|\Omega|} = m_W^2. \tag{7.5.1}$$

For the Ambjorn–Olesen equations (7.3.4), we can rewrite (7.2.19) as

$$\frac{4\pi(N - 2\sin^2\theta)}{|\Omega|\cos^2\theta} < \frac{g^2\varphi_0^2}{\cos^2\theta}. \tag{7.5.2}$$

It is clear that, for $N = 1, 2$, the condition (7.5.2) is covered in (7.2.18). Thus, (7.2.18) is a necessary and sufficient condition for the existence of a solution to the system (7.3.4) when $N = 1, 2$. Hence we propose the same problem as that for the massive non-Abelian equation.

Open Problem 7.5.2. For $N \geq 3$, find a necessary and sufficient condition for (7.3.4) to have a solution.

Similarly, it will be interesting to know whether (7.2.19) can be improved into the form

$$\frac{4\pi N}{|\Omega|} \leq \frac{8\pi\sin^2\theta}{|\Omega|} + g^2\varphi_0^2. \tag{7.5.3}$$

Recently, Bartolucci and Tarantello [26] made a remarkable progress concerning Open Problem 7.5.2. If we rewrite (7.2.19) as

$$\frac{4\pi N - g^2\varphi_0^2|\Omega|}{8\pi\sin^2\theta} < 1, \tag{7.5.4}$$

then their improved sufficiency condition reads

$$\frac{4\pi N - g^2 \varphi_0^2 |\Omega|}{8\pi \sin^2 \theta} < 2, \tag{7.5.5}$$

$$\frac{4\pi N - g^2 \varphi_0^2 |\Omega|}{8\pi \sin^2 \theta} \neq 1. \tag{7.5.6}$$

This result implies that (7.2.18) is a necessary and sufficient condition for the existence of an N-vortex solution for $N = 3, 4$ as well if (7.5.6) holds. Their proof is based on a study of a Liouville type equation with 'singular' data accomodating the Dirac sources, which extends a Brezis–Merle type analysis [55], as developed in [192, 193], for solution sequences admitting 'blow-up' points [26, 27]. Similar 'regular' problems have been previously studied in [92, 299]. It is reasonable to conjecture that, for existence, (7.5.6) is just a technicality and may yet be removed.

The comparison of the results obtained here for the W- and Higgs condensate equations in the classical Weinberg–Salam model found by Ambjorn and Olesen [4, 5, 6, 7] and its two-Higgs-doublet extension by Bimonte and Lozano [41] is interesting. In the doubly periodic setting, the conditions for existence for the former and latter are rather different. For the former, they imply that the area of the domain to confine multivortices with a given vortex number N must not be either too small or too large. While for the latter, there is no restriction to the upper bound of the area of a cell domain. Larger domains always accommodate more vortices. A common feature is that in both situations the solutions are minimizers of the energy functional and fluxes are quantized. Another is that the conditions for existence are independent of the locations of vortices but only depending on the total topological charge or the vortex number N. In the full plane setting, the former does not allow the existence of finite-energy Bogomol'nyi type vortices [293]. In order to get finite-energy solutions, it is important to put into consideration the effect of gravity through the coupling of the Einstein equations [348]. Now the metric noncompleteness leads to a sufficiently rapid decay of the metric at infinity and the existence of finite-energy multivortices may be ensured. Our conditions for existence involves the vortex number N again as in the doubly periodic case and nonuniqueness often occurs. Besides, the fluxes can assume continuous values from some explicitly given intervals. For the latter, there exists a unique finite-energy solution for any prescribed vortex distribution and total topological charge N. There is no freedom to designate the values of fluxes: They are all quantized and represented by the vortex number N. It is also remarkable that these values are of the same form as those from an N-vortex solution in a periodic lattice cell. It is hoped that this part of the work would also lend hints to more generalized or realistic electroweak models.

Finally, we mention that Bimonte and Lozano [40] have extended the work of Ambjorn and Olesen [4, 5, 6] on the W-condensed electroweak vortices to include a Higgs condensation in which the Higgs field is allowed to be complex-valued and vortices are generated from both the zeros of the W-field and the Higgs field. For this more general problem, the same mathematical analysis can be carried out to establish an existence theorem [348] and one would encounter the same obstacles arising from the Trudinger–Moser inequality.

8
Dyons

Dyons are hypothetical particles in high-energy physics that carry both electric and magnetic charges and are realized as a special class of static solutions of field equations. In this chapter, we shall present such static solutions representing dyons. In §8.1 and §8.2, respectively, we introduce Dirac's magnetic monopoles and Schwinger's dyons in the context of the Maxwell equations. In the rest of the chapter, we use a variational method to prove the existence of dyons in two important non-Abelian gauge field theory models for which the governing equations are nonlinear and the action functionals are indefinite: in §8.3 we present the Julia–Zee dyons and in §8.4–§8.6 we obtain dyons in the Weinberg–Salam electroweak theory.

8.1 Dirac Monopole

In this section we describe Dirac's magnetic monopole [94] and its consequences. We first remark on the duality between electricity and magnetism in the Maxwell equations. We next present Dirac's idea on the charge quantization and string singularity implied by the existence of a monopole. We then use the concept of a fiber bundle to rederive the quantization formula of Dirac and show that the string singularity may be removed and the Dirac monopole is indeed a point particle.

8.1.1 Electromagnetic duality

Let the vector fields $\mathbf{E}$ and $\mathbf{B}$ denote the electric and magnetic fields, respectively, which are induced from the presence of an electric charge density distribution, ρ, and a current density, $\mathbf{j}$. Then these fields are governed in the Heaviside–Lorentz rationalized units by the Maxwell equations,

$$\nabla \cdot \mathbf{E} \;=\; \rho, \tag{8.1.1}$$

$$\nabla \times \mathbf{B} - \frac{\partial \mathbf{E}}{\partial t} \;=\; \mathbf{j}, \tag{8.1.2}$$

$$\nabla \cdot \mathbf{B} \;=\; 0, \tag{8.1.3}$$

$$\nabla \times \mathbf{E} + \frac{\partial \mathbf{B}}{\partial t} \;=\; 0. \tag{8.1.4}$$

In vacuum where $\rho = 0$, $\mathbf{j} = 0$, these equations are invariant under the dual correspondence,

$$\mathbf{E} \mapsto \mathbf{B}, \quad \mathbf{B} \mapsto -\mathbf{E}. \tag{8.1.5}$$

In other words, in another world 'dual' to the original one, electricity and magnetism are seen as magnetism and electricity, and vice versa. This property is called electromagnetic duality or dual symmetry of electricity and magnetism. However, such a symmetry is broken in the presence of an external charge-current density, $(\rho, \mathbf{j})$. Dirac proposed a procedure that may be used to symmetrize (8.1.1)–(8.1.4), which starts from the extended equations,

$$\nabla \cdot \mathbf{E} \;=\; \rho_e, \tag{8.1.6}$$

$$\nabla \times \mathbf{B} - \frac{\partial \mathbf{E}}{\partial t} \;=\; \mathbf{j}_e, \tag{8.1.7}$$

$$\nabla \cdot \mathbf{B} \;=\; \rho_m, \tag{8.1.8}$$

$$\nabla \times \mathbf{E} + \frac{\partial \mathbf{B}}{\partial t} \;=\; \mathbf{j}_m, \tag{8.1.9}$$

where $\rho_e, \mathbf{j}_e$ and $\rho_m, \mathbf{j}_m$ denote the electric and magnetic source terms respectively. These equations are invariant under the transformation (8.1.5) if the source terms are transformed accordingly,

$$\rho_e \mapsto \rho_m, \quad \mathbf{j}_e \mapsto \mathbf{j}_m, \quad \rho_m \mapsto -\rho_e, \quad \mathbf{j}_m \mapsto -\mathbf{j}_e. \tag{8.1.10}$$

Hence, with the new magnetic source terms in (8.1.8) and (8.1.9), duality between electricity and magnetism is again achieved as in the vacuum case.

In (8.1.6)–(8.1.9), ρ_e and $\mathbf{j}_e$ are the usual electric charge and current density functions, but ρ_m and $\mathbf{j}_m$, which are the magnetic charge and current density functions, introduce completely new ideas, both technically and physically.

8.1.2   Dirac strings and charge quantization

In order to motivate our study, consider the classical situation that electromagnetism is generated from an ideal point electric charge q lying at the origin,

$$\rho_e = 4\pi q \delta(\mathbf{x}), \quad \mathbf{j}_e = \mathbf{0}, \quad \rho_m = 0, \quad \mathbf{j}_m = \mathbf{0}. \tag{8.1.11}$$

It is clear that, inserting (8.1.11), the system (8.1.6)–(8.1.9) can be solved to yield $\mathbf{B} = \mathbf{0}$ and

$$\mathbf{E} = \frac{q}{|\mathbf{x}|^3}\mathbf{x}, \tag{8.1.12}$$

which is the well-known Coulomb law in static electricity.

We now consider the case of a point magnetic charge g, or a monopole, at the origin,

$$\rho_e = 0, \quad \mathbf{j}_e = \mathbf{0}, \quad \rho_m = 4\pi g \, \delta(\mathbf{x}), \quad \mathbf{j}_m = \mathbf{0}. \tag{8.1.13}$$

Hence $\mathbf{E} = 0$ and

$$\mathbf{B} = \frac{g}{|\mathbf{x}|^3}\mathbf{x} = -g\nabla\left(\frac{1}{|\mathbf{x}|}\right). \tag{8.1.14}$$

Consequently, the magnetic flux through a sphere centered at the origin and of radius $r > 0$ is

$$\Phi = \int_{|\mathbf{x}|=r} \mathbf{B} \cdot d\mathbf{S} = 4\pi g, \tag{8.1.15}$$

which is independent of r and is identical to the Gauss law for static electricity. Nevertheless, we show below through quantum mechanics that the introduction of a magnetic charge yields drastically new physics because electric and magnetic fields are induced differently from a gauge vector potential.

Consider a point particle with mass m and electric charge q in the magnetic field of a monopole given in (8.1.14). The equation of motion is

$$m\ddot{\mathbf{x}} = q\dot{\mathbf{x}} \times \mathbf{B}. \tag{8.1.16}$$

The classical orbital momentum $\mathbf{x} \times \mathbf{p} = \mathbf{x} \times m\dot{\mathbf{x}}$ is no longer a conserved quantity due to (8.1.16) and (8.1.14),

$$\frac{d}{dt}(\mathbf{x} \times m\dot{\mathbf{x}}) = \mathbf{x} \times m\ddot{\mathbf{x}} = \frac{gq}{|\mathbf{x}|^3}\mathbf{x} \times (\dot{\mathbf{x}} \times \mathbf{x}) = \frac{d}{dt}\left(gq\frac{\mathbf{x}}{|\mathbf{x}|}\right),$$

but

$$\mathbf{J} = \mathbf{x} \times m\dot{\mathbf{x}} - gq\frac{\mathbf{x}}{|\mathbf{x}|} \tag{8.1.17}$$

is, which suggests that we should use (8.1.17) to define the total angular momentum when a monopole is present. Hence, in view of quantum mechanics, the eigenvalues of the operators J_i must be half-integers, which leads to the Dirac formula,

$$q = \frac{n}{2g}, \quad n = 0, \pm 1, \pm 2, \cdots. \tag{8.1.18}$$

One of the most important implication of (8.1.18) is that, if there is a monopole in the universe, then all electric charges are integral multiples of a basic unit quantity. In other words, the presence of a magnetic monopole implies electric charge quantization.

We now evaluate the energy of a monopole. Recall that the total energy of an electromagnetic field with electric component $\mathbf{E}$ and magnetic component $\mathbf{B}$ is given by

$$E = \int_{\mathbb{R}^3} (\mathbf{E}^2 + \mathbf{B}^2) \, dx. \tag{8.1.19}$$

Inserting (8.1.14) into (8.1.19) and using $r = |\mathbf{x}|$, we have

$$E = 4\pi g^2 \int_0^\infty \frac{1}{r^2} \, dr = \infty. \tag{8.1.20}$$

This energy blow-up seems to suggest that the idea of a magnetic monopole encounters an unacceptable obstacle. However, since the Coulomb law expressed in (8.1.12) for a point electric charge also leads to a divergent energy of the same form, (8.1.20), the infinite energy problem for a monopole is not a more serious one than that for a point electric charge which has been used effectively as good approximation for various particle models.

We now study the magnetic field generated from a monopole more closely. Since the quantum mechanical motion of a particle of mass m and charge q in an electromagnetic field is described by the gauged Schrödinger equation using a gauge vector potential of the electromagnetic field,

$$i\frac{\partial \psi}{\partial t} = \left(\frac{1}{2m} (\nabla - q\mathbf{A})^2 + q\phi \right)\psi, \tag{8.1.21}$$

where ϕ and $\mathbf{A}$ are potential fields for the electric and magnetic fields, respectively, for our purpose it will be essential to know how the magnetic field (8.1.14) of a monopole is induced from its gauge potentials. Note that (8.1.21) is invariant under the gauge transformation

$$\mathbf{A}(\mathbf{x}, t) \mapsto \mathbf{A}(\mathbf{x}, t) + \frac{1}{q}\nabla\omega(\mathbf{x}), \quad \psi(\mathbf{x}, t) \mapsto e^{i\omega(\mathbf{x})}\psi(\mathbf{x}, t). \tag{8.1.22}$$

Recall that for the electric field generated from a point electric charge q, the Coulomb law (8.1.12) gives us a scalar potential function $\phi = -q/|\mathbf{x}|$ such that $\mathbf{E} = \nabla\phi$ holds everywhere away from the point electric charge. Similarly, for the magnetic field (8.1.14) generated from a monopole, we study the existence of a vector field $\mathbf{A}$ such that

$$\mathbf{B} = \nabla \times \mathbf{A}. \tag{8.1.23}$$

In fact, using (8.1.15) and the Stokes theorem, it is easy to see that (8.1.23) cannot hold everywhere on any closed sphere centered at the origin. In other words, any such sphere would contain a singular point at which (8.1.23) fails. Intuitively, shrinking a sphere to the origin would give us a continuous family of singular points which is a string that links the origin to infinity. Such a string is called a Dirac string.

8.1.3 Fiber bundle device and removal of strings

In this subsection, we study the Dirac string more carefully and present a rigorous derivation of the charge quantization formula (8.1.18).

It may be checked directly that

$$\mathbf{A}^+ = (A_1^+, A_2^+, A_3^+),$$
$$A_1^+ = -\frac{x_2}{|\mathbf{x}|(|\mathbf{x}| + x_3)}g,$$
$$A_2^+ = \frac{x_1}{|\mathbf{x}|(|\mathbf{x}| + x_3)}g,$$
$$A_3^+ = 0 \tag{8.1.24}$$

satisfies (8.1.23) everywhere except on the negative x_3-axis, $x_1 = 0, x_2 = 0, x_3 \leq 0$. That is, with $\mathbf{A}^+$, the Dirac string S^- is the negative x_3-axis. Using spherical coordinates (r, θ, ϕ) where θ and ϕ are the polar and azimuthal angles, respectively, we have

$$A_r^+ = 0, \quad A_\theta^+ = \left(\frac{g}{r}\right)\frac{1 - \cos\phi}{\sin\phi}, \quad A_\phi^+ = 0. \tag{8.1.25}$$

Similarly, $\mathbf{A}^-$ satisfies (8.1.23) everywhere except on the positive x_3-axis, where

$$A_r^- = 0, \quad A_\theta^- = -\left(\frac{g}{r}\right)\frac{1 + \cos\phi}{\sin\phi}, \quad A_\phi^- = 0, \tag{8.1.26}$$

and the Dirac string S^+ is the positive x_3-axis.

From the point of view of physics, the magnetic field generated from a point monopole at the origin should be well defined everywhere except at the origin and should be independent of the choice of its gauge potentials. From the point of view of mathematics, electromagnetism induced from a monopole at the origin is simply a $U(1)$-bundle over the punctured space, $M = \mathbb{R}^3 - \{O\}$ so that the magnetic field $\mathbf{B}$ is the curvature 2-form of a connection 1-form $\mathbf{A}$,

$$\mathbf{B} = d\mathbf{A}, \tag{8.1.27}$$

which is a correct reformulation of (8.1.23) in terms of a fiber bundle. Then, with $U^+ = \mathbb{R}^3 - S^-$ and $U^- = \mathbb{R}^3 - S^+$, the bundle trivializes over U^+ and U^- and (8.1.27) or (8.1.23) holds individually with $\mathbf{A} = \mathbf{A}^+$ and

$\mathbf{A} = \mathbf{A}^-$ on U^+ and U^-, respectively. In order that $\mathbf{A}^+$ and $\mathbf{A}^-$ define a global connection 1-form $\mathbf{A}$ so that (8.1.27) holds on the entire M, the local representations $\mathbf{A}^+$ and $\mathbf{A}^-$ must be related to each other by the transition formula

$$\mathbf{A}^+ = \mathbf{A}^- + \frac{i}{q}\Omega d\Omega^{-1}, \quad \mathbf{x} \in U^+ \cap U^-, \quad \Omega = e^{i\omega} \in U(1), \qquad (8.1.28)$$

where the coefficient $1/q$ is introduced in order to be compatible with (8.1.22) for the quantum dynamics of a moving particle with mass m and electric charge q.

Since we need to consider gauge symmetry in terms of differential forms, we must correctly represent the vector potentials, $\mathbf{A}^+$ and $\mathbf{A}^-$, on their domains accordingly. The formula (8.1.24) determines $\mathbf{A}^+$ as a 1-form through

$$A_1^+ \, dx_1 + A_2^+ \, dx_2 + A_3^+ \, dx_3, \qquad (8.1.29)$$

where dx_1, dx_2, dx_3 form a basis of the cotangent fiber space. Inserting (8.1.24) into (8.1.29) and using the spherical coordinates, we see that (8.1.25) and (8.1.26) are representations of $\mathbf{A}^+$ and $\mathbf{A}^-$ as 1-forms under the basis

$$\mathbf{u}_r = dr, \quad \mathbf{u}_\theta = r \sin\phi \, d\theta, \quad \mathbf{u}_\phi = r \, d\phi. \qquad (8.1.30)$$

Therefore, we obtain from applying (8.1.30) in (8.1.25) and (8.1.26) that

$$\mathbf{A}^+ = g(1 - \cos\phi) \, d\theta, \quad \mathbf{A}^- = -g(1 + \cos\phi) \, d\theta. \qquad (8.1.31)$$

Substituting (8.1.31) into (8.1.28), we obtain

$$\frac{1}{q}\frac{\partial\omega}{\partial\theta} = 2g.$$

Consequently, $\omega = 2gq\theta$ and we have

$$\Omega = e^{i\omega} = \exp(2gqi\theta). \qquad (8.1.32)$$

Recall that we need $\Omega(\mathbf{x})$ to be a $U(1)$-valued element, which makes the transition relation (8.1.28) valid. Such a requirement imposes the necessary and sufficient condition that Ω be single-valued, which is equivalent to the property that (8.1.32) is invariant under the translation $\theta \mapsto \theta + 2\pi$. Hence, $2gq(2\pi) = 2n\pi$ or

$$gq = \frac{1}{2}n, \qquad (8.1.33)$$

where n is an integer. Thus we have arrived at the previously stated Dirac quantization formula, (8.1.18).

In conclusion, when the condition (8.1.18) or (8.1.33) holds, the magnetic field outside a monopole is well defined everywhere and is generated piecewise from suitable gauge potentials defined on their corresponding domains. In the language of fiber bundles, the magnetic field is in fact the curvature 2-form of a globally defined connection 1-form so that (8.1.23) is realized on local trivializations. Therefore the Dirac strings are seen to be artifacts and are removed.

8.2 Schwinger Theory

In the last section, we discussed electromagnetism of a point particle carrying either electric charge or magnetic charge, but not both. It will be interesting to extend our consideration to a particle simultaneously carrying electric and magnetic charges. Particles of this kind are called dyons by Schwinger [277] and are believed to be possible candidates for quarks and other elementary particles. A complete treatment of dyons would require some familiarity with quantum field theory and might distract our focus. Here we only satisfy ourselves with a short, self-contained, study. Zwanziger [357, 358] also investigated the quantum mechanical properties of dyons.

8.2.1 Rotation symmetry

We have seen that electromagnetic duality (8.1.5) leads us to a symmetrization of the Maxwell equations,

$$(8.1.1)-(8.1.4) \longrightarrow (8.1.6)-(8.1.9),$$

and the revised equations are now invariant under the full dual correspondence (8.1.5) and (8.1.10). It was Schwinger who first found that the invariance of (8.1.6)-(8.1.9) under (8.1.5) and (8.1.10) is a particular case of a uniform rotation by an angle θ defined by the rules,

$$
\begin{aligned}
\mathbf{E} &\mapsto \cos\theta\,\mathbf{E} + \sin\theta\,\mathbf{B}, & \mathbf{B} &\mapsto -\sin\theta\,\mathbf{E} + \cos\theta\,\mathbf{B}, \\
\rho_e &\mapsto \cos\theta\,\rho_e + \sin\theta\,\rho_m, & \rho_m &\mapsto -\sin\theta\,\rho_e + \cos\theta\,\rho_m, \\
\mathbf{j}_e &\mapsto \cos\theta\,\mathbf{j}_e + \sin\theta\,\mathbf{j}_m, & \mathbf{j}_m &\mapsto -\sin\theta\,\mathbf{j}_e + \cos\theta\,\mathbf{j}_m.
\end{aligned}
\tag{8.2.1}
$$

Thus, in view of (8.2.1), Schwinger pointed out that the observed absence of a magnetic charge in classical electromagnetism may well be described as the coexistence of electric and magnetic charges in the universal ratio indicated by

$$\frac{\rho_m}{\rho_e} = \tan\theta. \tag{8.2.2}$$

When (8.2.2) is violated, magnetic charge appears due to electromagnetic duality.

8.2.2 Charge quantization formula for dyons

We now extend our study of the motion of a charged particle in the magnetic field of a monopole to the case of dyons. We consider the motion of a dyon with mass m and electric and magnetic charges q_1 and g_1 in the electromagnetic field $(\mathbf{E}, \mathbf{B})$ of another dyon with electric and magnetic charges q_2 and g_2 at the origin. We assume that the second dyon is so heavy that it stays stationary throughout our study.

Solving (8.1.6)–(8.1.9) for the second dyon, we obtain

$$\mathbf{E} = q_2 \frac{\mathbf{x}}{|\mathbf{x}|^3}, \quad \mathbf{B} = g_2 \frac{\mathbf{x}}{|\mathbf{x}|^3}. \tag{8.2.3}$$

On the other hand, the nonrelativistic motion of the first dyon is governed by the Lorentz force,

$$m\ddot{\mathbf{x}} = q_1(\mathbf{E} + \dot{\mathbf{x}} \times \mathbf{B}) + g_1(\mathbf{B} - \dot{\mathbf{x}} \times \mathbf{E}). \tag{8.2.4}$$

Inserting (8.2.3) into (8.2.4), we arrive at

$$m\ddot{\mathbf{x}} = (q_1 q_2 + g_1 g_2) \frac{\mathbf{x}}{|\mathbf{x}|^3} + (q_1 g_2 - q_2 g_1)\dot{\mathbf{x}} \times \frac{\mathbf{x}}{|\mathbf{x}|^3}. \tag{8.2.5}$$

Using (8.2.5), we see that the previously studied conserved total angular momentum $\mathbf{J}$ defined in (8.1.17) is replaced by

$$\mathbf{J} = \mathbf{x} \times m\dot{\mathbf{x}} - (q_1 g_2 - q_2 g_1)\frac{\mathbf{x}}{|\mathbf{x}|}. \tag{8.2.6}$$

Comparing (8.2.6) with (8.1.17), we see that the Dirac charge quantization formula (8.1.18) for monopoles becomes the following Schwinger charge quantization formula for dyons,

$$g_2 q_1 - g_1 q_2 = \frac{n}{2}, \quad n = 0, \pm 1, \pm 2, \cdots. \tag{8.2.7}$$

When $g_1 = 0$ and $q_2 = 0$ (the motion of electrically charged particle in the magnetic field of a monopole), we recover from (8.2.7) the Dirac condition (8.1.18).

For later comparisons, we reformulate (8.1.6)–(8.1.9) in terms of the duality transformation via the Hodge star $*$.

With the notation $\mathbf{E} = (E^i)$, $\mathbf{B} = (B^i)$, we can define the skewsymmetric electromagnetic field tensor $(F^{\mu\nu})$ ($\mu, \nu = 0, 1, 2, 3$) by

$$F^{0i} = -E^i, \quad F^{ij} = -\epsilon_{ijk}B^k, \quad i, j, k = 1, 2, 3. \tag{8.2.8}$$

In fact we may also rewrite (8.2.8) in the following matrix form,

$$(F^{\mu\nu}) = \begin{pmatrix} 0 & -E^1 & -E^2 & -E^3 \\ E^1 & 0 & -B^3 & B^2 \\ E^2 & B^3 & 0 & -B^1 \\ E^3 & -B^2 & B^1 & 0 \end{pmatrix}. \tag{8.2.9}$$

Set $(j_e^\mu) = (\rho_e, \mathbf{j}_e)$ and $(j_m^\mu) = (\rho_m, \mathbf{j}_m)$. Recall that we may use the Hodge star $*$ to define the dual tensor $*F^{\mu\nu}$ of $F^{\mu\nu}$ by

$$*F^{\mu\nu} = \frac{1}{2}\epsilon^{\mu\nu\mu'\nu'} F_{\mu'\nu'}. \tag{8.2.10}$$

It can be seen that (8.2.10) corresponds to the transformation (8.1.5). Thus (8.1.6)–(8.1.9) become

$$\partial_\nu F^{\mu\nu} = -j_e^\mu, \quad \partial_\nu * F^{\mu\nu} = -j_m^\mu, \tag{8.2.11}$$

and the duality transformation, (8.1.5) and (8.1.10), takes the form

$$F^{\mu\nu} \mapsto *F^{\mu\nu}, \quad *F^{\mu\nu} \mapsto -F^{\mu\nu}, \quad j_e^\mu \mapsto j_m^\mu, \quad j_m^\mu \mapsto -j_e^\mu. \tag{8.2.12}$$

In a region where $F^{\mu\nu}$ is induced from a gauge vector potential, (A_μ), namely,

$$F_{\mu\nu} = \partial_\mu A_\nu - \partial_\nu A_\mu, \tag{8.2.13}$$

there holds the Bianchi identity,

$$\partial_\nu * F^{\mu\nu} = 0, \tag{8.2.14}$$

which forces the magnetic source current j_m^μ vanish, $j_m^\mu = 0$.

8.3   Julia–Zee Dyons

In the last section, we discussed monopoles and dyons in terms of the Maxwell equations for electromagnetism which is a theory of Abelian gauge fields. In fact it is more natural for monopoles and dyons to exist in non-Abelian gauge field-theoretical models because nonvanishing commutators themselves are now present as electric and magnetic source terms. In other words, non-Abelian monopoles and dyons are self-induced and inevitable. In this section, we shall present the simplest non-Abelian dyons known as the Julia–Zee dyons [162].

8.3.1   Field equations

Consider the simplest non-Abelian Lie group $SO(3)$, which has a set of generators $\{t_a\}$ $(a = 1, 2, 3)$ satisfying $[t_a, t_b] = \epsilon^{abc} t_c$. Consequently, the $so(3)$-valued quantities $A = A^a t_a$ and $B = B^a t_a$ give rise to a commutator,

$$[A, B] = \epsilon_{abc} A^a B^b t_c. \tag{8.3.1}$$

For convenience, we may view A and B as two 3-vectors, $\mathbf{A} = (A^a)$ and $\mathbf{B} = (B^a)$. Then, by (8.3.1), $[A, B]$ corresponds to the vector cross-product, $\mathbf{A} \times \mathbf{B}$. With these in mind, we make the following introduction to the $SO(3)$ (or $SU(2)$ since $SO(3)$ and $SU(2)$ have identical Lie algebras) Yang–Mills–Higgs theory.

Let $\mathbf{A}_\mu = (A_\mu^a)$ $(\mu = 0, 1, 2, 3)$ and $\phi = (\phi^a)$ $(a = 1, 2, 3)$ be a gauge and matter Higgs fields, respectively, interacting through the action density

$$\mathcal{L} = -\frac{1}{4} \mathbf{F}^{\mu\nu} \cdot \mathbf{F}_{\mu\nu} + \frac{1}{2} D^\mu \phi \cdot D_\mu \phi - \frac{\lambda}{4}(|\phi|^2 - 1)^2, \tag{8.3.2}$$

where the field strength tensor $\mathbf{F}_{\mu\nu}$ is defined by

$$\mathbf{F}_{\mu\nu} = \partial_\mu \mathbf{A}_\nu - \partial_\nu \mathbf{A}_\mu - e\mathbf{A}_\mu \times \mathbf{A}_\nu, \tag{8.3.3}$$

and the gauge-covariant derivative D_μ is defined by

$$D_\mu \phi = \partial_\mu \phi - e\mathbf{A}_\mu \times \phi. \tag{8.3.4}$$

Based on consideration on interactions [162, 310], it is recognized that the electromagnetic field $F_{\mu\nu}$ is defined by the formula

$$F_{\mu\nu} = \frac{1}{|\phi|}\phi \cdot \mathbf{F}_{\mu\nu} - \frac{1}{e|\phi|^3}\phi \cdot (D_\mu\phi \times D_\nu\phi). \tag{8.3.5}$$

The equations of motion of (8.3.2) can be derived as

$$\begin{aligned} D^\nu \mathbf{F}_{\mu\nu} &= -e\phi \times D_\mu\phi, \\ D^\mu D_\mu\phi &= -\lambda\phi(|\phi|^2 - 1). \end{aligned} \tag{8.3.6}$$

We are interested in static solutions of (8.3.6). In general, this is a difficult problem. Here we can only consider radially symmetric solutions.

Set $r = |\mathbf{x}|$. Following Julia and Zee [162], the most general radially symmetric solutions of (8.3.6) are of the form

$$\begin{aligned} A_0^a &= \frac{x^a}{er^2}J(r), \\ A_i^a &= \epsilon_{abi}\frac{x^b}{er^2}(K(r) - 1), \\ \phi^a &= \frac{x^a}{er^2}H(r), \quad a, b, c = 1, 2, 3. \end{aligned} \tag{8.3.7}$$

Inserting (8.3.7) into (8.3.6) and using prime to denote differentiation with respect to the radial variable r, we have

$$\begin{aligned} r^2 J'' &= 2JK^2, \\ r^2 H'' &= 2HK^2 - \lambda r^2 H\left(1 - \frac{1}{e^2 r^2}H^2\right), \\ r^2 K'' &= K(K^2 - J^2 + H^2 - 1). \end{aligned} \tag{8.3.8}$$

We need to specify boundary conditions for these equations. First, we see from (8.3.7) and regularity requirement that H, J, K must satisfy

$$\lim_{r\to 0}\left(\frac{H(r)}{er}, \frac{J(r)}{er}, K(r)\right) = (0, 0, 1). \tag{8.3.9}$$

Secondly, since the Hamiltonian density of (8.3.2) takes the form

$$\begin{aligned} \mathcal{H} &= \mathbf{F}_{0i} \cdot \mathbf{F}_{0i} + D_0\phi \cdot D_0\phi - \mathcal{L} \\ &= \frac{1}{e^2 r^2}(K')^2 + \frac{1}{2}(u')^2 + \frac{1}{2}(v')^2 \\ &\quad + \frac{1}{2e^2 r^4}(K^2 - 1)^2 + \frac{1}{r^2}K^2(u^2 + v^2) + \frac{\lambda}{4}(u^2 - 1)^2, \end{aligned} \tag{8.3.10}$$

where $u = H/er$ and $v = J/er$. Finite energy condition,

$$E = \int_{\mathbb{R}^3} \mathcal{H} \, dx < \infty, \tag{8.3.11}$$

and the formula (8.3.10) imply that $u(r) \to 1$ and $K(r) \to 0$ as $r \to \infty$. Besides, it is seen from (8.3.10) that $v(r) \to$ some constant C_0 as $r \to \infty$. However, C_0 cannot be determined completely. We record these results as follows,

$$\lim_{r \to \infty} \left(\frac{H(r)}{er}, \frac{J(r)}{er}, K(r) \right) = (1, C_0, 0). \tag{8.3.12}$$

8.3.2   Explicit solutions in BPS limit

Inserting (8.3.7) into (8.3.5), we find the electric and magnetic fields, $\mathbf{E} = (E^i)$ and $\mathbf{B} = (B^i)$, as follows,

$$E^i = -F^{0i} = \frac{x^i}{er} \frac{d}{dr} \left(\frac{J}{r} \right), \tag{8.3.13}$$

$$B^i = -\frac{1}{2} \epsilon_{ijk} F^{jk} = \frac{x^i}{er^3}. \tag{8.3.14}$$

From (8.3.13) we see that, if $J = 0$, $\mathbf{E} = \mathbf{0}$ and there is no electric field. The magnetic charge g may be obtained through integrating (8.3.14),

$$g = \frac{1}{4\pi} \lim_{r \to \infty} \oint_{|\mathbf{x}|=r} \mathbf{B} \cdot d\mathbf{S} = \frac{1}{e}, \tag{8.3.15}$$

which is similar to the Dirac quantization formula except that electric charge is doubled. Note that, in the above calculation, we have used the fact that the ith component of $d\mathbf{S}$ on the sphere $|\mathbf{x}| = r$ can be expressed as

$$dS^i = \frac{x^i}{r} \cdot r^2 \cdot d\sigma,$$

where $d\sigma$ is the standard area element of the unit sphere. Moreover, the equations of motion (8.3.8) are simplified into the form

$$r^2 H'' = 2HK^2 - \lambda r^2 H \left(1 - \frac{1}{e^2 r^2} H^2 \right),$$

$$r^2 K'' = K(K^2 + H^2 - 1). \tag{8.3.16}$$

These equations cannot be solved explicitly for general $\lambda > 0$ but an existence theorem has been established by using functional analysis [272]. Here we present a family of explicit solutions at the BPS limit $\lambda = 0$ due to Bogomol'nyi [42] and Prasad–Sommerfield [252].

When $\lambda = 0$, the system (8.3.16) becomes

$$
\begin{aligned}
r^2 H'' &= 2HK^2, \\
r^2 K'' &= K(K^2 + H^2 - 1),
\end{aligned}
\tag{8.3.17}
$$

with the associated energy

$$
\begin{aligned}
E &= \int_{\mathbb{R}^3} \mathcal{H}\, dx \\
&= \frac{4\pi}{e^2} \int_0^\infty \left\{ (K')^2 + \frac{1}{2r^2}(rH' - H)^2 + \frac{1}{2r^2}(K^2 - 1)^2 + \frac{1}{r^2} K^2 H^2 \right\} dr.
\end{aligned}
\tag{8.3.18}
$$

It is clear that the system (8.3.17) is the Euler–Lagrange equations of (8.3.18). Besides, using the boundary conditions (8.3.9) and (8.3.12), we have

$$
\begin{aligned}
E &= \frac{4\pi}{e^2} \int_0^\infty \left\{ \left(K' + \frac{1}{r} KH \right)^2 + \frac{1}{2r^2} \left(rH' - H + (K^2 - 1) \right)^2 \right. \\
&\quad \left. + \frac{d}{dr} \left(\frac{H}{r} - \frac{K^2 H}{r} \right) \right\} dr \\
&\geq \frac{4\pi}{e}.
\end{aligned}
\tag{8.3.19}
$$

Hence, we have the energy lower bound, $E \geq 4\pi/e$, which is attained when (H, K) satisfies

$$
\begin{aligned}
rK' &= -KH, \\
rH' &= H - (K^2 - 1).
\end{aligned}
\tag{8.3.20}
$$

Of course, any solution of (8.3.20) also satisfies (8.3.17). It was Maison [207] who first showed that (8.3.17) and (8.3.20) are actually equivalent, which is a topic we will not get into here.

We now obtain a family of explicit solutions of (8.3.20) (hence (8.3.17)). Introduce a change of variables from (H, K) to (U, V),

$$
-H = 1 + rU, \quad K = rV.
\tag{8.3.21}
$$

Then (8.3.20) becomes $U' = V^2, V' = UV$, which implies that $U^2 - V^2 = $ constant. Thus, by virtue of (8.3.12) and (8.3.21), we have

$$
U^2 - V^2 = e^2, \quad r > 0.
\tag{8.3.22}
$$

Inserting (8.3.22) into $U' = V^2$ and using (8.3.9), i.e., $U(r) \sim -1/r$ for $r > 0$ small, we obtain an explicit solution of (8.3.20),

$$
\begin{aligned}
H(r) &= er \coth(er) - 1, \\
K(r) &= \frac{er}{\sinh(er)},
\end{aligned}
\tag{8.3.23}
$$

which gives rise to a smooth, minimum energy (mass) $E = 4\pi/e$, monopole ($J = 0$) solution of the equations of motion (8.3.6) at the BPS limit $\lambda = 0$ through the radially symmetric prescription (8.3.7).

We next present a continuous family of explicit dyon solutions. At the BPS limit, $\lambda = 0$, the system (8.3.8) is

$$
\begin{aligned}
r^2 J'' &= 2JK^2, \\
r^2 H'' &= 2HK^2, \\
r^2 K'' &= K(K^2 - J^2 + H^2 - 1),
\end{aligned}
\tag{8.3.24}
$$

which becomes (8.3.17) when we compress H, J through

$$
H \mapsto (\cosh\alpha)H, \quad J \mapsto (\sinh\alpha)H,
$$

where α is a constant. Therefore, using (8.3.23), we have

$$
\begin{aligned}
H(r) &= \cosh\alpha(er\coth(er) - 1), \\
J(r) &= \sinh\alpha(er\coth(er) - 1), \\
K(r) &= \frac{er}{\sinh(er)}.
\end{aligned}
\tag{8.3.25}
$$

Consequently, in view of (8.3.7), we have obtained a family of explicit dyon solutions of (8.3.6). Note that all the boundary conditions stated in (8.3.9) and (8.3.12) are fulfilled except one, namely,

$$
\lim_{r\to\infty} \frac{H(r)}{er} = \cosh\alpha \neq 1.
\tag{8.3.26}
$$

However, since $\lambda = 0$, (8.3.26) is of no harm to the finite energy condition (8.3.11).

To compute the total electric charge, we use (8.3.13). We have

$$
\begin{aligned}
q &= \frac{1}{4\pi} \lim_{r\to\infty} \oint_{|\mathbf{x}|=r} \mathbf{E} \cdot d\mathbf{S} \\
&= \frac{1}{e} \lim_{r\to\infty} (rJ'(r) - J(r)) = \frac{\sinh\alpha}{e}.
\end{aligned}
\tag{8.3.27}
$$

It is a comfort to see that the solution becomes electrically neutral, $q = 0$, when $\alpha = 0$ and (8.3.25) reduces to the monopole solution (8.3.23).

8.3.3   Existence result in general

In the last subsection, we presented a family of explicit dyon solutions due to Bogomol'nyi and Prasad–Sommerfield at the BPS limit $\lambda = 0$. In this subsection, we aim at obtaining dyons for the general nonintegrable case $\lambda > 0$ by nonlinear analysis.

Theorem 8.3.1. *For any $\lambda > 0$ and $0 < C_0 < 1$, the non-Abelian gauge field equations (8.3.6) have a finite-energy static solution $(\mathbf{A}_\mu, \phi)$ defined by (8.3.7) so that*

$$K(r) > 0, \quad er > H(r) > J(r) > 0, \quad r > 0,$$

$$\lim_{r \to 0} \frac{H(r)}{r} = 0, \quad \lim_{r \to 0} \frac{J(r)}{r} = 0, \quad \lim_{r \to 0} K(r) = 1,$$

$$\frac{H(r)}{r} = e + O(r^{-1} e^{-\beta(1-\varepsilon)r}),$$

$$\frac{J(r)}{r} = C_0 e + O(r^{-1}), \quad K(r) = O(e^{-\alpha(1-\varepsilon)r})$$

as $r \to \infty$, where $\alpha = \sqrt{1 - C_0^2}, \beta = \min\{\alpha, \sqrt{2\lambda}\}$, and $0 < \varepsilon < 1$ is arbitrary. This solution carries a positive electric charge q,

$$q = \frac{1}{4\pi} \lim_{r \to \infty} \oint_{|x|=r} \mathbf{E} \cdot d\mathbf{S} = \frac{2}{e} \int_0^\infty \frac{J(r)}{r} K^2(r) \, dr,$$

and a magnetic charge $g = 1/e$.

We will prove the theorem by solving a variational problem with an indefinite action functional.

It will be convenient to use the substitution

$$\frac{H}{er} \mapsto u, \quad \frac{J}{er} \mapsto v, \quad K \to K, \quad er \mapsto r, \quad \frac{\lambda}{e^2} \mapsto \lambda. \qquad (8.3.28)$$

Then (8.3.8) and the boundary conditions (8.3.9), (8.3.12) take the form

$$(r^2 u')' = 2uK^2 - \lambda r^2 u(1 - u^2), \qquad (8.3.29)$$

$$(r^2 v')' = 2vK^2, \qquad (8.3.30)$$

$$r^2 K'' = K(K^2 + r^2[u^2 - v^2] - 1), \qquad (8.3.31)$$

$$\lim_{r \to 0}(u(r), v(r), K(r)) = (0, 0, 1), \qquad (8.3.32)$$

$$\lim_{r \to \infty}(u(r), v(r), K(r)) = (1, C_0, 0). \qquad (8.3.33)$$

We are to find a solution of (8.3.29)–(8.3.33) so that the finite energy condition (8.3.11) holds, or in terms of the transformed variables defined in (8.3.28), there holds

$$E(K, u, v) = \frac{4\pi}{e} \int_0^\infty \left\{ (K')^2 + \frac{r^2}{2}(u')^2 + \frac{r^2}{2}(v')^2 \right.$$

$$\left. + \frac{1}{2r^2}(K^2 - 1)^2 + K^2(u^2 + v^2) + \frac{\lambda}{4}r^2(u^2 - 1)^2 \right\} dr$$

$$< \infty. \qquad (8.3.34)$$

One of the most prominent features of this problem is that (8.3.29)–(8.3.31) are not the Euler–Lagrange equations of the positive definite energy functional (8.3.34), but the indefinite action functional

$$F(K, u, v) = \frac{4\pi}{e} \int_0^\infty \left\{ (K')^2 + \frac{r^2}{2}(u')^2 - \frac{r^2}{2}(v')^2 \right.$$
$$\left. + \frac{1}{2r^2}(K^2 - 1)^2 + K^2(u^2 - v^2) + \frac{\lambda}{4}r^2(u^2 - 1)^2 \right\} dr. \quad (8.3.35)$$

The functions under consideration will always be absolutely continuous on any compact subinterval of $r > 0$ or $(0, \infty)$ and $C_0 \neq 0$. We define the admissible space $\mathcal{A}$ to be

$$\left\{ (K, u, v) \,\middle|\, E(K, u, v) < \infty, K(0) = 1, K(\infty) = 0, u(\infty) = 1, v(\infty) = C_0 \right\}.$$

Note that we need K to be continuous for all $r \geq 0$ and we exclude the boundary conditions stated in (8.3.32) for u, v at $r = 0$ for the moment. It will be hard to find a critical point of F directly in $\mathcal{A}$ and a further restriction is to be imposed: we assume that (K, u, v) satisfies the constraint

$$\int_0^\infty dr \, (r^2 v' w' + 2K^2 vw) = 0, \quad \forall w, \quad (8.3.36)$$

where w satisfies $w(\infty) = 0$ and $J(K, w) < \infty$ with

$$J(K, w) = \int_0^\infty dr \left\{ \frac{r^2}{2}(w')^2 + K^2 w^2 \right\}. \quad (8.3.37)$$

Our restricted class $\mathcal{C}$ is defined as

$$\mathcal{C} = \left\{ (K, u, v) \in \mathcal{A} \,\middle|\, (K, u, v) \text{ satisfies (8.3.36)} \right\}.$$

Lemma 8.3.2. *We have $\mathcal{C} \neq \emptyset$. Besides, for any $(K, u, v) \in \mathcal{C}$, $v(r)$ is a monotone (either nonincreasing or nondecreasing) function of $r > 0$ and*

$$\lim_{r \to 0} v(r) = 0.$$

Proof. We can rewrite the action functional (8.3.35) as

$$F(K, u, v) = I(K, u) - J(K, v). \quad (8.3.38)$$

Then any element $(K, u, v) \in \mathcal{C}$ may be obtained by first choosing suitable K, u such that $I(K, u) < \infty$ and then choosing a unique v such that $v(\infty) = C_0$ and v solves the problem

$$\min \left\{ J(K, w) \,\middle|\, w(\infty) = C_0 \right\}. \quad (8.3.39)$$

In fact, the estimate

$$
\begin{aligned}
|w(r) - C_0| &\leq \int_r^\infty |w'(s)| \, ds \\
&\leq r^{-\frac{1}{2}} \left(\int_r^\infty s^2 (w'(s))^2 \, ds \right)^{\frac{1}{2}} \\
&\leq \sqrt{2} r^{-\frac{1}{2}} J^{\frac{1}{2}}(K, w) \quad\quad (8.3.40)
\end{aligned}
$$

ensures that the boundary condition $w(\infty) = C_0$ is preserved in the weak limit of any minimizing sequence of (8.3.39) and the convexity of the functional (8.3.37) with respect to w ensures the existence and uniqueness of a solution of (8.3.39).

Let $(K, u, v) \in C$. We only need to show that $v(r)$ is nondecreasing if $C_0 > 0$.

Since v is a minimizer of (8.3.39), it is clear that $0 < v < C_0$ because v also satisfies (8.3.30). Besides, we also have

$$
\liminf_{r \to 0} r^2 |v'(r)| = 0. \quad\quad (8.3.41)
$$

In fact, if (8.3.41) is false, there are $\delta > 0$ and $\varepsilon_0 > 0$ such that

$$
r^2 |v'(r)| \geq \varepsilon_0, \quad 0 < r < \delta, \quad\quad (8.3.42)
$$

which contradicts the convergence of the integral

$$
\int_0^\infty r^2 (v')^2 \, dr.
$$

Using (8.3.41), we obtain after integrating (8.3.30) that

$$
r^2 v'(r) = \int_0^r K^2(s) v(s) \, ds \geq 0 \qu\quad (8.3.43)
$$

and the monotonicity of v follows. In particular, there is a number $v_0 \geq 0$ such that

$$
\lim_{r \to 0} v(r) = v_0. \qu\quad (8.3.44)
$$

If $v_0 > 0$, we can use (8.3.30), $K(0) = 1$, $r^2 v'(r) \to 0$ as $r \to 0$ (see (8.3.43)), and the L'Hopital's rule to get

$$
2v_0 = 2 \lim_{r \to 0} K^2(r) v(r) = \lim_{r \to 0} (r^2 v')' = \lim_{r \to 0} \frac{r^2 v'(r)}{r} = \lim_{r \to 0} r v'(r). \qu\quad (8.3.45)
$$

Hence, there is a $\delta > 0$ such that

$$
v'(r) \geq \frac{1}{r} v_0, \quad 0 < r < \delta. \qu\quad (8.3.46)
$$

Integrating (8.3.46), we obtain

$$v(r_2) - v(r_1) \geq v_0 \ln\left(\frac{r_2}{r_1}\right),$$

which contradicts the existence of limit, (8.3.44).

Lemma 8.3.3. *If $|C_0| < 1$, there are numbers $\varepsilon_1, \varepsilon_2 > 0$ such that*

$$F(K, u, v) \geq$$
$$\int_0^\infty dr \left\{ (K')^2 + \varepsilon_1 r^2 (u')^2 + \varepsilon_2 K^2 u^2 + \frac{(K^2 - 1)^2}{2r^2} + \frac{\lambda}{4} r^2 (u^2 - 1)^2 \right\}$$

(8.3.47)

for any $(K, u, v) \in \mathcal{C}$. In other words, F is partially coercive over the constrained admissible class $\mathcal{C}$.

Proof. For any $(K, u, v) \in \mathcal{C}$, $v_1 = C_0 u$ satisfies $v_1(\infty) = C_0$. Therefore

$$C_0^2 J(K, u) = J(K, C_0 u) = J(K, v_1) \geq J(K, v). \tag{8.3.48}$$

Inserting (8.3.48) into (8.3.38), we arrive at

$$F(K, u, v) \geq I(K, u) - C_0^2 J(K, u)$$
$$= \int_0^\infty dr \left\{ (K')^2 + \frac{1}{2}(1 - C_0^2) r^2 (u')^2 + (1 - C_0^2) K^2 u^2 \right.$$
$$\left. + \frac{1}{2r^2}(K^2 - 1)^2 + \frac{\lambda}{4} r^2 (u^2 - 1)^2 \right\},$$

which gives us the lower estimate (8.3.47). $\qquad\qquad\qquad\square$

The above preparation allows us to consider the following constrained variational problem,

$$\min\{ F(K, u, v) \mid (K, u, v) \in \mathcal{C} \}. \tag{8.3.49}$$

Lemma 8.3.4. *The problem (8.3.49) has a solution if $|C_0| < 1$.*

Proof. Let $\{(K_n, u_n, v_n)\}$ be a minimizing sequence of (8.3.49). From (8.3.35), we have

$$|K_n(r) - 1| \leq r^{\frac{1}{2}} \left(\int_0^r K_n^2(s)\, ds \right)^{\frac{1}{2}}$$
$$\leq C r^{\frac{1}{2}} F(K_n, u_n, v_n), \tag{8.3.50}$$

$$|u_n(r) - 1| \leq r^{-\frac{1}{2}} \left(\int_r^\infty s^2 u_n^2(s)\, ds \right)^{\frac{1}{2}}$$
$$\leq C r^{-\frac{1}{2}} F(K_n, u_n, v_n). \tag{8.3.51}$$

Hence, $K_n(r) \to 1$ and $u_n(r) \to 1$ uniformly as $r \to 0$ and $r \to \infty$, respectively, for $n = 1, 2, \cdots$. In particular, $\{K_n\}$ and $\{u_n\}$ are both bounded sequences in $W^{1,2}(r_1, r_2)$ for any $0 < r_1 < r_2 < \infty$. By (8.3.48) and Lemma 8.3.3, we also have

$$J(K_n, v_n) \le C_0^2 J(K_n, u_n) \le CF(K_n, u_n, v_n),$$

which shows that $\{J(K_n, v_n)\}$ is bounded too.

Consider the Hilbert space $(X, (\cdot, \cdot))$ where the functions in X are all continuously defined on $r \ge 0$ and vanish at $r = 0$ and the inner product $(\cdot, \cdot)$ is defined by

$$(f_1, f_2) = \int_0^\infty f_1'(r) f_2'(r) \, dr, \quad f_1, f_2 \in X.$$

Since $\{K_n\}$ is bounded in $(X, (\cdot, \cdot))$, we may assume without loss of generality that $\{K_n\}$ has a weak limit, say, K, in the same space,

$$\int_0^\infty dr (K_n' \tilde{K}') \to \int_0^\infty dr (K' \tilde{K}') \quad \text{as } n \to \infty, \quad \forall \tilde{K} \in X. \tag{8.3.52}$$

Similarly, for the Hilbert space $(Y, (\cdot, \cdot))$ where the functions in Y are all continuously defined in $r > 0$ and vanish at infinity and the inner product $(\cdot, \cdot)$ is defined by

$$(f_1, f_2) = \int_0^\infty r^2 f_1'(r) f_2'(r) \, dr, \quad f_1, f_2 \in Y.$$

Since the sequences $\{u_n - 1\}$ and $\{v_n - C_0\}$ are bounded in $(Y, (\cdot, \cdot))$, then there are functions u, v with $u(\infty) = 1$, $v(\infty) = C_0$, and $u - 1, v - C_0 \in (Y, (\cdot, \cdot))$, such that

$$\int_0^\infty r^2 w_n' \tilde{w}' \, dr \to \int_0^\infty r^2 w' \tilde{w}' \, dr \quad \text{as } n \to \infty, \quad \forall \tilde{w} \in Y \tag{8.3.53}$$

for $w_n = u_n, w = u$ or $w_n = v_n, w = v$.

We now show that the constraint (8.3.36) will be preserved in the weak limit as $n \to \infty$.

First, the uniform bound (8.3.50) implies that there is a number $\delta > 0$ so that $K_n(r) \ge \frac{1}{2}$ for all $r \in (0, \delta)$. Thus

$$\int_0^\delta v_n^2 \, dr \le 4 \int_0^\infty K_n^2 v_n^2 \, dr \le CJ(K_n, v_n),$$

which leads to the weak convergence $v_n \rightharpoonup v$ in $L^2(0, \delta)$. Consequently, for the w in (8.3.36), we have

$$\left| \int_0^\delta (K_n^2 v_n - K^2 v) w \, dr \right|$$

$$\le \left| \int_0^\delta (K_n^2 v_n - K^2 v_n) w \, dr \right| + \left| \int_0^\delta K^2 (v_n - v) w \, dr \right| \to 0$$

because when we set

$$I(a, b; n) = \int_a^b (K_n^2 v_n - K^2 v_n) w \, dr,$$

there holds

$$|I(0, \delta; n)| \leq |I(0, \delta_1; n)| + |I(\delta_1, \delta; n)|$$

with

$$|I(0, \delta_1; n)| \leq \max_{0 < r < \delta_1} \{|K_n^2(r) - 1| + |K^2(r) - 1|\} \|v_n\|_{L^2(0,\delta)} \|w\|_{L^2(0,\delta)},$$

which, of course, can be made arbitrarily small by taking $\delta_1 > 0$ small, and $I(\delta_1, \delta; n) \to 0$ as $n \to \infty$. Besides, since $\{J(K_n, v_n)\}$ is bounded and $v_n(r) \to C_0$ uniformly and $w(r) = O(r^{-1/2})$, as $r \to \infty$ (see (8.3.40)), there holds

$$\left| \int_r^\infty K_n^2 v_n w \, dr \right| \leq C r^{-\frac{1}{2}} \int_0^\infty K_n^2 v_n^2 \, dr,$$

which vanishes at $r = \infty$ uniformly fast. Consequently, we arrive at

$$\lim_{n \to \infty} \int_0^\infty (r^2 v_n' w' + 2 K_n^2 v_n w) \, dr = \int_0^\infty (r^2 v' w' + 2 K^2 v w) \, dr, \quad (8.3.54)$$

which shows that the constraint (8.3.36) is indeed preserved.

We next show that (8.3.49) has a solution.

We claim that

$$r v_n' \to r v' \text{ strongly in } L^2(0, \infty) \text{ as } n \to \infty. \quad (8.3.55)$$

In fact, since $\{v_n\}$ satisfies (8.3.36), i.e.,

$$\int_0^\infty (r^2 v_n' w + 2 K_n^2 v_n w) \, dr = 0, \quad n = 1, 2, \cdots, \quad (8.3.56)$$

we get from setting $w = v - v_n$ in (8.3.36) and (8.3.56) and subtracting

$$\int_0^\infty r^2 (v_n' - v')^2 \, dr = 2 \int_0^\infty (K_n^2 v_n - K^2 v)(v_n - v) \, dr,$$

which goes to zero as $n \to \infty$ because $K_n(r) - K(r)$ and $v_n(r) - v(r)$ go to zero uniformly as $r \to 0$ and $r \to \infty$, respectively. Furthermore, it is clear that

$$\lim_{n \to \infty} \int_0^\infty K_n^2 v_n^2 \, dr = \int_0^\infty K^2 v^2 \, dr. \quad (8.3.57)$$

Inserting (8.3.55) and (8.3.57), we conclude that

$$F(K, u, v) \leq \liminf_{n \to \infty} F(K_n, u_n, v_n).$$

Therefore (8.3.49) is solved.

Lemma 8.3.5. *The solution (K, u, v) of the problem (8.3.49) obtained in Lemma 8.3.4 satisfies the equations (8.3.29)–(8.3.31) and the boundary conditions*

$$K(0) = 1, \quad K(\infty) = 0, \quad u(\infty) = 1, \quad v(\infty) = C_0. \tag{8.3.58}$$

Proof. We have already established (8.3.58) in the proof of the existence of a minimizer (K, u, v) of (8.3.49). We now show that (8.3.29)–(8.3.31) are all fulfilled. Since (8.3.36) is equivalent to (8.3.30), we are left with (8.3.29) and (8.3.31).

Let $\tilde{K} \in C_0^1(0, \infty)$ (set of functions with compact supports). For any $t \in \mathbb{R}$, there is a unique corresponding function v_t such that $(K + t\tilde{K}, u, v_t) \in C$ and that v_t smoothly depends on t. Set

$$v_t = v + \tilde{v}_t, \quad \tilde{w} = \left(\frac{\mathrm{d}}{\mathrm{d}t} \tilde{v}_t \right)_{t=0}.$$

Since v_t satisfies (8.3.36) when K is replaced by $K + t\tilde{K}$, we have with $w = \tilde{v}_t$ that

$$\int_0^\infty (r^2 (\tilde{v}_t')^2 + 2K^2 \tilde{v}_t^2) \, \mathrm{d}r = -2t \int_0^\infty (2K\tilde{K} + t\tilde{K}^2) v_t \tilde{v}_t \, \mathrm{d}r$$

$$\leq 2t \left(2t \int_0^\infty \tilde{K}^2 v_t^2 \, \mathrm{d}r + \frac{1}{2t} \int_0^\infty K^2 \tilde{v}_t^2 \, \mathrm{d}r \right) + 2t^2 \int_0^\infty \tilde{K}^2 |v_t| \, |\tilde{v}_t| \, \mathrm{d}r,$$

$$t \neq 0. \tag{8.3.59}$$

Applying the fact that both v and v_t are bounded functions, $0 \leq v, v_t \leq C_0$, we get

$$|\tilde{v}_t| \leq 2C_0. \tag{8.3.60}$$

Inserting (8.3.60) into (8.3.59), we arrive at

$$\int_0^\infty \left\{ r^2 \left(\frac{\tilde{v}_t'}{t} \right)^2 + K^2 \left(\frac{\tilde{v}_t}{t} \right)^2 \right\} \mathrm{d}r \leq C, \tag{8.3.61}$$

where $C > 0$ depends only on $\tilde{K}$ but is independent of t. As before (see the derivation of (8.3.51)), we have

$$\left| \frac{\tilde{v}_t(r)}{t} \right| \leq C^{\frac{1}{2}} r^{-\frac{1}{2}}. \tag{8.3.62}$$

Letting $t \to 0$ in (8.3.61) and (8.3.62), we see that $J(K, \tilde{w}) < \infty$ and $\tilde{w}(\infty) = 0$. Hence, $\tilde{w}$ can be used as a test function in (8.3.36),

$$\int_0^\infty (r^2 v' \tilde{w}' + 2K^2 v \tilde{w}) \, \mathrm{d}r = 0. \tag{8.3.63}$$

With the above preparation, we are ready to examine the equation (8.3.31). From

$$\left(\frac{\mathrm{d}}{\mathrm{d}t}F(K + t\tilde{K}, u, v_t)\right)\bigg|_{t=0} = 0,$$

we have in view of (8.3.63) that

$$\int_0^\infty \left\{K'\tilde{K}' + \frac{1}{r^2}(K^2 - 1)K\tilde{K} + (u^2 - v^2)K\tilde{K}\right\} \mathrm{d}r$$

$$= \frac{1}{2}\int_0^\infty (v'\tilde{w}' + 2K^2 v\tilde{w})\,\mathrm{d}r = 0, \quad \forall \tilde{K} \in C_0^1(0, \infty),$$

which is the weak form of (8.3.31). Using elliptic regularity theory, we see that (8.3.31) is verified.

It is direct to show that (8.3.29) holds because its weak form,

$$\int_0^\infty \{r^2 u'\tilde{u}' + 2K^2 u\tilde{u} + \lambda r^2(u^2 - 1)u\tilde{u}\}\,\mathrm{d}r = 0, \quad \forall \tilde{u} \in C_0^1(0, \infty),$$

is trivially valid, due to the independence of (8.3.36) from K, and is equivalent to (8.3.29).

The proof is thus complete.

Lemma 8.3.6. *Let (K, u, v) be the solution of (8.3.29)–(8.3.31) obtained in Lemma 8.3.5, with $C_0 > 0$. Then*

$$K(r) > 0, \quad 1 > u(r) > v(r) > 0, \quad \forall r > 0.$$

Proof. We have already shown $v > 0$. Since F is even in K and u, we have $K \geq 0$ and $u \geq 0$ since otherwise we may always achieve a lower value for the action functional F by using the substitutions $K \mapsto |K|$ and $u \mapsto |u|$. It remains to establish the strict inequalities.

Suppose otherwise that there is an $r_0 > 0$ such that $K(r_0) = 0$. Since $K \geq 0$, we have $K'(r_0) = 0$. Using the uniqueness theorem for the initial value problem of ordinary differential equations, we arrive at $K(r) = 0$ for all $r > 0$, which contradicts the property that $K(r) \to 1$ as $r \to 0$. Hence, $K > 0$.

The same argument is applicable to u to get $u > 0$. On the other hand, minimization of F over the constrained class $\mathcal{C}$ gives us the property $u \leq 1$. If there is a point $r_0 > 0$ such that $u(r_0) = 1$, then this point is a maximum point. Inserting this information into (8.3.29), we see a contradiction. Thus $u(r) < 1$ for all $r > 0$.

Finally, from (8.3.29) and (8.3.30), we have

$$(r^2(u - v)')' = 2K^2(u - v) + \lambda r^2 u(u^2 - 1), \quad r > 0. \tag{8.3.64}$$

Since $0 < u < 1$, there is a sequence $\{r_j\}$, $r_j \to 0$ as $j \to \infty$, such that the limit

$$\lim_{j \to \infty} u(r_j)$$

exists and is a nonnegative number, say, u_0. In particular, $u(r_j) - v(r_j) \to u_0 \geq 0$ as $j \to \infty$. On the other hand, we already know that $u(r) - v(r) \to 1 - C_0 > 0$ as $r \to \infty$. Thus, if there is a point $r_0 > 0$ such that $u(r_0) - v(r_0) \leq 0$, then this point can be assumed to be a local minimum point of $u - v$, which contradicts (8.3.64).

Lemma 8.3.7. *Let (K, u, v) be the solution of (8.3.29)–(8.3.31) obtained in Lemma 8.3.5. Then*

$$\lim_{r \to 0} u(r) = 0, \quad \lim_{r \to 0} v(r) = 0.$$

Proof. The limit for v was already established in Lemma 8.3.2. The proof of the limit for u is almost identical to that for v because the first term dominates over the second term on the right-hand side of (8.3.29) when $r > 0$ is small.

We next establish the decay estimates of the configuration functions, K, u, v, at infinity.

Lemma 8.3.8. *Let (K, u, v) be the solution of (8.3.29)–(8.3.33), with $1 > C_0 > 0$, obtained in the previous lemmas, and*

$$\alpha = \sqrt{1 - C_0^2}, \quad \beta = \min\{2\alpha, \sqrt{2\lambda}\}.$$

Then there hold the asymptotic decay estimates as $r \to \infty$,

$$K = O(e^{-\alpha(1-\varepsilon)r}), \quad u = 1 + O(r^{-1}e^{-\beta(1-\varepsilon)r}), \quad v = C_0 + O(r^{-1}),$$

where $\varepsilon \in (0, 1)$ is arbitrary.

Proof. Introduce a comparison function η,

$$\eta(r) = Ce^{-\alpha(1-\varepsilon)r}, \quad r > 0. \tag{8.3.65}$$

From (8.3.31) and the obtained asymptotic behavior of K, u, v, we see that there is a sufficiently large $r_\varepsilon > 0$ so that

$$(K - \eta)'' \geq \alpha^2(1 - \varepsilon)^2(K - \eta), \quad r > r_\varepsilon.$$

Choose the coefficient C in (8.3.65) large enough to make $K(r_\varepsilon) - \eta(r_\varepsilon) \leq 0$. Since $K - \eta$ vanishes at infinity, applying the maximum principle in the above inequality, we derive the bound $K < \eta$ for $r > r_\varepsilon$ as desired.

Set $U = r(u - 1)$. Then (8.3.29) and the decay estimate for K give us

$$U'' = \lambda u(u+1)U + \frac{2u}{r}K^2$$

$$\leq \beta^2\left(1 - \frac{\varepsilon}{2}\right)^2 U + C_1 e^{-2\alpha(1-\varepsilon)r}. \qquad (8.3.66)$$

Replacing α by β in (8.3.65) and using (8.3.66), we see that there exists an $r_\varepsilon > 0$ such that

$$(U + \eta)'' \leq \beta^2\left(1 - \frac{\varepsilon}{2}\right)^2 (U + \eta)$$

$$+ \beta^2\left\{(1-\varepsilon)^2 - \left(1 - \frac{\varepsilon}{2}\right)^2\right\}\eta + C_1 e^{-2\alpha(1-\varepsilon)r}$$

$$\leq \beta^2\left(1 - \frac{\varepsilon}{2}\right)^2 (U + \eta), \quad r > r_\varepsilon. \qquad (8.3.67)$$

As before, we can achieve $U(r_\varepsilon) + \eta(r_\varepsilon) > 0$ when the coefficient $C > 0$ in (8.3.65) (with $\alpha = \beta$) is chosen sufficiently large. On the other hand, finite energy implies the existence of a sequence $\{r_j\}$, $r_j \to \infty$ as $j \to \infty$, so that

$$U(r_j) = r_j(u(r_j) - 1) \to 0 \quad \text{as } j \to \infty. \qquad (8.3.68)$$

Using (8.3.68) as a boundary condition for U at infinity and applying the maximum principle in (8.3.67), we have $U + \eta > 0$ for all $r > r_\varepsilon$. In other words, we have obtained the estimate $0 < r(1 - u) < \eta$.

Finally, since the function v satisfies

$$(r[v - C_0])'' = \frac{2v}{r}K^2 \qquad (8.3.69)$$

and finite energy implies

$$\liminf_{r\to\infty} rv'(r) = 0, \qquad (8.3.70)$$

we may integrate (8.3.69) and use (8.3.70) as a boundary condition to get

$$(r[v(r) - C_0])' = -\int_r^\infty \frac{2v(s)}{s}K^2(s)\,ds.$$

Inserting the decay estimate for K in the above integral, we arrive at

$$(r[v(r) - C_0])' = O(e^{-2\alpha(1-\varepsilon)r}) \quad \text{as } r \to \infty. \qquad (8.3.71)$$

In particular, there exists a finite limit,

$$\lim_{r\to\infty} r(v(r) - C_0) \equiv V_0,$$

which implies the asymptotic estimate $v(r) = C_0 + O(r^{-1})$.

We may use (8.3.13) and (8.3.8) to calculate the electric charge of the solution in the original variables,

$$
\begin{aligned}
q &= \frac{1}{4\pi} \lim_{r\to\infty} \int_{|x|=r} \mathbf{E} \cdot d\mathbf{S} \\
&= \frac{1}{4\pi} \lim_{r\to\infty} \int_{|x|<r} \nabla \cdot \mathbf{E} \, dx \\
&= \frac{1}{4\pi e} \int_{\mathbb{R}^3} \partial_i \left\{ \frac{x_i}{r} \frac{d}{dr} \left(\frac{J(r)}{r} \right) \right\} dx \\
&= \frac{1}{e} \int_0^\infty \left(r^2 \left[\frac{J}{r} \right]' \right)' dr \\
&= \frac{1}{e} \int_0^\infty r J''(r) \, dr \\
&= \frac{2}{e} \int_0^\infty \frac{J(r)}{r} K^2(r) \, dr.
\end{aligned}
$$

Summarizing the above results, we see that Theorem 8.3.1 is established.

8.4   Weinberg–Salam Electroweak Dyons

We start from the bosonic Lagrangian action density of the Weinberg–Salam electroweak model

$$
\mathcal{L} = -\frac{1}{4} \mathbf{F}_{\mu\nu} \cdot \mathbf{F}^{\mu\nu} - \frac{1}{4} H_{\mu\nu} H^{\mu\nu} + (\hat{D}_\mu \phi) \cdot (\hat{D}^\mu \phi)^\dagger - \frac{\lambda}{2} \left(|\phi|^2 - \frac{\mu^2}{\lambda} \right)^2,
$$

where ϕ now is a Higgs complex doublet lying in the fundamental representation space of $SU(2)$, $\mathbf{F}_{\mu\nu}$ and $H_{\mu\nu}$ are the gauge fields of $SU(2)$ and $U(1)$ with the potential $\mathbf{A}_\mu$ and B_μ and the corresponding coupling constants g and g', respectively, for the generators of $SU(2)$ we use the conventional Pauli matrices τ^a ($a = 1, 2, 3$), $\mathbf{A}_\mu = (A_\mu^a)$, and the gauge-covariant derivatives are defined by the expressions

$$
\begin{aligned}
\hat{D}_\mu \Phi &= \left(\partial_\mu - i\frac{g}{2} \tau^a A_\mu^a - i\frac{g'}{2} B_\mu \right) \phi \\
&= D_\mu \Phi - i\frac{g'}{2} B_\mu \Phi.
\end{aligned}
$$

Therefore, within the above framework, the Euler–Lagrangian equations of $\mathcal{L}$ are

$$
\hat{D}_\mu \hat{D}_\mu \phi = \lambda \left(|\phi|^2 - \frac{\mu^2}{\lambda} \right) \phi,
$$

$$D_\mu \mathbf{F}_{\mu\nu} = i\frac{g}{2}(\phi^\dagger \tau [\hat{D}_\nu \phi] - [\hat{D}_\nu \phi]^\dagger \tau \phi),$$

$$\partial_\mu H_{\mu\nu} = i\frac{g'}{2}(\phi^\dagger [\hat{D}_\nu \phi] - [\hat{D}_\nu \phi]^\dagger \phi). \tag{8.4.1}$$

In order to pursue a static radially symmetric dyon solution, we follow Cho and Maison [83] to use the following general ansatz,

$$\phi = \frac{1}{\sqrt{2}}\rho\xi \quad (\rho^2 = 2|\phi|^2, \ \xi^\dagger\xi = 1),$$

$$\tilde{\phi} = \xi^\dagger \tau \xi,$$

$$A_\mu = \tilde{\phi} \cdot \mathbf{A}_\mu,$$

$$\rho = \rho(r),$$

$$\xi = i\begin{pmatrix} \sin(\theta/2)e^{-i\varphi} \\ -\cos(\theta/2) \end{pmatrix},$$

$$\mathbf{A}_\mu = \frac{1}{g}A(r)\partial_\mu t\tilde{\phi} + \frac{1}{g}(f(r)-1)\tilde{\phi} \times \partial_\mu\tilde{\phi},$$

$$B_\mu = -\frac{1}{g'}B(r)\partial_\mu t - \frac{1}{g'}(1-\cos\theta)\partial_\mu\varphi, \tag{8.4.2}$$

where (t, r, θ, φ) are the spherical coordinates. Note also that $\tilde{\phi}$ is now a real 3-vector.

The energy structure of the Weinberg–Salam dyon found by Cho and Maison [83] is that its energy is written in the decomposed form

$$E = E_0 + E_1,$$

where E_0 is an infinite part given by the divergent integral

$$E_0 = \frac{2\pi}{(g')^2}\int_0^\infty \frac{dr}{r^2}$$

and E_1 is a finite part and radially symmetric dyon solutions are those that make the 'truncated energy', E_1, finite. The existence result for the Weinberg–Salam dyon is stated as follows.

Theorem 8.4.1. *The Weinberg–Salam dyon equations (8.4.1) have a family of static finite E_1 energy smooth solutions which satisfy the radial symmetry properties given in (8.4.2). The obtained solution configuration functions f, ρ, A, B have the properties that*

$$f, \rho, A, B > 0, \quad B > A,$$

$$B(r) \quad and \quad \left(\frac{A}{g^2} + \frac{B}{g'^2}\right)(r) \quad increase \ for \ r > 0,$$

and are labelled by a real parameter $a_0 > 0$ lying in the range

$$a_0^2\left(\frac{1}{g^2} + \frac{1}{g'^2}\right) < 1, \quad \frac{a_0^2}{g^2} < \frac{1}{4}.$$

Furthermore, there hold the regular boundary conditions

$$\begin{aligned} f(0) &= 1, \quad \rho(0) = 0, \quad A(0) = 0, \quad B(0) = b_0, \\ f(\infty) &= 0, \quad \rho(\infty) = \rho_0, \quad A(\infty) = B(\infty) = \rho_0 a_0, \end{aligned}$$

where b_0 is a certain number satisfying

$$0 \le b_0 \le \rho_0 a_0\left(\frac{1}{g^2} + \frac{1}{g'^2}\right)$$

(hence, $b_0 \le \rho_0 a_0^{-1}$) and $\rho_0 = \mu\sqrt{2/\lambda}$. In fact these boundary conditions are realized at the asymptotic rates

$$\begin{aligned} f &= 1 + O(r^{2(1-\varepsilon)}), \quad \rho = O(r^{\delta(1-\varepsilon)}), \\ A &= O(r^{(1-\varepsilon)}), \quad B = b_0 + O(r^{(1+\sqrt{3})(1-\varepsilon)}) \quad \text{near } r = 0; \\ f &= O(e^{-\kappa(1-\varepsilon)r}), \quad \rho = \rho_0 + O(r^{-1}e^{-\sqrt{2}\mu_0(1-\varepsilon)r}), \\ A &= \rho_0 a_0 + O(r^{-1}), \quad B - A = O(r^{-1}e^{-\nu_0(1-\varepsilon)r}) \quad \text{near } r = \infty, \end{aligned}$$

where the decay exponents are defined by the expressions

$$\delta = \frac{-1 + \sqrt{3}}{2}, \quad \varepsilon \text{ is an arbitrary number lying in the interval } (0,1),$$

$$\mu_0 = \min\left\{\mu, \rho_0\sqrt{\frac{g^2}{2} - 2a_0^2}, \frac{\rho_0}{2}\sqrt{\frac{g^2 + g'^2}{2}}\right\},$$

$$\nu_0 = \min\left\{2\rho_0\sqrt{\frac{g^2}{4} - a_0^2}, \frac{\rho_0}{2}\sqrt{g^2 + g'^2}\right\}.$$

Besides, in the unitary gauge, the electromagnetic field carries an electric charge $q_e > 0$ and a magnetic charge $q_m > 0$ which can be calculated by the formulas

$$q_e = \frac{2}{e}\sin^2\theta_W \int_0^\infty f^2 A \, dr,$$

$$q_m = \frac{1}{e},$$

where θ_W is the classical Weinberg mixing angle and $e = g\sin\theta_W$ is the electric charge of the positron. Moreover, the Z boson does not carry any charges.

This existence theorem confirms the predictions and numerical results obtained by Cho and Maison in [83] and the solutions may also be called the Cho–Maison dyons.

8.5 Radial Equations and Action Principle

Within the radially symmetric ansatz (8.4.2), the Weinberg–Salam equations (8.4.1) become

$$f'' = \frac{1}{r^2}(f^2 - 1)f + \left(\frac{g^2}{4}\rho^2 - A^2\right)f,$$

$$(r^2\rho')' = \frac{1}{2}f^2\rho - \frac{1}{4}r^2(A - B)^2\rho + \lambda r^2\left(\frac{\rho^2}{2} - \frac{\mu^2}{\lambda}\right)\rho,$$

$$(r^2A')' = 2f^2A + \frac{g^2}{4}r^2\rho^2(A - B),$$

$$(r^2B')' = \frac{g'^2}{4}r^2\rho^2(B - A), \tag{8.5.1}$$

where $f' = df/dr$ but throughout this chapter we fix g' to denote as before a positive physical coupling constant which should not cause any confusion.

With $\rho_0 = \mu\sqrt{2/\lambda}$ and the substitution

$$r \mapsto \frac{2}{g\rho_0}r, \quad f \mapsto f, \quad \rho \mapsto \rho_0\rho, \quad A \mapsto \rho_0A, \quad B \mapsto \rho_0B,$$

the energy E_1 (up to a positive numerical factor, $2\sqrt{2}\mu\pi/g\sqrt{\lambda}$) takes the form

$$E_1(f, \rho, A, B)$$
$$= \int_0^\infty dr \left\{ (f')^2 + 2r^2(\rho')^2 + \frac{2}{g^2}r^2(A')^2 + \frac{2}{g'^2}r^2(B')^2 + \frac{(f^2 - 1)^2}{2r^2} \right.$$
$$\left. + f^2\rho^2 + \frac{4}{g^2}f^2A^2 + \frac{2}{g^2}r^2\rho^2(A - B)^2 + \frac{2}{g^2}\lambda r^2(\rho^2 - 1)^2 \right\} \tag{8.5.2}$$

and the governing equations (8.5.1) are

$$f'' = \frac{1}{r^2}(f^2 - 1)f + \left(\rho^2 - \frac{4}{g^2}A^2\right)f,$$

$$(r^2\rho')' = \frac{1}{2}f^2\rho - \frac{1}{g^2}r^2(A - B)^2\rho + \lambda r^2\frac{2}{g^2}(\rho^2 - 1)\rho,$$

$$(r^2A')' = 2f^2A + r^2\rho^2(A - B),$$

$$(r^2B')' = \frac{g'^2}{g^2}r^2\rho^2(B - A). \tag{8.5.3}$$

Regularity and finite-energy requirements imply that one needs to subject (8.5.3) to the boundary conditions

$$f(0) = 1, \tag{8.5.4}$$

$$\rho(0) = 0, \quad A(0) = 0, \quad B(0) = b_0, \tag{8.5.5}$$

$$f(\infty) = 0, \quad \rho(\infty) = 1, \quad A(\infty) = B(\infty) = a_0. \tag{8.5.6}$$

It will be seen that the parameters a_0 and b_0 play rather different roles. In fact a_0 may be prescribed, but b_0 is, so far, mysterious. Besides, (8.5.4) and (8.5.6) are direct consequences of the structure of the energy (8.5.2), but (8.5.5) is more delicate. These will become clearer later. For definiteness we will always assume that $a_0 > 0$.

The Weinberg–Salam dyon is a solution of the nonlinear system (8.5.3) subject to the boundary conditions (8.5.4)–(8.5.6) so that it makes the energy E_1 defined in (8.5.2) finite. To find such a solution, we will take a variational approach. The difficult feature here is that the system (8.5.3) is *not* the Euler–Lagrange equations of the positive definite energy (8.5.2) but is the Euler–Lagrange equations of the *indefinite* action functional

$$
F(f, \rho, A, B)
$$
$$
= \int_0^\infty dr \left\{ (f')^2 + 2r^2(\rho')^2 - \frac{2}{g^2} r^2 (A')^2 - \frac{2}{g'^2} r^2 (B')^2 + \frac{(f^2 - 1)^2}{2r^2} \right.
$$
$$
\left. + f^2 \rho^2 - \frac{4}{g^2} f^2 A^2 - \frac{2}{g^2} r^2 \rho^2 (A - B)^2 + \frac{2}{g^2} \lambda r^2 (\rho^2 - 1)^2 \right\}. \quad (8.5.7)
$$

Nevertheless, in the next two sections, we will use a constrained minimization method to find a solution of (8.5.3)–(8.5.6) as a critical point of (8.5.7). We will see that the difficulties arising from the negative terms in (8.5.7) may be circumvented by a suitable choice of constraint conditions.

8.6   Constrained Variational Method

In this section, we extend the constrained variational method for the proof of the existence of Julia–Zee dyons to establish our existence theorem for the Weinberg–Salam or Cho–Maison dyons. Since the solution is rather involved, we begin by a brief sketch of the variational formulation and a description of the method.

8.6.1   Admissible space

In the sequel we work on the set of functions depending on the radial variable r lying in the half-line $r \geq 0$. The form of the energy (8.5.2) or the action (8.5.7) says that the most general assumption on these functions is that they are absolutely continuous on any compact subinterval of $(0, \infty)$. This assumption will always be observed throughout.

The most natural admissible space $\mathcal{A}$ for our problem should be defined by

$$
\mathcal{A} = \left\{ (f, \rho, A, B) \,\middle|\, E(f, \rho, A, B) < \infty \text{ and } (8.5.4), (8.5.6) \text{ hold} \right\}.
$$

Note that we deliberately leave out the boundary condition (8.5.5) in the admissible space because it cannot be recovered simply from a finite energy requirement. On the other hand, the conditions (8.5.4) and (8.5.6) are direct consequences of our minimization procedure and may be used to derive sufficiently special properties of a solution so that (8.5.5) follows somewhat indirectly.

Our goal in this section is to find a critical point of the functional (8.5.7) in the admissible space $\mathcal{A}$. We will achieve this by going through the following steps.

(i) Devise a suitable set of constraints to restrict the consideration of (8.5.7) over a smaller admissible space, say, $\mathcal{C}$.

(ii) Find the correct conditions under which (8.5.7) becomes coercive on $\mathcal{C}$.

(iii) Minimize (8.5.7) on $\mathcal{C}$ and show that $\mathcal{C}$ can be preserved in the limit.

(iv) Show that the obtained minimizer is indeed a critical point of (8.5.7) over original admissible space $\mathcal{A}$ and thus is a solution of (8.5.3) subject to (8.5.4) and (8.5.6).

From the above steps we see that step (i), namely, a suitable choice of $\mathcal{C}$, affects all other steps and is the key to the entire analysis.

To motivate our choice of $\mathcal{C}$, recall that we need to make the indefinite functional (8.5.7) coercive. For this purpose, the only way to control the negative terms in (8.5.7) is to 'freeze' the unknowns A and B. Unfortunately, the pair A, B cannot be frozen arbitrarily because we are looking for a solution of (8.5.3) eventually. Hence, naturally, we have to require A, B satisfy the last two equations in (8.5.3) in a suitable sense for given f, ρ (which will lead to the weak form of the last two equations in (8.5.3). However, we will see later that these two equations are not enough for us to use as the full set of the constraints and an additional equation will have to be devised). Thus we see some clues for the definition of the constrained admissible space $\mathcal{C}$. Furthermore, the best convergence result we may expect to derive from the functional (8.5.7) is some kind of certain weak convergence for the minimizing sequence. Therefore we are led to assume, with each fixed pair f, ρ, that (A, B) is a critical point of the functional

$$J(A, B) = \int_0^\infty dr \left\{ \frac{2}{g^2} r^2 (A')^2 + \frac{2}{g'^2} r^2 (B')^2 + \frac{4}{g^2} f^2 A^2 + \frac{2}{g^2} r^2 \rho^2 (A - B)^2 \right\},$$
(8.6.1)

which is made from all the negative terms in (8.5.7). In fact, it is easily seen that the set all pairs (A, B) so that $J(A, B) < \infty$ and $A(\infty) = B(\infty) = a_0$ is an affine linear space. Besides, since J is strictly convex with respect to

(A, B), for each given pair (f, ρ), J can at most have one critical point. Of course, if (A, B) is a critical point, then there hold the necessary conditions

$$\int_0^\infty dr\{r^2 A' \tilde{A}' + 2f^2 A\tilde{A} + r^2\rho^2(A - B)\tilde{A}\} = 0,$$

$$\int_0^\infty dr\left\{\frac{1}{g'^2}r^2 B'\tilde{B}' - \frac{1}{g^2}r^2\rho^2(A - B)\tilde{B}\right\} = 0,$$

$$\int_0^\infty dr\left\{r^2 A'\tilde{A}_1' + \frac{g^2}{g'^2}r^2 B'\tilde{B}_1' + 2f^2 A\tilde{A}_1 + r^2\rho^2(A - B)(\tilde{A}_1 - \tilde{B}_1)\right\}$$
$$= 0, \tag{8.6.2}$$

for all $\tilde{A}, \tilde{B}, \tilde{A}_1, \tilde{B}_1$ such that

$$J(A + \tilde{A}, B) < \infty, \quad J(A, B + \tilde{B}) < \infty, \quad J(A + \tilde{A}_1, B + \tilde{B}_1) < \infty,$$

and $\tilde{A}(\infty) = \tilde{B}(\infty) = \tilde{A}_1(\infty) = \tilde{B}_1(\infty) = 0$. In fact $J(A+\tilde{A}_1, B+\tilde{B}_1) < \infty$ is equivalent to $J(\tilde{A}_1, \tilde{B}_1) < \infty$ and (8.6.2) follows from

$$\left(\frac{d}{dt}J(A + t\tilde{A}, B)\right)\bigg|_{t=0} = 0,$$

$$\left(\frac{d}{dt}J(A, B + t\tilde{B})\right)\bigg|_{t=0} = 0,$$

$$\left(\frac{d}{dt}J(A + t\tilde{A}_1, B + t\tilde{B}_1)\right)\bigg|_{t=0} = 0.$$

The first two equations in the above system or (8.6.2) form the weak form of the last two equations in (8.5.3) which may be viewed as the partial functional derivatives of J, but the last equation is somewhat nonstandard and may be viewed as a diagonal functional derivative. This requirement is due to the structure of the energy (8.6.1) in the problem here and will be important at the end of our construction to recover a classical solution of (8.5.1). Thus it is the full system (8.6.2) that will be used to define our constrained space $\mathcal{C}$. We will see indeed that the first two equations in the system (8.6.2) are sufficiently effective so that they make (8.5.7) coercive (partially) and that the full system (8.6.2) can be preserved in the weak limit of a minimizing sequence of (8.5.7) over $\mathcal{C}$. On the other hand, the last equation in (8.6.2), which cannot be derived from the first two equations, is crucial for us to arrive at (8.5.3) from a solution of our minimization problem.

Following the above observation, we now define the constrained admissible space

$$\mathcal{C} = \left\{(f, \rho, A, B) \in \mathcal{A} \,\middle|\, (f, \rho, A, B) \text{ satisfies } (8.6.2)\right\}.$$

We will show that, as promised, we can do minimization on $\mathcal{C}$ for the action functional F defined by (8.5.7) so that F becomes coercive there. For this purpose, the first preliminary result is to ensure that $\mathcal{C}$ is not an empty set.

Lemma 8.6.1. $\mathcal{C} \neq \emptyset$.

Proof. We will use a simple variational approach.

Let f and ρ fulfill the boundary conditions stated in (8.5.4) and (8.5.6) so that the positive part of the action (8.5.7) remains finite,

$$
\begin{aligned}
K(f,\rho) &= \\
&\int_0^\infty dr \left\{ (f')^2 + 2r^2(\rho')^2 + \frac{(f^2-1)^2}{2r^2} + f^2\rho^2 + \frac{2}{g^2}\lambda r^2(\rho^2-1)^2 \right\} \\
&< \infty.
\end{aligned}
\tag{8.6.3}
$$

For this fixed (f,ρ) and the functional J defined by (8.6.1), consider the optimization problem

$$
\min \left\{ J(A,B) \,\middle|\, (A,B) \in \mathcal{X} \right\},
$$
$$
\mathcal{X} = \left\{ (A,B) \,\middle|\, A, B \text{ satisfy } A(\infty) = B(\infty) = a_0 \right\}.
\tag{8.6.4}
$$

It can be seen that (8.6.4) has a unique solution as expected. In fact the uniqueness follows from the convexity of J as observed earlier. To see the existence, let $\{(A_n, B_n)\}$ be a minimizing sequence of (8.6.4). Then

$$
\begin{aligned}
|A_n(r) - a_0| &\leq \int_r^\infty |A'_n(s)|\,ds \\
&\leq r^{-\frac{1}{2}} \left(\int_0^\infty s^2 (A'_n)^2\,ds \right)^{\frac{1}{2}} \\
&\leq g r^{-\frac{1}{2}} J^{\frac{1}{2}}(A_n, B_n).
\end{aligned}
\tag{8.6.5}
$$

Hence, both A_n and B_n go to a_0 as $r \to \infty$ uniformly fast.

On the other hand, since $\|A| - |B\| \leq |A - B|$, we have $J(|A_n|, |B_n|) \leq J(A_n, B_n)$. In other words, we may assume that the minimizing sequence $\{(A_n, B_n)\}$ is nonnegative, $A_n \geq 0, B_n \geq 0$.

The inequality (8.6.5) also says that the sequence $\{(A_n, B_n)\}$ is pointwise bounded away from $r = 0$. Hence, $A_n, B_n \in W^{1,2}_{\text{loc}}(0,\infty)$ and for any pair $0 < r_1 < r_2 < \infty$, $\{A_n\}$ and $\{B_n\}$ are bounded sequences in $W^{1,2}(r_1, r_2)$. Using the compact embedding $W^{1,2}(r_1, r_2) \to C[r_1, r_2]$ and a diagonal subsequence argument, we may assume that there exists $(A,B) \in W^{1,2}_{\text{loc}}(0,\infty)$ so that

$$
\begin{aligned}
A_n \to A, \quad & B_n \to B \quad \text{weakly in } W^{1,2}(r_1, r_2), \\
A_n \to A, \quad & B_n \to B \quad \text{strongly in } C[r_1, r_2]
\end{aligned}
\tag{8.6.6}
$$

for each given pair $0 < r_1 < r_2 < \infty$.

Denote by $J_{\delta,R}$ the functional defined by the same integrand as that for J, but the integral is carried out over the bounded interval (δ, R) instead where $0 < \delta < R < \infty$. Then (8.6.6) implies

$$J_{\delta,R}(A, B) \leq \liminf_{n \to \infty} J_{\delta,R}(A_n, B_n) \leq \lim_{n \to \infty} J(A_n, B_n).$$

Hence, letting $\delta \to 0$, $R \to \infty$, we obtain $J(A, B) \leq \lim_{n \to \infty} J(A_n, B_n)$. However, $(A, B) \in \mathcal{X}$. Therefore (A, B) solves (8.6.4).

As a critical point of J, (A, B) of course satisfies (8.6.2). Hence $\mathcal{C} \neq \emptyset$ as claimed and the lemma follows.

From the above discussion, we understand the structure of $\mathcal{C}$ completely: for any pair (f, ρ) satisfying (8.5.4), (8.5.6), and (8.6.3), $(f, \rho, A, B) \in \mathcal{C}$ is the unique quartet so that (A, B) is the unique solution of (8.6.4). Thus (A, B) depends on (f, ρ),

$$(A, B) = (A_{(f,\rho)}, B_{(f,\rho)}),$$

and $\mathcal{C}$ looks like the graph of the map $(f, \rho) \mapsto (A_{(f,\rho)}, B_{(f,\rho)})$ in $\mathcal{A}$. Besides, we have also shown that $A \geq 0, B \geq 0$ everywhere. From this observation we can derive the uniform boundedness of A and B, which will be a crucial property for our later development. In fact, since (A, B) is a critical point of the functional (8.6.1), we obtain from the Euler–Lagrange equations of (8.6.1) that

$$\left(r^2 \left[\frac{1}{g^2} A' + \frac{1}{g'^2} B' \right] \right)' = \frac{2}{g^2} f^2 A \geq 0.$$

Hence, $r^2(A'/g^2 + B'/g'^2)$ is nondecreasing for $r > 0$. We may then argue as in the proof of Lemma 8.6.6 below that $r^2(A'/g^2 + B'/g'^2)$ goes to zero as $r \to 0$. Therefore $r^2(A'/g^2 + B'/g'^2) \geq 0$. In particular $A/g^2 + B/g'^2$ is nondecreasing and

$$\frac{1}{g^2} A(r) + \frac{1}{g'^2} B(r) \leq a_0 \left(\frac{1}{g^2} + \frac{1}{g'^2} \right)$$

because of the boundary condition (8.5.6). So the expected boundedness of A and B follows. These properties of A and B will be used in due course.

8.6.2 Partial coerciveness and minimization

We next derive a condition under which the functional (8.5.7) is positive definite and coercive with respect to f, ρ on $\mathcal{C}$.

In order to put various quantities on equal footing, we find it convenient to normalize A and B at infinity:

$$a = \frac{A}{a_0}, \quad b = \frac{B}{a_0}.$$

Then (8.6.1) becomes $a_0^2 J(a, b)$. We also consider

$$J_1(h) = 2a_0^2 \int_0^\infty dr \left\{ \left(\frac{1}{g^2} + \frac{1}{g'^2} \right) r^2 (h')^2 + \frac{2}{g^2} f^2 h^2 \right\},$$

namely, $J_1(h) = a_0^2 J(h, h)$. Of course J_1 has a unique minimizer within the class of functions satisfying $h(\infty) = 1$. We still use h to denote this minimizer.

Define

$$J_2(\rho) = \int_0^\infty dr \left\{ 2r^2 (\rho')^2 + f^2 \rho^2 \right\}$$

and assume

$$a_0^2 \left(\frac{1}{g^2} + \frac{1}{g'^2} \right) < 1, \qquad \frac{4a_0^2}{g^2} < 1. \tag{8.6.7}$$

Then, since (A, B) is a minimizer of J, we have

$$J(A, B) = a_0^2 J(a, b) \le a_0^2 J(h, h) = J_1(h) \le J_1(\rho) < J_2(\rho). \tag{8.6.8}$$

In fact, the above inequality can be strengthened. To see this, we set

$$\varepsilon_1 = 1 - a_0^2 \left(\frac{1}{g^2} + \frac{1}{g'^2} \right), \qquad \varepsilon_2 = 1 - \frac{4a_0^2}{g^2}. \tag{8.6.9}$$

Then $\varepsilon_1, \varepsilon_2 > 0$. We can rewrite (8.5.7) as

$$F(f, \rho, A, B) = \int_0^\infty dr \left\{ (f')^2 + \frac{(f^2 - 1)^2}{2r^2} + 2\varepsilon_1 r^2 (\rho')^2 + \varepsilon_2 f^2 \rho^2 \right\}$$
$$+ \int_0^\infty dr \left\{ \frac{2}{g^2} \lambda r^2 (\rho^2 - 1)^2 \right\} + J_1(\rho) - J(A, B). \tag{8.6.10}$$

This leads us to

Lemma 8.6.2. *Suppose that (8.6.7) holds and $\varepsilon_1, \varepsilon_2 > 0$ are defined in (8.6.9). Then we have the partial coerciveness*

$$F(f, \rho, A, B) \ge \int_0^\infty dr \left\{ (f')^2 + \frac{(f^2 - 1)^2}{2r^2} + 2\varepsilon_1 r^2 (\rho')^2 \right.$$
$$\left. + \varepsilon_2 f^2 \rho^2 + \frac{2}{g^2} \lambda r^2 (\rho^2 - 1)^2 \right\}. \tag{8.6.11}$$

Proof. The lemma follows from applying (8.6.8) in (8.6.10).

Lemma 8.6.3. *The optimization problem*

$$\min \left\{ F(f, \rho, A, B) \,\middle|\, (f, \rho, A, B) \in \mathcal{C} \right\} \tag{8.6.12}$$

has a solution provided that the condition (8.6.7) is fulfilled.

Proof. Let $\{(f_n, \rho_n, A_n, B_n)\}$ be a minimizing sequence of (8.6.12). Then applying (8.6.11) to $\{(f_n, \rho_n, A_n, B_n)\}$ we see that the following hold: (i) $f_n(r) \to 1$ as $r \to 0$ uniformly for $n = 1, 2, \cdots$; (ii) $\rho_n(r) \to 1$ as $r \to \infty$ uniformly for $n = 1, 2, \cdots$. Hence, $\{f_n\}$ and $\{\rho_n\}$ are bounded sequences in $W^{1,2}(r_1, r_2)$ for any $0 < r_1 < r_2 < \infty$. As before, we may assume that there exist $f, \rho \in W^{1,2}_{\text{loc}}(0, \infty)$ so that $f_n \to f$ and $\rho_n \to \rho$ as $n \to \infty$ weakly in $W^{1,2}(r_1, r_2)$ and strongly in $C[r_1, r_2]$ for any $0 < r_1 < r_2 < \infty$. Similarly both (8.6.5) and (8.6.6) are valid for A_n and B_n because (8.6.8) implies that $\{J(A_n, B_n)\}$ is a bounded sequence. We now prove that (8.6.2) holds in the limit.

First it is clear that the weak convergence under various suitable global inner products implies that

$$\int_0^\infty dr\{f_n' \tilde{f}'\} \;\to\; \int_0^\infty dr\{f' \tilde{f}'\},$$

$$\int_0^\infty dr\{r^2 \rho_n' \tilde{\rho}'\} \;\to\; \int_0^\infty dr\{r^2 \rho' \tilde{\rho}'\},$$

$$\int_0^\infty dr\{r^2 A_n' \tilde{A}'\} \;\to\; \int_0^\infty dr\{r^2 A' \tilde{A}'\},$$

$$\int_0^\infty dr\{r^2 B_n' \tilde{B}'\} \;\to\; \int_0^\infty dr\{r^2 B' \tilde{B}'\},$$

$$\int_0^\infty dr\{r^2 A_n' \tilde{A}_1'\} \;\to\; \int_0^\infty dr\{r^2 A' \tilde{A}_1'\},$$

$$\int_0^\infty dr\{r^2 B_n' \tilde{B}_1'\} \;\to\; \int_0^\infty dr\{r^2 B' \tilde{B}_1'\}$$

as $n \to \infty$, where $\tilde{f}, \tilde{\rho}, \tilde{A}, \tilde{B}, \tilde{A}_1, \tilde{B}_1$ all vanish at infinity and satisfy

$$\int_0^\infty (\tilde{f}')^2 dr < \infty, \quad \int_0^\infty r^2 (\tilde{\rho}')^2 dr < \infty, \quad \int_0^\infty r^2 (\tilde{A}')^2 dr < \infty,$$

$$\int_0^\infty r^2 (\tilde{B}')^2 dr < \infty, \quad \int_0^\infty r^2 (\tilde{A}_1')^2 dr < \infty, \quad \int_0^\infty r^2 (\tilde{B}_1')^2 dr < \infty.$$

It suffices to show that the first two equations in (8.6.2) will be preserved in the limit. The proof for the last equation is technically similar.

Since f_n satisfies the uniform decay estimate

$$|f_n(r) - 1| \leq r^{\frac{1}{2}} \left(\int_0^\infty (f_n'(s))^2 \, ds \right)^{\frac{1}{2}}, \tag{8.6.13}$$

there exists a $\delta > 0$ so that $|f_n(r)| \geq 1/2$ (say) for $0 < r < \delta$. Thus

$$\int_0^\delta A_n^2 dr \leq 4 \int_0^\infty f_n^2 A_n^2 \, dr \leq g^2 J(A_n, B_n)$$

and we may also assume that $A_n \to A$ weakly in $L^2(0,\delta)$. This result enables us to arrive at

$$\left| \int_0^\delta dr\{(f_n^2 A_n - f^2 A)\tilde{A}\} \right|$$

$$\leq \left| \int_0^\delta dr\{(f_n^2 A_n - f^2 A_n)\tilde{A}\} \right| + \left| \int_0^\delta dr\{f^2(A_n - A)\tilde{A}\} \right|$$

$$\to 0 \quad \text{as } n \to \infty. \tag{8.6.14}$$

In fact, the first term, say, T_1, on the right-hand side of (8.6.14) goes to zero because

$$\left| \int_0^{\delta_1} dr\{(f_n^2 A_n - f^2 A_n)\tilde{A}\} \right|$$

$$\leq \max_{0<r<\delta_1} \left\{ |f_n^2(r) - 1| + |f^2(r) - 1| \right\} \|A_n\|_{L^2(0,\delta)} \|\tilde{A}\|_{L^2(0,\delta)}$$

for any $0 < \delta_1 < \delta$ and (8.6.13) imply that T_1 can be written as the sum of two parts so that the first part is the integral over $(0, \delta_1)$ given as the left-hand side of the above inequality which may uniformly be made as small as we please with respect to $n = 1, 2, \cdots$ and the second part is the integral over (δ_1, δ) which goes to zero as $n \to \infty$ due to the pointwise convergence results mentioned earlier. Here we have used the property that $\tilde{A} \in L^2(0, \delta)$ which may easily be derived from the requirement $J(A_n + \tilde{A}, B_n) < \infty$ and the uniform limit $f(r) \to 1$ as $r \to 0$.

Besides, applying (8.6.5) to ρ_n, ρ and $A_n - B_n, A - B$, we find

$$\rho_n(r), \ \rho(r), \ A_n(r) - B_n(r), \ A(r) - B(r) = O(r^{-\frac{1}{2}}) \tag{8.6.15}$$

uniformly for $r \leq 1$ (say). Therefore, uniformly, we have

$$\lim_{\delta \to 0} \int_0^\delta r^2 \rho_n^2 (A_n - B_n)\tilde{A} dr = \lim_{\delta \to 0} \int_0^\delta r^2 \rho^2 (A - B)\tilde{A} dr = 0. \tag{8.6.16}$$

After the above control for the quantities near $r = 0$, we discuss what happens near infinity.

For $R > 0$ sufficiently large, we have $|\rho_n(r)| \geq 1/2$ for all $r \geq R$ (see (8.6.5)). Hence,

$$\int_R^\infty f_n^2 \, dr \leq 4K(f_n, \rho_n) \leq CF(f_n, \rho_n, A_n, B_n),$$

where $C > 0$ is a constant, and we may assume that $\{f_n\}$ is weakly convergent to f in $L^p(R, \infty)$ for any $p \geq 2$ in view of the embedding $W^{1,2}(R, \infty) \to L^p(R, \infty)$. Consequently,

$$\left| \int_R^\infty dr\{(f_n^2 A_n - f^2 A)\tilde{A}\} \right|$$

$$\leq \left| \int_R^\infty dr\{f_n^2(A_n - A)\tilde{A}\} \right| + \left| \int_R^\infty dr\{(f_n^2 - f^2)A\tilde{A}\} \right|$$

$$\to 0 \quad \text{as } n \to \infty, \tag{8.6.17}$$

where the first term, say, T_1, on the right-hand side of (8.6.17) goes to zero because

$$\left| \int_{R_1}^\infty dr\{f_n^2(A_n - A)\tilde{A}\} \right|$$

$$\leq C \max_{R_1 < r < \infty} \left\{ |A_n(r) - a_0| + |A(r) - a_0| \right\} \|f_n\|_{L^2(R,\infty)}^2$$

for any $R_1 > R$ where $C > 0$ is a suitable constant depending on $\tilde{A}$ and (8.6.5) imply that T_1 can be written as the sum of two parts so that the first part is the integral over (R, R_1) which goes to zero as $n \to \infty$ by the same pointwise convergence property and the second part is the integral over (R_1, ∞) given as the left-hand side of the above which vanishes uniformly fast with respect to $n = 1, 2, \cdots$ as $R_1 \to \infty$. The second term, say, T_2, goes to zero as well because

$$T_2 = \left| \int_R^{R_1} dr\{(f_n^2 - f^2)A\tilde{A}\} \right|$$

$$+ C \left(\max_{R_1 < r < \infty} \{|\tilde{A}(r)|\} \right) (\|f_n\|_{L^2(R_1,\infty)}^2 + \|f\|_{L^2(R_1,\infty)}^2).$$

The first term above goes to zero due to pointwise convergence and the second term may be made uniformly small for $n = 1, 2, \cdots$ since $\tilde{A} = O(r^{-1/2})$ for large r.

Furthermore, we have

$$\left| \int_R^\infty dr\{r^2[\rho_n^2(A_n - B_n) - \rho^2(A - B)]\tilde{A}\} \right| \leq I_1 + I_2, \tag{8.6.18}$$

where

$$I_1 = \left| \int_R^\infty dr\{r^2(\rho_n^2 - \rho^2)(A_n - B_n)\tilde{A}\} \right|$$

$$\leq \left(\max_{R < r < \infty} \{|\tilde{A}(r)|\} \right) \|r(A_n - B_n)\|_{L^2(R,\infty)}$$

$$\times \left(\|r(\rho_n^2 - 1)\|_{L^2(R,\infty)} + \|r(\rho^2 - 1)\|_{L^2(R,\infty)} \right),$$

$$I_2 = \left| \int_R^\infty dr\{r^2\rho^2[(A_n - B_n) - (A - B)]\tilde{A}\} \right|.$$

It is clear that $I_1 \to 0$ as $R \to \infty$. Besides, since $\{(A_n - B_n)\}$ is a bounded sequence in $L^2((R, \infty), r^2 dr)$, we may assume that $A_n - B_n$ weakly converges to $A - B$ in $L^2((R, \infty), r^2 dr)$. However, the condition $J(A + \tilde{A}, B) <$

∞ implies that $\tilde{A} \in L^2((R, \infty), r^2 dr)$, hence, so does $\rho^2 \tilde{A}$. Therefore, we see that $I_2 \to 0$ as $n \to \infty$. Thus we have proved that (8.6.18) goes to zero as $R, n \to \infty$.

We are now ready to show that the first constraint in (8.6.2) is preserved in the limit. In fact we have

$$
\begin{aligned}
I &= \left| \int_0^\infty dr \{ r^2 A' \tilde{A}' + 2 f^2 A \tilde{A} + r^2 \rho^2 (A - B) \tilde{A} \} \right| \\
&\leq \left| \int_0^\infty dr \{ r^2 (A' - A_n') \tilde{A} \} \right| + 2 \left| \int_0^\infty dr \{ (f^2 A - f_n^2 A_n) \tilde{A} \} \right| \\
&\quad + \left| \int_0^\infty dr \{ r^2 [\rho^2 (A - B) - \rho_n^2 (A_n - B_n)] \tilde{A} \} \right| \\
&= \left| \int_0^\infty dr \{ r^2 (A' - A_n') \tilde{A} \} \right| + T_{0,\delta}(n) + T_{\delta,R}(n) + T_{R,\infty}(n).
\end{aligned}
$$

$$(8.6.19)$$

Here, for example, $T_{0,\delta}(n)$ denotes the terms on the right-hand side of (8.6.19) that involve only integrals over the interval $(0, \delta)$. From (8.6.14) and (8.6.16) we see that for any $\varepsilon > 0$ there are $\delta > 0$ and $n_0 \geq 1$ so that for $n > n_0$ there holds $T_{0,\delta} < \varepsilon$. Similarly $T_{R,\infty}(n)$ denotes the terms on the right-hand side of (8.6.19) involving integrals over (R, ∞). By (8.6.17) and (8.6.19) we obtain $R > 0$ and $n_0 \geq 1$ so that $T_{R,\infty}(n) < \varepsilon$ whenever $n > n_0$. For these fixed $\delta > 0$ and $R > \delta$, the pointwise convergence $f_n, \rho_n, A_n, B_n \to f, \rho, A, B$ implies that $T_{\delta,R}(n) < \varepsilon$ when $n > n_0$. The first term on the right-hand side of (8.6.19) goes to zero as $n \to \infty$ by (8.6.13). This proves $I = 0$ and the first constraint in (8.6.2) is valid.

The same method shows that the second constraint in (8.6.2) is also preserved. To show that the last equation in (8.6.2) is preserved as well, we notice that the condition $J(A + \tilde{A}_1, B + \tilde{B}_1) < \infty$ implies that $\tilde{A}_1 - \tilde{B}_1 \in L^2((R, \infty), r^2 dr)$ for suitably large $R > 0$. The rest of the proof is then almost identical to the part for the first equation in (8.6.2) carried out earlier. Hence the full system (8.6.2) is valid for the limiting configuration (f, ρ, A, B) as expected.

Consider any fixed $n \geq 1$. The quartet (f_n, ρ_n, A_n, B_n) of course satisfies the first two equations in (8.6.2),

$$
\int_0^\infty dr \{ r^2 A_n' \tilde{A}' + 2 f_n^2 A_n \tilde{A} + r^2 \rho_n^2 (A_n - B_n) \tilde{A} \} = 0,
$$

$$
\int_0^\infty dr \left\{ \frac{1}{g'^2} r^2 B_n' \tilde{B}' - \frac{1}{g^2} r^2 \rho_n^2 (A_n - B_n) \tilde{B} \right\} = 0. \quad (8.6.20)
$$

In order to control the negative terms on the right-hand side of (8.5.7) when we consider the limiting behavior of F over the minimizing sequence (f_n, ρ_n, A_n, B_n), we will first need the property that

$$
r A_n' \to r A', \quad r B_n' \to r B', \quad r \rho_n (A_n - B_n) \to r \rho (A - B) \quad (8.6.21)
$$

strongly in $L^2(0,\infty)$ for $n \to \infty$. Indeed, subtracting (8.6.20) from (8.6.2) (correspondingly), setting $\tilde{A} = A - A_n$ and $\tilde{B} = B - B_n$ in the resulting expressions, and adding the two equations together, we easily obtain

$$
\int_0^\infty dr \left\{ r^2(A_n' - A')^2 + \frac{g^2}{g'^2} r^2(B_n' - B')^2 \right.
$$
$$
\left. + r^2(\rho_n[A_n - B_n] - \rho[A - B])^2 \right\}
$$
$$
= -2 \int_0^\infty dr \{ (f_n^2 A_n - f^2 A)(A_n - A) \}
$$
$$
+ \int_0^\infty dr \{ r^2(A_n - B_n)(A - B)(\rho_n - \rho)^2 \}. \qquad (8.6.22)
$$

Let I_1 and I_2 denote the first and second integrals on the right-hand side of (8.6.22), respectively. We also use $I_j(a,b)$ to denote the part of integral I_j that is carried out over the interval (a,b) $(j = 1,2)$. Then for any $0 < \delta < R < \infty$

$$
I_j = I_j(0,\delta) + I_j(\delta,R) + I_j(R,\infty), \quad j = 1,2.
$$

Since $\{f_n\}$ is a bounded sequence in $W^{1,2}(0,\infty)$ (see (8.6.11) and use the uniform limits $f_n(r) \to 1$ as $r \to 0$ and $\rho_n(r) \to 1$ as $r \to \infty$), it is also bounded in any $L^p(0,\infty)$ $(p \geq 2)$. Thus by the pointwise boundedness of $\{A_n\}$ (as well as A) we have

$$
|I_1(0,\delta)| \leq C \int_0^\delta (f_n^2 + f^2) \, dr \leq C\delta^{\frac{1}{2}} (\|f_n\|_{L^4(0,\delta)}^2 + \|f\|_{L^4(0,\delta)}^2),
$$

which implies that $I_1(0,\delta) \to 0$ uniformly for $n = 1,2,\cdots$.

The pointwise convergence $f_n \to f$, $A_n \to A$ already indicates that $I_1(\delta,R) \to 0$ as $n \to \infty$.

For $I_1(R,\infty)$, we have

$$
I_1(R,\infty) \leq
$$
$$
C \max_{R<r<\infty} \left\{ |A_n(r) - a_0| + |A(r) - a_0| \right\} \left(\|f_n\|_{L^2(R,\infty)}^2 + \|f\|_{L^2(R,\infty)}^2 \right),
$$

which goes to zero as $R \to \infty$ uniformly for $n = 1,2,\cdots$ because of the asymptotic estimate (8.6.5).

The above consideration enables us to arrive at the conclusion that $I_1 \to 0$ as $n \to \infty$.

For $I_2(0,\delta)$, since ρ_n and ρ obey the same estimate as A_n (see (8.6.5)), we have

$$
I_2(0,\delta) \leq C \int_0^\delta r^2(\rho_n - \rho)^2 \, dr \leq C_1 \delta^2,
$$

where, again, $C > 0$ is a constant independent of n. So $I_2(0, \delta) \to 0$ as $\delta \to 0$ uniformly too.

The convergence $I_2(\delta, R) \to 0$ for fixed $0 < \delta < R < \infty$ as $n \to \infty$ is obvious.

For $I_2(R, \infty)$ we notice that the inequality

$$|\rho_n(r) - 1| \leq r^{-\frac{1}{2}} \left(\int_r^\infty (s^2 \rho_n') \, ds \right)^{\frac{1}{2}}$$

$$\leq \frac{1}{\sqrt{2\varepsilon_1}} r^{-\frac{1}{2}} F^{\frac{1}{2}}(f_n, \rho_n, A_n, B_n)$$

implies the uniform boundedness of $\{\rho_n\}$ near $r = \infty$. Besides, since (8.5.7) is even in ρ, we see that we may assume that the sequence $\{\rho_n\}$ satisfies $\rho_n \geq 0$. Thus, by (8.6.11), we have

$$I_2(R, \infty) \leq$$
$$C \max_{R < r < \infty} \left\{ |A(r) - a_0| + |B(r) - a_0| \right\} \int_R^\infty r^2 ([\rho_n^2 - 1]^2 + [\rho^2 - 1]^2) dr,$$

which approaches zero uniformly fast as $R \to \infty$ by applying (8.6.5).

Consequently $I_2 \to$ as $n \to \infty$.

Taking $n \to \infty$ in (8.6.22), we immediately see that the right-hand side of (8.6.22) tends to zero as $n \to \infty$. Hence, (8.6.21) is proven.

Therefore, all the negative terms in the functional $F(f_n, \rho_n, A_n, B_n)$ (see (8.5.7)) are put under control, except the term of the form $-f_n^2 A_n^2$, which will have to be tackled separately and this is another place where the condition (8.6.7) is essential. We proceed as follows.

We will now need the property that

$$\lim_{n \to \infty} \int_0^\infty dr\{([A_n - a_0]^2 + 2a_0[A_n - a_0])f_n^2\}$$
$$= \int_0^\infty dr\{([A - a_0]^2 + 2a_0[A - a_0])f^2\}. \qquad (8.6.23)$$

The proof of (8.6.23) is similar to our earlier discussion on the other limits because the pieces of the integral on the left-hand side of (8.6.23) restricted over the intervals $(0, \delta)$ and (R, ∞) all approach zero uniformly fast as $\delta \to 0$ and $R \to \infty$, respectively.

We are now ready to show that the limit configuration (f, ρ, A, B) is a minimizer of the problem (8.6.12). To proceed, we rewrite the functional (8.5.7) evaluated over the minimizing sequence $\{(f_n, \rho_n, A_n, B_n)\}$ as

$$F(f_n, \rho_n, A_n, B_n) =$$
$$\int_0^\infty dr\left\{ (f_n')^2 + 2r^2(\rho_n')^2 + \frac{(f_n^2 - 1)^2}{2r^2} + \frac{2}{g^2}\lambda r^2(\rho_n^2 - 1)^2 \right.$$

$$+([\rho_n - 1]^2 + 2[\rho_n - 1])f_n^2 + \left(1 - \frac{4}{g^2}a_0^2\right)f_n^2$$

$$-\frac{2}{g^2}r^2(A_n')^2 - \frac{2}{g'^2}r^2(B_n')^2 - \frac{2}{g^2}r^2\rho_n^2(A_n - B_n)^2$$

$$-\frac{4}{g^2}([A_n - a_0]^2 + 2a_0[A_n - a_0])f_n^2\Big\}. \tag{8.6.24}$$

Taking $n \to \infty$ in (8.6.24) and applying (8.6.21), (8.6.23), and (8.6.7), we immediately arrive at the desired conclusion

$$F(f, \rho, A, B) \leq \lim_{n\to\infty} F(f_n, \rho_n, A_n, B_n).$$

Consequently, the quartet (f, ρ, A, B) solves (8.6.12) and the lemma is proven.

8.6.3 Weak solutions of governing equations

Lemma 8.6.4. *The action minimizing solution (f, ρ, A, B) of the problem (8.6.12) is a classical solution of the Cho–Maison dyon equations (8.5.3) subject to the partial boundary conditions (8.5.4) and (8.5.6).*

Proof. It will be sufficient to show that all the four equations in (8.5.3) are fulfilled in a weak sense.

We first fix a new admissible space,

$$\mathcal{S} = \left\{(f, \rho) \,\middle|\, f(0) = 1, f(\infty) = 0, \rho(\infty) = 1\right\}.$$

As observed earlier, the constrained class $\mathcal{C}$ is the image of a map $\chi : \mathcal{S} \to \mathcal{A}$, $(f, \rho) \mapsto (f, \rho, A_{(f,\rho)}, B_{(f,\rho)})$ with $A = A_{(f,\rho)}$ and $B = B_{(f,\rho)}$ being defined by (8.6.2), which is the weak form of the linear system consisting of the last two equations in (8.5.3). Consequently, χ is a smooth map in an obvious sense. Besides, the minimizer (f, ρ, A, B) of the constrained problem (8.6.12) obtained in Lemma 8.6.3 may simply be viewed as the image under χ of an absolute minimizer, (f, ρ), of the functional $F(f, \rho, A_{(f,\rho)}, B_{(f,\rho)})$ over the unconstrained class $\mathcal{S}$.

Let t be a real parameter confined in a small interval, say, $|t| < \frac{1}{2}$, and $\tilde{f} \in C_0^1(0, \infty)$ (functions with compact supports). Set $A_t = A_{(f+t\tilde{f}, \rho)}$ and $B_t = B_{(f+t\tilde{f}, \rho)}$. Furthermore, it will be convenient to use the notation

$$A_t = A + \tilde{A}_t, \quad B_t = B + \tilde{B}_t, \quad \tilde{A}_1 = \left(\frac{\mathrm{d}}{\mathrm{d}t}\tilde{A}_t\right)\Big|_{t=0}, \quad \tilde{B}_1 = \left(\frac{\mathrm{d}}{\mathrm{d}t}\tilde{B}_t\right)\Big|_{t=0}.$$

Note that $\tilde{A}_1 \neq \tilde{A}_t|_{t=1}$ and $\tilde{B}_1 \neq \tilde{B}_t|_{t=1}$. Then we may use $\tilde{A}_t$ and $\tilde{B}_t$ as test functions to rewrite the last equation in (8.6.2) as

$$\int_0^\infty dr \left\{ r^2 A' \tilde{A}_t' + \frac{g^2}{g'^2} r^2 B' \tilde{B}_t' + 2f^2 A \tilde{A}_t + r^2 \rho^2 (A - B)(\tilde{A}_t - \tilde{B}_t) \right\} = 0.$$
(8.6.25)

Thus, subtracting (8.6.25) from

$$\int_0^\infty dr \left\{ r^2 A_t' \tilde{A}_t' + \frac{g^2}{g'^2} r^2 B_t' \tilde{B}_t' + 2(f + t\tilde{f})^2 A_t \tilde{A}_t \right.$$

$$\left. + r^2 \rho^2 (A_t - B_t)(\tilde{A}_t - \tilde{B}_t) \right\} = 0,$$

we obtain the useful expression

$$\int_0^\infty dr \left\{ r^2 (\tilde{A}_t')^2 + \frac{g^2}{g'^2} r^2 (\tilde{B}_t')^2 + 2f^2 (\tilde{A}_t)^2 + r^2 \rho^2 (\tilde{A}_t - \tilde{B}_t)^2 \right\}$$

$$= -2t \int_0^\infty dr \{ 2f\tilde{f} A_t \tilde{A}_t + t\tilde{f}^2 A_t \tilde{A}_t \},$$

which, in view of the Schwartz inequality, enables us to arrive at the bound

$$\int_0^\infty dr \left\{ r^2 \left(\frac{\tilde{A}_t'}{t} \right)^2 + \frac{g^2}{g'^2} r^2 \left(\frac{\tilde{B}_t'}{t} \right)^2 + f^2 \left(\frac{\tilde{A}_t}{t} \right)^2 + r^2 \rho^2 \left(\frac{\tilde{A}_t}{t} - \frac{\tilde{B}_t}{t} \right)^2 \right\}$$

$$\leq 4 \int_0^\infty dr \{ \tilde{f}^2 A_t^2 + |\tilde{f}^2 A_t \tilde{A}_t| \} \leq C, \qquad t \neq 0,$$
(8.6.26)

where $C > 0$ depends on $\tilde{f}$ but is independent of t because of the uniform bound

$$0 \leq A(r), \; A_t(r) \leq a_0 \left(1 + \frac{g^2}{g'^2} \right)$$

derived in the remark following the proof of Lemma 8.6.1 and the fact that $|\tilde{A}_t| \leq |A| + |A_t|$.

From (8.6.26) and $\tilde{A}_t(\infty) = \tilde{B}_t(\infty) = 0$, we find as we did in (8.6.5) the uniform estimates for large $r > 0$,

$$\left| \frac{\tilde{A}_t(r)}{t} \right| \leq r^{-\frac{1}{2}} C^{\frac{1}{2}},$$

$$\left| \frac{\tilde{B}_t(r)}{t} \right| \leq r^{-\frac{1}{2}} C^{\frac{1}{2}} \frac{g'}{g}, \qquad t \neq 0.$$
(8.6.27)

As a consequence, we may take $t \to 0$ in (8.6.26) and (8.6.27) to deduce the properties

$$J(\tilde{A}_1, \tilde{B}_1) < \infty, \quad \tilde{A}_1(\infty) = 0, \quad \tilde{B}_1(\infty) = 0.$$

Therefore, for this pair, $\tilde{A}_1$ and $\tilde{B}_1$, the last equation in (8.6.2) holds.

Since (f, ρ) minimizes $F(f, \rho, A_{(f,\rho)}, B_{(f,\rho)})$, we have

$$\left(\frac{\mathrm{d}}{\mathrm{d}t} F(f + t\tilde{f}, \rho, A_t, B_t)\right)\bigg|_{t=0} = 0.$$

This equation gives us the following expression

$$\int_0^\infty \mathrm{d}r \left\{ f'\tilde{f}' + \frac{1}{r^2}(f^2 - 1)f\tilde{f} + \left(\rho^2 - \frac{4}{g^2}A^2\right)f\tilde{f}\right\}$$

$$= \frac{2}{g^2}\int_0^\infty \mathrm{d}r \left\{ r^2 A'\tilde{A}_1' + \frac{g^2}{g'^2}r^2 B'\tilde{B}_1'\right.$$

$$\left. + 2f^2 A\tilde{A}_1 + r^2\rho^2(A - B)(\tilde{A}_1 - \tilde{B}_1)\right\}$$

$$= 0 \tag{8.6.28}$$

in view of the last equation in (8.6.2). The left-hand side of (8.6.28) leads to the validity of a weak form of the first equation in (8.5.3).

Similarly, we fix a compactly supported test function $\tilde{\rho}$ and denote the functions $A_t = A_{(f,\rho+t\tilde{\rho})}$ and $B_t = B_{(f,\rho+t\tilde{\rho})}$ as before. We can show, almost verbatim as we did for (8.6.28), that

$$\int_0^\infty \mathrm{d}r \left\{ 2r^2\rho'\tilde{\rho}' + \left(f^2 - \frac{2}{g^2}(A - B)^2 + \frac{4}{g^2}\lambda r^2(\rho^2 - 1)\right)\rho\tilde{\rho}\right\} = 0. \tag{8.6.29}$$

Finally, supplementing (8.6.28) and (8.6.29) with the first two equations in (8.6.2), we see that, in standard sense, the quartet (f, ρ, A, B) is a weak solution of the original dyon equations (8.5.3). Thus, the well-known elliptic theory shows that (f, ρ, A, B) is a classical solution of (8.5.3) subject to the boundary conditions (8.5.4) and (8.5.6).

The proof of the lemma is complete.

The above proof indicates the necessity for the introduction of the last equation in (8.6.2) in the constraints. At a first glance, it seems that such an equation is already contained in the first two equations. However, the form of the functional J does not allow this passage. On the other hand, as seen in the proof of Lemma 8.6.4, the resolution of the optimization problem (8.6.12) does not rely on the last equation in (8.6.2). It is the final recovery of the dyon system (8.5.3) from a solution of (8.6.12) that needs such an additional equation in the constraint set. In fact the difficulty arises when one tries to derive (8.6.28) and (8.6.29) but only has the property $J(\tilde{A}_1, \tilde{B}_1) < \infty$ instead of the stronger property $J(A + \tilde{A}_1, B) < \infty$, $J(A, B + \tilde{B}_1) < \infty$, which is meant mainly for the first two equations in (8.6.2).

8.6.4 Full set of boundary conditions

In this section we show that the remaining boundary condition (8.5.5) will hold as well for the solution (f, ρ, A, B) obtained in the last section. As by-products, we will also derive some useful properties of the solution.

Lemma 8.6.5. *The minimizer* (ρ, f, A, B) *satisfies*

$$f(r) > 0, \quad \rho(r) > 0, \quad A(r) > 0, \quad B(r) > 0 \quad \forall r > 0. \qquad (8.6.30)$$

Proof. The structure of the functional (8.5.7) implies that for the action-minimizing sequence $\{(f_n, \rho_n, A_n, B_n)\}$ of the problem (8.6.12), we can always choose $f_n \geq 0, \rho_n \geq 0, n = 1, 2, \cdots$ (see also (8.6.24)). The discussion in the last section already gives us $A_n \geq 0, B_n \geq 0$. Thus this proves that we may assume that f, ρ, A, B already satisfy $f, \rho, A, B \geq 0$. It remains to confirm the strict inequalities.

We now prove the strict inequalities for A and B only because the proofs for f and ρ are actually simpler.

Suppose that there is an $r_0 > 0$ so that $A(r_0) = 0$. Then r_0 is a minimum point for $A(r)$ and $A'(r_0) = 0$, $A''(r_0) \geq 0$. Thus the third equation in (8.5.3) leads to

$$-r_0^2 \rho^2(r_0) B(r_0) \geq 0.$$

Consequently, we must have $B(r_0) = 0$. Thus $B'(r_0) = 0$ too. So (A, B) is a solution to the initial value problem

$$
\begin{aligned}
(r^2 A')' &= 2f^2 A + r^2 \rho^2 (A - B), \\
(r^2 B')' &= \frac{g'^2}{g^2} r^2 \rho^2 (B - A), \\
A(r_0) &= 0, \quad A'(r_0) = 0, \quad B(r_0) = 0, \quad B'(r_0) = 0.
\end{aligned}
$$

By the uniqueness theorem, we have $A(r) \equiv 0, B(r) \equiv 0$, which contradicts the boundary condition (8.5.6) for A and B. Thus we must have $A(r) > 0$ for all $r > 0$.

Similarly, $B(r_0) = 0$ for some $r_0 > 0$ also implies $A(r_0) = 0$, which is impossible. Hence, the proof is complete.

Lemma 8.6.6. *Let* (f, ρ, A, B) *be the solution of (8.5.3) obtained in the last section. Then* $B(r)$ *and* $(A/g^2 + B/g'^2)(r)$ *are increasing in* $r > 0$, $B(r) > A(r)$ *for all* $r > 0$, *and*

$$\lim_{r \to 0} \rho(r) = 0, \quad \lim_{r \to 0} A(r) = 0, \quad \lim_{r \to 0} B(r) = b_0$$

for a certain suitable number b_0 *satisfying* $0 \leq b_0 \leq a_0(1/g^2 + 1/g'^2)$. *In other words, the boundary condition (8.5.5) also holds.*

Proof. We concentrate on the pair A, B first.

As before, from the last two equations in (8.5.3), we have

$$\left(r^2\left[\frac{1}{g^2}A' + \frac{1}{g'^2}B'\right]\right)' = \frac{2}{g^2}f^2A, \quad r > 0. \tag{8.6.31}$$

Since $f, A > 0$, we see that $r^2(A/g^2 + B/g'^2)'$ is increasing for $r > 0$. In particular $r^2(A/g^2 + B/g'^2)'$ approaches a finite limit or $-\infty$ as $r \to 0$. We claim that

$$\lim_{r \to 0} r^2\left(\frac{1}{g^2}A + \frac{1}{g'^2}B\right)'(r) = 0. \tag{8.6.32}$$

Otherwise, there is an $r_0 > 0$ so that

$$\left|r^2\left(\frac{1}{g^2}A + \frac{1}{g'^2}B\right)'(r)\right| \geq c_0, \quad 0 < r < r_0$$

for some constant $c_0 > 0$, which implies that

$$r^2\left(\frac{1}{g^2}A' + \frac{1}{g'^2}B'\right)^2 \geq \frac{c_0}{r^2}, \quad 0 < r < r_0,$$

which contradicts the convergence of

$$\int_0^\infty dr\left\{r^2(A')^2 + r^2(B')^2\right\}.$$

Integrating (8.6.31) over $(0, r)$ with $0 < r < r_0$ and using (8.6.32), we obtain $r^2(A/g^2 + B/g'^2)'(r) > 0$. In other words, $(A/g^2 + B/g'^2)(r)$ increases for $r > 0$.

On the other hand, in view of $(A/g^2 + B/g'^2)(\infty) = a_0(1/g^2 + 1/g'^2)$, we see that $0 < (A/g^2 + B/g'^2)(r) < a_0(1/g^2 + 1/g'^2)$ for all $r > 0$. Hence, the limit

$$\lim_{r \to 0}\left(\frac{1}{g^2}A + \frac{1}{g'^2}B\right)(r) = b_0 \tag{8.6.33}$$

exists for some number $0 \leq b_0 \leq a_0(1/g^2 + 1/g'^2)$.

We first claim a weaker result for A, namely,

$$\liminf_{r \to 0} A(r) = 0. \tag{8.6.34}$$

Suppose that (8.6.34) is false. Then there exist $c_0 > 0$ and $r_0 > 0$ so that

$$A(r) \geq c_0, \quad 0 < r < r_0. \tag{8.6.35}$$

Inserting (8.6.35) into (8.6.31) and assuming r_0 to be small enough to make

$$2f^2(r) \geq 1, \quad 0 < r < r_0$$

valid, we obtain by the mean-value theorem the lower bound

$$r\left(\frac{1}{g^2}A'(r) + \frac{1}{g'^2}B'(r)\right) = \frac{1}{r}\left(r^2\left[\frac{1}{g^2}A'(r) + \frac{1}{g'^2}B'(r)\right]\right)$$

$$= \left(r^2\left[\frac{1}{g^2}A' + \frac{1}{g'^2}B'\right]\right)'(r_1) \quad (0 < r_1 < r < r_0)$$

$$\geq \frac{c_0}{g^2}, \quad 0 < r < r_0,$$

which implies that

$$\left(\frac{1}{g^2}A(r_2) + \frac{1}{g'^2}B(r_2)\right) - \left(\frac{1}{g^2}A(r_1) + \frac{1}{g'^2}B(r_1)\right)$$

$$\geq \frac{c_0}{g^2}\ln\left(\frac{r_2}{r_1}\right), \quad 0 < r_1, r_2 < r_0$$

and that the limit of $A(r)/g^2 + B(r)/g'^2$ as $r \to 0$ does not exist. This is in contradiction with (8.6.33). Hence, (8.6.34) is established.

By (8.6.34) we may find a sequence $\{r_j\}$, $r_j > 0$, $r_j \to 0$ as $j \to \infty$, so that

$$\lim_{j\to\infty} A(r_j) = 0, \quad \lim_{j\to\infty} B(r_j) = b_0. \tag{8.6.36}$$

We claim that

$$B(r) > A(r) \quad \text{for all } r > 0. \tag{8.6.37}$$

In fact, from the last two equations in (8.5.3), we obtain

$$(r^2[B' - A'])' = \left(\frac{g'^2}{g^2} + 1\right)r^2\rho^2(B - A) - 2f^2A. \tag{8.6.38}$$

We first show that a weaker form of (8.6.37) is true, namely,

$$B(r) \geq A(r), \quad r > 0. \tag{8.6.39}$$

Otherwise, let us assume that $B(r) < A(r)$ holds for some $r > 0$. Recall that $A(\infty) = B(\infty) = a_0$. Besides, using (8.6.36) and $b_0 \geq 0$, we have

$$\lim_{j\to\infty} A(r_j) \leq \lim_{j\to\infty} B(r_j).$$

These facts imply that $B - A$ has a negative local minimum at some point $r_0 > 0$. Thus $(r^2(B' - A'))'(r_0) \geq 0$, which violates (8.6.38). Consequently, the assertion (8.6.39) holds. To see why the stronger conclusion (8.6.37) is true, we again argue by contradiction. Suppose that there is a point $r_0 > 0$ so that $B(r_0) = A(r_0)$. Then (8.6.39) says that r_0 is a local minimum point for $B - A$. Such a property again violates the fact that $f^2(r_0)A(r_0) > 0$ (see (8.6.30)) by virtue of (8.6.38). Hence (8.6.37) is proven.

By the last equation in (8.5.3) and the result (8.6.37), we have $(r^2 B')' > 0$. So $r^2 B'(r)$ is increasing. As before, we can show that

$$\lim_{r \to 0} r^2 B'(r) = 0. \qquad (8.6.40)$$

Thus $r^2 B'(r) > 0$ for $r > 0$. This proves that $B(r)$ increases. In particular, $B(r) \to$ a limit as $r \to 0$ because $B(r) > 0$ for all $r > 0$. However, in view of the second limit in (8.6.36), we see that this limit must be b_0. Using (8.6.33) we arrive at the boundary condition (8.5.5) for A and B at the delicate end point $r = 0$.

For the boundary condition for ρ at $r = 0$, we note that the second equation in (8.5.3) gives us the property that

$$(r^2 \rho')' > \frac{1}{4} f^2 \rho > 0, \quad 0 < r < r_0 \qquad (8.6.41)$$

for some small $r_0 > 0$ because of the limit $f(r) \to 1$ as $r \to 0$, $\rho > 0$, $\rho(r) = O(r^{-1/2})$, and the fact that A, B are bounded functions. In particular, $r^2 \rho'(r)$ increases in $(0, r_0)$. Thus (8.6.40) is valid for $B = \rho$ and $r^2 \rho'(r) > 0$ in $(0, r_0)$. This proves that ρ increases in $(0, r_0)$ and

$$\lim_{r \to 0} \rho(r) = c_0 \qquad (8.6.42)$$

for some $c_0 \geq 0$. We claim that $c_0 = 0$. Otherwise, if $c_0 > 0$, we have $\rho(r) > c_0$ for $0 < r < r_0$ and (8.6.41) and the mean-value theorem would again give us

$$
\begin{aligned}
r\rho'(r) &= \frac{r^2 \rho'(r)}{r} \\
&= (r^2 \rho')'(r_1) \quad (0 < r_1 < r < r_0) \\
&\geq \frac{1}{4} c_0 \min_{0 \leq r \leq r_0} \left\{ f^2(r) \right\} > 0.
\end{aligned}
$$

It is readily seen as before that the above inequality contradicts the finite limit result stated in (8.6.42). Therefore $c_0 = 0$ in (8.6.42) and the proof of the lemma is complete.

8.6.5 Asymptotic estimates

We use the notation

$$\rho_0 = \mu \sqrt{\frac{2}{\lambda}}, \quad \kappa = \rho_0 \sqrt{\frac{g^2}{4} - a_0^2}, \quad \nu = \frac{\rho_0}{2} \sqrt{g^2 + g'^2}, \qquad (8.6.43)$$

and we consider the original dyon equations (8.5.1) without rescaling. Therefore we now have $f \to 0, \rho \to \rho_0, A \to \rho_0 a_0, B \to \rho_0 a_0$ as $r \to \infty$.

In order to calculate various charges, it will be useful to know some of the decay rates at $r = \infty$ as well as at $r = 0$. First, we have

Lemma 8.6.7. *For the dyon solution* (f, ρ, A, B) *of* $(8.5.1)$ *obtained in the last section, there hold the decay estimates*

$$f = O(e^{-\kappa(1-\varepsilon)r}), \quad \rho = \rho_0 + O(r^{-1}e^{-\sqrt{2}\mu_0(1-\varepsilon)r}),$$
$$A = \rho_0 a_0 + O(r^{-1}), \quad B - A = O(r^{-1}e^{-\nu_0(1-\varepsilon)r}), \quad (8.6.44)$$

for $r \to \infty$*, where* $0 < \varepsilon < 1$ *is arbitrary and the decay exponents* μ_0 *and* ν_0 *are defined by*

$$\mu_0 = \min\left\{\mu, \sqrt{2}\kappa, \frac{\nu}{\sqrt{2}}\right\}, \quad \nu_0 = \min\{2\kappa, \nu\}.$$

Proof. Define the comparison function $\eta(r) = Ce^{-\kappa(1-\varepsilon)r}$, where $C > 0$ is a constant to be chosen later. From the first equation in (8.5.1) and the property $f > 0$ we see that there is an $r_\varepsilon > 0$ so that

$$(f - \eta)'' \geq \kappa^2(1 - \varepsilon)^2(f - \eta), \quad r > r_\varepsilon. \quad (8.6.45)$$

Let $C > 0$ be large enough to make $(f - \eta)(r_\varepsilon) \leq 0$. Thus, in view of this and the boundary condition $(f - \eta)(r) \to 0$ as $r \to \infty$, we obtain by applying the maximum principle in (8.6.45) the result $f(r) < \eta(r)$ $(r > r_\varepsilon)$ as expected in (8.6.44).

Next, set $\sigma = r(B - A)$. Then Lemma 8.6.6 implies $\sigma > 0$. By virtue of the last two equations in (8.5.1), we have

$$\sigma'' = \frac{1}{4}\rho^2(g^2 + g'^2)\sigma - \frac{2}{r}f^2 A$$
$$\geq \nu_0^2\left(1 - \frac{\varepsilon}{2}\right)^2\sigma - C_1 e^{-2\kappa(1-\varepsilon/2)r} \quad (8.6.46)$$

for r sufficiently large. Now set $\eta = Ce^{-\nu_0(1-\varepsilon)r}$ and insert $\eta'' = \nu_0^2(1-\varepsilon)^2\eta$ into (8.6.46). We have

$$(\sigma - \eta)'' \geq \nu_0^2\left(1 - \frac{\varepsilon}{2}\right)^2(\sigma - \eta) + \nu_0^2\left\{\left(1 - \frac{\varepsilon}{2}\right)^2 - (1 - \varepsilon)^2\right\}\eta$$
$$\quad - C_1 e^{-2\kappa(1-\varepsilon/2)r}$$
$$\geq \nu_0^2\left(1 - \frac{\varepsilon}{2}\right)^2(\sigma - \eta), \quad r > r_\varepsilon, \quad (8.6.47)$$

where, again, $r_\varepsilon > 0$ is chosen to be sufficiently large. On the other hand, the energy being finite implies that there is a sequence $\{r_j\}$, $r_j \to \infty$ as $j \to \infty$ so that $\sigma(r_j) \to 0$ as $j \to \infty$. Using this in (8.6.47) and assuming that the constant C in the definition of η is large enough to make $(\sigma - \eta)(r_\varepsilon) \leq 0$, we arrive at $\sigma(r) < \eta(r)$ for all $r > r_\varepsilon$.

Then we consider the estimate for ρ stated in (8.6.44). For the new function $\tau = r(\rho - \rho_0)$, the second equation in (8.5.1) gives us

$$\tau'' = \frac{\lambda}{2}\rho(\rho + \rho_0)\tau + \frac{1}{2r}\left(f^2 - \frac{1}{2}r^2[A - B]^2\right)\rho. \tag{8.6.48}$$

It is seen that the coefficient of τ on the left-hand side of (8.6.48) goes to $\lambda\rho_0^2 = 2\mu^2$ as $r \to \infty$. Although this number is crucial for the exponential decay property stated in the lemma, the tail term of course also affects the speed of decay.

Set $\eta(r) = Ce^{-\sqrt{2}\mu_0(1-\varepsilon)r}$ with $C > 0$. Then (8.6.48) leads to

$$\begin{aligned}
(\tau - \eta)'' &= \frac{\lambda}{2}\rho(\rho + \rho_0)(\tau - \eta) \\
&\quad + C\left(\frac{\lambda}{2}\rho(\rho + \rho_0) - 2\mu_0^2(1 - \varepsilon)^2\right)e^{-\sqrt{2}\mu_0(1-\varepsilon)r} \\
&\quad + \frac{1}{2r}\left(f^2 - \frac{1}{2}r^2[A - B]^2\right)\rho \\
&\geq \frac{\lambda}{2}\rho(\rho + \rho_0)(\tau - \eta), \quad r > r_\varepsilon,
\end{aligned} \tag{8.6.49}$$

where $r_\varepsilon > 0$ is sufficiently large. Since the finite energy condition implies the existence of a sequence $\{r_j\}$ which goes to infinity so that $\tau(r_j) \to 0$ as $j \to \infty$, we can choose $C > 0$ large enough to make $(\tau - \eta)(r) < 0$, $r > r_\varepsilon$.

To get the other half of the estimate, we consider $\tau + \eta$ instead. We have, in place of (8.6.49), the inequality $(\tau + \eta)'' \leq (\lambda/2)\rho(\rho + \rho_0)(\tau + \eta)$ for $r >$ some large r_ε. Hence, there holds $(\tau + \eta)(r) > 0$, $r > r_\varepsilon$ when the coefficient C in the definition of η is made large enough. So the decay estimate for $\rho - \rho_0$ near infinity stated in (8.6.44) is established.

Finally, we rewrite the third equation in (8.5.1) as

$$(r[A - \rho_0 a_0])'' = \frac{2}{r}f^2 A + \frac{g^2}{4}r\rho^2(A - B). \tag{8.6.50}$$

In view of the finite energy condition, we have $r_j A'(r_j) \to 0$ for some sequence $r_j \to \infty$. Hence, (8.6.50) implies

$$(r[A(r) - \rho_0 a_0])' = -\int_r^\infty ds\left\{\frac{2}{s}f^2(s)A(s) + \frac{g^2}{4}s\rho^2(s)(A - B)(s)\right\}. \tag{8.6.51}$$

Inserting the obtained decay estimates for f and $A - B$ into (8.6.51), we see that the function $(r[A(r) - \rho_0 a_0])'$ also vanishes exponentially fast at infinity. In particular, we see that the stated decay estimate for $A - \rho_0 a_0$ near infinity holds.

The proof of the lemma is now complete.

Some other forms of decay estimates may be of use when we evaluate the associated electric and magnetic charges for a dyon solution. For example, we state

Lemma 8.6.8. *Let (f, ρ, A, B) be the solution of the dyon equations (8.5.1) obtained in the last section. Then $f', r^2\rho', r^2(A' - B')(r)$ all vanish at infinity exponentially fast.*

Proof. Since the proofs for these are similar, we only consider $r^2(A' - B')$. Indeed the argument following (8.6.51) in the proof of Lemma 8.6.7 shows that

$$(r[A - B])' = (r[A - \rho_0 a_0])' - (r[B - \rho_0 a_0])'$$

goes to zero exponentially fast as $r \to \infty$. Hence, in view of Lemma 8.6.7, so does $r(A' - B') = (r[A - B])' - (A - B)$. Consequently, the claim for $r^2(A' - B')$ follows.

We now study the decay estimates in the limit $r \to 0$.

Lemma 8.6.9. *The solution quartet (f, ρ, A, B) of (8.5.1) satisfies, near $r = 0$, the following asymptotic estimates,*

$$\begin{aligned} f &= 1 + O(r^{2(1-\varepsilon)}), \quad \rho = O(r^{\delta(1-\varepsilon)}), \\ A &= O(r^{(1-\varepsilon)}), \quad B = b_0 + O(r^{(1+\sqrt{3})(1-\varepsilon)}), \end{aligned} \tag{8.6.52}$$

where $\delta = (-1 + \sqrt{3})/2$ and ε is an arbitrary number lying in the interval $(0, 1)$.

Proof. Consider the second equation in (8.5.1) and introduce the new variable $s = \ln r$. Linearization at $\rho = 0, s = -\infty$ gives us $\rho_{ss} + \rho_s - \rho/2 = 0$ whose characteristic roots are $(-1 + \sqrt{3})/2$ and $(-1 - \sqrt{3})/2$. Thus, we may use the method in the proof of Lemma 8.6.7 to get the estimate for ρ (now at $s = -\infty$) stated in (8.6.52).

Next we study the decay rate of A. We rewrite the third equation in (8.5.1) as

$$A_{ss} + A_s = 2f^2 A + \frac{g^2}{4}e^{2s}\rho^2(A - B). \tag{8.6.53}$$

Since $f(s = -\infty) = 1$, $e^{2s}\rho^2(s) = O(e^{2(1+\delta[1-\varepsilon])s})$, and the characteristic roots of $A_{ss} + A_s - 2A = 0$ are -2 and 1, we see that the solution A of (8.6.53) verifies $A(s) = O(e^{(1-\varepsilon)s})$ as $s \to -\infty$ as claimed in (8.6.52).

For f we rewrite the first equation in (8.5.1) in the form

$$(f-1)_{ss} - (f-1)_s = (f+1)f(f-1) + e^{2s}\left(\frac{g^2}{4}\rho^2 - A^2\right)f. \tag{8.6.54}$$

Since $e^{2s}A^2(s) = O(e^{(4-\varepsilon)s})$, $(f+1)f$ goes to 2 as $s \to -\infty$, and the characteristic roots of $\tau_{ss} - \tau_s - 2\tau = 0$ are -1 and 2, we see that in (8.6.54) we have $f(s) - 1 = O(e^{2(1-\varepsilon)s})$ as $s \to -\infty$. Thus, the estimate for f also holds.

In view of the estimate for ρ, we see that the right-hand side of the last equation in (8.5.1) behaves like $r^{(1+\sqrt{3})(1-\varepsilon)}$. Integrating this equation and using $r^2 B'(r) \to 0$ as $r \to 0$ we arrive at $r^2 B'(r) = O(r^{(1+\sqrt{3})(1-\varepsilon)+1})$ or $B'(r) = O(r^{(1+\sqrt{3})(1-\varepsilon)-1})$ near $r = 0$. Integrating this last relation again gives $B(r) - b_0 = O(r^{(1+\sqrt{3})(1-\varepsilon)})$.

The proof of the lemma is thus complete.

8.6.6   Electric and magnetic charges

Finally we consider the model within the unitary gauge under which ξ defined in (8.4.2) is transformed into $(0,1)^\tau$, where $^\tau$ means taking transposition. Hence, the corresponding group element is

$$U = -\mathrm{i} \begin{pmatrix} \cos(\theta/2) & \sin(\theta/2)\mathrm{e}^{-\mathrm{i}\varphi} \\ \sin(\theta/2)\mathrm{e}^{\mathrm{i}\varphi} & -\cos(\theta/2) \end{pmatrix}.$$

Therefore, the vector field $\mathbf{A}_\mu$ is mapped to

$$\mathbf{A}_\mu \mapsto \frac{1}{g} \begin{pmatrix} (\sin\varphi\partial_\mu\theta + \sin\theta\cos\varphi\partial_\mu\varphi)f(r) \\ (-\cos\varphi\partial_\mu\theta + \sin\theta\sin\varphi\partial_\mu\varphi)f(r) \\ -A(r)\partial_\mu t - (1-\cos\theta)\partial_\mu\varphi \end{pmatrix}.$$

So in this gauge we have

$$A^3_\mu = -\frac{1}{g}A(r)\partial_\mu t - \frac{1}{g}(1-\cos\theta)\partial_\mu\varphi. \tag{8.6.55}$$

With A^3_μ defined in (8.6.55) and B_μ defined in (8.4.2), we can use the Weinberg angle θ_W determined by the relation

$$\cos\theta_W = \frac{g}{\sqrt{g^2 + g'^2}}$$

to designate the electromagnetic potential $\mathcal{A}_\mu$ and the neutral potential $\mathcal{Z}_\mu$ by

$$\begin{aligned} \mathcal{A}_\mu &= B_\mu\cos\theta_W + A^3_\mu\sin\theta_W, \\ \mathcal{Z}_\mu &= -B_\mu\sin\theta_W + A^3_\mu\cos\theta_W. \end{aligned}$$

From the above results we find

$$\begin{aligned} \mathcal{A}_\mu &= -e\left(\frac{1}{g^2}A + \frac{1}{g'^2}B\right)\partial_\mu t - \frac{1}{e}(1-\cos\theta)\partial_\mu\varphi, \\ \mathcal{Z}_\mu &= \frac{e}{gg'}(B - A)\partial_\mu t, \end{aligned} \tag{8.6.56}$$

where e is the electric charge of the positron: $e = g\sin\theta_W = gg'/\sqrt{g^2 + g'^2}$.

To calculate the associated electric and magnetic charges, we first get the field strengths

$$\mathcal{F}_{\mu\nu} = \partial_\mu \mathcal{A}_\nu - \partial_\nu \mathcal{A}_\mu, \quad \mathcal{Z}_{\mu\nu} = \partial_\mu \mathcal{Z}_\nu - \partial_\nu \mathcal{Z}_\mu$$

and then follow the standard prescription (taking $\mathcal{F}$, for example)

$$\mathbf{E} = (\mathcal{F}_{01}, \mathcal{F}_{02}, \mathcal{F}_{03})^\tau, \quad \mathbf{B} = (\mathcal{F}_{23}, \mathcal{F}_{31}, \mathcal{F}_{12})^\tau$$

to obtain the associated electric and magnetic fields. Thus, the corresponding charges are determined by the Gauss law,

$$q_e = \frac{1}{4\pi} \lim_{r \to \infty} \oint_{|x|=r} \mathbf{E} \cdot \mathrm{d}\mathbf{S},$$

$$q_m = \frac{1}{4\pi} \lim_{r \to \infty} \oint_{|x|=r} \mathbf{B} \cdot \mathrm{d}\mathbf{S}. \tag{8.6.57}$$

In the electromagnetic vector potential given in the first equation in (8.6.56), the second term on the right-hand side defines the classical Dirac string and gives rise to the magnetic field. Therefore, using the second formula in (8.6.57), we obtain the total magnetic charge,

$$q_m = \frac{1}{e}.$$

The first term on the right-hand side of the same equation in (8.6.56) defines the electric field. Thus the first formula in (8.6.57) yields

$$
\begin{aligned}
q_e &= \frac{1}{4\pi} \lim_{r \to \infty} \int_{|x|<r} \nabla \cdot \mathbf{E} \, \mathrm{d}x = e \int_0^\infty \mathrm{d}r \left\{ \left(r^2 \left[\frac{1}{g^2} A' + \frac{1}{g'^2} B' \right] \right)' \right\} \\
&= \frac{2}{e} \sin^2 \theta_W \int_0^\infty f^2 A \, \mathrm{d}r = e \left(r^2 \left[\frac{1}{g^2} A' + \frac{1}{g'^2} B' \right] \right) \Big|_{r=\infty} > 0,
\end{aligned}
$$

where, in deriving the last line above, we have applied (8.6.31) and (8.6.32).

For the neutral Z boson field, it is immediate from the second equation in (8.6.56) that the 'magnetic' field is absent and the 'electric' charge defined by the first expression in (8.6.57) is similarly given by

$$
\begin{aligned}
&-\frac{e}{gg'} \int_0^\infty \mathrm{d}r \{ (r^2[B' - A'])' \} \\
&= -\frac{e}{gg'} \left(\lim_{r \to \infty} (r^2[B' - A'])(r) - \lim_{r \to 0} (r^2[B' - A'])(r) \right) \\
&= -\frac{e}{4gg'} \int_0^\infty \mathrm{d}r \{ (g^2 + g'^2) r^2 \rho^2 (B - A) - 8 f^2 A \}, \tag{8.6.58}
\end{aligned}
$$

where we have used the equation (8.6.38). The exponential decay results stated in Lemma 8.6.7 imply the convergence of the last integral in (8.6.58).

Consequently the two limits in the middle of the equation (8.6.58) both exist. In fact the first limit is zero in view of Lemma 8.6.8. We also claim

$$\lim_{r \to 0} (r^2[B' - A'])(r) = 0.$$

To see this, we may argue again by contradiction: If the above is not true, then there is a positive number $\varepsilon_0 > 0$ and an $r_0 > 0$ so that

$$|r^2[B' - A']| \geq \varepsilon_0 \quad \text{for } 0 < r < r_0.$$

Thus

$$(r[B' - A'])^2 \geq \frac{\varepsilon_0^2}{r^2} \quad (r < r_0),$$

which violates the convergence

$$\int_0^\infty (r[B' - A'])^2 \, dr < \infty$$

derivable from the finite energy condition. Using this conclusion, we see that the quantity defined in (8.6.58) vanishes. In other words, we arrive at the zero 'electric' charge as well for the Z boson.

Summarizing our study in this section, we see that the proof of Theorem 8.4.1 is finally complete.

8.7 Remarks

In this chapter, we have seen that the existence of dyons in gauge field theory requires sophisticated analysis of some variational problems involving indefinite functionals. The original solution of the Julia–Zee dyon problem was due to Schechter and Weder [272], who used an abstract framework assisted with a few layers of the topological properties of the problem. In [351], the method of Schechter and Weder was modified to solve the more difficult electroweak dyon problem proposed in the work of Cho and Maison [83] and such a modified method may be simplified to reproduce directly the Julia–Zee dyons, as described in the previous sections. It is interesting to notice that in each situation dyons are a continuous family of classical solutions of the corresponding equations of motion and thus uniqueness does not hold. It is hoped that the methods presented in this chapter will be useful in solving other more challenging dyon problems also characterised by indefinite action functionals. For example, an important problem is

Open Problem 8.7.1. *Prove the existence of electrically and magnetically charged vortex solutions in the non-self-dual Chern–Simons–Higgs theory.*

For a description of the equations of motion for the above problem, see the work of Paul–Khare [243], de Vega–Schaposnik [87], and Kumar–Khare [174].

For a nice review on the history of monopoles and dyons, see the article by Goddard and Olive [130]. For the dynamics of the BPS monopoles, see [13, 209, 301].

Finally, we note that, at the sacrifice of minimal coupling and with a suitable design of nonlinearity, it is possible to have finite-energy Abelian monopoles [186, 350].

9
Ordinary Differential Equations

Many of the problems in two dimensions involve the Dirac function type singular source terms. The purpose of this chapter is to present a systematic study of one of the simplest such problems, namely, the radially symmetric solutions of a general scalar equation. In §9.1, we state our existence results. In §9.2, we present a dynamical analysis of all possible solutions. In §9.3, we show how to apply our results to obtain a complete understanding of symmetric Chern–Simons vortices. We shall later show that the results may be used to achieve a deep understanding of cosmic string solutions in terms of the string number and the universal gravitational constant.

9.1   Existence Results

The purpose of our study in this chapter is to achieve a thorough understanding of the radially symmetric solutions of the scalar equation

$$\Delta u + p(|x|)q(e^u) = 4\pi N\delta(x), \qquad x \in \mathbb{R}^2, \tag{9.1.1}$$

which arises in several important areas of field theory, including superconductivity and cosmology [77].

Since we shall concentrate on radially symmetric solutions, we see that, upon setting $r = |x|$ and $u = u(r)$, (9.1.1) becomes

$$u_{rr}(r) + \frac{1}{r}u_r(r) + p(r)q(e^{u(r)}) = 0, \qquad r > 0,$$

$$u(r) = 2N\ln r + O(1) \qquad \text{for small } r > 0.$$

Under the new variables

$$t = \ln r, \qquad U(t) = u(r), \qquad (9.1.2)$$

the problem becomes

$$U''(t) + f(t)g(U(t)) = 0, \qquad -\infty < t < \infty, \qquad (9.1.3)$$
$$U(t) = \alpha t + O(1) \qquad \text{as } t \to -\infty, \qquad (9.1.4)$$

where

$$f(t) = e^{2t}p(e^t), \quad g(u) = q(e^u), \quad \alpha = 2N.$$

Motivated by physical models, we shall assume that f and g satisfy the following conditions.

(H1) $f, g \in C^1(\mathbb{R})$ and

$$\int_{-\infty}^{0} |tf(t)| \, dt < \infty,$$

$$\sup_{u \in \mathbb{R}} \left\{ |g(u)| + |g'(u)| \right\} < \infty.$$

(H2) $f(\cdot) > 0$ in $\mathbb{R}$, $\lim_{t \to \infty} f(t) = \infty$, $g(0) = 0$, and $g(u) > 0$ for all $u < 0$.

(H3) $f'(t) \geq 0$ for all $t \in \mathbb{R}$.

(H4) There exists $M > 0$ such that $g'(u) > 0$ when $u < -M$ and

$$\int_{0}^{\infty} f(t)g(-Mt) \, dt < \infty.$$

(H5) If one defines

$$M_0 = \inf \left\{ M > 0 \, \middle| \, \int_{0}^{\infty} f(t)g(-Mt) \, dt < \infty \right\},$$

then

$$\int_{0}^{\infty} f(t)g(-M_0 t) \, dt = \infty.$$

In addition, for every $c > 0$,

$$\inf_{t > 0} \frac{f(t - c)}{f(t)} > 0.$$

(H6) Let $G_0(u) = \int_{-\infty}^{u} g(w) \, dw$. (Note that the assumptions (H2)–(H4) imply that

$$\int_{-\infty}^{0} g(u) \, du < \infty,$$

so that $G_0(u)$ is well defined.) Define

$$F_1(t) = \frac{f'(t)}{f(t)} \quad \text{and} \quad G_1(u) = \frac{G_0(u)}{g(u)}.$$

Then both $f_1 = \lim_{t \to \infty} F_1(t)$ and $g_1 = \lim_{u \to -\infty} G_1(u)$ exist and are finite.

(H7) The functions F_1 and G_1 defined in the assumption (H6) satisfy $F_1(t) \geq f_1$ for all $t \in \mathbb{R}$ and $G_1(u) \geq g_1$ for all $u \in (-\infty, 0)$.

(H8) There exists $\delta > 0$ such that $g'(u) \leq 0$ in $[-\delta, 0]$.

Our main result on (9.1.3), (9.1.4) is the following.

Theorem 9.1.1. *Consider the differential equation (9.1.3) with the boundary condition (9.1.4) where $\alpha \geq 0$ is a given constant and $f(\cdot)$ and $g(\cdot)$ satisfy (H1)–(H3). Then*

(i) There exists at least one solution of (9.1.3) and (9.1.4) such that $u \leq 0$, $u' \geq 0$, $u'' \leq 0$ in $\mathbb{R}$ (the equal signs hold only if $\alpha = 0$), and

$$\lim_{t \to \infty} u(t) = 0. \tag{9.1.5}$$

If in addition (H8) is fulfilled, then there exists a unique non-positive solution satisfying (9.1.5).

(ii) Assume also (H4)–(H6). Then, for every β in $(\alpha + 2f_1 g_1, \infty)$, there exists at least one solution u of (9.1.3) and (9.1.4), such that $u < 0$, $u'' < 0$ in $\mathbb{R}$ and

$$\lim_{t \to \infty} u'(t) = -\beta. \tag{9.1.6}$$

If in addition (H7) holds, then for any non-positive solution of (9.1.3) satisfying

$$\liminf_{t \to \infty} u(t) < 0,$$

there exists some $\beta \in (\alpha + 2f_1 g_1, \infty)$ to achieve (9.1.6).

The proof will be given in the next section.

9.2 Dynamical Analysis

In this section, we shall use a dynamical system approach to study (9.1.3) and (9.1.4) under the assumption (H1)–(H8). We first obtain local solutions near $-\infty$. We next analyze the initial data in relation to the global behavior of solutions. We then obtain a complete classification of the solutions.

9.2.1 Local solution via contractive mapping

First we establish the existence of the initial value problem for the equation (9.1.3).

Lemma 9.2.1. *Assume that (H1) holds. Then for any constants $\alpha \in \mathbb{R}$ and $a \in \mathbb{R}$, the equation (9.1.3) admits a unique solution U such that when $t \to -\infty$,*

$$U(t) = \alpha t + a + o(1). \tag{9.2.1}$$

Conversely, if $U(t)$ is a solution of (9.1.3) in some interval, then it can be uniquely extended to a global solution of (9.1.3) in $\mathbb{R}$ so that (9.2.1) holds for some $\alpha, a \in \mathbb{R}$.

Proof. One can directly verify that U is a solution of (9.1.3) satisfying (9.2.1) if and only if U verifies the integral equation

$$U(t) = \alpha t + a - \int_{-\infty}^{t} (t - s) f(s) g(U(s)) \, ds, \qquad t \in \mathbb{R}. \tag{9.2.2}$$

Let $T \in \mathbb{R}$ be a constant such that

$$\int_{-\infty}^{T} (T - s)|f(s)| \, ds = \int_{-\infty}^{T} \int_{-\infty}^{s_1} |f(s)| \, ds \, ds_1$$

$$< \frac{1}{2 \sup_{u \in \mathbb{R}} \{1 + |g(u)| + |g'(u)|\}}.$$

Then one can use the Picard successive iteration method, with the initial iteration $U^{(0)} = \alpha t + a$, to establish a solution in the interval $(-\infty, T]$. Since g is bounded, we can extend U to a solution of (9.1.3) in $\mathbb{R}$.

Next we prove the uniqueness. Assume that U^1 and U^2 are two solutions of (9.2.2) in the interval $(-\infty, T]$. Then their difference $U \equiv U^1 - U^2$ satisfies

$$|U(t)| = \left| \int_{-\infty}^{t} (t - s) f(s) (g(U^1(s)) - g(U^2(s))) \, ds \right|$$

$$\leq \sup_{u \in \mathbb{R}} |g'(u)| \int_{-\infty}^{T} (T - s)|f(s)| \, ds \sup_{(-\infty, T]} |U(\cdot)|$$

$$\leq \frac{1}{2} \sup_{(-\infty, T]} |U(\cdot)|, \qquad t < T,$$

by the assumption on T. Since the first equation implies that

$$\sup_{(-\infty, T]} |U(\cdot)| < \infty,$$

we obtain, upon taking the superum on the left-hand side of the above inequality, that $\sup_{(-\infty, T]} |U(\cdot)| = 0$; namely, $U^1 = U^2$ in $(-\infty, T]$. Hence, by the unique continuation, $U^1 = U^2$ in $\mathbb{R}$.

Finally, we prove the last assertion of the lemma. Assume that $U(t)$ is a solution of (9.1.3) in some interval. Then since $g(\cdot)$ is Lipschitz and bounded, U can be uniquely extended to a solution of (9.1.3) in $\mathbb{R}$. Noting that

$$\int_{-\infty}^{0} |f(s)g(U(s))|\,ds < \infty$$

and for any $t < 0$,

$$U'(t) = U'(0) + \int_{t}^{0} f(s)g(U(s))\,ds,$$

we know that

$$\alpha \equiv \lim_{t \to -\infty} U'(t)$$

exists and

$$\alpha = U'(0) + \int_{-\infty}^{0} f(s)g(U(s))\,ds.$$

Consequently, for any $t \in \mathbb{R}$,

$$U'(t) = \alpha - \int_{-\infty}^{t} f(s)g(U(s))\,ds$$

and

$$U(t) = U(0) + \alpha t - \int_{0}^{t} \int_{-\infty}^{s_1} f(s)g(U(s))\,ds\,ds_1. \tag{9.2.3}$$

Since

$$\int_{-\infty}^{t} \int_{-\infty}^{s_1} |f(s)g(U(s))|\,ds = \int_{-\infty}^{t} (t-s)|f(s)g(U(s))|\,ds < \infty,$$

we can write (9.2.3) as

$$U(t) = \alpha t + \left(U(0) + \int_{-\infty}^{0} \int_{-\infty}^{s_1} f(s)g(U(s))\,ds\,ds_1 \right)$$
$$- \int_{-\infty}^{t} \int_{-\infty}^{s_1} f(s)g(U(s))\,ds\,ds_1;$$

i.e., U satisfies (9.2.1) with

$$a = U(0) + \int_{-\infty}^{0} \int_{-\infty}^{s_1} f(s)g(U(s))\,ds\,ds_1.$$

This completes the proof of the lemma.

9.2.2 Parameter sets

In the sequel, we shall study the behavior of the solution as $t \to \infty$. To do this, we shall fix the constant $\alpha \geq 0$, and vary the parameter $a \in \mathbb{R}$. For convenience, we denote by $u(t, a)$ the solution given by Lemma 9.2.1 and denote by $'$ the derivative with respect to t and by a subscript $_a$ the derivative with respect to a.

Define

$$\mathcal{A}^+ = \left\{ a \in \mathbb{R} \,\middle|\, \text{there exists } t \in \mathbb{R} \text{ such that } u(t,a) > 0 \right\},$$

$$\mathcal{A}^0 = \left\{ a \in \mathbb{R} \,\middle|\, u(t,a) \leq 0,\ u'(t,a) \geq 0\ \forall t \in \mathbb{R} \right\},$$

$$\mathcal{A}^- = \left\{ a \in \mathbb{R} \,\middle|\, u(t,a) \leq 0\ \forall t \in \mathbb{R},\quad u'(t_0,a) < 0 \text{ for some } t_0 \in \mathbb{R} \right\}.$$

Clearly, the following relations hold,

$$\mathcal{A}^+ \cup \mathcal{A}^0 \cup \mathcal{A}^- = \mathbb{R}, \qquad \mathcal{A}^+ \cap \mathcal{A}^0 = \mathcal{A}^0 \cap \mathcal{A}^- = \mathcal{A}^+ \cap \mathcal{A}^- = \emptyset.$$

Lemma 9.2.2. *Assume (H1) and (H2). Then, the following holds.*
  (i) If $a \in \mathcal{A}^+$, then $u' > 0$ in the set $\{t \mid u(\tau, a) < 0\ \forall \tau \in (-\infty, t)\}$.
  (ii) If $a \in \mathcal{A}^0$, then $u'' \leq 0$ and $u' \geq 0$ in $\mathbb{R}$ and $\lim_{t \to \infty} u(t, a) = 0$.
  (iii) If $a \in \mathcal{A}^-$, then $u'' < 0$, $u < 0$ in $\mathbb{R}$ and $\lim_{t \to \infty} u(t, a) = -\infty$.
  (iv) $\mathcal{A}^+$ is open and if

$$a > M_1 \equiv \sup_{u \in \mathbb{R}} |g(u)| \int_{-\infty}^{0} |s f(s)|\, ds,$$

then $a \in \mathcal{A}^+$.
  (v) $\mathcal{A}^-$ is open.
  (vi) Let T be a positive constant such that

$$\left(\inf_{u \in [-2, -1]} g(u) \right) \inf_{t > T} \int_{t}^{t + \frac{1}{\alpha}} f(s)\, ds > 1 + \alpha.$$

Then $(-\infty, -M_1 - 2 - \alpha T) \subset \mathcal{A}^-$.
  (vii) $\mathcal{A}^0$ is non-empty, closed, and bounded.

Proof. (i) Let $a \in \mathcal{A}^+$ and t_0 be the first time at which $u(t, a)$ hits the t axis from below. Then $u(t, a) < 0$ for all $t \in (-\infty, t_0)$. Hence, by the assumption (H2) and the equation (9.1.3), $u'' < 0$ in $(-\infty, t_0)$, which implies that $u'(t, a) > 0$ in $(-\infty, t_0)$. The first assertion of the lemma thus follows.

(ii) If $a \in \mathcal{A}^0$, then by the assumption (H2), the equation (9.1.3), and the definition of $\mathcal{A}^0$, $u'' \leq 0$ in $\mathbb{R}$. In addition, $b \equiv \lim_{t \to \infty} u(t, a)$ exists and is non-positive. If $b < 0$, then $\lim_{t \to \infty} u''(t, a) = -g(b) \lim_{t \to \infty} f(t) = -\infty$, which is impossible. Hence, $b = 0$.

(iii) Since the only solution of (9.1.3) with $U(t_0) = U'(t_0) = 0$ is $U \equiv 0$, it follows that if $a \in \mathcal{A}^-$ then $u(\cdot, a) < 0$ in $\mathbb{R}$, and therefore $u''(\cdot, a) < 0$ in $\mathbb{R}$; that is, $u'(t, a)$ strictly decreases. Hence, $\limsup_{t \to \infty} u'(t, a) < 0$. Assertion (iii) of the lemma thus follows.

(iv) Since $u(t, a)$ is continuous in a (cf. the uniqueness proof of Lemma 9.2.1), if $u(t_0, a_0) > 0$, then $u(t_0, a) > 0$ when a is close to a_0; that is, $\mathcal{A}^+$ is open. From (9.2.2), $u(0, a) > a - M_1 > 0$ if $a > M_1$, so that $(M_1, \infty) \subset \mathcal{A}^+$.

(v) Assume $a_0 \in \mathcal{A}^-$. Then there exists $t_0 \in \mathbb{R}$ such that $u'(t_0, a_0) < 0$, and consequently, $u'(t_0, a) < 0$ when a is close to a_0. In addition, by the third assertion of the lemma, $u(t, a_0) < 0$ for all $t \leq t_0$, which also implies that $u(t, a) < 0$ for all $t \leq t_0$ and a close to a_0. (When t is negatively very large, use (9.2.1); when t is in a compact subset, use the continuity of the solution in a.) Furthermore, since the assumption (H2) implies that any solution of (9.1.3) cannot take a local negative minimum, $u'(t, a) \leq 0$ for all $t > t_0$ as long as $u(t_0, a) < 0$ and $u'(t_0, a) < 0$. Therefore, $u(t, a) < 0$ for all $t > t_0$ when a is close to a_0. That is, $\mathcal{A}^-$ is open.

(vi) We need only consider the case $\alpha > 0$ since when $\alpha = 0$, $\mathcal{A}^- = (-\infty, 0)$. Let $a < -M_1 - 2 - \alpha T$ be any constant. From (9.2.2), $u(a, t) \leq \alpha t + a + M_1 < -2$ for all $t \in (-\infty, 0]$. Since u cannot take a local negative minimum, it follows that if $a \notin \mathcal{A}^-$, then there exist positive constants T_1 and T_2 such that $T_2 < T_1$, $u(t, a) \leq -2$ in $(-\infty, T_2]$, $u(T_2, a) = -2$, $u'(T_2, a) \geq 0$, $u(t, a) \in [-2, -1]$ for all $t \in [T_2, T_1]$, $u(T_1, a) = -1$, and $u'(T_1, a) \geq 0$. It then follows that $u''(t, a) = -f(t)g(u(t)) \leq 0$ for all $t \leq T_1$, which implies that $u'(t, a) \leq \alpha$ for all $t \in (-\infty, T_1]$. Therefore, $T_2 \geq [u(T_2, a) - u(0, a)]/\alpha \geq [-2 - a - M_1]/\alpha > T$ and $T_1 - T_2 \geq 1/\alpha$. Consequently,

$$
\begin{aligned}
u'(T_1, a) &= u'(T_2, a) - \int_{T_2}^{T_1} f(s) g(u(s)) \, ds \\
&\leq \alpha - \left(\inf_{t \geq T} \int_t^{t + \frac{1}{\alpha}} f(s) \, ds \right) \left(\inf_{u \in [-2, -1]} g(u) \right) \\
&< -1,
\end{aligned}
$$

by the definition of T, which contradicts the assumption that $u'(T_1, a) \geq 0$. Hence $a \in \mathcal{A}^-$.

(vii) Since $\mathbb{R}$ cannot be decomposed into two disjoint non-empty open sets, the assertion follows from the conclusions (iv)–(vi).

The following lemma deals with the monotonicity of the solution with respect to the parameter a and will play an essential role in analyzing the three sets $\mathcal{A}^+$, $\mathcal{A}^0$, and $\mathcal{A}^-$.

Lemma 9.2.3. *Assume (H1)–(H3) and let $T_0(a) \in [-\infty, \infty]$ be the first time such that either $u'(t, a) > 0$ or $u(t, a) < 0$ is violated, namely,*

$$
T_0(a) = \sup \left\{ T \in [-\infty, \infty] \,\middle|\, u(t, a) < 0, u'(t, a) > 0 \quad \forall t \in (-\infty, T) \right\}.
$$

Then

$$u_a(t,a) \geq \frac{1}{\alpha} u'(t,a) > 0 \quad \forall t \in (-\infty, T_0(a)).$$

Proof. We need only consider the case $T_0(a) > -\infty$. From (9.2.2) and the standard ODE techniques on the continuous dependence of solutions on the the parameters, one can show that $v(t,a) \equiv u_a(t,a)$ exists, is smooth, and satisfies

$$v''(a,t) = -f(t)g'(u(t,a))v(t,a), \quad -\infty < t < \infty,$$
$$\lim_{t \to -\infty} v(t,a) = 1, \quad \lim_{t \to -\infty} v'(t,a) = 0.$$

Define $T_1(a) = \sup\{\tau \in \mathbb{R} \mid v(\cdot, a) > 0 \text{ in } (-\infty, \tau)\}$. Then, by the last two equations, $T_1(a) > -\infty$.

Set $w = u'$. Then, we see that $\lim_{t \to -\infty} w(t,a) = \alpha$, and by (9.1.3), $\lim_{t \to -\infty} w'(t,a) = 0$. It follows that the function $C(t,a) \equiv w(t,a)/v(t,a)$, $t \in (-\infty, T_1(a))$, satisfies $\lim_{t \to -\infty} C(t,a) = \alpha$ and $\lim_{t \to -\infty} C'(t,a) = 0$. Since the function w satisfies the equation

$$w'' = -f(t)g'(u)w - f'(t)g(u),$$

the method of variation of parameters yields

$$C'(t,a) = -\frac{1}{v^2(t,a)} \int_{-\infty}^{t} f'(s)u_a(t,a)g(u(s,a))\,ds,$$
$$\forall t \in (-\infty, T_1(a)). \tag{9.2.4}$$

Since $f' \geq 0$, it follows that $C' \leq 0$ and therefore $C \leq \alpha$ in the set $(-\infty, T_1(a))$; that is, $v(t,a) \geq \frac{1}{\alpha} w(t,a)$ in $(-\infty, T_1(a))$. Clearly this implies that $T_0(a) \leq T_1(a)$. The lemma thus follows.

The following statements characterize the sets $\mathcal{A}^+$, $\mathcal{A}^0$, and $\mathcal{A}^-$.

Lemma 9.2.4. *Assume (H1)–(H3). Then there exist constants a_1 and a_2 satisfying $a_1 \leq a_2$ such that*
(i) $\mathcal{A}^+ = (a_2, \infty)$;
(ii) $\mathcal{A}^- = (-\infty, a_1)$;
(iii) $\mathcal{A}^0 = [a_1, a_2]$;
(iv) if in addition (H8) holds, then $a_1 = a_2$.

Proof. We need only consider the case $\alpha > 0$ since in case $\alpha = 0$, one can directly verify that $\mathcal{A}^+ = (0, \infty)$, $\mathcal{A}^0 = \{0\}$, and $\mathcal{A}^- = (-\infty, 0)$.

(i) Since $\mathcal{A}^+$ is open, it suffices to show that if $(b_1, b_2) \subset \mathcal{A}^+$, then $b_2 \in \mathcal{A}^+$. For any $a \in (b_1, b_2)$, let $z_0(a)$ be the first time the solution crosses the t axis. (Since $\alpha > 0$, $\lim_{t \to -\infty} u(t,a) = -\infty$, so $z_0(a)$ is well defined.) Clearly, $u(z_0(a), a) = 0$, $u'(z_0(a), a) > 0$, and by Lemma 9.2.2 (i), $u' > 0$ in $(-\infty, z_0(a)]$. By Lemma 9.2.3, $u_a \geq \frac{1}{\alpha} u' > 0$ in $(-\infty, z_0(a)]$.

Applying the implicit function theorem to the equation $u(z_0(a), a) = 0$ then yields that $z_0(a)$ is a differentiable function of a in the set (b_1, b_2) and $\frac{d}{da} z_0(a) = -u_a(z_0(a), a)/u'(z_0(a), a) < 0$. Noting that (9.2.2) implies that $u \leq \alpha t + a$ in $(-\infty, z_0(a)]$, we then know that $z_0(a) \geq -a/\alpha$ for every $a \in (b_1, b_2)$. Thus $z_0(b_2) \equiv \lim_{a \nearrow b_2} z_0(a)$ exists and is finite. By continuity, $u(z_0(b_2), b_2) = 0$. Since $u'(z_0(b_2), b_2) = 0$ would result in $u(t, b_2) \equiv 0$, we also find that $u'(z_0(b_2), b_2) \neq 0$, which implies $u(t, b_2) > 0$ for t near $z_0(b_2)$. That is, $b_2 \in \mathcal{A}^+$. The first assertion of the lemma thus follows.

(ii) It is sufficient to show that $(b_1, b_2) \in \mathcal{A}^-$ implies $b_1 \in \mathcal{A}^-$. For every $a \in \mathcal{A}^-$, let $z_1(a)$ be the point where $u'(z_1(a), a) = 0$ and let $m(a) = u(z_1(a), a)$ be the maximum of $u(\cdot, a)$ in $\mathbb{R}$. Since $u''(z_1(a), a) < 0$, the implicit function theorem implies that $z_1(a)$ is a differentiable function on $\mathcal{A}^-$. Hence,

$$\frac{d}{da} m(a) = u'(z_1(a), a) \frac{d}{da} z_1(a) + u_a(z_1(a), a)$$
$$= u_a(z_1(a), a) \geq 0, \quad \forall a \in (b_1, b_2).$$

Consequently,

$$m(a) = \sup_{t \in \mathbb{R}} u(a, t) \leq m\left(\frac{b_1 + b_2}{2}\right), \quad \forall a \in \left(b_1, \frac{b_1 + b_2}{2}\right).$$

By continuity,

$$m(b_1) = \sup_{t \in \mathbb{R}} u(t, b_1) \leq m\left(\frac{b_1 + b_2}{2}\right) < 0.$$

This implies that $b_1 \in \mathcal{A}^0 \cup \mathcal{A}^-$. However, by Lemma 9.2.2 (ii), we can easily conclude that $b \notin \mathcal{A}^0$. Therefore, $b_1 \in \mathcal{A}^-$. This completes the proof of the second assertion.

(iii) Since $\mathcal{A}^0 = \mathbb{R} \setminus (\mathcal{A}^+ \cup \mathcal{A}^-)$, the assertion follows from the first two conclusions.

(iv) For every $a \in \mathcal{A}^0 = [a_1, a_2]$, we have $u' > 0$ in $\mathbb{R}$ and, by Lemma 9.2.3, $u_a(t, a) > 0$ in $\mathbb{R}$.

Moreover, since u is monotonic and $\lim_{t \to \infty} u(t, a) = 0$, for any $\delta > 0$ there is a continuous function $T_\delta(a)$ such that $u(T_\delta(a), a) = -\delta$ and $u(t, a) > -\delta$ in $(T_\delta(a), \infty)$. By the assumption (H8), $g'(u) \leq 0$ when $u \in [-\delta, 0]$. Therefore

$$u_a'' = -f(t) g'(u) u_a \geq 0 \qquad \forall t \geq T_\delta(a), \ a \in [a_1, a_2].$$

Hence, u_a is a non-negative convex function on $[T_\delta(a), \infty)$, so that

$$u_a(\infty, a) \equiv \lim_{t \to \infty} u_a(t, a)$$

exists and $u_a(\infty, a) \in [0, \infty]$. Suppose that we have shown $u_a(\infty, a) > 0$ for all $a \in [a_1, a_2]$. Then, by Fatou's lemma,

$$0 = \lim_{t \to \infty} (u(t, a_2) - u(t, a_1)) = \lim_{t \to \infty} \int_{a_1}^{a_2} u_a(t, a)\, da \geq \int_{a_1}^{a_2} u_a(\infty, a)\, da,$$

which implies that $a_1 = a_2$. It remains to show that $u_a(\infty, a) > 0$ for all $a \in [a_1, a_2]$.

Suppose, on the contrary, that $u_a(\infty, a) = 0$ for some $a \in [a_1, a_2]$. Then by (9.2.4), the function $C = u'/u_a$ satisfies

$$
\begin{aligned}
C'(t,a) &= -\frac{1}{u_a^2(t,a)} \cdot \int_{-\infty}^{t} f'(s) u_a(s,a) g(u(s,a)) ds \\
&\leq -\frac{1}{u_a^2(t,a)} \int_{-\infty}^{0} f'(s) u_a(s,a) g(u(s,a)) ds \\
&\to -\infty \quad \text{as } t \to \infty,
\end{aligned}
$$

which implies that $C(t,a) < 0$ when t is large enough. However, this is impossible since $C = u'/u_a > 0$ for all $t \in \mathbb{R}$. This proves that $u_a(\infty, a) > 0$ for all $a \in [a_1, a_2]$ and thus the last assertion of the lemma follows.

9.2.3 Asymptotic limits

Now we want to find more detailed behavior of the solution $u(t,a)$ when $t \to \infty$ and $a \in \mathcal{A}^-$.

Lemma 9.2.5. *Assume (H1)–(H4). Then for any $a \in \mathcal{A}^-$,*

$$
\beta(a) \equiv - \lim_{t \to \infty} u'(t,a)
$$

exists and is positive and finite.

Proof. Since for any $a \in \mathcal{A}^-$, $u'' < 0$ in $\mathbb{R}$, it follows that $\beta(a)$ exists and belongs to the interval $(0, \infty]$. We want to show that $\beta(a) < \infty$.

Assume that $\beta(a) > M$ where M is the constant in the assumption (H4). Then there exists a constant $T > 1$ such that $u(t,a) \leq -Mt$ for all $t > T$. Since $g'(u) > 0$ when $u < -M$, it follows that $g(u(t,a)) \leq g(-Mt)$ when $t \geq T$. Consequently, for all $t > T$,

$$
\begin{aligned}
u'(t,a) &= u'(T,a) - \int_{T}^{t} f(s) g(u(s,a)) \, ds \\
&\geq u'(T,a) - \int_{T}^{t} f(s) g(-Ms) \, ds \\
&\geq u'(T,a) - \int_{0}^{\infty} f(s) g(-Ms) \, ds.
\end{aligned}
$$

Therefore, $\beta(a) \leq -u'(T,a) + \int_{0}^{\infty} f(s) g(-Ms) \, ds < \infty$. The assertion of the lemma thus follows.

Finally, we would like to find the range of $\beta(a)$ when a runs over the set $\mathcal{A}^-$. Note that the assumptions (H2)–(H4) imply that $\int_{-\infty}^{0} g(u) \, du$ is finite, so that we can define

$$G_0(u) = \int_{-\infty}^{u} g(w)\, dw. \tag{9.2.5}$$

Lemma 9.2.6. *Assume (H1)–(H4) and let G_0 be defined as in (9.2.5). Then for any $a \in A^-$, both the function $f(\cdot)g(u(\cdot, a))$ and the function $f'(\cdot)G_0(u(\cdot, a))$ are in $L^1(\mathbb{R})$ and there hold the relations*

$$\beta(a) + \alpha = \int_{\mathbb{R}} f(t)g(u(t, a))\, dt, \tag{9.2.6}$$

$$\frac{1}{2}\beta^2(a) - \frac{1}{2}\alpha^2 = \int_{\mathbb{R}} f'(t)G_0(u(t, a))\, dt. \tag{9.2.7}$$

Proof. Since $u'(t, a) = \alpha - \int_{-\infty}^{t} f(t)g(u(t, a))\, dt$ and $fg \geq 0$, the identity (9.2.6) follows from Lemma 9.2.5.

To show (9.2.7), we use the identity

$$\frac{d}{dt}\left(\frac{u'^2(t, a)}{2} + f(t)G_0(u(t, a))\right) = f'(t)G_0(u(t, a)),$$

which follows by multiplying (9.1.3) by u'. Integrating both sides over $[-T, T]$ yields

$$\int_{-T}^{T} f'(t)G_0(u(t, a))\, dt = \frac{u'^2(t, a)}{2}\Big|_{t=-T}^{t=T} + f(t)G_0(u(t, a))\Big|_{t=-T}^{t=T}.$$

Since the integrand on the left-hand side is positive, to finish the proof, we need only show that

$$\lim_{t \to \pm\infty} f(t)G_0(u(t, a)) = 0.$$

Since $f(t) \to 0$ as $t \to -\infty$ and $G_0(u)$ is bounded, we have

$$f(t)G_0(u(t, a)) \to 0$$

as $t \to -\infty$. It remains to show that $f(t)G_0(u(t, a)) \to 0$ as $t \to \infty$.

Since $u' > -\beta(a)$ in $\mathbb{R}$, for T sufficiently large so that $u' < 0$ when $t > T$, we have

$$\begin{aligned}
G_0(u(t, a)) &= \int_{-\infty}^{u(t, a)} g(u(s, a))\, du(s, a) \\
&= \int_{\infty}^{t} g(u(s, a))u'(s, a)\, ds \leq \beta(a) \int_{t}^{\infty} g(u(s, a))\, ds.
\end{aligned}$$

Hence, using the facts that f monotonically increases and that

$$f(\cdot)g(u(\cdot, a)) \in L^1(\mathbb{R}),$$

we obtain that

$$0 \leq f(t)G_0(u(t,a)) \leq \beta(a) \int_t^\infty f(s)g(u(s,a))\,\mathrm{d}s \to 0$$

as $t \to \infty$, thereby completing the proof of the lemma.

9.2.4 Continuous dependence

Lemma 9.2.7. *Assume that (H1)–(H5) hold. Then the function $\beta(a)$ is continuous in $\mathcal{A}^-$.*

Proof. Let $a_0 \in \mathcal{A}^-$ be any point and M, M_0 be the constants in the assumptions (H4) and (H5). First we claim that $\beta(a_0) > M_0$. In fact, if $M_0 = 0$, there is nothing to prove. Thus it suffices to assume $M_0 > 0$. If the claim is not true, then since $u''(t,a) < 0$, $u'(t,a) > -M_0$ in $\mathbb{R}$, which implies that there exists a positive constant C such that $u(t,a) \geq -C - M_0 t$ for all $t > 0$. Let $T > 0$ be a time such that $u(t,a_0) < -M$ for all $t > T$. Then

$$\int_T^\infty f(t)g(u(t,a_0))\,\mathrm{d}t \;\geq\; \int_T^\infty f(t)g(-C - M_0 t)\,\mathrm{d}t$$

$$\geq\; \inf_{t>0} \frac{f(t - C/M_0)}{f(t)} \int_{T + \frac{C}{M_0}}^\infty f(t)g(-M_0 t)\,\mathrm{d}t$$

$$=\; \infty$$

by the assumption (H5), contradicting the finiteness of $\beta(a_0)$ and (9.2.6). This shows that $\beta(a_0) > M_0$.

Let $\delta = (\beta(a_0) - M_0)/4$. Then there exists a positive constant T_1 such that $(M_0 + \delta)T_1 \geq M$, $u'(T_1, a_0) < -(M_0 + 2\delta)$, and $u(T_1, a_0) \leq -(M_0 + 2\delta)T_1$. Since $u(a,t)$ and $u'(a,t)$ are continuous in a, $u'(T_1, a) \leq -(M_0 + \delta)$ and $u(T_1, a) \leq -(M_0 + \delta)T_1$ when a is close to a_0. It then follows that, since $u'' < 0$, $u(t,a) \leq -(M_0 + \delta)t$ in $[T_1, \infty)$ for all a close to a_0. Let $W(t)$ be the function defined by $W(t) = \sup_{u \in \mathbb{R}} |g(u)|f(t)$ for $t \leq T_1$ and $W(t) = f(t)g(-(M_0 + \delta)t)$ for $t > T_1$. Then by the definition of M_0, $W \in L^1(\mathbb{R})$. In addition, when a is close to a_0, $f(t)g(u(t,a)) \leq W(t)$ for all $t \in \mathbb{R}$. The assertion of the lemma then follows from the Lebesgue dominated convergence theorem and the formula (9.2.6).

9.2.5 Critical behavior and conclusion of proof

In order to find the range of $\beta(a)$, we need to find the behavior of $\beta(a)$ as $a \nearrow a_0 \equiv \sup_{a \in \mathcal{A}^-}\{a\}$ and as $a \to -\infty$.

Lemma 9.2.8. *Assume (H1)–(H4) and let $a_0 = \sup_{a \in A^-}\{a\}$. Then*

$$\lim_{a \nearrow a_0} \beta(a) = \infty.$$

Proof. Using the identity (9.2.7), we have

$$
\begin{aligned}
\liminf_{a \nearrow a_0} \frac{\beta^2(a)}{2} - \frac{\alpha^2}{2} &= \liminf_{a \nearrow a_0} \int_{\mathbb{R}} f'(t) G_0(u(t,a))\, dt \\
&\geq \liminf_{T \to \infty} \liminf_{a \nearrow a_0} \int_0^T f'(t) G_0(u(t,a))\, dt \\
&= \lim_{T \to \infty} \int_0^T f'(t) G_0(u(t,a_0))\, dt \\
&\geq \lim_{T \to \infty} G_0(u(0,a_0)) \int_0^T f'(t)\, dt \\
&= \infty,
\end{aligned}
$$

where, in the last inequality, we have used the property that both $G_0(\cdot)$ and $u(\cdot, a_0)$ are monotonic so that $G_0(u(t, a_0)) \geq G_0(u(0, a_0))$ for all $t > 0$. Since $\beta(a) > 0$, the assertion of the lemma thus follows.

To study the behavior of $\beta(a)$ as $a \to -\infty$, we need the following property of the solutions.

Lemma 9.2.9. *Assume (H1)–(H4) and for any $a \in A^-$, let $m(a) = \sup_{t \in \mathbb{R}} u(t,a)$. Then*

$$\lim_{a \to -\infty} m(a) = -\infty.$$

Proof. Since when $\alpha = 0$, $m(a) = a$, the assertion of the lemma is obviously true, so that we need only consider the case $\alpha > 0$.

Let $a \in A^-$ be any constant and let $z_1(a)$ be the point such that $u'(z_1(a), a) = 0$. Then (9.2.2) implies that $u(t, a) < \alpha t + a$ which, in turn, implies that $m(a) = u(z_1(a), a) < \alpha z_1(a) + a$; that is,

$$z_1(a) \geq \frac{m(a) - a}{\alpha}. \tag{9.2.8}$$

Since $0 \leq u' \leq \alpha$ in $(-\infty, z_1(a))$, there holds the inequality

$$m(a) - 1 < u(t, a) < m(a), \quad \forall t \in \left(z_1(a) - \frac{1}{\alpha}, z_1(a) \right),$$

so that

$$
\begin{aligned}
0 = u'(z_1(a), a) &= \alpha - \int_{-\infty}^{z_1(a)} f(s) g(u(s,a))\, ds \\
&\leq \alpha - \left(\inf_{u \in [m(a)-1, m(a)]} g(u) \right) \int_{z_1(a) - \frac{1}{\alpha}}^{z_1(a)} f(s)\, ds.
\end{aligned}
$$

Therefore, by (9.2.8) and the monotonicity of f,

$$\alpha \geq \left(\inf_{u \in [m(a)-1, m(a)]} g(u) \right) \int_{z_1(a)-\frac{1}{\alpha}}^{z_1(a)} f(s)\, ds$$

$$\geq \left(\inf_{u \in [m(a)-1, m(a)]} g(u) \right) \int_{\frac{m(a)-a-1}{\alpha}}^{\frac{m(a)-a}{\alpha}} f(s)\, ds.$$

Since $f(t) \to \infty$ as $t \to \infty$, the assertion of the lemma must hold.

Now we are in a position to find the behavior of $\beta(a)$ as $a \to -\infty$.

Lemma 9.2.10. *Assume (H1)–(H4) and (H6). Then*

$$\lim_{a \to -\infty} \beta(a) = \alpha + 2f_1 g_1. \tag{9.2.9}$$

If in addition (H5) holds, then $(\alpha + 2f_1 g_1, \infty) \subset \{\beta(a) \mid a \in \mathcal{A}^-\}$.

Proof. Let $a \ll -1$ and $T \gg 1$ be any fixed constants. Then the identity (9.2.7) implies that

$$\frac{\beta^2(a)}{2} - \frac{\alpha^2}{2} = \int_{\mathbb{R}} f'(t) G_0(u(t,a))\, dt$$

$$= \int_{-\infty}^{T} f'(t) G_0(u(t,a))\, dt + \int_{T}^{\infty} F_1(t) G_1(u(t,a)) f(t) g(u(t,a))\, dt$$

$$= \int_{-\infty}^{T} f'(t) G_0(u(t,a))\, dt + F_1(T^*) G_1(u(T^*,a)) \int_{T}^{\infty} f(t) g(u(t,a))\, dt$$

by the mean-value theorem, where $T^* \in [T, \infty)$. Using the identity (9.2.6) we have that

$$\frac{\beta^2(a)}{2} - \frac{\alpha^2}{2} = F_1(T^*) G_1(u(T^*,a))(\alpha + \beta(a)) + \Delta(T,a), \tag{9.2.10}$$

where

$$\Delta(T,a) = \int_{-\infty}^{T} \Big(f'(t) G_0(u(t,a)) - F_1(T^*) G_1(u(T^*,a)) f(t) g(u(t,a)) \Big)\, dt.$$

By Lemma 9.2.9, $\lim_{a \to -\infty} \Delta(T,a) = 0$. Solving $\beta(a)$ from the algebraic equation (9.2.10) yields that

$$\beta(a) = F_1(T^*) G_1(u(T^*,a)) + \sqrt{[\alpha + F_1(T^*) G_1(u(T^*,a))]^2 + 2\Delta(T,a)}.$$

Therefore,

$$\lim_{a \to -\infty} \beta(a) = \lim_{T \to \infty} \lim_{a \to -\infty} \Big(F_1(T^*) G_1(u(T^*,a)) \\ + \sqrt{[\alpha + F_1(T^*) G_1(u(T^*,a))]^2 + 2\Delta(T,a)} \Big)$$

$$= \lim_{T \to \infty} \lim_{a \to -\infty} \Big(\alpha + 2F_1(T^*) G_1(u(T^*,a)) \Big)$$

$$= \alpha + 2f_1 g_1$$

by the assumption (H6) and Lemma 9.2.9. This proves (9.2.9).

Since when (H5) holds, $\beta(a)$ is continuous in $\mathcal{A}^-$, so that the range of $\beta(a)$ when a runs over $\mathcal{A}^-$ contains the set $(\alpha + 2f_1g_1, \infty)$.

Lemma 9.2.11. *If (H1)–(H7) hold, then*

$$\left\{ \beta(a) \,\Big|\, a \in \mathcal{A}^- \right\} = (\alpha + 2f_1g_1, \infty).$$

Proof. We need only show that $\beta(a) > \alpha + 2f_1g_1$ for all $a \in \mathcal{A}^-$. In fact, if (H7) holds, then

$$
\begin{aligned}
\beta^2(a) - \alpha^2 &= 2\int_{\mathbb{R}} f'(t)G_0(u(t,a))\,\mathrm{d}t \\
&> 2f_1g_1 \int_{\mathbb{R}} f(t)g(u(t,a))\,\mathrm{d}t \\
&= 2f_1g_1(\beta(a) + \alpha),
\end{aligned}
$$

which implies that $\beta(a) > \alpha + 2f_1g_1$.

Clearly, Theorem 9.1.1 follows from Lemmas 9.2.1–9.2.11.

9.3 Applications

In Chapter 4, we studied a class of solutions of the Abelian Chern–Simons equations, called non-topological solutions. Due to limit on technicalities, we only presented an existence proof for those solutions. Theorem 9.1.1 here enables us to arrive at a complete understanding of the solutions with radial symmetry, topological or non-topological. In the next chapter, we will apply Theorem 9.1.1 to obtain a complete understanding of radially symmetric solutions of a rather complicated problem – cosmic strings arising from the coupled Einstein and Abelian Higgs equations in the theory of galaxy formation.

Consider the Chern–Simons equation with all vortices superimposed at the origin,

$$\Delta u = \frac{4}{\kappa^2}e^u(e^u - 1) + 4\pi N\delta(x), \quad x \in \mathbb{R}^2. \tag{9.3.1}$$

Using Theorem 9.1.1, we have

Theorem 9.3.1. *For $N \geq 0$, a radially symmetric solution of (9.3.1) is either trivial, $u \equiv 0$, or negative, $u < 0$. Corresponding to each given N, there exists a unique solution $u = u(r)$ $(r = |x|)$ satisfying*

$$\lim_{r \to \infty} u(r) = 0. \tag{9.3.2}$$

All other solutions observe the behavior $u(r) \to -\infty$ *as* $r \to \infty$ *and*

$$\lim_{r \to \infty} r u_r(r) = -\beta, \qquad \beta > 2N + 4. \tag{9.3.3}$$

More importantly, for any $\beta \in (2N+4, \infty)$, *there exists at least one solution* u *realizing the asymptote (9.3.3).*

Proof. Under the transformation (9.1.2), the equation (9.3.1) becomes (9.1.3) and (9.1.4) with

$$\alpha = 2N, \qquad f(t) = \frac{4}{\kappa^2} e^{2t}, \qquad g(u) = e^u(1 - e^u).$$

If u is a solution of (9.1.3) and (9.1.4) which becomes positive at some point $t = t_0$, then the maximum principle says that $u'(t_0) > 0$. Thus $u''(t) > 0$ and $u'(t) > 0$ for all $t > t_0$. In particular,

$$e^{u(t)} - 1 > e^{u(t_0)} - 1 > 0, \quad t > t_0.$$

Now the equation gives us the inequality

$$u'' > \delta e^u, \quad t > t_0,$$

where $\delta > 0$ is a constant depending on t_0. Clearly u blows up in finite time $t > t_0$.

To obtain non-positive solutions, we can modify $g(u)$ for $u \in (1, \infty)$ such that g and g' are uniformly bounded. Then, we can directly verify that such f and g satisfy (H1)–(H8), and hence the assertions of the theorem follow from Theorem 9.1.1.

Let u be an arbitrary radially symmetric solution produced in Theorem 9.3.1. Again use $z = x^1 + ix^2$ to denote a point in $\mathbb{R}^2$. Set

$$\phi(z) \;=\; \exp\left(\frac{1}{2} u(z) + iN \arg z\right),$$

$$A_1(z) \;=\; -2\mathrm{Re}\{i\bar{\partial} \ln \phi\}, \quad A_2(z) = -2\mathrm{Im}\{i\bar{\partial} \ln \phi\}. \tag{9.3.4}$$

From a solution u obtained in Theorem 9.3.1, we can use (9.3.4) to construct an N-vortex solution of the self-dual Chern–Simons equations (5.2.19), (5.2.20). For any given $\beta > 2N + 4$, let u be such a solution that (9.3.3) is fulfilled. Then it follows from (5.2.19), (5.2.20), (9.3.4), and (9.3.1) that the magnetic flux is

$$\begin{aligned}
\Phi \;&=\; \int_{\mathbb{R}^2} F_{12} \, dx = \frac{2}{\kappa^2} \int_{\mathbb{R}^2} e^u(1 - e^u) \, dx \\
&=\; \frac{4\pi}{\kappa^2} \int_0^\infty r e^{u(r)} (1 - e^{u(r)}) \, dr \\
&=\; \pi\left(\lim_{r \to 0} r u_r(r) - \lim_{r \to \infty} r u_r(r)\right) \\
&=\; 2\pi N + \pi \beta.
\end{aligned}$$

The first term on the right-hand side equals to the flux of a topological N-vortex solution. The electric charge is just $\kappa\Phi$. Furthermore, since the Chern–Simons energy density has the decomposition

$$\mathcal{H} = \frac{1}{4}\left[\frac{\kappa F_{12}}{|\phi|} + \frac{2}{\kappa}|\phi|(|\phi|^2 - 1)\right]^2 + |D_1\phi + iD_2\phi|^2$$
$$+F_{12} + \mathrm{Im}\left\{\partial_j\varepsilon_{jk}\phi^*(D_k\phi)\right\},$$

and, in view of (9.3.4), there holds

$$|D_j\phi|^2 = \frac{1}{2}u_r^2 e^u = O(r^{-(2+\beta)}), \quad \text{for large } r = |x|,$$

we have the total energy

$$E = \int_{\mathbb{R}^2} \mathcal{H}\,\mathrm{d}x = \int_{\mathbb{R}^2} F_{12}\,\mathrm{d}x = \Phi.$$

In summary, we can state

Theorem 9.3.2. *For given integer $N \geq 0$ and any $\beta > 2N+4$, the Chern–Simons system allows a non-topological N-vortex solution which realizes the following prescribed asymptotic decay properties,*

$$|\phi|^2 = O(r^{-\beta}), \quad |D_j\phi|^2 = O(r^{-(2+\beta)}), \quad F_{jk} = O(r^{-\beta})$$

for large $|x| = r > 0$ and the corresponding values of energy, electric charge, and magnetic flux

$$E = \Phi, \qquad Q = \kappa\Phi, \qquad \Phi = 2\pi N + \pi\beta.$$

Moreover, the radially symmetric topological N-vortex solution is uniquely determined by the vortex location.

9.4   Remarks

We have studied all possible radially symmetric negative solutions of the equation (9.1.1) under some conditions listed as a partial set of hypotheses (H1)–(H8). Although these conditions are quite general for applications in many physical models, there are some situations which cannot be covered by the results here. Hence it is natural to pursue a study when (H1)–(H8), or some of these hypotheses, are replaced by other hypotheses.

An important problem concerning (9.1.1) is whether all of its negative solutions are necessarily radially symmetric. More specifically, we state

Open Problem 8.4.1. *Are all the negative solutions of the equation (9.1.1) radially symmetric about the origin when the hypotheses (H1)–(H8), or a partial set of these hypotheses, are satisfied?*

An easier start would be the bare solution case when $N = 0$ as that for the Chern–Simons equation [295].

This problem is not merely of technical interest. An affirmative answer to it would imply, for cosmic strings, for example, which will be studied in the next chapter, that there is no finite-energy solution if all strings are clustered together in a compact setting.

10
Strings in Cosmology

In this chapter we present a series of results concerning the existence of cosmic strings. Such solutions arise from the Einstein equations coupled with a suitable matter and gauge field system and lead to local concentration of energy and gravitational curvature, a structure relevant in the theory of galaxy formation in the early-time cosmology [321, 322]. In §10.1 we discuss some basic notions in the study of cosmic strings. In particular, we present the elegant solution of Comtet and Gibbons [86] of the harmonic map string equations. In §10.2–§10.6 we present a study of the existence of multiple strings arising from the self-dual Abelian Higgs theory. In §10.7 we consider non-Abelian strings arising from the Weinberg–Salam electroweak theory.

10.1 Strings, Conical Geometry, and Deficit Angle

In this section, we first discuss some possible mathematical consequences from a concentrated distribution of the energy density over a conformally flat surface, a situation that gives rise to the simplest cosmic string solutions. We then present the explicit multiple string solutions of Comtet and Gibbons [86] for the sigma or harmonic map model.

10.1.1 Localized energy distribution and multiple strings

Let $g_{\mu\nu}$ be a general metric tensor with signature $(+ - - -)$, $R_{\mu\nu}$ the Ricci tensor, and R the scalar curvature. Then the Einstein tensor takes the form

$$G_{\mu\nu} = R_{\mu\nu} - \frac{1}{2} g_{\mu\nu} R. \tag{10.1.1}$$

The energy-momentum tensor of the matter sector is denoted by $T_{\mu\nu}$. Thus we can write down the Einstein equations,

$$G_{\mu\nu} = -8\pi G T_{\mu\nu}, \tag{10.1.2}$$

where G is Newton's gravitational constant. We first study some simplest string solutions of (10.1.2) which contain many important ingredients in our later study of the problem of the existence of straight, time independent cosmic string solutions.

Consider straight, time-independent cosmic strings in the Minkowski spacetime $\mathbb{R}^{3,1}$ as locally concentrated solutions of (10.1.2) generated from a suitably devised energy-momentum tensor $T_{\mu\nu}$. It is known that the most general metric over $\mathbb{R}^{3,1}$ compatible with reflection symmetry in the time variable $t = x^0$ and a specified vertical direction, say, x^3, is of the form

$$ds^2 = U dt^2 - V(dx^3)^2 - e^{\eta}((dx^1)^2 + (dx^2)^2), \tag{10.1.3}$$

where U, V, η are functions of the spatial variables x^1, x^2 only. A simple, idealistic situation, compatible with the metric (10.1.3), is that the energy-momentum $T_{\mu\nu}$ takes the diagonal form

$$T_0^0 = T_3^3 = \sum_{s=1}^{N} \sigma_s \delta_{p_s}; \qquad T_\mu^\nu = 0, \quad \text{for other } \mu, \nu = 0, 1, 2, 3, \tag{10.1.4}$$

where $\sigma_s > 0$ $(s = 1, 2, \cdots, N)$ are constants and δ_p is the Dirac distribution over the 2-surface $(S, \{g_{ij}\}) = (\mathbb{R}^2, \{e^{\eta} \delta_{ij}\})$ $(i, j = 1, 2)$ concentrated at the point $p \in \mathbb{R}^2$.

The special relation in (10.1.4), namely,

$$T_0^0 = T_3^3, \tag{10.1.5}$$

usually referred to as the boost invariance, implies that the metric (10.1.3) may be further simplified by imposing $U = V = $ constant. Therefore, after a rescaling, the metric (10.1.3) becomes

$$ds^2 = dt^2 - (dx^3)^2 - e^{\eta}((dx^1)^2 + (dx^2)^2), \tag{10.1.6}$$

which leads to a much simplified Einstein tensor,

$$\begin{aligned} G_{00} &= -G_{33} = -K_\eta = \text{ the Gauss curvature of } (\mathbb{R}^2, \{e^{\eta}\delta_{ij}\}), \\ G_{\mu\nu} &= 0 \quad \text{for other values of } \mu, \nu. \end{aligned} \tag{10.1.7}$$

Inserting (10.1.4) and (10.1.7) into the Einstein equations (10.1.2), we arrive at a single scalar equation,

$$K_\eta = 8\pi G \sum_{i=1}^{N} \sigma_s \delta_{p_s}, \quad x \in \mathbb{R}^2. \tag{10.1.8}$$

On the other hand, in terms of the conformal exponent η, the Gauss curvature K_η may be expressed as

$$K_\eta = -\frac{1}{2}e^{-\eta}\Delta\eta, \tag{10.1.9}$$

which tells us through (10.1.8) that η is a Green function,

$$\Delta\eta = -16\pi G \sum_{s=1}^{N} \sigma_s \delta_{p_s}(x), \tag{10.1.10}$$

where, now, $\delta_p(x)$ is the Dirac function over the Euclidean plane $\mathbb{R}^2$ at the point p.

Solving (10.1.10), we obtain

$$\eta(x) = -8G \sum_{s=1}^{N} \sigma_s \ln|x - p_s|, \tag{10.1.11}$$

which gives us the singular metric

$$ds^2 = dt^2 - (dx^3)^3 - \left(\prod_{s=1}^{N} |x - p_s|^{\sigma_s} \right)^{-8G} ((dx^1)^2 + (dx^2)^2), \tag{10.1.12}$$

representing a system of cosmic strings located at $p_1, p_2, \cdots, p_N \in \mathbb{R}^2$. Since $T_{00} = T_0^0$ is identified with energy density, $\mathcal{H}$, of the matter, (10.1.2), (10.1.4), and (10.1.8) give us the energy and curvature concentration formula

$$\mathcal{H} = (8\pi G)^{-1} K_\eta = \sum_{s=1}^{N} \sigma_s \delta_{p_s}(x). \tag{10.1.13}$$

It will be interesting to compare the metric (10.1.12) with the flat one,

$$ds^2 = dt^2 - (dx^3)^2 - (dx^1)^2 - (dx^2)^2, \tag{10.1.14}$$

near spatial infinity. For this purpose, let (r, θ) be the polar coordinates in the x^1x^2-plane. Then (10.1.14) becomes

$$ds^2 = dt^2 - (dx^3)^2 - dr^2 - r^2 d\theta^2, \tag{10.1.15}$$

On the other hand, asymptotically (for r large), the metric (10.1.12) looks like

$$ds^2 = dt^2 - (dx^3)^2 - r^{-8G\sigma}(dr^2 + r^2 d\theta^2), \tag{10.1.16}$$

where
$$\sigma = \sum_{s=1}^{N} \sigma_s. \tag{10.1.17}$$

Since the gravitational constant G is small, we assume that

$$4G\sigma < 1. \tag{10.1.18}$$

Hence, with the change of variables

$$R = (1 - 4G\sigma)^{-1} r^{1-4G\sigma}, \quad \Theta = (1 - 4G\sigma)\theta, \tag{10.1.19}$$

infinity is preserved and the metric (10.1.16) becomes a flat one of the form of (10.1.15),

$$ds^2 = dt^2 - (dx^3)^2 - dR^2 - R^2 d\Theta^2, \tag{10.1.20}$$

In fact (10.1.20) can be identified with (10.1.15) only locally and globally it is not flat because the second relation in (10.1.19) says that, as one travels around infinity, the polar angle Θ changes by

$$\Theta = 2\pi(1 - 4G\sigma) \tag{10.1.21}$$

instead of 2π recorded by the polar angle θ. The missing part,

$$\delta = 8\pi G\sigma = 8\pi G \sum_{s=1}^{N} \sigma_s, \tag{10.1.22}$$

is called the deficit angle which implies that infinity is a conical singularity. Thus the deficit angle measures the energy strength of cosmic strings.

In cosmology, cosmic strings realized as mixed states give rise to an array of concentration of curvature and energy which may serve as seeds for matter accretion for galaxy formation in the early universe [52, 118, 120, 134, 168, 169, 170, 180, 321, 322, 331, 332].

10.1.2 Harmonic map model

In the last subsection, we have seen that when the energy density of matter is distributed along a 2-surface as the Dirac functions, the gravitational metric defines a conical geometry and is singular at string locations. In this subsection, we present a family of multiple strings generated from a simplest field-theoretical model, the sigma model, known as harmonic maps in mathematics literature, which represent everywhere-regular metrics, although an analogous deficit angle formula holds.

Let $\phi = (\phi_1, \phi_2, \phi_3)$ be a map from the spacetime, with the metric $(g_{\mu\nu})$, to the unit sphere $S^2 \subset \mathbb{R}^3$. Consider the action density

$$\mathcal{L} = \frac{1}{2} g^{\mu\nu} \partial_\mu \phi \cdot \partial_\nu \phi. \tag{10.1.23}$$

As in Chapter 2, it will be more convenient to replace ϕ by its stereographic projection, in terms of a complex scalar field $u = u_1 + iu_2$ where

$$u_1 = \frac{\phi_1}{1 + \phi_3}, \quad u_2 = \frac{\phi_2}{1 + \phi_3}. \tag{10.1.24}$$

In view of (10.1.24), we see that (10.1.23) becomes

$$\mathcal{L} = \frac{2}{(1 + |u|^2)^2} g^{\mu\nu} \partial_\mu u \partial_\nu \bar{u}. \tag{10.1.25}$$

Variation of $g_{\mu\nu}$ in (10.1.25) gives us the energy-momentum tensor

$$T_{\mu\nu} = \frac{2}{(1 + |u|^2)^2}(\partial_\mu u \partial_\nu \bar{u} + \partial_\mu \bar{u} \partial_\nu u) - g_{\mu\nu}\mathcal{L}, \tag{10.1.26}$$

which leads to the following coupled Einstein and (harmonic map) matter equations,

$$G_{\mu\nu} = -8\pi G\, T_{\mu\nu},$$

$$\frac{1}{\sqrt{|g|}}\partial_\mu\left(\frac{\sqrt{|g|}g^{\mu\nu}}{(1 + |u|^2)^2}\partial_\nu u\right) = -\frac{2u}{(1 + |u|^2)^3}g^{\mu\nu}\partial_\mu u \partial_\nu \bar{u}. \tag{10.1.27}$$

We are interested in cosmic string solutions of (10.1.27) so that the metric takes the form (10.1.6) and u depends only on the spatial variables x^1 and x^2. Therefore, we have

$$\mathcal{H} = T_{00} = -T_{33} = \frac{2e^{-\eta}}{(1 + |u|^2)^2}(|\partial_1 u|^2 + |\partial_2 u|^2)$$

$$= \frac{2e^{-\eta}}{(1 + |u|^2)^2}|\partial_1 u \pm i\partial_2 u|^2 \pm i\frac{2e^{-\eta}}{(1 + |u|^2)^2}(\partial_1 u \partial_2 \bar{u} - \partial_1 \bar{u} \partial_2 u). \tag{10.1.28}$$

Using (10.1.28) and applying (2.1.24) and (2.1.25), we find as before the topological lower energy bound

$$E = \int_S \mathcal{H}\, d\Omega_g = \int_{\mathbb{R}^2} \mathcal{H}\, e^\eta\, dx$$

$$\geq \pm 2i\int_{\mathbb{R}^2}\frac{\partial_1 u \partial_2 \bar{u} - \partial_1 \bar{u} \partial_2 u}{(1 + |u|^2)^2}\, dx$$

$$= 4\pi|\deg(\phi)|, \tag{10.1.29}$$

with equality if and only if u satisfies the Cauchy–Riemann equations

$$\partial_1 u \pm i\partial_2 u = 0. \tag{10.1.30}$$

In view of (10.1.30), it is easily checked that T_{00} and T_{33} are the only nonvanishing components of the energy-momentum tensor. Besides, the second equation in (10.1.27) is automatically satisfied. Hence, applying (10.1.7), we see that the system (10.1.27) is reduced into a single equation,

$$K_\eta = 8\pi G \mathcal{H}, \tag{10.1.31}$$

where u satisfies (10.1.30) and the energy density $\mathcal{H}$ in (10.1.28) is given by

$$
\begin{aligned}
\mathcal{H} &= \pm J_{12} = \pm(\partial_1 J_2 - \partial_2 J_1), \\
J_k &= \frac{i}{1+|u|^2}(u\partial_k \overline{u} - \overline{u}\partial_k u), \quad k = 1, 2.
\end{aligned} \tag{10.1.32}
$$

For definiteness, we consider the case with the plus sign. Hence, from (10.1.30), we see that u is a meromorphic function in $z = x^1 + ix^2$, which may be chosen to be

$$u(z) = c\left(\prod_{s=1}^{N}(z - p_s)^{-1}\right)\left(\prod_{s=1}^{M}(z - q_s)\right), \tag{10.1.33}$$

where $c \neq 0$ is a complex parameter and p, q are points in the complex plane. Consequently, we can use the divergence theorem to arrive at

$$
\begin{aligned}
E &= 4\pi \deg(\phi) \\
&= \int_{\mathbb{R}^2} J_{12}\, dx \\
&= \lim_{r\to\infty} \oint_{|z|=r} J_k\, dx^k - \lim_{r\to 0}\sum_{s=1}^{N} \oint_{|z-p_s|=r} J_k\, dx^k,
\end{aligned} \tag{10.1.34}
$$

where the line integrals are all taken counterclockwise. Using (10.1.33), we easily see that, near infinity, J_k has the asymptotic estimate

$$
\begin{aligned}
J_k &= O(|x|^{-\delta}), \quad k = 1, 2. \\
\delta &= 1 + 2\max\{0, N - M\}.
\end{aligned} \tag{10.1.35}
$$

Hence, if the number of poles is larger than the number of zeros in (10.1.33), namely,

$$N > M, \tag{10.1.36}$$

then the first term on the right-hand side of (10.1.34) vanishes in view of (10.1.35). In fact, the condition (10.1.36) is exactly the one imposed by Belavin and Polyakov in absence of gravity (see Chapter 2).

In order to find the rest of the terms on the right-hand side of (10.1.34), we note that J_k has the estimate

$$J_k = -\epsilon_{jk}\frac{2|u|^2}{1+|u|^2}\left(\frac{y^j}{|z-p_s|^2} + O(1)\right), \quad k = 1, 2 \tag{10.1.37}$$

near the points p_s, where $y^1 = \mathrm{Re}\{z - p_s\}$ and $y^2 = \mathrm{Im}\{z - p_s\}$, $s = 1, 2, \cdots, N$. By virtue of (10.1.37) and the divergence theorem, we obtain

$$
\lim_{r \to 0} \oint_{|z-p_s|=r} J_k\, dx^k = \lim_{r \to 0} \left(\frac{2}{r^2} \oint_{|z-p_s|=r} y^2\, dy^1 - y^1\, dy^2 \right)
$$

$$
= -4\pi. \tag{10.1.38}
$$

Inserting (10.1.38) into (10.1.34), we see that $E = 4\pi N$ and $\deg(\phi) = N$ as in the model without gravity.

It remains to study the Einstein equation (10.1.31). Since, with $|u|^2 = e^v$, we have

$$
|\partial_1 u|^2 + |\partial_2 u|^2 = \frac{1}{2} e^v |\nabla v|^2,
$$

and the function v satisfies

$$
\Delta v = -4\pi \sum_{s=1}^{N} \delta_{p_s} + 4\pi \sum_{s=1}^{M} \delta_{q_s}. \tag{10.1.39}
$$

Therefore the energy density (10.1.28) becomes

$$
e^\eta \mathcal{H} = \frac{e^v}{(1+e^v)^2} |\nabla v|^2
$$

$$
= \Delta \ln(1 + e^v) + 4\pi \sum_{s=1}^{N} \delta_{p_s}. \tag{10.1.40}
$$

Inserting (10.1.19) and (10.1.40) into (10.1.31), we see that

$$
\eta + 16\pi G \left(\ln(1 + e^v) + \sum_{s=1}^{N} \ln|z - p_s|^2 \right)
$$

is a harmonic function, which may be assumed to be a constant for convenience. Applying this result and (10.1.33), we obtain the conformal factor for the gravitational metric explicitly as follows,

$$
e^\eta = g_0 \left(\prod_{s=1}^{N} |x - p_s|^2 + |\lambda|^2 \prod_{s=1}^{M} |x - q_s|^2 \right)^{-16\pi G}, \qquad x \in \mathbb{R}^2, \tag{10.1.41}
$$

where $g_0 > 0$ is a constant. This metric is everywhere regular. Near infinity, it has the estimate

$$
e^\eta = O(r^{-32N\pi G}) \tag{10.1.42}
$$

due to (10.1.36). Therefore, in view of the discussion made in the last subsection, we see that, when

$$
16\pi N G < 1, \tag{10.1.43}
$$

the deficit angle has the expression

$$\delta = 32\pi^2 NG. \tag{10.1.44}$$

Finally, in view of (10.1.42), the surface $(\mathbb{R}^2, e^\eta \delta_{ij})$ is geodesically complete (see a more general discussion in the last few paragraphs of §10.4) if and only if

$$N \le \frac{1}{16\pi G}. \tag{10.1.45}$$

10.2   Strings and Abelian Gauge Fields

In this section, we begin our study of the existence of cosmic strings generated as topological defects in the Abelian Higgs theory. We first write down the governing equations and remark on the important role played by defects. We then concentrate on the critical coupling limit first studied by Linet [198, 199] and Comtet and Gibbons [86]. We will establish our main obstruction conditions for existence of strings over a compact surface.

10.2.1   Governing equations over Riemann surfaces

In order to define the Abelian Higgs sector properly, we need to work in a framework in which the Higgs field ϕ is a cross-section on a $U(1)$-line bundle L [137] over the spacetime and the gauge field, say, A, is a connection 1-form. Suppose that h is a Hermitian metric of L and $\{(U_\alpha, e_\alpha)\}$ is such an atlas of local trivializations of L that e_α satisfies $h(e_\alpha, e_\alpha) = 1$, $\forall \alpha$. Let ϕ_α be the local representation of ϕ on $U_\alpha : \phi = \phi_\alpha e_\alpha$. Then we have $h(\phi, \phi) = |\phi_\alpha|^2$, which is obviously a local-chart-independent scalar field, and thus, may conveniently be denoted by $|\phi|^2$. Therefore, using local coordinates and local representations, we can write the Abelian Higgs action density, with a dimensionless coupling constant $\lambda > 0$, in the form

$$\mathcal{L} = -\frac{1}{4} g^{\mu\mu'} g^{\nu\nu'} F_{\mu\nu} F_{\mu'\nu'} + \frac{1}{2} g^{\mu\nu} (D_\mu \phi) \overline{(D_\nu \phi)} - \frac{\lambda}{8} (|\phi|^2 - \varepsilon^2)^2, \tag{10.2.1}$$

where and in the sequel we also allow the vanishing of the symmetry-breaking parameter, $\varepsilon = 0$, to include in the model the restoration of symmetry. Note that, in the above expression and subsequent discussion, we suppress the subscript "α" when there is no risk of confusion, and that D is the covariant derivative induced from the gauge connection A and $F = dA$ is the curvature of A or the Maxwell field. The presence of the gravitational metric $g_{\mu\nu}$ indicates the influence of gravity.

The Einstein equations coupled with the Abelian Higgs model are

$$G_{\mu\nu} = -8\pi G\, T_{\mu\nu},$$

$$\frac{1}{\sqrt{|g|}} D_\mu (g^{\mu\nu} \sqrt{|g|} [D_\nu \phi]) = \frac{\lambda}{2}(|\phi|^2 - \varepsilon^2)\phi,$$

$$\frac{1}{\sqrt{|g|}} \partial_{\mu'}(g^{\mu\nu} g^{\mu'\nu'} \sqrt{|g|} F_{\nu\nu'}) = \frac{i}{2} g^{\mu\nu}(\phi\overline{[D_\nu\phi]} - \overline{\phi}[D_\nu\phi]), \ (10.2.2)$$

where G is Newton's gravitational constant (or more precisely a dimensionless rescaling factor of the gravitational constant) and

$$T_{\mu\nu} = -g^{\mu'\nu'}F_{\mu\mu'}F_{\nu\nu'} + \frac{1}{2}([D_\mu\phi]\overline{[D_\nu\phi]} + \overline{[D_\mu\phi]}[D_\nu\phi]) - g_{\mu\nu}\mathcal{L} \quad (10.2.3)$$

is the energy-momentum tensor of the Abelian Higgs sector obtained by varying the gravitational metric in the action $L = \int \mathcal{L}$ with $\mathcal{L}$ being defined by (10.2.1).

We assume that the spacetime is uniform along the time axis $x^0 = t$ and the x^3-direction so that the line element takes the form

$$\begin{aligned} ds^2 &= g_{\mu\nu} dx^\mu dx^\nu \\ &= dt^2 - (dx^2)^2 - g_{jk} dx^j dx^k, \quad j,k = 1,2, \quad (10.2.4) \end{aligned}$$

where $\{g_{jk}\}$ is the Riemannian metric tensor of an orientable 2-surface S (without boundary), and that A_μ, ϕ depend only on the coordinates on S and

$$A_\mu = (0, A_1, A_2, 0).$$

Then $T_{\mu\nu}$ is simplified to

$$\begin{aligned} T_{00} &= \mathcal{H}, \ T_{33} = -\mathcal{H}, \ T_{03} = T_{0j} = T_{3j} = 0, \\ T_{jk} &= g^{j'k'} F_{jj'} F_{kk'} + \frac{1}{2}([D_j\phi]\overline{[D_k\phi]} + \overline{[D_j\phi]}[D_k\phi]) \\ &\quad - g_{jk}\mathcal{H}, \quad (10.2.5) \end{aligned}$$

where

$$\mathcal{H} = \frac{1}{4} g^{jj'} g^{kk'} F_{jk} F_{j'k'} + \frac{1}{2} g^{jk}(D_j\phi)\overline{(D_k\phi)} + \frac{\lambda}{8}(|\phi|^2 - \varepsilon^2)^2 \quad (10.2.6)$$

is the energy density of the Abelian Higgs sector which is now defined by the line bundle L restricted to the 2-surface S. The Maxwell field F_{jk} represents the first Chern class, of course. Besides, if we use K_g to denote the Gauss curvature of $(S, \{g_{jk}\})$, the Einstein tensor reduces under local isothermal coordinates into the form

$$\begin{aligned} -G_{00} &= G_{33} = K_g, \\ G_{\mu\nu} &= 0 \quad \text{for other values of } \mu, \nu \quad (10.2.7) \end{aligned}$$

as before. As a consequence, the system (10.2.2) becomes the following two-dimensional Einstein and Abelian Higgs equations on S,

$$K_g = 8\pi G \, \mathcal{H}, \quad T_{jk} = 0,$$

$$\frac{1}{\sqrt{g}} D_j(g^{jk}\sqrt{g}[D_k\phi]) = \frac{\lambda}{2}(|\phi|^2 - \varepsilon^2)\phi,$$

$$\frac{1}{\sqrt{g}} \partial_{j'}(g^{jk}g^{j'k'}\sqrt{g}F_{kk'}) = \frac{i}{2}g^{jk}(\phi\overline{[D_k\phi]} - \overline{\phi}[D_k\phi]), \quad (10.2.8)$$

where g also stands for the determinant formed from $\{g_{jk}\}$. Except in the next subsection, the coupling constant λ will always be assumed to be critical, $\lambda = 1$.

10.2.2 Role of defects

When the gravitational sector is ignored and the Abelian Higgs model is considered in a two-dimensional framework, the equations govern the electromagnetic properties of a planar superconductor so that the Higgs field ϕ appears as an order parameter. At the places where ϕ vanishes, superconductivity is destroyed and the magnetic field penetrates the sample in the form of vortex-lines. Thus, the zeros of ϕ are also called defects, which are indicators of spots of partial restoration of the symmetric normal phase. It is already well known in the Abelian Higgs model that the appearance of these defects is equivalent to the existence of mixed states characterized by $|\phi| \neq 1, F_{jk} \neq 0$ [1, 157]. Here we would like to know whether the same statement holds in the presence of gravity, i.e., whether the existence of strings or defects is crucial to producing gravity or a nonflat spacetime. For technical reasons, we will mostly concentrate on the compact case.

The basic result in this subsection says that the absence of string defects implies the absence of gravity. Such a fact may be seen intuitively as follows. When there are no string defects, the state is purely superconducting and the energy vacuum is attained. Thus the energy distribution of the matter-gauge sector is everywhere zero. However, the Einstein equations (i.e., the first equation in (10.2.8)) imply that the space curvature vanishes identically. As a consequence, we arrive at a flat space with trivial topology and there is no gravity.

Suppose that there are no defects, i.e., $\phi(x) \neq 0, \forall x \in S$. Then the line bundle L over S defined by the solution of (10.2.8) is trivial: $L = S \times \mathbb{C}$. We may view ϕ as a complex-valued function on S. Thus there is a real-valued function $f \in C^\infty(S)$ so that $\phi = \varphi e^{if}$ where $\varphi = |\phi| > 0$. Perform the gauge transformation

$$\phi \mapsto \phi e^{-if}, \qquad A_j \mapsto A_j - \partial_j f$$

in (10.2.8). We see that the last two equations in (10.2.8) take the form

$$\Delta_g \varphi = (g^{jk} A_j A_k)\varphi + \frac{\lambda}{2}(\varphi^2 - \varepsilon^2)\varphi,$$

$$\frac{1}{\sqrt{g}}\partial_j(g^{jk}\sqrt{g}A_k) = -2g^{jk}A_j\partial_k \ln\varphi,$$

$$\frac{1}{\sqrt{g}}\partial_{j'}(g^{jk}g^{j'k'}\sqrt{g}F_{kk'}) = -g^{jk}A_k\varphi^2, \tag{10.2.9}$$

where Δ_g is the Laplace–Beltrami operator induced from the metric $g = \{g_{jk}\}$ defined by

$$\Delta_g v = \frac{1}{\sqrt{g}}\partial_j(g^{jk}\sqrt{g}\partial_k v).$$

From the third equation in (10.2.9), we obtain

$$\frac{1}{\sqrt{g}}\partial_{j'}(g^{j'k'}\sqrt{g}[g^{jk}A_j F_{kk'}]) + \frac{1}{2}g^{j'k'}g^{jk}F_{jj'}F_{kk'} = -g^{jk}A_j A_k\varphi^2. \tag{10.2.10}$$

However, the first term on the left-hand side of (10.2.10) is a total divergence, thus an integration of (10.2.10) leads to

$$\int_S \left\{\frac{1}{2}g^{jj'}g^{kk'}F_{jk}F_{j'k'} + g^{jk}A_j A_k\varphi^2\right\} d\Omega_g = 0.$$

The two terms in the integrand are both positive-definite. As a consequence, we obtain $A_j = 0, j = 1, 2$. Inserting this information into the first equation in (10.2.9), we have

$$\Delta_g(\varphi - \varepsilon) = \frac{\lambda}{2}\varphi(\varphi + \varepsilon)(\varphi - \varepsilon), \qquad x \in S. \tag{10.2.11}$$

Using $\varphi > 0$ and the maximum principle in (10.2.11), we find $\varphi \equiv \varepsilon$. Consequently, $\mathcal{H} = 0$ everywhere. In view of the first equation in (10.2.8), we arrive at $K_g \equiv 0$, which indicates that S is a flat torus, gravity is absent, and the solution is trivial.

We summarize the above discussion as follows.

Theorem 10.2.1. *For a given solution triplet $(\phi, A, \{g_{jk}\})$ of the general Einstein and Abelian Higgs equations (10.2.8) with S being a compact 2-surface, the absence of string-like topological defects, i.e., $\phi \neq 0$ on S, implies that the solution is gauge-equivalent to a trivial solution so that S must be a flat torus, which is characterized by $K_g = 0, \phi = \varepsilon, A = 0$.*

A similar study may be carried out for the noncompact case so long as the field configurations decay fast enough at infinity so that the boundary terms resulting from integrating the first term on the left-hand side of (10.2.10) drop off. Thus, if such a property holds, the statement of Theorem 10.2.1 is also valid in general.

In the rest of this chapter, we will only consider the critical phase $\lambda = 1$.

10.2.3   Obstructions to existence

The system of equations (10.2.8) describes the interaction of the gravitational and gauge-matter sectors confined in a two-dimensional space. We will see that these two sectors are so strongly coupled that, topologically, they totally determine one another. In fact, the first Chern number,

$$c_1(L) = \frac{1}{4\pi} \int_S \epsilon^{jk} F_{jk} \, d\Omega_g,$$

classifies the line bundle L up to isomorphisms which clearly indicates the magnetic excitation pattern of the theory because the integer $N = |c_1(L)|$ is the number of magnetic strings through S, identified as the algebraic number of zeros of the order parameter ϕ. On the other hand, the Gauss curvature K_g reflects the topology of S and measures the gravitational strength, which, by the first equation in (10.2.8), is determined by the energy distribution of the matter and gauge fields. Thus it is clear that there must be a link between these structures through the coupling of gravity with the matter-gauge fields such as that given in (10.2.8).

More precisely, calling a solution of (10.2.8) with $|c_1(L)| = N$ an N-string, we have the following basic result.

Theorem 10.2.2. *Given an integer $N \geq 0$, there exists an N-string solution for the Einstein and Abelian Higgs equations (10.2.8) on a line bundle L equipped with a certain Hermitian structure over an appropriate compact Riemann surface S, only if the string number N, the first Chern number $c_1(L)$ of L, the Euler characteristic $\chi(S)$ of S, the symmetry-breaking parameter $\varepsilon > 0$, and the gravitational coupling factor G satisfy the exact relation*

$$\chi(S) = 4\pi\varepsilon^2 GN = 4\pi\varepsilon^2 G|c_1(L)|. \tag{10.2.12}$$

Furthermore, if $N \geq 2$, the condition (10.2.12) is necessary and sufficient for the existence of an N-string solution. In any case, the solutions of (10.2.8) with $\lambda = 1$ can all be obtained from a self-dual or anti-self-dual system in which the matter-gauge equations are all of the first order. Besides, the $N = 1$ may fail to exist.

First we would like to point out some of the interesting implications of this theorem.

Unique Topology of the Underlying Surface. It is well known that a compact orientable 2-surface S is topologically a sphere with n handles and the number n is called the genus of S. The Euler characteristic satisfies the equation $\chi(S) = 2 - 2n$. Thus the relation (10.2.12) implies that the only possible situation we may have so that the Einstein and Abelian Higgs system (10.2.8) has a cosmic string solution is given by $n = 0$. In other words, the 2-surface must be diffeomorphic to the Riemann sphere S^2 and all other geometries with $n \neq 0$ are ruled out. In particular, the surface S cannot be a torus ($n = 1$). This result implies the nonexistence of gravitational string condensation realized by the appearance of a periodic lattice structure.

Quantization of Symmetry Breaking Scale. Now we use the conclusion $\chi(S) = 2$ (or $n = 0$) arrived above and rewrite (10.2.12) in the form

$$\varepsilon = \varepsilon_N = \frac{1}{\sqrt{2\pi GN}}, \qquad N = 1, 2, \cdots. \tag{10.2.13}$$

We view the gravitational constant G as fixed. Equation (10.2.13) says that there are only countably many levels of the symmetry breaking scale ε for which there may exist cosmic string solutions, and, when ε is away from those quantized levels, there will be no strings. In particular, when $\varepsilon > \varepsilon_1 = 1/\sqrt{2\pi G}$, there is nonexistence. Such a fact seems to suggest that the existence of string solutions prefers lower values of the symmetry breaking scale ε. Indeed, in (10.2.12), the vanishing of ε implies that S is topologically a torus and (10.2.12) no longer presents a constraint to the string number N. However, we will see that in this case there are no nontrivial solutions. This simple fact will be established later. Thus, we observe that the existence of cosmic strings indeed requires symmetry breaking. Such a fact is also true for the noncompact case.

Effective Radius vs. Gravitational Attraction. Finally, since S is topologically a sphere, we may define the 'effective radius' of S, say, R_{eff}, by setting $4\pi R_{\text{eff}}^2 = |S|_g$ where $|S|_g$ is the total surface area of $(S, \{g_{jk}\})$. Then we find that R_{eff} has the lower bound

$$R_{\text{eff}} > \frac{1}{\varepsilon^2 \sqrt{2\pi G}}. \tag{10.2.14}$$

This simple inequality says that, as the gravitational constant, a smaller value of the symmetry-breaking scale leads to a larger R_{eff}. Thus it looks as if ε make an effective contribution to the attractive gravitational force.

10.2.4 Proof of equivalence and consequences

It is easily seen that the energy density $\mathcal{H}$ of the Abelian Higgs theory given in (10.2.6) can be rewritten in the form

$$\begin{aligned}
\mathcal{H} &= \frac{1}{4}g^{jj'}g^{kk'}(F_{jk} \pm \frac{1}{2}\epsilon_{jk}(|\phi|^2 - \varepsilon^2))(F_{j'k'} \pm \frac{1}{2}\epsilon_{j'k'}(|\phi|^2 - \varepsilon^2)) \\
&\quad + \frac{1}{4}g^{jk}(D_j\phi \pm \mathrm{i}\epsilon_j{}^{j'} D_{j'}\phi)\overline{(D_k\phi \pm \mathrm{i}\epsilon_k{}^{k'} D_{k'}\phi)} \\
&\quad \pm \frac{1}{4}\varepsilon^2\epsilon^{jk}F_{jk} \pm \nabla_j(\epsilon^{jk}J_k),
\end{aligned} \tag{10.2.15}$$

where ∇_j is the covariant derivative with respect to the metric $\{g_{jk}\}$ and J_k is the current vector defined by

$$J_k = \frac{\mathrm{i}}{4}(\phi\overline{[D_k\phi]} - \overline{\phi}[D_k\phi]). \tag{10.2.16}$$

From (10.2.15), we are led to the lower energy bound

$$\int_S \mathcal{H}\, d\Omega_g \geq \pi\varepsilon^2 |c_1(L)|$$

and the self-dual or anti-self-dual system

$$
\begin{aligned}
K_g - 8\pi G\mathcal{H} &= 0, \\
D_j\phi \pm i\epsilon_j{}^k D_k\phi &= 0, \\
\epsilon^{jk}F_{jk} \pm (|\phi|^2 - \varepsilon^2) &= 0.
\end{aligned}
\tag{10.2.17}
$$

That is, any solutions of (10.2.17) satisfy (10.2.8). Note that the last two equations in (10.2.17) imply immediately that $T_{jk} = 0$ ($j, k = 1, 2$). Here ϵ_{jk} is the standard Levi–Civita skew-symmetric 2-tensor satisfying $\epsilon_{12} = \sqrt{g}$. The system (10.2.17) was first derived in the work of Linet [198, 199] and Comtet and Gibbons [86]. Kim and Kim also studied the solutions of (10.2.17) based on some geometric and physical consideration [170]. Our subsequent discussion follows the main lines in [344, 345, 347].

Theorem 10.2.3. *The systems (10.2.8) and (10.2.17) are equivalent.*

Proof. Rewrite the last equation in (10.2.8) as

$$\partial_j(\epsilon^{j'k'}F_{j'k'}) + i\epsilon_{jj'}g^{j'k'}(\phi\overline{[D_{k'}\phi]} - \bar\phi[D_{k'}\phi]) = 0.$$

Therefore

$$\frac{1}{\sqrt{g}}\partial_j(\sqrt{g}g^{jk}\partial_k[\epsilon^{j'k'}F_{j'k'}]) + \frac{i}{\sqrt{g}}\partial_j(\sqrt{g}\epsilon^{jk}[\phi\overline{(D_k\phi)} - \bar\phi(D_k\phi)]) = 0.$$

Thus

$$\Delta_g(\epsilon^{jk}F_{jk}) - |\phi|^2(\epsilon^{jk}F_{jk}) + 2i\epsilon^{jk}(D_j\phi)\overline{(D_k\phi)} = 0.$$

Besides, we also have by using the second equation in (10.2.8) that

$$\Delta_g|\phi|^2 = |\phi|^2(|\phi|^2 - \varepsilon^2) + 2g^{jk}(D_j\phi)\overline{(D_k\phi)}.$$

These equations give us the useful expression

$$
\begin{aligned}
\Delta_g(\epsilon^{jk}F_{jk} \pm [|\phi|^2 - \varepsilon^2]) &= |\phi|^2(\epsilon^{jk}F_{jk} \pm [|\phi|^2 - \varepsilon^2]) \\
&\pm g^{jk}(D_j\phi \pm i\epsilon_j{}^{j'}D_{j'}\phi)\overline{(D_k\phi \pm i\epsilon_k{}^{k'}D_{k'}\phi)}.
\end{aligned}
\tag{10.2.18}
$$

Define for a solution triplet of (10.2.8) the quantities

$$
\begin{aligned}
P^+ &= \epsilon^{jk}F_{jk} + (|\phi|^2 - \varepsilon^2), \\
P^- &= \epsilon^{jk}F_{jk} - (|\phi|^2 - \varepsilon^2).
\end{aligned}
$$

Thus, according to the first equation in (10.2.8), we have, after a lengthy calculation,

$$P^+ P^- = 4g^{jk} T_{jk} \equiv 0. \tag{10.2.19}$$

The first part (with the plus sign) of (10.2.18) implies the elliptic inequality

$$\begin{aligned}
\Delta_g P^+ &= |\phi|^2 P^+ + g^{jk}(D_j\phi + i\epsilon_j{}^{j'} D_{j'}\phi)\overline{(D_k\phi + i\epsilon_k{}^{k'} D_{k'}\phi)} \\
&\geq |\phi|^2 P^+. \tag{10.2.20}
\end{aligned}$$

Since S is compact, the maximum principle implies that, either $P^+ \equiv 0$ or $P^+ < 0$ on S.

Similarly, the second part (with the minus sign) of (10.2.18) implies

$$\begin{aligned}
\Delta_g P^- &= |\phi|^2 P^- - g^{jk}(D_j\phi - i\epsilon_j{}^{j'} D_{j'}\phi)\overline{(D_k\phi - i\epsilon_k{}^{k'} D_{k'}\phi)} \\
&\leq |\phi|^2 P^-. \tag{10.2.21}
\end{aligned}$$

The maximum principle again says that, either $P^- \equiv 0$ or $P^- > 0$.

Using these observations in (10.2.19) we see that either $P^+ \equiv 0$ or $P^- \equiv 0$. Accordingly, we have either $D_j\phi + i\epsilon_j{}^k D_k\phi \equiv 0$ or $D_j\phi - i\epsilon_j{}^k D_k\phi \equiv 0$ due to (10.2.18). Thus (10.2.17) is fulfilled and the two systems (10.2.8) and (10.2.17) are equivalent.

As a consequence, we may concentrate on the system (10.2.17). Hence, in view of (10.2.15), the first equation in (10.2.17) becomes

$$K_g = \pm 2\pi\varepsilon^2 G\, \epsilon^{jk} F_{jk} \pm 8\pi G\, \nabla_j(\epsilon^{jk} J_k), \qquad x \in S. \tag{10.2.22}$$

We are now ready to determine the relation between the topology of S and the string number of a solution, or the topology of L.

For given $p \in S$, choose a specific isothermal coordinate system $(U, (x^j))$ so that $p \in U \subset S$ and $x^j(p) = 0$, $j = 1, 2$. Thus, around p, the second equation in (10.2.17) is simplified to

$$D_1\phi \pm iD_2\phi = 0,$$

which says in view of the $\bar{\partial}$-Poincaré lemma that, up to a nonvanishing factor, ϕ or $\bar{\phi}$ is holomorphic around p. In particular, if p is a zero of ϕ, then locally,

$$|\phi(x)| = |x|^n h(x), \tag{10.2.23}$$

where $h > 0$ and $n \geq 1$ is an integer. The zero is obviously isolated. In this case, there is a string passing through p with the winding number n. It is not hard to show that, for a nontrivial solution, $|\phi| < \varepsilon$ everywhere on S. Thus the third equation in (10.2.17) implies that the vorticity field acquires its maximal magnitude at the zeros of ϕ. In other words, the zeros of ϕ give the locations of the vortices or strings of a solution.

Suppose the zeros of the Higgs field ϕ are labelled by $p_1, \cdots, p_m \in S$ with multiplicities $n_1, \cdots, n_m$, respectively. $N = n_1 + \cdots + n_m$ is the total string or vortex number. The easily verified relation

$$N = \frac{1}{4\pi} \left| \int_S \epsilon^{jk} F_{jk} \, d\Omega_g \right|$$

says that $N = |c_1(L)|$ (the first Chern number). Since $|\phi|$ has the local representation (10.2.23) around each point $p = p_\ell$ (with $n = n_\ell$), $\ell = 1, \cdots, m$, we see that the substitution $u = \ln |\phi|^2$ renders the last two equations in (10.2.17) into the form

$$\Delta_g u = (e^u - \varepsilon^2) + 4\pi \sum_{\ell=1}^m n_\ell \delta_{p_\ell}, \qquad (10.2.24)$$

where δ_p is the Dirac distribution on $(S, \{g_{jk}\})$ concentrated at p.

Using the last two equations in (10.2.17) and (10.2.16), we obtain

$$
\begin{aligned}
\pm \frac{1}{4} \varepsilon^2 \epsilon^{jk} F_{jk} \pm \nabla_j (\epsilon^{jk} J_k) &= -\frac{1}{4} \varepsilon^2 (|\phi|^2 - \varepsilon^2) + \frac{1}{4} \Delta_g |\phi|^2 \\
&= -\frac{1}{4} \varepsilon^2 (e^u - \varepsilon^2) + \frac{1}{4} \Delta_g e^u.
\end{aligned}
$$

Hence the equation (10.2.22) becomes

$$K_g = -2\pi G(\varepsilon^2 [e^u - \varepsilon^2] - \Delta_g e^u). \qquad (10.2.25)$$

Let u_0 be a solution of the equation

$$\Delta_g u_0 = -\frac{4\pi N}{|S|_g} + 4\pi \sum_{\ell=1}^m n_\ell \delta_{p_\ell}. \qquad (10.2.26)$$

Thus (10.2.24) and (10.2.25) may be put into the form

$$\Delta_g(\varepsilon^2 u - e^u - \varepsilon^2 u_0) = -\frac{K_g}{2\pi G} + \frac{4\pi \varepsilon^2 N}{|S|_g}.$$

Since the function $w = \varepsilon^2 u - e^u - \varepsilon^2 u_0$ is smooth on S, the above equation leads to the consistency condition

$$\frac{1}{2\pi G} \int_S K_g \, d\Omega_g = 4\pi \varepsilon^2 N. \qquad (10.2.27)$$

As a consequence of the well-known Gauss–Bonnet theorem,

$$\int_S K_g \, d\Omega_g = 2\pi \chi(S),$$

we are led from the relation (10.2.27) to the equation (10.2.12).

In order to establish the lower bound (10.2.14), We subtract (10.2.26) from (10.2.24) to obtain the equation

$$\Delta_g(u - u_0) = e^u - \varepsilon^2 + \frac{4\pi N}{|S|_g}. \qquad (10.2.28)$$

Since $u - u_0$ and e^u are smooth functions, integrating (10.2.28) yields the inequality

$$4\pi N < \varepsilon^2 |S|_g. \qquad (10.2.29)$$

Combining (10.2.12), (10.2.29) and using $\chi(S) = 2$, we arrive at (10.2.14).

Now we consider the case $\varepsilon = 0$. We have seen that (10.2.8) and (10.2.17) are equivalent. Using the last equation in (10.2.17), we obtain

$$4\pi N = \pm \int_S \epsilon^{jk} F_{jk} \, d\Omega_g = - \int_S |\phi|^2 \, d\Omega_g.$$

Thus $N = 0$ and $\phi = 0$. The 2-surface S must be a flat torus, the energy density of the matter-gauge sector vanishes everywhere, and there is no gravity.

The proof of existence of string solutions will be presented in the next section.

10.3   Existence of Strings: Compact Case

In this section we prove the existence of multiple strings over a compact surface. It will be seen that both the total string number and local string strength are technically important.

10.3.1   Existence for $N \geq 3$

We now state and prove our existence theorem for N-string solutions over a compact Riemann surface S when $N \geq 3$.

Theorem 10.3.1. *Let the integer N satisfy the condition (10.2.12). For any $p_1, \cdots, p_m \in S$ and integers $n_1, \cdots, n_m$ with $n_1 + \cdots + n_m = N$ and*

$$n_1 < \frac{1}{2}N, \quad \cdots \cdots, \quad n_m < \frac{1}{2}N, \qquad (10.3.1)$$

the equations (10.2.8) or (10.2.17) have a smooth solution triplet (g, ϕ, A) so that it defines an appropriate Hermitian line bundle L over S with the first Chern number $c_1(L) = N$ and the zeros of the Higgs cross-section ϕ are exactly $p_1, \cdots, p_m$ with the respective multiplicities $n_1, \cdots, n_m$. In particular, when $N \geq 3$, the system (10.2.8) or (10.2.17) has an N-string solution so that the Higgs field has simple zeros at N distinct prescribed locations on S if and only if the condition (10.2.12) is fulfilled.

We present a proof of the theorem as follows.

Suppose that we can solve the coupled system of equations (10.2.24) and (10.2.25) for the unknown metric g and function u. Using u and applying the methods in [51, 116, 231, 232], we can construct a suitable line bundle L over (S, g) equipped with a Hermitian metric so that $c_1(L) = N$ and a solution pair (ϕ, A) of the last two equations in (10.2.17) is obtained. The validity of the first equation in (10.2.17) follows directly from (10.2.25). Thus we may only concentrate on the solvability of (10.2.24 and (10.2.25).

The proof splits into a few steps. We first combine the two equations, (10.2.24) and (10.2.25), into a single equation with a peculiar-looking nonlinearity. Then we perturb the resulting equation by a positive parameter so that a supersolution may be constructed in such a way that a monotone iterative scheme can be used to find a classical solution. Finally we take the zero parameter limit to recover a solution of the original equation.

Step 1. Reduction to an elliptic equation

We shall use the following standard device to get rid of the unknown gravitational background metric g on S.

Assume that g is conformal to a known metric g_0,

$$g = e^{\eta} g_0,$$

where η is an unknown conformal exponent, and that K_{g_0} is the Gauss curvature of (S, g_0). Then η, K_g, K_{g_0} are related through the equation

$$-\Delta_{g_0} \eta + 2K_{g_0} = 2K_g e^{\eta}.$$

Besides, we have $\Delta_g = e^{-\eta}\Delta_{g_0}$. Hence the equations (10.2.24) and (10.2.25) become

$$\Delta_{g_0} u = e^{\eta}(e^u - \varepsilon^2) + 4\pi \sum_{\ell=1}^{m} n_\ell \delta_{p_\ell}, \qquad (10.3.2)$$

$$\Delta_{g_0}(-\eta - 4\pi G e^u) = -2K_{g_0} - 4\pi G \varepsilon^2 e^{\eta}(e^u - \varepsilon^2), \qquad (10.3.3)$$

where now δ_p is the Dirac distribution at p on (S, g_0).

Let $v = u - u_0$, where u_0 is a solution of

$$\Delta_{g_0} u_0 = -\frac{4\pi N}{|S|_{g_0}} + 4\pi \sum_{\ell=1}^{m} n_\ell \delta_{p_\ell}. \qquad (10.3.4)$$

Then (10.3.2) and (10.3.3) take the form

$$\Delta_{g_0} v = e^{\eta}(e^{u_0+v} - \varepsilon^2) + \frac{4\pi N}{|S|_{g_0}}, \qquad (10.3.5)$$

$$\Delta_{g_0}(-\eta - 4\pi G e^{u_0+v} + 4\pi G \varepsilon^2 v) = -2K_{g_0} + \frac{16\pi^2 G \varepsilon^2 N}{|S|_{g_0}}. \qquad (10.3.6)$$

Using (10.2.12) and the Gauss–Bonnet theorem, we see that the right-hand side of (10.3.6) has zero integral. Consequently, there is a smooth function v_0 on S so that

$$\eta + 4\pi G e^{u_0+v} - 4\pi G \varepsilon^2 v = v_0 + c, \qquad (10.3.7)$$

where c is an arbitrary constant to be adjusted later.

We now insert (10.3.7) into (10.3.5). Use the notation $a = 4\pi G$. Then we arrive at the following equation,

$$\Delta_{g_0} v = \lambda e^{v_0 + a\varepsilon^2 v - a e^{u_0+v}}\left(e^{u_0+v} - \varepsilon^2\right) + \frac{4\pi N}{|S|_{g_0}} \qquad \text{on } S. \qquad (10.3.8)$$

Note that $\lambda = e^c$ is an adjustable parameter which should not be confused with the coupling parameter λ in the previous sections. We shall find in the rest of this section a solution of (10.3.8) which is obviously equivalent to (10.3.2) and (10.3.3) or (10.3.5) and (10.3.6).

Step 2. The perturbed problem

Let $(U_\ell, (x^j))$ be an isothermal coordinate chart near $p_\ell \in S$ for the surface (S, g_0) so that $x^j(p_\ell) = 0$ ($j = 1, 2$). According to [17], when U_ℓ is small the function u_0 (see (10.3.4)) has the property

$$u_0(x) = n_\ell \ln |x|^2 + w_\ell(x) \qquad \text{in } (U_\ell, (x^j)), \qquad (10.3.9)$$

where w_ℓ is a smooth function on U_ℓ.

We can use (10.3.9) to define a regular perturbation of u_0 as follows.

For any $\sigma > 0$ so that $p \in U_\ell$ whenever $|x(p)| < 3\sigma$, choose a function $\rho \in C^\infty(S)$ satisfying

$$0 \le \rho \le 1, \quad \rho(p) = 1 \quad \text{for } |x(p)| < \sigma, \quad \rho(p) = 0 \quad \text{for } |x(p)| > 2\sigma.$$

Take σ sufficiently small so that

$$\begin{aligned} u_0^\delta(x) &= n_\ell \ln(|x|^2 + \delta\rho(x)) + w_\ell(x) \quad \text{in } (U_\ell, (x^j)), \\ \ell &= 1, \cdots, m \end{aligned} \qquad (10.3.10)$$

($\delta > 0$) naturally extends to a smooth function on the full S and

$$u_0^\delta = u_0 \quad \text{in } S - \cup_{\ell=1}^m U_\ell; \quad u_0 \le u_0^\delta \quad \text{in } S.$$

It is more transparent to rewrite (10.3.8) in the form

$$\Delta_{g_0} v = \lambda e^{v_0 - a\varepsilon^2 u_0 + a\varepsilon^2(u_0+v) - a e^{u_0+v}}\left(e^{u_0+v} - \varepsilon^2\right) + \frac{4\pi N}{|S|_{g_0}} \qquad (10.3.11)$$

on S. The function $e^{-a\varepsilon^2 u_0}$ is a singular factor. We overcome this difficulty by introducing the perturbed equation

$$\Delta_{g_0} v = \lambda e^{v_0 - a\varepsilon^2 u_0^\delta + a\varepsilon^2(u_0+v) - a e^{u_0+v}}\left(e^{u_0+v} - \varepsilon^2\right) + \frac{4\pi N}{|S|_{g_0}} \qquad (10.3.12)$$

on S.

From now on, we assume $0 < \delta < 1$ (say).

Step 3. The sub/supersolution

Lemma 10.3.2. *There is a smooth function w on S independent of δ so that*

$$\Delta_{g_0} w > \lambda e^{v_0 - a\varepsilon^2 u_0^\delta + a\varepsilon^2 (u_0 + w) - a e^{u_0 + w}} \left(e^{u_0 + w} - \varepsilon^2 \right) + \frac{4\pi N}{|S|_{g_0}} \qquad (10.3.13)$$

on S for some suitable λ. In other words, w is a subsolution of (10.3.12) for all δ.

Proof. As before, let $(U_\ell, (x^j))$ be a coordinate chart near $p_\ell \in S$. Suppose that $\sigma > 0$ is a small number so that

$$\{x \in \mathbb{R}^2 \,|\, |x| < 3\sigma\} \subset \{x \in \mathbb{R}^2 \,|\, x = x(p) \text{ for some } p \in U_\ell\}, \quad \ell = 1, \cdots, m.$$

Define a function $f_\sigma \in C^\infty(S)$ so that

$$0 \le f_\sigma \le 1, \quad f_\sigma(p) = 1 \quad \text{for } |x(p)| \le \sigma \text{ and } p \in U_\ell, \quad \ell = 1, \cdots, m,$$

$f_\sigma(p) = 0$ for $|x(p)| \ge 2\sigma$ and $p \in U_\ell$, $\quad \ell = 1, \cdots, m$, or $p \in S - \cup_\ell^m U_\ell$.

Then the equation

$$\Delta_{g_0} w = \frac{8\pi N}{|S|_{g_0}} f_\sigma - C(\sigma) \qquad (10.3.14)$$

has a solution if $C(\sigma)$ satisfies

$$C(\sigma) = \frac{8\pi N}{|S|_{g_0}^2} \int_S f_\sigma \, d\Omega_{g_0}. \qquad (10.3.15)$$

The solution is unique up to an additive constant.

By (10.3.15) and the definition of f_σ, we see that $C(\sigma) \to 0$ as $\sigma \to 0$. Thus we can choose suitable $\sigma > 0$ to make

$$\frac{8\pi N}{|S|_{g_0}^2} - C(\sigma) > \frac{4\pi N}{|S|_{g_0}^2}. \qquad (10.3.16)$$

Define

$$U_\ell^\sigma = \{p \in S \,|\, p \in U_\ell, \, |x(p)| < \sigma\}, \qquad \ell = 1, \cdots, m.$$

Hence, in view of (10.3.14) and (10.3.16), we have

$$\Delta_{g_0} w > \frac{4\pi N}{|S|_{g_0}^2} \quad \text{in } \cup_{\ell=1}^m U_\ell^\sigma.$$

Of course, we can choose w such that $e^{u_0 + w} - \varepsilon^2 < 0$ on S. Consequently, the inequality (10.3.13) holds in $\cup_{\ell=1}^m U_\ell$ for any λ, δ.

Recall that w is independent of δ. Besides, by the definition of u_0^δ, we have $u_0^\delta \leq u_0^1$. Thus

$$e^{-a\varepsilon^2 u_0^\delta + a\varepsilon^2(u_0+w) - ae^{u_0+w}} \geq e^{a\varepsilon^2(u_0-u_0^1) + a\varepsilon^2 w - ae^{u_0+w}}$$

$$\geq C_0 > 0 \quad \text{in } S - \cup_{\ell=1}^m U_\ell^\sigma. \tag{10.3.17}$$

Of course,

$$\sup_S(e^{u_0+w} - \varepsilon^2) = -C_1, \quad C_1 > 0. \tag{10.3.18}$$

The constants C_0, C_1 are independent of δ. From (10.3.17) and (10.3.18), we see that when λ is large enough, the inequality (41) holds in $S - \cup_{\ell=1}^m U_\ell^\sigma$ as well.

In summary, w satisfies (10.3.13) for all δ.

Lemma 10.3.3. *Define $v_1 = -u_0 + \ln \varepsilon^2$. Then $v_1 > w$ on S.*

Proof. Use the notation of Lemma 10.3.2. Suppose $\sigma > 0$ is small so that $U_\ell^\sigma \cap U_{\ell'}^\sigma = \emptyset$ for $\ell \neq \ell'$. Of course,

$$w < v_1 \quad \text{in } \overline{U}_\ell^\sigma, \quad \ell = 1, \cdots, m$$

when σ is sufficiently small. Using (10.3.4), we can rewrite (10.3.13) in the form

$$\Delta_{g_0}(u_0 + w - \ln \varepsilon^2) > \lambda \varepsilon^2 e^{v_0 - a\varepsilon^2 u_0^\delta + a\varepsilon^2(u_0+w) - ae^{u_0+w}}(e^{u_0+w-\ln \varepsilon^2} - 1)$$

$$\text{in } S - \cup_{\ell=1}^m U_\ell^\sigma. \tag{10.3.19}$$

On ∂U_ℓ^σ $(\ell = 1, \cdots, m)$, we already have $u_0 + w - \ln \varepsilon^2 < 0$. If there is a point p so that $(u_0 + w - \ln \varepsilon^2)(p) \geq 0$, then the function $u_0 + w - \ln \varepsilon^2$ has a nonnegative interior maximum in $S - \cup_{\ell=1}^m U_\ell^\sigma$ which is false due to (10.3.19) and the maximum principle.

The lemma is proven.

Step 4. Solution of the perturbed equation

We shall use v_1 as a supersolution to find a solution of (10.3.12). Note that v_1 is singular at the points $p_1, \cdots, p_m$. Nevertheless, we can apply the following iterative scheme. We define

$$(\Delta_{g_0} - C_\delta)v_n = \lambda e^{v_0 - a\varepsilon^2 u_0^\delta + a\varepsilon^2(u_0+v_{n-1}) - ae^{u_0+v_{n-1}}}(e^{u_0+v_{n-1}} - \varepsilon^2)$$

$$-C_\delta v_{n-1} + \frac{4\pi N}{|S|_{g_0}} \quad \text{on } S, \quad n = 2, 3, \cdots, \tag{10.3.20}$$

where $C_\delta > 0$ is a constant to be determined as follows.

Consider the function

$$f(t) = e^{a\varepsilon^2 t - ae^t}(e^t - \varepsilon^2). \tag{10.3.21}$$

It is clear that the derivative $f'(t)$ is bounded for $t \in \mathbb{R}$.

Set

$$C_\delta = 1 + \lambda \sup_{x \in S}\{e^{v_0(x) - a\varepsilon^2 u_0^\delta(x)}\} \cdot \sup_{t \in \mathbb{R}}\{f'(t)\}$$

in (10.3.20). We have

Lemma 10.3.4. *There holds on S the inequality*

$$v_1 > v_2 > \cdots > v_n > \cdots > w. \qquad (10.3.22)$$

Proof. We have seen in Lemma 10.3.3 that $v_1 > w$. Using (10.3.20), we obtain

$$(\Delta_{g_0} - C_\delta)v_2 = -C_\delta v_1 + \frac{4\pi N}{|S|_{g_0}} \quad \text{on } S. \qquad (10.3.23)$$

Thus $(\Delta_{g_0} - C_\delta)(v_2 - v_1) = 0$ in $S - \{p_1, \cdots, p_m\}$. Since $v_1 \in L^p(S)$ for any $p > 1$, we have $v_2 \in W^{2,p}(S)$ and $v_2 \in C^{1,\alpha}(S)$ for any $0 < \alpha < 1$. In particular v_2 is bounded. Using the maximum principle, we get $v_1 > v_2$.

Besides, using the notation (10.3.21), the inequality (10.3.13), i.e.,

$$\Delta_{g_0} w > \lambda e^{v_0 - a\varepsilon^2 u_0^\delta} f(u_0 + w) + \frac{4\pi N}{|S|_{g_0}},$$

Lemma 10.3.3, and (10.3.23), we find

$$(\Delta_{g_0} - C_\delta)(w - v_2)$$
$$> \lambda e^{v_0 - a\varepsilon^2 u_0^\delta}(f(u_0 + w) - f(u_0 + v_1)) - C_\delta(w - v_1)$$
$$= (\lambda e^{v_0 - a\varepsilon^2 u_0^\delta} f'(u_0 + \xi) - C_\delta)(w - v_1)$$
$$> 0 \qquad (w < \xi < v_1).$$

Thus the maximum principle implies that $w < v_2$.

Suppose $v_{k-1} > v_k > w$ with $k \geq 2$. Then (10.3.20) says that

$$(\Delta_{g_0} - C_\delta)(v_{k+1} - v_k)$$
$$= \lambda e^{v_0 - a\varepsilon^2 u_0^\delta}(f(u_0 + v_k) - f(u_0 + v_{k-1})) - C_\delta(v_k - v_{k-1})$$
$$= (\lambda e^{v_0 - a\varepsilon^2 u_0^\delta} f'(u_0 + \xi_k) - C_\delta)(v_k - v_{k-1}) > 0,$$

where ξ_k is between v_{k-1} and v_k. Hence, $v_{k+1} < v_k$.

Moreover, we have

$$(\Delta_{g_0} - C_\delta)(w - v_{k+1})$$
$$> \lambda e^{v_0 - a\varepsilon^2 u_0^\delta}(f(u_0 + w) - f(u_0 + v_k)) - C_\delta(w - v_k)$$
$$= (\lambda e^{v_0 - a\varepsilon^2 u_0^\delta} f'(u_0 + \xi_k) - C_\delta)(w - v_k) > 0.$$

Consequently, $w < v_{k+1}$. The proof of the lemma is finished.

Taking the limit

$$\lim_{n \to \infty} v_n \equiv v^\delta$$

in (10.3.20) and using (10.3.22), we see that v^δ is a solution of the equation (10.3.12) satisfying

$$v_1 > v^\delta \geq w \qquad \text{on } S. \tag{10.3.24}$$

In order to find a solution of the original equation (10.3.11) or (10.3.8), we need to consider the $\delta \to 0$ limit in the following step.

Step 5. Passage to limit

First, observe that (10.3.24) implies that, for any $p > 1$, there is a constant $C > 0$ independent of δ so that

$$\|v^\delta\|_{L^p(S)} \leq C. \tag{10.3.25}$$

Next, it is easy to see that the one-parameter function $f(u_0 + v^\delta)(x)$ is uniformly bounded.

Besides, since $e^{-a\varepsilon^2 u_0^\delta} \leq e^{-a\varepsilon^2 u_0}$ and in the local isothermal coordinates $(U_\ell, (x^j))$ around p_ℓ we have

$$e^{-a\varepsilon^2 u_0(x)} = e^{-a\varepsilon^2 w_\ell(x)} |x|^{-2a\varepsilon^2 n_\ell}$$

due to (10.3.9), we see that there is a constant $C > 0$ independent of δ so that

$$\|e^{-a\varepsilon^2 u_0^\delta}\|_{L^p(S)} \leq C \tag{10.3.26}$$

provided that

$$2a\varepsilon^2 n_\ell p < 2. \tag{10.3.27}$$

From the condition (10.2.12) and $\chi(S) = 2$, we have

$$a\varepsilon^2 N = 2. \tag{10.3.28}$$

Combining (10.3.27) and (10.3.28), we arrive at the condition

$$n_\ell p < \frac{1}{2} N, \qquad \ell = 1, \cdots, m. \tag{10.3.29}$$

Suppose that n_ℓ's satisfy the restriction (10.3.1). Then there is a $p > 1$ so that (10.3.29) is valid. Using (10.3.25) and (10.3.26) in (10.3.12), we see that there is a constant $C > 0$ independent of δ to confine the $W^{2,p}$-norm of v^δ,

$$\|v^\delta\|_{W^{2,p}(S)} \leq C, \qquad \forall 0 < \delta < 1. \tag{10.3.30}$$

From the embedding $W^{k,p}(S) \to C^m(S)$ for $0 \leq m < k - 2/p$ with $k = 2$ and $p > 1$, we infer in view of (10.3.30) that

$$|v^\delta|_{C(S)} \leq C \tag{10.3.31}$$

for any δ, where $C > 0$ is a δ-independent constant.

It is now useful to rewrite (10.3.12) with $v = v^\delta$ in the form

$$\Delta_{g_0} v^\delta = \lambda e^{v_0 + a\varepsilon^2(u_0 - u_0^\delta) + a\varepsilon^2 v^\delta - ae^{u_0 + v^\delta}} (e^{u_0 + v^\delta} - \varepsilon^2) + \frac{4\pi N}{|S|_{g_0}}. \quad (10.3.32)$$

By the definition of u_0^δ, we know that the factor $e^{a\varepsilon^2(u_0 - u_0^\delta)}$ is bounded. Using (10.3.31), we see in particular that the right-hand side of (10.3.32) has uniformly bounded L^p-norm for any $p > 1$. Thus (10.3.30) holds for any $p > 1$. As a consequence, $\{v^\delta\}$ is bounded in $C^{1,\alpha}$ for any $0 < \alpha < 1$. Applying this fact in (10.3.32), we conclude that $\{v^\delta\}$ is bounded in $C^{2,\alpha}(S)$. However, the compact embedding $C^{2,\alpha}(S) \to C^2(S)$ enables us to get a convergent subsequence $\{v^{\delta_n}\}$ ($\delta_n \to 0$ as $n \to \infty$) so that

$$v^{\delta_n} \to \text{ some element } v \text{ in } C^2(S). \quad (10.3.33)$$

Inserting (10.3.33) into (10.3.32), we find that v is a solution of (10.3.8).

The proof of Theorem 10.3.1 is complete.

10.3.2 Existence for $N = 2$ and nonexistence for $N = 1$

We continue our study of cosmic strings over a compact Riemann surface S. Since, topologically, S must be a sphere, we see that $\chi(S) = 2$ and the string number N and the symmetry-breaking parameter $\varepsilon > 0$ must satisfy the constraint

$$N = \frac{1}{2\pi\varepsilon^2 G}. \quad (10.3.34)$$

In the previous subsection, we established an existence theorem for $N \geq 3$. In this subsection, we state our results for the remaining cases, $N = 1, 2$.

Theorem 10.3.5. *Consider the cosmic string equations (10.2.17) with the prescribed string number N over the standard 2-sphere S^2 and assume that the symmetry-breaking scale $\varepsilon > 0$ already satisfies the necessary condition (10.3.34).*

(i) When $N = 2N_0$ ($N_0 \geq 1$) is an even integer, the system (10.2.17) has an N-string solution so that the centers of the strings are at the north and south poles and there are exactly N_0 strings at each of these two poles. In particular, there exists a 2-string solution with strings located at the opposite poles.

(ii) The system (10.2.17) does not have any N-string solution so that the strings are all superimposed at one point on S^2 and the field configuration is symmetric about this point. In particular, there does not exist any symmetric 1-string solution at all.

We postpone the proof of this theorem until §10.6.

10.4   Existence of Strings: Noncompact Case

The sufficient conditions stated in Theorems 10.3.1 and 10.3.5 impose some restrictions to the local winding numbers n_ℓ's of the strings. Such a barrier comes from the topological type of the base space and from the technicalities involved in the existence proof. We shall now see that, when (S, g) is conformally flat, the constraint from topology disappears and one is able to get solutions realizing an arbitrarily prescribed string distribution and the respectively designated local winding numbers n_ℓ's.

10.4.1   Existence results

Our existence theorem below concerning multiple strings is valid under a sufficient condition imposed only on the total string number N.

Theorem 10.4.1. *Consider the coupled Einstein and Abelian Higgs equations (10.2.8) over an open Riemann surface (S, g). For any $p_1, \cdots, p_m \in S$ and $n_1, \cdots, n_m \in \mathbb{N}$, the system (10.2.8) has a finite-energy solution so that $S = \mathbb{R}^2$, g is conformal to the standard metric of $\mathbb{R}^2$: $g_{jk} = e^\eta \delta_{jk}$, the zeros of ϕ are exactly $p_1, \cdots, p_m$ with the corresponding multiplicities $n_1, \cdots, n_m$, the conformal factor e^η verifies the sharp decay estimate*

$$e^{\eta(x)} = O(|x|^{-4\pi\varepsilon^2 GN}) \qquad as\ |x| \to \infty$$

and there hold the following quantized values of the total gravitational curvature, the magnetic flux, and the energy of the matter-gauge sector

$$\int_{\mathbb{R}^2} K_g e^\eta\, dx = 8\pi^2\varepsilon^2 GN, \qquad \int_{\mathbb{R}^2} F_{12}\, dx = 2\pi N, \qquad \int_{\mathbb{R}^2} \mathcal{H} e^\eta\, dx = \pi\varepsilon^2 N,$$

provided that the total string number $N = n_1 + \cdots + n_m$ satisfies the bound

$$N \le \frac{1}{4\pi\varepsilon^2 G}. \tag{10.4.1}$$

Besides, the Gauss curvature K_g, and the physical energy density terms F_{jk}^2, $|D_j\phi|^2$, $(|\phi|^2 - \varepsilon^2)^2$ obey the sharp decay estimates

$$K_g, \quad F_{jk}^2, \quad |D_j\phi|^2, \quad 0 < \varepsilon^2 - |\phi|^2 = O(|x|^{-b}) \quad as\ |x| \to \infty$$

for any $b > 0$ if $N < 1/4\pi\varepsilon^2 G$. When $N = 1/4\pi\varepsilon^2 G$ but $m \ge 2$, there hold instead the asymptotic decay rates at $r = |x| = \infty$:

$$K_g = O(r^{-1}), \qquad F_{jk} = O(r^{-4}),$$
$$|D_j\phi|^2 = O(r^{-3}), \qquad 0 < \varepsilon^2 - |\phi|^2 = O(r^{-2}).$$

While, when $N = 1/4\pi\varepsilon^2 G$ and $m = 1$, the radially symmetric solution satisfies

$$K_g = O(r^{-2\sqrt{2N}}), \qquad F_{jk} = O(r^{-(2+\sqrt{2N})}),$$
$$|D_j\phi| = O(r^{-(1+\sqrt{2N})}), \qquad 0 < \varepsilon^2 - |\phi|^2 = O(r^{-\sqrt{2N}}).$$

Furthermore, in the same category of solutions, the obtained conformally flat surface $(\mathbb{R}^2, e^{\eta}\delta_{jk})$ *is geodesically complete if and only if the integer* N *fulfills the condition (10.4.1). When* $N > 1/4\pi\varepsilon^2 G$, *although an* N-*string solution may exist and the corresponding Gauss curvature is the curvature function for some conformal metric which is complete, the obtained gravitational metric itself in the solution is not complete.*

The solutions stated in Theorem 10.4.1 are most interesting because they belong to the category that the Higgs field tends to the asymmetric vacuum at infinity in spite of the decay of the gravitational metric and are crucial in the Higgs mechanism.

In the next subsection, we prove the existence of multiple strings for a given string distribution and obtain some preliminary properties of the solutions which will be useful in deriving the desired asymptotic estimates. We then establish the decay estimates of the solutions and identify in terms of the total string number N the criterion for completeness of the gravitational metric.

10.4.2 Construction of solutions

We now study the problem on the full plane. We shall look for a solution so that it is such that the Higgs field ϕ vanishes precisely at the given points $p_1, \cdots, p_m \in \mathbb{R}^2$ at the respective orders $n_1, \cdots, n_m$. We can use the same device to reduce this problem to a system of nonlinear elliptic equations of the form (10.3.2) and (10.3.3). It will be convenient to rescale (10.3.2) and (10.3.3) by the translation

$$u \mapsto u + \ln \varepsilon^2.$$

We again split the proof into a few steps.

Step 1. The elliptic equation and its perturbation

Set $a = 4\pi G\varepsilon^2$ as before. Choose g_0 to be the standard flat metric. Hence, $\Delta_{g_0} = \Delta$, $K_{g_0} = 0$, and we end up with the following equations on $\mathbb{R}^2$,

$$\Delta u = \varepsilon^2 e^{\eta}(e^u - 1) + 4\pi \sum_{\ell=1}^{m} n_\ell \delta_{p_\ell}, \qquad (10.4.2)$$

$$\Delta(\eta + ae^u) = a\varepsilon^2 e^{\eta}(e^u - 1). \qquad (10.4.3)$$

Define the background functions

$$u_0 = \sum_{\ell=1}^{m} n_\ell \ln\left(\frac{|x - p_\ell|^2}{1 + |x - p_\ell|^2}\right),$$

$$w_0 = \sum_{\ell=1}^{m} n_\ell \ln(1 + |x - p_\ell|^2).$$

Then it is seen that $u_0 < 0$, $w_0 \geq 0$, and

$$
\begin{aligned}
\Delta u_0 &= 4\pi \sum_{\ell=1}^{m} n_\ell \delta_{p_\ell} - \Delta w_0 \\
&= 4\pi \sum_{\ell=1}^{m} n_\ell \delta_{p_\ell} - g,
\end{aligned}
$$

where, in this subsection and part of the next subsection,

$$
g = \Delta w_0 = 4 \sum_{\ell=1}^{m} \frac{n_\ell}{(1 + |x - p_\ell|^2)^2} > 0,
$$

which should not be confused with our notation for the Riemannian metric $g = \{g_{jk}\}$.

Let $u = u_0 + v$. Then (10.4.2) becomes

$$
\Delta v = \varepsilon^2 e^\eta (e^{u_0 + v} - 1) + g. \tag{10.4.4}
$$

By (10.4.3) and (10.4.4), we easily infer that

$$
h = \frac{\eta}{a} + e^{u_0 + v} - v + w_0
$$

is an entire harmonic function, which clearly defines a background for the gravitational metric. The choice of h is often crucial in establishing existence in some situations. In the radially symmetric case, h is a constant. We now take the point of view that, far away from local regions, the solutions look radially symmetric, thus, multiple string solutions and radially symmetric string solutions should reside in the same metric background. Therefore we are led to assuming that $h = c =$const. Of course, such a choice may restrict the range of non-symmetric solutions we are searching for, and hence, is only a technical convenience. In this section, we again use the notation $\lambda = e^{ac}$ as in the compact case, which should not be confused with the coupling parameter λ considered in the Abelian Higgs action density. Note that this parameter is adjustable as in the compact case. Finally, inserting the above expression into (10.4.4), we find the resulting equation

$$
\Delta v = \lambda e^{-aw_0 + a(v - e^{u_0 + v})}(e^{u_0 + v} - 1) + g \quad \text{in } \mathbb{R}^2. \tag{10.4.5}
$$

As in the last section, the equation (10.4.5) is not convenient to work with. We can avoid the difficulty by introducing a δ-regularization of the equation (10.4.5) as follows,

$$
\Delta v = \lambda e^{-aw_0 + a(v - e^{u_0^\delta + v})}(e^{u_0^\delta + v} - 1) + g \quad \text{in } \mathbb{R}^2, \tag{10.4.6}
$$

where

$$
u_0^\delta = \sum_{\ell=1}^{m} n_\ell \ln(\delta + |x - p_\ell|^2) - w_0.
$$

Note that

$$\Delta u_0^\delta = 4\delta \sum_{\ell=1}^{m} \frac{n_\ell}{(\delta + |x - p_\ell|^2)^2} - g. \tag{10.4.7}$$

Step 2. The solution of the perturbed equation via sub/supersolutions
We first find a supersolution of (10.4.6).

Lemma 10.4.2. *The function $v_1^\delta = -u_0^\delta$ $(0 < \delta < 1)$ is a supersolution of (10.4.6).*

Proof. By virtue of (10.4.7),

$$\Delta v_1^\delta = -4\delta \sum_{\ell=1}^{m} \frac{n_\ell}{(\delta + |x - p_\ell|^2)^2} + g$$

$$< \quad g = \lambda e^{-aw_0 + a(v_1^\delta - e^{u_0^\delta + v_1^\delta})}(e^{u_0^\delta + v_1^\delta} - 1) + g.$$

Thus, v_1^δ is a supersolution as expected.

We now turn to the recognition of a suitable subsolution.

Lemma 10.4.3. *There is a λ_0 independent of $0 < \delta < 1/2$ (say) so that for $aN \leq 1$, there holds*

$$0 > \lambda e^{-aw_0 - ae^{u_0^\delta}}(e^{u_0^\delta} - 1) + g \qquad in \ \mathbb{R}^2 \tag{10.4.8}$$

whenever $\lambda > \lambda_0$. In other words, $v = 0$ is a subsolution of (10.4.6) for all δ.

Proof. We rewrite u_0^δ as

$$u_0^\delta = -\sum_{\ell=1}^{m} n_\ell \ln \left(1 + \frac{1 - \delta}{\delta + |x - p_\ell|^2} \right) < 0 \tag{10.4.9}$$

and

$$e^{u_0^\delta} - 1 = e^{\xi(u_0^\delta)u_0^\delta} u_0^\delta \qquad (0 < \xi(u_0^\delta) < 1).$$

Note that, since $u_0^\delta < 0$,

$$e^{\xi(u_0^\delta)u_0^\delta} = \frac{e^{u_0^\delta} - 1}{u_0^\delta} < 1.$$

Besides, by (10.4.9), we know, for $0 < \delta < \frac{1}{2}$, that $u_0^\delta \to 0$ uniformly as $r = |x| \to \infty$. Thus $e^{\xi(u_0^\delta)u_0^\delta} \to 1$ uniformly as $r \to \infty$. Hence,

$$e^{-aw_0 - ae^{u_0^\delta}}(e^{u_0^\delta} - 1) \leq e^{-aw_0 - a}(e^{u_0^\delta} - 1)$$

$$= e^{-aw_0 - a + \xi(u_0^\delta)u_0^\delta} u_0^\delta$$

$$= e^{-a+\xi(u_0^\delta)u_0^\delta} u_0^\delta \left(\prod_{\ell=1}^{m} \frac{1}{(1+|x-p_\ell|^2)^{an_\ell}} \right)$$

$$\text{(by (10.4.9))} = -e^{-a+\xi(u_0^\delta)u_0^\delta} \left(\sum_{\ell=1}^{m} \frac{n_\ell}{1+\xi_\ell} \cdot \frac{1-\delta}{\delta+|x-p_\ell|^2} \right)$$

$$\times \left(\prod_{\ell=1}^{m} \frac{1}{(1+|x-p_\ell|^2)^{an_\ell}} \right)$$

$$\equiv -h_\delta,$$

where

$$\xi_\ell = \theta_\ell \left(\frac{1-\delta}{\delta+|x-p_\ell|^2} \right) \quad \text{with } 0 < \theta_\ell < 1, \quad \ell = 1, \cdots, m.$$

By virtue of the assumption $aN \le 1$ or $2aN + 2 \le 4$, we obtain

$$r^4 h_\delta(x) \to \infty \quad (\text{if } aN < 1) \quad \text{as } r = |x| \to \infty$$

uniformly with respect to $0 < \delta < 1/2$ or

$$r^4 h_\delta(x) \to \text{some number } c_\delta > 0 \quad (\text{if } aN = 1) \quad \text{as } r = |x| \to \infty.$$

Using the definition of h_δ, we can see that there is a suitable c_0 so that $c_\delta \ge c_0$ for all $0 < \delta < \frac{1}{2}$.

The above observation enables us to conclude that there is an $r_0 > 0$ and $\lambda_1 > 0$ so that

$$\lambda e^{-aw_0 - ae^{u_0^\delta}} (e^{u_0^\delta} - 1) + g \le -\frac{1}{r^4}(\lambda r^4 h_\delta - r^4 g) < 0 \qquad (10.4.10)$$

whenever $r = |x| \ge r_0$ and $\lambda > \lambda_1$ because $r^4 g = O(1)$ at infinity.

On the other hand, we see by the definition of u_0^δ that $u_0^\delta \le u_0^{\frac{1}{2}}$ ($\delta \le \frac{1}{2}$). Thus,

$$e^{-aw_0 - ae^{u_0^\delta}} (e^{u_0^\delta} - 1) \le e^{-aw_0 - a} (e^{u_0^{\frac{1}{2}}} - 1). \qquad (10.4.11)$$

By (10.4.11), we can find $\lambda_0 \ge \lambda_1$ so that (10.4.8) holds in $\{x \in \mathbb{R}^2 \mid |x| < r_0\}$ for all $0 < \delta < 1/2$ whenever $\lambda > \lambda_0$.

Step 3. Solution of the original equation – the nonradial case

Since $v_1^\delta > 0$ is a supersolution, $v = 0$ is a subsolution, and all involved are regular, so we see that (10.4.6) has a smooth solution v^δ in $\mathbb{R}^2$ satisfying

$$-u_0^\delta = v_1^\delta \ge v^\delta \ge 0 \quad \text{in } \mathbb{R}^2. \qquad (10.4.12)$$

Such a step may be based on a general lemma established by Ni [226].

We now study the passage $\delta \to 0$ of the family $\{v^\delta\}$.

For $\delta_1 < \delta_2$, we have $u_0^{\delta_1} < u_0^{\delta_2}$. Hence $v_1^{\delta_1} > v_1^{\delta_2}$. In particular, $v_0 \equiv -u_0 = v_1^0 > v_1^\delta$ for all $\delta > 0$. Thus a weaker form of (10.4.12) is

$$v_0 > v^\delta \geq 0 \qquad \text{in } \mathbb{R}^2. \tag{10.4.13}$$

Let us consider the right-hand side of (10.4.6). It is clear that

$$e^{-ae^{u_0^\delta+v^\delta}}\left(e^{u_0^\delta+v^\delta} - 1\right)$$

is a bounded function with an upper bound independent of δ. Besides, we have in view of (10.4.13) the bound

$$e^{-aw_0+av^\delta} \leq e^{-aw_0+av_0} = e^{-a(u_0+w_0)}$$
$$= \prod_{\ell=1}^{m} |x - p_\ell|^{-2an_\ell} = f \tag{10.4.14}$$

for the other factor on the right-hand side of (10.4.6). So the δ-independent upper bound function f has singularities at $x = p_\ell$, $\ell = 1, \cdots, m$. We hope to apply L^p-estimates to control the sequence $\{v^\delta\}$. For this purpose, we require

$$f \in L^p_{\text{loc}}(\mathbb{R}^2) \qquad \text{for some } p > 1. \tag{10.4.15}$$

By (10.4.14) and $n_\ell \leq N$, we see that (10.4.15) is ensured provided that $aN < 1$. When $aN = 1$, we only consider the nonradial case where $m \geq 2$ (there are more than two centers of strings). Thus $n_\ell < N$ for $\ell = 1, \cdots, m$, and (10.4.15) is still ensured. Roughly speaking, the condition $aN \leq 1$ is sufficient to give us (10.4.15). It is interesting to note that this local regularity condition is the same as the condition stated in Lemma 10.4.3 where we need to control the behavior of the nonlinearity at infinity in order to obtain a subsolution.

As a consequence, we conclude that the right-hand side of (10.4.6) has uniform L^p bound on any given compact domain in $\mathbb{R}^2$. In other words, for any bounded domain $\mathcal{O} \subset \mathbb{R}^2$, there is a constant $C(p, \mathcal{O}) > 0$ independent of δ, so that

$$\|\Delta v^\delta\|_{L^p(\mathcal{O})} \leq C(p, \mathcal{O}). \tag{10.4.16}$$

Applying the interior L^p-estimates [3, 38] and using (10.4.13) and (10.4.16), we see that

$$\|v^\delta\|_{W^{2,p}(\mathcal{O})} \leq C(p, \mathcal{O}) \tag{10.4.17}$$

for some δ-independent constant $C(p, \mathcal{O}) > 0$. Using the continuous embedding

$$W^{k,p}(\mathcal{O}) \to C^m(\overline{\mathcal{O}}) \qquad \text{for } 0 \leq m < k - \frac{2}{p} \tag{10.4.18}$$

with $k = 2$, we see that $\{v^\delta\}$ is bounded in $C(\overline{\mathcal{O}})$. From this fact and (10.4.13), we conclude that $\{v^\delta\}$ is uniformly bounded over the full $\mathbb{R}^2$.

In view of (10.4.6), the boundedness of $\{v^\delta\}$ implies that $\{\Delta v^\delta\}$ is also bounded in $\mathbb{R}^2$. Thus the interior L^p-estimates say that (10.4.17) holds for any $p > 1$ and any given bounded domain $\mathcal{O}$. Take $p > 2$. The embedding (10.4.18) gives us the bound

$$|v^\delta|_{C^1(\overline{\mathcal{O}})} \le C(p, \mathcal{O}), \qquad \forall \delta > 0. \tag{10.4.19}$$

Using the properties of the right-hand side of (10.4.6), we see in view of (10.4.19) that

$$|\Delta v^\delta|_{C^1(\overline{\mathcal{O}})} \le C(p, \mathcal{O})$$

for any δ. Since $\mathcal{O}$ is arbitrary, the above results and the interior Schauder estimates enable us to conclude that for each $\alpha \in (0, 1)$ there is a constant $C(\alpha, p, \mathcal{O})$ independent of $\delta > 0$ so that

$$|v^\delta|_{C^{2,\alpha}(\overline{\mathcal{O}})} \le C(\alpha, p, \mathcal{O}). \tag{10.4.20}$$

We are now ready to use a standard diagonal subsequence argument to obtain a solution of (10.4.5) on $\mathbb{R}^2$ in the limit $\delta \to 0$.

Let $r_1 < r_2 < \cdots < r_i < \cdots$, $r_i \to \infty$ (as $i \to \infty$) be a sequence of positive numbers and

$$B_i = \{x \in \mathbb{R}^2 \mid |x| < r_i\}.$$

Applying the estimate (10.4.20) on $\{v^\delta\}$ with $\mathcal{O} = B_i$, $i = 1, 2, \cdots$, and the compact embedding $C^{2,\alpha}(\overline{B_i}) \to C^2(\overline{B_i})$, we can extract on each B_i a convergent subsequence of $\{v^\delta\}$ in $C^2(\overline{B_i})$. We start from B_1. Choose δ_n^1, $\delta_n^1 \to 0$ as $n \to \infty$ and $v_1 \in C^2(\overline{B_1})$ so that $v^{\delta_n^1} \to v_1$ in $C^2(\overline{B_1})$ as $n \to \infty$. Then there is a subsequence $\{\delta_n^2\}$ (say) of $\{\delta_n^1\}$ and an element $v_2 \in C^2(\overline{B_2})$ satisfying $\delta_n^2 \to \infty$ and $v^{\delta_n^2} \to v_2$ in $C^2(\overline{B_2})$ as $n \to \infty$. Of course, $v_1 = v_2$ in B_1. We can repeat this procedure to get sequences $\{\delta_n^i\}$, $i = 1, 2, \cdots$, so that

(i) $\{\delta_n^i\} \subset \{\delta_n^{i-1}\}$, $i = 2, 3, \cdots$;
(ii) for each fixed $i = 1, 2, \cdots$, $\delta_n^i \to 0$ as $n \to \infty$;
(iii) for each fixed $i = 1, 2, \cdots$, there is an element $v_i \in C^2(\overline{B_i})$ satisfying $v^{\delta_n^i} \to v_i$ as $n \to \infty$;
(iv) there holds $v_i = v_{i-1}$ on B_{i-1}, $i = 2, 3, \cdots$.

Set $v(x) = v_i(x)$ for $x \in B_i$ and $i = 1, 2, \cdots$. The property (iv) above says that v is a well-defined C^2-function on $\mathbb{R}^2$. By virtue of (i) and (iii), we see that $v^{\delta_n^n} \to v$ as $n \to \infty$ in $C^2(\overline{\mathcal{O}})$-norm for any given bounded domain $\mathcal{O}$ in $\mathbb{R}^2$. Take $\delta = \delta_n^n$ in (10.4.6). Letting $n \to \infty$ and using (i) above, we find that v is a smooth solution of (10.4.5). Moreover, the inequality (10.4.13) implies that v verifies the same bounds

$$0 \le v \le -u_0 \quad \text{in } \mathbb{R}^2. \tag{10.4.21}$$

In particular, v vanishes at infinity.

Step 4. Solution of the radial case

We now deal with the case $aN = 1$ and $m = 1$ (clustered strings) individually. Without loss of generality, we assume that the single center of the N strings is at the origin. It suffices to find a radially symmetric solution. The equations (10.4.2) and (10.4.3) give us

$$\frac{1}{a}\eta + e^u - u + 2N \ln r = c = \text{constant}.$$

Thus, we come up with a single equation, replacing (10.4.5) with $\lambda = e^{ac}$ as before,

$$
\begin{aligned}
u_{rr} + \frac{1}{r}u_r &= \lambda r^{-2aN} e^{a(u-e^u)}(e^u - 1), && r > 0, \\
\lim_{r\to 0} ru_r(r) &= 2N, \quad \lim_{r\to\infty} u(r) = 0. && (10.4.22)
\end{aligned}
$$

The boundary condition at $r = 0$ is important when we use the radial solution defined in the punctured plane $r > 0$ to get a classical solution of the original problem over the entire $\mathbb{R}^2$.

We now introduce the new variables

$$t = \ln r, \qquad U(t) = u(e^t).$$

Then the system (10.4.22) becomes

$$
\begin{aligned}
U'' &= \lambda e^{a(U-e^U)}(e^U - 1), && -\infty < t < \infty, \\
\lim_{t\to -\infty} U'(t) &= 2N, \quad \lim_{t\to\infty} U(t) = 0. && (10.4.23)
\end{aligned}
$$

In Chapter 9, we saw that the equation (10.4.23) has a negative solution in the neighborhood of $t = -\infty$ and $\lim_{t\to -\infty} U(t) = -\infty$. Besides, note that the right-hand side of the equation, $g(U)$, can be written in the form

$$g(U) = \lambda e^{a(U-e^U)}(e^U - 1) = -\frac{\lambda}{a}\frac{d}{dU}\left[e^{a(U-e^U)}\right].$$

Therefore, multiplying the equation (10.4.23) by U' and integrating over $(-\infty, t)$, we find the reduced equation

$$U'^2(t) = 4N^2 - \frac{2\lambda}{a}e^{a(U-e^U)} \equiv F(U). \qquad (10.4.24)$$

It will be useful to study the critical points of this equation first. Suppose $\underline{U}$ is a number that $F(\underline{U}) = 0$. In order to ensure the uniqueness property at the equilibrium $U = \underline{U}$ for (10.4.24), we need to require that

$$F'(\underline{U}) = 2\lambda(e^{\underline{U}} - 1)e^{a(\underline{U}-e^{\underline{U}})} = 0.$$

The only choice is $\underline{U} = 0$. Inserting this result into $F(\underline{U}) = 0$, we find that

$$\lambda = 2aN^2e^a = 2Ne^{\frac{1}{N}}, \tag{10.4.25}$$

where we have used the condition $aN = 1$. In the sequel, we will always assume (10.4.25). Hence, for t in a neighborhood of $t = -\infty$, we can rewrite (10.4.24) in the following form so that the derivative is explicit:

$$U'(t) = 2N\sqrt{1 - e^{\frac{1}{N}(1+U-e^U)}}, \tag{10.4.26}$$

where we have chosen the positive radical root because, according to the boundary condition at $t = -\infty$ in (10.4.23), $U' > 0$ initially. Since, in (10.4.26), $U' > 0$ and the uniqueness holds at the equilibrium $\underline{U} = 0$, we can use the fact that $F(U)$ decreases in $U < 0$ to conclude that $U = U(t)$ (the local solution of (10.4.23) near $t = -\infty$) solves (10.4.26) in the entire interval $-\infty < t < \infty$ and $U(t) < 0 = \underline{U}$ for all t.

Besides, (10.4.26) can be rewritten in the integral form

$$\int_{U(0)}^{U(t)} \frac{dU}{\sqrt{F(U)}} = t. \tag{10.4.27}$$

Consequently, we must have $U(t) \to \underline{U} = 0$ as $t \to \infty$. In fact, since

$$F''(0) = 2\lambda e^{-a} = 4N,$$

we can derive from (10.4.27) the sharp estimate

$$|U| = O(e^{-\sqrt{2N}t}) \qquad \text{as } t \to \infty. \tag{10.4.28}$$

Returning to the original variable $r = e^t, u(r) = U(\ln r)$, we can see that a desired solution of (10.4.2) and (10.4.3) is obtained. However, we do not have a bound like (10.4.21). Instead, from (10.4.28), we have

$$|u(r)| = O(r^{-\sqrt{2N}}) \qquad \text{as } r \to \infty. \tag{10.4.29}$$

Note also that (10.4.26) implies $U'(t)/|U(t)| \to \sqrt{F''(0)/2} = \sqrt{2N}$ as $t \to \infty$. Hence,

$$|u_r(r)| = O(r^{-(1+\sqrt{2N})}) \qquad \text{as } r \to \infty. \tag{10.4.30}$$

In order to see that the solutions just obtained carry finite energies, we need to study the asymptotic behavior of v in (10.4.21) or u itself.

10.4.3 Asymptotic decay estimates

In this subsection, we complete the proof of Theorem 10.4.1 by obtaining the asymptotic properties of the solutions and the condition that is crucial for the gravitational metric to be complete.

Decay estimates

Let v be a solution of (10.4.5) satisfying (10.4.21). Then $u = u_0 + v$ fulfills the equation (10.4.2) with the function η given by the expression

$$\eta = -aw_0 + a(v - e^{u_0 + v}) + c. \tag{10.4.31}$$

Hence, by the definition of w_0, we have

$$e^\eta = e^{c + a(v - e^{u_0 + v})} \prod_{\ell=1}^{m} \frac{1}{(1 + |x - p_\ell|^2)^{an_\ell}}. \tag{10.4.32}$$

Choose $r_0 > 0$ sufficiently large so that

$$\{p_1, \cdots, p_m\} \subset B(r_0) = \{x \in \mathbb{R}^2 \mid |x| < r_0\}.$$

Then (10.4.2) becomes

$$\Delta u = e^\eta(e^u - 1) \quad \text{in } \mathbb{R}^2 - \overline{B(r_0)}. \tag{10.4.33}$$

The decay property (10.4.32) implies the following.

Lemma 10.4.4. *Suppose that $aN < 1$. Then the solution u of (10.4.33) has the bounds*

$$-C_b|x|^{-b} < u(x) < 0, \qquad |x| > r_0 \tag{10.4.34}$$

for any $b > 0$. Here $C_b > 0$ is a constant depending on b. If $aN = 1$ and $m \geq 2$ (there are at least two string centers), (10.4.34) holds for $b = 2$. If $aN = 1$ but $m = 1$ (superimposed strings), then the radial solution satisfies (10.4.34) with $b = \sqrt{2N}$.

Proof. Assume $aN < 1$ first. Introduce the comparison function

$$w(x) = C|x|^{-b}. \tag{10.4.35}$$

Then

$$\Delta w = b^2 r^{-2} w, \qquad |x| = r > r_0. \tag{10.4.36}$$

Choose $\xi \in [0, 1]$ so that $e^u - 1 = e^{\xi u} u$. Thus, by (10.4.33), (10.4.36), and setting $\sigma = e^{\eta + \xi u}$, we have for $r_0 > 0$ sufficiently large,

$$\begin{aligned}
\Delta(u + w) &= \sigma u + b^2 r^{-2} w \\
&< b^2 r^{-2}(u + w), \qquad |x| = r > r_0,
\end{aligned} \tag{10.4.37}$$

since e^η satisfies (10.4.32) and $2aN < 2$. For such fixed r_0, we can take the constant C in (10.4.35) large to make

$$(u(x) + w(x))\Big|_{|x| = r_0} > 0.$$

Applying the maximum principle in (10.4.37), we obtain $u + w > 0$ in $\mathbb{R}^2 - \overline{B(r_0)}$ as expected.

If $aN = 1$ but $m \geq 2$, the estimate comes from (10.4.21) and the definition of u_0 (see the proof of Lemma 10.4.5 below). If $aN = 1$ and $m = 1$, the estimate follows from (10.4.29).

We next show that $\partial_j u$ satisfies similar decay properties as u stated in (10.4.34). For this purpose, we first prove

Lemma 10.4.5. *Suppose $aN < 1$ or $aN = 1$ but $m \geq 2$. There holds $\partial_j v = 0$ at infinity, $j = 1, 2$.*

Proof. By (10.4.21) and

$$-u_0(x) = \sum_{\ell=1}^{m} n_\ell \ln \left(1 + \frac{1}{|x - p_\ell|^2} \right) = O(r^{-2})$$

as $|x| = r \to \infty$, we see that $v \in L^2(\mathbb{R}^2)$. Besides, using

$$0 \leq 1 - e^{u_0 + v} \leq 1 - e^{u_0}$$
$$= 1 - \prod_{\ell=1}^{m} \frac{|x - p_\ell|^{2n_\ell}}{(1 + |x - p_\ell|^2)^{n_\ell}} = O(r^{-2})$$

and

$$e^{-aw_0} = \prod_{\ell=1}^{m} \frac{1}{(1 + |x - p_\ell|^2)^{an_\ell}} = O(r^{-2aN}) \tag{10.4.38}$$

(at $|x| = r = \infty$), we conclude that the right-hand side of (10.4.5) lies in $L^2(\mathbb{R}^2)$ as well. Hence the L^2-estimates for (10.4.5) enable us to get $v \in W^{2,2}(\mathbb{R}^2)$.

Furthermore, differentiating (10.4.5) gives us

$$\begin{aligned}
\Delta(\partial_j v) &= \lambda e^{-aw_0 + a(v - e^{u_0 + v})}(e^{u_0 + v} - a[e^{u_0 + v} - 1]^2)(\partial_j v) \\
&\quad + \lambda e^{-aw_0 + a(v - e^{u_0 + v})}(1 + a - ae^{u_0 + v})e^v(\partial_j e^{u_0}) \\
&\quad + \lambda e^{a(v - e^{u_0 + v})}(e^{u_0 + v} - 1)(\partial_j e^{-aw_0}) + (\partial_j g).
\end{aligned} \tag{10.4.39}$$

Of course, $\partial_j g \in L^2(\mathbb{R}^2)$. Using

$$\partial_j e^{u_0} = \partial_j \left(\prod_{\ell=1}^{m} \frac{|x - p_\ell|^{2n_\ell}}{(1 + |x - p_\ell|^2)^{n_\ell}} \right) = O(r^{-1})$$

and (10.4.38), we find

$$e^{-aw_0}(\partial_j e^{u_0}) = O(r^{-(2aN+1)}) \in L^2(\mathbb{R}^2).$$

Differentiating (10.4.38) gives us

$$\partial_j e^{-aw_0} = O(r^{-(2aN+1)}).$$

Inserting the above information into (10.4.39), we see that the right-hand side of (10.4.39) lies in $L^2(\mathbb{R}^2)$. Thus the elliptic L^2-estimates lead us to the conclusion $\partial_j v \in W^{2,2}(\mathbb{R}^2)$. Consequently, $\partial_j v \to 0$ as $|x| \to \infty$.

We are now ready to derive the decay estimates for $|\nabla u|$.

Lemma 10.4.6. *For the solution u of (10.4.33), we have*

$$|\nabla u|^2 \le C_b |x|^{-b}, \qquad |x| > r_0, \tag{10.4.40}$$

where $b > 0$ is again arbitrary for $aN < 1$ and $C_b > 0$ is a constant, while, for $aN = 1$ but $m \ge 2$, the solution of (10.4.33) can be so obtained that (10.4.40) holds for $b = 3$.

Proof. From Lemma 10.4.5 and the definition of u_0, we see that $|\nabla u| \to 0$ as $|x| \to \infty$.

Differentiating (10.4.33), we obtain

$$
\begin{aligned}
\Delta(\partial_j u) &= (\partial_j \eta) e^{\eta}(e^u - 1) + e^{\eta+u}(\partial_j u) \\
&= f_j + (e^{\eta+u} - ae^{\eta}(e^u - 1)^2)(\partial_j u). \tag{10.4.41}
\end{aligned}
$$

In view of (10.4.31) and Lemma 10.4.4, we see that, at infinity,

$$f_j = \begin{cases} O(r^{-b_1}), & \forall b_1 > 0, \quad \text{when } aN < 1, \\ O(r^{-5}), & \text{when } aN = 1, \, m \ge 2. \end{cases} \tag{10.4.42}$$

Set $h = |\nabla u|^2$. Thus, as a consequence of (10.4.41), there holds

$$
\begin{aligned}
\Delta h &\ge 2e^{\eta+u}h + 2(\nabla u \cdot \nabla \eta)e^{\eta}(e^u - 1) \\
&\ge e^{\eta+u}h + q(x), \qquad |x| > r_0, \tag{10.4.43}
\end{aligned}
$$

where $q(x)$ satisfies the same decay estimate as f_j in (10.4.42).

Suppose that w is given by (10.4.35). From (10.4.36) and (10.4.43), we have

$$\Delta(h - w) \ge e^{\eta+u}h - b^2 r^{-2}w + q(x), \qquad |x| = r > r_0. \tag{10.4.44}$$

Assume that $b_1 > 2 + b$ when $aN < 1$ or $5 \ge 2 + b$ when $aN = 1$ but $m \ge 2$. Then there is some $C > 0$ in (10.4.35) so that

$$q(x) > -b^2 r^{-2}w(x) \qquad \text{for } |x| = r > r_0. \tag{10.4.45}$$

By (10.4.32), if $aN < 1$, we may also assume that

$$e^{\eta+u} > 2b^2 r^{-2} \qquad \text{for } |x| = r > r_0; \tag{10.4.46}$$

while, if $aN = 1$ but $m \geq 2$, by the uniform bound (10.4.21) and the arbitrariness of the constant c in (10.4.31) or (10.4.32) so that $(e^{\eta(x)})|_{|x|=r_0}$ may be made sufficiently large, we still have the validity of (10.4.46).

Inserting (10.4.45) and (10.4.46) into (10.4.44), we find the inequality

$$\Delta(h-w) \geq e^{\eta+u}h - 2b^2r^{-2}w$$
$$\geq 2b^2r^{-2}(h-w), \qquad |x| = r > r_0. \qquad (10.4.47)$$

Of course, we can adjust the constant C in (10.4.35) to make

$$(h(x) - w(x))\Big|_{|x|=r_0} \leq 0. \qquad (10.4.48)$$

Using the boundary conditions (10.4.48) and $h - w \to 0$ (as $|x| \to \infty$) in (10.4.47), we arrive at $h \leq w$ for $|x| > r_0$ as desired.

Let $u = u_0 + v$ and η be defined in (10.4.31). Then $g_{jk} = e^{\eta}\delta_{jk}$ and (ϕ, A) give rise to a solution triplet of the coupled Einstein and Abelian Higgs equations (10.2.8) or (10.2.17) on the conformally flat surface $(\mathbb{R}^2, \{g_{jk}\})$, where

$$\phi(z) = \exp\left(\frac{1}{2}u(z) + i\sum_{\ell=1}^{m} n_\ell \arg(z - p_\ell)\right),$$

$$A_1(z) = -\mathrm{Re}\{2i\overline{\partial}\ln\phi(z)\}, \quad A_2(z) = -\mathrm{Im}\{2i\overline{\partial}\ln\phi(z)\}. \qquad (10.4.49)$$

In general, if ϕ has the local representation $\phi = e^{\sigma+i\omega}$ where σ and ω are real functions, by $\partial_1 = \partial + \overline{\partial}$, $\partial_2 = i(\partial - \overline{\partial})$ and the second line in (10.4.49), we have

$$D_1\phi = (\partial + \overline{\partial})\phi - \left\{\frac{\overline{\partial}\phi}{\phi} - \frac{\overline{\partial\overline{\phi}}}{\overline{\phi}}\right\}\phi = 2\phi\partial\sigma,$$

$$D_2\phi = i(\partial - \overline{\partial})\phi + i\left\{\frac{\overline{\partial}\phi}{\phi} + \frac{\overline{\partial\overline{\phi}}}{\overline{\phi}}\right\}\phi = 2i\phi\partial\sigma.$$

Using these formulas and (10.4.49), we have the relation $|D_1\phi|^2 + |D_2\phi|^2 = \frac{1}{2}e^u|\nabla u|^2$.

Let b be the exponent described in Lemma 10.4.4. Then by the third equation in (10.2.17) and (10.4.32), we have

$$F_{12} = O(r^{-(b+2aN)}),$$
$$1 - |\phi|^2 = O(r^{-b})$$

for $r = |x|$ large. By Lemma 10.4.6, we have

$$|D_j\phi|^2 = \begin{cases} O(r^{-b}), & \forall b > 0, \text{ where } aN < 1, \\ O(r^{-3}), & \text{where } aN = 1, m \geq 2, \\ O(r^{-2(1+\sqrt{2N})}), & \text{where } aN = 1, m = 1. \end{cases} \qquad (10.4.50)$$

These estimates imply immediately that

$$\int_{\mathbb{R}^2} F_{12} dx = 2\pi N.$$

Consequently, by the first equation in (10.2.17) and (10.4.32), we find that the Gauss curvature satisfies

$$K_g = \begin{cases} O(r^{-b}), & \forall b > 0, \text{ where } aN < 1, \\ O(r^{-1}), & \text{where } aN = 1, m \geq 2, \\ O(r^{-2\sqrt{2N}}), & \text{where } aN = 1, m = 1. \end{cases}$$

Thus the decay estimates stated in Theorem 10.4.1 are obtained.

Furthermore, since we can rewrite the energy density of the matter-gauge sector in the form

$$\begin{aligned} \mathcal{H} &= \frac{1}{2} e^{-2\eta} F_{12}^2 + \frac{1}{2} e^{-\eta} (|D_1\phi|^2 + |D_2\phi|^2) + \frac{1}{8}(|\phi|^2 - 1)^2 \\ &= \frac{1}{2}\left(e^{-\eta} F_{12} + \frac{1}{2}(|\phi|^2 - 1)\right)^2 + \frac{1}{2} e^{-\eta}|D_1\phi + iD_2\phi|^2 \\ &\quad - \frac{1}{2} e^{-\eta} F_{12}(|\phi|^2 - 1) + \frac{i}{2} e^{-\eta}([D_1\phi]\overline{[D_2\phi]} - \overline{[D_1\phi]}[D_2\phi]) \\ &= \frac{1}{2} e^{-\eta} F_{12} + \frac{1}{2} e^{-\eta} \text{Im}\{\partial_j e^{jk} \overline{\phi}(D_k\phi)\}; \end{aligned} \qquad (10.4.51)$$

hence, (10.4.50) and (10.4.51) lead us to the following total energy of the matter-gauge coupling,

$$\int_{\mathbb{R}^2} \mathcal{H} e^\eta \, dx = \pi N$$

and the energy of the gravitational sector which is realized by the total Gauss curvature

$$\int_{\mathbb{R}^2} K_g e^\eta \, dx = 8\pi^2 GN,$$

where we have used the first equation in (10.2.17) to relate the total curvature to the matter-gauge energy. Returning to the original system with the symmetry-breaking parameter $\varepsilon > 0$, we see that the decay estimates and the quantization identities are established.

Completeness of metric

Let p, q be two arbitrarily given points in the conformally flat surface $(\mathbb{R}^2, e^\eta \delta_{jk})$ where the conformal exponent η is determined by the expression (10.4.31). Suppose that $\mathcal{C}$ is the set of piecewise differentiable curves of the form $x = x(t) \in \mathbb{R}^2$, $0 \leq t \leq T$ connecting p, q, i.e., $x(0) = p, x(T) = q$. The geodesic metric d on $(\mathbb{R}^2, e^\eta \delta_{jk})$ is defined by

$$\begin{aligned} d(p,q) &= \inf_{x \in \mathcal{C}} \int_0^T \sqrt{g_{jk} \dot{x}^j \dot{x}^k} \, dt \\ &= \inf_{x \in \mathcal{C}} \int_0^T e^{\frac{1}{2}\eta(x(t))} |\dot{x}(t)| \, dt, \qquad p, q \in \mathbb{R}^2. \quad (10.4.52) \end{aligned}$$

Recall that $(\mathbb{R}^2, e^\eta \delta_{jk})$ is called complete if it is a complete metric space with respect to the metric d and, by the Hopf–Rinow–de Rham theorem, this latter property is equivalent to the statement that each geodesic on $(\mathbb{R}^2, e^\eta \delta_{jk})$ can be extended to a global geodesic defined on the entire real line $\mathbb{R}$ (see, e.g., [296]). In this case, the infimum in (10.4.52) may always be attained by minimizing geodesics, which is obviously an important feature.

We now proceed to show that the metric completeness for the solutions obtained is equivalent to the restriction $N \leq 1/4\pi\varepsilon^2 G$.

In fact, for the obtained solutions, the conformal factor e^η satisfies the estimates in $\mathbb{R}^2$,

$$C_1(1+r)^{-2aN} \leq e^{\eta(x)} \leq C_2(1+r)^{-2aN}, \qquad r = |x|, \qquad (10.4.53)$$

where $C_1, C_2 > 0$ are constants. By (10.4.53), it suffices to consider the radially symmetric metric $g_0 = (1+r)^{-2aN} \delta_{jk}$.

Considering rays starting from the origin as geodesics, it is straightforward to see that the completeness of g_0 is equivalent to the property that the geodesic distance from the origin to infinity is infinite. That is, the integral

$$\int_0^\infty (1+r)^{-aN} dr.$$

is divergent, which clearly requires that $aN \leq 1$ as expected.

Thus we see that the obtained gravitational metric is complete if and only if the total string or vortex number N satisfies the upper bound $N \leq 1/4\pi\varepsilon^2 G$, which coincides with the existence condition stated in Theorem 10.4.1.

The proof of Theorem 10.4.1 is complete.

10.5   Symmetric Solutions

The goal of this section is to achieve a complete understanding of finite energy symmetric solutions of (10.2.8) when the underlying surface is conformally a plane. We first state our main existence theorem under a simple necessary and sufficient condition. We then present our analysis based on the existence results established in Chapter 9.

10.5.1   Necessary and sufficient condition for existence

In the last section, we have studied the existence and properties of multi-string solutions, in the broken symmetry category $\lim_{|x|\to\infty} |\phi(x)| = \varepsilon > 0$, of the coupled Einstein and Abelian Higgs equations (10.2.8) governing gravitational curvature and energy concentration over a conformally

Euclidean surface under the condition

$$N \leq \frac{1}{4\pi\varepsilon^2 G}, \tag{10.5.1}$$

which ensures the completeness of the induced gravitational metric. It will be important to know whether (10.5.1) may be improved for the more general category of finite energy solutions.

The most general radially symmetric N-string solutions centered at the origin of the system (10.2.8) takes the form

$$\eta(x) = \eta(r), \quad \phi(x) = U(r)e^{iN\theta},$$
$$A_1(x) = -NV(r)\frac{x^2}{r^2}, \quad A_2(x) = NV(r)\frac{x^1}{r^2}, \tag{10.5.2}$$

where r, θ are the polar coordinates of the point $x \in \mathbb{R}^2$ and the integer N is the winding number of the Higgs field ϕ near infinity.

Theorem 10.5.1. *Under the cylindrical symmetry assumption (10.5.2) and the finite energy condition*

$$\int_{\mathbb{R}^2} \mathcal{H} e^\eta dx < \infty, \qquad \int_{\mathbb{R}^2} K e^\eta dx < \infty,$$

the equations (10.2.8) have a solution for $\varepsilon > 0$ if and only if the topological charge N satisfies the condition

$$N < \frac{1}{2\pi\varepsilon^2 G}. \tag{10.5.3}$$

In addition, the solution gives rise to a geodesically complete metric if and only if it fulfills the condition (10.5.1).

If $\varepsilon = 0$, the system (10.2.8) has no solution except the trivial ones.

It is remarkable that the finite energy condition (10.5.3) is similar to the completeness condition (10.5.1): both conditions impose constraints on the total string number N and place the gravitational constant G and the symmetry-breaking scale $\varepsilon > 0$ at equal positions.

The proof of Theorem 10.5.1 is based on a study of (10.2.17).

10.5.2   Equivalence theorem

Our starting point in the proof of Theorem 10.5.1 is the following result.

Theorem 10.5.2. *The coupled Einstein and Abelian Higgs system of equations (10.2.8) and the self-dual system (10.2.17) are equivalent in the category of radially symmetric solutions of the form (10.5.2).*

It is interesting to note that the equivalence assertion of Theorem 10.5.2 is valid for *any* symmetric solutions, regardless of the asymptotic behavior. In particular, it says that for solutions of infinite energy, the equations (10.2.8) and (10.2.17) are also equivalent. Thus we can describe, as well, all infinite energy solutions of (10.2.8) with the symmetry (10.5.2) by studying (10.2.17).

The proof of the theorem is similar to that in §10.2. We use P^+ and P^- to denote the same quantities. It suffices to show either $P^+ \equiv 0$ or $P^- \equiv 0$.

Suppose otherwise that $P^- \not\equiv 0$. Then (10.2.19) says that $P^+ = 0$ somewhere. Using the radial ansatz (10.5.2), we see that there exists an $r \geq 0$ so that

$$P^+(r) = 2Ne^{-\eta(r)}\frac{V_r(r)}{r} + (U^2(r) - \varepsilon^2) = 0.$$

Set

$$Z = \{r \geq 0 \mid P^+(r) = 0\}.$$

We assert that the number of points in Z, $\#Z$, does not exceed 3 unless $P^+ \equiv 0$. In fact, if $\#Z \geq 4$ but $P^+ \not\equiv 0$, then there were $r_1, r_2, r_3 \in Z$ so that $0 < r_1 < r_2 < r_3$.

Inserting the radial ansatz (10.5.2), we reduce (10.2.20) to the form

$$P^+_{rr} + \frac{1}{r}P^+_r \geq U^2e^\eta P^+. \tag{10.5.4}$$

Consider (10.5.4) over (r_1, r_3). Since $P^+(r_2) = 0$, P^+ would have a nonnegative maximum inside the interval (r_1, r_3), contradicting the strong maximum principle. Consequently $\#Z \leq 3$. Hence, $P^+(r) \neq 0$ for all $r > 0$, with at most three exceptions. Therefore we must have $P^- \equiv 0$, another contradiction.

Assume now $P^+ \not\equiv 0$. Then (10.2.19) says that $P^- = 0$ somewhere. The radial version of (10.2.21) reads

$$P^-_{rr} + \frac{1}{r}P^-_r \leq U^2e^\eta P^-. \tag{10.5.5}$$

Again, according to the strong maximum principle, P^- cannot have a nonpositive local minimum unless it is a constant. Thus we obtain $P^- \equiv 0$.

10.5.3 N-strings

Based on the equivalence theorem of the last subsection, in this subsection, we prove Theorem 10.5.1 by a complete study of the symmetric solutions of the system (10.2.17).

If all N strings are superimposed at the origin, then we are led by (10.4.2) and (10.4.3) to the system

$$\Delta u = \varepsilon^2 e^\eta(e^u - 1) + 4\pi N\delta(x), \tag{10.5.6}$$

$$\Delta(\eta + ae^u) = a\varepsilon^2 e^\eta(e^u - 1). \tag{10.5.7}$$

From (10.5.6) and (10.5.7), we see that

$$h = \frac{\eta}{a} + e^u - u + 2N \ln |x|$$

is a harmonic function over $\mathbb{R}^2$. Since we are interested only in radially symmetric solutions, h must be a constant. Using this result, we conclude that the system of equations (10.5.6) and (10.5.7) becomes a scalar equation,

$$\Delta u = \lambda |x|^{-2aN} e^{a(e^u - u)}(e^u - 1) + 4\pi N \delta(x), \quad x \in \mathbb{R}^2, \qquad (10.5.8)$$

where $\lambda > 0$ is an arbitrary constant.

We now concentrate on radially symmetric solutions of (10.5.8) for which the equation reduces to an ordinary differential equation contained as a special case of the equation (9.1.1), well studied in Chapter 9. Thus we may directly apply our results there.

We shall show that the only physically interesting solutions are those verifying

$$u \leq 0 \qquad (10.5.9)$$

(see Theorem 10.5.4). The following result gives us a necessary and sufficient condition for the existence of solutions of (10.5.8) subject to (10.5.9).

Theorem 10.5.3. *For non-positive valued radially symmetric solutions of (10.5.8), a necessary and sufficient condition for existence is*

$$N < \frac{2}{a} = \frac{1}{2\pi\varepsilon^2 G}. \qquad (10.5.10)$$

More precisely, we have
(i) When $aN < 1$, for each given $\lambda > 0$ there is a unique solution $u = u(r)$ ($r = |x|$) satisfying

$$\lim_{r \to \infty} u(r) = 0. \qquad (10.5.11)$$

All other solutions satisfy the property that $u(r) \to -\infty$ as $r \to \infty$ and

$$\lim_{r \to \infty} ru_r(r) = -\beta \qquad (10.5.12)$$

for some number

$$\beta > \frac{4}{a} - 2N = \frac{2(1 - 2\pi\varepsilon^2 GN)}{2\pi\varepsilon^2 G}.$$

More importantly, for any

$$\beta \in \left(\frac{4}{a} - 2N, \infty \right) = \left(\frac{2(1 - 2\pi\varepsilon^2 GN)}{2\pi\varepsilon^2 G}, \infty \right),$$

there exists at least one solution u realizing the asymptote (10.5.12).

(ii) When $aN > 1$, for each given $\lambda > 0$ there exists a non-positive solution if and only if (10.5.10) holds.

(iii) When $aN = 1$, there is a solution for any sufficiently large value of $\lambda > 0$.

Proof. Under the transformation (9.1.2), the equation (10.5.8) (of the form (9.1.1)) becomes (9.1.3) and (9.1.4) with

$$\alpha = 2N,$$
$$f(t) = e^{2(1-aN)t},$$
$$g(u) = e^{a(u-e^u)}(1 - e^u) = \frac{1}{a}\frac{d}{du}\left[e^{a(u-e^u)}\right].$$

(i) Clearly, when $aN < 1$, (H1)–(H8) hold with $f_1 = 2(1 - aN)$ and $g_1 = 1/a$, and therefore the first assertion of the theorem follows from Theorem 9.1.1.

(ii) When $aN > 1$, we make a transformation from t to $-t$, obtaining an equation of the form (9.1.3) with $f(t) = e^{2(aN-1)t}$, whereas (9.1.4) is replaced by $\lim_{t\to\infty} u'(t) = -2N$. In this case (H1)–(H8) are valid. Hence, if there exists a non-positive radially symmetric solution of (10.5.8), then $\alpha = \lim_{t\to-\infty} u'(t)$ exists and is non-negative, where $u = u(t)$ is the corresponding solution of (9.1.3) and (9.1.4). In fact, the existence and finiteness of α is trivial from (H1)–(H8). If $\alpha < 0$, then u gives rise to a symmetric solution of (10.5.8) which becomes positive when $r > 0$ is sufficiently large. Hence, there exists a non-positive solution of (10.5.8) if and only if $2N \in (\alpha + 4(aN - 1)/a, \infty)$ for some $\alpha \geq 0$, or equivalently, $aN < 2$.

(iii) This case was already covered in the previous section. However, there exists another type of solution satisfying $u(r) \to -\infty$ as $r \to \infty$ which will be studied in the next section.

We now characterize all finite-energy solutions.

Theorem 10.5.4. *Consider the radially symmetric solutions of the system (10.2.17) so that the 2-surface is conformally flat,*

$$(S, \{g_{jk}\}) = (\mathbb{R}^2, \{e^\eta \delta_{jk}\}),$$

and that the zero of ϕ is the origin of $\mathbb{R}^2$ with multiplicity $N \geq 1$. Let K be the Gaussian curvature of the surface. Then the finite energy condition

$$\int_{\mathbb{R}^2} \mathcal{H}\, e^\eta\, dx < \infty, \qquad \int_{\mathbb{R}^2} K\, e^\eta\, dx < \infty \qquad (10.5.13)$$

is equivalent to the bound $|\phi| < \varepsilon$ or $u < 0$ where $|\phi|^2 = \varepsilon^2 e^u$.

Proof. With $u = \ln(|\phi|^2/\varepsilon^2)$, we can start from the (10.5.8) or the system

$$u_{rr} + \frac{1}{r}u_r = \lambda r^{-2aN} e^{a(u-e^u)}(e^u - 1), \qquad r > 0,$$
$$\lim_{r\to 0} r u_r(r) = 2N. \qquad (10.5.14)$$

Assume first that (10.5.13) is true. Let us verify $u < 0$.

Suppose otherwise that there is some $r_0 > 0$ to make $u(r_0) \geq 0$. Since $u(r) < 0$ for $r > 0$ small, we may assume r_0 to be the smallest such number at which $u(r_0) \geq 0$. Obviously $u(r_0) = 0$. Because $u(r_0)$ cannot be a local minimum of u and $r = r_0$ is an isolated zero of u, we see that there exists some $\delta > 0$ so that $u(r) > 0$ for $r \in (r_0, r_0 + \delta)$. The maximum principle prohibits the existence of an $r_1 > r_0$ to make $u(r_1) = 0$. Thus $u(r) > 0$ for all $r > r_0$.

Consequently, we can strengthen the above observation by the statement $u_r(r) > 0$ $(r > 0)$. In fact, if there were some $r_1 > 0$ so that $u_r(r_1) = 0$, then $r_1 \neq r_0$. Thus $u(r_1) < 0$ if $r_1 < r_0$ or $u(r_1) > 0$ if $r_1 > r_0$. However, either case would violate the maximum principle applied to (10.5.14).

Thus the equation in (10.5.14) says that $(ru_r(r))_r > 0$ when $r > r_0$. Therefore $ru_r(r) > r_0 u_r(r_0) \equiv \sigma > 0$ for all $r > r_0$, which implies that

$$u(r) > \sigma[\ln r - \ln r_0], \qquad r > r_0.$$

Using the expression

$$|D_j\phi|^2 = \frac{1}{2}\varepsilon^2 u_r^2 e^u, \tag{10.5.15}$$

we find

$$|D_j\phi|^2 > \frac{1}{2}\varepsilon^2 \sigma^2 r_0^{-\sigma} r^{\sigma-2}, \qquad |x| = r > r_0.$$

Since $\mathcal{E} > \frac{1}{2}e^{-\eta}|D_j\phi|^2$, we arrive at

$$\int_{\mathbb{R}^2} \mathcal{H} \, e^\eta \, dx > \frac{\pi}{2}\varepsilon^2 \int_{r_0}^\infty \sigma^2 r_0^{-\sigma} r^{\sigma-1} \, dr = \infty.$$

That is, the solution does not carry a finite energy.

Suppose next that $u < 0$ holds. Then, according to Theorem 10.5.3, $aN < 2$. We consider here the case $1/a < N < 2/a$. Other cases are similar.

Let u be a solution satisfying (10.5.11). Then (10.5.12) holds with $\beta \geq 0$. If $\beta = 0$, then $u_r(r) > 0$, $r > 0$, because $ru_r(r)$ is a decreasing function in view of the property $u < 0$ and the equation in (10.5.14). Thus $\lim_{r\to\infty} u(r)$ exists and is non-positive. Applying this fact to

$$u_r(r) = -\frac{1}{r}\int_r^\infty \rho^{1-2aN} e^{a(u-e^u)}(e^u - 1) \, d\rho$$

yields directly the bound

$$ru_r(r) = O(r^{2(1-aN)}) \qquad \text{for large } r > 0.$$

On the other hand, in terms of u, we easily obtain the expressions

$$\eta = -2aN \ln r + a(u - e^u), \quad K = \frac{a}{2}[u_r^2 e^{u-\eta} + \varepsilon^2(e^u - 1)^2],$$

$$|\phi|^2 = \varepsilon^2 e^u, \quad |D_j\phi|^2 = \frac{1}{2}\varepsilon^2 u_r^2 e^u, \quad F_{12} = \frac{1}{2}\varepsilon^2 e^\eta(1 - e^u).$$

Therefore we arrive at the estimates at $r = \infty$,

$$e^{\eta}, \ F_{jk}, \ Ke^{\eta} = O(r^{-8\pi NG}), \quad |\phi|^2 = O(1),$$
$$|D_j\phi|^2 = O(r^{-2(8\pi NG-1)}). \tag{10.5.16}$$

If $\beta > 0$, we easily derive the estimates at $r = \infty$ as follows,

$$e^{\eta}, \ F_{jk} = O(r^{-4\pi G(2N+\beta)}), \quad |\phi|^2 = O(r^{-\beta}),$$
$$|D_j\phi|^2, \ Ke^{\eta} = O(r^{-(2+\beta)}). \tag{10.5.17}$$

Both (10.5.16) and (10.5.17) lead to (10.5.13).

Assume next $\varepsilon = 0$. Since the equations (10.2.8) and (10.2.17) are equivalent, we will use both of them to argue freely depending on convenience.

First, we observe that for a nontrivial solution of the form (10.5.2), we have $U(r) \neq 0$ at any $r > 0$. In fact, if there is an $r_0 > 0$ so that $U(r_0) = 0$, then, applying the boundary condition $\phi = 0$ on $|x| = r_0$ in the inequality $\Delta_g|\phi|^2 \geq |\phi|^4$ over $|x| < r_0$, we have $\phi = 0$ for $|x| < r_0$. That is, $U(r) = 0$ for $r < r_0$.

With the radial symmetry (10.5.2), we have

$$F_{12} = N\frac{V_r(r)}{r}.$$

Inserting this expression into the last equation in (10.2.17), i.e., $2e^{\eta}F_{12} \pm U^2 = 0$, we obtain $V(r) = $ const. for $r < r_0$. However, the regularity of A_j requires $V(0) = 0$. Thus $V(r) = 0$ for $r < r_0$. As a consequence, we find that $\mathcal{H}(\phi, A)(r) = 0$, $r < r_0$. Using these facts in the radial version of (10.2.8), namely,

$$\eta_{rr} + \frac{1}{r}\eta_r + 8\pi Ge^{\eta}\mathcal{H} = 0,$$
$$U_{rr} + \frac{1}{r}U_r - \frac{N^2}{r^2}(V-1)^2U - \frac{1}{2}e^{\eta}U^3 = 0,$$
$$V_{rr} - \frac{1}{r}(1+r\eta_r)V_r - e^{\eta}(V-1)U^2 = 0, \tag{10.5.18}$$

and applying in (10.5.18) the uniqueness theorem for the initial value problem of ordinary differential equations, we conclude that $U(r) = 0, V(r) = 0$, and $\eta =$ const. for all $r > 0$. Thus, we arrive at a trivial solution.

Consequently, we may assume in the following that $U(r) \neq 0$ for $r > 0$. Since (10.5.18) is invariant under the reflection $U \mapsto -U$, we may assume also that $U(r) > 0$, $r > 0$.

Suppose in (10.5.2), $N \geq 1$. The regularity of ϕ at the origin requires that $U(0) = 0$ and $U(r) = r^N f(r)$ near $r = 0$, where $f(0) \neq 0$. Hence the substitution $u = \ln U^2$ reduces the last two equations in (10.2.17) to

$$u_{rr} + \frac{1}{r}u_r = e^{\eta+u}, \qquad r > 0,$$

$$\lim_{r \to 0} \frac{u(r)}{\ln r} = 2N.$$

The boundary condition above implies $\lim_{r \to 0} ru_r(r) = 2N$. Thus

$$ru_r(r) = 2N + \int_0^r \rho e^{\eta(\rho)+u(\rho)} \, d\rho, \qquad r > 0.$$

In particular, $ru_r(r) > 2N$, $r > 0$, which implies, say,

$$u(r) > 2N \ln r + u(1), \qquad r > 1.$$

Therefore we obtain the lower bound

$$|D_1\phi|^2 + |D_2\phi|^2 \geq \frac{4N^2}{r^2} \cdot r^{2N} e^{u(1)} = 4N^2 e^{u(1)} r^{2N-2}, \quad |x| = r > 1.$$

Consequently,

$$\int_{\mathbb{R}^2} \mathcal{H} e^{\eta} \, dx \geq 8\pi N^2 e^{u(1)} \int_1^\infty r^{2N-1} \, dr = \infty.$$

In other words, we have energy blow-up.

If $N = 0$, then by (10.5.2), $A_j \equiv 0$, $j = 1, 2$. Thus the last equation in (10.2.17) implies that $U \equiv 0$, contradicting the assumption that $U(r) > 0$, $r > 0$.

In conclusion, there is no nontrivial solution when $\varepsilon = 0$ except those carrying infinite energy and the last statement in Theorem 10.5.1 is proven.

In the compact case, the nonexistence result at $\varepsilon = 0$ may also be established easily by a similar method. In fact, from the second equation in (10.2.8), we have as before $\Delta_g |\phi|^2 \geq |\phi|^4$. Thus the maximum principle implies that $\phi = 0$ everywhere. Inserting this fact into the last equation in (10.2.17), we find $F_{jk} = 0$. Thus $\mathcal{H} = 0$ and $K_g = 0$, which imply that S is a 2-torus and the solution triplet $(\phi, A, \{g_{jk}\})$ is trivial.

10.6 Symmetric Solutions on S^2

In this section we construct symmetric strings on S^2. In particular, we establish the existence of an $N = 2$ string solution and nonexistence of any $N = 1$ string solution. Our method is partially based on the complete understanding achieved in the last section of finite-energy symmetric solutions on a plane.

10.6.1 Balanced strings at opposite poles

First we consider again cosmic strings of the Abelian Higgs model living on a compact surface S. In this case we have shown earlier that topologically S

must be the sphere S^2 and the string number N and the symmetry-breaking parameter $\varepsilon > 0$ must satisfy the constraint

$$N = \frac{1}{2\pi\varepsilon^2 G}. \tag{10.6.1}$$

Furthermore, under (10.6.1) and the condition $N \geq 3$, we proved that the equations (10.2.17) have an N-string solution. The main goal of the present section is to settle the case $N = 2$ and give a nonexistence result for the case $N = 1$.

The main existence result for symmetric strings on S^2 below states that there must be a perfect balance between strings.

Theorem 10.6.1. *Consider the cosmic string equations (10.2.17) with the prescribed string number N over the standard 2-sphere S^2 and assume that the symmetry-breaking scale $\varepsilon > 0$ already satisfies the necessary condition (10.6.1). For any $N \geq 2$ and $N_1, N_2 \geq 1$ so that $N_1 + N_2 = N$, there exists a symmetric N-string solution realizing N_1 and N_2 strings located at the north and south poles, respectively, if and only if N is even and $N_1 = N_2 = N/2$. In particular, there exists a 2-string solution with strings located at the opposite poles but there does not exist any symmetric 1-string solution.*

In fact these S^2 solutions are in one-to-one correspondence with radially symmetric solutions on conformally Euclidean surfaces studied in the last section.

In the following development, we first set up the differential equation to be solved which is defined in the complement of one of the poles on S^2 and may conveniently be regarded as an equation over $\mathbb{R}^2$. Next, we concentrate on the balanced case $N_1 = N_2 = N/2$ and construct a special family of solutions of the resulting equation to be used to produce a solution later on the full S^2. We then consider the unbalanced case, $N_1 \neq N_2$, which concludes our proof.

10.6.2 Differential equation

Consider (10.2.17) defined over the base manifold (S^2, g) where g is an unknown gravitational metric. Suppose that the zeros of ϕ are $p_1, \cdots, p_N \in S^2$. Then the substitution $u = \ln|\phi|^2$ transforms (10.2.17) into the equivalent form,

$$
\begin{aligned}
K_g &= -2\pi G(\varepsilon^2[e^u - \varepsilon^2] - \Delta_g e^u), \\
\Delta_g u &= (e^u - \varepsilon^2) + 4\pi \sum_{\ell=1}^{N} \delta_{p_\ell}.
\end{aligned}
\tag{10.6.2}
$$

$N_1 = N_2 \equiv N_0$ or $N = 2N_0$, $N_0 \geq 1$.

Suppose that $p_1 = \cdots = p_{N_0} = n = $ the north pole of S^2 and $p_{N_0+1} = \cdots = p_N = s = $ the south pole of S^2. We use $\mathcal{P} = (\mathbb{R}^2, (x)) = S^2 - \{s\}$ and $\mathcal{P}' = (\mathbb{R}^2, (x')) = S^2 - \{n\}$ to cover S^2 through stereographical projections from the south and north poles, respectively. Assume that, in $\mathcal{P}$, the unknown metric g is conformal to the Euclidean metric, namely, $g_{jk} = e^\eta \delta_{jk}$. Then, on $S^2 - \{s\} = \mathbb{R}^2$ under $\mathcal{P}$, the system (10.6.2) becomes

$$\Delta(\eta + 4\pi G e^u) = 4\pi\varepsilon^2 G e^\eta (e^u - \varepsilon^2),$$
$$\Delta u = e^\eta (e^u - \varepsilon^2) + 4\pi N_0 \delta(x), \qquad (10.6.3)$$

where $x \in \mathbb{R}^2$ and $\delta(x)$ is the Dirac distribution on $\mathbb{R}^2$ concentrated at the origin. For simplicity, we introduce a rescaling or translation

$$u \mapsto u + \ln\varepsilon^2 \qquad (10.6.4)$$

and set $a = 4\pi\varepsilon^2 G$. Thus the system (10.6.3) gives us

$$\frac{\eta}{a} = u - e^u - 2N_0 \ln|x| + h, \qquad (10.6.5)$$

where h is a harmonic function over $\mathbb{R}^2$ which may be assumed to be an undetermined constant. Using (10.6.5), we obtain from (10.6.3) a single scalar equation,

$$\Delta u = \lambda |x|^{-2aN_0} e^{a(u-e^u)}(e^u - 1) + 4\pi N_0 \delta(x), \qquad (10.6.6)$$

where $\lambda > 0$ is a parameter which comes from $h = $ constant in (10.6.5) and may be adjusted according to our need.

10.6.3 Solution on $\mathcal{P}$

In this section, we establish the existence of a solution of (10.6.6). Such a solution is only local for the S^2 problem because it is only defined on the coordinate patch $\mathcal{P}$ around the north pole. In the next subsection, we will show that our solution obtained here can be used to produce a solution with strings sitting at both poles.

Since we look for radially symmetric solutions of (10.6.6), the problem is equivalent to solving the following ordinary differential equation subject to a singular boundary condition at $r = 0$,

$$u_{rr} + \frac{1}{r}u_r = \lambda r^{-2aN_0} e^{a(u-e^u)}(e^u - 1), \quad r > 0,$$

$$\lim_{r\to 0} \frac{u(r)}{\ln r} = \lim_{r\to 0} r u_r(r) = 2N_0. \qquad (10.6.7)$$

It will again be convenient to use the new variable $t = \ln r$. Thus, by virtue of (10.6.1) or $aN_0 = 1$, (10.6.7) becomes

$$u_{tt} = \lambda e^{a(u-e^u)}(e^u - 1), \quad -\infty < t < \infty,$$

$$\lim_{t\to -\infty} \frac{u(t)}{t} = \lim_{t\to -\infty} u_t(t) = 2N_0. \qquad (10.6.8)$$

We will look for a special subclass of solutions of (10.6.8) so that the asymptotic property

$$\lim_{t \to \infty} \frac{u(t)}{t} = \lim_{t \to \infty} u_t(t) = -2N_0 \qquad (10.6.9)$$

holds true. Using the maximum principle in the differential equation in (10.6.8), we see that a solution of such type must be globally concave down and has a unique negative maximum. Since the equation is autonomous, we may assume that the maximum is attained at $t = 0$. This observation motives our study of the following initial value problem as in the Chern–Simons case

$$\begin{aligned} u_{tt} &= \lambda e^{a(u-e^u)}(e^u - 1), \quad -\infty < t < \infty, \\ u(0) &= -\alpha, \quad \alpha > 0, \quad u_t(0) = 0. \end{aligned} \qquad (10.6.10)$$

Simple arguments show that for any $\alpha > 0$, the problem (10.6.10) has a unique global solution. For each given α, we will find a suitable λ so that the solution of (10.6.10) fulfills the boundary condition stated in (10.6.8) and (10.6.9). Therefore we are to carry out a shooting procedure to match the 'two-point' boundary value condition stated in (10.6.8) and (10.6.9) at $t = \pm\infty$. Furthermore, we will show as a remark later that α, λ may be determined explicitly. Such a property makes (10.6.10) practical in providing a numerical solution of (10.6.8) and (10.6.9).

Let u be the solution of (10.6.10). Of course $u_{tt} < 0$ and $u_t < 0$ ($t > 0$). Integrating the differential equation in (10.6.10), we have

$$u_t(t) = \int_0^t \lambda e^{a(u(\tau)-e^{u(\tau)})}(e^{u(\tau)} - 1) \, d\tau.$$

We set formally

$$f(\lambda, \alpha) = \int_0^\infty \lambda e^{a(u(\tau)-e^{u(\tau)})}(e^{u(\tau)} - 1) \, d\tau. \qquad (10.6.11)$$

Lemma 10.6.2. *The function $f(\lambda, \alpha)$ is continuous in $\lambda > 0, \alpha > 0$.*

Proof. *Step 1. $f(\lambda, \alpha)$ is finite.*

In fact, since $u_{tt} < 0$, we see that $f(\lambda, \alpha) = u_t(\infty)$ for each pair $\lambda > 0, \alpha > 0$ is either a negative number or $-\infty$. However, the latter does not happen because

$$u_t(t) < u_t(1) = -|u_t(1)|, \qquad t > 1 \qquad (10.6.12)$$

implies that $u(t) < -|u_t(1)|(t - 1) + u(1)$ ($t > 1$), which gives the convergence of (10.6.11).

Step 2. $f(\lambda, \alpha)$ is continuous.

To see this property, we first notice that the continuous dependence of u on λ and α implies that the quantity

$$u_t(1) = \int_0^1 \lambda e^{a(u(\tau) - e^{u(\tau)})}(e^{u(\tau)} - 1)\,d\tau$$

satisfies the bound

$$u_t(1) \leq -C_0 \quad \text{for } \lambda \in [\lambda_1, \lambda_2], \ \alpha \in [\alpha_1, \alpha_2], \tag{10.6.13}$$

where $\lambda_j, \alpha_j > 0$ and $C_0 > 0$ only depends on $\lambda_j, \alpha_j, j = 1, 2$. Inserting (10.6.13) into (10.6.12), we have

$$u(t) < -C_0(t-1) + u(1) < -C_0(t-1), \quad t > 1. \tag{10.6.14}$$

Using (10.6.14) in (10.6.11), we see that (10.6.11) is uniformly convergent in $[\lambda_1, \lambda_2] \times [\alpha_1, \alpha_2]$. Consequently, (10.6.11) is continuous in $\lambda > 0, \alpha > 0$ as expected. This proves the lemma.

We are now ready to invoke a shooting argument to prove

Lemma 10.6.3. *For any $\alpha > 0$, there is a $\lambda = \lambda(\alpha) > 0$ so that the solution $u(t)$ of (10.6.10) satisfies*

$$\lim_{t \to \infty} u_t(t) = -2N_0. \tag{10.6.15}$$

Proof. By the property $-\infty < u \leq -\alpha < 0$, we have $e^{-a} < e^{-ae^u} < 1$. Hence the equation in (10.6.10) gives us the inequality

$$\lambda e^{-a} e^{au}(e^u - 1) > u_{tt} > \lambda e^{au}(e^u - 1), \quad t > 0. \tag{10.6.16}$$

The right-hand side of (10.6.16) leads us to

$$u_{tt} > -\lambda e^{au}, \quad t > 0. \tag{10.6.17}$$

Since $u_t < 0$ for $t > 0$, we obtain by multiplying (10.6.17) by u_t and integrating in $t > 0$ the upper bound,

$$\frac{1}{2}(u_t(t))^2 < -\frac{\lambda}{a}e^{au(t)}\bigg|_0^t$$

$$= \frac{\lambda}{a}(e^{-a\alpha} - e^{au(t)}).$$

Therefore,

$$0 > u_t(t) > -\sqrt{\frac{2\lambda}{a}(e^{-a\alpha} - e^{au(t)})}. \tag{10.6.18}$$

On the other hand, the left-hand side of (10.6.16) implies

$$u_{tt} < \lambda e^{-a}(e^{-\alpha} - 1)e^{au}, \quad t > 0. \tag{10.6.19}$$

Multiplying (10.6.19) by $u_t < 0$ and integrating in $t > 0$, we obtain the lower bound,

$$\frac{1}{2}(u_t(t))^2 > \frac{\lambda}{a}e^{-a}(e^{-\alpha} - 1)(e^{au} - e^{-a\alpha})$$

$$= \frac{\lambda}{a}e^{-a}(1 - e^{-\alpha})(e^{-a\alpha} - e^{au}), \quad t > 0.$$

Thus,

$$u_t(t) < -\sqrt{\frac{2\lambda}{a}e^{-a}(1 - e^{-\alpha})(e^{-a\alpha} - e^{au(t)})}. \tag{10.6.20}$$

From (10.6.18) and (10.6.20), we find after setting $t = \infty$ the inequalities

$$0 \geq f(\lambda, \alpha) \geq -\sqrt{\frac{2\lambda}{a}e^{-a\alpha}},$$

$$f(\lambda, \alpha) \leq -\sqrt{\frac{2\lambda}{a}e^{-a}(1 - e^{-\alpha})e^{-a\alpha}}. \tag{10.6.21}$$

By (10.6.21) we can find $\lambda_1 > 0$ and $\lambda_2 > 0$ so that

$$f(\lambda_1, \alpha) < -2N_0 < f(\lambda_2, \alpha).$$

Since $f(\cdot, \alpha)$ is continuous by Lemma 10.6.2, we conclude from the above that there exists some $\lambda = \lambda(\alpha)$ between λ_1 and λ_2 so that $f(\lambda(\alpha), \alpha) = -2N_0$. The proof of the lemma is complete.

Lemma 10.6.4. *Any solution of (10.6.10) must be an even function.*

Proof. Given $\alpha > 0$, let $u(t)$ be the unique solution of (10.6.2). Define a new function $\tilde{u}$ by setting

$$\tilde{u}(t) = \begin{cases} u(t), & t \geq 0, \\ u(-t), & t < 0. \end{cases}$$

It is easily checked that $\tilde{u}$ is an even function which also solves (10.6.10). By uniqueness, $u = \tilde{u}$ and the lemma follows.

In view of Lemmas 10.6.3 and 10.6.4, we see that for any $\alpha > 0$ there is a $\lambda = \lambda(\alpha) > 0$ so that the problem (10.6.10) has a unique solution which also satisfies (10.6.8) and (10.6.9). Returning to the original variable r, we obtain a radially symmetric solution of (10.6.6) which gives rise to a solution of (10.2.17) in the coordinate patch $\mathcal{P}$ so that the first N_0 strings are all located at the north pole. The rest of the problem is to show that this solution also gives rise to N_0 strings at the south pole.

10.6.4 Solutions on full S^2

Let u be a radially symmetric solution of (10.6.6) found in the last subsection as a solution of (10.6.8) and (10.6.9). Then, as a function of $r = |x|$, u satisfies both the boundary condition in (10.6.7) and the additional property

$$\lim_{r \to \infty} r u_r(r) = -2N_0. \tag{10.6.22}$$

We will show in this section that the property (10.6.22) enables us to extend u originally constructed over $\mathcal{P} = S^2 - \{s\}$ to the entire S^2 to give rise to a full S^2-solution with N_0 strings sitting at both north and south poles. To this end, we recall that the system (10.6.2) under the coordinate patch $\mathcal{P}' = (\mathbb{R}^2, (x'))$ around the south pole may be reduced into an equation similar to (10.6.6),

$$\Delta u = \lambda |x'|^{-2aN_0} e^{a(u-e^u)}(e^u - 1) + 4\pi N_0 \delta(x'), \quad x' \in \mathbb{R}^2, \tag{10.6.23}$$

so that, with $r' = |x'|$, its radially symmetric version is of the form

$$u_{r'r'} + \frac{1}{r'} u_{r'} = \lambda r'^{-2aN_0} e^{a(u-e^u)}(e^u - 1), \quad r' > 0,$$

$$\lim_{r' \to 0} \frac{u(r')}{\ln r'} = \lim_{r' \to 0} r' u_{r'}(r') = 2N_0. \tag{10.6.24}$$

On the other hand, since $\mathcal{P} = (\mathbb{R}^2, (x))$ and $\mathcal{P}' = (\mathbb{R}^2, (x'))$ are represented by the stereographical projections with respect to the north and south poles, respectively, we have the relation

$$rr' = 1, \tag{10.6.25}$$

where the fact that S^2 has unit radius has been used. With (10.6.25) in mind, we deduce the simple identities

$$r' u_{r'} = -r u_r, \quad r'^2 u_{r'r'} + r' u_{r'} = r^2 u_{rr} + r u_r, \quad \text{in } \mathcal{P} \cap \mathcal{P}'. \tag{10.6.26}$$

Using $aN_0 = 1$, (10.6.22), (10.6.25), and (10.6.26), we see immediately that u obtained on $\mathcal{P}$ in the last subsection satisfies (10.6.24). In other words, we have produced an $N = 2N_0$-string solution realizing precisely N_0 strings superimposed at each of the two poles.

10.6.5 Nonexistence of unbalanced solutions

Case (i): $1 \le N_1 \ne N_2$.

For definiteness, we assume $N_1 < N_2$. Therefore $u = \ln(|\phi|^2/\varepsilon^2)$ satisfies the boundary value problem

$$u_{rr} + \frac{1}{r} u_r = \lambda r^{-2aN_1} e^{a(u-e^u)}(e^u - 1), \quad r > 0,$$

$$\lim_{r \to 0} \frac{u(r)}{\ln r} = \lim_{r \to 0} r u_r(r) = 2N_1,$$

$$\lim_{r \to \infty} \frac{u(r)}{\ln r} = \lim_{r \to \infty} r u_r(r) = -2N_2.$$

On the other hand, in view of (10.6.1), N_1 satisfies

$$N_1 < \frac{1}{4\pi\varepsilon^2 G} = \frac{1}{a}.$$

Therefore, according to Theorem 10.5.3, N_2 lies in the interval

$$\left(\frac{2}{a} - N_1, \infty \right) = \left(\frac{1}{2\pi\varepsilon^2 G} - N_1, \infty \right),$$

which contradicts the condition (10.6.1) with $N = N_1 + N_2$.

It remains to show that (10.6.2) does not have any symmetric solution when the strings are all superimposed at a single point. That is, $p_1 = \cdots = p_N = n =$ the north pole (say). To this end, we first state a simple criterion.

Lemma 10.6.5. *Let (u, g) be a solution of (10.6.2). Then $u < \ln \varepsilon^2$.*

Proof. Let S_δ^2 be the complement of the δ-neighborhoods of the points $p_1, \cdots, p_N$ in S^2. Thus, for $\delta > 0$ sufficiently small, we have $u(x) < \ln \varepsilon^2$ when $x \in \partial S_\delta^2$. Using this property and the maximum principle to the equation $\Delta_g u = (e^u - \varepsilon^2)$, $x \in S_\delta^2$, we have $\max_{x \in S_\delta^2} u(x) < \ln \varepsilon^2$ as expected.

Lemma 10.6.6. *In (10.6.2), let $p_1 = \cdots = p_N = n =$ the north pole. Under this circumstance the system (10.6.2) has no solution on the full S^2 which is symmetric with respect to the string point n.*

Proof. Consider (10.6.2) in the open set $S^2 - \{s\} = U$. The metric g on $U = \mathbb{R}^2$ is globally conformal to the standard Euclidean metric of $\mathbb{R}^2$. Hence $g_{jk} = e^\eta \delta_{jk}$ and (10.6.3) holds with $N_0 = N$. We again use the translation (10.6.4) to simplify the notation. Since we are only interested in symmetric solutions, the unknown pair (u, η) may be assumed to be radially symmetric with respect to the origin of $\mathbb{R}^2$ which corresponds to the north pole n of S^2. Therefore the harmonic function h in (10.6.5) (with $N_0 = N$) must be a constant by virtue of the radial symmetry. Hence, we arrive consecutively at (10.6.6) and then (10.6.7) with $N_0 = N$. Thus we obtain the ordinary differential equation

$$u_{tt} = \lambda e^{2(1-aN)t} e^{a(u-e^u)}(e^u - 1), \quad -\infty < t < \infty,$$

$$\lim_{t \to -\infty} \frac{u(t)}{t} = \lim_{t \to -\infty} u_t(t) = 2N. \tag{10.6.27}$$

Recall the condition (10.6.1), namely, $aN = 2$. Then the new variable $\tau = -t$ in (10.6.27) gives us the system

$$u_{\tau\tau} = \lambda e^{2\tau} e^{a(u-e^u)}(e^u - 1), \quad -\infty < \tau < \infty,$$

$$\lim_{\tau \to \infty} \frac{u(\tau)}{\tau} = \lim_{\tau \to \infty} u_\tau(\tau) = -2N. \tag{10.6.28}$$

If there is a solution for the original string problem, then, by Lemma 10.6.5, the corresponding solution u of (10.6.28) satisfies $u < 0$ everywhere. Consequently, the integral

$$\int_{-\infty}^{0} e^{2\tau} e^{a(u-e^u)}(e^u - 1)\, d\tau$$

is convergent. This result implies that the number

$$\alpha = \lim_{\tau \to -\infty} u_\tau(\tau) \tag{10.6.29}$$

is a finite number. Using the relation

$$u\Big|_{\tau \to -\infty} = u\Big|_{x=s \in S^2} = \text{a finite number},$$

we see that the only situation we can have in (10.6.29) is $\alpha = 0$. Therefore, using Theorem 10.5.3 or 9.1.1 directly, we find

$$2N \in (4/a, \infty)$$

or $N > 1/2\pi\varepsilon^2 G$, which contradicts the condition (10.6.1). The lemma is proven.

In conclusion, the proof of Theorem 10.6.1 is complete.

The study of this section also establishes the existence of a finite-energy symmetric N-string solution on $\mathbb{R}^2$ under the condition

$$N = \frac{1}{4\pi\varepsilon^2 G}, \tag{10.6.30}$$

which satisfies

$$\lim_{r \to \infty} \frac{u(r)}{\ln r} = \lim_{r \to \infty} r u_r(r) = -2N. \tag{10.6.31}$$

In the proof, we have seen that, to ensure the solvability of (10.6.8) and (10.6.9) from the initial value problem (10.6.10), the parameter λ depends on the value of u at $t = 0$, $u(0) = -\alpha$. We remark that such a dependence can explicitly be determined. In fact, multiplying the differential equation in (10.6.8) by u_t and integrating over the interval $(-\infty, t)$, we have

$$u_t^2(t) = 4N_0^2 - \frac{2\lambda}{a} e^{a(u(t) - e^{u(t)})}, \quad -\infty < t < \infty. \tag{10.6.32}$$

Using $u(0) = -\alpha$ and $u_t(0) = 0$ in (10.6.32) we obtain

$$\lambda = 4\pi\varepsilon^2 G N_0^2 e^{\alpha + e^{-\alpha}}. \tag{10.6.33}$$

Thus we can state that a solution of (10.6.10) gives rise to a solution of (10.6.8) and (10.6.9) (strings concentrated at two opposite poles) if and only if (10.6.33) is fulfilled. The condition (10.6.33) is of obvious importance when one wants to obtain numerical solutions of (10.6.8) and (10.6.9).

The formula (10.6.32) and the concavity of u have yet another interesting mathematical implication. To see it, we rewrite (10.6.32) as

$$u_t(t) = \begin{cases} \sqrt{4N_0^2 - \frac{2\lambda}{a} e^{a(u(t) - e^{u(t)})}}, & t \leq 0, \\ -\sqrt{4N_0^2 - \frac{2\lambda}{a} e^{a(u(t) - e^{u(t)})}}, & t > 0, \end{cases} \tag{10.6.34}$$

where λ satisfies (10.6.33) and $u(0) = -\alpha$. In view of (10.6.33), we see that $u \equiv -\alpha$ is an equilibrium solution of (10.6.34). On the other hand, our solution to (10.6.8) and (10.6.9) also solves (10.6.34) subject to the same initial condition. Therefore the solutions of (10.6.34) under the initial condition $u(0) = -\alpha$ suffer nonuniqueness. For this reason the original second-order problem (10.6.10) is more useful than its first-order reduction (10.6.34).

We have seen that cosmic strings on S^2 are closely related via stereographical projections to strings on the full plane. When all the strings are superimposed at one point on S^2, the condition (10.5.10) for the existence of planar finite-energy solutions prohibits the existence of symmetric solutions on S^2 because the topology of S^2 requires (10.6.1) which makes (10.5.10) invalid.

10.7   Non-Abelian Cosmic Strings

In this section we consider the Einstein theory coupled with the non-Abelian electroweak gauge theory. It will be seen that in this case it is necessary to consider the Einstein equation with a cosmological term. Note that such a cosmological term has an interesting history [329] in theoretical physics. We shall first discuss the simplified massive W-string model along the line of [4]. We then study the full bosonic Weinberg–Salam model [5, 6, 7]. We show that the structure of these models leads to significantly positive values of the cosmological constant and new problems challenging mathematical analysts.

10.7.1 Massive W-boson and strings

In the presence of a cosmological constant, Λ, the Einstein equations governing a gravitational metric tensor $g_{\mu\nu}$ of signature $(+ - --)$ are

$$G_{\mu\nu} - \Lambda g_{\mu\nu} = -8\pi G T_{\mu\nu}, \qquad (10.7.1)$$

where the energy-momentum tensor $T_{\mu\nu}$ originates from matter and is to be specified later.

Let $\{t_a\}_{a=1,2,3}$ be a set of generators of $SO(3)$ satisfying the commutation relation

$$[t_a, t_b] = i\epsilon_{abc}t_c, \qquad a, b, c = 1, 2, 3.$$

Then the $SO(3)$ gauge potential A_μ can be expressed in the matrix form

$$A_\mu = A_\mu^a t_a.$$

As in Chapter 7, introduce the complex W-vector boson by setting

$$W_\mu = \frac{1}{\sqrt{2}}[A_\mu^1 + iA_\mu^2].$$

Then the Lagrangian of the $SO(3)$ matter-gauge sector in the presence of a gravitational metric $ds^2 = g_{\mu\nu}dx^\mu dx^\nu$ under consideration is

$$\mathcal{L} = -\frac{1}{4}g^{\mu\mu'}g^{\nu\nu'}F_{\mu\nu}^a F_{\mu'\nu'}^a + m_W^2 g^{\mu\nu}\overline{W}_\mu W_\nu,$$

where

$$F_{\mu\nu} = \partial_\mu A_\nu - \partial_\nu A_\mu + ie[A_\mu, A_\nu],$$

$m_W > 0$ is the mass of the W-particle, and $-e$ is the electron charge.

Put $A_\mu^3 = P_\mu$ and

$$P_{\mu\nu} = \partial_\mu P_\nu - \partial_\nu P_\mu.$$

Thus $\mathcal{L}$ is reduced after a calculation to

$$
\begin{aligned}
\mathcal{L} = {} & -\frac{1}{4}g^{\mu\mu'}g^{\nu\nu'}P_{\mu\nu}P_{\mu'\nu'} \\
& -\frac{1}{2}g^{\mu\mu'}g^{\nu\nu'}(D_\mu W_\nu - D_\nu W_\mu)\overline{(D_{\mu'}W_{\nu'} - D_{\nu'}W_{\mu'})} \\
& +m_W^2 g^{\mu\nu}\overline{W}_\mu W_\nu + ieg^{\mu\mu'}g^{\nu\nu'}P_{\mu\nu}\overline{W}_{\mu'}W_{\nu'} \\
& +\frac{e^2}{2}\left((g^{\mu\mu'}\overline{W}_\mu\overline{W}_{\mu'})g^{\nu\nu'}W_\nu W_{\nu'} - (g^{\mu\mu'}\overline{W}_\mu W_{\mu'})^2\right),
\end{aligned}
$$

where $D_\mu = \partial_\mu - ieP_\mu$. Varying the metric $\{g_{\mu\nu}\}$ in $\mathcal{L}$ leads to the following expression of the energy-momentum tensor,

$$T_{\mu\nu} = -g^{\mu'\nu'}P_{\mu\mu'}P_{\nu\nu'}$$

$$-2\mathrm{Re}\{g^{\mu'\nu'}(D_\mu W_{\mu'} - D_{\mu'}W_\mu)\overline{(D_\nu W_{\nu'} - D_{\nu'}W_\nu)}\}$$
$$+m_W^2(W_\mu\overline{W}_\nu + W_\nu\overline{W}_\mu)$$
$$+ieg^{\mu'\nu'}(P_{\mu'\mu}[\overline{W}_{\nu'}W_\nu - W_{\nu'}\overline{W}_\nu] + P_{\mu'\nu}[\overline{W}_{\nu'}W_\mu - W_{\nu'}\overline{W}_\mu])$$
$$-e^2\Big((g^{\mu'\nu'}\overline{W}_{\mu'}W_{\nu'})(\overline{W}_\mu W_\nu + \overline{W}_\nu W_\mu)$$
$$-(\overline{W}_\mu\overline{W}_\nu)(g^{\mu'\nu'}W_{\mu'}W_{\nu'}) - (g^{\mu'\nu'}\overline{W}_{\mu'}\overline{W}_{\nu'})(W_\mu W_\nu)\Big)$$
$$-g_{\mu\nu}\mathcal{L}. \tag{10.7.2}$$

Besides, the equations of motion of the matter-gauge Lagrangian $\mathcal{L}$ are

$$\frac{1}{\sqrt{-g}}D_\mu[g^{\mu\mu'}g^{\nu\nu'}\sqrt{-g}(D_{\mu'}W_{\nu'} - D_{\nu'}W_{\mu'})]$$
$$= \quad ieg^{\mu\mu'}g^{\nu\nu'}P_{\mu'\nu'}W_\mu - m_W^2 g^{\mu\nu}W_\mu$$
$$+e^2(g^{\mu\nu}\overline{W}_\mu(g^{\mu'\nu'}W_{\mu'}W_{\nu'}) - (g^{\mu'\nu'}W_{\mu'}\overline{W}_{\nu'})g^{\mu\nu}W_\mu),$$
$$\frac{1}{\sqrt{-g}}\partial_\mu[g^{\mu'\nu}g^{\mu\nu'}\sqrt{-g}P_{\mu'\nu'}]$$
$$= \quad ieg^{\mu\nu}g^{\mu'\nu'}\overline{(D_\mu W_{\nu'} - D_{\nu'}W_\mu)}W_{\mu'}$$
$$-ieg^{\mu\nu}g^{\mu'\nu'}(D_\mu W_{\nu'} - D_{\nu'}W_\mu)\overline{W}_{\mu'}$$
$$+\frac{ie}{\sqrt{-g}}\partial_\mu(g^{\mu'\nu}g^{\mu\nu'}\sqrt{-g}(\overline{W}_{\mu'}W_{\nu'} - \overline{W}_{\nu'}W_{\mu'})). \tag{10.7.3}$$

The string metric is still of the form (10.1.6). A symmetry consideration shows that it may be consistent to assume that

$$W_0 = W_3 = 0, \quad P_0 = P_3 = 0,$$

W_j, P_j ($j = 1, 2$) depend only on x^k ($k = 1, 2$), and there is a complex scalar field W so that

$$W_1 = W, \quad W_2 = iW.$$

Thus, in view of the expression (10.7.2), we have

$$T_{\mu\nu} = 0, \quad \mu \neq \nu.$$

Moreover,

$$T_{00} = -T_{33} = \mathcal{H},$$

where $\mathcal{H} = -\mathcal{L}$ can be written

$$\mathcal{H} = \frac{1}{2}e^{-2\eta}P_{12}^2 + e^{-2\eta}|D_1 W + iD_2 W|^2 + 2m_W^2 e^{-\eta}|W|^2$$
$$-2ee^{-2\eta}P_{12}|W|^2 + 2e^2 e^{-2\eta}|W|^4$$
$$= \quad e^{-2\eta}|D_1 W + iD_2 W|^2 + \frac{1}{2}e^{-2\eta}\left[P_{12} - \left(\frac{m_W^2}{e}e^\eta + 2e|W|^2\right)\right]^2$$
$$-\frac{1}{2}\frac{m_W^4}{e^2} + \frac{m_W^2}{e}e^{-\eta}P_{12}. \tag{10.7.4}$$

Besides, the other two nonvanishing components of $T_{\mu\nu}$ are

$$
\begin{aligned}
T_{11} &= T_{22} \\
&= e^{-\eta}P_{12}^2 + 2e^{-\eta}|D_1W + iD_2W|^2 + 2m_W^2|W|^2 \\
&\quad -4ee^{-\eta}P_{12}|W|^2 + 4e^2e^{-\eta}|W|^4 - e^{\eta}\mathcal{H} \\
&= e^{-\eta}|D_1W + iD_2W|^2 + \frac{1}{2}e^{-\eta}\left[P_{12} - \left(\frac{m_W^2}{e}e^{\eta} + 2e|W|^2\right)\right]^2 \\
&\quad + \frac{m_W^2}{e}\left[P_{12} - \left(\frac{m_W^2}{e}e^{\eta} + 2e|W|^2\right)\right] + \frac{1}{2}\frac{m_W^4}{e^2}e^{\eta}.
\end{aligned}
$$

The form of the Hamiltonian energy density $\mathcal{H}$ suggests the following curved-space version of the self-dual equations

$$
\begin{aligned}
D_1W + iD_2W &= 0, \\
P_{12} &= \frac{m_W^2}{e}e^{\eta} + 2e|W|^2.
\end{aligned}
\tag{10.7.5}
$$

It can be verified directly that (10.7.5) implies the equations of motion, (10.7.3). Furthermore, since in view of (10.1.7), the full Einstein equations (10.7.1) are reduced to

$$
K_\eta = \Lambda + 8\pi G\mathcal{H}, \quad \Lambda e^{\eta}\delta_{jk} = 8\pi GT_{jk}, \quad j,k = 1,2, \tag{10.7.6}
$$

we see that the consistency in (10.7.6) requires that the cosmological constant take the unique value

$$
\Lambda = 4\pi G\frac{m_W^4}{e^2}. \tag{10.7.7}
$$

Inserting (10.7.7) into (10.7.6) and using (10.7.5) in (10.7.4), we see that the Einstein system (10.7.6) is simplified into the single equation

$$
K_\eta = 8\pi G\frac{m_W^2}{e}e^{-\eta}P_{12}. \tag{10.7.8}
$$

Therefore, by (10.1.9), the equation (10.7.8) becomes

$$
\Delta\eta + 16\pi G\frac{m_W^2}{e}P_{12} = 0. \tag{10.7.9}
$$

Thus the governing equations are reduced to (10.7.5) and (10.7.9).

We finish this subsection by condensing the coupled equations (10.7.5) and (10.7.9) into a second-order elliptic partial differential equation.

Let the zeros of W be denoted by $p_1, \cdots, p_N$ (a zero of multiplicity m is counted as m zeros). Then the substitution $u = \ln|W|^2$ reduces (10.7.5) into the form

$$
\Delta u = -2m_W^2 e^{\eta} - 4e^2 e^u + 4\pi\sum_{n=1}^{N}\delta_{p_n} \quad \text{in } \mathbb{R}^2. \tag{10.7.10}
$$

Furthermore, using (10.7.5) in (10.7.9), we obtain

$$\Delta\left(\frac{\eta}{8\pi G}\right) = -\frac{2m_W^4}{e^2}e^\eta - 4m_W^2 e^u. \tag{10.7.11}$$

Hence, we have seen that the original system of equations, (10.7.5) and (10.7.9), is equivalent to the system of equations (10.7.10) and (10.7.11).

The special form of (10.7.11) allows a further simplification of the system. In fact, inserting (10.7.11) into (10.7.10), we get

$$\Delta u = \frac{e^2}{m_W^2}\Delta\left(\frac{\eta}{8\pi G}\right) + 4\pi\sum_{n=1}^{N}\delta_{p_n}.$$

That is to say,

$$w = u - \frac{e^2}{8\pi G m_W^2}\eta - 2\sum_{n=1}^{N}\ln|x - p_n|$$

is a harmonic function in $\mathbb{R}^2$. For simplicity we assume w to be a constant. As a consequence, if we set

$$v = \frac{\eta}{8\pi G}, \quad a = \frac{2m_W^4}{e^2}, \quad b = 8\pi G, \quad c = \frac{e^2}{m_W^2},$$

the system of equations (10.7.10) and (10.7.11) is reduced to the single equation

$$\Delta v = -ae^{bv} - \gamma\left(\prod_{n=1}^{N}|x - p_n|^2\right)e^{cv}, \tag{10.7.12}$$

where $\gamma > 0$ is an arbitrary, adjustable, constant.

The existence of solutions of this interesting nonlinear equation is not difficult to establish when $p_1, \cdots, p_n$ coincide.

10.7.2 Einstein–Weinberg–Salam system

Let t_a ($a = 1, 2, 3$) be the generators of $SU(2)$ used in the last subsection (note that $SU(2)$ and $SO(3)$ have the same Lie algebra) and set

$$t_0 = \frac{1}{2}\begin{pmatrix} 1 & 0 \\ 0 & 1 \end{pmatrix}.$$

Then the gauge group $SU(2) \times U(1)$ in the Weinberg–Salam electroweak theory transforms a complex doublet ϕ according to the rules

$$\phi \mapsto \exp(-i\omega_a t_a)\phi, \quad \omega_a \in \mathbb{R}, \quad a = 1, 2, 3,$$
$$\phi \mapsto \exp(-i\xi t_0)\phi, \quad \xi \in \mathbb{R}.$$

The $SU(2)$ and $U(1)$ gauge fields are denoted by $A_\mu = A_\mu^a t_a$ (or $A_\mu = (A_\mu^a)$) and B_μ respectively, where both A_μ^a and B_μ are real 4-vectors. Besides, the field strength tensors and the $SU(2) \times U(1)$ gauge-covariant derivative are

$$
\begin{aligned}
F_{\mu\nu} &= \partial_\mu A_\nu - \partial_\nu A_\mu + ig_1[A_\mu, A_\nu], \\
H_{\mu\nu} &= \partial_\mu B_\nu - \partial_\nu B_\mu, \\
D_\mu \phi &= \partial_\mu \phi + ig_1 A_\mu^a t_a \phi + ig_2 B_\mu t_0 \phi,
\end{aligned}
$$

where $g_1, g_2 > 0$ are coupling constants.

In the presence of the gravitational metric $ds^2 = g_{\mu\nu} dx^\mu dx^\nu$, the Lagrangian density of the bosonic sector of the Weinberg–Salam theory is

$$
\begin{aligned}
\mathcal{L} = {} & -\frac{1}{4} g^{\mu\mu'} g^{\nu\nu'} F_{\mu\nu}^a F_{\mu'\nu'}^a - \frac{1}{4} g^{\mu\mu'} g^{\nu\nu'} H_{\mu\nu} H_{\mu'\nu'} \\
& + g^{\mu\nu} (D_\mu \phi)(D_\nu \phi)^\dagger - \lambda (\varphi_0^2 - \phi^\dagger \phi)^2,
\end{aligned}
\tag{10.7.13}
$$

where $\lambda > 0$ is a constant and $\varphi_0 > 0$ is the vacuum expectation value of the Higgs field ϕ.

We now go to the standard unitary gauge. We introduce the new vector fields P_μ and Z_μ as a rotation of the pair A_μ^3 and B_μ,

$$
\begin{aligned}
P_\mu &= B_\mu \cos\theta + A_\mu^3 \sin\theta, \\
Z_\mu &= -B_\mu \sin\theta + A_\mu^3 \cos\theta.
\end{aligned}
$$

Thus D_μ becomes

$$
\begin{aligned}
D_\mu = {} & \partial_\mu + ig_1(A_\mu^1 t_1 + A_\mu^2 t_2) \\
& + iP_\mu(g_1 \sin\theta t_3 + g_2 \cos\theta t_0) \\
& + iZ_\mu(g_1 \cos\theta t_3 - g_2 \sin\theta t_0).
\end{aligned}
$$

As usual, if the coupling constants $g_1, g_2 > 0$ are so chosen that the electron charge satisfies

$$
e = \frac{g_1 g_2}{\sqrt{g_1^2 + g_2^2}},
$$

then there is an angle θ, $0 < \theta < \pi/2$ (the Weinberg angle), so that

$$
e = g_1 \sin\theta = g_2 \cos\theta.
$$

In this situation, the operator D_μ has the expression

$$
D_\mu = \partial_\mu + ig_1(A_\mu^1 t_1 + A_\mu^2 t_2) + iP_\mu eQ + iZ_\mu eQ'.
$$

Here $eQ = e(t_3 + t_0)$ and $Q' = \cot\theta t_3 - \tan\theta t_0$ are charge and neutral charge operators, respectively.

Assume now

$$
\phi = \begin{pmatrix} 0 \\ \varphi \end{pmatrix},
$$

where φ is a real scalar field. Then

$$D_\mu\phi = \begin{pmatrix} \dfrac{ig_1}{2}[A_\mu^1 - iA_\mu^2]\varphi \\ \partial_\mu\varphi - \dfrac{ig_1}{2\cos\theta}Z_\mu\varphi \end{pmatrix}.$$

As before, define the complex vector field

$$W_\mu = \frac{1}{\sqrt{2}}(A_\mu^1 + iA_\mu^2)$$

and set $\mathcal{D} = \partial_\mu - ig_1 A_\mu^3$. With the notation $P_{\mu\nu} = \partial_\mu P_\nu - \partial_\nu P_\mu$ and $Z_{\mu\nu} = \partial_\mu Z_\nu - \partial_\nu Z_\mu$, the Lagrangian density takes the form

$$\begin{aligned}
\mathcal{L} &= -\frac{1}{4}g^{\mu\mu'}g^{\nu\nu'}P_{\mu\nu}P_{\mu'\nu'} - \frac{1}{4}g^{\mu\mu'}g^{\nu\nu'}Z_{\mu\nu}Z_{\mu'\nu'} \\
&\quad -\frac{1}{2}g^{\mu\mu'}g^{\nu\nu'}(\mathcal{D}_\mu W_\nu - \mathcal{D}_\nu W_\mu)\overline{(\mathcal{D}_{\mu'}W_{\nu'} - \mathcal{D}_{\nu'}W_{\mu'})} \\
&\quad +g^{\mu\nu}\partial_\mu\varphi\partial_\nu\varphi + \frac{g_1^2}{4\cos^2\theta}\varphi^2 g^{\mu\nu}Z_\mu Z_\nu + \frac{g_1^2}{2}\varphi^2 g^{\mu\nu}\overline{W}_\mu W_\nu \\
&\quad -\frac{1}{2}g_1^2((g^{\mu\nu}\overline{W}_\mu W_\nu)^2 - (g^{\mu\mu'}W_\mu W_{\mu'})\overline{(g^{\nu\nu'}W_\nu W_{\nu'})}) \\
&\quad -ig_1 g^{\mu\mu'}g^{\nu\nu'}(Z_{\mu'\nu'}\cos\theta + P_{\mu'\nu'}\sin\theta)\overline{W}_\mu W_\nu \\
&\quad -\lambda(\varphi^2 - \varphi_0^2)^2.
\end{aligned} \tag{10.7.14}$$

Moreover, the equations of motion of the Lagrangian (10.7.14) are

$$\begin{aligned}
&\frac{1}{\sqrt{-g}}\mathcal{D}_\mu\left(g^{\mu\mu'}g^{\nu\nu'}\sqrt{-g}(\mathcal{D}_{\mu'}W_{\nu'} - \mathcal{D}_{\nu'}W_{\mu'})\right) \\
&= -ig_1 g^{\mu\mu'}g^{\nu\nu'}(Z_{\mu'\nu'}\cos\theta + P_{\mu'\nu'}\sin\theta)W_\mu - \frac{1}{2}g_1^2\varphi^2 g^{\mu\nu}W_\mu \\
&\quad +g_1^2([g^{\mu'\nu'}\overline{W}_{\mu'}W_{\nu'}]g^{\mu\nu}W_\mu - [g^{\mu'\nu'}W_{\mu'}W_{\nu'}]g^{\mu\nu}\overline{W}_\mu), \\[1em]
&\frac{1}{\sqrt{-g}}\partial_\mu\left(g^{\mu'\nu}g^{\mu\nu'}\sqrt{-g}P_{\mu'\nu'}\right) \\
&= ig_1\sin\theta g^{\mu'\nu}g^{\mu\nu'}\overline{(\mathcal{D}_{\mu'}W_{\nu'} - \mathcal{D}_{\nu'}W_{\mu'})}W_\mu \\
&\quad -ig_1\sin\theta g^{\mu'\nu}g^{\mu\nu'}(\mathcal{D}_{\mu'}W_{\nu'} - \mathcal{D}_{\nu'}W_{\mu'})\overline{W}_\mu \\
&\quad -\frac{i}{\sqrt{-g}}g_1\sin\theta\partial_\mu(g^{\mu'\nu}g^{\mu\nu'}\sqrt{-g}[\overline{W}_{\mu'}W_{\nu'} - \overline{W}_{\nu'}W_{\mu'}]), \\[1em]
&\frac{1}{\sqrt{-g}}\partial_\mu[g^{\mu'\nu}g^{\mu\nu'}\sqrt{-g}Z_{\mu'\nu'}] \\
&= ig_1\cos\theta g^{\mu'\nu}g^{\mu\nu'}\overline{(\mathcal{D}_{\mu'}W_{\nu'} - \mathcal{D}_{\nu'}W_{\mu'})}W_\mu \\
&\quad -ig_1\cos\theta g^{\mu'\nu}g^{\mu\nu'}(\mathcal{D}_{\mu'}W_{\nu'} - \mathcal{D}_{\nu'}W_{\mu'})\overline{W}_\mu
\end{aligned}$$

$$-\frac{i}{\sqrt{-g}}g_1\cos\theta\partial_\mu\left(g^{\mu'\nu}g^{\mu\nu'}\sqrt{-g}(\overline{W}_{\mu'}W_{\nu'}-\overline{W}_{\nu'}W_{\mu'})\right)$$

$$+\frac{g_1^2}{2\cos^2\theta}\varphi^2 g^{\mu\nu}Z_\mu,$$

$$\frac{1}{\sqrt{-g}}\partial_\mu(g^{\mu\nu}\sqrt{-g}\partial_\nu\varphi)$$

$$=\frac{g_1^2}{4\cos^2\theta}\left(g^{\mu\nu}Z_\mu Z_\nu\right)\varphi+\frac{g_1^2}{2}(g^{\mu\nu}\overline{W}_\mu W_\nu)\varphi$$

$$-2\lambda(\varphi^2-\varphi_0^2)\varphi. \tag{10.7.15}$$

We shall show that, when λ satisfies a specific condition, (10.7.15) allows a reduction into a first-order system.

Varying the metric $\{g_{\mu\nu}\}$, we obtain from (10.7.14) the following expression for the energy-momentum tensor of the electroweak matter-gauge sector,

$$T_{\mu\nu}=-g^{\mu'\nu'}P_{\mu\mu'}P_{\nu\nu'}-g^{\mu'\nu'}Z_{\mu\mu'}Z_{\nu\nu'}$$

$$-2\mathrm{Re}\{g^{\mu'\nu'}(\mathcal{D}_\mu W_{\mu'}-\mathcal{D}_{\mu'}W_\mu)\overline{(\mathcal{D}_\nu W_{\nu'}-\mathcal{D}_{\nu'}W_\nu)}\}$$

$$+2\partial_\mu\varphi\partial_\nu\varphi+\frac{g_1^2}{2\cos^2\theta}\varphi^2 Z_\mu Z_\nu+g_1^2\varphi^2\mathrm{Re}\{W_\mu\overline{W}_\nu\}$$

$$-2g_1^2(g^{\mu'\nu'}\overline{W}_{\mu'}W_{\nu'})\mathrm{Re}\{W_\mu\overline{W}_\nu\}+2g_1^2\mathrm{Re}\{(g^{\mu'\nu'}W_{\mu'}W_{\nu'})\overline{W}_\mu\overline{W}_\nu\}$$

$$-ig_1\cos\theta g^{\mu'\nu'}(Z_{\mu'\mu}[\overline{W}_{\nu'}W_\nu-\overline{W}_\nu W_{\nu'}]+Z_{\mu'\nu}[\overline{W}_{\nu'}W_\mu-W_{\nu'}\overline{W}_\mu])$$

$$-ig_1\sin\theta g^{\mu'\nu'}(P_{\mu'\mu}[\overline{W}_{\nu'}W_\nu-\overline{W}_\nu W_{\nu'}]+P_{\mu'\nu}[\overline{W}_{\nu'}W_\mu-W_{\nu'}\overline{W}_\mu])$$

$$-g_{\mu\nu}\mathcal{L}.$$

In the sequel, we impose again the conformally flat string metric (10.1.6) and assume that

$$P_0=P_3=Z_0=Z_3=0,$$

$$W_0=W_3=0,\quad W_1=W,\quad W_2=iW,$$

and that P_j, Z_j $(j=1,2)$, W, and φ depend only on x^k $(k=1,2)$. Then (10.7.6) still holds with the electroweak Hamiltonian density $\mathcal{H}=-\mathcal{L}$ being given by

$$\mathcal{H}=\frac{1}{2}e^{-2\eta}P_{12}^2+\frac{1}{2}e^{-2\eta}Z_{12}^2+e^{-2\eta}|\mathcal{D}_1 W+i\mathcal{D}_2 W|^2+e^{-\eta}|\nabla\varphi|^2$$

$$+g_1^2 e^{-\eta}\varphi^2|W|^2+\frac{g_1^2}{4\cos^2\theta}e^{-\eta}\varphi^2[Z_1^2+Z_2^2]+2g_1^2 e^{-2\eta}|W|^4$$

$$-2g_1\cos\theta e^{-2\eta}Z_{12}|W|^2-2g_1\sin\theta e^{-2\eta}P_{12}|W|^2+\lambda(\varphi^2-\varphi_0^2)^2$$

$$=\frac{1}{2}e^{-2\eta}\left[P_{12}-\frac{g_1}{2\sin\theta}\varphi_0^2 e^\eta-2g_1\sin\theta|W|^2\right]^2$$

$$+\frac{1}{2}e^{-2\eta}\left[Z_{12} - \frac{g_1}{2\cos\theta}(\varphi^2 - \varphi_0^2)e^{\eta} - 2g_1\cos\theta|W|^2\right]^2$$

$$+e^{-2\eta}|\mathcal{D}_1 W + i\mathcal{D}_2 W|^2 + e^{-\eta}\left[\frac{g_1}{2\cos\theta}\varphi Z_j + \epsilon_j^k \partial_k\varphi\right]^2$$

$$+\left[\lambda - \frac{g_1^2}{8\cos^2\theta}\right][\varphi^2 - \varphi_0^2]^2 - \frac{g_1^2\varphi_0^4}{8\sin^2\theta} + \frac{g_1\varphi_0^2}{2\sin\theta}e^{-\eta}P_{12}$$

$$-\frac{g_1\varphi_0^2}{2\cos\theta}e^{-\eta}Z_{12} - \frac{g_1}{2\cos\theta}e^{-\eta}\partial_k(\epsilon_j^k Z_j\varphi^2). \qquad (10.7.16)$$

The form of (10.7.16) suggests that we may impose the critical condition

$$\lambda = \frac{g_1^2}{8\cos^2\theta} \qquad (10.7.17)$$

and the Bogomol'nyi–Ambjorn–Olesen equations

$$\mathcal{D}_1 W + i\mathcal{D}_2 W = 0,$$
$$P_{12} = \frac{g_1}{2\sin\theta}\varphi_0^2 e^{\eta} + 2g_1\sin\theta|W|^2,$$
$$Z_{12} = \frac{g_1}{2\cos\theta}(\varphi^2 - \varphi_0^2)e^{\eta} + 2g_1\cos\theta|W|^2,$$
$$Z_j = -\frac{2\cos\theta}{g_1}\epsilon_j^k\partial_k\ln\varphi. \qquad (10.7.18)$$

In fact, we can examine directly that any solution of (10.7.18) also satisfies the full equations of motion (10.7.15) when (10.7.17) is fulfilled.

We now simplify the Einstein equations. In view of (10.7.16), (10.7.17), and (10.7.18), $\mathcal{H}$ may be rewritten in the form

$$\mathcal{H} = -\frac{g_1^2\varphi_0^4}{8\sin^2\theta} + \frac{g_1\varphi_0^2}{2\sin\theta}e^{-\eta}P_{12}$$

$$+\frac{g_1}{2\cos\theta}e^{-\eta}Z_{12}(\varphi^2 - \varphi_0^2) + 2e^{-\eta}|\nabla\varphi|^2. \qquad (10.7.19)$$

Furthermore, it is straightforward to check using the last equation in (10.7.18) that

$$T_{12} = T_{21}$$
$$= 2\partial_1\varphi\partial_2\varphi + \frac{g_1^2}{2\cos^2\theta}\varphi^2 Z_1 Z_2 = 0.$$

The other off-diagonal components $T_{\mu\nu} = 0$ ($\mu \neq \nu$) are direct consequences of the string ansatz. Besides, we have

$$T_{11} = \frac{1}{2}e^{-\eta}P_{12}^2 + \frac{1}{2}e^{-\eta}Z_{12}^2 + [\partial_1\varphi]^2 - [\partial_2\varphi]^2$$

$$+2g_1^2 e^{-\eta}|W|^4 + \frac{g_1^2}{4\cos^2\theta}\varphi^2[Z_1^2 - Z_2^2]$$

$$-2g_1 \sin\theta e^{-\eta} P_{12}|W|^2 - 2g_1 \cos\theta e^{-\eta} Z_{12}|W|^2$$

$$-\frac{g_1^2}{8\cos^2\theta} e^{\eta}[\varphi^2 - \varphi_0^2]^2$$

$$= \frac{1}{2}e^{-\eta}\left[(P_{12} - 2g_1\sin\theta|W|^2) - \frac{g_1\varphi_0^2}{2\sin\theta}e^{\eta}\right]$$

$$\times\left[(P_{12} - 2g_1\sin\theta|W|^2) + \frac{g_1\varphi_0^2}{2\sin\theta}e^{\eta}\right]$$

$$+\frac{1}{2}e^{-\eta}\left[(Z_{12} - 2g_1\cos\theta|W|^2) - \frac{g_1}{2\cos\theta}e^{\eta}(\varphi^2 - \varphi_0^2)\right]$$

$$\times\left[(Z_{12} - 2g_1\cos\theta|W|^2) + \frac{g_1}{2\cos\theta}e^{\eta}(\varphi^2 - \varphi_0^2)\right]$$

$$+\left[\partial_1\varphi - \frac{g_1}{2\cos\theta}\varphi Z_2\right]\left[\partial_1\varphi + \frac{g_1}{2\cos\theta}\varphi Z_2\right]$$

$$+\left[\frac{g_1}{2\cos\theta}\varphi Z_1 - \partial_2\varphi\right]\left[\frac{g_1}{2\cos\theta}\varphi Z_1 + \partial_2\varphi\right] + \frac{g_1^2\varphi_0^4}{8\sin^2\theta}e^{\eta}.$$

Thus in view of (10.7.18) again, we find $T_{11} = g_1^2\varphi_0^4 e^{\eta}/8\sin^2\theta$. Similarly, we can show that T_{22} takes the same value as T_{11}.

Inserting the energy-momentum tensor $T_{\mu\nu}$ just obtained into the Einstein equations (10.7.1) (or (10.7.6)), we see that there holds the condition

$$\Lambda = \frac{\pi G g_1^2\varphi_0^4}{\sin^2\theta}. \tag{10.7.20}$$

Thus, now, in view of (10.7.19), (10.7.1) or (10.7.6) is equivalent to the single equation

$$\Delta\left(\frac{\eta}{8\pi G}\right) + \frac{g_1\varphi_0^2}{\sin\theta}P_{12} + \frac{g_1}{\cos\theta}[\varphi^2 - \varphi_0^2]Z_{12} + 4|\nabla\varphi|^2 = 0. \tag{10.7.21}$$

Thus, we have seen that, under the critical coupling condition (10.7.17), the Einstein–Weinberg–Salam system of equations, (10.7.1) and (10.7.15), is reduced to the simpler equations (10.7.18) and (10.7.21). The presence of such solutions requires that the cosmological constant Λ verify the unique condition (10.7.20).

We conclude this section by writing (10.7.18) and (10.7.21) as a system of second-order nonlinear elliptic equations. Assume that the strings are at $p_1, \cdots, p_N$. Let u, v be such that

$$|W|^2 = e^u, \quad \varphi^2 = e^v.$$

Then it is straightforward to show that (10.7.18) and (10.7.20) become

$$\Delta u = -g_1^2 e^{v+\eta} - 4g_1^2 e^u + 4\pi\sum_{n=1}^N \delta_{p_n},$$

$$\Delta v = \frac{g_1^2}{2\cos^2\theta}[e^v - \varphi_0^2]e^\eta + 2g_1^2 e^u,$$

$$\Delta\left(\frac{\eta}{8\pi G}\right) = -\frac{g_1^2}{2}e^\eta\left[\frac{(e^v - \varphi_0^2)^2}{\cos^2\theta} + \frac{\varphi_0^4}{\sin^2\theta}\right] - 2g_1^2 e^{u+v} - |\nabla v|^2 e^v.$$

$$(10.7.22)$$

The solutions of (10.7.22) give rise to N-string solutions of the original Einstein–Weinberg–Salam theory.

It is important to note that the system (10.7.22) can be compressed further.

Multiplying the second equation by the factor $e^v - \varphi_0^2$ and adding it to the last equation in the system (10.7.22), we have

$$\Delta\left(e^v + \frac{\eta}{8\pi G}\right) = \frac{g_1^2\varphi_0^2}{2\cos^2\theta}\left(e^v - \frac{\varphi_0^2}{\sin^2\theta}\right)e^\eta. \qquad (10.7.23)$$

Besides, the first two equations in (10.7.22) may be put together to form

$$\Delta(u + 2v) = \frac{g_1^2}{\cos^2\theta}(\sin^2\theta e^v - \varphi_0^2)e^\eta + 4\pi\sum_{n=1}^{N}\delta_{p_n}. \qquad (10.7.24)$$

Combining (10.7.23) and (10.7.24), we see that

$$e^v + \frac{\eta}{8\pi G} - \frac{\varphi_0^2}{2\sin^2\theta}\left(u + 2v - 2\sum_{n=1}^{N}\ln|x - p_n|\right) \qquad (10.7.25)$$

is a harmonic function on $\mathbb{R}^2$, which may be assumed to be an arbitrary constant. Thus we can express the conformal exponent η in terms of u, v as

$$\frac{\eta}{8\pi G} = c + \frac{\varphi_0^2}{2\sin^2\theta}(u + 2v) - e^v - \frac{\varphi_0^2}{\sin^2\theta}\sum_{n=1}^{N}\ln|x - p_n|, \qquad (10.7.26)$$

which eliminates the last equation in (10.7.22). Setting

$$u = w - 2\ln g_1 - \ln 2 + 2\sum_{n=1}^{N}\ln|x - p_n|, \quad a = \frac{4\pi G\varphi_0^2}{\sin^2\theta}, \quad b = 8\pi G, \qquad (10.7.27)$$

we have

$$\eta = C + a(w + 2v) - be^v \qquad (10.7.28)$$

and we see that the original quasilinear system (10.7.22) becomes a complicated semilinear system,

$$\Delta w = -\lambda e^{a(w+2v)+v-be^v} - 2\left(\prod_{n=1}^{N}|x - p_n|\right)^2 e^w,$$

$$\Delta v \;=\; \frac{\lambda}{2\cos^2\theta}(e^v - \varphi_0^2)e^{a(w+2v)-be^v} + \left(\prod_{n=1}^{N}|x-p_n|\right)^2 e^w,$$

$$(10.7.29)$$

where $\lambda > 0$ is an arbitrary, adjustable, constant. This system governs electroweak cosmic strings which give rise to a positive cosmological constant uniquely determined by the formula (10.7.20).

10.8   Remarks

In this chapter, we have presented a series of static solutions of the Einstein gravitational equations coupled with various field theoretical models and we have seen that the existence problems require a deep understanding of the corresponding highly nonlinear equations, except in the sigma model case completely solved by Comtet and Gibbons [86]. Our analysis leads us to many surprises: (i) the presence of string-like defects is necessary to gravitation, (ii) gravitation manifests its presence by imposing an explicit obstruction to the total string number, (iii) a perfect balance is important for the existence of strings over a compact surface, and, (iv) the existence of self-dual non-Abelian strings such as those in the electroweak theory of Weinberg–Salam developed by Ambjorn and Olesen [4, 5, 6, 7] leads to a uniquely determined *positive* value of the cosmological constant expressed in terms of some fundamental parameters in particle physics, such as the electron charge, the W-particle mass, and the Weinberg mixing angle.

Our methods here may be used directly to obtain semilocal strings [121].

It will be interesting to recall other related studies concerning the existence of static solutions of the Einstein equations. It is well known that the Schwarzschild blackhole solution is the only solution of the vacuum Einstein equations, which has singularities. If the Einstein equations are coupled with the pure sourceless Maxwell equations, the only solution is the Reissner–Nordström solution, which is again singular somewhere. However, the recent numerical work of Bartnik and McKinnon [25] and the analytic work of Smoller, Wasserman, Yau, and McLeod [288] established the existence of everywhere regular static solutions of the Einstein equations coupled with a non-Abelian gauge theory, the pure $SU(2)$ Yang–Mills equations, which indicates that a suitable repulsive force or a current source is necessary to counter-balance the gravitational attraction and to prevent the appearance of a singular structure. The results of this chapter show that, in order to obtain static regular cosmic string solutions, a minimal presence of matter source is sufficient which may be in the form of the sigma model (without a Maxwell field), the Abelian Higgs theory (with a Maxwell field), or a non-Abelian gauge field theory (with some nuclear force fields such as those in the electroweak theory).

We conclude this chapter with a few important open problems.

Open Problem 10.8.1. *Obtain arbitrarily distributed N-string solutions for the coupled Einstein and Abelian Higgs equations (10.2.17) on $\mathbb{R}^2$ under the condition*

$$N > \frac{1}{4\pi\varepsilon^2 G}. \tag{10.8.1}$$

We have constructed multiple strings (Theorem 10.4.1) for

$$N \leq 1/4\pi\varepsilon^2 G \tag{10.8.2}$$

and symmetric superimposed strings (Theorem 10.5.1) for

$$N < 1/2\pi\varepsilon^2 G. \tag{10.8.3}$$

However, for the range

$$\frac{1}{4\pi\varepsilon^2 G} < N < \frac{1}{2\pi\varepsilon^2 G}, \tag{10.8.4}$$

the existence of multiple strings is an open question which involves a deeper understanding of the nonlinear elliptic equation (10.4.5). In our study we need (10.8.2) in a crucial way for the control of both local and global behavior of an approximation sequence. In order to establish an existence theorem under (10.8.4), some new techniques are to be developed.

Open Problem 10.8.2. *Does there exist a finite-energy N-string solution when (10.8.3) is violated?*

Our study indicates that the presence of gravitation introduces drastically different phenomena from the case in absence of gravitation. For example, on a sphere, although there does not exist any symmetric solution representing superimposed strings, there exist a broad class of solutions representing suitably balanced multiple strings. Such an observation suggests that multiple strings *may* exist under the condition

$$N > \frac{1}{2\pi\varepsilon^2 G}, \tag{10.8.5}$$

although we know from Theorem 10.5.1 that finite-energy symmetric solutions no longer exist.

Open Problem 10.8.3. *Study whether all solutions of the coupled system*

$$\Delta_g e^u - (e^u - 1) = NK_g,$$
$$\Delta_g u = e^u - 1 + 4\pi N\delta_p \tag{10.8.6}$$

over the sphere S^2 are necessarily symmetric about the point $p \in S^2$ where K_g is the Gauss curvature of an unknown metric g over S^2 and δ_p is the Dirac distribution on (S^2, g) concentrated at the point p.

This problem is related to the problem of whether there exists a solution over a compact surface representing N superimposed strings in the Abelian Higgs theory. If the answer to Open Problem 10.8.3 is affirmative, it would mean that there are no superimposed strings at all since it has been shown that there is no superimposed symmetric solution on S^2 (Theorem 10.6.1). The system (10.8.6) comes from setting $p_1 = p_2 = \cdots = p_N = p$, putting $\varepsilon = 1$ by rescaling, and using (10.6.1), in (10.6.2).

Open Problem 10.8.4. *Develop an existence theory for the W-string equation (10.7.12).*

As mentioned earlier, for any $\gamma > 0$, it is trivial to show the existence of solutions when the points $p_1, p_2, \cdots, p_N$ coincide because it becomes a simple ODE problem. When these points are different from each other, no existence result has been obtained whatsoever.

Open Problem 10.8.5. *Develop an existence theory for the electroweak cosmic string equations (10.7.29).*

Nothing is known about this problem even for radially symmetric solutions when all the points $p_1, p_2, \cdots, p_N$ coincide. It seems that the system (10.7.29) defies all available techniques and is a true challenge to analysts.

11
Vortices and Antivortices

In this chapter we consider the coexistence of vortices (strings) and antivortices (antistrings) in an Abelian gauge theory. In §11.1, we introduce the gauge field model and state our main existence theorems. In §11.2, we calculate various components of the energy-momentum tensor and reduce the equations of motion into a self-dual system and boil the problem down to a nonlinear elliptic equation with sources through an integration of the Einstein equations. In §11.3, we prove the existence of a unique solution for the elliptic equation governing vortices, in absence of gravity, and, we prove the existence of a solution for the equation governing strings. In §11.4, we calculate the precise value of the quantized energy, and total curvature, of a solution possessing M vortices (strings) and N antivortices (antistrings) and observe the roles played by these two different types of vortices (strings), energetically. In §11.5, we consider the problem over a closed Riemann surface.

11.1 Gauge Field Theory and Coexisting Strings

In this section we introduce an Abelian gauge field theory and state our main results concerning the coexistence of vortices and antivortices and strings and antistrings.

11.1.1 Action density

To motivate our study, it will be instructive to review first the classical $O(3)$ sigma model for a planar ferromagnet where the field configuration is a spin vector $\phi = (\phi_1, \phi_2, \phi_3)$ which maps $\mathbb{R}^2$ into the unit sphere, S^2, in $\mathbb{R}^3$, namely, $\phi_1^2 + \phi_2^2 + \phi_3^2 = 1$. The energy is defined by the expression

$$E(\phi) = \frac{1}{2} \int (\partial_1 \phi)^2 + (\partial_2 \phi)^2, \qquad (11.1.1)$$

where and in the sequel, the integral is taken over the full $\mathbb{R}^2$ with respect to the Lebesgue measure dx, unless stated otherwise. The finite-energy condition implies that one may compactify $\mathbb{R}^2$ so that ϕ belongs to a second homotopy class on S^2 characterized by its corresponding Brouwer degree, $\deg(\phi)$, which may be represented by the formula

$$\deg(\phi) = \frac{1}{4\pi} \int \phi \cdot (\partial_1 \phi \times \partial_2 \phi). \qquad (11.1.2)$$

In view of (11.1.2), we have seen in Chapter 2 that there holds the topological lower energy bound

$$E(\phi) \geq 4\pi |\deg(\phi)| \qquad (11.1.3)$$

and that, in each $\deg(\phi) = N$ class, solutions saturating the energy lower bound (11.1.3) could be constructed explicitly via meromorphic functions with N prescribed poles and M prescribed zeros at the points

$$p_1, p_2, \cdots, p_N \quad \text{and} \quad q_1, q_2, \cdots, q_M, \qquad (11.1.4)$$

respectively. In particular, only the N poles make their contributions to the total energy, $E = 4\pi N$ but the M zeros are phantoms.

In the gauged $O(3)$ sigma model studied in Chapter 2, the energy (11.1.1) is extended to take the form

$$E(\phi, A) = \frac{1}{2} \int F_{12}^2 + (D_1 \phi)^2 + (D_2 \phi)^2 + (1 - \mathbf{n} \cdot \phi)^2, \qquad (11.1.5)$$

where A_j $(j = 1, 2)$ is a vector field, $F_{12} = \partial_1 A_2 - \partial_2 A_1$ is the induced magnetic curvature, $D_j \phi = \partial_j + A_j (\mathbf{n} \times \phi)$ are covariant derivatives which replace the conventional ones, and $\mathbf{n} = (0, 0, 1)$ is the north pole on S^2, and the degree formula (11.1.2) may be rewritten as

$$\deg(\phi) = \frac{1}{4\pi} \int \phi \cdot (D_1 \phi \times D_2 \phi) + F_{12}(1 - \mathbf{n} \cdot \phi). \qquad (11.1.6)$$

We have seen the validity of the same lower energy bound (11.1.3) shown that it may be saturated if and only if, after projecting ϕ into a complex scalar field which is again a meromorphic function up to a locally nonvanishing factor, the numbers of poles and zeros, N and M, satisfy $N \geq M+2$. Hence, energetically, the M zeros are still phantoms, although they manifest themselves magnetically in a peculiar, and partial, way. This special result may be viewed as an early indication of a possibility to find a gauge field theory model in which the points of poles and zeros of a complex scalar field can serve as the prescribed locations of vortices (strings) and antivortices (antistrings) which will play *equal* roles, both energetically and magnetically. This chapter confirms such a possibility. The main idea is to explore an important modification of (11.1.5) in which the symmetric vacuum is maximally broken so that the potential density is replaced by $(\mathbf{n} \cdot \phi)^2$. Such a framework enables one to recognize vortices and antivortices or cosmic strings and antistrings as energetically and magnetically indistinguishable solitons as described above.

In our study to follow, we mainly concentrate on the cosmic string problem for which the gauge field model will be coupled with Einstein's gravitational equations through Hilbert's principle.

Let $g_{\mu\nu}$ be the gravitational metric of our spacetime of signature $(+ - -)$. Thus the action density of the matter-gauge sector that modifies (11.1.5) into a form with a maximally broken vacuum symmetry as stated above is written

$$\mathcal{L} = -\frac{1}{4}g^{\mu\mu'}g^{\nu\nu'}F_{\mu\nu}F_{\mu'\nu'} + \frac{1}{2}g^{\mu\nu}(D_\mu\phi)\cdot(D_\nu\phi) - \frac{1}{2}(\mathbf{n}\cdot\phi)^2, \quad (11.1.7)$$

where $F_{\mu\nu} = \partial_\mu A_\nu - \partial_\nu A_\mu$ is the electromagnetic curvature induced from the 4-vector connection A_μ ($\mu, \nu = 0, 1, 2, 3$ with $t = x_0$) and $D_\mu\phi = \partial_\mu\phi + A_\mu(\mathbf{n} \times \phi)$. To see how this field theory model produces a system of strings and antistrings in equilibrium with very interesting physical properties, it is more transparent to go to its equivalent stereographic representation. In doing so, we introduce a complex scalar field $u = u_1 + iu_2$ where

$$u_1 = \frac{\phi_1}{1 + \phi_3}, \quad u_2 = \frac{\phi_2}{1 + \phi_3}. \quad (11.1.8)$$

That is, we project S^2 onto the complex plane through the south pole $-\mathbf{n}$, which corresponds to infinity of u. For convenience, we replace the original vector field A_μ by $-A_\mu$.

Thus, with the new gauge-covariant derivatives $D_\mu u = \partial_\mu u - iA_\mu u$ ($\mu = 0, 1, 2, 3$), the action density (11.1.7) becomes

$$\mathcal{L} = -\frac{1}{4}g^{\mu\mu'}g^{\nu\nu'}F_{\mu\nu}F_{\mu'\nu'} + \frac{2}{(1 + |u|^2)^2}g^{\mu\nu}(D_\mu u)\overline{(D_\nu u)} - \frac{1}{2}\left(\frac{1 - |u|^2}{1 + |u|^2}\right)^2.$$
$$(11.1.9)$$

There are three interesting facts.

(i) Like that in the classical Yang–Mills–Higgs theory, the potential density function for the complex scalar field u also has a double-well profile.

In particular, when we take the $|u|^2 \to 1$ limit in the denominators of the second and third terms in (11.1.9), we see that the model approaches that of the classical Abelian Higgs theory,

$$\mathcal{L} = -\frac{1}{4}g^{\mu\mu'}g^{\nu\nu'}F_{\mu\nu}F_{\mu'\nu'} + \frac{1}{2}g^{\mu\nu}(D_\mu u)\overline{(D_\nu u)} - \frac{1}{8}(1 - |u|^2)^2. \quad (11.1.10)$$

(ii) The preimages of the north pole under the original spin vector ϕ become the zeros of the complex field u and those of the south pole become the poles of u.

(iii) The action density (11.1.9) is invariant under the transformation

$$(u, A_\mu) \mapsto \left(\frac{1}{u}, -A_\mu\right) \quad (11.1.11)$$

in addition to its $U(1)$ gauge invariance. This important feature indicates that the poles and zeros will play equal roles. In this chapter, we regard (11.1.9) as an *independent* gauge field theory.

Denote by R and $R_{\mu\nu}$ the scalar curvature and Ricci tensor, respectively, induced from the metric $g_{\mu\nu}$ and $G_{\mu\nu} = R_{\mu\nu} - \frac{1}{2}g_{\mu\nu}R$ the Einstein tensor. From (11.1.9) and the Hilbert principle, the coupled system of equations of motion are

$$G_{\mu\nu} + \Lambda g_{\mu\nu} = -8\pi G T_{\mu\nu}, \quad (11.1.12)$$

$$\frac{1}{\sqrt{|g|}}D_\mu\left(\frac{\sqrt{|g|}g^{\mu\nu}}{(1+|u|^2)^2}D_\mu u\right) = f, \quad (11.1.13)$$

$$\frac{1}{\sqrt{|g|}}\partial_{\mu'}(g^{\mu\nu}g^{\mu'\nu'}\sqrt{|g|}F_{\nu\nu'}) = j^\mu, \quad (11.1.14)$$

where, in the Einstein equations (11.1.12), G is Newton's gravitational constant, Λ is the cosmological constant, $T_{\mu\nu}$ is the energy-momentum tensor of the matter-gauge sector defined by

$$T_{\mu\nu} = -g^{\mu'\nu'}F_{\mu\mu'}F_{\nu\nu'} + \frac{2}{(1+|u|^2)^2}(D_\mu u\overline{D_\nu u} + \overline{D_\mu u}D_\nu u) - g_{\mu\nu}\mathcal{L}, \quad (11.1.15)$$

and the force and current densities in the wave and Maxwell equations, (11.1.13) and (11.1.14), are defined by

$$f = \frac{(1 - |u|^2 - 2g^{\mu\nu}D_\mu u\overline{D_\nu u})}{(1+|u|^2)^3}u, \quad (11.1.16)$$

$$j^\mu = \frac{2}{(1+|u|^2)^2}ig^{\mu\nu}(\overline{u}D_\nu u - u\overline{D_\nu u}), \quad (11.1.17)$$

Our solutions that represent systems of cosmic strings and antistrings in equilibrium will be obtained for the coupled equations (11.1.12)–(11.1.14).

11.1.2 Existence theorems

Let (x^μ) $(\mu = 0, 1, 2, 3)$ be the coordinates of a point in the Minkowski spacetime $\mathbb{R}^{1,3}$ for which $x^0 = t$ is the temporal component and (x^a) $(a = 1, 2, 3)$ are the spatial coordinates. Recall that the cosmic string solutions of (11.1.12)–(11.1.17) we are looking for are a special class of static field configurations which depend only on (x^j) $(j = 1, 2)$, $A_0 = A_3 = 0$, and the metric is of the form

$$g_{\mu\nu} = \operatorname{diag} \left\{ 1, -e^\eta, -e^\eta, -1 \right\}. \tag{11.1.18}$$

Here is our main result concerning the coexistence of cosmic strings and antistrings.

Theorem 11.1.1. *For given prescribed points*

$$p_1, p_2, \cdots, p_N \quad and \quad q_1, q_2, \cdots, q_M$$

in $\mathbb{R}^2$ satisfying

$$M + N \leq \frac{1}{8\pi G}, \tag{11.1.19}$$

the equations (11.1.12)–(11.1.17) have a static energy minimizing solution (η, u, A_j) $(A_0 = A_3 = 0)$ so that the gravitational metric is of the form (11.1.18), the field configurations η, u, A_j depend only on $x = (x^1, x^2) \in \mathbb{R}^2$, the points $p_1, p_2, \cdots, p_N$ are poles and $q_1, q_2, \cdots, q_M$ are zeros of the complex scalar function u, the magnetic field $F_{12} = \partial_1 A_2 - \partial_2 A_1$ has opposite vorticities at these poles and zeros,

$$F_{12}(p_s) < 0, \quad s = 1, 2, \cdots, N; \quad F_{12}(q_s) > 0, \quad s = 1, 2, \cdots, M, \tag{11.1.20}$$

the total magnetic flux Φ and matter-gauge energy E have the following quantized values,

$$\Phi = 2\pi(M - N), \tag{11.1.21}$$

$$E = 2\pi(M + N), \tag{11.1.22}$$

and, as $|x| \to \infty$, the solution approaches the vacuum with broken symmetry at the rate

$$\begin{aligned}
|u|^2 - 1 &= O(|x|^{-b}), \\
|D_1 u| + |D_2 u| &= O(|x|^{-b_1}), \\
F_{12} &= O(|x|^{-b}), \tag{11.1.23}
\end{aligned}$$

where $b, b_1 > 0$ are arbitrary if $8\pi G(M + N) < 1$; $b = 2, b_1 = 2 + \varepsilon$ if $8\pi G(M + N) = 1$ and there are at least two distinct points among those listed in (11.1.4); $b = (1-\varepsilon)\sqrt{M}, b_1 = (1-\varepsilon)\sqrt{M}+1$ if $N = 0, 8\pi G M = 1$, and all q's are identical; $b = (1 - \varepsilon)\sqrt{N}, b_1 = (1 - \varepsilon)\sqrt{N} + 1$ if $M = 0$,

$8\pi G N = 1$, and all p's are identical. Here $\varepsilon > 0$ may be made arbitrarily small. Moreover, the associated two-dimensional gravitational metric $g_{jk} = e^{\eta}\delta_{jk}$ is geodesically complete and is determined explicitly by the formula

$$e^{\eta} = g_0 \left(\frac{|u|^2}{(1+|u|^2)^2} \prod_{s=1}^{N} |x - p_s|^{-2} \prod_{s=1}^{M} |x - q_s|^{-2} \right)^{8\pi G}, \qquad (11.1.24)$$

where $g_0 > 0$ is a constant. In fact, the condition for existence, namely, the inequality (11.1.19), is a necessary and sufficient condition for the metric g_{jk} determined by (11.1.24) to be complete. Furthermore, the total curvature is also quantized according to the formula

$$\int_{\mathbb{R}^2} K_g \, d\Omega_g = 16\pi^2 G(M + N), \qquad (11.1.25)$$

where $d\Omega_g$ is the canonical area element and K_g is the Gauss curvature of the two-surface $(\mathbb{R}^2, \{g_{jk}\})$ on which the cosmic strings reside. Consequently, the M cosmic strings located at $q_1, q_2, \cdots, q_M$ and the N antistrings located at $p_1, p_2, \cdots, p_N$ magnetically annihilate each other (as stated in (11.1.21)) like charged particles but make indistinguishable equal contributions to the total energy (as stated in (11.1.22)) and global geometry (as stated in (11.1.19), (11.1.24), and (11.1.25)). However, such a symmetry between strings and antistrings will be broken when an external magnetic field is switched on. In other words, in the presence of a weak applied field, one type of strings will be energetically preferred over another in such a way that the excited magnetic field chooses to be aligned along the same direction of the applied field.

In absence of gravity, the model (11.1.9) takes the special form that $g_{\mu\nu}$ is the flat Minkowski metric,

$$g_{\mu\nu} = \mathrm{diag}\left\{1, -1, -1, -1\right\} \qquad (11.1.26)$$

and the equations of motion are simply (11.1.13), (11.1.14), (11.1.16), (11.1.17), subject to (11.1.26). Our existence result can be stated as follows.

Theorem 11.1.2. *Suppose that gravity is absent. Given the points*

$$q_1, q_2, \cdots, q_M \quad and \quad p_1, p_2, \cdots, p_N$$

in $\mathbb{R}^2$, there is a unique energy-minimizing solution (u, A_j) $(A_0 = A_3 = 0$ and u, A_j depend only on $x = (x^j)$ $(j = 1, 2) \in \mathbb{R}^2)$ representing M vortices at the points q's and N antivortices at the points p's so that the q's and p's are zeros and poles of u, respectively, and, as $|x| \to \infty$, the solution approaches the vacuum state with broken symmetry exponentially fast at the rate

$$|u|^2 - 1, \quad |D_1 u| + |D_2 u|, \quad F_{12} = O(e^{-(1-\varepsilon)|x|}), \qquad (11.1.27)$$

where $\varepsilon > 0$ is an arbitrarily small number. Besides, the same flux and energy formulas (11.1.21) and (11.1.22) hold and the opposite vorticity values for strings and antistrings, stated in (11.1.20), are replaced by the following explicit values

$$
\begin{aligned}
F_{12}(p_s) &= -1, \quad s = 1, 2, \cdots, N; \\
F_{12}(q_s) &= 1, \quad s = 1, 2, \cdots, M,
\end{aligned}
\tag{11.1.28}
$$

at the vortex and antivortex points. In fact the values given in (11.1.28) are also the global minimum and maximum of F_{12} on the full plane. Moreover, the same symmetry-breaking mechanism for strings and antistrings in the presence of an applied magnetic field takes place for vortices and antivortices here.

These results will be established in subsequent sections.

11.2 Simplification of Equations

In its original form, the system (11.1.12)–(11.1.17) is hard to approach. However, as in the classical $O(3)$ sigma model and Abelian Higgs model, we also have a self-dual structure to explore so that the level of difficulty in the existence problem is greatly reduced. To proceed, we first introduce a new current density

$$
J_k = \frac{i}{1 + |u|^2}(u\overline{D_k u} - \overline{u}D_k u), \quad j = 1, 2.
\tag{11.2.1}
$$

Using the following easily verified commutation relation for gauge-covariant derivatives,

$$
[D_j, D_k]u = (D_j D_k - D_k D_j)u = -iF_{jk}u, \quad j, k = 1, 2,
\tag{11.2.2}
$$

we obtain by differentiating (11.2.1) and applying $\partial_j(u_1\overline{u_2}) = (D_j u_1)\overline{u_2} + u_1\overline{(D_j u_2)}$ that

$$
J_{12} = \partial_1 J_2 - \partial_2 J_1 = -\frac{2|u|^2}{1 + |u|^2}F_{12} + 2i\frac{D_1 u\overline{D_2 u} - \overline{D_1 u}D_2 u}{(1 + |u|^2)^2}.
\tag{11.2.3}
$$

Besides, it can be shown that there holds the identity

$$
|D_1 u + iD_2 u|^2 = |D_1 u|^2 + |D_2 u|^2 - i(D_1 u\overline{D_2 u} - \overline{D_1 u}D_2 u).
\tag{11.2.4}
$$

Thus, in order to derive the self-dual reduction of the system (11.1.13) and (11.1.14), we may apply (11.2.1), (11.2.3), and (11.2.4) to note that the energy density $\mathcal{H} = T_{00}$ induced from the energy-momentum tensor (11.1.15) of the matter-gauge sector has the representation

$$
\begin{aligned}
\mathcal{H} &= \frac{1}{2}e^{-2\eta}F_{12}^2 + \frac{2e^{-\eta}}{(1+|u|^2)^2}(|D_1 u|^2 + |D_2 u|^2) + \frac{1}{2}\left(\frac{1-|u|^2}{1+|u|^2}\right)^2 \\
&= \frac{1}{2}\left(e^{-\eta}F_{12} - \frac{1-|u|^2}{1+|u|^2}\right)^2 + e^{-\eta}F_{12}\frac{1-|u|^2}{1+|u|^2} \\
&\quad + \frac{2e^{-\eta}}{(1+|u|^2)^2}(|D_1 u + iD_2 u|^2 + i[D_1 u\overline{D_2 u} - \overline{D_1 u}D_2 u]) \\
&= e^{-\eta}(F_{12} + J_{12}) \\
&\quad + \frac{1}{2}\left(e^{-\eta}F_{12} - \frac{1-|u|^2}{1+|u|^2}\right)^2 + \frac{2e^{-\eta}}{(1+|u|^2)^2}|D_1 u + iD_2 u|^2 . (11.2.5)
\end{aligned}
$$

As a consequence of (11.2.5), we obtain the self-dual system

$$
D_1 u + iD_2 u = 0, \tag{11.2.6}
$$

$$
F_{12} = e^{\eta}\frac{1-|u|^2}{1+|u|^2}. \tag{11.2.7}
$$

It is straightforward to examine that any solution of (11.2.6) and (11.2.7) necessarily satisfies (11.1.13) and (11.1.14) where the metric tensor is defined by (11.1.18).

We next consider what form the Einstein equations (11.1.12) will be reduced to. Recall that the metric (11.1.18) first implies that the Einstein tensor $G_{\mu\nu}$ assumes the form

$$
G_{00} = -G_{33} = -K_g, \quad G_{\mu\nu} = 0 \quad \text{for other values of } \mu, \nu, \tag{11.2.8}
$$

where K_g is the Gauss curvature of the two-surface $(\mathbb{R}^2, e^{\eta}\delta_{jk})$ which can calculated by the expression

$$
K_g = -\frac{1}{2}e^{-\eta}\Delta\eta \tag{11.2.9}
$$

with $\Delta = \partial_1^2 + \partial_2^2$ being the Laplace operator on $\mathbb{R}^2$.

To achieve consistency in (11.1.12), we consider the energy-momentum tensor (11.1.15). By (11.1.18) and the assumption that all fields depend only on $x = (x^j)$ $(j = 1, 2)$, we easily find that

$$
T_{00} = -T_{33} = \mathcal{H}; \quad T_{0\mu} = 0, \quad \mu \neq 0; \quad T_{3\mu} = 0, \quad \mu \neq 3.
$$

Furthermore, in view of (11.2.6) and (11.2.7), we have from (11.1.15) by setting $\mu = \nu = 1$ that

$$
T_{11} = e^{-\eta}F_{12}^2 + \frac{4|D_1 u|^2}{(1+|u|^2)^2}
$$

$$-e^{\eta}\left(\frac{1}{2}e^{-2\eta}F_{12}^2 + \frac{2e^{-\eta}}{(1+|u|^2)^2}(|D_1u|^2+|D_2u|^2) + \frac{1}{2}\left[\frac{1-|u|^2}{1+|u|^2}\right]^2\right)$$

$$= \frac{1}{2}e^{\eta}\left(e^{-\eta}F_{12} - \frac{1-|u|^2}{1+|u|^2}\right)\left(e^{-\eta}F_{12} + \frac{1-|u|^2}{1+|u|^2}\right) = 0. \qquad (11.2.10)$$

Similarly, $T_{22} = 0$. Besides, by (11.1.18) and (11.2.6), we have

$$T_{12} = \frac{2}{(1+|u|^2)^2}(D_1u\overline{D_2u} + \overline{D_1u}D_2u) = 0.$$

Thus, inserting these results into (11.1.12) and using (11.2.8), we see that the cosmological constant must vanish, $\Lambda = 0$, and the Einstein equations are boiled down to a single scalar equation relating the Gauss curvature to the energy density,

$$K_g = 8\pi G\mathcal{H}. \qquad (11.2.11)$$

It turns out that the sets of zeros and poles, say, Q and P, of u, where (u, A_j, η) is a solution triplet of (11.2.6), (11.2.7), and (11.2.11), are the sites for strings and antistrings to reside. To begin our study, we now analyze the behavior of u near P and Q. In fact, the equation (11.2.6) is the same as that in the classical Abelian Higgs model. Therefore, any point $q \in Q$ obeys the characterization that there are a locally well-defined nonvanishing function h_q and an integer $n_q \geq 1$ so that, in a neighborhood around q,

$$u(z) = h_q(z)(z-q)^{n_q}, \quad z = x^1 + ix^2. \qquad (11.2.12)$$

To study the behavior near a pole $p \in P$, we use the substitution $U = 1/u$ and define $d_j = \partial_j + iA_j$, $(j = 1, 2)$. There holds $d_jU = -U^2D_ju$. By (11.2.6) we have

$$d_1U = -id_2U, \quad \text{away from } P.$$

Therefore, using the same argument as before and the removable singularity theorem, we see that U has a similar representation near a given $p \in P$ as u near q expressed in (11.2.12). Reinterpreting this result for u itself, we obtain

$$u(z) = h_p(z)(z-p)^{-n_p}, \quad z = x^1 + ix^2, \qquad (11.2.13)$$

where h_p is nonvanishing near $z = p$ and $n_p \geq 1$ is an integer.

Thus, we see that the winding number of u,

$$\frac{1}{2\pi i}\oint_C \frac{du}{u},$$

along a circle C near q is positive, which is n_q, but is negative, $-n_p$, near p. In the flat case, $\eta = 0$, the equation (11.2.7) says that the vorticity field F_{12} attains its positive maximum value, 1, at q, and its negative minimum value, -1, at p. In the nonflat case, $\eta \neq 0$, we can derive similar conclusions. It will also be seen later that the solution configuration around such pair of points, q and p, behave magnetically like particles of opposite charges with flux $2\pi(n_q - n_p)$ and energy $2\pi(n_q + n_p)$. In this way we obtain a string and an antistring at q and p with opposite magnetic charges n_q and $-n_p$, respectively.

In conclusion, the governing system (11.1.12)–(11.1.17) is reduced to the system of equations (11.2.6), (11.2.7), and (11.2.11), from which we will produce a general class of solutions representing multiply centered cosmic strings and antistrings located at the sets Q and P, in equilibrium, as stated in Theorem 11.1.1.

In absence of gravity, the metric takes the flat form (11.1.26) and the equations of motion, governing vortices and antivortices, are reduced to (11.2.6), (11.2.7) over $\mathbb{R}^2$ with $\eta = 0$. In the rest of this section, we derive from (11.2.6), (11.2.7), and (11.2.11) a scalar elliptic equation for string solutions. The elliptic equation for vortices will be viewed as a special case of that for strings.

We first use the substitution $v = \ln|u|^2$ and assume that the sets

$$Q = \left\{q_1, q_2, \cdots, q_M\right\}, \quad P = \left\{p_1, p_2, \cdots, p_N\right\}$$

determine the locations of strings and antistrings of unit charge (we can always accommodate strings of higher charges by allowing arbitrary multiplicities of those points in Q or P). Then, in view of (11.2.12), (11.2.13), and the relation (according to (11.2.6))

$$F_{12} = -\frac{1}{2}\Delta\ln|u|^2, \quad \text{away from the set } P \cup Q, \tag{11.2.14}$$

the system of equations (11.2.6) and (11.2.7) is reduced via $v = \ln|u|^2$ to

$$\Delta v = 2e^\eta\left(\frac{e^v - 1}{e^v + 1}\right) - 4\pi\sum_{s=1}^{N}\delta_{p_s} + 4\pi\sum_{s=1}^{M}\delta_{q_s}. \tag{11.2.15}$$

Here the unknown metric conformal factor e^η is to be determined from (11.2.11). A solution pair (u, A_j) of (11.2.6) and (11.2.7) may be constructed from a solution v of (11.2.15) by the following standard prescription

$$u(z) = \exp\left(\frac{1}{2}v(z) + i\theta(z)\right),$$

$$\theta(z) = -\sum_{s=1}^{N}\arg(z - p_s) + \sum_{s=1}^{M}\arg(z - q_s),$$

$$A_1(z) = -\operatorname{Re}\{2i\overline{\partial}\ln u(z)\}, \quad A_2(z) = -\operatorname{Im}\{2i\overline{\partial}\ln u(z)\}. \tag{11.2.16}$$

Hence, we may use the above to write down the two gauge-covariant derivatives as follows,

$$D_1 u = (\partial + \overline{\partial})u - \left(\frac{\overline{\partial}u}{u} - \frac{\partial\overline{u}}{\overline{u}}\right)u = u\partial v,$$

$$D_2 u = i(\partial - \overline{\partial})u + i\left(\frac{\overline{\partial}u}{u} + \frac{\partial\overline{u}}{\overline{u}}\right)u = iu\partial v. \qquad (11.2.17)$$

A straightforward consequence of the above formulas is

$$|D_1 u|^2 + |D_2 u|^2 = \frac{1}{2}e^v|\nabla v|^2. \qquad (11.2.18)$$

We are now ready to resolve the Einstein equation (11.2.11).

In view of (11.2.7) and (11.2.14), we rewrite the energy density $\mathcal{H}$ in the form

$$
\begin{aligned}
\mathcal{H} &= e^{-\eta}F_{12}\left(\frac{1-|u|^2}{1+|u|^2}\right) + \frac{2e^{-\eta}}{(1+|u|^2)^2}(|D_1 u|^2 + |D_2 u|^2) \\
&= e^{-\eta}\left(\frac{\Delta v}{2}\left[\frac{e^v-1}{e^v+1}\right] + \frac{e^v|\nabla v|^2}{(e^v+1)^2}\right) \\
&= e^{-\eta}\Delta\left(\ln(1+e^v) - \frac{1}{2}v\right), \quad \text{away from } P \cup Q.
\end{aligned}
$$

Since $\mathcal{H}$ is a smooth function, the above expression indicates that we can compensate the singular sources at the points p's and q's to arrive at the relation

$$e^{\eta}\mathcal{H} = \Delta\left(\ln(1+e^v) - \frac{1}{2}v\right) + 2\pi\sum_{s=1}^{N}\delta_{p_s} + 2\pi\sum_{s=1}^{M}\delta_{q_s},$$

which is now valid in the full $\mathbb{R}^2$. Inserting the above and (11.2.9) into (11.2.11), we see that

$$\frac{\eta}{16\pi G} + \ln(1+e^v) - \frac{1}{2}v + \sum_{s=1}^{N}\ln|x-p_s| + \sum_{s=1}^{M}\ln|x-q_s|$$

is a harmonic function, which we assume to be a constant. Consequently, we obtain explicitly the conformal factor e^{η} of the metric as follows,

$$e^{\eta} = g_0\left(\frac{e^v}{(1+e^v)^2}\prod_{s=1}^{N}|x-p_s|^{-2}\prod_{s=1}^{M}|x-q_s|^{-2}\right)^{8\pi G}. \qquad (11.2.19)$$

Here $g_0 > 0$ is an arbitrary constant. Note that, since v satisfies (11.2.15), the metric (11.2.19) is everywhere regular. Thus, only infinity is to be concerned, and, at the string and antistring points, q's and p's, respectively, we have, in view of (11.2.7), opposite associated vorticities as expected,

$$
\begin{aligned}
F_{12}(q_s) &= e^{\eta(q_s)} > 0, \quad 1 \le s \le M; \\
F_{12}(p_s) &= -e^{\eta(p_s)} < 0, \quad 1 \le s \le N.
\end{aligned}
\tag{11.2.20}
$$

Recall that we are interested in solutions in the broken symmetry category so that $v = 0$ or $|u|^2 = 1$ at infinity. This fact and (11.2.19) imply the validity of the following global inequality in $\mathbb{R}^2$,

$$
C_1(1 + |x|)^{-16\pi G(M+N)} \le e^{\eta(x)} \le C_2(1 + |x|)^{-16\pi G(M+N)}, \tag{11.2.21}
$$

where $C_1, C_2 > 0$ are suitable constants. The inequality (11.2.21) enables us to draw the immediate conclusion that a solution leads to a geodesically complete metric if and only if the condition (11.1.19) holds. Thus, in the sense of a complete metric, the numbers of strings and antistrings play equal parts and a large number of strings of either type will make the metric incomplete.

Inserting (11.2.19) into (11.2.15), we arrive at the final governing equation

$$
\begin{aligned}
\Delta v = {} & 2g_0 \left(\frac{e^v}{(1+e^v)^2} \prod_{s=1}^{N} |x - p_s|^{-2} \prod_{s=1}^{M} |x - q_s|^{-2} \right)^{8\pi G} \left(\frac{e^v - 1}{e^v + 1} \right) \\
& -4\pi \sum_{s=1}^{N} \delta_{p_s} + 4\pi \sum_{s=1}^{M} \delta_{q_s}.
\end{aligned}
\tag{11.2.22}
$$

Equation (11.2.22) and its flat-space limit when gravity is absent, $G = 0$, $g_0 = 1$, will be solved in the next section.

11.3   Proof of Existence

In this section we prove the coexistence theorems for vortices and antivortices and strings and antistrings stated earlier. We first study the easy case, vortices and antivortices. We next use an approximation method to construct strings and antistrings. We then establish asymptotic estimates for the solutions obtained.

11.3.1   Vortices and antivortices

We start from the simpler case that gravity is absent. Hence $G = 0$ and the metric is the flat one given by (11.1.26) so that $g_0 = 1$ in (11.2.19). Therefore, the governing equation (11.2.22) is reduced to

$$
\Delta v = 2 \left(\frac{e^v - 1}{e^v + 1} \right) - 4\pi \sum_{s=1}^{N} \delta_{p_s} + 4\pi \sum_{s=1}^{M} \delta_{q_s}. \tag{11.3.1}
$$

Theorem 11.3.1. *For any prescribed vortex and antivortex points,*

$$q_1, q_2, \cdots, q_M \quad and \quad p_1, p_2, \cdots, p_N,$$

the equation (11.3.1) has a unique solution which vanishes at infinity according to the exponential decay rate

$$|v(x)| = O(e^{-(1-\varepsilon)|x|}), \quad |\nabla v(x)| = O(e^{-(1-\varepsilon)|x|}),$$

where $\varepsilon > 0$ is an arbitrary small number.

Proof. We will use the method of monotone iterations to solve (11.3.1). While such a method is elementary, it has two interesting features. The first is that we are able to use the solution of the classical Abelian Higgs equation to get a comparable pair of (distributional) sub- and supersolutions, which greatly simplifies our work. The second is that, although the sub- and supersolutions are not those in classical sense, the structure of the equation (11.3.1) allows us to control the limit of our iterative sequence in a suitable norm, which turns out to be strong enough to yield a smooth solution of the original problem. Besides, the idea of the proof will also be useful for the study of the original string equation (11.2.22). These special features made us decide to present in some detail the method here.

In order to solve (11.3.1), instead, we consider the equation

$$\Delta v_1 = (e^{v_1} - 1) + 4\pi \sum_{s=1}^{N} \delta_{p_s}. \tag{11.3.2}$$

It is well known [157] that the equation (11.3.2) has a unique solution that vanishes exponentially fast at infinity. Furthermore, the solution also satisfies $v_1 < 0$, which can easily be seen from using the maximum principle. Hence, from (11.3.2), we arrive at

$$\Delta(-v_1) = (1 - e^{v_1}) - 4\pi \sum_{s=1}^{N} \delta_{p_s}$$

$$\leq 2\left(\frac{1 - e^{v_1}}{1 + e^{v_1}}\right) - 4\pi \sum_{s=1}^{N} \delta_{p_s}$$

$$\leq 2\left(\frac{e^{-v_1} - 1}{e^{-v_1} + 1}\right) - 4\pi \sum_{s=1}^{N} \delta_{p_s} + 4\pi \sum_{s=1}^{M} \delta_{q_s}, \tag{11.3.3}$$

which says that $-v_1$ is a supersolution of (11.3.1) in sense of distribution. Similarly, let v_2 be the unique solution of

$$\Delta v_2 = (e^{v_2} - 1) + 4\pi \sum_{s=1}^{M} \delta_{q_s} \tag{11.3.4}$$

that vanishes exponentially fast at infinity. Then $v_2 < 0$. Consequently, from (11.3.4), we obtain

$$\Delta v_2 > 2\left(\frac{e^{v_2} - 1}{e^{v_2} + 1}\right) - 4\pi \sum_{j=1}^{N_1} \delta_{p_j} + 4\pi \sum_{j=1}^{N_2} \delta_{q_j}, \qquad (11.3.5)$$

which implies that v_2 is a subsolution of (11.3.1) in sense of distribution.

It is clear that the desired comparison $-v_1 > v_2$ (the supersolution is above the subsolution) holds. However, since our pair of supersolution $-v_1$ and subsolution v_2 of (11.3.1) are not in the classical sense, we need to elaborate more to obtain a solution of (11.3.1) between this pair.

Let us introduce the background functions

$$v_{0,1} = -\sum_{s=1}^{N} \ln(1 + |x - p_s|^{-2}), \ v_{0,2} = -\sum_{s=1}^{M} \ln(1 + |x - q_s|^{-2});$$

$$g_1 = 4\sum_{s=1}^{N}(1 + |x - p_s|^2)^{-2}, \ g_2 = 4\sum_{s=1}^{M}(1 + |x - q_s|^2)^{-2}. \quad (11.3.6)$$

We use the substitutions $v = -v_{0,1} + v_{0,2} + w$ in (11.3.1), $-v_1 = -v_{0,1} + v_{0,2} + w_+$ in (11.3.3), and $v_2 = -v_{0,1} + v_{0,2} + w_-$ in (11.3.5). We have

$$\Delta w = 2\left(\frac{e^{v_{0,2}+w} - e^{v_{0,1}}}{e^{v_{0,2}+w} + e^{v_{0,1}}}\right) - g_1 + g_2, \qquad (11.3.7)$$

and w_+ and w_- are super- and subsolutions of (11.3.7) in the sense of distribution. We show that (11.3.7) has a solution, w, satisfies $w_- < w < w_+$. For this purpose, we define the following iterative scheme,

$$\begin{aligned}
\Delta w_n - C_0 w_n &= 2\left(\frac{e^{v_{0,2}+w_{n-1}} - e^{v_{0,1}}}{e^{v_{0,2}+w_{n-1}} + e^{v_{0,1}}}\right) - C_0 w_{n-1} - g_1 + g_2 \\
&\equiv F(x, w_{n-1}), \\
w_n &\to 0 \quad \text{as } |x| \to \infty, \quad n = 2, 3, \cdots, \\
w_1 &= w_+,
\end{aligned}$$

where $C_0 > 0$ is to be determined. The definition of w_1 implies that, as a function of x, $F(x, w_1(x)) \in L^2(\mathbb{R}^2)$. Hence, by L^2-theory of elliptic equations, we have $w_2 \in W^{2,2}(\mathbb{R}^2)$. In particular, $w_2 = 0$ at infinity. Using induction, we see easily that we can define $w_n \in W^{2,2}(\mathbb{R}^2)$ for all $n \geq 2$ by the above scheme.

We next establish the monotone property

$$w_- < \cdots < w_n < \cdots < w_2 < w_1 = w_+. \qquad (11.3.8)$$

for a sufficiently large number C_0. In fact, since w_2 is bounded, we see that $w_2 < w_1 = w_+$ near $Q = \{q_1, q_2, \cdots, q_M\}$. In $\mathbb{R}^2 - Q$, we have $\Delta(w_2 - w_1) \geq$

$C_0(w_2 - w_1)$. Hence, the maximum principle gives us $w_2 < w_1$ everywhere. Besides, since $w_- < w_+$, we have

$$(\Delta - C_0)(w_- - w_2) \geq \left(\frac{4e^{v_{0,1}+v_{0,2}+\xi}}{(e^{v_{0,2}+\xi} + e^{v_{0,1}})^2} - C_0 \right)(w_- - w_+)$$

$$(w_- < \xi < w_+)$$

$$\geq (2 - C_0)(w_- - w_+) \geq 0$$

if $C_0 \geq 2$. Thus $w_- < w_2$. We will fix the assumption $C_0 \geq 2$ in the rest of the proof.

Suppose we have already proved that $w_- < w_n$ and $w_n < w_{n-1}$ for some $n \geq 2$. Then, for some ξ lying between w_{n-1} and w_n, we have

$$(\Delta - C_0)(w_{n+1} - w_n) = \left(\frac{4e^{v_{0,1}+v_{0,2}+\xi}}{(e^{v_{0,2}+\xi} + e^{v_{0,1}})^2} - C_0 \right)(w_n - w_{n-1}) \geq 0.$$

Hence, $w_{n+1} < w_n$. Moreover, for some ξ lying between w_- and w_n, we have

$$(\Delta - C_0)(w_- - w_{n+1}) \geq \left(\frac{4e^{v_{0,1}+v_{0,2}+\xi}}{(e^{v_{0,2}+\xi} + e^{v_{0,1}})^2} - C_0 \right)(w_- - w_n) \geq 0.$$

Thus, again, $w_- < w_{n+1}$. Consequently, the general property (11.3.8) holds.

Since both w_- and w_+ are in $L^2(\mathbb{R}^2)$, the inequality (11.3.8) implies that the sequence $\{w_n\}$ converges in $L^2(\mathbb{R}^2)$. Using a standard bootstrap argument, we see that the convergence also holds in $W^{2,2}(\mathbb{R}^2)$. Of course, its limit in $W^{2,2}(\mathbb{R}^2)$, say, w, is a classical solution of (11.3.7).

Returning to the original variable, $v = -v_{0,1} + v_{0,2} + w$, we get a solution of (11.3.1) which vanishes at infinity.

The uniqueness follows from the fact that the nonlinearity of (11.3.1) is monotone increasing and a use of the maximum principle.

The exponential decay estimates will be established later in this section.

11.3.2   Strings and antistrings

In this subsection, we will find a solution of (11.2.22) subject to the boundary condition $v(x) \to 0$ as $|x| \to \infty$. Since the coefficient, say, $F_0(x)$, in the nonlinear term on the right-hand side of (11.2.22) involves singularities of a power type, the direct method employed in the last section for the vortex equation (11.3.1) fails and it becomes necessary to consider a few separate cases. In fact, it can be seen that a major obstacle comes from the orders of singularities at the strings. Our partial differential equation technique needs to keep F_0 in L^p_{loc} for some $p > 1$. In view of the condition (11.1.19), we need to assume that there are at least two distinct points among the set of centers of strings, p's and q's, given in (11.1.4). When all these points coincide, we have $F_0 \in L_{\mathrm{loc}}$ $(p = 1)$ and the partial differential equation method no longer works. Fortunately, we can still pursue a solution in its ordinary differential equation version within a radial symmetry assumption.

Here is our existence result for multiple strings.

Theorem 11.3.2. *Under the condition* $8\pi G(M+N) < 1$ *or* $8\pi G(M+N) = 1$ *but at least two of the points listed in (11.1.4) are distinct, the equation (11.2.22) has a solution which vanishes at infinity.*

Because of some technical difficulties, we need to consider a regularized version of the equation (11.2.22),

$$
\Delta v = 2g_0 F_\delta(x) \left(\frac{e^v}{(1+e^v)^2} \right)^{8\pi G} \left(\frac{e^v - 1}{e^v + 1} \right)
$$

$$
- \sum_{s=1}^{N} \frac{4\delta}{(\delta + |x - p_s|^2)^2} + \sum_{s=1}^{M} \frac{4\delta}{(\delta + |x - q_s|^2)^2}, \quad (11.3.9)
$$

where the coefficient function F_δ is defined by

$$
F_\delta(x) = \left(\prod_{s=1}^{N} (\delta + |x - p_s|^2)^{-1} \prod_{s=1}^{M} (\delta + |x - q_s|^2)^{-1} \right)^{8\pi G}
$$

and $0 < \delta < 1$ is a parameter. Note that the original equation (11.2.22) may be recovered from (11.3.9) from taking the $\delta \to 0$ limit.

We first look at a modified version of (11.3.9) as follows,

$$
\Delta v = g_0 F_\delta(x) \left(\frac{e^v}{(1+e^v)^2} \right)^{8\pi G} (e^v - 1) + \sum_{s=1}^{N} \frac{4\delta}{(\delta + |x - p_s|^2)^2}. \quad (11.3.10)
$$

We will solve (11.3.10) under the same boundary condition as stated in Theorem 11.3.2. The solution will be used as a supersolution of (11.3.9).

Lemma 11.3.3. *The function* $v_+ = 0$ *is a supersolution of (11.3.10).*

Proof. In fact, we have

$$
0 = \Delta v_+ < g_0 F_\delta(x) \left(\frac{e^{v_+}}{(1+e^{v_+})^2} \right)^{8\pi G} (e^{v_+} - 1) + \sum_{s=1}^{N} \frac{4\delta}{(\delta + |x - p_s|^2)^2}
$$

and the lemma follows immediately.

We now introduce the functions

$$
v_{\delta,1} = \sum_{s=1}^{N} \ln \left(\frac{\delta + |x - p_s|^2}{1 + |x - p_s|^2} \right), \quad v_{\delta,2} = \sum_{s=1}^{M} \ln \left(\frac{\delta + |x - q_s|^2}{1 + |x - q_s|^2} \right). \quad (11.3.11)
$$

Of course, $v_{0,j} < v_{\delta,j} < 0$ and $v_{\delta,j} \to 0$ as $|x| \to \infty$, $j = 1, 2$.

Lemma 11.3.4. *If $8\pi G(M+N) \le 1$, then there is a constant $C_0 > 0$ so that, whenever $g_0 \ge C_0$, $v_- = v_{\delta,1}$ is a subsolution of (11.3.10) for any $0 < \delta < 1/2$ (say).*

Proof. Since

$$\Delta v_{\delta,1} = \sum_{s=1}^{N} \frac{4\delta}{(\delta + |x-p_s|^2)^2} - g_1,$$

where g_1 has been defined in (11.3.6), it suffices to show that

$$0 > g_0 F_\delta(x) \left(\frac{e^{v_{\delta,1}}}{[1+e^{v_{\delta,1}}]^2} \right)^{8\pi G} (e^{v_{\delta,1}} - 1) + g_1 \tag{11.3.12}$$

everywhere in $\mathbb{R}^2$ for any $0 < \delta < 1/2$ and $g_0 \ge$ some $C_0 > 0$. For this purpose, we rewrite $v_{\delta,1}$ as

$$v_{\delta,1} = -\sum_{s=1}^{N} \ln\left(1 + \frac{1-\delta}{\delta + |x-p_s|^2}\right). \tag{11.3.13}$$

The expression (11.3.13) says that $v_{\delta,1} \to 0$ uniformly as $|x| \to \infty$. Thus, so does $e^{v_{\delta,1}} \to 1$. Note that

$$e^{v_{\delta,1}} - 1 = e^{\xi(v_{\delta,1})} v_{\delta,1} \qquad (v_{\delta,1} < \xi(v_{\delta,1}) < 0).$$

Hence, $e^{\xi(v_{\delta,1})} \to 1$ uniformly as $|x| \to \infty$ as well. On the other hand,

$$F_\delta(x)\left(\frac{e^{v_{\delta,1}}}{[1+e^{v_{\delta,1}}]^2}\right)^{8\pi G} (e^{v_{\delta,1}} - 1)$$

$$= -\left(\sum_{s=1}^{N} \ln\left(1 + \frac{1-\delta}{\delta + |x-p_s|^2}\right)\right)$$

$$\times \frac{e^{\xi(v_{\delta,1})}}{(1+e^{v_{\delta,1}})^{16\pi G}} \left(\prod_{s=1}^{N}(1+|x-p_s|^2)\prod_{s=1}^{M}(\delta+|x-q_s|^2)\right)^{-8\pi G}$$

$$= -\left(\sum_{s=1}^{N} \frac{1}{1+\xi_s} \cdot \frac{1-\delta}{\delta + |x-p_s|^2}\right)\frac{e^{\xi(v_{\delta,1})}}{(1+e^{v_{\delta,1}})^{16\pi G}}$$

$$\times \left(\prod_{s=1}^{N}(1+|x-p_s|^2)\prod_{s=1}^{M}(\delta+|x-q_s|^2)\right)^{-8\pi G}$$

$$\equiv -h_\delta(x), \tag{11.3.14}$$

where

$$\xi_s = \theta_s\left(\frac{1-\delta}{\delta + |x-p_s|^2}\right) \quad \text{with } 0 < \theta_s < 1, \quad s = 1, 2, \cdots, N.$$

Hence, setting $r = |x|$, we have

$$\lim_{r \to \infty} r^4 h_\delta(x) = \begin{cases} \infty & \text{uniformly w. r. t. } \delta \text{ if } 16\pi G(M+N) + 2 < 4, \\ a_\delta & \geq \text{ some } a_0 > 0 \text{ if } 16\pi G(M+N) + 2 = 4. \end{cases}$$

These conditions are contained in the requirement $8\pi G(M+N) \leq 1$ stated in the lemma. Consequently, we can find C_0 and r_0 so that

$$-\frac{1}{r^4}(g_0 r^4 h_\delta - r^4 g_1) < 0, \quad \forall r \geq r_0, \quad g_0 \geq C_0. \tag{11.3.15}$$

That is to say, (11.3.12) is valid for $|x| \geq r_0$, $g_0 \geq C_0$.

We next choose r_0 sufficiently large so that

$$P \cup Q \subset \left\{ x \,\middle|\, |x| < r_0 \right\} \equiv D.$$

We study the equation in the disk D. Since $v_{\delta,1} \leq v_{1/2,1}$ and $v_{\delta,j} < 0$ $(j = 1, 2)$, we have

$$e^{v_{\delta,1}} - 1 \leq e^{v_{1/2,1}} - 1,$$

$$F_\delta(x) \left(\frac{e^{v_{\delta,1}}}{[1 + e^{v_{\delta,1}}]^2} \right)^{8\pi G} \geq 2^{-16\pi G} \left(\prod_{s=1}^N (1+|x-p_s|^2) \prod_{s=1}^M (1+|x-q_s|^2) \right)^{-8\pi G}.$$

Hence, by making C_0 sufficiently large, we see that (11.3.12) is valid on D as well.

Lemma 11.3.5. *Equation (11.3.9) has a positive supersolution, say, v_δ^+, which vanishes at infinity.*

Proof. Use v_+ and v_- to denote the super- and subsolutions of (11.3.10). Then Lemmas 11.3.3 and 11.3.4 give us $v_- < v_+$. Thus, (11.3.10) has a solution, say, v_δ, satisfying $v_- < v_\delta < v_+$. Since v_- and v_+ both vanish at infinity, so does v_δ.

We now show that $v_\delta^+ \equiv -v_\delta > 0$ is a supersolution of (11.3.9). In fact, we have

$$\begin{aligned} \Delta v_\delta^+ &= g_0 F_\delta(x) \left(\frac{e^{v_\delta}}{[1 + e^{v_\delta}]^2} \right)^{8\pi G} (1 - e^{v_\delta}) - \sum_{s=1}^N \frac{4\delta}{(\delta + |x - p_s|^2)^2} \\ &< 2g_0 F_\delta(x) \left(\frac{e^{v_\delta^+}}{[1 + e^{v_\delta^+}]^2} \right)^{8\pi G} \left(\frac{1 - e^{v_\delta}}{1 + e^{v_\delta}} \right) - \sum_{s=1}^N \frac{4\delta}{(\delta + |x - p_s|^2)^2} \\ &< 2g_0 F_\delta(x) \left(\frac{e^{v_\delta^+}}{[1 + e^{v_\delta^+}]^2} \right)^{8\pi G} \left(\frac{e^{v_\delta^+} - 1}{e^{v_\delta^+} + 1} \right) \\ &\quad - \sum_{s=1}^N \frac{4\delta}{(\delta + |x - p_s|^2)^2} + \sum_{s=1}^M \frac{4\delta}{(\delta + |x - q_s|^2)^2} \end{aligned}$$

as expected. The lemma follows.

In order to get a suitable subsolution of (11.3.9), we turn to the equation

$$\Delta v = g_0 F_\delta(x) \left(\frac{e^v}{(1 + e^v)^2} \right)^{8\pi G} (e^v - 1) + \sum_{s=1}^{M} \frac{4\delta}{(\delta + |x - q_s|^2)^2}$$

(11.3.16)

which is identical to (11.3.10) after replacing N by M and the points p's by q's. Hence, for g_0 sufficiently large, it has a solution v_δ satisfying $v_{\delta,2} < v_\delta < 0$. Define $v_\delta^- = v_\delta$. We have

Lemma 11.3.6. *The function v_δ^- satisfies $v_{\delta,2} < v_\delta^- < 0$ and is a subsolution of the equation (11.3.9).*

Proof. From (11.3.16), we have, as before,

$$
\begin{aligned}
\Delta v_\delta^- &= g_0 F_\delta(x) \left(\frac{e^{v_\delta^-}}{[1 + e^{v_\delta^-}]^2} \right)^{8\pi G} (e^{v_\delta^-} - 1) + \sum_{s=1}^{M} \frac{4\delta}{(\delta + |x - q_s|^2)^2} \\
&> 2g_0 F_\delta(x) \left(\frac{e^{v_\delta^-}}{[1 + e^{v_\delta^-}]^2} \right)^{8\pi G} \left(\frac{e^{v_\delta^-} - 1}{e^{v_\delta^-} + 1} \right) \\
&\quad - \sum_{s=1}^{N} \frac{4\delta}{(\delta + |x - p_s|^2)^2} + \sum_{s=1}^{M} \frac{4\delta}{(\delta + |x - q_s|^2)^2}.
\end{aligned}
$$

Therefore, the proof is complete.

We can now prove Theorem 11.3.2. By Lemmas 11.3.5 and 11.3.6, the equation (11.3.9) has a positive supersolution v_δ^+ and a negative subsolution v_δ^-. Hence, it has a solution, say, v_δ, satisfying

$$-v_{0,1} > -v_{\delta,1} > v_\delta^+ > v_\delta > v_\delta^- > v_{\delta,2} > v_{0,2}, \quad 0 < \delta < \frac{1}{2}, \quad (11.3.17)$$

$$\Delta v_\delta = 2g_0 F_\delta(x) \left(\frac{e^{v_\delta}}{(1 + e^{v_\delta})^2} \right)^{8\pi G} \left(\frac{e^{v_\delta} - 1}{e^{v_\delta} + 1} \right)$$

$$-4\pi \sum_{s=1}^{N} \frac{\delta}{(\delta + |x - p_s|^2)^2} + 4\pi \sum_{s=1}^{M} \frac{\delta}{(\delta + |x - q_s|^2)^2} \quad \text{in } \mathbb{R}^2. \quad (11.3.18)$$

Using (11.3.17) in (11.3.18) and a diagonal subsequence argument along with an elliptic bootstrap iteration, it is straightforward to show that there is a sequence $\{\delta_n\}$, $\delta_n \to 0$ as $n \to \infty$, such that $\{v_{\delta_n}\}$ converges to a function $v \in C^\infty(\mathbb{R}^2 - P \cup Q)$ in any $C^m(K)$ norm for each arbitrarily given compact subset K of $\mathbb{R}^2 - P \cup Q$ and $m \in \mathbb{N}$ (the set of positive integers). Furthermore, using the Green function to rewrite (11.3.18) with

$\delta = \delta_n$ over an open disk containing $P \cup Q$ in an integral form and setting $n \to \infty$, we see in view of the condition stated in Theorem 11.3.2 which ensures the property $F_0 = F_\delta|_{\delta=0} \in L^p_{\text{loc}}(\mathbb{R}^2)$ that we can take $n \to \infty$ to show that v is the solution of the equation (11.2.22). Consequently, the proof of Theorem 11.3.2 is complete.

We now deal with the exceptional cases of (11.2.22) which are not covered in the somewhat restrictive condition stated in Theorem 11.3.2.

Theorem 11.3.7. *Suppose that $M \geq 1, N = 0, 8\pi GM = 1$, and all the points q_s's are identical: $q_s = q_0$, $s = 1, 2, \cdots, M$. Then, with the choice $g_0 = 2^{16\pi G} M$, (11.2.22) has a solution that vanishes at infinity and is symmetric about the point q_0.*

Proof. Without loss of generality, let $q_s = q_0 =$ the origin of $\mathbb{R}^2$, $s = 1, 2, \cdots, M$. Set

$$a = 8\pi G \tag{11.3.19}$$

and use $r = |x|$. Then (11.2.22) becomes

$$\Delta v = 2g_0 r^{-2aM} \frac{e^{av}}{(1 + e^v)^{2a+1}} (e^v - 1) + 4\pi M \delta(x). \tag{11.3.20}$$

An application of the removable singularity theorem indicates that, in the context of radially symmetric solutions, the equation (11.3.20) is equivalent to

$$(rv_r)_r = 2g_0 r^{(1-2aM)} \frac{e^{av}}{(1 + e^v)^{2a+1}} (e^v - 1), \quad r > 0, \tag{11.3.21}$$

$$\lim_{r \to 0} \frac{v(r)}{\ln r} = \lim_{r \to 0} rv_r(r) = 2M. \tag{11.3.22}$$

We now use the variable $t = \ln r$ and the condition $aM = 1$ to transform (11.3.21), (11.3.22) into an autonomous problem,

$$v'' = 2g_0 \frac{e^{av}}{(1 + e^v)^{2a+1}} (e^v - 1), \quad -\infty < t < \infty, \tag{11.3.23}$$

$$\lim_{t \to -\infty} \frac{v(t)}{t} = \lim_{t \to -\infty} v'(t) = 2M, \tag{11.3.24}$$

where $v' = dv/dt$. We denote the right-hand side of the equation (11.3.23) by $h(v)$. Then the system (11.3.23), (11.3.24) is recast into the integral equation

$$v(t) = 2Mt + \int_{-\infty}^{t} (t - \tau) h(v(\tau)) \, d\tau, \quad t \in \mathbb{R}. \tag{11.3.25}$$

For convenience, set $w = v - 2Mt$ and rewrite (11.3.25) as

$$w(t) = \int_{-\infty}^{t} (t - \tau) h(2M\tau + w(\tau)) \, d\tau, \quad t \in \mathbb{R}. \tag{11.3.26}$$

We will look for a solution of (11.3.26) which vanishes at $t = -\infty$. We denote the right-hand side of (11.3.26) by $T(w)$ and we arrive at a fixed-point problem, $w = T(w)$. We will consider the function space

$$\mathcal{X} = \left\{ w \in C(-\infty, t_0] \,\middle|\, \lim_{t \to -\infty} w(t) = 0, \ \sup_{t \le t_0} |w(t)| \le 1 \right\},$$

where the time-span t_0, $-\infty < t_0 < \infty$, is to be specified. We first show that we may choose t_0 to make T a map from $\mathcal{X}$ into itself. In fact, take any $w \in \mathcal{X}$, we have by the definition of h that

$$|T(w)| \le 2g_0 \sup_{t \le t_0} \int_{-\infty}^{t} |t - \tau| e^{a(2M\tau + 1)} \, d\tau \le 1$$

provided that t_0 is properly chosen. We next show that T is a contraction. To this end, we calculate the derivative of h to get

$$|h'(v)| = 2g_0 \left(\frac{e^v}{[1 + e^v]^2} \right)^a \frac{|2e^v - a(1 - e^v)^2|}{(1 + e^v)^2} \le C_1 e^{av}, \quad \forall v,$$

where $C_1 > 0$ is a constant depending only on g_0. Therefore, for any $w_1, w_2 \in \mathcal{X}$, we have

$$|(T(w_1) - T(w_2))(t)|$$
$$= \left| \int_{-\infty}^{t} (t - \tau) h'(2M\tau + \tilde{w}(\tau))(w_1(t) - w_2(\tau)) \, d\tau \right|$$
$$\le C_1 \sup_{t \le t_0} |w_1(t) - w_2(t)| \int_{-\infty}^{t_0} |t_0 - \tau| e^{a(2M\tau + 1)} \, d\tau,$$

where $\tilde{w}(t)$ lies between $w_1(t)$ and $w_2(t)$. The above inequality indicates that, when t_0 is properly chosen, $T : \mathcal{X} \to \mathcal{X}$ is a contraction.

Hence we have shown that (11.3.26) has a unique solution in a neighborhood of $t = -\infty$. From (11.3.26) it is straightforward to see that $w'(t) \to 0$ as $t \to -\infty$. Furthermore, using (11.3.26) and a standard continuation argument, we can extend w to a solution over the entire $\mathbb{R}$. Thus we have solved (11.3.23), (11.3.24). It remains to achieve the boundary condition

$$\lim_{t \to \infty} v(t) = 0. \tag{11.3.27}$$

Define the function $H(v)$ by

$$H(v) = -\int_{-\infty}^{v} h(w) \, dw = \frac{2g_0}{a} \frac{e^{av}}{(1 + e^v)^{2a}} > 0.$$

Multiply the equation (11.3.23) by $v'(t)$ and integrate over $(-\infty, t)$. We get

$$(v'(t))^2 = 4M^2 - 2H(v(t)). \tag{11.3.28}$$

We will need to use a suitable critical point of the equation (11.3.28), say, $\hat{v}$. Thus, we arrive at

$$2M^2 - H(\hat{v}) = 2M^2 - \frac{2g_0}{a} \frac{e^{a\hat{v}}}{(1+e^{\hat{v}})^{2a}} = 0. \tag{11.3.29}$$

On the other hand, in order to ensure uniqueness at the equilibrium $\hat{v}$, we are motivated to require

$$H'(\hat{v}) = -h(\hat{v}) = 2g_0 \frac{e^{a\hat{v}}}{(1+e^{\hat{v}})^{2a+1}}(1 - e^{\hat{v}}) = 0.$$

Thus we are left with the only choice, $\hat{v} = 0$. As a consequence, the relation (11.3.29) allows us to determine the parameter g_0,

$$g_0 = 2^{2a}aM^2 = 2^{2a}M, \tag{11.3.30}$$

where we have inserted in the above the condition $aM = 1$. The value (11.3.30) for the conformal factor g_0 is the one stated in the theorem.

Since $v'(t) > 0$ when t is sufficiently negative, we have from (11.3.28) that

$$\frac{dv}{dt} = \sqrt{4M^2 - 2H(v)} \equiv \sqrt{F(v)} \tag{11.3.31}$$

and $v(t)$ is an increasing function. Using the definition (11.3.30), we note that there holds the limit

$$\lim_{v \to 0^-} \frac{\sqrt{F(v)}}{v} = -\frac{\sqrt{g_0}}{2a} = -\sqrt{M} < 0. \tag{11.3.32}$$

Furthermore, (11.3.31) gives us

$$\int_{v(\tau)}^{v(t)} \frac{dv}{\sqrt{F(v)}} = t - \tau. \tag{11.3.33}$$

Combining (11.3.32) and (11.3.33), we can deduce the conclusion that v satisfies the desired property (11.3.27). In fact, since the right-hand side of (11.3.28), $F(v)$, decreases as a function of $v \leq 0$, and $F(0) = 0$, we have $F(v) > 0$ when $v < 0$. Using this and (11.3.24), we see that $v'(t) > 0$ whenever $v < 0$. We claim that $v(t) < 0$ for all t. Otherwise, if there were a finite t_0 so that $v(t_0) = 0$, then, due to (11.3.32), there would exist a number $\delta > 0$ so that

$$\frac{\sqrt{F(v(t))}}{v(t)} < -\frac{\sqrt{M}}{2}, \quad t_0 - \delta \leq t < t_0.$$

In view of this and (11.3.33), we would arrive at the contradiction

$$t - (t_0 - \delta) > -\frac{2}{\sqrt{M}} \int_{v(t_0-\delta)}^{v(t)} \frac{dv}{v} = \frac{2}{\sqrt{M}} \ln\left|\frac{v(t_0 - \delta)}{v(t)}\right|, \quad t_0 > t > t_0 - \delta,$$

because the right-hand side of the above inequality approaches infinity as $t \to t_0$ while its left-hand side remain finite. Hence, $v(t) < 0$ and $v'(t) > 0$ for all t. In particular, $v(t)$ approaches its limiting value $v_\infty \leq 0$ as $t \to \infty$. We now show that the only possibility is $v_\infty = 0$. Otherwise, suppose $v_\infty < 0$. Since $F(v)$ is bounded from below by $F(v_\infty) > 0$, we see that the left-hand side of (11.3.33) remains bounded as $t \to \infty$, which contradicts the right-hand side of (11.3.33) when taking $t \to \infty$.

For any $0 < \varepsilon < 1$, let $\tau > 0$ be large enough so that

$$\frac{\sqrt{F(v(t))}}{v(t)} < -(1 - \varepsilon)\sqrt{M}, \quad t \geq \tau.$$

Inserting this result into (11.3.33), we obtain

$$t - \tau > -\frac{1}{(1-\varepsilon)\sqrt{M}} \int_{v(\tau)}^{v(t)} \frac{dv}{v} = \frac{1}{(1-\varepsilon)\sqrt{M}} \ln \left| \frac{v(\tau)}{v(t)} \right|, \quad t > \tau,$$

which leads us to the decay estimate

$$|v| = O(e^{-(1-\varepsilon)\sqrt{M}t}) \quad \text{for } t \to \infty. \tag{11.3.34}$$

Returning to the original variable $r = e^t$, we see from (11.3.34) that

$$|v(r)| = O(r^{-(1-\varepsilon)\sqrt{M}}) \quad \text{as } r \to \infty. \tag{11.3.35}$$

Besides, since $v'(t)/v(t) \to -\sqrt{M}$ as $t \to \infty$, we have from (11.3.35) that

$$|v_r(r)| = O(r^{-([1-\varepsilon]\sqrt{M}+1)}) \quad \text{as } r \to \infty. \tag{11.3.36}$$

The theorem is thus proven.

Theorem 11.3.8. *Suppose that $M = 0, N \geq 1, 8\pi GN = 1$, and all the points p_s's are identical: $p_s = p_0$, $s = 1, 2, \cdots, N$. Then, with the choice $g_0 = 2^{16\pi G}N$, the equation (11.2.22) has a solution that vanishes at infinity and is symmetric about the point p_0.*

Proof. Let $p_0 = $ the origin of $\mathbb{R}^2$, $s = 1, 2, \cdots, N$. Set $V = -v$. Then (11.2.22) becomes

$$\Delta V = 2g_0 r^{-2aN} \left(\frac{e^V}{(1 + e^V)^2} \right)^a \left(\frac{e^V - 1}{e^V + 1} \right) + 4\pi N \delta(x), \tag{11.3.37}$$

which takes the same form as (11.3.20). Hence, we may get a solution V for (11.3.37) that fulfills the decay properties (11.3.35) and (11.3.36) with $M = N$. Therefore $v = -V$ satisfies the same decay estimates

$$|v(r)| = O(r^{-(1-\varepsilon)\sqrt{N}}), \quad |v_r(r)| = O(r^{-([1-\varepsilon]\sqrt{N}+1)}), \tag{11.3.38}$$

as $r \to \infty$, and the theorem follows.

11.3.3 Asymptotic estimates

In this subsection, our goal is to obtain the decay estimates for multistring solutions in the general case $8\pi G(M + N) < 1$ or $8\pi G(M + N) = 1$ but (11.1.4) contains at least two distinct points.

Lemma 11.3.9. *Suppose that $a(M + N) < 1$. Then for any $b > 0$, there are suitable constants $C_b > 0$ so that the solution v of (11.2.22) obtained in Theorem 11.3.2 obeys the asymptotic estimates*

$$-C_b r^{-b} < v(x) < C_b r^{-b}, \quad |x| = r > r_0,$$

where r_0 satisfies

$$r_0 > \max\left\{|p_s| \,\Big|\, s = 1, 2, \cdots, N\right\}, \quad r_0 > \max\left\{|q_s| \,\Big|\, s = 1, 2, \cdots, M\right\}.$$

If $a(M + N) = 1$ and there are at least two distinct points among those listed in (11.1.4), then the above estimates hold with $b = 2$; if $N = 0$, and all q's are identical, or $M = 0$, and all p's are identical, then in the above estimates, $b = (1-\varepsilon)\sqrt{M}$ or $b = (1-\varepsilon)\sqrt{N}$, respectively, for any $\varepsilon \in (0, 1)$.

Proof. Introduce the comparison function

$$w(x) = C|x|^{-2b}. \tag{11.3.39}$$

Then

$$\Delta w = 4b^2 r^{-2} w, \quad |x| = r > r_0 > 0. \tag{11.3.40}$$

On the other hand, since in (11.2.22), $2a(M + N) < 2$ and $v \to 0$ (as $r \to \infty$), we have $\Delta v^2 > 4b^2 r^{-2} v^2$ for $r > r_0$ when r_0 is sufficiently large. Using this result and (11.3.40), we get the inequality $\Delta(v^2 - w) > 4b^2 r^{-2}(v^2 - w)$, $r > r_0$. For such fixed r_0, we can choose constant C in (11.3.39) large to make $v^2 - w < 0$ at $r = r_0$. Applying the maximum principle and the boundary condition $v = w = 0$ at infinity, we have $v^2 - w < 0$ for $r > r_0$ as claimed.

When $a(M + N) = 1$ and at least two p's or two q's are distinct, the estimates with $b = 2$ comes directly from using the crude inequalities in (11.3.17) and taking the $\delta \to 0$ limit.

In the case $a(M + N) = 1$ and all q's or p's are identical, the estimates with $b = (1 - \varepsilon)\sqrt{M}$ or $b = (1 - \varepsilon)\sqrt{N}$ are already obtained as in (11.3.35) or (11.3.38), respectively.

The proof is complete.

Lemma 11.3.10. *For the solution v of (11.2.22) obtained in the last section, we have for sufficiently large r_0 the decay estimate*

$$|\nabla v| \le C_b |x|^{-b}, \quad |x| > r_0, \tag{11.3.41}$$

where $b > 0$ is again arbitrary if $a(M+N) < 1$ and $C_b > 0$ is a constant. If $a(M+N) = 1$ but there are at least two distinct points among those listed in (11.1.4), the above estimate (11.3.41) holds with $b = 3(1-\varepsilon)$ where $0 < \varepsilon < 1$ is arbitrary; if $M = 0$ and all p's or $N = 0$ and all q's are identical, then $b = (1-\varepsilon)\sqrt{N} + 1$ or $b = (1-\varepsilon)\sqrt{M} + 1$, respectively.

Proof. In the region $\Omega = \{x \,|\, |x| > r_0\}$, we write down the equation (11.2.22),

$$\Delta v =$$
$$2g_0 \left(\frac{e^v}{(1+e^v)^2} \prod_{s=1}^{N} |x - p_s|^{-2} \prod_{s=1}^{M} |x - q_s|^{-2} \right)^{8\pi G} \left(\frac{e^v - 1}{e^v + 1} \right). \quad (11.3.42)$$

Since $v \in L^2(\Omega)$, we have $v \in W^{2,2}(\Omega)$. Besides, differentiating (11.3.42) and using the L^2-estimates for elliptic equations again, we get the result $\partial_j v \in W^{2,2}(\Omega)$, $j = 1, 2$. Consequently, $|\nabla v|$ vanishes at infinity as well. For the function $V = \partial_j v$, (11.3.42) implies that, asymptotically as $|x| = r \to \infty$, there holds

$$\Delta V = H(v)r^{-2a(M+N)}V + \mathrm{O}(r^{-2a(M+N)-b-1}), \quad (11.3.43)$$

where $H(v) \to$ some constant $H_0 > 0$ as $v \to 0$ and the meaning of the exponent $b > 0$ in the tail term above is as given in Lemma 11.3.9. Hence the rest of the argument follows from using (11.3.43) and suitable comparison functions as in the proof of Lemma 11.3.9. In fact, if $a(M+N) < 1$, then the exponent b in (11.3.43) may be arbitrarily large. For the function w defined by

$$w(x) = C|x|^{-b_1}, \quad C > 1, \quad b_1 > 1, \quad (11.3.44)$$

we have

$$\Delta w = b_1^2 r^{-2} w = (b_1^2 + 1)r^{-2}w - Cr^{-2-b_1}$$
$$< H(v)r^{-2a(M+N)}w - Cr^{-2-b_1}, \quad r \geq r_0,$$

where r_0 is sufficiently large. Thus

$$\Delta(V - w) \geq H(v)r^{-2a(M+N)}(V - w) + (Cr^{-2-b_1} + \mathrm{O}(r^{-2a(M+N)-b-1})),$$

where b can be made as large as we please due to Lemma 11.3.9. Thus, for $2a(M+N) + b + 1 > 2 + b_1$, the above inequality gives us

$$\Delta(V - w) > H(v)r^{-2a(M+N)}(V - w), \quad |x| > r_0$$

for a large enough $r_0 > 0$ uniformly with respect to the choice of the constant C in (11.3.44) in view of the uniform bound given in (11.3.17). Therefore, we can let C be large that $V - w \leq 0$ for $|x| = r_0$. Hence, an application of the maximum principle gives $V \leq w$ for $|x| > r_0$.

If $a(M + N) = 1$ and there are at least two distinct p's or q's, we have $b = 2$ in Lemma 11.3.9. Thus, V satisfies

$$\Delta V = H(v)r^{-2}V + O(r^{-5}). \tag{11.3.45}$$

Therefore, we have as before

$$\Delta(V - w) \geq H(v)r^{-2}(V - w) + (Cr^{-2-3(1-\varepsilon)} + O(r^{-5})),$$

where $b_1 = 3(1 - \varepsilon)$ in (11.3.44). Here we have used the fact that $H(v)$ may be made arbitrarily large due to the uniform bound (11.3.17) and the adjustability of g_0. For fixed ε, we can find $r_\varepsilon > 0$ independent of $C \geq 1$ such that $Cr^{-2-3(1-\varepsilon)} + O(r^{-5}) > 0$ for $r > r_\varepsilon$. Namely, $\Delta(V - w) > H(v)r^{-2}(V - w)$. The maximum principle again gives us $V < w$ when C in (11.3.44) is chosen large enough.

On the other hand, for $|x|$ sufficiently and $a(M + N) < 1$, we have

$$\Delta(V + w) \leq H(v)r^{-2a(M+N)}(V + w) - (Cr^{-2-b_1} + O(r^{-2a(M+N)-b-1})),$$

which leads to

$$\Delta(V + w) < H(v)r^{-2a(M+N)}(V + w)$$

for $2 + b_1 < 2a(M + N) + b + 1$; for the case $a(M + N) = 1$ and there at least two distinct p's or q's, we choose $b_1 = 3(1 - \varepsilon)$ again and we have for large g_0 and $r_\varepsilon > 0$,

$$\Delta(V + w) \leq H(v)r^{-2}(V + w) - (Cr^{-2-3(1-\varepsilon)} + O(r^{-5})), \quad |x| > r_\varepsilon,$$

which implies $\Delta(V+w) < H(v)r^{-2}(V+w)$ as before. Consequently, in both cases, the maximum principle gives us $V > -w$ for $|x| > r_\varepsilon$ and sufficiently large C defined in (11.3.44).

The last case, $a(M + N) = 1$, $M = 0$ or $N = 0$ and all p's or q's are identical, is already obtained in (11.3.36) or (11.3.38), respectively.

In the absence of gravity, $\eta = 0$ and $g_0 = 1$. We have stronger decay rate.

Lemma 11.3.11. *Let v be the solution of (11.3.1) obtained in Theorem 11.3.1. Then the exponential decay estimates stated there hold.*

Proof. Linearizing (11.3.1) near $v = 0$ in a neighborhood of infinity, we have $\Delta v = f(x)u$ where $f(x) \to 1$ as $|x| \to \infty$. Thus for any $0 < \varepsilon < 1$ there are constants $C(\varepsilon) > 0$ and $R(\varepsilon) > 0$ such that

$$|v(x)| \leq C(\varepsilon)e^{-(1-\varepsilon)|x|}, \quad |x| > R(\varepsilon). \tag{11.3.46}$$

Next, fix $j = 1, 2$ and set $V = \partial_j v$. Then V satisfies by view of (11.3.1) the equation

$$\Delta V = 2\frac{e^v}{1 + e^v}V - 2\frac{(e^v - 1)e^v}{(1 + e^v)^2}V \tag{11.3.47}$$

away from a sufficiently large local neighborhood. The decay estimate (11.3.46) says that v is an L^2-function. Using this, the fact that the right-hand side of (11.3.1) is L^2 (neglecting a local region), and the L^2-estimates in (11.3.1), we see that $v \in W^{2,2}$ (again, neglecting a local region). Applying this result in (11.3.47) and using the L^2-estimates, we obtain $V \in W^{2,2}$ (in the same sense). In particular, $V \to 0$ as $|x| \to \infty$. Asymptotically, V satisfies $\Delta V = f(x)V$ where $f(x) \to 1$ as $|x| \to \infty$. Hence V also verifies (11.3.46) as expected.

11.4   Quantized Flux, Total Curvature, and Topology

In view of (11.2.5)–(11.2.7), the energy E is written

$$E = \int_{\mathbb{R}^2} \mathcal{H} \, d\Omega_g = \int_{\mathbb{R}^2} (F_{12} + J_{12}), \qquad (11.4.1)$$

where $d\Omega_g = e^\eta d^2 x$ is the canonical surface element of $(\mathbb{R}^2, e^\eta \delta_{jk})$. On the other hand, the equation (11.2.7) indicates that the magnetic flux is given by

$$\Phi = \int_{\mathbb{R}^2} F_{12} = \int_{\mathbb{R}^2} e^\eta \left(\frac{1 - |u|^2}{1 + |u|^2} \right) = \int_{\mathbb{R}^2} e^\eta \left(\frac{1 - e^v}{1 + e^v} \right). \qquad (11.4.2)$$

First we calculate Φ. To this end, we use the substitution $v = -v_{0,1} + v_{0,2} + w$ to transform (11.2.15) into the regular form

$$\Delta w = 2e^\eta \left(\frac{e^{v_{0,2}+w} - e^{v_{0,1}}}{e^{v_{0,2}+w} + e^{v_{0,1}}} \right) - g_1 + g_2. \qquad (11.4.3)$$

Recall that the decay estimates stated in Lemmas 11.3.10 and 11.3.11 imply that, as $|x| \to \infty$, the quantity $|\nabla v|$ vanishes at least according to the rate

$$|\nabla v| = O(r^{-\alpha}), \quad \forall \alpha \in (2, 3). \qquad (11.4.4)$$

Since $|\nabla v_{0,j}| = O(r^{-3})$ $(j = 1, 2)$ asymptotically, we see that $|\nabla w|$ also satisfies (11.4.4). Consequently,

$$\int_{\mathbb{R}^2} \Delta w = \lim_{r \to \infty} \oint_{|x|=r} \frac{\partial w}{\partial n} \, ds = 0. \qquad (11.4.5)$$

Integrating (11.4.3), we find by (11.4.2) and (11.4.5) that

$$\Phi = -\int_{\mathbb{R}^2} e^\eta \left(\frac{e^{v_{0,2}+w} - e^{v_{0,1}}}{e^{v_{0,2}+w} + e^{v_{0,1}}} \right) = -\frac{1}{2} \int_{\mathbb{R}^2} (g_1 - g_2) = 2\pi(M - N). \qquad (11.4.6)$$

We next calculate the flux generated from J_{12} which contributes to the energy as an additional term (see (11.4.1)). By (11.2.1), the current density J_k is regular at the vortex points q's. Thus, we have by the divergence theorem,

$$\int_{\mathbb{R}^2} J_{12} = \lim_{r \to \infty} \oint_{|x|=r} J_k \, dx^k - \lim_{r \to 0} \sum_{\ell=1}^{n} \oint_{|x-p_\ell|=r} J_k \, dx^k$$

$$\equiv \lim_{r \to \infty} I(r) - \lim_{r \to 0} \sum_{\ell=1}^{n} I_\ell(r), \tag{11.4.7}$$

where, only in this section, the integer n denotes the number of distinct antivortex (or antistring) points p_ℓ's with the corresponding multiplicities n_ℓ, $\ell = 1, 2, \cdots, n$, and

$$N = \sum_{\ell=1}^{n} n_\ell \tag{11.4.8}$$

still denotes the total number of antivortices (or antistrings). Note also that, in (11.4.7), all path integrals are taken counter-clockwise.

Using (11.2.17), and the definition of J_k stated in (11.2.1), we have

$$I_\ell(r) = i \oint_{|x-p_\ell|=r} \frac{1}{1+|u|^2} (u \overline{D_k u} - \overline{u} D_k u) \, dx^k$$

$$= i \oint_{|x-p_\ell|=r} \frac{|u|^2}{1+|u|^2} ([\overline{\partial} - \partial]v \, dx^1 - i[\overline{\partial} + \partial]v \, dx^2)$$

$$= \oint_{|x-p_\ell|=r} \frac{e^v}{1+e^v} (-\partial_2 v \, dx^1 + \partial_1 v \, dx^2). \tag{11.4.9}$$

Besides, recall that, near p_ℓ ($\ell = 1, 2, \cdots, n$), v has the local representation

$$v(x) = -n_\ell \ln |x - p_\ell|^2 + w_\ell(x) \tag{11.4.10}$$

where w_ℓ is a smooth function (see also (11.2.13)). Therefore, near $x = p_\ell$, we may write

$$\nabla v = -2n_\ell \frac{x - p_\ell}{|x - p_\ell|^2} + \nabla w_\ell, \quad \ell = 1, 2, \cdots, n. \tag{11.4.11}$$

Inserting (11.4.10) and (11.4.11) into (11.4.9) and taking the $r \to 0$ limit, we obtain immediately

$$\lim_{r \to 0} I_\ell(r) = -4\pi n_\ell, \quad \ell = 1, 2, \cdots, n. \tag{11.4.12}$$

Furthermore, applying the decay estimates for $|\nabla v|$ stated in Lemmas 11.3.10 and 11.3.11, we have the vanishing result for the other path integral,

$$\lim_{r \to \infty} I(r) = \lim_{r \to \infty} \oint_{|x|=r} \frac{e^v}{1+e^v} (-\partial_2 v \, dx^1 + \partial_1 v \, dx^2) = 0. \tag{11.4.13}$$

Inserting (11.4.12) (noting (11.4.8)) and (11.4.13) into (11.4.7), we get

$$\int_{\mathbb{R}^2} J_{12} = 4\pi N, \tag{11.4.14}$$

which only counts the total number of antistrings or antivortices. In view of (11.4.14), we arrive from (11.4.1), (11.4.6), and the Einstein equation (11.2.11) at the expected quantization formulas (11.1.22) and (11.1.25) for the matter-gauge energy and total curvature for the obtained system of M strings and N antistrings. In absence of gravity, the same energy formula holds for a system of M vortices and N antivortices.

Like that in the classical Abelian Higgs model, the magnetic flux Φ generated from M vortices (strings) and N antivortices (antistrings) given by the quantization formula

$$\Phi = \int_{\mathbb{R}^2} F_{12} = 2\pi(M - N), \tag{11.4.15}$$

is topological and resembles the first Chern class which determines only the difference of the numbers of vortices (strings) and antivortices (antistrings). It will be interesting to find another topological quantity which may be used, along with (11.4.15), to determine completely both numbers of vortices (strings) and antivortices (antistrings) simultaneously. To this end, we recall that our model originates from the sigma model, whose solutions are all characterized by the associated Hopf degrees. Hence, it will be natural for us to explore this notion. We will recover a map $\phi : \mathbb{R}^2 \to S^2$ from a solution configuration (u, A) representing M vortices (strings) and N antivortices (antistrings) by using the inverse of the transformation (11.1.8) directly, i.e., namely,

$$\phi_1 = \frac{2}{1 + |u|^2}\text{Re}(u), \quad \phi_2 = \frac{2}{1 + |u|^2}\text{Im}(u), \quad \phi_3 = \frac{1 - |u|^2}{1 + |u|^2}, \tag{11.4.16}$$

and see what implications can be found in terms of Hopf's topological degrees. In fact, it is easy to agree that such a topological invariant is not well defined because u at infinity does not have a definite value in the extended complex plane which means that the map given in (11.4.16) cannot be viewed as an element in the homotopy class $\pi_2(S^2)$. Nevertheless, it may be seen that the integral form of the degree formula (11.1.2) can still yield interesting insight for us. In the rest of this section, we continue to use 'deg(ϕ)' to denote the integral given on the right-hand side of (11.1.2), although we already know that it *no longer* gives us the (Brower) degree of a map ϕ.

To proceed, we obtain after a somewhat lengthy computation, the result

$$\phi \cdot (\partial_1 \phi \times \partial_2 \phi) = \frac{2i}{(1 + |u|^2)^2}(\partial_1 u \partial_2 \overline{u} - \partial_1 \overline{u} \partial_2 u)$$

$$= \frac{2i}{(1+|u|^2)^2}([D_1u\overline{D_2u} - \overline{D_1u}D_2u] - iA_2\partial_1|u|^2 + iA_1\partial_2|u|^2)$$

$$= 2\left(-F_{12}\frac{|u|^2}{1+|u|^2} + i\frac{D_1u\overline{D_2u} - \overline{D_1u}D_2u}{(1+|u|^2)^2}\right)$$

$$+2\left(\frac{1}{(1+|u|^2)^2}(A_2\partial_1|u|^2 - A_1\partial_2|u|^2) + F_{12}\frac{|u|^2}{1+|u|^2}\right)$$

$$= J_{12} + Q_{12}, \tag{11.4.17}$$

where $Q_{12} = \partial_1 Q_2 - \partial_2 Q_1$ and

$$Q_j = 2A_j\frac{|u|^2}{1+|u|^2}, \quad j = 1, 2,$$

which allows us to write in view of the decay estimates stated in Theorem 11.1.1 or 11.1.2 (see also Lemma 11.3.10 or 11.3.11) the relation

$$Q_j = A_j - A_j\frac{1-|u|^2}{1+|u|^2} = A_j + O(|x|^{-\alpha}) \quad \text{as } |x| \to \infty, \quad j = 1, 2. \tag{11.4.18}$$

Note that, in (11.4.18), $\alpha \geq 2$ at least. Therefore, again by the divergence theorem, we have

$$\int_{\mathbb{R}^2} Q_{12} = \lim_{r \to \infty} \oint_{|x|=r} A_j \, dx^j = \int_{\mathbb{R}^2} F_{12} = 2\pi(M - N). \tag{11.4.19}$$

As a consequence of (11.4.17) and (11.4.19), we see that the degree formula (11.1.2) gives us

$$\deg(\phi) = \frac{1}{4\pi}\int_{\mathbb{R}^2}\phi\cdot(\partial_1\phi \times \partial_2\phi)$$

$$= \frac{1}{4\pi}\int_{\mathbb{R}^2}(F_{12} + J_{12})$$

$$= \frac{1}{2}(M + N) = \frac{1}{4\pi}(\text{energy } E), \tag{11.4.20}$$

which takes half-integer values unless $M = N$ mod(2). Here again the numbers of vortices (strings) and antivortices (antistrings) play equal parts.

In general, (11.4.20) implies the energy inequality $E(u, A) \geq 4\pi|\deg(\phi)|$ in place of the classical one, (11.1.3).

The obtained quantization formulas for flux, energy, and total curvature indicate a perfect symmetry between vortices (strings) and antivortices (antistrings). Here we observe an interesting phenomenon that such a symmetry can be broken by an external field, no matter how weak it appears to be. To see this, we switch on a constant magnetic field along the x_3 axis, say, $B = (0, 0, H)$ and consider the model in absence of gravity. The energy density is now

$$\mathcal{H}' = \mathcal{H} - F_{12}H$$

$$= \frac{1}{2}F_{12}^2 - F_{12}H + \frac{2}{(1 + |u|^2)^2}(|D_1u|^2 + |D_2u|^2) + \frac{1}{2}\left(\frac{1 - |u|^2}{1 + |u|^2}\right)^2.$$

$$(11.4.21)$$

Since H is constant, the equations of motion of (11.4.21) are the same as those without the H term. Hence, the energy minimizers are those of the solutions of the self-dual equations (11.2.6) and (11.2.7) with $\eta \equiv 0$ and the energy of a solution representing M vortices and N antivortices is given by

$$E' = \int_{\mathbb{R}^2} \mathcal{H}' \, d\Omega_\eta = 2\pi M(1 - H) + 2\pi N(1 + H). \qquad (11.4.22)$$

If $0 < H < 1$ (subcritical), the expression (11.4.22) says that vortices are energetically preferred over antivortices, i.e., $N = 0$ yields a lower energy; similarly, if $-1 < H < 0$, antivortices are preferred over vortices. Consequently, in either case, the excited magnetic field F_{12} chooses to be aligned everywhere along the same direction of the applied field B. In other words, no matter how weak the external field is, its presence breaks the symmetry between vortices and antivortices.

For strings, a similar argument leads to the same symmetry-breaking mechanism.

The proofs of Theorems 11.1.1 and 11.1.2 are now complete.

In conclusion, we have obtained the coexistence of a system of vortices (strings) and antivortices (antistrings) in an Abelian gauge field model (coupled with Einstein's gravity) which originates naturally from a gauged sigma model with broken symmetry. The induced magnetic flux is proportional to the difference of the numbers of vortices (strings) and antivortices (antistrings) but the energy (as well as the total curvature in presence of gravity) is proportional to the sum of these numbers. In absence of an external field, vortices (strings) and antivortices (antistrings) are indistinguishable either energetically or magnetically. However, the presence of a weak external field breaks such a symmetry so that only one of the two types of the vortices (strings) is preferred in order to comply with the applied field.

11.5   Unique Solutions on Compact Surfaces

In this section we construct coexisting vortices and antivortices on a compact Riemann surface. We first formulate the problem in the framework of a complex line bundle. We next consider the number count for the energy and flux in terms of vortices and antivortices parallel to the planar solution case in the previous sections. We then use a fixed-point theory argument to establish an existence theorem under a necessary and sufficient condition.

11.5.1 Formulation on line bundles

Let (ξ, h) be a complex line bundle over a compact Riemann surface (S, g) where h is a Hermitian metric on ξ and g is a Riemannian metric on S. We choose an atlas $\{(U_\alpha, e_\alpha)\}$ which trivializes ξ so that $h(e_\alpha, e_\alpha) = 1$ for all α. Thus, if u is a section of $\xi \to S$ with $u = u_\alpha e_\alpha$, we have $h(u, u) = |u_\alpha|^2$, which is a local-chart-independent scalar field and may conveniently be written as $|u|^2$. We use $A = (A_j)$ $(j = 1, 2)$ to denote a (real) connection 1-form, $D_A u = du - iAu = (D_j u) = (\partial_j u - iA_j u)dx^j$ $(j = 1, 2)$ the connection, and $F = F_A = dA = \frac{1}{2}F_{jk}dx^j \wedge dx^k$ $(j, k = 1, 2)$ its induced curvature. Then the energy density of our Abelian gauge field theory still reads

$$\mathcal{H} = \frac{1}{4}g^{jj'}g^{kk'}F_{jk}F_{j'k'} + \frac{2}{(1+|u|^2)^2}g^{jk}D_j u\overline{D_k u} + \frac{1}{2}\left(\frac{1-|u|^2}{1+|u|^2}\right)^2. \quad (11.5.1)$$

The equations of motion of (11.5.1) are

$$\frac{1}{\sqrt{|g|}}D_j\left(\frac{\sqrt{|g|}g^{jk}}{(1+|u|^2)^2}D_k u\right) = \frac{(|u|^2 - 1 - 2g^{jk}D_j u\overline{D_k u})}{(1+|u|^2)^3}u, \quad (11.5.2)$$

$$\frac{1}{\sqrt{|g|}}\partial_{j'}(g^{jk}g^{j'k'}\sqrt{|g|}F_{kk'}) = \frac{2}{(1+|u|^2)^2}ig^{jk}(u\overline{D_k u} - \overline{u}D_k u). \quad (11.5.3)$$

We will again find a first integral of the system of equations, (11.5.2) and (11.5.3). For this purpose, we use as before ϵ_{jk} to denote the standard Levi-Civita skewsymmetric 2-tensor satisfying $\epsilon_{12} = \sqrt{g}$. Then, after a tedious calculation, we can rewrite the energy (11.5.1) into the form

$$\begin{aligned}
\mathcal{H} = {} & \pm\frac{1}{2}\epsilon^{jk}F_{jk} \pm \frac{1}{2}\epsilon^{jk}J_{jk} \\
& + \frac{1}{4}g^{jj'}g^{kk'}\left(F_{jk} \mp \epsilon_{jk}\frac{1-|u|^2}{1+|u|^2}\right)\left(F_{j'k'} \mp \epsilon_{j'k'}\frac{1-|u|^2}{1+|u|^2}\right) \\
& + \frac{1}{(1+|u|^2)^2}g^{jk}(D_j u \pm i\epsilon_j^{j'}D_{j'}u)\overline{(D_k u \pm i\epsilon_k^{k'}D_{k'}u)}, \quad (11.5.4)
\end{aligned}$$

where the new 'vorticity' field J_{jk} $(j, k = 1, 2)$ is defined by

$$\begin{aligned}
J_{jk} &= (2dJ)_{jk} = -\frac{2|u|^2}{1+|u|^2}F_{jk} + 2i\frac{D_j u\overline{D_k u} - \overline{D_j u}D_k u}{(1+|u|^2)^2}, \\
(J)_k &= J_k = \frac{i}{1+|u|^2}(u\overline{D_k u} - \overline{u}D_k u). \quad (11.5.5)
\end{aligned}$$

Use $d\Omega_g$ to denote the canonical area element of the surface (S, g). Then (11.5.4) implies the energy lower bound

$$E(A, u) = \int_S \mathcal{H} \, d\Omega_g \geq \frac{1}{2} \left| \int_S \epsilon^{jk} F_{jk} \, d\Omega_g + \int_S \epsilon^{jk} J_{jk} \, d\Omega_g \right|. \qquad (11.5.6)$$

The right-hand side of (11.5.6) is a topological invariant. Hence, such a lower bound is saturated if and only if the configuration pair (A, u) satisfies the self-dual or anti-self-dual equations

$$D_j u = \mp i \epsilon_j^k D_k u, \qquad (11.5.7)$$

$$\epsilon^{jk} F_{jk} = \pm 2 \frac{1 - |u|^2}{1 + |u|^2}. \qquad (11.5.8)$$

It is straightforward to examine that the system (11.5.7) and (11.5.8) implies (11.5.2) and (11.5.3). It is clear that the self-dual case and anti-self-dual case are related through the correspondence

$$(A, u) \rightleftharpoons (-A, \overline{u}).$$

In the rest of this section, we concentrate on the self-dual case.

Solutions of the equation (11.5.7) are sections u which are meromorphic with respect to the connection D_A. Suppose that the set of zeros and the set of poles of u are

$$\left\{ q_1, q_2, \cdots, q_M \right\} \quad \text{and} \quad \left\{ p_1, p_2, \cdots, p_N \right\}, \qquad (11.5.9)$$

where arbitrary multiplicities of points are allowed to accommodate possible higher order zeros and poles. Consequently, if we use

$$B = \frac{1}{2} \epsilon^{jk} F_{jk}$$

to represent a 'magnetic' field across the surface S, then (11.5.8) gives us opposite magnetic orientations at the zeros and poles,

$$B(q_s) = 1, \quad s = 1, 2, \cdots, M; \quad B(p_s) = -1, \quad s = 1, 2, \cdots, N.$$

In other words, when B is interpreted as a vorticity field, the points q's and p's represent vortices and antivortices, respectively.

In the next subsection, we show by a direct calculation that the numbers of vortices and antivortices can again be used to determine global physical quantities such as magnetic flux and energy.

11.5.2   Number count

With the prescribed data (11.5.9) and the substitution $v = \ln |u|^2$, it can easily be shown that the system of equations, (11.5.7) and (11.5.8), becomes

$$\Delta v = 2\left(\frac{e^v - 1}{e^v + 1}\right) + 4\pi \sum_{s=1}^{M} \delta_{q_s} - 4\pi \sum_{s=1}^{N} \delta_{p_s}, \tag{11.5.10}$$

where δ_p is the Dirac distribution concentrated at the point $p \in (S, g)$ and Δ is the Laplace–Beltrami operator on (S, g) defined by

$$\Delta v = \frac{1}{\sqrt{g}} \partial_j (g^{jk} \sqrt{g} \partial_k v).$$

We use $|S|$ to denote the total volume (area) of (S, g). Then there are functions v'_0 and v''_0 which satisfy

$$\Delta v'_0 = -\frac{4\pi M}{|S|} + 4\pi \sum_{s=1}^{M} \delta_{q_s}, \quad \Delta v''_0 = -\frac{4\pi N}{|S|} + 4\pi \sum_{s=1}^{N} \delta_{p_s}. \tag{11.5.11}$$

From (11.5.10), (11.5.11), and the substitution $v = v'_0 - v''_0 + w$, we have

$$\Delta w = 2\left(\frac{e^{v'_0 + w} - e^{v''_0}}{e^{v'_0 + w} + e^{v''_0}}\right) + \frac{4\pi}{|S|}(M - N), \tag{11.5.12}$$

which is now in a regular (singularity-free) form.

Let w be a solution of (11.5.12). Then $v = v'_0 - v''_0 + w$ solves (11.5.10). As in the Abelian Higgs model, we can use such a v to recover a solution pair (A, u) of the self-dual system (11.5.7) and (11.5.8). Thus, from (11.5.8), we can calculate the total magnetic flux

$$\begin{aligned}
\Phi &= \frac{1}{2}\int_S \epsilon^{jk} F_{jk} \, d\Omega_g = \int_S \frac{1 - |u|^2}{1 + |u|^2} \, d\Omega_g \\
&= \int_S \frac{1 - e^v}{1 + e^v} \, d\Omega_g = \int_S \frac{1 - e^{v'_0 - v''_0 + w}}{1 + e^{v'_0 - v''_0 + w}} \, d\Omega_g \\
&= \int_S \frac{e^{v''_0} - e^{v'_0 + w}}{e^{v''_0} + e^{v'_0 + w}} \, d\Omega_g = 2\pi(M - N). \tag{11.5.13}
\end{aligned}$$

We next calculate the flux induced from the new current density J_k. From (11.5.5) and the divergence theorem, we see that we only need to concentrate on a neighborhood of the points $p_1, p_2, \cdots, p_N$. For convenience, we use positively oriented isothermal coordinates near these points. Then we have in view of the similar calculation carried out in §11.4 that

$$\frac{1}{2}\int_S \epsilon^{jk} J_{jk} \, d\Omega_g = \sum_{s=1}^{N} \lim_{r \to 0} \oint_{\partial B(p_s; r)} J_k \, dx^k = 4\pi N, \tag{11.5.14}$$

where $B(p_s; r)$ is the circular region centered at p_s with radius $r > 0$, under the isothermal coordinate representation in $\mathbb{R}^2$, and the path integrals are all taken clockwise.

Therefore, from (11.5.13) and (11.5.14), we see in view of (11.5.6) that the minimum energy is indeed represented correctly in terms of the number of vortices, M, and the number of antivortices, N, by the formula

$$E = 2\pi(M + N). \tag{11.5.15}$$

11.5.3 Solution and fixed-point method

In this subsection, we obtain *all* possible self-dual solutions realizing the topological minimum energy levels we have just described. The fact that the problem is formulated over a closed surface S introduces a few additional difficulties if one tries to obtain an iterative existence proof as in the last section for the problem over $\mathbb{R}^2$. Here we shall use the compactness of S to solve the problem by a fixed-point method via the Leray–Schauder theorem.

We first derive a necessary condition for the solvability of (11.5.12). For this purpose, we rewrite (11.5.12) as

$$\Delta w = \frac{4e^{v_0' + w}}{e^{v_0' + w} + e^{v_0''}} - C_0, \quad C_0 = 2 - \frac{4\pi}{|S|}(M - N). \tag{11.5.16}$$

An integration of the above equation leads to the necessary condition $0 < C_0 < 4$, or equivalently,

$$|M - N| < \frac{|S|}{2\pi}. \tag{11.5.17}$$

Throughout this section, we shall assume that (11.5.17) is fulfilled.

We now proceed to prove that (11.5.17) is also sufficient for the existence of a solution to the equation (11.5.16). We will use a fixed-point argument over the Sobolev space $W^{1,2}(S)$ (the space of L^2-functions on S, with L^2 distributional derivatives, equipped with the standard inner product). Consider a proper subspace of $W^{1,2}(S)$ defined by

$$X = \left\{ w \in W^{1,2}(S) \,\middle|\, \int_S w \, d\Omega_g = 0 \right\}$$

which is the orthogonal complement of the set of constant functions in $W^{1,2}(S)$. Of course, we have the direct sum $W^{1,2}(S) = \mathbb{R} \oplus X$ and the Poincaré inequality

$$\int_S w^2 \, d\Omega_g \leq c_0 \int_S g^{jk} \partial_j w \partial_k w \, d\Omega_g \equiv c_0 \int_S |\nabla w|^2 \, d\Omega_g, \quad w \in X, \tag{11.5.18}$$

where $c_0 > 0$ is a constant. Besides, we also recall the following Trudinger–Moser inequality,

$$\int_S e^w \, d\Omega_g \leq c_1 \exp\left(c_2 \int_S |\nabla w|^2 \, d\Omega_g\right), \quad w \in X, \qquad (11.5.19)$$

where the ranges of the positive constants c_1 and c_2 are of no concern in our treatment of the problem.

We now introduce an operator $T : X \to X$ so that its fixed point is a solution of the governing equation (11.5.16).

Lemma 11.5.1. *For given $w \in X$, there is a unique number $c(w) \in \mathbb{R}$ so that*

$$\int_S \frac{4e^{v_0' + c(w) + w}}{e^{v_0' + c(w) + w} + e^{v_0''}} \, d\Omega_g = C_0 |S|. \qquad (11.5.20)$$

Proof. Consider the function

$$f(t) = \int_S \frac{4e^{v_0' + t + w}}{e^{v_0' + t + w} + e^{v_0''}} \, d\Omega_g.$$

Then it is easily seen that

$$\lim_{t \to -\infty} f(t) = 0, \quad \lim_{t \to \infty} f(t) = 4|S|.$$

Using (11.5.17) (namely, $0 < C_0 < 4$) and the continuity of $f(\cdot)$, we see that there is a point, $t_w : -\infty < t_w < \infty$, such that $f(t_w) = C_0 |S|$. On the other hand, since

$$\frac{df}{dt}(t) = \int_S \frac{4e^{v_0' + v_0'' + t + w}}{(e^{v_0' + t + w} + e^{v_0''})^2} \, d\Omega_g > 0,$$

we conclude that such a t_w must be unique as claimed.

Lemma 11.5.2. *For given $w \in X$, let $c(w)$ be defined as in Lemma 11.5.1. Then, viewed as a function, $c : X \to \mathbb{R}$ is continuous with respect to the weak topology of X.*

Proof. Let $\{w_n\}$ be a weakly convergent sequence in X such that $w_n \rightharpoonup w_0 \in X$ weakly as $n \to \infty$. The compact embedding $W^{1,2}(S) \to L^p(S)$ for $p \geq 1$ indicates that $w_n \to w_0$ strongly in $L^p(S)$ ($\forall p \geq 1$). We need to show that $c(w_n) \to c(w_0)$ as $n \to \infty$. Without loss of generality, we may assume otherwise that there is some number $\varepsilon_0 > 0$ such that

$$|c(w_n) - c(w_0)| \geq \varepsilon_0 > 0, \quad \forall n. \qquad (11.5.21)$$

We will then show after a few steps that (11.5.21) leads to a contradiction.

(i) The sequence $\{c(w_n)\}$ is bounded from above.

Otherwise, going to a subsequence if necessary, we may assume that $c(w_n) \to \infty$ as $n \to \infty$. The strong convergence $w_n \to w_0$ in $L^p(S)$ and the Egorov theorem imply that for any $\varepsilon > 0$ there is a sufficiently large number $K_\varepsilon > 0$ and a subset $S_\varepsilon \subset S$ such that

$$|w_n(x)| \leq K_\varepsilon, \quad x \in S - S_\varepsilon; \quad |S_\varepsilon| < \varepsilon.$$

Hence, we have

$$
\begin{aligned}
C_0|S| &= \int_S \frac{4e^{v_0' + c(w_n) + w_n}}{e^{v_0' + c(w_n) + w_n} + e^{v_0''}} \, d\Omega_g \\
&\geq \int_{S-S_\varepsilon} \frac{4e^{v_0' + c(w_n) + w_n}}{e^{v_0' + c(w_n) + w_n} + e^{v_0''}} \, d\Omega_g \\
&\geq \int_{S-S_\varepsilon} \frac{4e^{v_0' + c(w_n) - K_\varepsilon}}{e^{v_0' + c(w_n) - K_\varepsilon} + e^{v_0''}} \, d\Omega_g.
\end{aligned}
\tag{11.5.22}
$$

Letting $n \to \infty$ on the right-hand side of the above inequality, we arrive at

$$C_0|S| > 4(|S| - \varepsilon).$$

Since $\varepsilon > 0$ is arbitrary, we see that the condition $C_0 < 4$ is violated.

(ii) The sequence $\{c(w_n)\}$ is bounded from below.

For any $\varepsilon > 0$, let S_ε be a neighborhood of the points $p_1, p_2, \cdots, p_N$ so that $p_s \in S_\varepsilon$ $(\forall s)$ and $|S_\varepsilon| < \varepsilon$.

Note that the boundedness of $\{w_n\}$ in X and (11.5.19) imply that

$$\sup_n \int_S e^{w_n} \, d\Omega_g \leq C < \infty. \tag{11.5.23}$$

Therefore, using the equality statement in (11.5.22), and (11.5.23), we have

$$
\begin{aligned}
C_0|S| &\leq \int_{S-S_\varepsilon} \frac{4e^{v_0' + c(w_n) + w_n}}{e^{v_0''}} \, d\Omega_g + 4 \int_{S_\varepsilon} d\Omega_g \\
&\leq 4 \left(\sup_{S-S_\varepsilon} e^{-v_0''} \right) \int_{S-S_\varepsilon} e^{v_0' + c(w_n) + w_n} \, d\Omega_g + 4 \int_{S_\varepsilon} d\Omega_g.
\end{aligned}
$$

To proceed further from the above estimate, we assume that ε satisfies $4\varepsilon < C_0|S|$. Hence, we are led from (11.5.23) to the lower boundedness of the sequence $\{c(w_n)\}$ as claimed,

$$c(w_n) \geq \ln(C_0|S| - 4\varepsilon) - \ln\left(\int_S e^{w_n} \, d\Omega_g \right) - C_1 \geq -C_2 > -\infty.$$

(iii) Since $\{c(w_n)\}$ is bounded, by going to a subsequence if necessary, we may assume that $c(w_n) \to$ some $\tilde{c} \in \mathbb{R}$ as $n \to \infty$. The assumption (11.5.21) gives us

$$\tilde{c} \neq c(w_0). \tag{11.5.24}$$

On the other hand, since there is a constant $\varepsilon_0 > 0$ such that $e^{v_0''(x)} \geq \varepsilon_0$ for all $x \in S - S_\varepsilon$, we have for some $\tilde{w}_n$ lying between $\tilde{c} + w_0$ and $c(w_n) + w_n$, the estimate

$$
\begin{aligned}
\alpha &\equiv \left| \int_S \frac{4e^{v_0' + \tilde{c} + w_0}}{e^{v_0' + \tilde{c} + w_0} + e^{v_0''}} \, d\Omega_g - C_0 |S| \right| \\
&= \left| \int_S \frac{4e^{v_0' + \tilde{c} + w_0}}{e^{v_0' + \tilde{c} + w_0} + e^{v_0''}} \, d\Omega_g - \int_S \frac{4e^{v_0' + c(w_n) + w_n}}{e^{v_0' + c(w_n) + w_n} + e^{v_0''}} \, d\Omega_g \right| \\
&\leq 8|S_\varepsilon| + 4 \int_{S - S_\varepsilon} \left| \frac{e^{v_0' + \tilde{c} + w_0}}{e^{v_0' + \tilde{c} + w_0} + e^{v_0''}} - \frac{e^{v_0' + c(w_n) + w_n}}{e^{v_0' + c(w_n) + w_n} + e^{v_0''}} \right| d\Omega_g \\
&< 8\varepsilon + 4 \int_{S - S_\varepsilon} \frac{e^{v_0' + v_0'' + \tilde{w}_n}}{(e^{v_0' + \tilde{w}_n} + e^{v_0''})^2} |(\tilde{c} + w_0) - (c(w_n) + w_n)| \, d\Omega_g \\
&< 8\varepsilon + C_\varepsilon \int_S e^{|w_0| + |w_n|} (|\tilde{c} - c(w_n)| + |w_0 - w_n|) \, d\Omega_g.
\end{aligned}
$$

Applying the Schwartz inequality and (11.5.19), we see that the integral on the right-hand side of the above goes to zero as $n \to \infty$. Hence, $\alpha \leq 8\varepsilon$. But $\varepsilon > 0$ is arbitrary, we obtain $\alpha = 0$. Therefore, in view of Lemma 11.5.1, we must have $\tilde{c} = c(w_0)$, contradicting (11.5.24).

The proof of the lemma thus follows.

We can now formulate a fixed-point theory argument to prove the existence of a solution of the equation (11.5.16).

For each given $w \in X$, consider the equation

$$
\Delta W = \frac{4e^{v_0' + c(w) + w}}{e^{v_0' + c(w) + w} + e^{v_0''}} - C_0. \tag{11.5.25}
$$

By (11.5.20), we see that the right-hand side of (11.5.25) has zero average value on S. Therefore the equation (11.5.25) has a unique solution $W \in X$. This correspondence, $w \to W$, gives us a well-defined operator T that maps X into itself, $W = T(w)$.

Lemma 11.5.3. *The operator $T : X \to X$ is completely continuous.*

Proof. Let $w_n \to w_0$ weakly in X as $n \to \infty$. Then $w_n \to w_0$ strongly in any $L^p(S)$ ($p \geq 1$). Set $W_n = T(w_n)$ and $W_0 = T(w_0)$. Then

$$
\Delta(W_n - W_0) = \frac{4e^{v_0' + c(w_n) + w_n}}{e^{v_0' + c(w_n) + w_n} + e^{v_0''}} - \frac{4e^{v_0' + c(w_0) + w_0}}{e^{v_0' + c(w_0) + w_0} + e^{v_0''}}. \tag{11.5.26}
$$

Multiplying (11.5.26) by $W_n - W_0$ and integrating by parts, we obtain

$$
\int_S |\nabla(W_n - W_0)|^2 \, d\Omega_g
$$

$$\leq \int_S \frac{4e^{v_0' + v_0'' + \tilde{c}_n + \tilde{w}_n}}{(e^{v_0' + \tilde{c}_n + \tilde{w}_n} + e^{v_0''})^2} (|c(w_n) - c(w_0)| + |w_n - w_0|) |W_n - W_0| \, d\Omega_g$$

(where $\tilde{c}_n$ and $\tilde{w}_n$ lie between $c(w_n), c(w_0)$ and w_n, w_0, respectively)

$$\leq 2|c(w_n) - c(w_0)| \, \|S\|^{\frac{1}{2}} \|W_n - W_0\|_2 + 2\|w_n - w_0\|_2 \|W_n - W_0\|_2.$$

Applying the Poincaré inequality (11.5.18) and a suitable interpolation, we get from the above the estimate

$$\int_S |\nabla(W_n - W_0)|^2 \, d\Omega_g \leq C_1 (|c(w_n) - c(w_0)|^2 + \|w_n - w_0\|_2^2),$$

which proves that $W_n \to W_0$ strongly in X and the lemma follows.

We now study the fixed-point equation labelled by a parameter t,

$$w_t = t \, T(w_t), \quad 0 \leq t \leq 1. \tag{11.5.27}$$

Lemma 11.5.4. *There is a constant $C > 0$ independent of $t \in [0, 1]$ so that*

$$\|w_t\|_X \leq C, \quad 0 \leq t \leq 1. \tag{11.5.28}$$

Consequently, T has a fixed point in X.

Proof. Of course, for $t > 0$, the function w_t satisfies the equation

$$\Delta w_t = 4t \frac{e^{v_0' + c(w_t) + w_t}}{e^{v_0' + c(w_t) + w_t} + e^{v_0''}} - tC_0. \tag{11.5.29}$$

Multiplying (11.5.29) by w_t and integrating by parts, we have

$$\int_S |\nabla w_t|^2 \, d\Omega_g \leq 4 \int_S |w_t| \, d\Omega_g.$$

In view of (11.5.18) and the above inequality, we see immediately that (11.5.28) holds.

The existence of a fixed point is a consequence of Lemma 11.5.3, the apriori estimate (11.5.28), and the Leray–Schauder theory. See Gilbarg and Trudinger [123].

Lemma 11.5.5. *The elliptic governing equation (11.5.12) or (11.5.16) has a solution if and only if the condition (11.5.17) is valid. Furthermore, if there exists a solution, it must be unique.*

Proof. The existence part follows from the fact that, if $\tilde{w}$ is a fixed of T, then $w = c(\tilde{w}) + \tilde{w}$ is a solution of (11.5.16). The uniqueness part follows from the monotonicity of the right-hand side of the equation (11.5.16) as a function of w.

We note that the above existence result may also be proved by a variational method. However, it appears to us that such an approach is less as elegant and greatly complicates the analysis.

We can now summarize our results obtained in this section as follows.

Theorem 11.5.6. *Let $\xi \to S$ be a complex line bundle over a closed Riemann surface S Then the Abelian gauge field energy*

$$E(A, u) = \int_S \mathcal{H}(A, u) \, d\Omega_g$$

with $\mathcal{H}$ defined in (11.5.1) has the topological lower bound

$$E(A, u) \geq 2\pi|\tau_1 + \tau_2|, \tag{11.5.30}$$

where the topological invariants τ_1 and τ_2 are represented by

$$\tau_1 = \frac{1}{4\pi} \int_S \epsilon^{jk} F_{jk} \, d\Omega_g, \tag{11.5.31}$$

$$\tau_2 = \frac{1}{4\pi} \int_S \epsilon^{jk} J_{jk} \, d\Omega_g. \tag{11.5.32}$$

Furthermore, if a section u has M zeros and N poles, then the flux formulas

$$\tau_1 = M - N, \quad \tau_2 = 2N$$

hold, and for any prescribed M zeros and N poles, counting possible multiplicities, stated in (11.5.9), there exists an energy-minimizing pair (A, u) realizing these zeros and poles and the quantized energy minimum,

$$E(A, u) = 2\pi|\tau_1 + \tau_2| = 2\pi(M + N), \tag{11.5.33}$$

if and only if the condition

$$|M - N| < \frac{|S|}{2\pi} \tag{11.5.34}$$

is satisfied. Finally, the pair (A, u) is uniquely determined by the prescribed zeros and poles given in (11.5.9).

In fact the topological invariant τ_1 is the first Chern class, $c_1(\xi)$, of the line bundle $\xi \to S$. On the other hand, it may be shown [287] that the topological invariant τ_2 corresponds to the Thom class [49] of the associated bundle ξ^*.

A few additional comments are in order.

First, it will be interesting to compare our results here with those of the classical Abelian Higgs model, also framed on a complex line bundle $\xi \to S$, and defined by the energy functional

$$\mathcal{H} = \frac{1}{4} g^{jj'} g^{kk'} F_{jk} F_{j'k'} + \frac{1}{2} g^{jk} D_j u \overline{D_k u} + \frac{1}{8}(1 - |u|^2)^2. \tag{11.5.35}$$

In this case the section u of a finite-energy field configuration pair (A, u) can only have M zeros, $|c_1(\xi)| = M$, and there holds the quantized energy lower bound

$$E(A, u) \geq \pi |c_1(\xi)| = \pi M.$$

This lower bound is saturated if and only if [51, 231, 232]

$$|c_1(\xi)| = M < \frac{|S|}{4\pi}, \tag{11.5.36}$$

which implies that the number of vortices has an upper limit. On the other hand, (11.5.34) does not give a restriction on either M or N but on their difference $|M - N|$. Since $c_1(\xi) = M - N$ is a measure of the non-triviality of ξ, (11.5.34) defines a range in which the bundle may be non-trivial. In this sense, (11.5.34) and (11.5.36) are similar.

Secondly, in the limiting case $c_1(\xi) = 0$, we have $\xi = \mathbb{C} \times S$ and we may set $A = 0$ in the model (11.5.1) and drop the potential term, which reduces the energy density into the form

$$
\begin{aligned}
\mathcal{H}(u) &= \frac{2}{(1+|u|^2)^2} g^{jk} \partial_j u \partial_k \overline{u} \\
&= \pm \frac{1}{2} \epsilon^{jk} J_{jk} + \frac{1}{(1+|u|^2)^2} g^{jk} (\partial_j u \pm \mathrm{i}\epsilon_j^{j'} \partial_{j'} u) \overline{(\partial_k u \pm \mathrm{i}\epsilon_k^{k'} \partial_{k'} u)},
\end{aligned}
\tag{11.5.37}
$$

where, now,

$$J_{jk} = (2\mathrm{d}J)_{jk} = 2\mathrm{i} \frac{\partial_j u \partial_k \overline{u} - \partial_j \overline{u} \partial_k u}{(1+|u|^2)^2}, \quad j, k = 1, 2,$$

$$(J)_k = J_k = \frac{\mathrm{i}}{1+|u|^2} (u \partial_k \overline{u} - \overline{u} \partial_k u), \quad k = 1, 2.$$

As before, we have the quantized integral,

$$\frac{1}{2} \int_S \epsilon^{jk} J_{jk} \, \mathrm{d}\Omega_g = \pm 4\pi N, \tag{11.5.38}$$

where N is the number of poles of the complex function u over S and the $\pm$ sign is determined by the orientation of S. Inserting (11.5.38) into (11.5.37), we obtain

$$E(u) = \int_S \mathcal{H}(u) \, \mathrm{d}\Omega_g \geq 4\pi N,$$

and this lower bound is saturated if and only u (or $\overline{u}$) is a meromorphic function over S with N poles and (necessarily) *identical* number of zeros. Since now $E = 2\pi(N + N)$, we recover (11.5.33).

Thirdly, the symmetry between M vortices and N antivortices as stated in Theorem 11.5.6 will be broken when the model (11.5.1) is replaced by the one with a symmetric vacuum, $u = 0, A = 0$, defined by

$$\mathcal{H} = \frac{1}{4}g^{jj'}g^{kk'}F_{jk}F_{j'k'} + \frac{2}{(1+|u|^2)^2}g^{jk}D_ju\overline{D_ku} + \frac{2|u|^4}{(1+|u|^2)^2}. \quad (11.5.39)$$

As before, we can rewrite (11.5.39) in the form

$$\begin{aligned}
\mathcal{H} &= \pm\frac{1}{2}\epsilon^{jk}J_{jk} \\
&\quad + \frac{1}{4}g^{jj'}g^{kk'}\left(F_{jk} \pm 2\epsilon_{jk}\frac{|u|^2}{1+|u|^2}\right)\left(F_{j'k'} \pm 2\epsilon_{j'k'}\frac{|u|^2}{1+|u|^2}\right) \\
&\quad + \frac{1}{(1+|u|^2)^2}g^{jk}(D_ju \pm i\epsilon_j^{j'}D_{j'}u)\overline{(D_ku \pm i\epsilon_k^{k'}D_{k'}u)}, \quad (11.5.40)
\end{aligned}$$

which implies the topological lower bound

$$E(A, u) \geq \frac{1}{2}\left|\int_S \epsilon^{jk}J_{jk}\, d\Omega_g\right|$$

and the self-dual (anti-self-dual) equations

$$\begin{aligned}
D_ju &= \mp i\epsilon_j^k D_ku, & (11.5.41) \\
\epsilon^{jk}F_{jk} &= \mp\frac{4|u|^2}{1+|u|^2}. & (11.5.42)
\end{aligned}$$

Again, we are interested in solutions of (11.5.41) and (11.5.42) with designated zeros and poles stated in (11.5.9). The governing equation becomes

$$\Delta v = \frac{4e^v}{e^v + 1} + 4\pi\sum_{s=1}^{M}\delta_{q_s} - 4\pi\sum_{s=1}^{N}\delta_{p_s}. \quad (11.5.43)$$

The equation (11.5.43) is similar to (11.5.10). Thus we can establish the following result.

Theorem 11.5.7. *Let the energy E be defined in (11.5.39). Then there holds the topological energy lower bound,*

$$E(A, u) \geq 2\pi|\tau_2| = 4\pi N,$$

and, this lower bound is saturated if and only if

$$0 < \tau_1 = c_1(\xi) < \frac{|S|}{\pi}.$$

In other words, the quantized minimum energy level $E(A, u) = 4\pi N$ is realized by a solution of (11.5.41) and (11.5.42), with M zeros and N poles given in (11.5.9) if and only if

$$0 < N - M < \frac{|S|}{\pi} \quad (11.5.44)$$

is satisfied. Moreover, for the prescribed zeros and poles, the solution is unique.

Comparing Theorem 11.5.6 with Theorem 11.5.7, we see that in the model (11.5.39) vortices and antivortices play rather uneven roles. In particular, only antivortices make contributions to the total energy but vortices are phantoms.

11.6   Remarks

In this chapter, we have developed a field theory for coexisting vortices and antivortices and cosmic strings and antistrings with opposite magnetic properties, a theory originally proposed by Schroers [275] and elaborated in more detail in [352, 353]. It will be interesting to extend this theory to include coexisting vortices and antivortices and strings and antistrings of opposite electric properties.

Open Problem 11.6.1.   *Develop a gauge field theory to allow the coexistence of static vortices and antivortices of opposite electric charges, both locally and globally.*

It is not clear whether the theory may be extended to produce coexisting monopoles and antimonopoles.

Open Problem 11.6.2.   *Develop a gauge field theory to allow the coexistence of static monopoles and antimonopoles of opposite magnetic or electric charges, both locally and globally.*

It will also be interesting to study coexisting cosmic strings and antistrings on a compact gravitational surface as was done for the Einstein and Abelian Higgs equations in Chapter 10.

12
Born–Infeld Solutions

In this chapter we study field equations arising from the classical Born–Infeld electromagnetic theory, a topic of current research activities in theoretical physics. In §12.1, we introduce the Born–Infeld theory and use the Bernstein type theorems for minimal surface equations to study its electrostatic solutions. In §12.2, we study electrostatic and magnetostatic solutions in view of finite energy, obtain a generalized Bernstein problem, and find a connection between the minimal surface equations in Euclidean spaces and the maximal surface equations in Minkowskian spaces. In §12.3, we study the Born–Infeld wave equations. In particular, we solve the one-dimensional equations explicitly. We shall also illustrate the connection between the Born–Infeld theory and the Nambu–Goto string theory. In §12.4, we study string-like solutions arising from an Abelian Higgs theory within the framework of the Born–Infeld electromagnetism.

12.1   Nonlinear Electromagnetism

In this section, we consider the electromagnetic theory of Born and Infeld and its mathematical structure. We first review the point charge problem and its resolution in the Born–Infeld theory. We then introduce the Bernstein theorem for minimal surface equations and show how to use it in the study of electrostatic solutions.

12.1.1   Point charge problem

One of the major motivations for the introduction of the Born–Infeld electromagnetic field theory is to overcome the infinity problem associated with a point charge source in the original Maxwell theory. It is observed that, since the Einstein mechanics of special relativity may be obtained from the Newton mechanics by replacing the classical action function $\mathcal{L} = \frac{1}{2}mv^2$ by the relativistic expression

$$\mathcal{L} = mc^2\left(1 - \sqrt{1 - \frac{v^2}{c^2}}\right) = b^2\left(1 - \sqrt{1 - \frac{1}{b^2}mv^2}\right), \quad b^2 = mc^2, \quad (12.1.1)$$

so that no physical particle of a positive rest mass m can move at a speed v greater than the speed of light c, it will be acceptable to replace the action function of the Maxwell theory,

$$\mathcal{L} = \frac{1}{2}(\mathbf{E}^2 - \mathbf{B}^2), \quad (12.1.2)$$

where $\mathbf{E}$ and $\mathbf{B}$ are electric and magnetic fields, respectively, by a corresponding expression of the form

$$\mathcal{L} = b^2\left(1 - \sqrt{1 - \frac{1}{b^2}(\mathbf{E}^2 - \mathbf{B}^2)}\right), \quad (12.1.3)$$

where $b > 0$ is a suitable scaling parameter, often called the Born–Infeld parameter. It is clear that (12.1.3) defines a nonlinear theory of electromagnetism and the Maxwell theory, (12.1.2), may be recovered in the weak field limit $\mathbf{E}, \mathbf{B} \to \mathbf{0}$. Note that the choice of sign in front of the Lagrangian density (12.1.2) is the opposite of that of Born–Infeld [48] and is widely adopted in contemporary literature. This convention will be observed throughout the chapter.

Intrinsically, if (12.1.2) is replaced by $\mathcal{L} = -\frac{1}{4}F_{\mu\nu}F^{\mu\nu}$, then (12.1.3) takes the form

$$\mathcal{L} = b^2\left(1 - \sqrt{1 + \frac{1}{2b^2}F_{\mu\nu}F^{\mu\nu}}\right), \quad (12.1.4)$$

where

$$F_{\mu\nu} = \partial_\mu A_\nu - \partial_\nu A_\mu \quad (12.1.5)$$

is the electromagnetic field strength curvature induced from a gauge vector potential A_μ. More precisely, if we use

$$\mathbf{E} = (E^1, E^2, E^3), \quad \mathbf{B} = (B^1, B^2, B^3) \quad (12.1.6)$$

to denote the electric and magnetic fields, respectively, then there holds the standard identification

$$F^{0i} = -E^i, \quad F^{ij} = -\epsilon^{ijk}B^k, \quad i, j, k = 1, 2, 3, \quad (12.1.7)$$

which has the following matrix form,

$$(F^{\mu\nu}) = \begin{pmatrix} 0 & -E^1 & -E^2 & -E^3 \\ E^1 & 0 & -B^3 & B^2 \\ E^2 & B^3 & 0 & -B^1 \\ E^3 & -B^2 & B^1 & 0 \end{pmatrix}. \tag{12.1.8}$$

The dual of $F^{\mu\nu}$ reads

$$*F^{\mu\nu} = \tilde{F}^{\mu\nu} = \frac{1}{2}\epsilon^{\mu\nu\alpha\beta}F_{\alpha\beta}. \tag{12.1.9}$$

From (12.1.5), there holds again the Bianchi identity

$$\partial_\mu \tilde{F}^{\mu\nu} = 0. \tag{12.1.10}$$

On the other hand, it is easy to find that the Euler–Lagrange equations of (12.1.4) are

$$\partial_\mu P^{\mu\nu} = 0, \tag{12.1.11}$$

$$P^{\mu\nu} = \frac{F^{\mu\nu}}{\sqrt{1 + \frac{1}{2b^2}F_{\alpha\beta}F^{\alpha\beta}}}. \tag{12.1.12}$$

Corresponding to the electric field $\mathbf{E}$ and magnetic field $\mathbf{B}$, we introduce the electric displacement $\mathbf{D}$ and magnetic intensity $\mathbf{H}$,

$$\mathbf{D} = (D^1, D^2, D^3), \quad \mathbf{H} = (H^1, H^2, H^3), \tag{12.1.13}$$

and make the identification

$$P^{0i} = -D^i, \quad P^{ij} = -\epsilon^{ijk}H^k, \quad i, j, k = 1, 2, 3, \tag{12.1.14}$$

which has the following matrix form,

$$(P^{\mu\nu}) = \begin{pmatrix} 0 & -D^1 & -D^2 & -D^3 \\ D^1 & 0 & -H^3 & H^2 \\ D^2 & H^3 & 0 & -H^1 \\ D^3 & -H^2 & H^1 & 0 \end{pmatrix}. \tag{12.1.15}$$

Inserting (12.1.8) into (12.1.10) and (12.1.15) into (12.1.11), we obtain the fundamental governing equations of the Born–Infeld electromagnetic theory,

$$\frac{\partial \mathbf{B}}{\partial t} + \nabla \times \mathbf{E} = \mathbf{0}, \quad \nabla \cdot \mathbf{B} = 0, \tag{12.1.16}$$

$$-\frac{\partial \mathbf{D}}{\partial t} + \nabla \times \mathbf{H} = \mathbf{0}, \quad \nabla \cdot \mathbf{D} = 0, \tag{12.1.17}$$

which look exactly like the vacuum Maxwell equations, except that, in view of the relations (12.1.8), (12.1.12), and (12.1.15), the fields $\mathbf{E}, \mathbf{B}$ and $\mathbf{D}, \mathbf{H}$ are related nonlinearly,

$$\mathbf{D} = \frac{\mathbf{E}}{\sqrt{1 + \frac{1}{b^2}(\mathbf{B}^2 - \mathbf{E}^2)}}, \tag{12.1.18}$$

$$\mathbf{H} = \frac{\mathbf{B}}{\sqrt{1 + \frac{1}{b^2}(\mathbf{B}^2 - \mathbf{E}^2)}}. \tag{12.1.19}$$

Hence, the Born–Infeld electromagnetism introduces $\mathbf{E}$- and $\mathbf{B}$-dependent dielectrics and permeability,

$$\mathbf{D} = \varepsilon(\mathbf{E}, \mathbf{B})\mathbf{E}, \quad \mathbf{B} = \mu(\mathbf{E}, \mathbf{B})\mathbf{H}. \tag{12.1.20}$$

If there is an external current source, $(j^\mu) = (\rho, \mathbf{j})$, the equation (12.1.11) will be replaced by

$$\partial_\mu G^{\mu\nu} = j^\nu \tag{12.1.21}$$

and equivalently, the equations in (12.1.17) become

$$-\frac{\partial \mathbf{D}}{\partial t} + \nabla \times \mathbf{H} = \mathbf{j}, \quad \nabla \cdot \mathbf{D} = \rho, \tag{12.1.22}$$

We now examine the point charge problem.

Consider the electrostatic field generated from a point particle of electric charge q placed at the origin. Then $\mathbf{B} = \mathbf{0}, \mathbf{H} = \mathbf{0}$, and the Born–Infeld equations become a single one,

$$\nabla \cdot \mathbf{D} = 4\pi q \delta(x), \tag{12.1.23}$$

which can be solved to give us

$$\mathbf{D} = \frac{qx}{|x|^3}, \tag{12.1.24}$$

which is singular at the origin. However, from (12.1.18), we have

$$\mathbf{D} = \frac{\mathbf{E}}{\sqrt{1 - \frac{1}{b^2}\mathbf{E}^2}}, \tag{12.1.25}$$

which implies that

$$\mathbf{E} = \frac{\mathbf{D}}{\sqrt{1 + \frac{1}{b^2}\mathbf{D}^2}}$$

$$= \frac{qx}{|x|\sqrt{|x|^4 + \left(\frac{q}{b}\right)^2}}. \tag{12.1.26}$$

In particular, the electric field $\mathbf{E}$ is globally bounded. It is interesting to see that, when $|x|$ is sufficiently large, $\mathbf{E}$ given in (12.1.26) approximates that given by the Coulomb law, a consequence of the Maxwell equations.

As for the energy, we obtain from the Lagrange density (12.1.4) the energy-momentum tensor

$$T_\mu^\nu = -\frac{F^{\gamma\nu}F_{\gamma\mu}}{\sqrt{1 + \frac{1}{2b^2}F_{\alpha\beta}F^{\alpha\beta}}} - \delta_\mu^\nu \mathcal{L}, \qquad (12.1.27)$$

which gives us in the electrostatic case the Hamiltonian energy density

$$
\begin{aligned}
\mathcal{H} \;=\; T_0^0 &= b^2\left(\frac{1}{\sqrt{1 - \frac{1}{b^2}\mathbf{E}^2}} - 1\right) \\[2mm]
&= b^2\left(\sqrt{1 + \frac{1}{b^2}\mathbf{D}^2} - 1\right) \\[2mm]
&= b^2\left(\sqrt{1 + \left(\frac{q}{b}\right)^2 \frac{1}{|x|^4}} - 1\right).
\end{aligned}
\qquad (12.1.28)
$$

From (12.1.28), it is seen that the total energy of a point electric charge is now finite,

$$E = \int_{\mathbb{R}^3} \mathcal{H}\,dx < \infty. \qquad (12.1.29)$$

Similarly, we can consider the magnetostatic field generated from a point magnetic charge g placed at the origin of $\mathbb{R}^3$. In this case, $\mathbf{D} = 0$, $\mathbf{E} = 0$, and the Born–Infeld equations become

$$\nabla \cdot \mathbf{B} = 4\pi g\delta(x). \qquad (12.1.30)$$

From (12.1.30), we have as before,

$$\mathbf{B} = \frac{gx}{|x|^3}, \qquad (12.1.31)$$

$$\mathbf{H} = \frac{gx}{|x|\sqrt{|x|^4 + \left(\frac{g}{b}\right)^2}}. \qquad (12.1.32)$$

Thus $\mathbf{H}$ is a bounded vector field. In view of (12.1.27), the Hamiltonian density of a magnetostatic field takes the form,

$$\mathcal{H} = b^2\left(\sqrt{1 + \frac{1}{b^2}\mathbf{B}^2} - 1\right). \qquad (12.1.33)$$

Inserting (12.1.31) into (12.1.33), we see that the total energy of a point magnetic charge is also finite in the Born–Infeld theory.

12.1.2   Bernstein theorems

Since the equations of motion for the Born–Infeld field are nonlinear, a natural question may be raised as to whether there exists a solution that represents a self-induced electrostatic field or a magnetic monopole, either in the limit $\mathbf{B} = 0$ or $\mathbf{E} = 0$, without an external source.

Consider the electrostatic case. The Born–Infeld equations (12.1.16) and (12.1.17) become

$$\nabla \times \mathbf{E} = 0, \quad \nabla \cdot \mathbf{D} = 0. \tag{12.1.34}$$

The first equation in (12.1.34) says that $\mathbf{E}$ is conservative. Hence, there is a real scalar field f such that $\mathbf{E} = b\nabla f$. Inserting this into (12.1.25) and using the second equation in (12.1.34), we arrive at the following equivalent reduction of (12.1.34),

$$\nabla \cdot \left(\frac{\nabla f}{\sqrt{1 - |\nabla f|^2}} \right) = 0. \tag{12.1.35}$$

The equation (12.1.35) over $\mathbb{R}^n$ is the maximal space-like hypersurface equation in the Minkowski spacetime $\mathbb{R}^{n,1}$ which is closely related to the following non-parametric minimal hypersurface equation in the Euclidean space $\mathbb{R}^{n+1}$,

$$\nabla \cdot \left(\frac{\nabla f}{\sqrt{1 + |\nabla f|^2}} \right) = 0, \quad x \in \mathbb{R}^n. \tag{12.1.36}$$

The Bernstein theorem for (12.1.35) established by Calabi [64] for $n = 2, 3, 4$ and by Cheng and Yau [81] for all n states that any entire space-like solution f of (12.1.35), satisfying $|\nabla f| < 1$, must be a linear function. Hence, $\mathbf{E} = b\nabla f$ is a constant vector. The finite energy condition then implies that $\mathbf{E} = 0$ everywhere. This proves that in the Born–Infeld theory there is no self-induced electrostatic field.

Motivated by the above and other applications to the Born–Infeld theory, we present below a discussion of the Bernstein theorems for (12.1.35) and (12.1.36).

In 1927, S. Bernstein published his work on the minimal surface ($n = 2$) equation (12.1.36) which states that any entire solution of this equation must be linear in its two variables. Geometrically, this means that a non-parametric minimal surface in $\mathbb{R}^3$ must be a plane. This result is similar to the Liouville theorem and has been extended to higher dimensions by many authors [241]. See also [173]. It is known that the same result is true for $n \leq 7$ but not true for $n > 7$. Here we present the simplest case when $n = 2$ and the elementary proof of Nitsche [229, 230, 290].

We use x, y to denote the coordinates of $\mathbb{R}^2$. The following preliminary theorem is due to Jörgens [161].

Theorem 12.1.1. *If $u : \mathbb{R}^2 \to \mathbb{R}$ is a function satisfying the equation*

$$\frac{\partial^2 u}{\partial x^2} \frac{\partial^2 u}{\partial y^2} - \left(\frac{\partial^2 u}{\partial x \partial y}\right)^2 = 1 \qquad (12.1.37)$$

on the full $\mathbb{R}^2$, then u is a quadratic polynomial in x, y.

Proof. We use the notation

$$p = \frac{\partial u}{\partial x}, \quad q = \frac{\partial u}{\partial y}, \quad r = \frac{\partial^2 u}{\partial x^2}, \quad s = \frac{\partial^2 u}{\partial x \partial y}, \quad t = \frac{\partial^2 u}{\partial y^2}$$

to rewrite (12.1.37) as

$$rt - s^2 = 1. \qquad (12.1.38)$$

Since $rt > 0$, we may assume without loss of generality that $r, t > 0$ because otherwise we may replace u by $-u$.

We will use the Lewy transformation [187] to simplify (12.1.38),

$$
\begin{aligned}
T : (x, y) &\to (\xi, \eta), \\
T(x, y) &= (\xi(x, y), \eta(x, y)) \\
&= (x + p(x, y), y + q(x, y)).
\end{aligned} \qquad (12.1.39)
$$

To show that T is 1-1, we fix two points (x_1, y_1) and (x_2, y_2) in $\mathbb{R}^2$ and consider their images under T. For this purpose, we use τ to denote a real parameter in the unit interval $[0, 1]$ and define the function

$$h(\tau) = u(x_1 + \tau(x_2 - x_1), y_1 + \tau(y_2 - y_1)).$$

Then

$$
\begin{aligned}
h'(\tau) &= (x_2 - x_1)p + (y_2 - y_1)q, \\
h''(\tau) &= (x_2 - x_1)^2 r + 2(x_2 - x_1)(y_2 - y_1)s + (y_2 - y_1)^2 t,
\end{aligned} \qquad (12.1.40)
$$

where p, q, r, s, t are evaluated at the point $(x_1 + \tau(x_2 - x_1), y_1 + \tau(y_2 - y_1))$. Using (12.1.38), we can see that the quadratic form (12.1.40) in $x_2 - x_1$ and $y_2 - y_1$ is positive definite. Hence, $h'(\tau)$ increases as a function of τ. In particular $h'(1) \geq h'(0)$. That is,

$$(x_2 - x_1)(p_2 - p_1) + (y_2 - y_1)(q_2 - q_1) \geq 0, \qquad (12.1.41)$$

where

$$p_i = p(x_i, y_i), \quad q_i = q(x_i, y_i), \quad i = 1, 2.$$

Using $\|\cdot\|$ to denote the norm of $\mathbb{R}^2$ and (12.1.39), (12.1.41), we obtain the estimate

$$\|T(x_1, y_1) - T(x_2, y_2)\| \geq \|(x_1, y_1) - (x_2, y_2)\|. \qquad (12.1.42)$$

An immediate and standard consequence of (12.1.42) is that the mapping T is 1-1 and closed.

On the other hand, in view of (12.1.38) and $r, t > 0$, it is straightforward to see that the Jacobian of T is positive everywhere,

$$
\begin{vmatrix} \frac{\partial \xi}{\partial x} & \frac{\partial \xi}{\partial y} \\ \frac{\partial \eta}{\partial x} & \frac{\partial \eta}{\partial y} \end{vmatrix} = \begin{vmatrix} 1 + r & s \\ s & 1 + t \end{vmatrix}
$$

$$
= 1 + r + t + rt - s^2
$$

$$
\geq 2.
$$

Hence T is also open. Since $\mathbb{R}^2$ is connected, we conclude that T is onto. Therefore T is a diffeomorphism of $\mathbb{R}^2$ onto itself.

The inverse of T is denoted by T^{-1}. Thus the derivative of $T^{-1}(\xi, \eta) = (x, y)$ may be calculated by the formula

$$
\begin{aligned}
dT^{-1} &= \begin{pmatrix} \frac{\partial x}{\partial \xi} & \frac{\partial x}{\partial \eta} \\ \frac{\partial y}{\partial \xi} & \frac{\partial y}{\partial \eta} \end{pmatrix} \\
&= (dT)^{-1} \\
&= \begin{pmatrix} 1 + r & s \\ s & 1 + t \end{pmatrix}^{-1} \\
&= \frac{1}{2 + r + t} \begin{pmatrix} 1 + t & -s \\ -s & 1 + r \end{pmatrix},
\end{aligned} \tag{12.1.43}
$$

which enables us to read off the partial derivatives of x, y with respect to ξ, η quickly.

Define a complex function F by

$$
\begin{aligned}
F(\xi + i\eta) &= U(\xi, \eta) + iV(\xi, \eta), \\
U(\xi, \eta) &= x - p \\
&= x(\xi, \eta) - p(x(\xi, \eta), y(\xi, \eta)), \\
V(\xi, \eta) &= -y + q \\
&= -y(\xi, \eta) + q(x(\xi, \eta), y(\xi, \eta)).
\end{aligned}
$$

Then, by (12.1.43), we have

$$
\begin{aligned}
\frac{\partial U}{\partial \xi} &= \frac{t - r}{2 + r + t} = \frac{\partial V}{\partial \eta}, \\
\frac{\partial V}{\partial \xi} &= \frac{2s}{2 + r + t} = -\frac{\partial U}{\partial \eta}.
\end{aligned}
$$

In other words, the pair U, V satisfy the Cauchy–Riemann equations. Hence F is an entire function (holomorphic on the full complex plane). Besides, we can compute the derivative of F as

$$
F'(\xi + i\eta) = \frac{\partial U}{\partial \xi} + i\frac{\partial V}{\partial \xi} = \frac{t - r + 2is}{2 + r + t}. \tag{12.1.44}
$$

By virtue of (12.1.38), we have

$$|F'(\xi + i\eta)|^2 = \frac{r + t - 2}{r + t + 2},\tag{12.1.45}$$

which is seen to be bounded. Consequently, F' must be a constant in view of the Liouville theorem.

It will be useful to rewrite (12.1.45) as

$$1 - |F'(\xi + i\eta)|^2 = \frac{4}{r + t + 2},\tag{12.1.46}$$

Hence, from (12.1.44)–(12.1.46), we can solve for r, s, t to obtain

$$
\begin{aligned}
r &= \frac{2}{1 - |F'|^2} - 1 - \frac{2\mathrm{Re}(F')}{1 - |F'|^2}, \\
s &= \frac{2\mathrm{Im}(F')}{1 - |F'|^2}, \\
t &= \frac{2}{1 - |F'|^2} - 1 + \frac{2\mathrm{Re}(F')}{1 - |F'|^2}.
\end{aligned}
$$

In particular, r, s, t are all constants.

We are now prepared to prove the original Bernstein theorem.

Theorem 12.1.2. *For $n = 2$, any entire solution f of (12.1.36) must be linear in its two variables, x, y. In other words, planes are the only minimal surfaces in $\mathbb{R}^3$ which are the graph of a function $f : \mathbb{R}^2 \to \mathbb{R}$.*

Proof. We continue to use the letters p, q, r, s, t to denote the respective partial derivatives of a solution f of the minimal surface equation (12.1.36) in the xy-plane, which may now be rewritten as

$$(1 + q^2)r - 2pqs + (1 + p^2)t = 0.\tag{12.1.47}$$

With the notation $w = \sqrt{1 + p^2 + q^2}$, we may apply (12.1.47) to arrive at the integrability conditions

$$\frac{\partial}{\partial x}\left(\frac{1 + q^2}{w}\right) = \frac{\partial}{\partial y}\left(\frac{pq}{w}\right), \qquad \frac{\partial}{\partial x}\left(\frac{pq}{w}\right) = \frac{\partial}{\partial y}\left(\frac{1 + p^2}{w}\right).$$

More precisely, there are globally defined functions g, h so that

$$\nabla g = \left(\frac{1 + p^2}{w}, \frac{pq}{w}\right), \qquad \nabla h = \left(\frac{pq}{w}, \frac{1 + q^2}{w}\right).\tag{12.1.48}$$

Besides, since we can read off

$$\frac{\partial g}{\partial y} = \frac{\partial h}{\partial x}$$

from (12.1.48), we see that there exists a globally defined function u over $\mathbb{R}^2$ such that

$$\frac{\partial u}{\partial x} = g, \quad \frac{\partial u}{\partial y} = h. \tag{12.1.49}$$

Combining (12.1.48) and (12.1.49), we obtain

$$\frac{\partial^2 u}{\partial x^2} = \frac{1+p^2}{w}, \quad \frac{\partial^2 u}{\partial x \partial y} = \frac{pq}{w}, \quad \frac{\partial^2 u}{\partial y^2} = \frac{1+q^2}{w}. \tag{12.1.50}$$

Consequently u satisfies (12.1.37).

By Jörgens' theorem (Theorem 12.1.1), the quantities listed in (12.1.50) are all constants, from which we can solve for p and q to show that p and q are also constants.

We now turn our attention to maximal hypersurface equation (12.1.35). We show that the Bernstein theorem is valid for $n = 2$ by observing that (12.1.35) and (12.1.36) are equivalent when $n = 2$. See also [64].

Theorem 12.1.3. *If $f : \mathbb{R}^2 \to \mathbb{R}$ is a function satisfying the condition of being space-like, $|\nabla f|^2 < 1$, and (12.1.35) with $n = 2$, then f must be a linear function in its variables. In fact, there is a 1-1 correspondence between (12.1.35) and (12.1.36) when $n = 2$.*

Proof. Let u be a solution of (12.1.36) and p, q be defined as in the proof of Theorem 12.1.1. Set $w = \sqrt{1 + p^2 + q^2}$. Then (12.1.36) reads

$$\frac{\partial}{\partial x}\left(\frac{p}{w}\right) + \frac{\partial}{\partial y}\left(\frac{q}{w}\right) = 0.$$

Hence, there is a real-valued function U such that

$$P \equiv \frac{\partial U}{\partial x} = -\frac{q}{w}, \quad Q \equiv \frac{\partial U}{\partial y} = \frac{p}{w}.$$

Therefore, we have

$$1 - P^2 - Q^2 = \frac{1}{w^2} > 0$$

and U is space-like. Inserting the relations $p = Qw$, $q = -Pw$, and $w = 1/W$ where $W = \sqrt{1 - P^2 - Q^2}$ into the identity $\partial p/\partial y = \partial q/\partial x$, we arrive at

$$\frac{\partial}{\partial x}\left(\frac{P}{W}\right) + \frac{\partial}{\partial y}\left(\frac{Q}{W}\right) = 0.$$

Thus, U solves (12.1.35).

The inverse correspondence from (12.1.35) to (12.1.36) may be established similarly.

12.2 Relation of Electrostatic and Magnetostatic Fields

In the last section, we see that the Bernstein theorems may be used to deduce the conclusion that there does not exist a self-induced electrostatic field. The purpose of this section is to find an equivalence relation between electrostatic and magnetostatic fields. As a by-product, we will see that there does not exist a self-induced magnetostatic field either. Such a study motivates a natural extension of the Bernstein or Born–Infeld type equations.

12.2.1 Electrostatic fields

We begin by considering the energy of an electrostatic field. Hence, (12.1.5) gives rise to a time-independent purely electric field $\mathbf{E}$ with $\mathbf{B} = 0$, $\partial_0 A_\mu = 0$ and there is a real-valued potential function ϕ such that

$$A_0 = -\phi, \quad \mathbf{E} = \nabla\phi,$$

and the Born–Infeld equations (12.1.16) and (12.1.17) become

$$\nabla \cdot \left(\frac{\nabla\phi}{\sqrt{1 - \frac{1}{b^2}|\nabla\phi|^2}} \right) = 0, \tag{12.2.1}$$

which is the Euler–Lagrange equation of the action

$$I(\phi) = \int_{\mathbb{R}^3} b^2 \left(1 - \sqrt{1 - \frac{1}{b^2}|\nabla\phi|^2} \right) \, dx. \tag{12.2.2}$$

However, in the current situation, the Hamiltonian $\mathcal{H} = T_0^0$ defined by (12.1.27) takes the form

$$
\begin{aligned}
\mathcal{H} &= \frac{|\nabla\phi|^2}{\sqrt{1 - \frac{1}{b^2}|\nabla\phi|^2}} - b^2 \left(1 - \sqrt{1 - \frac{1}{b^2}|\nabla\phi|^2} \right) \\
&= b^2 \left(\frac{1}{\sqrt{1 - \frac{1}{b^2}|\nabla\phi|^2}} - 1 \right).
\end{aligned}
\tag{12.2.3}
$$

Therefore, by the simple inequality

$$1 - s \leq \frac{1}{s} - 1, \quad \forall s \in (0, 1],$$

we see that a finite-energy solution of (12.2.1) satisfying

$$\int_{\mathbb{R}^3} \mathcal{H} \, dx < \infty$$

is a finite-energy critical point of the functional (12.2.2). In other words, the problem of obtaining an electrostatic field may be reduced to the more general problem of obtaining a finite-energy critical point ϕ of the functional (12.2.2).

If ϕ is a critical point of (12.2.2), we see by rescaling ϕ and the fact that $\phi_\lambda(x) = \phi(\lambda x)$ is a critical point at $\lambda = 1$ that

$$\frac{\mathrm{d}}{\mathrm{d}\lambda}\left\{\int_{\mathbb{R}^3} b^2\left(1 - \sqrt{1 - \frac{1}{b^2}|\nabla\phi_\lambda|^2}\right)\mathrm{d}x\right\}\bigg|_{\lambda=1} = 0. \tag{12.2.4}$$

However, in the integrand in (12.2.4), we may absorb the parameter λ correspondingly in the integral by $x \mapsto x_\lambda = \lambda x$, $\mathrm{d}x \mapsto \lambda^{-3}\mathrm{d}x_\lambda$ and returning to the original notation $x_\lambda \to x$. Hence, we arrive from (12.2.4) at the identity

$$\int_{\mathbb{R}^3} \frac{|\nabla\phi|^2}{\sqrt{1 - \frac{1}{b^2}|\nabla\phi|^2}}\,\mathrm{d}x = 3b^2 \int_{\mathbb{R}^3}\left(1 - \sqrt{1 - \frac{1}{b^2}|\nabla\phi|^2}\right)\mathrm{d}x. \tag{12.2.5}$$

In particular, we find that the left-hand side of (12.2.5) is finite. Consequently, we can integrate (12.2.3) and insert (12.2.5) to obtain the relation

$$2I(\phi) = \int_{\mathbb{R}^3} \mathcal{H}\,\mathrm{d}x. \tag{12.2.6}$$

Therefore, we conclude that the problem of obtaining an electrostatic field is in fact equivalent to the problem of obtaining a finite-energy critical point ϕ of the functional (12.2.2).

12.2.2 Magnetostatic fields

We next consider a pure static magnetic field or self-induced magnetostatic field case, $A_0 = 0$, $\mathbf{E} = 0$, $\partial_0\mathbf{B} = 0$ and we prove similarly that $\mathbf{B} = 0$.

In view of (12.1.27), the Hamiltonian density takes the simple form

$$\mathcal{H} = b^2\left(\sqrt{1 + \frac{1}{b^2}\mathbf{B}^2} - 1\right). \tag{12.2.7}$$

Besides, recall that we can represent $\mathbf{B}$ by its vector gauge potential $\mathbf{A} = -(A_1, A_2, A_3)$: $\mathbf{B} = \nabla \times \mathbf{A}$. Hence the Born–Infeld equations of motion become

$$\nabla \times \left(\frac{\nabla \times \mathbf{A}}{\sqrt{1 + \frac{1}{b^2}|\nabla \times \mathbf{A}|^2}}\right) = \mathbf{0}, \tag{12.2.8}$$

which are the Euler–Lagrange equations of the energy functional

$$E(\mathbf{A}) = \int_{\mathbb{R}^3} b^2\left(\sqrt{1 + \frac{1}{b^2}|\nabla \times \mathbf{A}|^2} - 1\right)\mathrm{d}x \tag{12.2.9}$$

defined by (12.2.7).

The same rescaling argument gives us the identity for a critical point of (12.2.9) as follows,

$$\int_{\mathbb{R}^3} \frac{|\nabla \times \mathbf{A}|^2}{\sqrt{1 + \frac{1}{b^2}|\nabla \times \mathbf{A}|^2}}\, d\mathbf{x} = 3b^2 \int_{\mathbb{R}^3} \left(\sqrt{1 + \frac{1}{b^2}|\nabla \times \mathbf{A}|^2} - 1 \right) d\mathbf{x}. \quad (12.2.10)$$

We can show that a Bernstein theorem *also* holds for the vector field equation (12.2.8). To prove it, we first note from (12.2.8) that the vector field

$$\mathbf{H} = \frac{\nabla \times \mathbf{A}}{\sqrt{1 + \frac{1}{b^2}|\nabla \times \mathbf{A}|^2}} \quad (12.2.11)$$

is conservative. Thus there is a real scalar field ϕ such that

$$\mathbf{H} = \nabla \phi. \quad (12.2.12)$$

Using (12.2.11) and (12.2.12), we have

$$|\nabla \times \mathbf{A}|^2 = \frac{|\nabla \phi|^2}{1 - \frac{1}{b^2}|\nabla \phi|^2}. \quad (12.2.13)$$

Therefore, we can invert (12.2.11) and (12.2.12) to obtain the useful relation

$$\begin{aligned} \nabla \times \mathbf{A} &= \mathbf{H}\sqrt{1 + \frac{1}{b^2}|\nabla \times \mathbf{A}|^2} \\ &= \frac{\nabla \phi}{\sqrt{1 - \frac{1}{b^2}|\nabla \phi|^2}}. \end{aligned} \quad (12.2.14)$$

Inserting (12.2.14) into the Bianchi identity $\nabla \cdot (\nabla \times \mathbf{A}) = 0$, we arrive at the equation (12.2.1) and we see that $\nabla \phi$ must be a constant vector. Hence so is $\mathbf{B} = \nabla \times \mathbf{A}$ as claimed. The finite-energy condition then implies that $\mathbf{B} = 0$ as expected.

The above calculation indicates that there is a relation between the electrostatic equation (12.2.1) and magnetostatic equation (12.2.8). Here we remark that these two equations are in fact equivalent. To see this, we observe from (12.2.1) that the vector field

$$\mathbf{H} = \frac{\nabla \phi}{\sqrt{1 - \frac{1}{b^2}|\nabla \phi|^2}} \quad (12.2.15)$$

is solenoidal. Therefore, there is a vector field $\mathbf{A}$ such that

$$\mathbf{H} = \nabla \times \mathbf{A}. \quad (12.2.16)$$

Inserting (12.2.15) into (12.2.16), we have

$$|\nabla\phi|^2 = \frac{|\nabla \times \mathbf{A}|^2}{1 + \frac{1}{b^2}|\nabla \times \mathbf{A}|^2}, \tag{12.2.17}$$

which inverts the relation (12.2.13) obtained earlier. Using the expressions (12.2.15), (12.2.16), and (12.2.17), we obtain

$$\begin{aligned} \nabla\phi &= \mathbf{H}\sqrt{1 - \frac{1}{b^2}|\nabla\phi|^2} \\ &= \frac{\nabla \times \mathbf{A}}{\sqrt{1 + \frac{1}{b^2}|\nabla \times \mathbf{A}|^2}}. \end{aligned} \tag{12.2.18}$$

In view of the identity $\nabla \times (\nabla\phi) = 0$, we arrive at (12.2.8), which proves the claimed equivalence. It is interesting to note that the finite-energy conditions for the electrostatic and magnetostatic cases are also equivalent as can been seen from rewriting (12.2.17) as

$$\frac{|\nabla\phi|^2}{\sqrt{1 - \frac{1}{b^2}|\nabla\phi|^2}} = \frac{|\nabla \times \mathbf{A}|^2}{\sqrt{1 + \frac{1}{b^2}|\nabla \times \mathbf{A}|^2}} \tag{12.2.19}$$

and applying the energy identities (12.2.5) and (12.2.10).

12.2.3 Generalized Bernstein problem

The above study on the equivalence of the electrostatic equation (12.2.1) and magnetostatic equation (12.2.8) may be extended to reveal an interesting link between minimal hypersurface equations over Euclidean and maximal hypersurface equations over Minkowskian spaces. To see this, we use Λ^p to denote the space of p-forms over $\mathbb{R}^n$ equipped with the usual inner product and consider the differential equation

$$d * \left(\frac{d\omega}{\sqrt{1 + |d\omega|^2}}\right) = 0, \qquad \omega \in \Lambda^p, \tag{12.2.20}$$

where d is the exterior derivative and $*$ is the Hodge isometry between Λ^p and Λ^{n-p} satisfying

$$** = (-1)^{p(n-p)}. \tag{12.2.21}$$

Let ω be a solution of (12.2.20) and set

$$\tau = * \left(-\frac{d\omega}{\sqrt{1 + |d\omega|^2}}\right). \tag{12.2.22}$$

Since τ is a closed $(n - p - 1)$-form and the de Rham cohomology over $\mathbb{R}^n$ is trivial, there is an $(n - p - 2)$-form σ such that $\tau = d\sigma$, namely,

$$d\sigma = * \frac{d\omega}{\sqrt{1 + |d\omega|^2}}, \tag{12.2.23}$$

which gives us $|d\sigma|^2 < 1$ and

$$|d\omega|^2 = \frac{|d\sigma|^2}{1 - |d\sigma|^2}, \tag{12.2.24}$$

because $*$ is an isometry. In view of (12.2.21) and (12.2.23), we have

$$
\begin{aligned}
d\omega &= (-1)^{p(n-p)+n-1} * d\sigma\sqrt{1 + |d\omega|^2} \\
&= (-1)^{p(n-p)+n-1} * \frac{d\sigma}{\sqrt{1 - |d\sigma|^2}}.
\end{aligned} \tag{12.2.25}
$$

Applying $d^2\omega = 0$ in (12.2.25), we arrive at the equation

$$d * \left(\frac{d\sigma}{\sqrt{1 - |d\sigma|^2}}\right) = 0, \qquad \sigma \in \Lambda^{n-p-2}. \tag{12.2.26}$$

Conversely, we can follow the same path to show that (12.2.26) implies (12.2.20).

Hence, we see that, via a 'dual correspondence', the equations (12.2.20) and (12.2.26) are equivalent and that, in terms of differential forms, we have obtained a generalized version of Theorem 12.1.3 in any dimensions. We remark that the method of dual correspondence is well known in the classical Hodge theory. This technique has also been used by Sibner and Sibner [285] in a nonlinear Hodge theory.

Besides, in view of (12.2.24), there holds the 'energy' identity

$$\frac{|d\omega|^2}{\sqrt{1 + |d\omega|^2}} = \frac{|d\sigma|^2}{\sqrt{1 - |d\sigma|^2}}, \tag{12.2.27}$$

which generalizes (12.2.19).

To make sense out of the equivalent pair of equations (12.2.20) and (12.2.26), we must have $0 \le p \le n - 2$. Thus there are two borderline situations, $p = 0$ and $p = n - 2$. In fact, we have seen that these two situations bear clear geometric significance and are well understood. When $p = 0$, the equation (12.2.20) is the classical non-parametric minimal surface equation. When $p = n - 2$, the equation (12.2.26) is a scalar equation and defines maximal space-like hypersurfaces in the Minkowski space $\mathbb{R}^{n,1}$. Except for these two cases, there is a large open area for the equation (12.2.20) or (12.2.26) for p in the interval

$$0 < p < n - 2 \tag{12.2.28}$$

regarding its geometric or physical meaning and the Bernstein type properties.

In the rest of this subsection, we consider two simplest situations for the equation (12.2.20), under which we derive the triviality property, $d\omega = 0$.

Firstly, we consider (12.2.20) over a closed (compact without boundary) oriented Riemannian manifold, (M, g), and we use $\Lambda^p(M)$ to denote the space (fiber bundle) of p-forms on M.

The natural inner product on any fiber $\Lambda^p_x(M)$ ($x \in M$) induced from the metric g is denoted by $< \cdot, \cdot >_x$ and the canonical volume element of (M, g) is simply $*1$. Recall also that one can define on $\Lambda^p(M)$ a global inner product $(\cdot, \cdot)$ as

$$(\sigma, \tau) = \int_M \sigma \wedge (*\tau) = \int_M <\sigma, \tau>_x *1, \quad \sigma, \tau \in \Lambda^p(M).$$

Use δ here to denote the co-differential operator,

$$\delta : \Lambda^p(M) \to \Lambda^{p-1}(M),$$
$$\delta = (-1)^{n(p+1)+1} * d * .$$

Then the following well-known Green formula holds,

$$(d\sigma, \tau) = (\sigma, \delta\tau), \quad \sigma \in \Lambda^{p-1}(M), \tau \in \Lambda^p(M). \tag{12.2.29}$$

Therefore, if ω is a p-form on M which is a solution of the equation (12.2.20), then, using (12.2.29) and the fact that $*$ is an isometry, we have

$$
\begin{aligned}
0 &= \left(d * \frac{d\omega}{\sqrt{1 + |d\omega|^2}}, *\omega \right) \\
&= \left(* \frac{d\omega}{\sqrt{1 + |d\omega|^2}}, \delta * \omega \right) \\
&= (-1)^{n(p+1)+1} \left(* \frac{d\omega}{\sqrt{1 + |d\omega|^2}}, *d * *\omega \right) \\
&= (-1)^{n-p^2+1} \left(* \frac{d\omega}{\sqrt{1 + |d\omega|^2}}, *d\omega \right) \\
&= \pm \left(\frac{d\omega}{\sqrt{1 + |d\omega|^2}}, d\omega \right) \\
&= \pm \int_M \frac{|d\omega|^2}{\sqrt{1 + |d\omega|^2}} * 1.
\end{aligned}
$$

Thus $d\omega = 0$ everywhere on M.

Secondly, we consider (12.2.20) on $\mathbb{R}^n$ assuming for convenience that $|\omega|$ is bounded and

$$|d\omega| \in L^1(\mathbb{R}^n) \cap L^2(\mathbb{R}^n). \tag{12.2.30}$$

Let $\xi(x)$ be a smooth function on $\mathbb{R}^n$ satisfying

$$\xi(x) = 1, \quad |x| \le 1; \quad \xi(x) = 0, \quad |x| \ge 2; \quad 0 \le \xi(x) \le 1, \quad \forall x \in \mathbb{R}^n.$$

For $\rho > 0$, construct the cut-off function $\xi_\rho(x) = \xi(x/\rho)$. Then

$$|\nabla \xi_\rho| \leq \frac{K}{\rho} \quad \text{on } \mathbb{R}^n, \tag{12.2.31}$$

where $K > 0$ is a constant independent of ρ.

We have

$$d\left(* \frac{d\omega}{\sqrt{1 + |d\omega|^2}} \wedge \xi_\rho \omega \right) = d\left(* \frac{d\omega}{\sqrt{1 + |d\omega|^2}} \right) \wedge (\xi_\rho \omega)$$

$$+ (-1)^{n-p-1} \xi_\rho \left(* \frac{d\omega}{\sqrt{1 + |d\omega|^2}} \right) \wedge d\omega$$

$$+ (-1)^{n-p-1} \frac{*d\omega}{\sqrt{1 + |d\omega|^2}} \wedge d\xi_\rho \wedge \omega. \tag{12.2.32}$$

Integrating (12.2.32), neglecting the boundary terms at infinity, and using (12.2.20), we obtain the estimate

$$\int_{\mathbb{R}^n} \xi_\rho \frac{|d\omega|^2}{\sqrt{1 + |d\omega|^2}} \, dx \leq$$

$$C \left(\int_{\rho < |x| < 2\rho} |\nabla \xi_\rho|^m \, dx \right)^{\frac{1}{m}} \left(\int_{\mathbb{R}^n} |d\omega|^{\frac{m}{m-1}} \, dx \right)^{\frac{m-1}{m}}, \tag{12.2.33}$$

where m is any number satisfying

$$m > \max\{n, 2\}. \tag{12.2.34}$$

Set $u = |d\omega|$ and $s + t = m/(m-1)$. Let $p, q > 1$ be conjugate exponents such that

$$\frac{1}{p} + \frac{1}{q} = 1, \quad ps = 1, \quad qt = 2.$$

Then we may solve to get

$$s = \frac{m-2}{m-1}, \quad t = \frac{2}{m-1}, \quad p = \frac{m-1}{m-2}, \quad q = m - 1.$$

With these numbers, we can apply the Schwarz inequality to obtain

$$\int_{\mathbb{R}^n} u^{\frac{m}{m-1}} \, dx \leq \left(\int_{\mathbb{R}^n} u \, dx \right)^{\frac{m-2}{m-1}} \left(\int_{\mathbb{R}^n} u^2 \, dx \right)^{\frac{1}{m-1}}. \tag{12.2.35}$$

Inserting (12.2.35) into (12.2.33), we have

$$\int_{\mathbb{R}^n} \xi_\rho \frac{|d\omega|^2}{\sqrt{1 + |d\omega|^2}} \, dx \leq C(\|\omega\|_\infty, \|d\omega\|_{L^1}, \|d\omega\|_{L^2}) \rho^{-(1 - \frac{n}{m})}. \tag{12.2.36}$$

Letting $\rho \to \infty$ and using (12.2.34), we find that

$$\int_{\mathbb{R}^n} \frac{|d\omega|^2}{\sqrt{1 + |d\omega|^2}} \, dx = 0.$$

That is, we have obtained $d\omega = 0$ everywhere on $\mathbb{R}^n$ as expected.

Of course, other convenient conditions may also be imposed to ensure $d\omega = 0$. For example, one can assume that there are conjugate exponents $p, q > 1$ so that $\omega \in L^p(\mathbb{R}^n)$ and $d\omega \in L^q(\mathbb{R}^n)$ to arrive at the conclusion.

12.2.4 Mixed interaction case

It will also be interesting to study sourceless static solutions without assuming either $\mathbf{E}$ or $\mathbf{B}$ to be zero. With $\mathbf{E} = \nabla\phi$ and $\mathbf{B} = \nabla \times \mathbf{A}$, the equations of motion are

$$\nabla \cdot \left(\frac{\nabla\phi}{\sqrt{1 + \frac{1}{b^2}(|\nabla \times \mathbf{A}|^2 - |\nabla\phi|^2)}} \right) = 0, \qquad (12.2.37)$$

$$\nabla \times \left(\frac{\nabla \times \mathbf{A}}{\sqrt{1 + \frac{1}{b^2}(|\nabla \times \mathbf{A}|^2 - |\nabla\phi|^2)}} \right) = 0. \qquad (12.2.38)$$

From (12.2.38), we see that there is a real scalar function ψ such that

$$\frac{\nabla \times \mathbf{A}}{\sqrt{1 + \frac{1}{b^2}(|\nabla \times \mathbf{A}|^2 - |\nabla\phi|^2)}} = \nabla\psi, \qquad (12.2.39)$$

which leads us to the relation

$$|\nabla \times \mathbf{A}|^2 = \left(1 - \frac{1}{b^2}|\nabla\phi|^2\right) \frac{|\nabla\psi|^2}{\left(1 - \frac{1}{b^2}|\nabla\psi|^2\right)}. \qquad (12.2.40)$$

Inserting (12.2.40) into (12.2.39), we obtain

$$\nabla \times \mathbf{A} = \nabla\psi \frac{\sqrt{1 - \frac{1}{b^2}|\nabla\phi|^2}}{\sqrt{1 - \frac{1}{b^2}|\nabla\psi|^2}}. \qquad (12.2.41)$$

In view of (12.2.40) and (12.2.41), we see that the static Born–Infeld equations (12.2.37) and (12.2.38) are equivalent to the following coupled system of two scalar equations,

$$\nabla \cdot \left(\nabla\phi \frac{\sqrt{1 - \frac{1}{b^2}|\nabla\psi|^2}}{\sqrt{1 - \frac{1}{b^2}|\nabla\phi|^2}} \right) = 0, \qquad (12.2.42)$$

$$\nabla \cdot \left(\nabla\psi \frac{\sqrt{1 - \frac{1}{b^2}|\nabla\phi|^2}}{\sqrt{1 - \frac{1}{b^2}|\nabla\psi|^2}} \right) = 0. \qquad (12.2.43)$$

This system generalizes the minimal or maximal surface equations and is the system of the Euler–Lagrange equations of the action functional

$$\mathcal{A}(\phi, \psi) = \int_{\mathbb{R}^3} \left(1 - \sqrt{1 - \frac{1}{b^2}|\nabla\phi|^2}\sqrt{1 - \frac{1}{b^2}|\nabla\psi|^2} \right) dx. \qquad (12.2.44)$$

It is interesting to ask whether the Bernstein property holds for (12.2.42) and (12.2.43) or whether any entire solution (ϕ, ψ) of these equations must be such that ϕ and ψ are affine linear functions. However, it is easily seen that an entire solution of (12.2.42) and (12.2.43) is not necessarily affine linear. In fact, a counterexample is obtained by setting $\phi = \psi = h$ in these equations where h is an arbitrary harmonic function over the full space. Thus we have yet to identify the most general trivial solutions of these coupled equations. Here we will be satisfied with a proof that all solutions of (12.2.42) and (12.2.43) are constant if the magnetic and electric fields $\mathbf{E}$ and $\mathbf{B}$ decay to their vacuum states $\mathbf{E} = \mathbf{0}$ and $\mathbf{B} = \mathbf{0}$ at infinity at least at a rate of the form $O(r^{-1})$, which is a natural physical (finite energy) requirement. Indeed, such a condition implies that both $\nabla\phi$ and $\nabla\psi$ decay at infinity at the same rate (see (12.2.39)). Therefore, ϕ and ψ are bounded functions. Applying the Harnack inequality for strictly elliptic equations to (12.2.42) and (12.2.43) separately, we conclude that ϕ and ψ must be constant. Therefore, we have proved the fact that $\mathbf{E} = \mathbf{0}$ and $\mathbf{B} = \mathbf{0}$ everywhere as expected. This result again excludes the possibility of a self-induced static solution carrying either an electric or magnetic charge.

12.3   Nonlinear Wave Equations

In this section, we study the nonlinear wave equations in view of the Born–Infeld theory. We shall first present a general discussion of static solutions and solve the simplest one-dimensional equations. We then derive the Born–Infeld dynamics from the Nambu–Goto string theory.

12.3.1   Static solutions

The calculation in the previous section indicates that the Born–Infeld action corresponds to the rescaling of variables differently. In particular, it will be interesting to reconsider the Derrick theorem in our case of an $(n+1)$-dimensional spacetime with the Lorentz metric $\text{diag}(1, -1, \cdots, -1)$ where in view of (12.1.1) the Lagrangian function for a scalar (real or vector-valued) field u takes the form

$$\mathcal{L} = b^2 \left(1 - \sqrt{1 - \frac{1}{b^2}\partial_\mu u \partial^\mu u} \right) - V(u), \qquad (12.3.1)$$

where $V \geq 0$ is a potential density function.

We are interested in static configurations, $\partial_0 u = 0$. The energy from (12.3.1) is

$$\int_{\mathbb{R}^n} \left\{ b^2 \left(\sqrt{1 + \frac{1}{b^2} |\nabla u|^2} - 1 \right) + V(u) \right\} dx. \qquad (12.3.2)$$

The equation of motion of (12.3.1) is

$$\nabla \cdot \left(\frac{\nabla u}{\sqrt{1 + \frac{1}{b^2} |\nabla u|^2}} \right) = \frac{\delta V}{\delta u}, \qquad (12.3.3)$$

which is also the Euler–Lagrange equation of (12.3.2). Hence, we can focus on the critical points of (12.3.2) for which a rescaling argument like the one used in the last section leads us to the identity

$$\int_{\mathbb{R}^n} \frac{|\nabla u|^2}{\sqrt{1 + \frac{1}{b^2} |\nabla u|^2}} \, dx$$

$$= n \int_{\mathbb{R}^n} \left\{ b^2 \left(\sqrt{1 + \frac{1}{b^2} |\nabla u|^2} - 1 \right) + V(u) \right\} dx. \qquad (12.3.4)$$

It is clear that no value of $n = 1, 2, \cdots$ is excluded, which is in sharp contrast with the conclusion for the Klein–Gordon wave theory model with the Lagrangian function

$$\mathcal{L} = \frac{1}{2} \partial_\mu u \partial^\mu u - V(u)$$

for which only the numbers of dimensions, $n = 1, 2$, are possible, known as Derrick's theorem [89], presented in §1.3.3.

It will be interesting to develop an existence theory for solutions of (12.3.3) of nontrivial topology characterized by their asymptotic properties at infinity. For example, if u is a map from $\mathbb{R}^n$ to $\mathbb{R}^n$ and the potential function V singles out a ground state that breaks the global $SO(n)$ symmetry of the model such as the one of the form $|u| = 1$, u induces a map from S^{n-1} to itself and is thus characterized by the homotopy group $\pi_{n-1}(S^{n-1}) = \mathbb{Z}$ $(n \geq 2)$. Since (12.3.3) is not uniformly elliptic (ellipticity fails at $|\nabla u| = \infty$), an analytic study of it would require some tools such as Shiffman's regularization [281, 283, 285] or BV-spaces [126]. In this section we restrain ourselves from the general existence problem $(n \geq 2)$ but only consider the simplest case, $n = 1$ and u is real-valued (Born–Infeld kinks or domain walls).

When $n = 1$, the spatial variable is denoted by x and $\partial u / \partial x = u'$. The Hamiltonian density can then be decomposed as follows,

$$\mathcal{H} = b^2 \left(\sqrt{1 + \frac{1}{b^2} (u')^2} - 1 \right) + V(u)$$

$$= \frac{1}{2}\left(u' \pm b\sqrt{1 + \frac{1}{b^2}(u')^2} \sqrt{P(u)} \right)^2 \frac{1}{\sqrt{1 + \frac{1}{b^2}(u')^2}}$$

$$+ \frac{1}{2} \frac{b^2}{\sqrt{1 + \frac{1}{b^2}(u')^2}} \left(\sqrt{1 + \frac{1}{b^2}(u')^2} \sqrt{1 - P(u)} - 1 \right)^2$$

$$\mp bu'\sqrt{P(u)}, \tag{12.3.5}$$

where P and V are related through

$$V = b^2(1 - \sqrt{1 - P}), \quad P = 1 - \left(1 - \frac{V}{b^2}\right)^2. \tag{12.3.6}$$

Hence, $V = 0$ if and only if $P = 0$, which is a crucial property for analyzing ground states.

Recall that the equation of motion, (12.3.3), becomes

$$\left(\frac{u'}{\sqrt{1 + \frac{1}{b^2}(u')^2}} \right)' = V'(u), \tag{12.3.7}$$

and we are looking for a solution of (12.3.7) of finite energy, $\int \mathcal{H}\, dx < \infty$. Assume symmetry breaking so that V has at least two local minima, u_1 and u_2, satisfying $V(u_1) = V(u_2) = 0$, and set

$$Q = \int_{u_1}^{u_2} \sqrt{P(u)}\, du. \tag{12.3.8}$$

Then (12.3.5) indicates that any configuration u subject to the boundary condition

$$\lim_{x \to -\infty} u = u_1, \quad \lim_{x \to \infty} u = u_2, \tag{12.3.9}$$

has the energy lower bound

$$E(u) = \int_{-\infty}^{\infty} \mathcal{H}\, dx \geq b|Q|, \tag{12.3.10}$$

which is attained if and only if u makes the two quadratic terms on the right-hand side of (12.3.5) vanish. This latter requirement is equivalent to

$$u' = \mp \frac{b\sqrt{P(u)}}{\sqrt{1 - P(u)}}. \tag{12.3.11}$$

In other words, (12.3.11) subject to (12.3.9) gives us energy-minimizing solutions of the original equation, (12.3.7), with the minimum energy value

$E = b|Q|$. In particular, (12.3.11) implies (12.3.7). In fact, the two equations, (12.3.7) and (12.3.11), are equivalent. To see this, we use the substitution

$$v = \frac{u'}{\sqrt{1 + \frac{1}{b^2}(u')^2}} \tag{12.3.12}$$

to recast (12.3.7) into an autonomous first-order system,

$$u' = \frac{v}{\sqrt{1 - \frac{1}{b^2}v^2}},$$

$$v' = V'(u),$$

which leads to a separable equation,

$$\frac{du}{dv} = \frac{v}{V'(u)\sqrt{1 - \frac{1}{b^2}v^2}}.$$

Integrating the above equation and using the boundary condition $V(u) \to 0$, $v \to 0$, as $x \to \infty$, we obtain

$$V(u) = b^2 \left(1 - \sqrt{1 - \frac{v^2}{b^2}}\right).$$

Inserting (12.3.12) and using (12.3.6), we arrive at (12.3.11).

It is elementary to prove that, if $u_1 < u_2$, (12.3.11) subject to (12.3.9) has a solution with $u' > 0$ (lower sign in the equation) whenever $P \neq 0$ in the interval (u_1, u_2), and if $u_1 > u_2$, there is a solution with $u' < 0$ (upper sign in the equation) whenever $P \neq 0$ in (u_2, u_1). Translational invariance of (12.3.11) then gives us two continuous families of solutions with prescribed locations of maxima for u'. Consequently, we obtain the following integral for the solutions,

$$\int \sqrt{\frac{1}{P(u)} - 1}\, du = \mp bx + C. \tag{12.3.13}$$

We examine a special 'sine-Gordon' case where

$$V(u) = \sin^2 u, \quad u_1 = 0, \quad u_2 = \pi. \tag{12.3.14}$$

Using (12.3.14) in (12.3.6), we see that (12.3.11) becomes

$$u' = \frac{b \sin u \sqrt{2b^2 - \sin^2 u}}{b^2 - \sin^2 u}, \tag{12.3.15}$$

which can be solved implicitly.

12.3.2   In view of Nambu–Goto string theory

The relativistic motion of a free point particle of mass m described by the spatial coordinates $(x^i(t)) = \mathbf{x}(t)$ at any time $t = x^0$ is governed by the variational equation of the action,

$$L = \int_{t_1}^{t_2} m(1 - \sqrt{1 - \mathbf{v}^2})\, dt, \qquad (12.3.16)$$

where the velocity of light is set to unit, $c = 1$, and $\mathbf{v}(t) = d\mathbf{x}(t)/dt$ is the velocity of the particle. It will be sufficient to concentrate on the part of the action that depends on $\mathbf{v}$,

$$\mathcal{A} = -m \int_{t_1}^{t_2} \sqrt{1 - \mathbf{v}^2}\, dt. \qquad (12.3.17)$$

We now consider the motion of a free string of a uniform mass density, ρ_0, parametrized by a real parameter, s, with the spatial coordinates given as a parametrized curve,

$$\mathbf{x} = \mathbf{x}(s, t), \quad s_1 \le s \le s_2,$$

at any fixed time t. Following (12.3.17), the action for the motion of the infinitesimal portion

$$d\ell = \left| \frac{\partial \mathbf{x}}{\partial s} \right| ds, \quad s_1 \le s \le s_2,$$

is given by

$$-\rho_0\, d\ell \int_{t_1}^{t_2} \sqrt{1 - \mathbf{v}^2}\, dt. \qquad (12.3.18)$$

In the Nambu–Goto theory [133], the internal forces between neighboring points along a string do not contribute to the action. Hence, $\mathbf{v}$ in (12.3.18) should be taken to be the normal component of the velocity vector of the string according to the orthogonal decomposition

$$\frac{\partial \mathbf{x}}{\partial t} = \mathbf{v} + a(s, t) \frac{\partial \mathbf{x}}{\partial s}, \quad \mathbf{v} \cdot \frac{\partial \mathbf{x}}{\partial s} = 0. \qquad (12.3.19)$$

From (12.3.19), we can determine the scalar factor $a(s, t)$ and obtain $\mathbf{v}$ as follows,

$$\mathbf{v} = \frac{\partial \mathbf{x}}{\partial t} - \frac{\left(\frac{\partial \mathbf{x}}{\partial t} \cdot \frac{\partial \mathbf{x}}{\partial s} \right)}{\left(\frac{\partial \mathbf{x}}{\partial s} \right)^2} \frac{\partial \mathbf{x}}{\partial s}. \qquad (12.3.20)$$

Integrating (12.3.18), we obtain the total action for a Nambu–Goto string,

$$\mathcal{A} = -\rho_0 \int_{t_1}^{t_2} \int_{s_1}^{s_2} \sqrt{\left(\frac{\partial \mathbf{x}}{\partial s} \right)^2 (1 - \mathbf{v}^2)}\, ds\, dt.$$

Replacing the time variable t by the invariant proper time τ, renaming the parameter s,

$$t = t(\tau, \sigma), \quad s = \sigma,$$

and using the abbreviations

$$\dot{f} = \frac{\partial f}{\partial \tau}, \quad f' = \frac{\partial f}{\partial \sigma}$$

for partial derivatives, then

$$\frac{\partial \tau}{\partial t} = \frac{1}{\dot{t}},$$

$$\frac{\partial \tau}{\partial s} = -\frac{t'}{\dot{t}},$$

$$\frac{\partial(t, s)}{\partial(\tau, \sigma)} = \dot{t}.$$

As a consequence, we have

$$\frac{\partial \mathbf{x}}{\partial t} = \frac{\dot{\mathbf{x}}}{\dot{t}},$$

$$\frac{\partial \mathbf{x}}{\partial s} = \mathbf{x}' - \frac{t'}{\dot{t}}\dot{\mathbf{x}}.$$

Based on the above relations and (12.3.20), we have, after a lengthy calculation,

$$\left(\frac{\partial \mathbf{x}}{\partial s}\right)^2 (1 - \mathbf{v}^2) = \dot{t}^2((\dot{x}x')^2 - \dot{x}^2 x'^2), \tag{12.3.21}$$

where, for any two vectors, $x = (x^\mu) = (x^0, \mathbf{x})$ and $y = (y^\mu) = (y^0, \mathbf{y})$, in the flat $(3+1)$-dimensional Minkowski space with metric

$$(\eta_{\mu\nu}) = \mathrm{diag}\{1, -1, -1, -1\},$$

xy stands for the inner product

$$xy = x^\mu \eta_{\mu\nu} y^\mu = x^0 y^0 - x^i y^i = x^0 y^0 - \mathbf{x} \cdot \mathbf{y}.$$

The physical assumption that the motion is space-like, $|\mathbf{v}| < 1$, implies that the quantity expressed in (12.3.21) is positive,

$$(\dot{x}x')^2 - \dot{x}^2 x'^2 > 0. \tag{12.3.22}$$

Hence we arrive at the Nambu–Goto action

$$\mathcal{A} = -\rho_0 \int_{\sigma_1}^{\sigma_2} d\sigma \int_{\tau_1(\sigma)}^{\tau_2(\sigma)} d\tau \sqrt{(\dot{x}x')^2 - \dot{x}^2 x'^2}. \tag{12.3.23}$$

In order to find a geometric meaning of the action (12.3.23), we calculate the line element of the embedded 2-surface $x^\mu = x^\mu(\tau, \sigma)$, in the flat Minkowski spacetime $\mathbb{R}^{3,1}$, parametrized by the parameters τ and σ as follows,

$$
\begin{aligned}
ds^2 &= dx^\mu \eta_{\mu\nu} dx^\nu \\
&= (\dot{x}^\mu d\tau + x'^\mu d\sigma)\eta_{\mu\nu}(\dot{x}^\nu d\tau + x'^\nu d\sigma) \\
&= \dot{x}^2 d\tau^2 + 2\dot{x}x' d\tau d\sigma + x'^2 d\sigma^2 \\
&= g_{\alpha\beta} du^\alpha du^\beta, \quad \alpha, \beta = 0, 1,
\end{aligned}
$$

where $u^0 = \tau, u^1 = \sigma$, and

$$
(g_{\alpha\beta}) = \begin{pmatrix} \dot{x}^2 & \dot{x}x' \\ \dot{x}x' & x'^2 \end{pmatrix}. \tag{12.3.24}
$$

From (12.3.22)–(12.3.24), we see that the Nambu–Goto string action is simply a surface integral,

$$
A = -\rho_0 \int_\Omega \sqrt{|g|}\, d\tau d\sigma = -\rho_0 \int_S dS, \tag{12.3.25}
$$

where $|g|$ is the absolute value of the determinant of the matrix (12.3.24) and dS is the canonical area element of the embedded 2-surface, $(S, \{g_{\alpha\beta}\})$, in the Minkowski spacetime.

As a comparison, the action (12.3.17) for a point particle is simply a path integral,

$$
A = -m \int_{\tau_1}^{\tau_2} \sqrt{\dot{x}^2}\, d\tau = -m \int_C dC,
$$

where dC is the line element of the path C parametrized by $x^\mu = x^\mu(\tau)$, $\tau_1 \leq \tau \leq \tau_2$, in the Minkowski spacetime.

Therefore the motion of a point particle follows an extremized path, the world line, and the motion of a Nambu–Goto string follows an extremized surface, the world sheet.

To see the connection between the Nambu–Goto string action (12.3.25) and the Born–Infeld theory, we follow [24] to consider the motion of a string confined in the plane $x^3 = $ constant and parametrized by

$$
x^0 = t, \quad x^1 = u, \quad x^2 = v = v(t, u). \tag{12.3.26}
$$

In view of (12.3.26), the line element of the 2-surface is reduced to

$$
\begin{aligned}
ds^2 &= \eta_{\mu\nu} dx^\mu dx^\nu \\
&= dt^2 - du^2 - \left(\frac{\partial v}{\partial t} dt + \frac{\partial v}{\partial u} du\right)^2 \\
&\equiv g_{\alpha\beta} du^\alpha du^\beta, \quad u^0 = t, \quad u^1 = u, \quad \alpha, \beta = 0, 1,
\end{aligned}
$$

where

$$(g_{\alpha\beta}) = \begin{pmatrix} 1 - \left(\frac{\partial v}{\partial t}\right)^2 & -\frac{\partial v}{\partial t}\frac{\partial v}{\partial u} \\ -\frac{\partial v}{\partial t}\frac{\partial v}{\partial u} & -1 - \left(\frac{\partial v}{\partial u}\right)^2 \end{pmatrix}.$$

Hence the Nambu–Goto string action reads

$$\mathcal{A} = -\rho_0 \int_\Omega \sqrt{1 - \left(\left[\frac{\partial v}{\partial t}\right]^2 - \left[\frac{\partial v}{\partial u}\right]^2\right)}\, dtdu.$$

Therefore, the Lagrangian action L corresponding to (12.3.16) now takes the form

$$L = \rho_0 \int (1 - \sqrt{1 - \partial^\mu v \partial_\mu v})\, dtdu, \qquad (12.3.27)$$

which is of the Born–Infeld form as expected.

It is immediate to extend the Nambu–Goto string theory to include higher dimensional geometric entities such as relativistic membranes [85]. The Born–Infeld electromagnetism also arises from the quantized open string theory [111].

12.4   Abelian Strings

In this section, we study the classical Abelian Higgs model in which electromagnetism obeys the Born–Infeld geometric dynamics. We first formulate the problem and state our main existence and uniqueness theorems concerning multiple strings or vortices. We then present our analysis on the compact and noncompact cases separately. It will be seen that the Born–Infeld parameter plays an interesting and distinctive role in the solutions obtained.

12.4.1   Existence and uniqueness theorems

Suppose that the Minkowski spacetime metric is $\{g_{\mu\nu}\}$. The Abelian Higgs model in which the electrodynamics is governed by the Born–Infeld theory is defined by the action density

$$\mathcal{L} = b^2 \left(1 - \sqrt{1 + \frac{1}{2b^2} g^{\mu\mu'} g^{\nu\nu'} F_{\mu\nu} F_{\mu'\nu'}}\right)$$
$$+ g^{\mu\nu}(D_\mu\phi)(\overline{D_\nu\phi}) - V(|\phi|^2), \qquad (12.4.1)$$

where $F_{\mu\nu} = \partial_\mu A_\nu - \partial_\nu A_\mu$ is the electromagnetic field induced from the gauge field A_μ, ϕ is a complex scalar field, $D_\mu\phi = \partial_\mu\phi - iA_\mu\phi$ is the gauge-covariant derivative, and V is a Higgs potential density function which spontaneously breaks vacuum symmetry with (say) $V(1) = 0$.

The equations of motion of (12.4.1) are

$$\frac{1}{\sqrt{|g|}} D_\mu(g^{\mu\nu}\sqrt{|g|}[D_\nu\phi]) = -V'(|\phi|^2)\phi, \qquad (12.4.2)$$

$$\frac{1}{\sqrt{|g|}}\partial_{\mu'}\left(\frac{g^{\mu\nu}g^{\mu'\nu'}\sqrt{|g|}F_{\nu\nu'}}{\sqrt{1+\frac{1}{2b^2}g^{\alpha\alpha'}g^{\beta\beta'}F_{\alpha\beta}F_{\alpha'\beta'}}}\right) = ig^{\mu\nu}(\phi\overline{D_\nu\phi} - \overline{\phi}D_\nu\phi).$$

$$(12.4.3)$$

These equations govern a Born–Infeld type Abelian Higgs dynamics over a gravitational background. If we use this model to generate gravitation, the Einstein equations

$$G_{\mu\nu} = -8\pi G T_{\mu\nu} \qquad (12.4.4)$$

should also be considered. Here $G_{\mu\nu}$ is the Einstein tensor, G is Newton's constant, and $T_{\mu\nu}$ is the energy-momentum tensor of (12.4.1) given by

$$T_{\mu\nu} = -\frac{g^{\mu'\nu'}F_{\mu\mu'}F_{\nu\nu'}}{\sqrt{1+\frac{1}{2b^2}g^{\alpha\alpha'}g^{\beta\beta'}F_{\alpha\beta}F_{\alpha'\beta'}}}$$
$$+(D_\mu\phi\overline{D_\nu\phi} + \overline{D_\mu\phi}D_\nu\phi) - g_{\mu\nu}\mathcal{L}. \qquad (12.4.5)$$

We shall look for string-like solutions for which the spacetime is of the form $\mathbb{R}^{1,1}\times S$ where S is a Riemann surface. In order to simplify calculation, we use locally conformally flat (isothermal) coordinate patches on S so that the line element becomes diagonal,

$$(g_{\mu\nu}) = \text{diag}\{1, -e^\eta, -e^\eta, -1\}, \qquad (12.4.6)$$

where η depends on local coordinates x^1, x^2 on S only. Accordingly, if we assume that the fields ϕ and A_μ depend only on x^1, x^2 and $A_0 = A_3 = 0$, we obtain from (12.4.5) the components of $T_{\mu\nu}$,

$$\begin{aligned}
T_{03} &= T_{0j} = T_{3j} = 0, \quad j = 1, 2,\\
T_{00} &= -T_{33} = \mathcal{H}\\
T_{12} &= D_1\phi\overline{D_2\phi} + \overline{D_1\phi}D_2\phi,\\
T_{11} &= (|D_1\phi|^2 - |D_2\phi|^2) - e^\eta V(|\phi|^2)\\
&\quad +e^\eta b^2\left(1 - \left[1 + \frac{1}{b^2}e^{-2\eta}F_{12}^2\right]^{-\frac{1}{2}}\right),\\
T_{22} &= (|D_2\phi|^2 - |D_1\phi|^2) - e^\eta V(|\phi|^2)\\
&\quad +e^\eta b^2\left(1 - \left[1 + \frac{1}{b^2}e^{-2\eta}F_{12}^2\right]^{-\frac{1}{2}}\right), \qquad (12.4.7)
\end{aligned}$$

where $\mathcal{H}$ is the Hamiltonian density of the matter-gauge sector defined by

$$\mathcal{H} = -\mathcal{L} = b^2\left(\sqrt{1 + \frac{1}{b^2}e^{-2\eta}F_{12}^2} - 1\right)$$

$$+e^{-\eta}(|D_1\phi|^2 + |D_2\phi|^2) + V(|\phi|^2). \qquad (12.4.8)$$

With the identity

$$|D_1\phi|^2 + |D_2\phi|^2 = |D_1\phi \pm iD_2\phi|^2 \pm i(D_1\phi\overline{D_2\phi} - \overline{D_1\phi}D_2\phi), \qquad (12.4.9)$$

we may rewrite (12.4.8) as

$$\mathcal{H} = \frac{1}{2}\left(e^{-\eta}F_{12} \pm \sqrt{1 + \frac{1}{b^2}e^{-2\eta}F_{12}^2}(|\phi|^2 - 1)\right)^2 \left(1 + \frac{1}{b^2}e^{-2\eta}F_{12}^2\right)^{-\frac{1}{2}}$$

$$+ \frac{b^2}{2}\left(\sqrt{1 + \frac{1}{b^2}e^{-2\eta}F_{12}^2}\sqrt{1 - \frac{1}{b^2}(|\phi|^2 - 1)^2} - 1\right)^2 \left(1 + \frac{1}{b^2}e^{-2\eta}F_{12}^2\right)^{-\frac{1}{2}}$$

$$- b^2 \mp e^{-\eta}F_{12}(|\phi|^2 - 1) + b^2\sqrt{1 - \frac{1}{b^2}(|\phi|^2 - 1)^2}$$

$$+ e^{-\eta}|D_1\phi \pm iD_2\phi|^2 \pm e^{-\eta}\,i(D_1\phi\overline{D_2\phi} - \overline{D_1\phi}D_2\phi) + V(|\phi|^2). \qquad (12.4.10)$$

Recall that $(D_1D_2 - D_2D_1)\phi = -iF_{12}\phi$. Hence the 'current' density

$$J_k = \frac{1}{2}i(\phi\overline{D_k\phi} - \bar{\phi}D_k\phi), \quad k = 1,2 \qquad (12.4.11)$$

gives rise to the following 'vorticity' field,

$$J_{12} = \partial_1 J_2 - \partial_2 J_1 = i(D_1\phi\overline{D_2\phi} - \overline{D_1\phi}D_2\phi) - |\phi|^2 F_{12}. \qquad (12.4.12)$$

Inserting (12.4.12) into (12.4.8), we have the inequality

$$\mathcal{H} \geq \pm e^{-\eta}(F_{12} + J_{12}) + V(|\phi|^2) - b^2\left(1 - \sqrt{1 - \frac{1}{b^2}(|\phi|^2 - 1)^2}\right), \qquad (12.4.13)$$

and, the equality

$$\mathcal{H} = \pm e^{-\eta}(F_{12} + J_{12}) \qquad (12.4.14)$$

holds if the following are satisfied simultaneously,

$$V(|\phi|^2) = b^2\left(1 - \sqrt{1 - \frac{1}{b^2}(|\phi|^2 - 1)^2}\right), \qquad (12.4.15)$$

$$D_1\phi \pm iD_2\phi = 0, \qquad (12.4.16)$$

$$\frac{e^{-\eta}F_{12}}{\sqrt{1 + \frac{1}{b^2}e^{-2\eta}F_{12}^2}} = \pm(1 - |\phi|^2), \qquad (12.4.17)$$

$$\sqrt{1 + \frac{1}{b^2}e^{-2\eta}F_{12}^2}\sqrt{1 - \frac{1}{b^2}(|\phi|^2 - 1)^2} = 1. \qquad (12.4.18)$$

The expression (12.4.15) specifies the form of the potential density V. In order to accommodate the symmetric (normal) vacuum, $\phi = 0$, we must require that

$$b > 1. \qquad (12.4.19)$$

This condition and (12.4.15) will be observed throughout the rest of the chapter.

A short calculation shows that the system of equations, (12.4.16) through (12.4.18), is equivalent to the system

$$D_j\phi \pm i\epsilon_j^k D_k\phi = 0, \tag{12.4.20}$$

$$\frac{1}{2}\epsilon^{jk}F_{jk} = \pm\frac{(1-|\phi|^2)}{\sqrt{1-\frac{1}{b^2}(1-|\phi|^2)^2}}, \tag{12.4.21}$$

where we have used the skew-symmetric 2-tensor ϵ_{jk} on $(S,\{g_{jk}\})$ with $\epsilon_{12} = \sqrt{g}$ to express the equations intrinsically (globally independent of specification of local coordinates) on S. We have used g to denote the determinant of the metric $\{g_{jk}\}$. Sometimes we also use g to suppress the notation $\{g_{jk}\}$.

The definition of the current J_k gives us

$$\int_S dJ = \int_S \epsilon^{jk}J_{jk}\,d\Omega_g = 0.$$

Besides, if we use a complex line bundle L over S to describe the model in which ϕ is a section, $A = (A_j)$ is a connection 1-form, $(F_{jk}) = dA$ its curvature, and $D_j\phi = (D_A\phi)_j$ denotes the jth component of the connection, then

$$\frac{1}{4\pi}\int_S \epsilon^{jk}F_{jk}\,d\Omega_g = c_1(L) \tag{12.4.22}$$

is the first Chern class of $L \to S$. The inequality (12.4.13) implies the validity of the following topological energy lower bound

$$E = \int_S \mathcal{H}\,d\Omega_g \geq 2\pi|c_1(L)| \tag{12.4.23}$$

and the solutions of (12.4.20) and (12.4.21) are energy minimizers in the same topological class.

In fact, it is straightforward to show that, on S, any solution of (12.4.20) and (12.4.21) satisfies the equations of motion, (12.4.2) and (12.4.3). In other words, the system of equations (12.4.20) and (12.4.21) is a reduction of the system of equations (12.4.2) and (12.4.3). Through these reduced equations, we are able to establish the following existence theorems.

Theorem 12.4.1. *Suppose that S is compact and $|S|$ is the total surface area of (S,g). For any integer N, the system of equations (12.4.2) and (12.4.3) has an energy-minimizing solution to saturate the lower bound (12.4.23) with $c_1(L) = \pm N$ if and only if*

$$N < \frac{|S|}{2\pi\sqrt{1-\frac{1}{b^2}}}. \tag{12.4.24}$$

Moreover, if N satisfies (12.4.24), then for any given N points on S, the system of equations (12.4.2) and (12.4.3) has a unique solution (ϕ, A) so that ϕ vanishes exactly at these points and (ϕ, A) realizes the quantized minimum energy value

$$E = 2\pi N. \tag{12.4.25}$$

If S is noncompact, say, $(S, g) = (\mathbb{R}^2, \delta_{jk})$, we have

Theorem 12.4.2. *For any prescribed N points in $\mathbb{R}^2$, the system of equations (12.4.2) and (12.4.3) has a unique energy-minimizing solution (ϕ, A) so that ϕ vanishes at these points, the energy of the solution is given by (12.4.25), and the total magnetic flux Φ has the expression*

$$\Phi = \int_{\mathbb{R}^2} F_{12} \, dx = 2\pi N \ or \ -2\pi N. \tag{12.4.26}$$

Besides, the quantities $|\phi|^2 - 1$, $|D_j\phi|^2$, and F_{12} all vanish at infinity exponentially fast according to the sharp decay rate

$$|\phi|^2 - 1, \quad |D_1\phi|^2 + |D_2\phi|^2, \quad F_{12} = O(e^{-\sqrt{2}|x|}).$$

It is interesting to notice that the decay estimates above are independent of the Born–Infeld parameter $b > 0$. On the other hand, however, in view of (12.4.21) the maximum penetration of the magnetic field $B = \pm F_{12}$ appears at the zeros of ϕ,

$$B_{\max} = \frac{1}{\sqrt{1 - \frac{1}{b^2}}},$$

which may achieve arbitrarily high values as $b \to 1$. In other words, b may be adjusted to give rise to arbitrarily localized solutions with sharp high peaks for the magnetic field. Such a phenomenon is also true for the solutions over a compact surface. In fact, (12.4.24) implies in addition that, as b gets closer and closer to 1, there exist arbitrarily localized solutions of more and more magnetic lumps.

We next consider the coupling of the Einstein equations (12.4.4) in the problem.

Inserting (12.4.15)–(12.4.18) into (12.4.7), we see that, except for T_{00} and T_{33}, all $T_{\mu\nu}$ vanish. On the other hand, in view of the metric (12.4.6), the Einstein tensor has the reduction

$$G_{00} = -G_{33} = -K_g, \quad G_{\mu\nu} = 0 \quad \text{for other values of } \mu, \nu, \tag{12.4.27}$$

where K_g is the Gauss curvature of the surface $(S, \{g_{jk}\})$. Hence, the Einstein equations (12.4.4) are reduced to a scalar one,

$$K_g = 8\pi G \mathcal{H}. \tag{12.4.28}$$

This equation gives us a precise relation between geometry and energy.

When S is compact, the Gauss–Bonnet theorem constrains the total curvature of $(S, \{g_{jk}\})$ to the topology of S,

$$\int_S K_g \, d\Omega_g = 2\pi\chi(S), \quad \chi(S) = 2 - 2n, \qquad (12.4.29)$$

where $\chi(S)$ is the Euler characteristic and n is the genus of S. For nontrivial solutions, we must have

$$\int_S \mathcal{H} \, d\Omega_g > 0. \qquad (12.4.30)$$

Hence, (12.4.28)–(12.4.30) imply that $\chi(S) > 0$ or

$$n = 0. \qquad (12.4.31)$$

In other words, topologically, S must be a sphere. This conclusion is identical to the case in the Abelian Higgs theory. Besides, since the equations of motion are reduced to (12.4.20), (12.4.21), and (12.4.28), the energy is quantized according to (12.4.25). Therefore, integrating (12.4.28) and using (12.4.25), (12.4.29), and (12.4.31), we obtain the quantization of Newton's constant,

$$G = G_N \equiv \frac{1}{4\pi N}. \qquad (12.4.32)$$

Such a rigidity in fact comes from the specification of the energy scale of the spontaneous symmetry-breaking characterized by the vacuum state $|\phi|^2 = 1$. If we replace it by an arbitrary energy scale, $|\phi|^2 = \varepsilon^2$ ($\varepsilon > 0$), the Higgs potential density assumes the form

$$V(|\phi|^2) = b^2 \left(1 - \sqrt{1 - \frac{1}{b^2}(|\phi|^2 - \varepsilon^2)^2} \right). \qquad (12.4.33)$$

It can be shown that a similar procedure may be worked out to arrive at a quantization formula for ε instead of G (for fixed G). Here we only study the problem defined by the fixed vacuum $|\phi|^2 = 1$.

Using again local isothermal coordinates and the complex differentiation,

$$\partial = \frac{1}{2}(\partial_1 - i\partial_2), \quad \bar{\partial} = \frac{1}{2}(\partial_1 + i\partial_2),$$

we can express A_1 and A_2 via (12.4.16) as

$$A_1 = -\mathrm{Re}\{2i\bar{\partial}\ln\phi\}, \quad A_2 = -\mathrm{Im}\{2i\bar{\partial}\ln\phi\}. \qquad (12.4.34)$$

Thus, by the (local) relation $\phi = e^{\frac{1}{2}u + i\omega}$ where ω is a real-valued function, we have

$$D_1\phi = (\partial + \bar{\partial})\phi - \left(\frac{\bar{\partial}\phi}{\phi} - \frac{\partial\bar{\phi}}{\bar{\phi}}\right)\phi = \phi\partial u,$$

$$D_2\phi = i(\partial - \bar{\partial})\phi + i\left(\frac{\bar{\partial}\phi}{\phi} + \frac{\partial\bar{\phi}}{\bar{\phi}}\right)\phi = i\phi\partial u.$$

Hence, we have

$$|D_1\phi|^2 + |D_2\phi|^2 = \frac{1}{2}e^u|\nabla u|^2. \tag{12.4.35}$$

In view of (12.4.34), we see that, away from zeros of ϕ, there holds

$$F_{12} = -\frac{1}{2}\Delta \ln|\phi|^2. \tag{12.4.36}$$

Since the zeros of ϕ are discrete and of integer multiplicities, we can apply (12.4.36) to arrive at the global relation

$$|\phi|^2 F_{12} = -\frac{1}{2}e^u\Delta u. \tag{12.4.37}$$

Using (12.4.12), (12.4.16), (12.4.21), (12.4.35), and (12.4.37) in (12.4.14) and return to general coordinates, we have

$$\mathcal{H} = -\frac{(e^u-1)}{\sqrt{1 - \frac{1}{b^2}(e^u-1)^2}} + \frac{1}{2}\Delta_g(e^u), \tag{12.4.38}$$

where Δ_g is the Laplace–Beltrami operator with the sign convention $\Delta_g = \Delta$ when the metric $g = \{g_{jk}\}$ is Euclidean.

On the other hand, suppose that the unknown gravitational metric g is a conformal deformation of a known one, g_0. Then there is a function η such that $g = e^\eta g_0$ and $\Delta_g = e^{-\eta}\Delta_{g_0}$. Besides, the Gauss curvatures K_g on (S, g) and K_{g_0} on (S, g_0) are related through

$$-\Delta_{g_0}\eta + 2K_{g_0} = 2K_g e^\eta. \tag{12.4.39}$$

Inserting (12.4.38) and (12.4.39) into (12.4.28), we have

$$\Delta_{g_0}(\eta + 8\pi G e^u) = 2K_{g_0} + 16\pi G e^\eta \frac{(e^u-1)}{\sqrt{1 - \frac{1}{b^2}(e^u-1)^2}}. \tag{12.4.40}$$

Furthermore, with $u = \ln|\phi|^2$, we can compress the equations (12.4.20) and (12.4.21) into a scalar one,

$$\Delta_{g_0}u = 2e^\eta \frac{(e^u-1)}{\sqrt{1 - \frac{1}{b^2}(e^u-1)^2}} + 4\pi\sum_{s=1}^{N}\delta_{p_s}, \tag{12.4.41}$$

where δ_p is the Dirac distribution on (S, g_0) concentrated at the point p.

Combining (12.4.40) and (12.4.41), we arrive at

$$\Delta_{g_0}(\eta + 8\pi G[e^u - u]) = 2K_{g_0} - 32\pi^2 G\sum_{s=1}^{N}\delta_{p_s}. \tag{12.4.42}$$

Using the condition (12.4.32) in (12.4.42), which now reads

$$4\pi = \int_S K_{g_0}\, d\Omega_{g_0} = 16\pi^2 GN,$$

we see that there is a function u_0 which is smooth away from p_s, behaves like $\ln|x - p_s|^2$ near p_s ($s = 1, 2, \cdots, N$), and resolves (12.4.42),

$$\eta + 8\pi G(e^u - u) = -8\pi G u_0 + c, \qquad (12.4.43)$$

where c is an arbitrary constant. From (12.4.43), we see that the governing equations (12.4.40) and (12.4.41) become a single equation in the new variable $v = u - u_0$,

$$\Delta_{g_0} v = \lambda e^{8\pi G(v - e^{u_0 + v})} \frac{(e^{u_0 + v} - 1)}{\sqrt{1 - \frac{1}{b^2}(e^{u_0 + v} - 1)^2}} + \frac{K_{g_0}}{4\pi G}, \qquad (12.4.44)$$

where $\lambda > 0$ is an arbitrarily adjustable parameter.

We next consider the noncompact case when (S, g) is conformal to $\mathbb{R}^2$, namely, $(S, \{g_{jk}\}) = (\mathbb{R}^2, \{e^\eta \delta_{jk}\})$. Hence g_0 is Euclidean, $K_{g_0} = 0$, and (12.4.43) may be rewritten

$$\eta = 8\pi G(u - e^u) - 8\pi G \sum_{s=1}^{N} \ln|x - p_s|^2 + c.$$

Consequently, the conformal factor of the gravitational metric takes the form

$$e^\eta = c_0 \left(\prod_{s=1}^{N} |x - p_s| \right)^{-16\pi G} e^{8\pi G(u - e^u)} \qquad (12.4.45)$$

where $c_0 > 0$ is arbitrary. Inserting (12.4.45) into (12.4.41) and writing $\lambda = 2c_0$, we obtain

$$\Delta u = \lambda \left(\prod_{s=1}^{N} |x - p_s| \right)^{-16\pi G} e^{8\pi G(u - e^u)} \frac{(e^u - 1)}{\sqrt{1 - \frac{1}{b^2}(e^u - 1)^2}} + 4\pi \sum_{s=1}^{N} \delta_{p_s}, \qquad (12.4.46)$$

Since u satisfies (12.4.46), it has logarithmic singularities at the points $p_1, p_2, \cdots, p_N$ as well. These singularities cancel those power-function-type singularities seen in the expression of e^η given in (12.4.45). Therefore e^η is everywhere regular. On the other hand, within the category of solutions of our interest (solutions with broken vacuum symmetry), $u = 0$ at infinity. Applying this information in (12.4.46), we see the validity of the following sharp asymptotic estimate,

$$e^{\eta(x)} = O(|x|^{-16\pi GN}), \quad x \in \mathbb{R}^2, \quad |x| \to \infty. \qquad (12.4.47)$$

This result implies that the metric $g = \{e^\eta \delta_{jk}\}$ is complete if and only if the string number N is bounded by

$$N \leq \frac{1}{8\pi G}. \tag{12.4.48}$$

This conclusion is similar to that for the classical Abelian Higgs cosmic strings studied earlier. In fact, the equations (12.4.44) and (12.4.46) are technically analogous to the corresponding equations in the Abelian Higgs theory and the mathematical analysis there may be carried over to the Born–Infeld theory to establish existence and nonexistence results both over a compact surface and a noncompact surface conformal to $\mathbb{R}^2$. For example, we state without proof our existence theorem for multiple strings over $\mathbb{R}^2$ as follows.

Theorem 12.4.3. *For any prescribed points $p_1, p_2, \cdots, p_N$ in $\mathbb{R}^2$ satisfying the necessary and sufficient condition (12.4.48) for completeness of the gravitational metric, the Born–Infeld–Higgs equations (12.4.2) and (12.4.3) coupled with the Einstein equations (12.4.4) have a solution so that the metric is of the form (12.4.6), $A_0 = A_3 = 0$, ϕ and A_j ($j = 1, 2$) depend only on coordinates on $S = \mathbb{R}^2$, the points $p_1, p_2, \cdots, p_N$ are the zeros of ϕ. Moreover, the Hamiltonian density $\mathcal{H}$ and the Gauss curvature K_g of S both vanish at infinity sufficiently fast so that the total energy and magnetic flux are quantized according to (12.4.25) and (12.4.26), respectively, and the total curvature takes the value*

$$\int_{\mathbb{R}^2} K_g \, e^\eta \, dx = 16\pi^2 GN.$$

In the next two subsections, we present our proofs of Theorems 12.4.1 and 12.4.2.

12.4.2   Analysis of compact surface case

Consider the system of equations (12.4.20) and (12.4.21) when the background manifold (S, g) is a compact surface. The prescribed zeros of ϕ are $p_1, p_2, \cdots, p_N$. Then, with the substitution $u = \ln |\phi|^2$, we arrive at the equivalent nonlinear elliptic equation

$$\Delta_g u = \frac{2(e^u - 1)}{\sqrt{1 - \frac{1}{b^2}(e^u - 1)^2}} + 4\pi \sum_{s=1}^N \delta_{p_s}, \quad x \in S. \tag{12.4.49}$$

It is easily seen from the maximum principle that $u < 0$ everywhere.

Let u_0 be a solution of

$$\Delta_g u_0 = -\frac{4\pi N}{|S|} + 4\pi \sum_{s=1}^N \delta_{p_s}, \quad x \in S. \tag{12.4.50}$$

Then $v = u - u_0$ satisfies

$$\Delta_g v = \frac{2(e^{u_0+v} - 1)}{\sqrt{1 - \frac{1}{b^2}(e^{u_0+v} - 1)^2}} + \frac{4\pi N}{|S|} \qquad (12.4.51)$$

and

$$u_0 + v < 0, \quad x \in S. \qquad (12.4.52)$$

On the other hand, since the function

$$\sigma(t) = \frac{2(t - 1)}{\sqrt{1 - \frac{1}{b^2}(t - 1)^2}}, \quad 0 \le t \le 1, \qquad (12.4.53)$$

is increasing, we have, in view of (12.4.52), that

$$-\frac{2}{\sqrt{1 - \frac{1}{b^2}}} \le \frac{2(e^{u_0+v} - 1)}{\sqrt{1 - \frac{1}{b^2}(e^{u_0+v} - 1)^2}}. \qquad (12.4.54)$$

Therefore, integrating (12.4.51) and using (12.4.54), we get

$$a \equiv \frac{2}{\sqrt{1 - \frac{1}{b^2}}} > \frac{4\pi N}{|S|}, \qquad (12.4.55)$$

which is the condition stated in Theorem 12.4.1.

We now show that (12.4.55) is also sufficient. To this end, we consider the modified equation

$$\Delta_g w = a(e^{u_0+w} - 1) + \frac{4\pi N}{|S|}, \quad x \in S. \qquad (12.4.56)$$

We solve (12.4.56) by the following constrained optimization problem,

$$\min \left\{ I(w) \,\middle|\, w \in W^{1,2}(S), \; J(w) = a|S| - 4\pi N \right\}, \qquad (12.4.57)$$

where I and J are functionals defined by

$$I(w) = \int_S \frac{1}{2}|\nabla w|^2 \, d\Omega_g - (a|S| - 4\pi N)\underline{w},$$

$$J(w) = a \int_S e^{u_0+w} \, d\Omega_g,$$

$$w \in W^{1,2}(S), \quad w = \underline{w} + w', \quad \underline{w} \in \mathbb{R}, \quad \int_S w' \, d\Omega_g = 0.$$

It follows from the Trudinger–Moser inequality,

$$\int_S e^{w'} \, d\Omega_g \le C_1 \int_S e^{C_2 \|\nabla w'\|_2^2} \, d\Omega_g,$$

$$w' \in W^{1,2}(S), \quad \int_S w' \, d\Omega_g = 0, \qquad (12.4.58)$$

where C_1 and C_2 are some positive constants and $\|\cdot\|$ denotes the L^2 norm for a function over S, that the functional J is well-defined and is weakly continuous.

Let w lie in the described admissible class. Inserting the decomposition $w = \underline{w} + w'$ into the constraint $J(w) = a|S| - 4\pi N > 0$, we obtain

$$\underline{w} = \ln\left(|S| - \frac{4\pi N}{a}\right) - \ln\left(\int_S e^{u_0 + w'}\, d\Omega_g\right). \tag{12.4.59}$$

In view of Jensen's inequality, we have

$$\frac{1}{|S|}\int_S e^{u_0 + w'}\, d\Omega_g \geq \exp\left(\frac{1}{|S|}\int_S u_0\, d\Omega_g\right). \tag{12.4.60}$$

Using (12.4.60) in (12.4.59), we see that $\underline{w}$ is uniformly bounded from above.

Furthermore, substituting (12.4.59) into $I(w)$, we have

$$\begin{aligned}
I(w) &= \frac{1}{2}\|\nabla w'\|_2^2 - (a|S| - 4\pi N)\ln\left(|S| - \frac{4\pi N}{a}\right) \\
&\quad + (a|S| - 4\pi N)\ln\left(\int_S e^{u_0 + w'}\, d\Omega_g\right).
\end{aligned} \tag{12.4.61}$$

Applying (12.4.55) and (12.4.60) in (12.4.61), we see that there is a constant $C > 0$ such that

$$I(w) \geq \frac{1}{2}\|\nabla w'\|_2^2 - C. \tag{12.4.62}$$

Thus, if w belongs to a minimizing sequence of the problem (12.4.57), then (12.4.62) implies that $\|\nabla w'\|_2^2$ is uniformly bounded. Hence, by (12.4.58) and (12.4.59), we see that $\underline{w}$ is also uniformly bounded from below.

In summary, a minimizing sequence of (12.4.57) is bounded in $W^{1,2}(S)$, which must have a weakly convergent subsequence. Since I is weakly lower semicontinuous and J is weakly continuous, the weak limit in $W^{1,2}(S)$, say, w, solves (12.4.57).

Let w be a solution of (12.4.57). Then there is a number λ, the Lagrange multiplier, such that

$$\int_S \nabla w \cdot \nabla f\, d\Omega_g - \left(a - \frac{4\pi N}{|S|}\right)\int_S f\, d\Omega_g$$

$$= \lambda \int_S e^{u_0 + w} f\, d\Omega_g, \quad \forall f \in W^{1,2}(S). \tag{12.4.63}$$

Choose $f = 1$ in (12.4.63). Using the constraint $J(w) = a|S| - 4\pi N$ again, we get $\lambda = -a$. Inserting this result into (12.4.63), we see that w is a weak solution of (12.4.56). The elliptic theory then indicates that w is a smooth solution of (12.4.56).

With the above preparation, we can find a pair of sub- and supersolutions of (12.4.51), say, v_- and v_+ satisfying $v_- < v_+$.

To this end, let w be a solution of (12.4.56) just obtained. It is clear that $u_0 + w < 0$. Consequently,

$$1 - \frac{1}{b^2}(e^{u_0+w} - 1)^2 \geq 1 - \frac{1}{b^2},$$

which immediately leads us to

$$\Delta_g w \leq \frac{2(e^{u_0+w} - 1)}{\sqrt{1 - \frac{1}{b^2}(e^{u_0+w} - 1)^2}} + \frac{4\pi N}{|S|}, \quad x \in S.$$

In other words, $v_+ = w$ is a supersolution of (12.4.51).

On the other hand, since for $\varepsilon > 0$ sufficiently small, we have in view of (12.4.55) that

$$a_\varepsilon \equiv \frac{2}{\sqrt{1 - \frac{(1-\varepsilon)}{b^2}}} > \frac{4\pi N}{|S|}. \tag{12.4.64}$$

Therefore, the equation

$$\Delta_g w_\varepsilon = a_\varepsilon(e^{u_0+w_\varepsilon} - 1) + \frac{4\pi N}{|S|} \tag{12.4.65}$$

also has a solution w_ε satisfying $u_0 + w_\varepsilon < 0$. Choose $c > 0$ sufficiently large so that

$$0 < 1 - \frac{1}{b^2}(e^{u_0+w_\varepsilon-c} - 1)^2 < 1 - \frac{(1-\varepsilon)}{b^2} \quad \text{on } S.$$

Hence, from (12.4.65), we have

$$\Delta_g w_\varepsilon \geq a_\varepsilon(e^{u_0+w_\varepsilon-c} - 1) + \frac{4\pi N}{|S|}$$

$$> \frac{2(e^{u_0+w_\varepsilon-c} - 1)}{\sqrt{1 - \frac{1}{b^2}(e^{u_0+w_\varepsilon-c} - 1)^2}} + \frac{4\pi N}{|S|}.$$

That is, $w_- = w_\varepsilon - c$ is a subsolution of (12.4.51). We may of course choose c large to make $w_- < w_+$.

Using the pair $w_- < w_+$, we can show that (12.4.51) has a solution w: $w_- \leq w \leq w_+$.

Since the function (12.4.53) is increasing, it is standard to show by the maximum principle that (12.4.51) may only have one solution.

Theorem 12.4.1 is proven.

12.4.3 Solutions on noncompact surfaces

Motivated by the study of the last subsection, we continue to resort to sub- and supersolutions. Equation (12.4.49) with $S = \mathbb{R}^2$ becomes

$$\Delta u = \frac{2(e^u - 1)}{\sqrt{1 - \frac{1}{b^2}(e^u - 1)^2}} + 4\pi \sum_{s=1}^{N} \delta_{p_s}, \quad x \in \mathbb{R}^2. \tag{12.4.66}$$

We first recall that the modified equation

$$\Delta v = a(e^v - 1) + 4\pi \sum_{s=1}^{N} \delta_{p_s}, \quad x \in \mathbb{R}^2 \tag{12.4.67}$$

occurs in the classical Abelian Higgs model and is known [157] to have a unique solution, say, v_a, that vanishes at infinity. Here $a > 0$ is an arbitrary parameter.

Let $a = a_1$ in (12.4.67) be defined by

$$a_1 = \frac{2}{\sqrt{1 - \frac{1}{b^2}}}. \tag{12.4.68}$$

Since $v_{a_1} < 0$, we see that $u_+ = v_{a_1}$ is a supersolution of (12.4.66) because

$$a_1(e^{v_{a_1}} - 1) \le \frac{2(e^{v_{a_1}} - 1)}{\sqrt{1 - \frac{1}{b^2}(e^{v_{a_1}} - 1)^2}}.$$

On the other hand, let $a = a_2 > 0$ in (12.4.67) be small enough so that

$$a_2 < \frac{2}{\sqrt{1 - \frac{1}{b^2}(e^{v_{a_1}} - 1)^2}}.$$

Since $a_2 < a_1$ and $v_{a_2} < 0$, it is clear that v_{a_2} is a subsolution of (12.4.67) when $a = a_1$. The maximum principle implies that $v_{a_1} > v_{a_2}$. Consequently,

$$a_2 < \frac{2}{\sqrt{1 - \frac{1}{b^2}(e^{v_{a_1}} - 1)^2}} < \frac{2}{\sqrt{1 - \frac{1}{b^2}(e^{v_{a_2}} - 1)^2}},$$

which says that $u_- = v_{a_2}$ is a subsolution of (12.4.66).

Since $u_- < u_+$, the existence of a solution u to (12.4.66) such that $u_- \le u \le u_+$ follows. Using the decay estimates for u_- and u_+, we can see that u also vanishes at infinity exponentially fast. The uniqueness of a solution follows easily from the property that the nonlinear term in (12.4.66) has a positive derivative.

We now consider the decay rate of a solution near infinity.

It is easily seen that there is a function $\xi(x)$ satisfying $\xi(x) \to 0$ as $|x| \to \infty$ such that

$$\frac{2(e^u - 1)}{\sqrt{1 - \frac{1}{b^2}(e^u - 1)^2}} = \frac{2e^\xi}{\left(1 - \frac{1}{b^2}(e^\xi - 1)^2\right)^{3/2}} u. \tag{12.4.69}$$

Using (12.4.69) in (12.4.66), we find

$$u(x) = (e^{-\sqrt{2}|x|}) \quad \text{as } |x| \to \infty. \tag{12.4.70}$$

On the other hand, in view of (12.4.70) and L^p-estimates, it is not hard to conclude that $u \in W^{2,p}$ ($p > 2$) near infinity of $\mathbb{R}^2$. In particular, $U_j = \partial_j u \to 0$ as $|x| \to \infty$ ($j = 1, 2$). By (12.4.66), we have, away from the points $p_1, p_2, \cdots, p_N$, that

$$\Delta U_j = \frac{2e^u}{\left(1 - \frac{1}{b^2}(e^u - 1)^2\right)^{3/2}} U_j. \qquad (12.4.71)$$

Since the coefficient of U_j on the right-hand side of (12.4.71) goes to 2 as $|x| \to \infty$, we arrive at the same exponential decay estimate for U_j as that for u stated in (12.4.70).

Finally, using the relation $|\phi|^2 = e^u$, (12.4.21), and (12.4.35), we see that the decay estimates for $|\phi|^2 - 1$, F_{12}, and $|D_j\phi|^2$ ($j = 1, 2$) follow immediately. Quantized magnetic flux and energy are standard consequences of these decay estimates.

The proof of Theorem 12.4.2 is complete.

We can also observe the stated b-independent decay result from the point of view of field theory.

With the fixed vacuum state

$$\langle 0|\phi|0\rangle = 1,$$

we represent the Higgs field ϕ as a perturbation from the vacuum state by a pair of real scalar fields ψ_1 and ψ_2,

$$\phi(x) = 1 + \frac{1}{\sqrt{2}}(\psi_1(x) + i\psi_2(x)). \qquad (12.4.72)$$

Substituting (12.4.72) into the Born–Infeld action density

$$\mathcal{L} = b^2\left(1 - \sqrt{1 + \frac{1}{2b^2}F_{\mu\nu}F^{\mu\nu}}\right) + (D_\mu\phi)\overline{(D^\mu\phi)} - V(|\phi|^2), \qquad (12.4.73)$$

where V is as defined by (12.4.15), and neglecting cubic and higher-order terms, we obtain

$$\begin{aligned}
\mathcal{L} = {}& -\frac{1}{4}F_{\mu\nu}F^{\mu\nu} + A_\mu A^\mu + \frac{1}{2}\partial_\mu\psi_1\partial^\mu\psi_1 \\
& + \frac{1}{2}\partial_\mu\psi_2\partial^\mu\psi_2 - (\psi_1)^2 + \sqrt{2}A^\mu\partial_\mu\psi_2 + \cdots. \qquad (12.4.74)
\end{aligned}$$

It is seen that both the gauge field A_μ and the real scalar field ψ_1 are massive, carrying the same mass $\sqrt{2}$, but the real scalar field ψ_2 is massless. In fact, the mixed term involving A^μ and $\partial_\mu\psi_2$ indicates that a propagating 'photon' could turn into a ψ_2 field and is thus not a physical field. Indeed, it may be removed using a gauge transformation,

$$1 + \frac{1}{\sqrt{2}}(\psi_1 + i\psi_2) \mapsto \left(1 + \frac{1}{\sqrt{2}}(\psi_1 + i\psi_2)\right)e^{i\Omega}, \quad A_\mu \mapsto A_\mu + \partial_\mu\Omega, \quad (12.4.75)$$

and one is left with a gauge field and a real scalar field, both carrying the same mass $\sqrt{2}$ as before. Consequently, both fields obey the same exponential decay rate, $O(e^{-\sqrt{2}|x|})$, at infinity as expected.

12.5 Remarks

In this chapter, we have made a general study and presented a series of locally concentrated solutions in the Born–Infeld theory [47, 48, 355]. In particular, existence and uniqueness theorems are established for multiply centered magnetic string solutions induced from a Higgs field over a closed Riemann surface or an Euclidean plane. We have seen a few interesting new phenomena. For example, on any given compact surface, the Born–Infeld parameter may be adjusted under a necessary and sufficient condition to allow the existence of an arbitrarily large number of strings.

The action density (12.1.4) is the main, and simplest, ingredient of the Born–Infeld geometric theory of electromagnetism and can also be derived from an invariant principle consideration [47, 48]. In fact, (12.1.4) was proposed earlier by Born himself [45, 46] and reconsidered later based on the invariance principle by Born and Infeld [47, 48], which eventually led them to write down the more elegant action density,

$$
\begin{aligned}
\mathcal{L} &= b^2 \left(1 - \sqrt{ -\det \left(\eta_{\mu\nu} + \frac{1}{b} F_{\mu\nu} \right) } \right) \\
&= b^2 \left(1 - \sqrt{ 1 + \frac{1}{2b^2} F_{\mu\nu} F^{\mu\nu} - \frac{1}{16 b^4} (F_{\mu\nu} \tilde{F}^{\mu\nu})^2 } \right), \quad (12.5.1)
\end{aligned}
$$

where $(\eta_{\mu\nu}) = \text{diag}\,\{1, -1, -1, -1\}$ is the Minkowski spacetime metric and

$$
\tilde{F}^{\mu\nu} = \frac{1}{2} \epsilon^{\mu\nu\alpha\beta} F_{\alpha\beta}
$$

is the dual of $F_{\mu\nu}$. Clearly, (12.5.1) introduces some higher-order interaction terms, although both (12.1.4) and (12.5.1) satisfy the invariance principle and take the Maxwell theory as their weak-field limit. For this reason, Born and Infeld stated: 'which of these action principles is the right one can only be decided by their consequences'. In recent years, due to its relevance in the theory of superstrings and membranes, the Born–Infeld nonlinear theory of electromagnetism has received much attention from theoretical physicists [53, 65, 90, 122, 132, 219, 222, 247, 250, 312, 314, 315]. Since in our study, we mainly focus on magnetostatic solutions (static and in the temporal gauge), (12.1.4) and (12.5.1) are identical and our corresponding solutions are the solutions for both (12.1.4)- and (12.5.1)-based theories. In the original work of Born–Infeld [48], a point-charge electrostatic solution in space was obtained. Shortly after, Pryce [256, 257] studied planar solutions and their uniqueness. Most recently, Gibbons [122] has conducted a systematic study of the Born–Infeld theory and has obtained exact solutions in numerous situations. In particular, his work links two exciting areas of physics (particles and fields) and mathematics (differential geometry) and motivated the further development here. The self-dual structure of the Abelian Higgs theory with a Born–Infeld electromagnetism was originally discovered in the work of Shiraishi and Hirenzaki [282] in the context of radially symmetric configurations.

We now propose a few open problems.

Open Problem 12.5.1. *Identify and investigate the Bernstein type property for the equation (12.2.20) under the condition (12.2.28).*

A weaker problem associated with our discussion in §12.2.3 is

Open Problem 12.5.2. *Identify a characteristic condition under which the equation (12.2.20) would only allow closed solution, $d\omega = 0$.*

For example, one of such conditions could be the finite-energy condition,

$$\int_{\mathbb{R}^n} (\sqrt{1 + |d\omega|^2} - 1) \, dx < \infty. \tag{12.5.2}$$

We may ask whether a solution of (12.2.20) satisfying (12.5.2) must also satisfy $d\omega = 0$. It is clear that low-dimensional cases are trivial.

Due to the nonlinear interaction introduced by the Born–Infeld electromagnetism, it will be more interesting to study sourceless static solutions without assuming a pure electrostatic or magnetostatic limit. In this connection, it is important to know whether there is self-induced electromagnetism which leads us to the coupled system of equations (12.2.42) and (12.2.43) and

Open Problem 12.5.3. *Identify a Bernstein type property for the system of equations (12.2.42) and (12.2.43). In other words, identify the most general space-like 'trivial' solutions of this system over $\mathbb{R}^3$ or over any $\mathbb{R}^n$.*

As seen in §12.3, there is no Derrick's theorem type obstruction to the existence of finite-energy static solutions of the Born–Infeld wave theory. Thus we state

Open Problem 12.5.4. *For the specific potential function*

$$V(u) = \lambda(|u|^2 - 1)^2,$$

where u maps $\mathbb{R}^n$ into itself, develop an existence theory for the critical points of the energy (12.3.2) characterized by the homotopy group $\pi_{n-1}(S^{n-1}) = \mathbb{Z}$, $n > 1$.

The Born–Infeld theory contains an ambiguous parameter, b. It will be interesting to find a mechanism that fixes b or a modification that does not require such a parameter.

Note that the Cauchy problems of the Born–Infeld wave equations and nonlinear electromagnetic field equations in $(1 + 1)$ dimensions have been studied in [22, 23].

Open Problem 12.5.5. *Develop an existence theory for the static vortex solutions in the Born–Infeld–Higgs model defined by, say, the energy*

$$E(A, \phi) = \int_{\mathbb{R}^2} \{(\sqrt{1 + F_{12}^2} - 1) + |D_1\phi|^2 + |D_2\phi|^2 + \lambda(|\phi|^2 - 1)^2\} \, dx, \quad (12.5.3)$$

where $\lambda > 0$ is a constant.

A first step would be to obtain the existence of radially symmetric solutions as was done for the Abelian Higgs model or the Ginzburg–Landau theory [249].

References

[1] A. A. Abrikosov, On the magnetic properties of superconductors of the second group, *Sov. Phys. JETP* **5**, 1174–1182 (1957).

[2] A. Actor, Classical solutions of $SU(2)$ Yang–Mills theories, *Rev. Modern Phys.* **51**, 461–525 (1979).

[3] S. Agmon, A. Douglis, and L. Nirenberg, Estimates near the boundary of solutions of elliptic partial differential equations satisfying general boundary conditions. I, *Comm. Pure Appl. Math.* **12**, 623–727 (1959).

[4] J. Ambjorn and P. Olesen, Anti-screening of large magnetic fields by vector bosons, *Phys. Lett.* B **214**, 565–569 (1988).

[5] J. Ambjorn and P. Olesen, On electroweak magnetism, *Nucl. Phys.* B **315**, 606–614 (1989).

[6] J. Ambjorn and P. Olesen, A magnetic condensate solution of the classical electroweak theory, *Phys. Lett.* B **218**, 67–71 (1989).

[7] J. Ambjorn and P. Olesen, A condensate solution of the classical electroweak theory which interpolates between the broken and the symmetric phase, *Nucl. Phys.* B **330**, 193–204 (1990).

[8] A. Ambrosetti and P. Rabinowitz, Dual variational methods in critical point theory and applications, *J. Funct. Anal.* **14**, 349–381 (1973).

[9] V. A. Andreev, Application of the inverse scattering method to the equation $\sigma_{xt} = e^{\sigma}$, *Theoret. Math. Phys.* **29**, 1027–1035 (1976).

[10] M. M. Ansourian and F. R. Ore, Jr., Pseudoparticle solutions on the $O(5,1)$ light cone, *Phys. Rev.* D **16**, 2662–2665 (1977).

[11] K. Arthur, D. H. Tchrakian, and Y. Yang, Topological and non-topological self-dual Chern–Simons solitons in a gauged $O(3)$ sigma model, *Phys. Rev.* D **54**, 5245–5258 (1996).

[12] M. F. Atiyah, V. G. Drinfeld, N. J. Hitchin and Yu. I. Manin, Construction of instantons, *Phys. Lett.* A **65**, 185–187 (1978).

[13] M. F. Atiyah and H. J. Hitchin, *The Geometry and Dynamics of Magnetic Monopoles*, Princeton U. Press, Princeton, NJ, 1988.

[14] M. F. Atiyah, N. J. Hitchin, and I. M. Singer, Deformation of instantons, *Proc. Natl. Acad. Sci. USA* **74**, 2662–2663 (1977).

[15] M. F. Atiyah, N. J. Hitchin, and I. M. Singer, Self-duality in four-dimensional Riemannian geometry, *Proc. Roy. Soc.* A **362**, 425–461 (1978).

[16] M. F. Atiyah and I. M. Singer, The index of elliptic operators. I, *Ann. of Math.* **87**, 484–530 (1968).

[17] T. Aubin, *Nonlinear Analysis on Manifolds: Monge–Ampére Equations*, Springer, Berlin and New York, 1982.

[18] T. Aubin, Meilleures constantes dans le théorème d'inclusion de Sobolev et un théorème de Fredholm non linéaire pour la transformation conforme de courburne scalaire, *J. Funct. Anal.* **32**, 148–174 (1979).

[19] M. Audin, *The Topology of Torus Actions on Symplectic Manifolds*, Birkhäuser, Basel, 1991.

[20] P. Aviles, Conformal complete metrics with prescribed non-negative Gaussian curvature in $\mathbb{R}^2$, *Invent. Math.* **83**, 519–544 (1986).

[21] A. Bahri and J. Coron, Une theorie des points critiques a l'infini pour l'equation de Yamabe et le probleme de Kazdan–Warner, *C. R. Acad. Sci. Paris*, Ser. I **15**, 513–516 (1985).

[22] B. M. Barbashov and N. A. Chernikov, Solution and quantization of a nonlinear two-dimensional model for a Born–Infeld type field, *Soviet. Phys. JETP* **23**, 861–868 (1966).

[23] B. M. Barbashov and N. A. Chernikov, Scattering of two plane electromagnetic waves in the nonlinear Born–Infeld electrodynamics, *Commun. Math. Phys.* **3**, 313–322 (1966).

[24] B. M. Barbashov and V. V. Nesterenko, *Introduction to the Relativistic String Theory*, World Scientific, Singapore, 1990.

[25] R. Bartnik and J. McKinnon, Particle-like solutions of the Einstein–Yang–Mills equations, *Phys. Rev. Lett.* **61**, 141–144 (1988).

[26] D. Bartolucci and G. Tarantello, Liouville type equations with singular data and their applications to the electroweak theory, *Preprint*, 2000.

[27] D. Bartolucci and G. Tarantello, The Liouville equation with singular data: a concentration-compactness principle via a local representation formula, *Preprint*, 2000.

[28] A. O. Barut, *Electrodynamics and Classical Theory of Fields and Particles*, Dover, New York, 1980.

[29] P. Bauman, N. N. Carlson, and D. Phillips, On the zeros of solutions to Ginzburg–Landau type systems, *SIAM J. Math. Anal.* **24**, 1283–1293 (1993).

[30] P. Bauman, C.-N. Chen, D. Phillips, and P. Sternberg, Vortex annihilation in nonlinear heat flow for Ginzburg–Landau systems, *European J. Appl. Math.* **6**, 115–126 (1995).

[31] P. Bauman, D. Phillips, and Q. Tang, Stable nucleation for the Ginzburg–Landau system with an applied magnetic field, *Arch. Rat. Mech. Anal.* **142**, 1–43 (1998).

[32] R. Becker, *Electromagnetic Fields and Interactions*, Dover, New York, 1982.

[33] A. A. Belavin and A. M. Polyakov, Metastable states of two-dimensional isotropic ferromagnets, *JETP Lett.* **22**, 245–247 (1975).

[34] A. A. Belavin, A. M. Polyakov, A. S. Schwartz and Yu. S. Tyupkin, Pseudoparticle solutions of the Yang–Mills equations, *Phys. Lett.* B **59**, 85–87 (1975).

[35] M. S. Berger, On Riemannian structures of prescribed Gaussian curvature for compact 2-manifolds, *J. Diff. Geom.* **5**, 325–332 (1971).

[36] M. S. Berger, *Nonlinearity and Functional Analysis*, Acad. Press, New York, 1977.

528   References

[37] M. S. Berger and Y. Y. Chen, Symmetric vortices for the Ginzburg–Landau equations of superconductivity and the nonlinear desingularization phenomenon, *J. Funct. Anal.* **82**, 259–295 (1989).

[38] L. Bers, F. John, and M. Schechter, *Partial Differential Equations*, Amer. Math. Soc., Providence, 1964.

[39] F. Bethuel, H. Brezis, and F. Helein, *Ginzburg–Landau Vortices*, Birkhäuser, Boston, 1994.

[40] G. Bimonte and G. Lozano, Z flux-line lattices and self-dual equations in the standard model, *Phys. Rev.* D **50**, 6046–6050 (1994).

[41] G. Bimonte and G. Lozano, Vortex solutions in two-Higgs-doublet systems, *Phys. Lett.* B **326**, 270–275 (1994).

[42] E. B. Bogomol'nyi, The stability of classical solutions, *Sov. J. Nucl. Phys.* **24**, 449–454 (1976).

[43] G. Bor, Yang–Mills fields which are not self-dual, *Commun. Math. Phys.* **145**, 393–410 (1992).

[44] G. Bor and R. Montgomery, $SO(3)$ invariant Yang–Mills fields which are not self-dual, in *"Hamiltonian Systems, Transformation Groups and Spectral Transform Method"*, pp. 191–198, University of Montréal, 1990.

[45] M. Born, Modified field equations with a finite radius of the electron, *Nature* **132**, 282 (1933).

[46] M. Born, On the quantum theory of the electromagnetic field, *Proc. Roy. Soc.* A **143**, 410–437 (1934).

[47] M. Born and L. Infeld, Foundation of the new field theory, *Nature* **132**, 1004 (1933).

[48] M. Born and L. Infeld, Foundation of the new field theory, *Proc. Roy. Soc.* A **144**, 425–451 (1934).

[49] R. Bott and L. W. Tu, Differential Forms in Algebraic Topology, Springer, Berlin and New York, 1982.

[50] J. P. Bourguignon and H. B. Lawson, Jr., Stability and isolation phenomena for Yang–Mills fields, *Commun. Math. Phys.* **79**, 189–230 (1981).

[51] S. Bradlow, Vortices in holomorphic line bundles over closed Kähler manifolds, *Commun. Math. Phys.* **135**, 1–17 (1990).

[52] R. H. Brandenberger, Cosmic strings and the large-scale structure of the universe, *Phys. Scripta* **T36**, 114–126 (1991).

[53] D. Brecher, BPS states of the non-Abelian Born–Infeld action, *Phys. Lett.* B **442**, 117–124 (1998).

[54] H. Brezis, Some variational problems of the Thomas–Fermi type, in *Variational Inequalities*, Cottle, Gianessi and Lions, ed., Wiley, 1980, pp. 53–73.

[55] H. Brezis and F. Merle, Uniform estimates and blow-up behavior for solutions of $-\Delta u = V(x)e^u$ in two dimensions, *Comm. P.D.E.* **16**, 1223-1253 (1991).

[56] H. Brezis and L. Nirenberg, Minima locaux relatifs a C^1 et H^1, *C. R. Acad. Sci. Paris*, Ser. I, **317**, 465–472 (1993).

[57] L. E. J. Brouwer, Uber Abbildung von Mannigfaltigkeiten, *Math. Ann.* **71**, 97–115 (1912).

[58] L. S. Brown, R. D. Carlitz, and C. Lee, Massless excitations in pseudoparticle fields, *Phys. Rev.* D **16**, 417–422 (1977).

[59] J. Burzlaff, Non-self dual solutions of $SU(3)$ Yang–Mills theory and a two-dimensional Abelian Higgs model, *Phys. Rev.* D **24**, 546–547 (1981).

[60] J. Burzlaff, A finite-energy $SU(3)$ solution which does not satisfy the Bogomol'nyi equation, *Czech. J. Phys.* B **32**, 624 (1982).

[61] J. Burzlaff, A. Chakrabarti and D. H. Tchrakian, Axially symmetric instantons in generalized Yang–Mills theory in $4p$ dimensions, *J. Math. Phys.* **34**, 1665–1680 (1993).

[62] L. Caffarelli, B. Gidas and J. Spruck, On multimeron solutions of the Yang–Mills equation, *Commun. Math. Phys.* **87**, 485–495 (1983).

[63] L. Caffarelli and Y. Yang, Vortex condensation in the Chern–Simons Higgs model: an existence theorem, *Commun. Math. Phys.* **168**, 321–336 (1995).

[64] E. Calabi, Examples of Bernstein Problems for some nonlinear equations, *Global Analysis* (Proc. Sympos. Pure Math. **15**, Berkeley, ed. S.-S. Chern and S. Smale, AMS, 1968), pp. 223–230.

[65] C. G. Callan, Jr. and J. M. Maldacena, Brane dynamics from the Born–Infeld action, *Nucl. Phys.* B **513**, 198–212 (1998).

[66] R. W. Carter, Simple groups and simple Lie algebras, *J. London Math. Soc.* **40**, 193–240 (1965).

530   References

[67] R. W. Carter, *Simple Groups of Lie Type*, Wiley, New York, 1972.

[68] D. Chae and O. Yu. Imanuvilov, The existence of nontopological multivortex solutions in the relativistic self-dual Chern–Simons theory, *Commun. Math. Phys.* **215**, 119–142 (2000).

[69] D. Chae and N. Kim, Topological multivortex solutions of the self-dual Maxwell–Chern–Simons–Higgs system, *J. Diff. Eq.* **134**, 154–182 (1997).

[70] M. Chaichian and N. F. Nelipa, *Introduction to Gauge Field Theory*, Springer, Berlin and New York, 1984.

[71] A. Chakrabarti, T. N. Sherry and D. H. Tchrakian, On axially symmetric self-dual field configurations in $4p$ dimensions, *Phys. Lett.* B **162**, 340–344 (1985).

[72] S. Chakravarty and Y. Hosotani, Anyon model on a cylinder, *Phys. Rev.* D **44**, 441–451 (1991).

[73] S. Y. A. Chang and P. Yang, Prescribing Gaussian curvature on S^2, *Acta Math.* **159**, 215–259 (1987).

[74] S. Chanillo and M. Kiessling, Rotational symmetry of solutions of some nonlinear problems in statistical mechanics and geometry, *Commun. Math, Phys.* **160**, 217–238 (1994).

[75] S. Chanillo and M. Kiessling, Conformally invariant systems of nonlinear PDE of Liouville type, *Geom. Funct. Anal.* **5**, 924–947 (1995).

[76] W. Chen and W. Ding, Scalar curvature on S^2, *Trans. Amer. Math. Soc.* **303**, 365–382 (1987).

[77] X. Chen, S. Hastings, J. B. McLeod, and Y. Yang, A nonlinear elliptic equation arising from gauge field theory and cosmology, *Proc. Roy. Soc.* A **446**, 453–478 (1994).

[78] W. Chen and C. Li, Classification of solutions of some nonlinear elliptic equations, *Duke Math. J.* **63**, 615–622 (1991).

[79] W. Chen and C. Li, Qualitative properties of solutions to some nonlinear elliptic equations in $\mathbb{R}^2$, *Duke Math. J.* **71**, 427–439 (1993).

[80] K. Cheng and W.-M. Ni, One the structure of the conformal Gaussian curvature equation on $\mathbb{R}^2$, *Duke Math. J.* **62**, 721–737 (1991).

[81] S.-Y. Cheng and S.-T. Yau 1976, Maximal space-like hypersurfaces in the Lorentz–Minkowski spaces, *Ann. of Math.* **104**, 407–419.

[82] C. Chevalley, Sur certains groupes de transformations finis et continus, *Tôhoku Math. J.* **7**, 14–66 (1955).

[83] Y. M. Cho and D. Maison, Monopole configurations in Weinberg–Salam model, *Phys. Lett.* B **391**, 360–365 (1997).

[84] Y. Choquet-Bruhat, S. M. Paneitz, and I. E. Segal, The Yang–Mills equations on the universal cosmos, *J. Funct. Anal.* **53**, 112–150 (1983).

[85] P. A. Collins and R. W. Tucker, Classical and quantum mechanics of free relativistic membranes, *Nucl. Phys.* B **112**, 150–176 (1976).

[86] A. Comtet and G. W. Gibbons, Bogomol'nyi bounds for cosmic strings, *Nucl. Phys.* B **299**, 719–733 (1988).

[87] H. J. de Vega and F. Schaposnik, Electrically charged vortices in non-Abelian gauge theories with Chern–Simons term, *Phys. Rev. Lett.* **56**, 2564–2566 (1986).

[88] J. Deang, Q. Du, M. Gunzburger, and J. Peterson, Vortices in superconductors: modelling and computer simulations, *Philos. Trans. Roy. Soc.* A **355**, 1957–1968 (1997).

[89] G. H. Derrick, Comments on nonlinear wave equations as models for elementary particles, *J. Math. Phys.* **5**, 1252–1254 (1964).

[90] S. Deser and G. W. Gibbons, Born–Infeld–Einstein actions? *Class. Quant. Grav.* **15**, L35–L39 (1998).

[91] W. Ding, J. Jost, J. Li, and G. Wang, An analysis of the two-vortex case in the Chern–Simons Higgs model, *Calc. Var.* **7**, 87–97 (1998).

[92] W. Ding, J. Jost, J. Li, and G. Wang, Existence results for mean field equations, *Ann. Inst. H. Poincarè - Anal. non linéaire* **16**, 653–666 (1999).

[93] P. A. M. Dirac, *General Theory of Relativity*, Princeton U. Press, Princeton, NJ, 1996.

[94] P. A. M. Dirac, Quantized singularities in the electromagnetic field, *Proc. Roy. Soc.* A **133**, 60–72 (1931).

[95] Q. Du, M. Gunzburger, and J. Peterson, Analysis and approximation of the Ginzburg–Landau model of superconductivity, *SIAM Rev.* **34**, 54–81 (1992).

[96] G. Dunne, *Self-Dual Chern–Simons Theories*, Lecture Notes in Physics, vol. m **36**, Springer, Berlin, 1995.

[97] G. Dunne, Mass degeneracies in self-dual models, *Phys. Lett.* B **345**, 452–457 (1995).

[98] G. Dunne, R. Jackiw, S.-Y. Pi and C. Trugenberger, Self-dual Chern–Simons solitons and two-dimensional nonlinear equations, *Phys. Rev.* D **43**, 1332–1345 (1991).

[99] Weinan E, Dynamics of vortices in Ginzburg–Landau theories with applications to superconductivity, *Physica* D **77**, 383–404 (1994).

[100] D. M. Eardley and V. Moncrief, The global existence of Yang–Mills–Higgs fields in 4-dimensional Minkowski space, Parts I and II, **83**, 171–212 (1982).

[101] J. D. Edelstein and C. Nunez, Supersymmetric electroweak cosmic strings, *Phys. Rev.* D **55**, 3811–3819 (1997).

[102] T. Eguchi, P. Gilkey, and A. Hanson, Gravitation, gauge theories, and differential geometry, *Phys. Rep.* **66**, 213–393 (1980).

[103] J. Ellis, S. Kelley, and D. V. Nanopoulos, Precision LEP data, supersymmetric GUTs and string unification, *Phys. Lett.* B **249**, 441–448 (1990).

[104] M. J. Esteban, A direct variational approach to Skyrme's model for meson fields, *Commun. Math. Phys.* **105**, 571–591 (1986).

[105] M. J. Esteban, A new setting for Skyrme's problem, *Variational Methods* (Paris, 1988), pp. 77–93, Progress in Nonlin. Diff. Eqs. Appl. **4**, Birkhäuser, Boston, 1990.

[106] M. J. Esteban and S. Müller, Sobolev maps with integer degree and applications to Skyrme's problem, *Proc. Roy. Soc.* A **436**, 197–201 (1992).

[107] P. Fayet and J. Iliopoulos, Spontaneously broken supergauge symmetries and Goldstone spinors, *Phys. Lett.* B **51**, 461–464 (1974).

[108] B. Felsager, *Geometry, Particles, and Fields*, Springer, Berlin and New York, 1998.

[109] L. Fontana, Sharp borderline Sobolev inequalities on compact Riemannian manifolds, *Comment. Math. Helv.* **68**, 415–454 (1993).

[110] J. Foster and J. D. Nightingale, *A Short Course in General Relativity*, Longman, London and New York, 1979.

[111] E. S. Fradkin and A. A. Tseytlin, Nonlinear electrodynamics from quantized strings, *Phys. Lett.* B **163**, 123–130 (1985).

[112] J. Fröhlich and P. Marchetti, Quantum field theory of anyons, *Lett. Math. Phys.* **16**, 347–358 (1988).

[113] J. Fröhlich and P. Marchetti, Quantum field theory of vortices and anyons, *Commun. Math. Phys.* **121**, 177–223 (1989).

[114] J. Fröhlich, The fractional quantum Hall effect, Chern–Simons theory, and integral lattices, *Proc. Internat. Congr. Math.*, pp. 75–105, Birkhäuser, Basel, 1995.

[115] N. Ganoulis, P. Goddard and D. Olive, Self-dual monopoles and Toda molecules, *Nucl. Phys.* B **205**, 601–636 (1982).

[116] O. Garcia-Prada, A direct existence proof for the vortex equations over a compact Riemann surface, *Bull. London Math. Soc.* **26**, 88–96 (1994).

[117] C. S. Gardner, G. Green, M. Kruskal, and R. Miura, Method for solving the Korteweg–de Vries equation, *Phys. Rev. Lett.* **19**, 1095–1097 (1967).

[118] D. Garfinkle, General relativistic strings, *Phys. Rev.* D **32**, 1323–1329 (1985).

[119] P. K. Ghcsh and S. K. Ghosh, Topological and non-topological solitons in a gauged $O(3)$ sigma model with Chern–Simons term, *Phys. Lett.* B **366**, 199 (1996).

[120] G. W. Gibbons, M. E. Ortiz, and F. R. Ruiz, Existence of global strings coupled to gravity, *Phys. Rev.* D **39**, 1546–1551 (1989).

[121] G. W. Gibbons, M. E. Ortiz, F. R. Ruiz, and T. M. Samols, Semilocal strings and monopoles, *Nucl. Phys.* B **385**, 127–144 (1992).

[122] G. W. Gibbons, Born–Infeld particles and Dirichlet p-branes, *Nucl. Phys.* B **514**, 603–639 (1998).

[123] D. Gilbarg and N. Trudinger, *Elliptic Partial Differential Equations of Second Order*, Springer, Berlin and New York, 1977.

[124] J. Ginibre and G. Velo, The Cauchy problem for coupled Yang–Mills and scalar fields in the temporal gauge, *Commun. Math. Phys.* **82**, 1–28 (1981).

[125] V. L. Ginzburg and L. D. Landau, On the theory of superconductivity, in *Collected Papers of L. D. Landau* (edited by D. Ter Haar), pp. 546–568, Pergamon, New York, 1965.

[126] E. Giusti 1984, *Minimal Surfaces and Functions of Bounded Variation*, Birkhäuser, Boston.

[127] J. Gladikowski, Topological Chern–Simons vortices in the $O(3)$ sigma model, *Z. Phys.* C **73**, 181–188 (1996).

[128] R. T. Glassey and W. A. Strauss, Some global solutions of the Yang–Mills equations in Minkowski space, *Commun. Math. Phys.* **81**, 171–188 (1981).

[129] J. Glimm and A. Jaffe, Multiple meron solution of the classical Yang–Mills equation, *Phys. Lett.* B **73**, 167–170 (1978).

[130] P. Goddard and D. I. Olive 1978, Magnetic monopoles in gauge field theories, *Rep. Prog. Phys.* **41**, 1360–1437.

[131] G. H. Golub and J. M. Ortega, *Scientific Computing and Differential Equations*, Academic, San Diego, 1992.

[132] S. Gonorazky, C. Nunez, F. A. Schaposnik, and G. Silva, Bogomol'nyi bounds and the supersymmetric Born–Infeld theory, *Nucl. Phys.* B **531**, 168–184 (1998).

[133] T. Goto, Relativistic quantum mechanics of one-dimensional mechanical continuum and subsidiary condition of dual resonance model, *Prog. Theoret. Phys.* **46**, 1560–1569 (1971).

[134] R. Gregory, Gravitational stability of local strings, *Phys. Rev. Lett.* **59**, 740–743 (1987).

[135] W. Greiner and B. Müller, *Quantum Mechanics – Symmetries*, 2nd ed., Springer, Berlin and New York, 1994.

[136] W. Greiner and J. Reinhardt, *Field Quantization*, Springer, Berlin and New York, 1986.

[137] P. A. Griffiths and J. Harris, *Principle of Algebraic Geometry*, Wiley, New York, 1978.

[138] B. Grossman, T.W. Kephart and J. D. Stasheff, Solutions to the Yang–Mills field equations in 8 dimensions and the last Hopf map, *Commun. Math. Phys.* **96** 431–437 (1984). Erratum, **100**, 311 (1985).

[139] V. Guilleman and A. Pollack, *Differential Topology*, Prentice-Hall, Englewood Cliffs, NJ, 1974.

[140] Z.-C. Han, Prescribing Gaussian curvature on S^2, *Duke Math. J.* **61**, 679–703 (1990).

[141] N. J. Hitchin, The self-duality equations on a Riemann surface, *Proc. London Math. Soc.* **55**, 59–126 (1987).

[142] M.-C. Hong, J. Jost, and M. Struwe, Asymptotic limits of a Ginzburg–Landau type functional, in *Geometry, Analysis and the Calculus of Variations*, pp. 99–123, Internat. Press, Cambridge, MA, 1996.

[143] J. Hong, Y. Kim and P.-Y. Pac, Multivortex solutions of the Abelian Chern–Simons–Higgs theory, *Phys. Rev. Lett.* **64**, 2330–2333 (1990).

[144] R. A. Horn and C. R. Johnson, *Matrix Analysis*, Cambridge U. Press, Cambridge, U. K., 1990.

[145] Y. Hosotani, Gauge invariance in Chern–Simons theory on a torus, *Phys. Rev. Lett.* **62**, 2785–2788 (1989).

[146] K. Huang and R. Tipton, Vortex excitations in the Weinberg–Salam theory, *Phys. Rev.* D **23**, 3050–3057 (1981).

[147] J. E. Humphreys, *Introduction to Lie Algebras and Representation Theory*, Springer, New York and Heidelberg, 1972.

[148] R. Iengo and K. Lechner, Quantum mechanics of anyons on a torus, *Nucl. Phys.* B **346**, 551–575 (1990).

[149] R. Jackiw, Quantum meaning of classical field theory, *Rev. Modern Phys.* **49**, 681–707 (1977).

[150] R. Jackiw, K. Lee, and E. J. Weinberg, Self-dual Chern–Simons solitons, *Phys. Rev.* D **42**, 3488–3499 (1990).

[151] R. Jackiw, C. Nohl, and C. Rebbi, Conformal properties of pseudoparticle configurations, *Phys. Rev.* D **15**, 1642–1646 (1977).

[152] R. Jackiw and S.-Y. Pi, Soliton solutions to the gauged nonlinear Schrödinger equation on the plane, *Phys. Rev. Lett.* **64**, 2969–2972 (1990).

[153] R. Jackiw and S.-Y. Pi, Classical and quantum nonrelativistic Chern–Simons theory, *Phys. Rev.* D **42**, 3500–3513 (1990).

[154] R. Jackiw, S.-Y. Pi, and E. J. Weinberg, Topological and non-topological solitons in relativistic and non-relativistic Chern–Simons theory, *Particles, Strings and Cosmology* (Boston, 1990), pp. 573–588, World Sci. Pub., River Edge, NJ, 1991.

[155] R. Jackiw and C. Rebbi, Degrees of freedom in pseudoparticle systems, *Phys. Lett.* B **67**, 189–192 (1977).

[156] R. Jackiw and E. J. Weinberg, Self-dual Chern–Simons vortices, *Phys. Rev. Lett.* **64**, 2334–2337 (1990).

536   References

[157] A. Jaffe and C. H. Taubes, *Vortices and Monopoles*, Birkhäuser, Boston, 1980.

[158] M. James, L. Perivolaropoulos, and T. Vachaspati, Stability of electroweak strings, *Phys. Rev.* D **46**, 5232–5235 (1992).

[159] M. James, L. Perivolaropoulos, and T. Vachaspati, Detailed stability analysis of electroweak strings, *Nucl. Phys.* B **395**, 534–546 (1993).

[160] T. Jonsson, O. McBryan, F. Zirilli and J. Hubbard, An existence theorem for multimeron solutions to classical Yang–Mills field equations, *Commun. Math. Phys.* **68**, 259–273 (1979).

[161] K. Jörgens, Uber die Lösungen der Differentialgleichung $rt - s^2 = 1$, *Math. Ann.* **127**, 130–134 (1954).

[162] B. Julia and A. Zee, Poles with both magnetic and electric charges in non-Abelian gauge theory, *Phys. Rev.* D **11**, 2227–1232 (1975).

[163] H. C. Kao and K. Lee, Self-dual $SU(3)$ Chern–Simons Higgs systems, *Phys. Rev.* D **50**, 6626–6632 (1994).

[164] J. L. Kazdan and F. W. Warner, Integrability conditions for $\Delta u = k - Ke^{2u}$ with applications to Riemannian geometry, *Bull. Amer. Math. Soc.* **77**, 819–823 (1971).

[165] J. L. Kazdan and F. W. Warner, Curvature functions for compact 2-manifolds, *Ann. Math.* **99**, 14–47 (1974).

[166] J. L. Kazdan and F. W. Warner, Curvature functions for open 2-manifolds, *Ann. Math.* **99**, 203–219 (1974).

[167] I. R. Kenyon, *General Relativity*, Oxford U. Press, London, 1990.

[168] T. W. B. Kibble, Some implications of a cosmological phase transition, *Phys. Rep.* **69**, 183–199 (1980).

[169] T. W. B. Kibble, Cosmic strings – an overview, in *The Formation and Evolution of Cosmic Strings*, ed. G. Gibbons, S. Hawking, and T. Vachaspati, Cambridge University Press, pp. 3–34, 1990.

[170] C. Kim and Y. Kim 1994, Vortices in the Bogomol'nyi limit of the Einstein–Maxwell–Higgs theory with or without external sources, *Phys. Rev.* D **50**, 1040–1050.

[171] K. Kimm, K. Lee, and T. Lee, Anyonic Bogomol'nyi solitons in a gauged $O(3)$ sigma model, *Phys. Rev.* D **53**, 4436–4440 (1996).

[172] B. Kostant, The solution to a generalized Toda lattice and representation theory, *Adv. Math.* **34**, 195–338 (1979).

[173] H.-T. Ku and M.-K. Ku, *Vector Bundles, Connections, Minimal Submanifolds and Gauge Theory*, Lecture Notes in Math. Sci., Hubei Acad. Press, Wuhan, 1986.

[174] C. Kumar and A. Khare, Charged vortex of finite energy in non-Abelian gauge theories with Chern–Simons term, *Phys. Lett.* B **178**, 395–399 (1986).

[175] O. Ladyzhenskaya, *Mathematical Theory of Viscous and Compressible Flow*, Gordon & Breach, New York 1969.

[176] O. Ladyzhenskaya and N. Ural'tseva, *Linear and Quasilinear Elliptic Equations*, Academic Press, New York, 1968.

[177] C. H. Lai (ed.), *Selected Papers on Gauge Theory of Weak and Electromagnetic Interactions*, World Scientific, Singapore, 1981.

[178] L. Landau and E. Lifshitz, *The Classical Theory of Fields*, Addison-Wesley, Cambridge, MA, 1951.

[179] P. Langacker and M. Luo, Implications of precision electroweak experiments for $m_t, \rho_0, \sin^2 \theta_W$, and grand unification, *Phys. Rev.* D **44**, 817–822 (1991).

[180] P. Laguna-Castillo and R. A. Matzner, Coupled field solutions for $U(1)$-gauge cosmic strings, *Phys. Rev.* D **36**, 3663–3673 (1987).

[181] P. D. Lax, Integrals of nonlinear equations of evolution and solitary waves, *Commun. Pure Appl. Math.* **21**, 467–490 (1968).

[182] O. Lechtenfeld, W. Nahm and D. H. Tchrakian, Dirac equations in $4p$-dimensions, *Phys. Lett.* B **162**, 143–147 (1985).

[183] C. Lee, K. Lee, and H. Min, Self-dual Maxwell Chern–Simons solitons, *Phys. Lett.* B **252**, 79–83 (1990).

[184] K. Lee, Relativistic nonabelian self-dual Chern–Simons systems, *Phys. Lett.* B **255**, 381–384 (1991).

[185] K. Lee, Self-dual nonabelian Chern–Simons solitons, *Phys. Rev. Lett.* **66**, 553–355 (1991).

[186] K. Lee and E. J. Weinberg, Nontopological magnetic monopoles and new magnetically charged black holes, *Phys. Rev. Lett.* **73**, 1203–1206 (1994).

[187] H. Lewy, A priori limitations for solutions of Monge–Ampére equations: II, *Trans. Amer. Math. Soc.* **41**, 365–374 (1937).

538   References

[188] A. N. Leznov, On the complete integrability of a nonlinear system of partial differential equations in two-dimensional space, *Theoret. Math. Phys.* **42**, 225–229 (1980).

[189] A. N. Leznov and M. V. Saveliev, Representation of zero curvature for the system of nonlinear partial differential equations $x_{\alpha,z\bar{z}} = \exp(kx)_\alpha$ and its integrability, *Lett. Math. Phys.* **3**, 489–494 (1979).

[190] A. N. Leznov and M. V. Saveliev, Representation theory and integration of nonlinear spherically symmetric equations to gauge theories, *Commun. Math. Phys.* **74**, 111–118 (1980).

[191] Y. Li, On Nirenberg's problem and related topics, *Topo. Methods Nonlin. Anal.* **3**, 221–233 (1994).

[192] Y. Li, Harnack-type inequality: the method of moving planes, *Comm. Math. Phys.* **200**, 421-444 (1999).

[193] Y. Li and I. Shafrir Blow-up analysis for solutions of $-\Delta u = V(x)e^u$ in dimension two, *Ind. Univ. Math. J.* **43**, 1255-1270 (1994).

[194] E. H. Lieb, Remarks on the Skyrme model, *Proc. Sympos. Pure Math.* **54**, Part 2, pp. 379–384, Amer. Math. Soc., Providence, RI, 1993.

[195] F.-H. Lin, Complex Ginzburg–Landau equations and dynamics of vortices, filaments, and codimension-2 submanifolds, *Comm. Pure Appl. Math.* **51**, 385–441 (1998).

[196] F.-H. Lin, Mixed vortex-antivortex solutions of Ginzburg–Landau equations, *Arch. Rat. Mech. Anal.* **133**, 103–127 (1995).

[197] S.-Y. Lin and Y. Yang, Computation of superconductivity in thin films, *J. Comput. Phys.* **89**, 257–275 (1990).

[198] B. Linet, A vortex-line model for a system of cosmic strings in equilibrium, *General Relat. Grav.* **20**, 451–456 (1988).

[199] B. Linet, On the supermassive $U(1)$ gauge cosmic strings, *Class. Quantum Grav.* **7**, L75–L79 (1990).

[200] J. Liouville, Sur l'équation aux différences partielles $\frac{d^2 \log \lambda}{du\,dv} \pm \frac{\lambda}{2a^2} = 0$, *Journal de Mathématiques Pures et Appl.* **18**, 71–72 (1853).

[201] M. S. Longair, *Theoretical Concepts in Physics*, Cambridge U. Press, London, 1984.

[202] M. Loss, The Skyrme model on Riemannian manifold, *Lett. Math. Phys.* **14**, 149–156 (1987).

[203] J. Lykken, J. Sonnenschein, and N. Weiss, The theory of anyonic superconductivity, *Inter. J. Mod. Phys.* A **6**, 5155 (1991).

[204] Zhong-Qi Ma, G.M. O'Brien and D.H. Tchrakian, Dimensional reduction and higher-order topological invariants: descent by even steps and applications, *Phys. Rev.* D **33**, 1177–1180 (1986).

[205] Zhong-Qi Ma and D.H. Tchrakian, Gauge Field Systems on $\mathbb{C}P^n$, *J. Math. Phys.* **31**, 1506–1512 (1990).

[206] S. W. MacDowell and O. Törnkvist, Structure of the ground state of the electroweak gauge theory in a strong magnetic field, *Phys. Rev.* D **45**, 3833–3844 (1992).

[207] D. Maison, Uniqueness of the Prasad–Sommerfield monopole solution, *Nucl. Phys.* B **182**, 144–150 (1981).

[208] P. Mansfield, Solutions of Toda systems, *Nucl. Phys.* B **208**, 277–300 (1982).

[209] N. S. Manton, A remark on the scattering of BPS monopoles, *Phys. Lett.* B **110**, 54–56 (1982).

[210] N. S. Manton, Geometry of Skyrmions, *Commun. Math. Phys.* **111**, 469–478 (1987).

[211] N. S. Manton and P. J. Ruback, Skyrmions in flat space and curved space, *Phys. Lett.* B **181**, 137–140 (1986).

[212] P. J. McCarthy, Bäcklund transformations as nonlinear Dirac equations, *Lett. Math. Phys.* **2**, 167–170 (1977).

[213] R. McOwen, Conformal metrics in $\mathbb{R}^2$ with prescribed Gaussian curvature and positive total curvature, *Indiana U. Math. J.* **34**, 97–104 (1984).

[214] R. McOwen, On the equation $\Delta u + K e^{2u} = f$ and prescribed negative curvature in $\mathbb{R}^2$, *J. Math. Anal. Appl.* **103**, 365–370 (1984).

[215] A. Mikhailov, M. Olshanetsky and A. Perelomov, Two-dimensional generalized Toda lattice, *Commun. Math. Phys.* **79**, 473–488 (1981).

[216] J. W. Milnor, *Topology from the Differentiable Viewpoint*, Princeton U. Press, Princeton, 1965.

[217] C. W. Misner, K. S. Thorne, and J. A. Wheeler, *Gravitation*, Freeman, New York, 1973.

[218] M. Monastyrsky, *Topology of Gauge Fields and Condensed Matter*, Plenum, New York and London, 1993.

[219] E. Moreno, C. Nunez, and F. A. Schaposnik, Electrically charged vortex solution in Born–Infeld theory, *Phys. Rev.* D **58**, 025015 (1998).

[220] J. Moser, On a nonlinear problem in differential geometry, in *Dynamical Systems* (M. Peixoto ed.), Academic, New York, 1973.

[221] J. Moser, A sharp form of an inequality of N. Trudinger, *Indiana U. Math. J.* **20**, 1077–1092 (1971).

[222] A. Nakamura and K. Shiraishi, Born–Infeld monopoles and instantons, *Hadronic J.* **14**, 369–375 (1991).

[223] C. Nash and S. Sen, *Topology and Geometry for Physicists*, Academic, London and New York, 1983.

[224] J. C. Neu, Vortices in complex scalar fields, *Physica* D **43**, 385–406 (1990).

[225] W.-M. Ni, On the elliptic equation $\Delta u + K(x)e^{2u} = 0$ and conformal metrics with prescribed Gaussian curvatures, *Invent. Math.* **66**, 343–352 (1982).

[226] W.-M. Ni, On the elliptic equation $\Delta u + K(x)u^{(n+2)/(n-2)} = 0$, its generalizations, and applications in geometry, *Indiana U. Math. J.* **31**, 493–529 (1982).

[227] H. Nielsen and P. Olesen, Vortex-line models for dual strings, *Nucl. Phys.* B **61**, 45–61 (1973).

[228] L. Nirenberg, Variational and topological methods in nonlinear problems, *Bull. Amer. Math. Soc.* **4**, 267–302 (1981).

[229] J. C. C. Nitsche, Elementary proof of Bernstein's theorem on minimal surfaces, *Ann. of Math.* **66**, 543–544 (1957).

[230] J. C. C. Nitsche, *Lectures on Minimal Surfaces*, Cambridge U. Press, London, 1989.

[231] M. Noguchi, Abelian Higgs theory on Riemann surfaces, Thesis, Duke University, 1985.

[232] M. Noguchi, Yang–Mills–Higgs theory on a compact Riemann surface, *J. Math. Phys.* **28**, 2343–2346 (1987).

[233] M. Nolasco and G. Tarantello, On a sharp Sobolev type inequality on two dimensional compact manifolds, *Arch. Rat. Mech. Anal.* **145**, 161–195 (1998).

[234] M. Nolasco and G. Tarantello, Double vortex condensates in the Chern–Simons–Higgs theory, *Calc. Var. & PDE's* **9**, 31–91 (1999).

[235] G.M. O'Brien and D.H. Tchrakian, Spin-connection self-dual GYM fields on double-self-dual GEC backgrounds, *J. Math. Phys.* **29**, 1212–1219 (1988).

[236] O. A. Oleinik, On the equation $\Delta u + k(x)e^u = 0$, *Russian Math. Surveys* **33**, 243–244 (1978).

[237] D. Olive and N. Turok, The symmetries of Dynkin diagrams and the reduction of Toda field equations, *Nucl. Phys.* B **215**, 470–494 (1983).

[238] P. Olesen, Soliton condensation in some self-dual Chern–Simons theories, *Phys. Lett.* B **265**, 361–365 (1991); Erratum, B **267**, 541 (1991).

[239] D. O'Sè and D.H. Tchrakian, Conformal properties of the BPST instantons of the generalised Yang–Mills system, *Lett. Math. Phys.* **13**, 211–218 (1987).

[240] R. Osserman, On the inequality $\Delta u \geq f(u)$, *Pacific J. Math.* **7**, 1641–1647 (1957).

[241] R. Osserman, *A Survey of Minimal Surfaces*, Dover, New York, NY, 1986.

[242] T. H. Parker, Nonminimal Yang–Mills fields and dynamics, *Invent. Math.* **107**, 397–420 (1992).

[243] S. Paul and A. Khare, Charged vortices in an Abelian Higgs model with Chern–Simons term, *Phys. Lett.* B **174**, 420–422 (1986).

[244] P. J. E. Peebles, *Principles of Physical Cosmology*, Princeton U. Press, Princeton, NJ, 1993.

[245] L. Perivolaropoulos, Existence of double vortex solutions, *Phys. Lett.* B **316**, 528–533 (1993).

[246] W. B. Perkins, W-condensation in electroweak strings, *Phys. Rev.* D **47**, 5224–5227 (1993).

[247] M. Perry and J. H. Schwarz, Interacting chiral gauge fields in six dimensions and Born–Infeld theory, *Nucl. Phys.* B **489**, 47–64 (1997).

[248] B. Piette and W. J. Zakrzewski, Skyrmion model in 2+1 dimensions with soliton bound states, *Nucl. Phys.* B **393**, 65–78 (1993).

[249] B. J. Plohr, The existence, regularity, and behavior of isotropic solutions of classical gauge field theories, Thesis, Princeton University, 1980.

[250] J. Polchinski, *Tasi Lectures on D-Branes*, hep-th/9611050, 1996.

[251] A. Polychronakos, Abelian Chern–Simons theories in $2 + 1$ dimensions, *Ann. Phys.* **203**, 231–254 (1990).

[252] M. K. Prasad and C. M. Sommerfield, Exact classical solutions for the 't Hooft monopole and the Julia–Zee dyon, *Phys. Rev. Lett.* **35**, 760–762 (1975).

[253] J. Preskill, Vortices and monopoles, *Architecture of Fundamental Interactions at Short Distances*, edited by P. Ramond and R. Stora, Elsevier, Amsterdam, pp. 235–337, 1987.

[254] J. Preskill, Semilocal defects, *Phys. Rev.* D **46**, 4218–4231 (1992).

[255] M. H. Protter and H. F. Weinberger, *Maximum Principles in Differential Equations*, Prentice-Hall, Englewood, NJ, 1967.

[256] M. H. L. Pryce, The two-dimensional electrostatic solutions of Born's new field equations, *Proc. Cambridge Phil. Soc.* **31**, 50–57 (1935).

[257] M. H. L. Pryce, On a uniqueness theorem, *Proc. Cambridge Phil. Soc.* **31**, 625–628 (1935).

[258] J. Qing, Renormalized energy for Ginzburg–Landau vortices on closed surfaces, *Math. Z.* **225**, 1–34 (1997).

[259] R. Rajaraman, *Solitons and Instantons*, North Holland, Amsterdam, 1982.

[260] S. Randjbar, A. Salam, and J. A. Strathdee, Anyons and Chern–Simons theory on compact spaces of finite genus, *Phys. Lett.* B **240**, 121–126 (1990).

[261] J. H. Rawnsley, Spherically symmetric monopoles are smooth, *J. Phys.* A **10**, L139–L141 (1977).

[262] C. Rebbi and G. Soliani (eds.), *Solitons and Particles*, World Scientific, Singapore, 1984.

[263] M. Renardy, On bounded solutions of a classical Yang–Mills equation, *Commun. Math. Phys.* **76**, 277–287 (1980).

[264] T. Ricciardi and G. Tarantello, Vortices in the Maxwell–Chern–Simons theory, *Comm. Pure Appl. Math.* **53**, 811–851 (2000).

[265] W. Rindler, *Essential Relativity: Special, General and Cosmological*, 2nd edition, Springer, Berlin and New York, 1977.

[266] T. Riviere, Line vortices in the $U(1)$-Higgs model, *ESAIM Controle Optim. Calc. Var.* **1**, 77–167 (1995/1996).

[267] T. Riviere and F. Bethuel, Vortices for a variational problem related to superconductivity, *Ann. Inst. H. Poincaré – Anal. Nonlinéaire* **12**, 243–303 (1995).

[268] J. Rubinstein and P. Sternberg, On the slow motion of vortices in the Ginzburg–Landau heat flow, *SIAM J. Math. Anal.* **26**, 1452–1466 (1995).

[269] L. H. Ryder, *Quantum Field Theory*, 2nd edition, Cambridge U. Press, London, 1996.

[270] L. Sadun and J. Segert, Non-self-dual Yang–Mills connections with quadrupole symmetry, *Commun. Math. Phys.* **145**, 362–391 (1992).

[271] D. H. Sattinger, Conformal metrics in $\mathbb{R}^2$ with prescribed curvature, *Indiana U. Math. J.* **22**, 1–4 (1972).

[272] M. Schechter and R. Weder, A theorem on the existence of dyon solutions, *Ann. Phys.* **132**, 292–327 (1981).

[273] J. Schiff, Integrability of Chern–Simons Higgs and Abelian Higgs vortex equations in a background metric, *J. Math. Phys.* **32**, 753 (1991).

[274] B. J. Schroers, Bogomol'nyi solitons in a gauged $O(3)$ sigma model, *Phys. Lett.* B **356**, 291–296 (1995).

[275] B. J. Schroers, The spectrum of Bogomol'nyi solitons in gauged linear sigma models, *Nucl. Phys.* B **475**, 440–468 (1996).

[276] A. S. Schwartz, On regular solutions of Euclidean Yang–Mills equations, *Phys. Lett.* B **67**, 172–174 (1977).

[277] J. Schwinger, A magnetic model of matter, *Science* **165**, 757–761 (1969).

[278] I. E. Segal, The Cauchy problem for the Yang–Mills equations, *J. Funct. Anal.* **33**, 175–194 (1979).

[279] J. Shatah, Weak solutions and development of singularities of the $SU(2)$ σ-model, *Commun. Pure Appl. Math.* **41**, 459–469 (1988).

[280] T.N. Sherry and D.H. Tchrakian, Dimensional reduction and higher order topological invariants, *Phys. Lett.* B **147** 121-126 (1984).

[281] M. Shiffman, On the existence of subsonic flows of a compressible fluid, *J. Rat. Mech. Anal.* **1**, 605–652 (1952).

[282] K. Shiraishi and S. Hirenzaki 1991, Bogomol'nyi equations for vortices in Born–Infeld Higgs systems, *Int. J. Mod. Phys.* A **6**, 2635–2647.

[283] L. M. Sibner, An existence theorem for a non-regular variational problem, *Manuscripta Math.* **43**, 45–73 (1983).

[284] L. M. Sibner, The isolated point singularity problem for the coupled Yang–Mills equations in higher dimensions, *Math. Ann.* **271**, 125–131 (1985).

[285] L. M. Sibner and R. J. Sibner, A nonlinear Hodge–De Rham theorem, *Acta Math.* **125**, 57–73 (1970).

[286] L. M. Sibner, R. J. Sibner, and K. Uhlenbeck, Solutions to Yang–Mills equations that are not self-dual, *Proc. Nat. Acad. Sci. USA* **86**, 8610–8613 (1989).

[287] L. M. Sibner, R. J. Sibner, and Y. Yang, Abelian gauge theory on Riemann surfaces and new topological invariants, *Proc. Roy. Soc.* A **456**, 593–613 (2000).

[288] J. A. Smoller, A. G. Wasserman, S. T. Yau, and J. B. McLeod, Smooth static solutions of the Einstein/Yang–Mills equations, *Commun. Math. Phys.* **143**, 115–147 (1991).

[289] A. A. Sokolov, I. M. Ternov, V. Ch. Zhukovskii, and A. V. Borisov, *Quantum Electrodynamics*, Mir Pub., Moscow, 1988.

[290] M. Spivak, *A Comprehensive Introduction to Differential Geometry*, IV, Publish or Perish, Berkeley, CA, 1979.

[291] J. Spruck, D. H. Tchrakian, and Y. Yang, Multiple instantons representing higher-order Chern–Pontryagin classes, *Commun. Math. Phys.* **188**, 737–751 (1997).

[292] J. Spruck and Y. Yang, On multivortices in the electroweak theory I: existence of periodic solutions, *Commun. Math. Phys.* **144**, 1–16 (1992).

[293] J. Spruck and Y. Yang, On multivortices in the electroweak theory II: Existence of Bogomol'nyi solutions in $\mathbb{R}^2$, *Commun. Math. Phys.* **144**, 215–234 (1992).

[294] J. Spruck and Y. Yang, Topological solutions in the self-dual Chern–Simons theory: existence and approximation, *Ann. Inst. H. Poincaré – Anal. non linéaire*, **12**, 75–97 (1995).

[295] J. Spruck and Y. Yang, The existence of non-topological solitons in the self-dual Chern–Simons theory, *Commun. Math. Phys.* **149**, 361–376 (1992).

[296] S. Sternberg, *Lectures on Differential Geometry*, Prentice-Hall, NJ, 1964.

[297] J. Stoer and R. Burlirsch, *Introduction to Numerical Analysis*, Springer, New York, 1983.

[298] W. A. Strauss, *Nonlinear Wave Equations*, Amer. Math. Soc., Providence, RI, 1989.

[299] M. Struwe and G. Tarantello, On multivortex solutions in Chern–Simons gauge theory, *Boll. U.M.I.* **8**, 109-121 (1998).

[300] D. Stuart, Dynamics of Abelian Higgs vortices in the near Bogomol'nyi regime, *Commun. Math. Phys.* **159**, 51–91 (1994).

[301] D. Stuart, The geodesic approximation for the Yang–Mills–Higgs equations, *Commun. Math. Phys.* **166**, 149–190 (1994).

[302] G. Tarantello, Multiple condensate solutions for the Chern–Simons–Higgs theory, *J. Math. Phys.* **37**, 3769–3796 (1996).

[303] C. H. Taubes, Arbitrary N-vortex solutions to the first order Ginzburg–Landau equations, *Commun. Math. Phys.* **72**, 277–292 (1980).

[304] C. H. Taubes, On the equivalence of the first and second order equations for gauge theories, *Commun. Math. Phys.* **75**, 207–227 (1980).

[305] C. H. Taubes, The existence of a non-minimal solution to the $SU(2)$ Yang–Mills–Higgs equations on $\mathbb{R}^3$, Parts I, II, *Commun. Math. Phys.* **86**, 257–320 (1982).

[306] D. H. Tchrakian, N-dimensional instantons and monopoles, *J. Math. Phys.* **21**, 166–169 (1980).

[307] D.H. Tchrakian, Spherically symmetric gauge field configurations in $4p$ dimensions, *Phys. Lett.* B **150**, 360–362 (1985).

[308] D.H. Tchrakian, Yang-Mills hierarchy, *Int. J. Mod. Phys.* (Proc. Suppl.) A **3**, 584–587 (1993).

[309] D.H. Tchrakian and A. Chakrabarti, How overdetermined are the generalised self-duality relations?, *J. Math. Phys.* **32**, 2532–2539 (1991).

546   References

[310] G. 't Hooft, Computation of the quantum effects due to a four-dimensional pseudoparticle, *Phys. Rev.* D **14**, 3432–3450 (1976).

[311] G. 't Hooft, A property of electric and magnetic flux in nonabelian gauge theories, *Nucl. Phys.* B **153**, 141–160 (1979).

[312] L. Thorlacius, Born–Infeld strings as a boundary conformal field theory, *Phys. Rev. Lett.* **80**, 1588–1590 (1998).

[313] N. Trudinger, On embedding into Orlitz spaces and some applications, *J. Math. Phys.* **17**, 473–484 (1967).

[314] A. A. Tseytlin, Self-duality of Born–Infeld action and Dirichlet 3-brane of type II B superstring theory, *Nucl. Phys.* B **469**, 51–67 (1996).

[315] A. A. Tseytlin, On non-abelian generalisation of Born–Infeld action in string theory, *Nucl. Phys.* B **501**, 41–52 (1997).

[316] Yu. S. Tyupkin, V. A. Fateev, and A. S. Shvarts, Particle-like solutions of the equations of gauge theories, *Theoret. Math. Phys.* **26**, 270–273 (1976).

[317] K. K. Uhlenbeck, Removable singularities in Yang–Mills fields, *Bull. Amer. Math. Soc.* **1**, 579–581 (1979).

[318] K. K. Uhlenbeck, Removable singularities in Yang–Mills fields, *Commun. Math. Phys.* **83**, 11-29 (1982).

[319] T. Vachaspati, Vortex solutions in the Weinberg–Salam model, *Phys. Rev. Lett.* **68**, 1977–1980 (1992).

[320] T. Vachaspati, Electroweak strings, *Nucl. Phys.* B **397**, 648–671 (1993).

[321] A. Vilenkin, Cosmic strings and domain walls, *Phys. Rep.* **121**, 263-315 (1985).

[322] A. Vilenkin and E. P. S. Shellard, *Cosmic Strings and Other Topological Defects*, Cambridge University Press, 1994.

[323] C. von Westenholz, *Differential Forms in Mathematical Physics*, North-Holland Pub., Amsterdam, New York, and Oxford, 1978.

[324] R. Wang, The existence of Chern–Simons vortices, *Commun. Math. Phys.* **137**, 587–597 (1991).

[325] S. Wang and Y. Yang, Abrikosov's vortices in the critical coupling, *SIAM J. Math. Anal.* **23**, 1125–1140 (1992).

[326] S. Wang and Y. Yang, Symmetric superconducting states in thin films, *SIAM J. Appl. Math.* **52**, 614–629 (1992).

[327] E. J. Weinberg, Multivortex solutions of the Ginzburg–Landau equations, *Phys. Rev.* D **19**, 3008–3012 (1979).

[328] S. Weinberg, *Gravitation and Cosmology*, Wiley, New York, 1972.

[329] S. Weinberg, The cosmological constant problem, *Rev. Mod. Phys.* **61**, 1–23 (1989).

[330] F. Wilczek, *Fractional Statistics and Anyonic Superconductivity*, World Scientific, Singapore, 1990.

[331] E. Witten, Some exact multipseudoparticle solutions of classical Yang–Mills theory, *Phys. Rev. Lett.* **38**, 121–124 (1977).

[332] E. Witten, Superconducting strings, *Nucl. Phys.* B **249**, 557–592 (1985).

[333] E. Witten, Superconducting cosmic strings, *Preprint*, 1987 (unpublished).

[334] E. Witten, Phase of $N = 2$ theories in two dimensions, *Nucl. Phys.* B **403**, 159–222 (1993).

[335] H. Wittich, Ganze Lösungen der Differentialgleichung $\Delta u = e^u$, *Math. Z.* **49**, 579–582 (1944).

[336] C. N. Yang and R. Mills, Conservation of isotopic spin and isotopic gauge invariance, *Phys. Rev.* **96**, 191–195 (1954).

[337] Y. Yang, Generalized Skyrme model on higher-dimensional Riemannian manifolds, *J. Math. Phys.* **30**, 824–828 (1989).

[338] Y. Yang, Existence, regularity, and asymptotic behavior of the Ginzburg–Landau equations on $\mathbb{R}^3$, *Commun. Math. Phys.* **123**, 147–161 (1989).

[339] Y. Yang, Blow-up of the $SU(2, \mathbb{C})$ Yang–Mills fields, *J. Math. Phys.* **31**, 1237–1239 (1990).

[340] Y. Yang, Ginzburg–Landau equations for superconducting films and the Meissner effect, *J. Math. Phys.* **31**, 1284–1289 (1990).

[341] Y. Yang, Existence of massive $SO(3)$ vortices, *J. Math. Phys.* **32**, 1395–1399 (1991).

[342] Y. Yang, Self duality of the gauge field equations and the cosmological constant, *Commun. Math. Phys.* **162**, 481–498 (1994).

[343] Y. Yang, The existence of Ginzburg–Landau solution on the plane by a direct variational method, *Ann. Inst. H. Poincaré - Anal. non linéaire* **11**, 517–536 (1994).

[344] Y. Yang, Obstructions to the existence of static cosmic strings in an Abelian Higgs model, *Phys. Rev. Lett.* **73**, 10–13 (1994).

[345] Y. Yang, Prescribing topological defects for the coupled Einstein and Abelian Higgs equations, *Commun. Math. Phys.* **170**, 541–582 (1995).

[346] Y. Yang, A necessary and sufficient condition for the existence of multisolitons in a self-dual gauged sigma model, *Commun. Math. Phys.* **181**, 485–506 (1996).

[347] Y. Yang, Static cosmic strings on S^2 and criticality, *Proc. Roy. Soc. A* **453**, 581–591 (1997).

[348] Y. Yang, Topological solitons in the Weinberg–Salam theory, *Physica D* **101**, 55–94 (1997).

[349] Y. Yang, The relativistic non-Abelian Chern–Simons equations, *Commun. Math. Phys.* **186**, 199–218 (1997).

[350] Y. Yang, The Lee–Weinberg magnetic monopole of unit charge: existence and uniqueness, *Physica D* **117**, 215–240 (1998).

[351] Y. Yang, Dually charged particle-like solutions in the Weinberg–Salam theory, *Proc. Roy. Soc. A* **454**, 155–178 (1998).

[352] Y. Yang, Coexistence of vortices and antivortices in an Abelian gauge theory, *Phys. Rev. Lett.* **80**, 26–29 (1998).

[353] Y. Yang, Strings of opposite magnetic charges in a gauge field theory, *Proc. Roy. Soc. A* **455**, 601–629 (1999).

[354] Y. Yang, On a system of nonlinear elliptic equations arising in theoretical physics, *J. Funct. Anal.* **170**, 1–36 (2000).

[355] Y. Yang, Classical solutions in the Born–Infeld theory, *Proc. Roy. Soc. A* **456**, 615–640 (2000).

[356] V. E. Zakharov and A. B. Shabat, Interaction between solitons in a stable medium, *Sov. Phys. JETP* **37**, 823–828 (1973).

[357] D. Zwanziger, Exactly soluble nonrelativistic model of particles with both electric and magnetic charges, *Phys. Rev.* **176**, 1480–1488 (1968).

[358] D. Zwanziger, Quantum field theory of particles with both electric and magnetic charges, *Phys. Rev.* **176**, 1489–1495 (1968).

Index